# INHIBITION OF MATRIX METALLOPROTEINASES

## THERAPEUTIC APPLICATIONS

ANNALS OF THE NEW YORK ACADEMY OF SCIENCES
Volume 878

# INHIBITION OF MATRIX METALLOPROTEINASES

## THERAPEUTIC APPLICATIONS

*Edited by Robert A. Greenwald, Stanley Zucker,
and Lorne M. Golub*

*The New York Academy of Sciences*
*New York, New York*
*1999*

The cover of the paper-bound edition of this volume shows the catalytic domain of a crystal structure of truncated MMP-3 bound to an MMP inhibitor. The inhibitor shown is PGE-116611, the chemical name of which is: (2R)-isobutyl-(3S)-[N-hydroxycarboxamido]-6-hydroxyhexanoic acid amide of (1N)-2-[methoxyethyl]-caprolactam-(3S)-amine. The cover illustration was generously provided by Drs. Biswanath De and Glen Mieling of Procter and Gamble Pharmaceuticals.

### Library of Congress Cataloging-in-Publication Data

Inhibition of matrix metalloproteinases : therapeutic applications
/ edited by Robert A. Greenwald, Stanley Zucker, and Lorne M. Golub.
    p.  cm. — (Annals of the New York Academy of Sciences,
0077-8923 ; v. 878)
    Includes bibliographical references and index.
    ISBN 1-57331-180-4 (cloth : alk. paper)
    ISBN 1-57331-181-2 (pbk : alk. paper)
    1. Metalloproteinases—Inhibitors—Therapeutic use Congresses. 2.
Extracellular matrix proteins Congresses. I. Greenwald, Robert A.,
1943– II. Zucker, Stanley. III. Golub, Lorne M. IV. Series.
    Q11 .N5   vol. 878 RM666.M512
    500 s—dc21
    [615' .35]
                        99-30122
                        CIP

K-M Research/PCP
*Printed in the United States of America*
**ISBN** 1-57331-180-4 (cloth)
**ISBN** 1-57331-181-2 (paper)
**ISSN** 0077-8923

ANNALS OF THE NEW YORK ACADEMY OF SCIENCES

Volume 878

June 30, 1999

# INHIBITION OF MATRIX METALLOPROTEINASES
## THERAPEUTIC APPLICATIONS[a]

*Editors and Conference Chairs*
ROBERT A. GREENWALD, STANLEY ZUCKER, AND LORNE M. GOLUB

## CONTENTS

[a]This volume is the result of a conference entitled **Inhibition of Matrix Metalloproteinases: Therapeutic Applications,** which was sponsored jointly by Long Island Jewish Medical Center, North Shore–Long Island Jewish Health System, the NIH, and the New York Academy of Sciences and held on October 21–24, 1998 in Tampa, Florida.

# Part VI. Clinical Studies—Cancer

# Part VII. Clinical Studies—Dental

# Part VIII. Angiogenesis and Related Phenomena

**Financial assistance was received from:**

*Joint Sponsors*
• LONG ISLAND JEWISH MEDICAL CENTER, NORTH SHORE–LONG ISLAND JEWISH HEALTH SYSTEM
• NATIONAL INSTITUTES OF HEALTH
    • NATIONAL INSTITUTE OF DENTAL RESEARCH
    • NATIONAL CANCER INSTITUTE
    • NATIONAL INSTITUTE OF ARTHRITIS AND MUSCULOSKELETAL AND SKIN DISEASES

*Major Funder*
• COLLAGENEX PHARMACEUTICALS, INC.

*Supporter*
• AMERSHAM PHARMACIA BIOTECH

*Contributors*
• BLOCK DRUG COMPANY, INC.
• BRISTOL-MEYERS SQUIBB PHARMACEUTICAL RESEARCH INSTITUTE
• GLAXO WELLCOME, INC
• NOVARTIS
• PARKE-DAVIS
• TNO PHARMA, PHARMACEUTICAL RESEARCH DIVISION
• WARNER LAMBERT, INC.

# Preface

ROBERT A. GREENWALD

*Division of Rheumatology, Long Island Jewish Medical Center,*
*New Hyde Park, New York 11040, USA*

From October 21 to October 25, 1998, 275 scientists from 18 countries, representing over 30 scientific disciplines, gathered in Tampa, Florida for a meeting known colloquially as Tampa-II. The idea for the Tampa MMP inhibitor meetings had been spawned in 1991 as a local gathering of perhaps a half-dozen persons who might want to share data on the use of certain compounds for inhibition of pathologically excessive collagenase activity, as seen in rheumatoid arthritis, periodontal disease, etc. With the help of the New York Academy of Sciences, the "local gathering" became an international conference, which was held in Tampa in January 1994. That meeting, subtitled Therapeutic Potential, explored ways in which inhibitors of matrix metalloproteinases (MMPIs) might be medically useful in arthritis, cancer, periodontal disease, vascular diseases, eye and skin conditions, and osteoporosis.

By 1998, with clinical applications more clearly defined and with human trials well under way, the subtitle was updated to Therapeutic Applications. The timing of Tampa-II was prophetic: just three weeks earlier, the United States Food and Drug Administration approved low-dose doxycycline (Periostat, CollaGenex Pharmaceuticals, Inc.) as adjunctive treatment for periodontal disease. This is the first agent to be marketed with FDA-approved labeling as an inhibitor of collagenase. The 1998 audience, which included virtually all of the major names in the field, comprised academic and industrial researchers in roughly equal numbers. Over 100 posters were presented.

For Tampa-II, my cochairman, Larry Golub, from the Department of Oral Biology & Pathology at the State University of New York at Stony Brook, was joined by Dr. Stanley Zucker, an internationally known oncologist with wide connections within the MMP field.

Two themes from Tampa-I carried over to Tampa-II. First, it still has not been decided whether a specific inhibitor of just one or two MMPs (e.g., MMP-13) would be better than a broad-spectrum agent as treatment for a given disease. Some workers believe that the musculoskeletal toxicity of drugs such as batimistat may relate to their broad-spectrum action—e.g., inhibition of fibroblast collagenase. Nevertheless, the papers presented at the 1998 meeting were much less concentrated on single-enzyme inhibition than they had been in 1994. Second, everyone recognizes that inhibition is assay dependent, and there is no substitute for animal-model and eventually human testing.

Several sessions expanded the scope of the field into new areas. One paper reported that MMPs were involved in the antigen presentation required for immune reactivity, extending the MMPs into immunology. Three excellent papers dealt with the role of MMPs in acute coronary plaque rupture and in aneurysm expansion. Finally, the ADAMs family made its way into the proceedings because of their relationship to TACE. If there is another such meeting (*vide infra*), many new MMPs and areas of human or animal importance can be expected.

## ACKNOWLEDGMENTS

Like Tampa-I, Tampa-II was cosponsored by the New York Academy of Sciences, the NIH, and Long Island Jewish Medical Center.

My cochairmen, Larry Golub and Stan Zucker, were irreplaceably important in selecting and inviting the speakers and in fund raising. They labored long and hard over many a broiled salmon dinner going over draft after draft of the program. Sherryl Usmani from the Science & Technology Meetings Department of the New York Academy of Sciences was absolutely stupendous: she took care of thousands of details, never missed a beat, and remembered all the special-request phone calls from six months before. Rashid Shaikh and Sue Davies, also of the Academy's Science & Technology Meetings Department, provided important support as well. We are also grateful to Richard Stiefel of the Academy's editorial department and to John W. Kennedy (K-M Research) for so efficiently seeing this volume through the press.

Long Island Jewish Medical Center, my home institution, and the National Institutes of Health (dental, arthritis, cancer) provided critical financial support, as did a variety of pharmaceutical firms as listed separately in this volume.

## THE NEXT MMP MEETING

Before Tampa-II ended, at least 20 people had asked me when the next meeting would be held. Since this is an ad hoc event, there being no society behind this topic, an exact answer is not possible as of this writing (April, 1999). The Gordon Research conferences on MMPs meet every other summer in odd-numbered years; although the GRCs do not emphasize inhibition, the topic always comes up. The GRCs, however, are limited to 135 participants; over half of the Tampa-II audience would have to be turned away. It is clear that interest is high and that there are new areas to explore (e.g., immunology, cardiology). If interest is sustained and industry support remains available, I may try to convince the New York Academy of Sciences to try it again in 2002. One thing, however, is certain: there will be no Tampa-III. As of this writing, pending years of site research yet to be done, Tampa-II will be followed by Bermuda-I (or equivalent). And perhaps the subtitle for the next meeting will be Inhibition of Matrix Metalloproteinases: Therapeutic Triumph!

# Common Names of Matrix Metalloproteinases[a]

COMPILED BY ROBERT A. GREENWALD AND J. FRED WOESSNER, JR.

| MMP No. | Common Name | MMP No. | Common Name |
|---------|-------------|---------|-------------|
| 1 | Collagenase 1<br>Fibroblast collagenase<br>Interstitial collagenase | 10<br>11 | Stromelysin 2<br>Stromelysin 3 |
| 2 | Gelatinase A<br>72-kDa Gelatinase | 12 | Macrophage elastase |
| 3 | Stromelysin 1 | 13 | Collagenase 3<br>Rat osteoblast collagenase |
| 4 | Not used | 14 | MT1-MMP |
| 5 | Not used | 15 | MT2-MMP |
| 6 | Not used | 16 | MT3-MMP |
| 7 | Matrilysin | 17 | MT4-MMP |
| 8 | Collagenase 2<br>Neutrophil collagenase | 18<br>19 | Collagenase 4<br>No trivial name |
| 9 | Gelatinase B<br>92-kDa Gelatinase | 20 | Enamelysin |

[a]As of February 1999.

# Engineering of Selective TIMPs

HIDEAKI NAGASE,[a,d] QI MENG,[b] VLADIMIR MALINOVSKII,[b] WEN HUANG,[b]
LINDA CHUNG,[a] WOLFRAM BODE,[c] KLAUS MASKOS,[c] AND KEITH BREW[b]

[a]*Department of Biochemistry and Molecular Biology,
University of Kansas Medical Center, Kansas City, Kansas 66160, USA*

[b]*Department of Biochemistry and Molecular Biology,
University of Miami School of Medicine, P.O. Box 016129, Miami, Florida 33101, USA*

[c]*Max-Planck-Institut für Biochemie, Abteilung für Strukturforschung,
Martinsried, Germany*

ABSTRACT: Differences in proteinase susceptibility between free TIMP-1 and the TIMP-1–MMP-3 complex and mutagenesis studies suggested that the residues around the disulfide bond between Cys1 and Cys70 in TIMP-1 may interact with MMPs. The crystal structure of the complex between TIMP-1 and the catalytic domain of MMP-3 has revealed that the $\alpha$-amino group of Cys1 bidentately chelates the catalytic zinc of MMP-3 and the Thr2 side chain occupies the $S_1{}'$ pocket. Generation of the N-terminal domain of TIMP-1 (N-TIMP-1) variants with 15 different amino acid substitutions for Thr2 has indicated that the nature of the side chain of residue 2 has a major effect on the affinity of N-TIMP-1 for three different MMPs (MMPs-1, -2 and -3). The results also demonstrate that the mode of binding of N-TIMP-1 residue 2 differs from the binding of the $P_1{}'$ residue of a peptide substrate.

Matrix metalloproteinases (MMPs), also called matrixins, play major roles in extracellular matrix turnover during biological processes such as morphogenesis, development, wound healing, and tissue resorption.[1,2] The expression of most matrixins is transcriptionally regulated by growth factors, hormones, inflammatory cytokines, cell–matrix interactions, and cellular transformation.[1,2] Matrixins function primarily on the cell surface or in the extracellular space, and their activities are controlled through zymogen activation and inhibition by endogenous protein inhibitors, $\alpha_2$-macroglobulin, and tissue inhibitors of metalloproteinases (TIMPs). $\alpha_2$-Macroglobulin and related inhibitors, being general proteinase inhibitors in plasma, regulate matrixins primarily in the fluid phase, whereas TIMPs are key inhibitors of matrixins in the tissue. Currently four TIMPs (TIMPs -1 to -4) have been identified in vertebrates, and their expression in tissues and different cell types are also controlled to maintain balanced extracellular matrix catabolism during tissue remodeling under physiological conditions.[3,4] Disruption of this balance may result in diseases associated with uncontrolled proteolysis of connective tissue matrices such as arthritis, tumor cell invasion and metastasis, atherosclerosis, and fibrosis.[1,2,4]

[d]Address for correspondence: H. Nagase, Department of Biochemistry and Molecular Biology, University of Kansas Medical Center, 3901 Rainbow Blvd., Kansas City, KS 66160 USA. Phone, 913/588-7079; fax, 913/588-7111; e-mail, hnagase@kumc.edu

1

The four TIMPs share about 40% identity in sequence including 12 conserved cysteines and they consist of N-terminal and C-terminal domains, each stabilized by three disulfide bonds.[5] Human TIMP-1 is a 29–30-kDa glycoprotein of 184 amino acids, whereas human TIMP-2 is an unglycosylated protein of 194 amino acids. TIMP-3 was first found as a 21-kDa protein produced by Rous sarcoma virus–transfected chick embryonic fibroblasts.[6] While chicken TIMP-3 is not glycosylated, recombinant human and mouse TIMP-3s, expressed in mammalian cells, are mainly glycoproteins.[7] TIMP-3 has unique properties: it tightly binds to the extracellular matrix[6]; mutations in the C-terminal domain of TIMP-3 are responsible for Sorsby's fundus dystrophy, an autosomal dominant macular disorder[8]; and overexpression of TIMP-3 in smooth muscle cell enhances apoptosis.[9,10] TIMP-4 was first identified by cDNA cloning from a human heart cDNA library.[11] All four TIMPs form tight 1:1 stoichiometric complexes with MMPs, and they show little selectivity against different types of matrixins except that TIMP-1 is an ineffective inhibitor of membrane-type MMPs (MT1-MMP and MT2-MMP).[12] Studies with the truncated amino-terminal domains of TIMP-1 (N-TIMP-1) and TIMP-2 (N-TIMP-2) showed that MMP inhibitory activity is located in those domains.[13–15] TIMP-1 binds to the precursor of MMP-9 (proMMP-9), and TIMP-2 and TIMP-4 bind to proMMP-2, but their interactions are through the C-terminal domains of the inhibitor and the proenzyme.[16,17] Thus, TIMPs bound to progelatinases can inhibit other MMPs, and the activation of progelatinase in the complex requires blocking of the free N-terminal domain of TIMP by MMPs, otherwise activated gelatinase is readily inhibited by the cognate TIMP.[18,19] The specific interaction of proMMP-2 and TIMP-2 through their C-terminal domains is important for the cell surface activation process of proMMP-2 by MT1-MMP.[20] Recent studies by Butler *et al.*[21] suggest that a TIMP-2-MT1-MMP complex on the cell surface acts as a receptor for proMMP-2 and that the bound proMMP-2 is activated by the action of a second MT1-MMP molecule. Once activated on the cell surface, MMP-2 is inhibited by TIMP-2 bound to the membrane.[22] The latter mode of interaction between TIMP-2 and the membrane is not inhibited by a synthetic hydroxamate inhibitor of MMPs, whereas the interaction of TIMP-2 and MT1-MMP is sensitive to a hydroxamate inhibitor.[22] These observations indicate that TIMP-2 plays a critical role in controlling the pericellular activity of MMP-2.

## IDENTIFICATION OF THE MMP INTERACTION SITE IN TIMP-1

The inhibitory site of TIMP-1 was first investigated by differential proteolytic susceptibility of free TIMP-1 and the TIMP-1-MMP complex and by mutagenesis of N-TIMP-1. Okada *et al.*[23] reported that trypsin, chymotrypsin, and human neutrophil elastase (HNE) destroy the inhibitory activity of TIMP-1. While trypsin and chymotrypsin digest TIMP-1 into several fragments, HNE initially cleaves the inhibitor into 16-kDa and 20-kDa fragments, which are then degraded into smaller pieces. The initial cleavage of TIMP-1 by HNE correlated with the inactivation of the inhibitor. We, therefore, postulated that the site cleaved by HNE is critical for the association of TIMP-1 and MMPs or for the integrity of the inhibitor structure. We also postulated that if this site is directly involved in binding to MMPs, the cleavage of TIMP-1 by HNE should be prevented when TIMP-1 is bound to MMPs.

The protection of TIMP-1 from HNE cleavage was demonstrated by forming a complex between TIMP-1 and either the 28-kDa or 45-kDa form of MMP-3.[24] Full TIMP-1 activity was recovered from the HNE-treated TIMP-1-MMP-3 complex after dissociation of the complex at pH 3.0 in the presence of 25 mM EDTA and dialysis against a neutral buffer. HNE was shown to hydrolyze the Val69-Cys70 bond of free TIMP-1. These results suggest that the region around Val69-Cys70 bond is close to the site involved in MMP binding. Specific inactivation of TIMP-1 by HNE also takes place in a complex of proMMP-9 and TIMP-1, and this action of HNE is considered to be important in activation of proMMP-9 in the complex.[25] While free proMMP-9 is readily activated by a catalytic amount of MMP-3, when it is bound to TIMP-1, MMP-3 is inhibited. For the activation of MMP-9 it is therefore necessary to either saturate the reactive site of TIMP-1 with MMPs or inactivate TIMP-1. Specific inactivation of TIMP-1 by HNE renders proMMP-9 in the complex to be readily activatable, and this action of HNE is probably significant in tissue damage at sites of inflammation.

Informed by the results of proteinase protection experiments and chemical modification studies, a series of mutagenesis studies were conducted to investigate the role of individual residues in TIMP inhibition. N-TIMP-1 and its mutants were expressed as inclusion bodies in *E. coli* using the pET3a expression vector and folded to produce active native protein.[14] Twenty recombinant N-TIMP-1 mutants were expressed and characterized for their inhibitory activities against MMP-1, MMP-2, and MMP-3.[26] The mutation of Thr2 to Ala (T2A) resulted in more than 100-fold decrease in affinity for the catalytic domain of MMP-3 [MMP-3($\Delta$C)]. This mutant exhibited significant discriminatory inhibition activity against three MMPs. The inhibition constants, $K_i$ values, of T2A against MMP-1, MMP-2 and MMP-3 were $2090 \pm 180$ nM, $308 \pm 17$ nM, and $126 \pm 4$ nM, respectively, whereas those for wild-type N-TIMP-1 were $1.5 \pm 0.4$ nM, $1.1 \pm 0.1$ nM, and $1.9 \pm 0.1$ nM.[26] Mutation of either Cys1 or Cys70, which are disulfide-bonded in native TIMP-1, decreases the affinity for MMP-3($\Delta$C) by more than three orders of magnitude. Other mutants that exhibited more than 10-fold decreases in affinity for MMP-3($\Delta$C) were M66A, V69I, V69T, and the double mutant, V69A/A103V. A Y38A mutant had the $K_i$ value of 470 nM against MMP-3($\Delta$C), but near- and far-UV CD spectra of this mutant indicated losses in fixed tertiary structure. Mutations at other positions had relatively low effects on activity.

Taken together, these studies suggested that the residues around the disulfide bond between Cys1 and Cys70 in TIMP-1 interact with MMPs. This region was noted to form an exposed ridge on the surface of a low-resolution NMR structure of N-TIMP-2.[27]

## INHIBITION MECHANISMS OF MMPs BY TIMPs

The crystal structure of the TIMP-1-MMP-3($\Delta$C) complex, determined by Gomis-Rüth *et al.*,[28] revealed the principle interactions of the two molecules and the mechanism by which TIMP inhibits MMPs. Six sequentially separate polypeptide segments of TIMP-1 interact with MMP-3($\Delta$C) (four are in the N-terminal domain and two in the C-terminal domain), and solvent-accessible surface areas of about

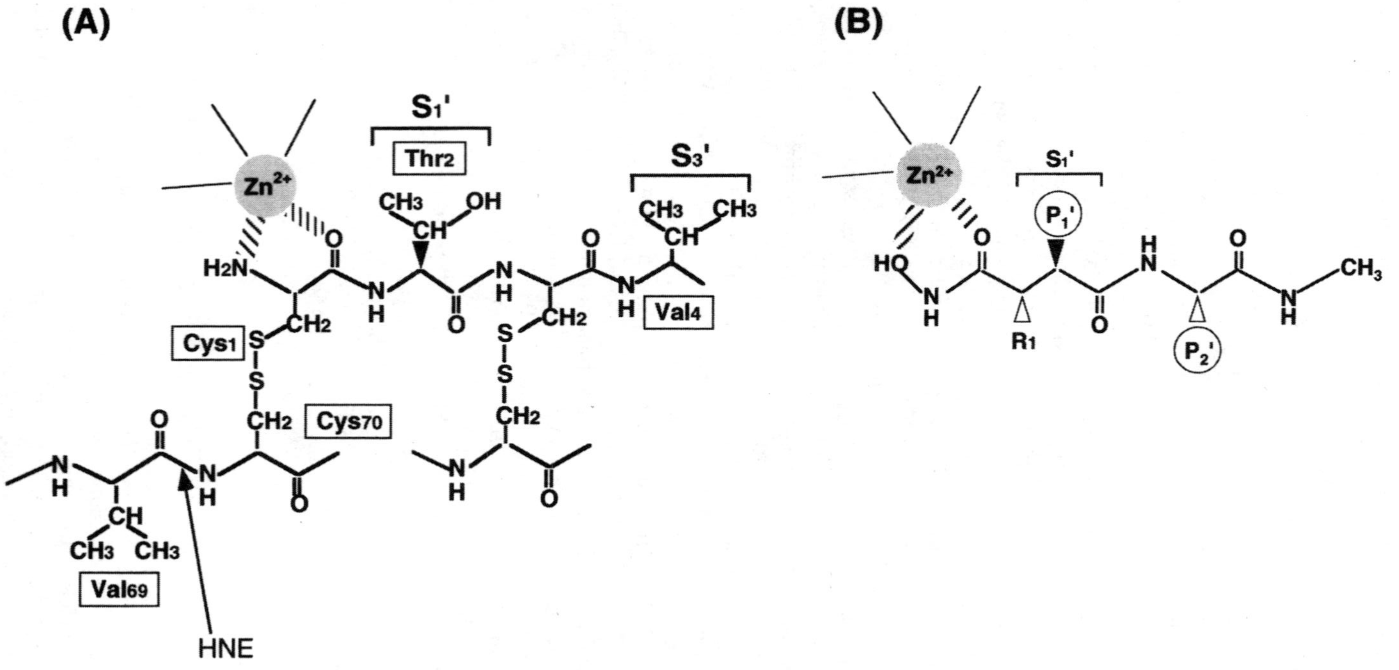

**FIGURE 1.** Schematic representation of the NH$_2$-terminal regions of TIMP-1 (**A**) and a peptidyl hydroxamate inhibitor (**B**). The scheme of the TIMP-1 is based on the crystal structure of the TIMP-1-MMP-3(ΔC) complex.[28] The peptide bond cleaved by human neutrophil elastase (HNE) is indicated.

1,300 $\text{Å}^2$ of each molecule are buried on formation of the complex. Three-quarters of all intermolecular contacts are located around the disulfide bond between Cys1 and Cys70. The N-terminal Cys1-Val4 and Ala65-Cys70 occupies the active site of MMP-3. The N-terminal Cys1 sits on top of the catalytic $Zn^{2+}$ of the metalloproteinase and coordinates the metal ion through its $\alpha$-amino group and peptide carboxyl group (FIG. 1A). The Cys1-Val4 segment binds to subsites $S_1$ to $S_3{}'$ in a manner similar to a peptide substrate ($P_1$ to $P_3{}'$), and the side chain of Thr2 extends into the large $S_1{}'$ specificity pocket of MMP-3. The $\alpha$-amino group of Cys1 and the Thr2 side chain partially superimpose with the hydroxamic acid moiety and the $P_1{}'$ side chain of synthetic MMP inhibitors, respectively (FIG. 1B). Ser68 and Val69 occupy the part of the active site ($S_2$ and $S_3$ subsites, respectively), but they are arranged in a nearly opposite orientation to the $P_3$-$P_2$ segment of a bound peptide substrate.

## INFLUENCE OF THE SIDE CHAIN OF POSITION 2 ON MMP INHIBITION

The $S_1{}'$ specificity pockets of MMPs differ in size. Residue 2 in currently known TIMPs is Thr or Ser. Thus, TIMP variants with larger side chains at position 2 may be more selective for different MMPs. We therefore investigated the effect of the side chain of this position in N-TIMP-1 by substituting it with fifteen different amino acids.

N-TIMP-1 and its variants were expressed in *E. coli* and they were refolded from the inclusion bodies and purified as described previously.[14] The His2 mutant did not fold *in vitro*, but other mutants were compatible with native folding, as indicated by their near-UV CD spectra. In some mutants a misfolded conformer was also generated by *in vitro* folding, but this form was separated from the native protein by cation exchange chromatography. The $K_i$ values of N-TIMP-1 variants were measured against MMP-1, MMP-2, and MMP-3 (FIG. 2). Inhibition studies with higher-affinity mutants were analyzed using a treatment for tight-binding inhibitors, according to Morrison and Walsh.[29]

Among fourteen variants of N-TIMP-1 at position 2, the Gly mutant was the weakest inhibitor for all three MMPs. This mutant was available in limited amount and as the highest concentration used for inhibition studies (8 μM) inhibited MMP-1 and MMP-2 only 31% and 7.2%, respectively, implying that the $K_i$ value must be much higher than this concentration. The affinity for MMP-3 was slightly higher ($K_i = 1.4$ μM). These results indicate that the removal of the side chain of position 2 results in practically inactive inhibitor. The reduced affinity of the Ala2 mutant for three MMPs in the previous mutagenesis studies is consistent with this concept.

Negatively charged side chains at position 2 (Asp2 and Glu2 mutants) are unfavorable for all three MMPs as compared with the Asn2 and Gln2, which exhibited significantly higher affinities towards three MMPs than Asp2 and Glu2 mutants. The Arg2 mutant showed most striking discrimination between MMP-1 and MMP-2 with $K_i$ values of 5 μM and 12 nM, respectively. The affinity for MMP-3 was similar to that for MMP-2. The reasonably high affinity of the Arg2 mutant and MMP-2 with MMP-3 was surprising since peptide substrates with Arg at the $P_1{}'$ position are extremely poor substrates for these enzymes (TABLE 1). The unfavorable interaction of

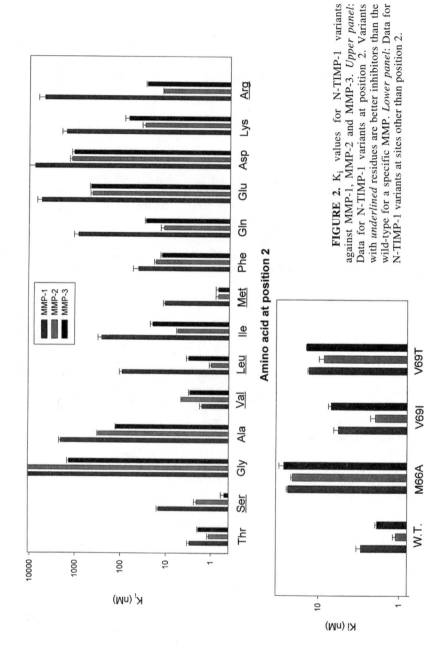

**FIGURE 2.** $K_i$ values for N-TIMP-1 variants against MMP-1, MMP-2 and MMP-3. *Upper panel:* Data for N-TIMP-1 variants at position 2. Variants with *underlined* residues are better inhibitors than the wild-type for a specific MMP. *Lower panel:* Data for N-TIMP-1 variants at sites other than position 2.

the Arg2 mutant and MMP-1 may reflect a structural feature of the $S_1{}'$ pocket of MMP-1, where Arg195, which replaces Leu in MMP-2 and MMP-3, projects into the pocket and creates a less deep and more cationic environment compared with those of MMP-2 and MMP-3. Nonpolar side chains of increasing size increase the affinity for MMP-2 and MMP-3, except Ile and Phe, whereas longer aliphatic chains at position 2 reduce the affinity for MMP-1. This also contrasts with the substrate specificity of MMP-1. The best inhibitor for MMP-1 was the Val2 variant and the best inhibitors for MMP-2 and MMP-3 are the Met2 and Ser2 variants, respectively.

One of the striking features of reside 2 in N-TIMP-1 is that mutation at this site significantly alters the affinity for different MMPs. Log $K_i$ values of the position 2 mutants for the three MMPs show large differences, particularly for MMP-1 vs. MMP-2 and MMP-1 vs. MMP-3 (FIG. 2A). In comparison for MMP-2 vs. MMP-3, Ser, Gly, Ala, and Leu deviate notably. By contrast, mutations at sites that are known to be involved in interaction, other than position 2, with MMP-3 (i.e., M66A, Val69I, V69T) only had small effects on relative affinity for the three MMPs (FIG. 2B).

Although residue 2 of TIMP-1 binds to the $S_1{}'$ specificity pocket, the interaction of this side chain appears to differ from that of the $P_1{}'$ residue of a peptide substrate. As summarized in TABLE I, MMPs favor substrates with large aliphatic or automatic side chains at the $P_1{}'$ subsite. Substrates with Val, Ser or a charged group at the $P_1{}'$ position are very poor substrates for MMPs. Plots of $-\log K_i$ for N-TIMP-1 mutants and $\log(k_{cat}/K_m)$ for octapeptide substrates with substitutions at the $P_1{}'$ site show very poor correlations. As illustrated in FIGURE 3, the correlation for MMP-3 was negative with $r^2$ of 0.08. This discrepancy may be due to a greater loss of conformational entropy associated with peptide substrate–MMP interaction than TIMP-1-MMP interaction. The orientation of residue 2 of TIMP-1 may also be influenced by the relatively rigid structure around the disulfide-bonded Cys1-Cys70 region and by the interaction of Cys1 with the active site $Zn^{2+}$.

**TABLE 1. Relative sequence specificities of matrixins[a]**

| | Relative rate of hydrolysis | | |
| --- | --- | --- | --- |
| | MMP-1 | MMP-2 | MMP-3 |
| P4 -P3 -P2- P1 ~ P1´-P2´- P3´- P4´ | | | |
| Gly-Pro-Gln-Gly~ **Ile**-Ala-Gly-Gln | 100 | 100 | 100 |
| Gly-Pro-Gln-Gly~ **Leu**-Ala-Gly-Gln | 130 | 88 | 110 |
| Gly-Pro-Gln-Gly~**Val**-Ala-Gly-Gln | 9.1 | 30 | 53 |
| Gly-Pro-Gln-Gly~**Ser**-Ala-Gly-Gln | 5.9 | 15 | 45 |
| Gly-Pro-Gln-Gly~**Phe**-Ala-Gly-Gln | 20 | 55 | 140 |
| Gly-Pro-Gln-Gly~**Met**-Ala-Gly-Gln | 110 | 230 | 60 |
| Gly-Pro-Gln-Gly~**Gln**-Ala-Gly-Gln | 28 | 34 | 38 |
| Gly-Pro-Gln-Gly~**Glu**-Ala-Gly-Gln | < 0.5 | < 0.5 | < 0.002 |
| Gly-Pro-Gln-Gly~**Arg**-Ala-Gly-Gln | < 0.5 | < 0.5 | < 4.9 |

[a]Data for MMP-1, MMP-2 and MMP-3 are from Refs. 30, 31, and 32, respectively.

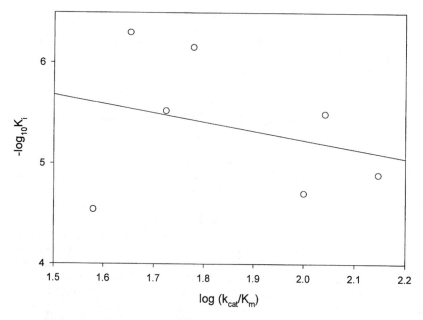

**FIGURE** 3. Effects of the residues that interact with the $S_1{}'$ pocket of MMP-3 on N-TIMP-1 affinity ($-\log K_i$) and efficiency of substrates $\log(k_{cat}/K_m)$. Data are from Leu, Phe, Met, Val, Gln, Ser, and Ile substituted for Thr2 in N-TIMP-1 and for X in the substrate Gly-Pro-Gln-Gly~X-Ala-Gly-Gln.[30–32]

## CONCLUSIONS

Our studies have demonstrated that the specificity of TIMPs for matrixins can be altered by site-directed mutagenesis to generate variants that are more selective towards certain MMPs or a subgroup of MMPs. The side chain of Thr2 in TIMP-1 extends into the large $S_1{}'$ specificity pocket of matrixins, but the mode of interaction of this side chain significantly differs from that of $P_1{}'$ side chain of a peptide substrate. The nature of the side chain of residue 2 in TIMPs thus also has a major influence on the affinity for MMPs. The removal of this side chain results in an essentially inactive inhibitor. The crystal structure of the TIMP-1-MMP-3($\Delta$C) complex indicates that six sequentially separated polypeptide segments of TIMP-1 interact with the enzyme.[28] Thus, combinations of mutations have the potential to generate TIMPs that are highly selective inhibitors of individual MMPs. Such mutants may be utilized in conjunction with gene therapy technologies for tissue- or cell-targeted inhibition of specific MMPs to intervene in disease processes resulting from unbalanced breakdown of the extracellular matrix.

## ACKNOWLEDGMENTS

This work was supported by NIH Grant AR40994.

## REFERENCES

1. NAGASE, H. 1996. Matrix metalloproteinases. *In* Zinc Metalloproteinases in Health and Disease. N.M. Hooper, Ed.: 153–204, Taylor & Francis Ltd., London.
2. PARKS, W.C & R.P. MECHAM. 1998. Matrix metalloproteinases. Academic Press. San Diego, CA.
3. DOUGLAS, D.A., Y.E. SHI & G.A. SANG. 1997. Computational sequence analysis of the tissue inhibitor of metalloproteinase family. J. Protein Chem. **16:** 237–255.
4. GOMEZ, D.E., D.F. ALONSO, H. YOSHIJI & U.P. THORGEIRSSON. 1997. Tissue inhibitors of metalloproteinases: structure, regulation and biological functions. Eur. J. Cell Biol. **74:** 111–122.
5. WILLIAMSON, R.A., F.A.O. MARSTON, S. ANGAL, P. KOKLITIS, M. PANICO, H.R. MORRIS, A.F. CARNE, B.J. SMITH, T.J.R. HARRIS & R.B. FREEDMAN. 1990. Disulphide bond assignment in human tissue inhibitor of metalloproteinases (TIMP). Biochem. J. **268:** 267–274.
6. PAVLOFF, N., P.W. STASKUS, N.S. KISHNANI & S.P. HAWKES. 1992. A new inhibitor of metalloproteinases from chicken: ChIMP-3. A third member of the TIMP family. J. Biol. Chem. **267:** 17321–17326.
7. APTE, S.S., B.R. OLSEN & G. MURPHY. 1995. The gene structure of tissue inhibitor of metalloproteinases (TIMP)-3 and its inhibitory activities define the distinct TIMP gene family. J. Biol. Chem. **270:** 14313–14318.
8. FELBOR, U., C. BEUKWITZ, M.L. KLEIN, J. GREENBERG, C.Y. GREGORY & B.H.F. WEBER. 1997. Sorsby fundus dystrophy. Reevaluation of variable expressivity in patients carrying a TIMP-3 founder mutation. Arch. Ophthalmol. **115:** 1569–1571.
9. AHONEN, M., A.H. BAKER & V.-M. KÄHÄRI. 1998. Adenovirus-mediated gene delivery of tissue inhibitor of metalloproteinases-3 inhibits invasion and induces apoptosis in melanoma cells. Cancer Res. **58:** 2310–2315.
10. BAKER, A.H., A.B. ZALTSMAN, S.J. GEORGE & A.C. NEWBY. 1998. Divergent effects of tissue inhibitor of metalloproteinase-1, -2, or -3 overexpression on rat vascular smooth muscle cell invasion, proliferation, and death in vitro. TIMP-3 promotes apoptosis. J. Clin. Invest. **101:** 1478–1487.
11. GREENE, J., M. WANG, Y.E. LIU, L.A. RAYMOND, C. ROSEN & Y.E. SHI. 1996. Molecular cloning and characterization of human tissue inhibitor of metalloproteinase 4. J. Biol. Chem. **271:** 30375–30380.
12. WILL, H., S.J. ATKINSON, G.S. BUTLER, B. SMITH & G. MURPHY. 1996. The soluble catalytic domain of membrane type 1 matrix metalloproteinase cleaves the propeptide of progelatinase A and initiates autoproteolytic activation. Regulation by TIMP-2 and TIMP-3. J. Biol. Chem. **271:** 17119–17123.
13. MURPHY, G., A. HOUBRECHTS, M.I. COCKETT, R.A.WILLIAMSON, M. O'SHEA & A.J.P. DOCHERTY. 1991. The N-terminal domain of human tissue inhibitor of metalloproteinases 1 retains metalloproteinase inhibitory activity. Biochemistry **30:** 8097–8102.
14. HUANG, W., K. SUZUKI, H. NAGASE, S. ARUMUGAM, S.R. VAN DOREN & K. BREW. 1996. Folding and characterization of the amino-terminal domain of human tissue inhibitor of metalloproteinases-1 (TIMP-1) expressed at high yield in E. coli. FEBS Lett. **384:** 155–161.
15. WILLIAMSON, R.A., D. NATALIA, C.K. GEE, G. MURPHY, M.D. CARR & R.B. FREEDMAN. 1996. Chemically and conformationally authentic active domain of human tissue inhibitor of metalloproteinases-2 refolded from bacterial inclusion bodies. Eur. J. Biochem. **241:** 476–483.
16. MURPHY, G. & F. WILLENBROCK. 1995. Tissue inhibitors of matrix metalloproteinases. Methods Enzymol. **248:** 496–510.

17. BIGG, H.F., Y.E. SHI, Y.E. LIU, B. STEFFENSEN & C.M. OVERALL. 1997. Specific, high affinity binding of tissue inhibitor of metalloproteinases-4 (TIMP-4) to the COOH-terminal hemopexin-like domain of human gelatinase A. TIMP-4 binds progelatinase A and the COOH-terminal domain in a similar manner to TIMP-2. J. Biol. Chem. **272:** 15496–15500.

18. ITOH, Y., S. BINNER & H. NAGASE. 1995. Steps involved in activation of the complex of pro-matrix metalloproteinase 2 (Progelatinase A) and tissue inhibitor of metallo-proteinases (TIMP)-2 by 4-aminophenylmercuric acetate. Biochem. J. **308:** 645–651.

19. OGATA, Y., Y. ITOH & H. NAGASE. 1995. Steps involved in activation of the pro-matrix metalloproteinase 9 (Progelatinase B)-tissue inhibitor of metalloproteinases-1 complex by 4-aminophenylmercuric acetate and proteinases. J. Biol. Chem. **270:** 18506–18511.

20. STROGIN, A.Y., I. COLLIER, G. BANNIKOV, B.L. MARMER, G.A. GRANT & G.I. GOLDBERG. 1995. Mechanism of cell surface activation of 72-kDa type IV colla-genase. Isolation of the activated form of the membrane metalloproteinase. J. Biol. Chem. **270:** 5331–5338.

21. BUTLER, G.S., H. WILL, S.J. ATKINSON & G. MURPHY. 1997. Membrane-type-2 matrix metalloproteinase can initiate the processing of progelatinase A and is regulated by the tissue inhibitors of metalloproteinases. Eur. J. Biochem. **244:** 653–657.

22. ITOH, Y., A. ITO, K. IWATA, K. TANZAWA, Y. MORI & H. NAGASE. 1998. Plasma mem-brane-bound tissue inhibitor of metalloproteinases (TIMP)-2 specificity inhibits matrix metalloproteinase 2 (Gelatinase A) activated on the cell surface. J. Biol. Chem. **273:** 24360–24367.

23. OKADA, Y., S. WATANABE, I. NAKANISHI, J. KISHI, T. HAYAKAWA, W. WATOREK, J. TRAVIS & H. NAGASE. 1988. Inactivation of tissue inhibitor of metalloproteinases by neutrophil elastase and other serine proteinases. FEBS Lett. **229:** 157–160.

24. NAGASE, H., K. SUZUKI, T.E. CAWSTON & K. BREW. 1997. Involvement of a region near valine-69 of tissue inhibitor of metalloproteinases (TIMP) -1 in the interaction with matrix metalloproteinase 3 (stromelysin 1). Biochem. J. **325:** 163–167.

25. ITOH, Y. & NAGASE, H. 1995. Preferential inactivation of tissue inhibitor of metallo-proteinases-1 that is bound to the precursor of matrix metalloproteinase 9 (progelati-nase B) by human neutrophil elastase. J. Biol. Chem. **270:** 16518–16521.

26. HUANG, W., Q. MENG, K. SUZUKI, H. NAGASE & K. BREW. 1997. Mutational study of the amino-terminal domain of human tissue inhibitor of metalloproteinases 1 (TIMP-1) locates an inhibitory region for matrix metalloproteinases. J. Biol. Chem. **272:** 22086–22091.

27. WILLIAMSON, R.A., G. MARTORELL, M.D. CARR, G. MURPHY, A.J.P. DOCHERTY, R.B. FREEDMAN & J. FEENEY. 1994. Solution structure of the active domain of tissue inhibitor of metalloproteinases-2. A new member of the OB fold protein family. Bio-chemistry. **33:** 11745–11759.

28. GOMIS-RÜTH, F.-X., K. MASKOS, M. BETZ, A. BERGER, R. HUBER, K. SUZUKI, N. YOSHIDA, H. NAGASE, K. BREW, G.B. BOURENKOW, H. BARTUNIK & W. BODE. 1997. Mechanism of inhibition of the human metalloproteinase stromelysin-1 by TIMP-1. Nature **389:** 77–81.

29. MORRISON, J.F., & C.T. WALSH. 1988 The behavior and significance of slow-binding enzyme inhibitors. Adv. Enzymol. Relat. Areas Mol. Biol. **61:** 201–301.

30. NETZEL-ARNETT, S., G. FIELDS, H. BIRKEDAL-HANSEN & H.E. VAN WART. 1991. Sequence specificities of human fibroblast and neutrophil collagenases. J. Biol. Chem. **266:** 6747–6755.

31. NETZEL-ARNETT, S., Q.-X. SANG, W.G.I. MOORE, M. NAVRE, H. BERKEDAL-HANSEN & H.E. VAN WART. 1993. Comparative sequence specificities of human 72- and 92-kDa gelatinases (Type IV collagenases) and PUMP (Matrilysin). Biochemistry. **32:** 6427–6432.
32. FIELDS, G.B. 1988. The application of solid phase peptide synthesis to the study of structure-function relationships in the collagen/collagenase system. Ph.D. thesis, Florida State University, Tallahassee, Florida.

# Role of Matrix Metalloproteinases and Their Inhibition in Cutaneous Wound Healing and Allergic Contact Hypersensitivity

BRIAN K. PILCHER,[a] MIN WANG, XUE-JIN QIN, WILLIAM C. PARKS, ROBERT M. SENIOR, AND HOWARD G. WELGUS[b,c]

*Department of Medicine, Washington University School of Medicine, St. Louis, Missouri 63110, USA*

ABSTRACT: Normal wounds can heal by secondary intention (epidermal migration to cover a denuded surface) or by approximation of the wound edges (e.g., suturing). In healing by secondary intention, epidermis-derived MMPs are important. Keratinocyte migration begins within 3–6 hr post injury, as basal cells detach from underlying basal lamina and encounter a dermal substratum rich in type I collagen. Cell contact with type I collagen *in vitro* stimulates collagenase-1 expression, which is mediated by the $\alpha2\beta1$ integrin, the major keratinocyte collagen-binding receptor. Collagenase-1 activity alone is necessary and sufficient for keratinocyte migration over a collagen subsurface. Stromelysins-1 and -2 are also found in the epidermis of normal acute wounds. Stromelysin-2 co-localizes with collagenase-1 and may facilitate cell migration over non-collagenous matrices of the dermis. In contrast, stromelysin-1 is expressed by keratinocytes behind the migrating front and which remain on basal lamina, i.e., the proliferative cell population. Studies with stromelysin-1-deficient mice that suggest this MMP plays a role in keratinocyte detachment from underlying basement membrane to initiate cell migration. In chronic ulcers, MMP levels are markedly elevated, in contrast to their precise temporal and spatial expression in acute wounds. Both collagenase-1 and stromelysin-1 are found in fibroblasts underlying the nonhealing epithelium, and stromelysin-1 expression is especially prominent. Two key questions underlie the use of MMP inhibitors and wound healing: (1) will these agents impair normal reepithelialization in wounds that heal by secondary intention; and (2) can MMP inhibitors be effective therapy for chronic ulcers? The answer to neither is known. Batimastat and marimastat appear not to interfere with normal wound healing, but only in sutured surgical wounds, a situation in which MMP expression has practically no role. We also show the first example of an *in vivo* immune response, contact hypersensitivity, which is dependent upon MMP activity. Using gene-deficient mice, we demonstrate that stromylysin-1 (MMP-3) is required for sensitization, whereas gelatinase B (MMP-9) is required for timely resolution of the reaction to antigenic challenge.

[a]Present address: Department of Cell Biology and Neuroscience, UT Southwestern Medical Center, Dallas, Texas 75235-9039.

[b]Present address: Parke-Davis Pharmaceutical Research, 2800 Plymouth Road, Ann Arbor, Michigan 48105.

[c]Corresponding author. Phone, 734/622-5086; fax, 734/622-4333; e-mail, howard.welgus@wl.com

## INTRODUCTION

Matrix metalloproteinases (MMPs) are a gene family of at least 15 members which collectively can degrade all components of the extracellular matrix (ECM). Recently, it has become apparent that these enzymes can function in other proteolytic capacities, such as the degradation of non-matrix macromolecules (e.g., myelin basic protein[1]), α1-proteinase inhibitor,[2] release of latent growth factors from ECM, and cleavage of bioactive cell surface molecules. This manuscript will discuss two major functions of MMPs in the skin. The first has been considered a "traditional" function—wound healing. Our data have shown that MMPs have an important role in the normal healing of acute wounds as well as in the pathogenesis of chronic ulcers. The second is a novel function and the first *in vivo* evidence for a role of MMPs in the immune response—cutaneous contact hypersensitivity.

## WOUND HEALING

Prior to the early 1990s, many studies had shown that MMPs are present in wound fluid of acute wounds and at increased levels in wound fluid of chronic ulcers.[3–5] However, the cellular sources and precise tissue localization of these enzymes had not been determined. Knowledge of the cell types responsible for MMP expression and the localization of MMP protein during wound healing is critical to understanding enzyme function in repair. Such information provides clues as to which ECM macromolecules are being targeted and which cell populations are actively involved in tissue remodeling. To address these questions, we[6–9] and others[10–12] have used *in situ* hybridization and immunohistochemistry of human tissues. In general, since these enzymes are rapidly secreted from cells after their synthesis and are regulated pretranslationally,[13] *in situ* hybridization provides the most appropriate and sensitive

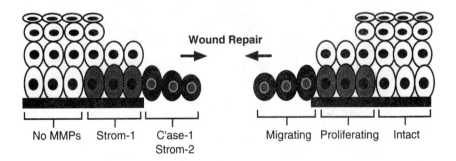

**FIGURE 1.** Schematic of MMP production in wound healing. Uninjured normal skin does not express MMPs detectable by *in situ* hybridization. After injury, keratinocytes migrate off basement membrane (*thick line* under keratinocytes) and onto dermal matrix. Migrating keratinocytes express collagenase-1 and stromelysin-2. Hyperproliferative or premigratory keratinocytes produce stromelysin-1.

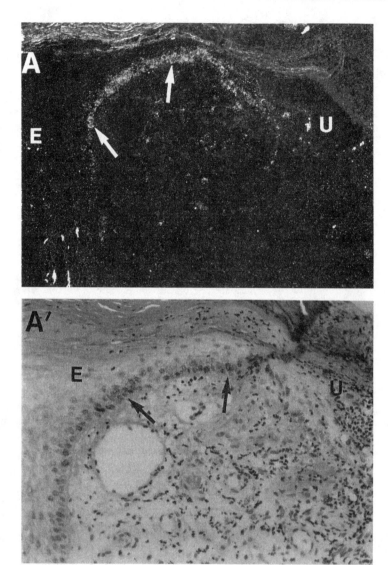

**FIGURE 2(A).** Localization of collagenase-1 mRNA to the migrating epithelial front. Sections of pyogenic granuloma with overt ulceration were hybridized with a [35]S-labeled antisense RNA probe for collagenase-1. Autoradiographic exposure was for 10 days, and sections were stained with hematoxylin and eosin. (A) Dark-field photomicrograph reveals strong autoradiographic signal (*arrows*) for collagenase mRNA in the epidermis (*E*) surrounding the ulceration (*U*), and this signal diminishes as the epidermis becomes more intact. Under dark-field illumination silver grains appear white and indicate the collagenase-1-producing cells. Original magnification ×200. (A´) Bright-field photomicrograph of the migrating epithelial tip shown in panel (**A**). Under bright-field illumination silver grains appear black and collagenase-expressing cells are seen to be limited to basal keratinocytes. Original magnification ×400.

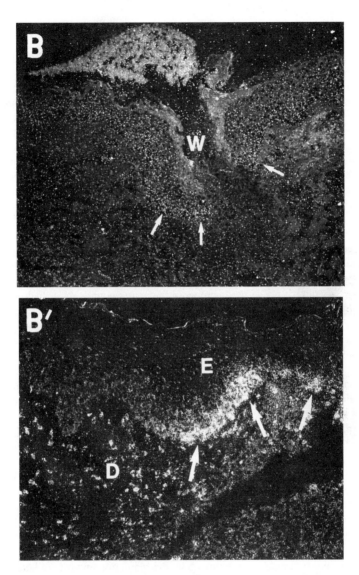

**FIGURE 2(B).** Collagenase-1 expression in a normally healing acute wound *versus* a chronic ulcer. **(B)** Dark-field photomicrograph of a *normally healing acute wound* hybridized as described above. This section of biopsy taken 3 days after wounding reveals signal for collagenase-1 mRNA (*arrows*) in basal keratinocytes of the epidermal front migrating below the incisional wound (W). Original magnification ×120. **(B´)** Dark-field photomicrograph of a *chronic nonspecific ulcer*. The ulcerated area (U) is indicated, and markedly positive collagenase-1-expressing basal keratinocytes are seen at the leading edge of migration (*arrows*). In this specimen, collagenase-1 mRNA is also expressed by underlying dermal fibroblasts. Original magnification ×200. Autoradiographic exposure for both **(B)** and **(B´)** is 10 days. (From Saarialho-Kere *et al.*[8] Reproduced by permission.)

technique to determine the site of expression of MMPs in the *in vivo* wound environment.

The most comprehensive report of MMP expression in normal wound healing comes from Saarialho-Kere *et al.*[14] These investigators analyzed 10-mm punch donor graft sites and found strong signal for collagenase-1 and stromelysin-2 in basal keratinocytes at the migrating epithelial front and stromelysin-1 in hyperproliferative basal keratinocytes lagging behind the migrating front. Cells expressing collagenase-1 and stromelysin-2 were migratory and contacted dermal matrix. Cells expressing stromelysin-1 were premigratory and rested on basement membrane. These were distinct keratinocyte populations without spatial overlap. These relationships are shown in the schematic of FIGURE 1. We analyzed keratome samples of human volunteers and pyogenic granulomas, and found an identical distribution of these enzymes (FIGS. 2A and 2B).[6] Stricklin *et al.*[11] have also observed collagenase-1 expression in migrating epithelial cells of acute burn wounds. In addition, they found considerable dermal fibroblast collagenase-1 production adjacent to sites of destroyed tissue in this more severe type of injury. Expression of 92-kDa gelatinase (gelatinase B, MMP-9) has been reported in epidermal cells of acute wounds in rodents,[15] but thus far not in human wounds.

When Saarialho-Kere *et al.*[14] compared the acute wounds described above to a variety of chronic ulcers, they found MMP levels to be greater in the ulcers, but the same enzymes were expressed by the same spatial populations of epidermal cells. In earlier studies, more than 100 chronic ulcers of diverse etiology were examined, nearly all exhibiting *prominent and temporally unrestricted*, but *spatially confined* epithelial expression of MMPs (FIG. 2B).[7,8] In addition, in chronic ulcers, collagenase-1, and especially stromelysin-1, were frequently produced by dermal fibroblasts (FIG. 2B), whereas MMP expression by these cells in simple acute healing wounds was sporadic.

MMPs have evolved to fulfill many normal biological functions, including ovulation, blastocyst implantation,[16] postuterine involution,[17] branching morphogenesis,[18] axonal growth,[19] and re-epithelialization after wounding.[20] Yet abundant data indicate that overproduction of MMPs is associated with tissue destruction in chronic inflammatory diseases such as rheumatoid and osteoarthritis,[21] tumor invasion and metastasis,[22,23] aneurysms,[24] atherosclerosis,[25] and skin ulcers.[26] Moreover, for each tissue involved, proteinases which are overexpressed in disease are generally those required for normal physiologic function, which is exactly the case in skin for normal wound healing and chronic ulcers.

We have considerable information about the regulation and role of keratinocyte-derived collagenase-1 in wound healing. Post wounding, keratinocytes rapidly migrate off basement membrane and onto dermal matrix, contacting type I collagen via their $\alpha 2\beta 1$ integrin receptors and triggering collagenase-1 transcription.[20] Collagenase-1 expression during re-epithelialization is sustained by an autocrine HB-EGF/ EGFR loop (FIG. 3)[27] and the enzyme's production is terminated by contact with laminin-1 of reforming basement membrane.[20] Collagenase is uniquely capable of cleaving the triple helix of native type I collagen, and such activity is *required* for movement of keratinocytes over a type I collagen substratum.[28] As shown in FIGURE 4, we have demonstrated a requirement for collagenase-1 activity for keratinocyte migration over type I collagen using three strategies: (1) inhibition with anti-

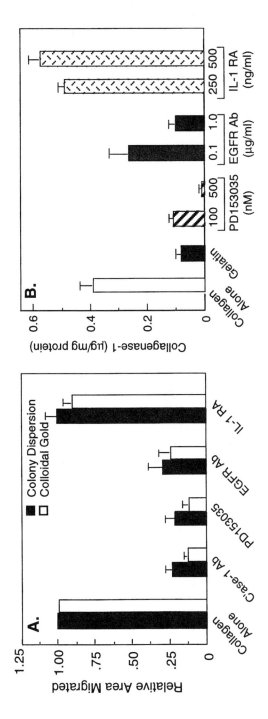

**FIGURE 3.** Inhibition of EGFR signaling markedly inhibits keratinocyte migration on type I collagen and collagen-mediated collagenase-1 production. (**A**) Primary human keratinocytes were subjected to colony dispersion and colloidal gold migration assays in the presence of collagen alone (1.0 mg/ml), affinity-purified collagenase-1 antisera (1:4 dilution), PD 153035 (500 nM, a specific EGFR kinase inhibitor), EGFR blocking antibody 528 (1.0 μg/ml), or IL-1 RA (500 ng/ml). The data presented are means ± SD of values from three separate wells per treatment group, and migration is expressed as arbitrary units relative to 0-hr controls. (**B**) Similar experiment to (**A**) with indicated concentrations of reagents, but with collagenase-1 content of conditioned media measured by ELISA. (From Pilcher *et al.*[27] Reproduced by permission.)

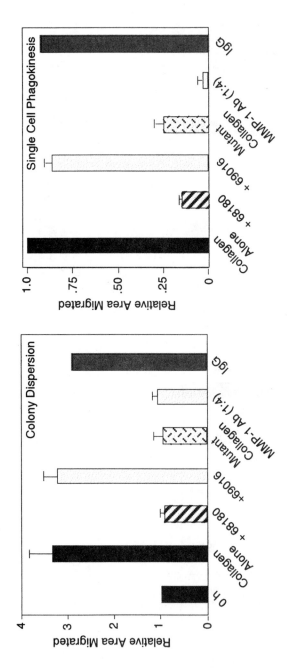

**FIGURE 4.** Keratinocyte migration on type I collagen is dependent upon collagenase-1 activity. Migration of primary human skin keratinocytes was determined in either colony dispersion (migration outward from a cloning cylinder) or single-cell phagokinesis (gold particle) assays over a type I collagen substratum. In the case of colony dispersion, migration was determined at 96 hr and compared to that at 0 hr (area of the cloning chamber). Migration was essentially abolished by the broad-spectrum MMP hydroxamate SC-68180 (10 μM), specific anti-collagenase-1 (MMP-1) antiserum, and when the substratum contained a type I collagen mutated at the collagenase cleavage site. Migration was normal in the presence of the collagenase-1-sparing hydroxamate inhibitor SC-69016 (10 μM) and nonimmune IgG. Essentially identical results were obtained with the single-cell phagokinesis assay. (From Pilcher et al.[28] Reproduced by permission.)

collagenase-1 antiserum; (2) inhibition using a collagen mutated at the collagenase cleavage site[29,30] and (3) inhibition by broad-spectrum but not collagenase-1-sparing hydroxamate inhibitors. The hydroxamate inhibitors provide perhaps the most compelling conclusions. Blocking of keratinocyte migration with the broad-spectrum hydroxamate but retention of full migration with the collagenase-1-sparing agent (which inhibits all MMPs except for collagenase-1) demonstrates that collagenase-1 is *both* necessary and sufficient for keratinocyte migration over type I collagen. These data also suggest that catalytic activity of MMPs may be necessary for cell migration in general, even when such migration involves movement *over* a matrix rather than *through* it.

Unfortunately, there are no data on the effects of MMP inhibitors on either chronic skin ulcers or normal wound healing. However, since it is clear that collagenase-1, stromelysin-1, and stromelysin-2 are expressed in the healing of normal acute wounds in a temporally and spatially defined manner, this question must be addressed during future tests of MMP-inhibiting drugs. In this regard, two types of wound healing must be distinguished. Healing by primary intention involves the sutured closing of a vertical cut or an elliptical incision. In this scenario, tensile strength must be regained, but very little re-epithelialization must occur because the sutures approximate the wounded margins. From our knowledge of the role of MMPs in wound repair, one would predict that healing by primary intention would not be affected adversely by systemically administered MMP inhibitors. In contrast, healing by secondary intention involves the granulating in of an open wound in which the wound edges are not approximated. Here, cutaneous integrity and function must be re-established by wound contraction, acquisition of tensile strength, and re-epithelialization of the opposing epidermal margins. From the observations just discussed, there is justifiable concern that wounding by secondary intention would be affected adversely by MMP inhibitors. Such concern at present is only hypothetical, but data need to be generated to either support or negate the concept.

## CONTACT HYPERSENSITIVITY

MMPs are expressed in T cells and macrophages, but there is a paucity of evidence for their role in immune responses. We studied stromelysin-1 and gelatinase B "knockout" mice in a DNFB-induced model of contact hypersensitivity (CHS). DNFB was applied as a 0.5% suspension to truncal skin of mice on days 1 and 2 of the experiment. On day 7, the dorsal surfaces of mouse ears were challenged with 0.2% DNFB and ear thickness was measured with a micrometer on subsequent days as an index of inflammation. This system is based upon antigen recognition and processing by Langerhans cells, located suprabasally in the epidermis. Processed antigen is then taken by the Langerhans cells from epidermis into dermal lymphatics and into T- cell-rich areas of lymph nodes. The antigen is presented by the Langerhans cell to naive T cells, which then proliferate into specific T cells recognizing the antigen. Upon subsequent challenge, these T cells detect the antigen and migrate into the challenge site, releasing proinflammatory cytokines and causing the release of chemokines that attract other inflammatory cell types. Delayed hypersensitivity is the basis for reactions to poison ivy. Approximately 60% of individuals exposed to

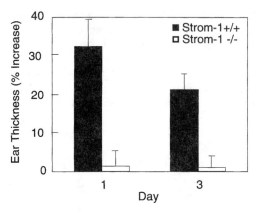

**FIGURE 5.** Contact hypersensitivity is impaired in stromelysin-1-deficient mice. Stromelysin-1-deficient and wild-type mice were sensitized with 0.5% DNFB to the trunk on successive days and challenged 5 days later with 0.2% DNFB applied to the dorsal surface of the ear. Ear thickness, an index of inflammation resulting from the CHS response, was determined by a digital micrometer 1 and 3 days after challenge. (From Wang et al.[34] Reproduced by permission.)

poison ivy repeatedly will become sensitized in this manner to the chemical coating the leaf.

As shown in FIGURE 5, stromelysin-1-deficient mice exhibited a markedly impaired CHS response to topical DNFB, although they responded normally to cutaneously applied phenol, an acute irritant that generates nonimmune mediated inflammation (not shown). This result was striking both in its magnitude and also in its consistency. Lymphocytes from lymph nodes of DNFB-sensitized stromelysin-1-deficient mice did not proliferate in response to specific soluble antigen DNBS, but did proliferate identically to lymph node lymphocytes from wild-type mice presented with the mitogen concanavalin-A (not shown). Most importantly, intradermal injection of stromelysin-1 just before the time of DNFB sensitization restored a normal contact hypersensitivity response to subsequent challenge with DNFB in stromleysin-1-deficient mice.

These results indicate that stromelysin-1 catalytic activity is essential to the afferent limb of the CHS response. The ability of exogenously administered stromelysin-1 to restore normal CHS in stromelysin-1-deficient mice when given just prior

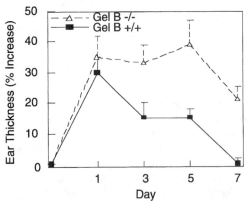

**FIGURE 6.** Contact hypersensitivity fails to resolve in gelatinase B-deficient mice. Gelatinase B-deficient and wild-type mice were sensitized and challenged as described in FIGURE 5. Ear thickness was determined at days 1, 3, 5 and 7 after challenge. (From Wang et al.[34] Reproduced by permission.)

to sensitization and at the sensitization site suggests that stromelysin-1 functions "early" in this afferent limb. Such potential functions include the following: (1) antigen recognition and/or processing; (2) breaking of Langherans cell:cell contacts with adjacent keratinocytes; (3) crossing the dermo-epidermal basement membrane; (4) transit through dermal matrix; or (5) entry into lymphatic vessels. While our data do not identify which of these potential roles is subserved by stromelysin-1, it is interesting that Bissell and colleagues[31] have reported that mammary epithelial cells transfected to overexpress spontaneously active stromelysin-1 appeared to detach from one another. This cell:cell detachment was apparently mediated by stromelysin-1 cleavage of E-cadherin and potentially other cell:cell contact proteins.

Interestingly, gelatinase B-deficient mice exhibited a CHS response very different from that of the stromelysin-1–deficient mice. Gelatinase B-deficient mice displayed an inflammatory reaction comparable to that of their wild-type controls at 1 day, but the response persisted to 7 days in contrast to the complete resolution observed in wild-type mice by 7 days (FIG. 6). Persistent inflammation was observed only with CHS reactions, as gelatinase B-deficient mice had a normal rate of resolution of inflammation elicited by cutaneous phenol. The prolonged inflammatory response to DNFB in gelatinase B-deficient mice was accompanied by a lack of IL-10 production at the site of ear swelling, an essential part of resolution that occurred in control mice (not shown).

Interleukin-10 is produced by several cell types, including TH2 lymphocytes, keratinocytes, and macrophages, and is a powerful inhibitor of delayed type hypersensitivity reactions.[32] Our data show that gelatinase B-deficient mice exhibit prolonged inflammation after antigen challenge and that this is associated with failure to induce IL-10 production. We measured IL-10 protein levels by ELISA (R & D Systems) using polyclonal antiserum generated against the entire IL-10 protein. Significantly, the tissue extract supernatants we examined had been subjected to multiple homogenizations and freeze-thawings in the presence of detergent, so that we are confident all of the tissue IL-10 was released for assay. Furthermore, the N-terminal 18 amino acids of IL-10 indicate the presence of a secretion-leader sequence, mouse and human cDNA clones are readily expressed as secreted proteins in monkey COS cells, and there is no evidence for protein processing of IL-10 in any system.[33] Thus, we believe that our observations indicate that gelatinase B-deficient mice fail to induce IL-10 *production,* and this leads to an exaggerated and prolonged CHS response. Importantly, we have determined that spleen cells from naive gelatinase B-deficient and wild-type mice provide similar induction of IL-10 upon exposure to LPS (not shown). Therefore, gelatinase B-deficient mice are able to produce IL-10 outside the context of a CHS response.

In the course of these studies, we also examined CHS in matrilysin-deficient mice and in metalloelastase-deficient mice. Both of these MMP knockouts exhibited responses identical to those of their wild-type controls.

## SUMMARY

Our findings indicate that MMPs are important to cutaneous wound healing and the development of chronic ulcers. We also show that stromelysin-1 and gelatinase

B serve important functions in CHS. Stromelysin-1 is required for initiation of the response, while gelatinase B plays a critical role in its resolution. To our knowledge, this is the first demonstration for an *in vivo* role of MMPs in immunologic responses. Our findings thus extend even further the potential role of MMPs outside of the traditionally thought of degradation of ECM.

## ACKNOWLEDGMENT

This work was supported by grants from the National Institutes of Health.

## REFERENCES

1. CHANDLER, S., R. COATES, A. GEARING, J. LURY, G. WELLS & E. BONE. 1995. Matrix metalloproteinases degrade myelin basic protein. Neuroscience Lett. **201:** 223–226.
2. SIRES, U.I., G. MURPHY, H.G. WELGUS & R.M. SENIOR. 1994. Matrilysin is much more efficient than other metalloproteinases in the proteolytic inactivation of a1-antitrypsin. Biochem. Biophys. Res. Commun. **204:** 613–620.
3. AGREN, M., C. TAPLIN, J. WOESSNER, W. EAGLESTEIN & P. MERTZ. 1992. Collagenase in wound healing: effect of wound age and type. J. Invest. Dermatol. **99:** 709–714.
4. BUCKLEY-STURROCK, A., S.C. WOODWARD, R.M. SENIOR, G.L. GRIFFIN, M. KLAGSBRUN & J.M. DAVIDSON. 1989. Differential stimulation of collagenase and chemotatic activity in fibroblasts derived from rat wound repair tissue and human skin by growth factors. J. Cell. Physiol. **138:** 70–78.
5. WYSOCKI, A.B., L. STAIANO-COICO & F. GRINNELL. 1993. Wound fluid from chronic leg ulcers contains elevated levels of metalloproteinases MMP-2 and MMP-9. J. Invest. Dermatol. **101:** 64–68.
6. SAARIALHO-KERE, U.K., S.O. KOVACS, A.P. PENTLAND, J. OLERUD, H.G. WELGUS & W.C. PARKS. 1993. Cell-matrix interactions influence interstitial collagenase expression by human keratinocytes actively involved in wound healing. J. Clin. Invest. **92:** 2858–2866.
7. SAARIALHO-KERE, U.K., S.O. KOVACS, A.P. PENTLAND, W.C. PARKS & H.G. WELGUS. 1994. Distinct populations of keratinocytes express stromelysin-1 and -2 in chronic wounds. J. Clin. Invest. **94:** 79–88.
8. SAARIALHO-KERE, U.K., E.S. CHANG, H.G. WELGUS & W.C. PARKS. 1992. Distinct localization of collagenase and TIMP expression in wound healing associated with ulcerative pyogenic granuloma. J. Clin. Invest. **90:** 1952–1957.
9. SAARIALHO-KERE, U., M. VAALAMO, K. AIROLA, K.-M. NIEMI, A.I. OIKARINEN & W.C. PARKS. 1995. Interstitial collagenase is expressed by keratinocytes that are actively involved in reepithelialization in blistering skin diseases. J. Invest. Dermatol. **104:** 982–988.
10. STRICKLIN, G.P., L. LI & L.B. NANNEY. 1994. Localization of mRNAs representing interstitial collagenase, 72 kDa gelatinase, and TIMP in healing porcine burn wounds. J. Invest. Dermatol. **103:** 352–358.
11. STRICKLIN, G.P. AND L.B. NANNEY. 1994. Immunolocalization of collagenase and TIMP in healing human burn wounds. J. Invest. Dermatol. **103:** 488–492.
12. STRICKLIN, G.P., L. LI, V. JANCIC, B.A. WENCZAK & L.B. NANNEY. 1993. Localization of mRNAs representing collagenase and TIMP in sections of healing human burn wounds. Am. J. Path. **143:** 1657–1666.
13. BIRKEDAL-HANSEN, H. 1995. Proteolytic remodeling of extracellular matrix. Curr. Opin. Cell Biol. **7:** 728–735.

14. VAALAMO, M,, M. WECKROTH, P. PUOLAKKAINEN, J. KERE, P. SAARINEN, J. LAUHA-RANTA & V. SAARIALHO-KERE. 1996. Patterns of matrix metalloproteinase and TIMP-1 expression in chronic and normally healing human cutaneous wounds. Br. J. Dermatol. **135:** 52–59.

15. OKADA, A., C. TOMASETTO, Y. LUTZ, J.-P. BELLOCQ, M.-C. RIO & P. BASSET. 1997. Expression of matrix metalloproteinases during rat skin wound healing: evidence that membrane type-1 matrix metalloproteinase is a stromal activator of progelatinase A. J. Cell Biol. **137:** 67–77.

16. LIBRACH, C.L., Z. WERB, M.L. FIZTGERALD, K. CHIU, N.M. CORWIN, R.A. ESTEVES, D. GROBELNY, R. GALARDY, C.H. DAMSKY & S.J. FISHER. 1991. 92-kD type IV collagenase mediates invasion of human cytotrophoblasts. J. Cell Biol. **113:** 437–449.

17. WILCOX, B.D., J.A. DUMIN & J.J. JEFFREY. 1994. Serotonin regulation of IL-1 mRNA in rat uterine smooth muscle cells. J. Biol. Chem. **269:** 29658–29664.

18. SYMPSON, C.J., R.S. TALHOUK, C.M. ALEXANDER, J.R. CHIN, S.M. CLIFT, M.J. BISSELL & Z. WERB. 1994. Targeted expression of stromelysin-1 in mammary gland provides evidence for a role of proteinases in branching morphogenesis and the requirement for an intact basement membrane for tissue-specific gene expression. J. Cell Biol. **125:** 681–693.

19. DESOUZA, S., J. LOCHNER, C.M. MACHIDA, L.M. MATRISIAN & G. CIMENT. 1995. A novel nerve growth factor-responsive element in the stromelysin-1 (transin) gene that is necessary and sufficient for gene expression in PC12 cells. J. Biol. Chem. **270:** 9106–9114.

20. SUDBECK, B.D., B.K. PILCHER, H.G. WELGUS & W.C. PARKS. 1997. Induction and repression of collagenase-1 by keratinocytes is controlled by distinct components of different extracellular matrix compartments. J. Biol. Chem. **272:** 22103–22110.

21. REBOUL, P., J.-P. PELLETIER, G. TARDIF, J.-M. CLOUTIER & J. MARTEL-PELLETIER. 1996. The new collagenase, collagenase-3, is expressed and synthesized by human chondrocytes but not by synoviocytes: a role in osteoarthritis. J. Clin. Invest. **97:** 2011–2019.

22. YAMAMOTO, M., S. MOHANAM, R. SAWAYA, G.N. FULLER, M. SEIKI, H. SATO, Z.L. GOKASLAN, L.A. LIOTTA, G.L. NICOLSON & J.S. RAO. 1996. Differential expression of membrane-type matrix metalloproteinase and its correlation with gelatinase A activation in human malignant brain tumors in vivo and in vitro. Cancer Res. **56:** 384–392.

23. NEUMUNAITIS, J., C. POOLE, J. PRIMROSE, A. ROSEMURGY, J. MALFETANO, P. BROWN, A. BERRINGTON, A. CORNISH, K. LYNCH, H. RASMUSSEN, D. KERR, D. COX & A. MILLAR. 1998. Combined analysis of studies of the effect of the MMP inhibitor marimastat on serum tumor markers in advanced cancer: selection of a biologically active and tolerable dose for longer-term studies. Clin. Cancer Res. **4:** 1101–1109.

24. THOMPSON, R.W., D.R. HOLMES, R.A. MERTENS, S. LIAO, M.D. BOTNEY, R.P. MECHAM, H.G. WELGUS & W.C. PARKS. 1995. Production and localization of 92 kD gelatinase in abdominal aortic aneurysms: an elastolytic metalloproteinase expressed by aneurysm-infiltrating macrophages. J. Clin. Invest. **96:** 318–326.

25. GALIS, Z.S., G.K. SUKHOVA, R. KRANZHOFER, S. CLARK & P. LIBBY. 1995. Macrophage foam cells from experimental atheroma constitutively produce matrix-degrading proteinases. Proc. Natl. Acad. Sci. USA. **92:** 402–406.

26. VAALAMO, M., M. WECKROTH, P. PUOLAKKAINEN, J. KERE, P. SAARINEN, J. LAUHARANTA & U. SAARIALHO-KERE. 1996. Patterns of matrix metalloproteinase and TIMP-1 expression in chronic and normally healing human cutaneous wounds. Brit. J. Dermatol. **135:** 52–59.

27. PILCHER, B.K., J. DUMIN, M.J. SCHWARTZ, B.A. MAST, G.S. SCHULTZ, W.C. PERKS & H.G. WELGUS. 1999. Keratinocyte collagenase-1 expression requires an epidermal growth factor receptor autocrine mechanism. J. Biol. Chem. **274:** 10372–10381.
28. PILCHER, B.K., B.D. SUDBECK, J. DUMIN, S.M. KRANE, H.G. WELGUS & W.C. PARKS. 1997. The activity of collagenase-1 is required for keratinocyte migration on type I collagen. J. Cell Biol. **137:** 1445–1457.
29. LIU, X., H. WU, M. BYRNE, J. JEFFREY, S. KRANE & R. JAENISCH. 1995. A targeted mutation at the known collagenase cleavage site in mouse type I collagen impairs tissue remodeling. J. Cell Biol. **130:** 227–237.
30. KRANE, S.M., M.H. BYRNE, V. LEMAITRE, P. HENRIET, J.J. JEFFREY, J.P. WITTER, X. LIU, H. WU, R. JAENISCH & Y. EECKHOUT. 1996. Different collagenase gene products have different roles in degradation of type I collagen. J. Biol. Chem. **271:** 28509–28515.
31. LOCHTER, A., S. GALOSY, J. MUSCHLER, N. FREEDMAN, Z. WERB & M.J. BISSELL. 1997. Matrix metalloproteinase stromelysin-1 triggers a cascade of molecular alterations that leads to stable epithelial-to-mesenchymal conversion and a premalignant phenotype in mammary epithelial cells. J. Cell Biol. **139:** 1861–1872.
32. FERGUSON, T.A., P. DUBE & T.S. GRIFFITH. 1994. Regulation of contact hypersensitivity by interleukin 10. J. Exp. Med. **179:** 1597–1604.
33. MOSMANN, T.R. 1994. Properties and functions of interleukin-10. Adv. Immunol. **56:** 1–26.
34. WANG, M., X. QIN, J.S. MUDGETT, T.A. FERGUSON, R.M. SENIOR, AND H.G. WELGUS. 1999. Metalloproteinase deficiencies alter contact hypersensitivity: stromelysin-1 deficiency prevents the response and gelatinase B deficiency prolongs the response. Proc. Natl. Acad. Sci. (USA). In press.

# Evaluation of Some Newer Matrix Metalloproteinases

GILLIAN MURPHY,[a,b] VERA KNÄUPER,[a] SUSAN COWELL,[c]
ROSALIND HEMBRY,[a] HEATHER STANTON,[a] GEORGINA BUTLER,[a]
JOSÉ FREIJE,[d] ALBERTO M. PENDÁS,[d] AND CARLOS LÓPEZ-OTÍN[d]

[a]School of Biological Sciences, University of East Anglia, Norwich NR4 7TJ, UK

[c]Strangeways Research Laboratory, Cambridge CB1 4RN, UK

[d]Departamento de Bioquimica y Biologia Molecular, Facultad de Medicina,
Universidad de Oviedo, Oviedo, Spain

ABSTRACT: Recombinant protein expression techniques have been utilized
to facilitate the biochemical and cell biological characterization of human matrix metalloproteinases (MMPs). The importance of the membrane type 1
MMP (MMP 14) in the regulation of pericellular proteolysis, either directly or
through the activation of MMP-2, MMP-9, and MMP-13 has been identified.
Studies on an *in vitro* chondrocyte-like cell and an *in vivo* cartilage repair model indicated that such MT1 MMP–regulated activation cascades are physiologically feasible. MMP19 shows a limited sequence identity with other MMPs
and may represent a novel subclass. However, analysis of the recombinant protein identified a number of biochemical properties typical of the MMP family.

Proteolysis of the extracellular matrix (ECM) is a key component of the inflammatory response, acting as a key effector in the modulation of the cell–cell and cell–
ECM interactions underlying both disease and repair activities. The role of matrix
metalloproteinases (MMPs) in matrix turnover has long been under scrutiny, and it
has become evident that there is a bewildering array of these enzymes with apparently overlapping substrate specificities and expression patterns. A number of the
MMPs that have been identified more recently have been biochemically and biologically characterized in our laboratories in order to relate their function to that of the
more established members of the MMP family.

## MEMBRANE-TYPE MATRIX METALLOPROTEINASES, MT-MMPs

Four MT-MMPs have been cloned, MT1-MMP (MMP-14), MT2-MMP (MMP-
15), MT3-MMP (MMP-16), and MT4-MMP (MMP-17).[1–4] These membrane-associated members of the MMP family are expressed at low levels by many cell types.
MT1-MMP is the most predominant and the most clearly regulated by cytokines and
growth factors (M. Lafleur and D. Edwards, personal communication[5]). However,

[b]Address for correspondence: Gillian Murphy, School of Biological Sciences, University of
East Anglia, Norwich, NR4 7TJ, UK. Phone, +44 1603 593811; fax, +44 1603 593874; e-mail:
g.murphy@uea.ac.uk

both MT2-MMP and MT3-MMP share the ability to initiate the activation of proMMP-2 with MT1-MMP. In contrast, MT4-MMP has negligible proMMP-2 processing activity (W. English, A. Merryweather, and G. Murphy, unpublished material). Although there are a number of reports on the regulation of MT1-MMP mRNA by cytokines, this does not necessarily correlate with an increase in proMMP-2 activation. Indeed, in our hands, the modulation of this MMP, in terms of the expression of activity on proMMP-2, can occur at many levels other than the transcriptional one, including the extent of cell surface expression and aggregation, the extent of propeptide processing and further proteolytic inactivation, and the levels of TIMP-2 present. The majority of these events are still not well understood and require clarification. As part of our studies on the potential for ECM regulation of MMP activity we looked at the effect of fibronectin and laminin-1 on the proMMP-2 activation cascade in HT1080 cells.[6] Fibronectin, but not laminin, caused the activation of MMP-2 produced by these cells. Culture of the cells on peptide fragments of fibronectin derived from the cell-binding domain had a similar effect, as did culture on immobilized antibodies to the $\alpha_5$ and $\beta_1$ integrin subunits. The data indicated that the activation of MMP-2 is regulated by the nature of the ECM and that signals via the $\alpha_5\beta_1$ receptor may be involved. Interestingly, another ligand for this integrin, vitronectin, did not have the same effect as fibronectin. Levels of TIMP-2 expressed by the HT1080 cells did not vary between cells cultured on fibronectin or laminin, but the amount of MT1-MMP protein (but not mRNA) was increased almost two-fold. This increase was largely attributed to an increase in levels of a truncated 45-kDa form of MT1-MMP rather than the 60-kDa active form (very little of the 63-kDa proform was detectable in any case). Parallel studies using gelatin zymography demonstrated that the upregulation of the production of the 45-kDa form was concomitant with proMMP-2 activation. Inhibition studies revealed that the truncation of MT1-MMP to a 45-kDa form is MMP-mediated, although not inhibited by TIMP-1. This implies that either MMP-2 or MT1-MMP could effect this processing, presenting the potential for a self-regulatory termination to the activation cascade.

It is known that soluble forms of MT1-MMP and MT2-MMP are relatively efficient proteinases and degrade denatured interstitial collagens, cartilage aggrecan, perlecan, fibulin-1 and -2, fibronectin, vitronectin, nidogen, large tenascin-C, and laminin-1, as well as a pro-tumor necrosis factor $\alpha$-GST fusion protein.[7-9] In the case of MT1-MMP, cleavage of native triple helical collagens I, II and III into the characteristic three-quarter, one-quarter fragments has been demonstrated, suggesting that this enzyme shares some characteristics with the collagenases.[8,9] C-terminal deletion mutants of MT1-MMP are not able to cleave triple helical collagen—that is, the mechanism of action involves the noncatalytic hemopexin domain, as described for the traditional human collagenases, MMPs 1, 8, and 13.

In order to assess whether MT1-MMP was functional at the surface of cells we transiently transfected CHO cells, which make negligible amounts of this MMP, with an HCMV-driven expression vector expressing the cDNA.[10] Indirect immunofluorescence showed that MT1-MMP was expressed at the surface of the cells. Transfected cells were cultured on thin films of Texas Red–labeled gelatin which had been fixed to the culture vessel. Over 24 hours the cells degraded the gelatin, and lysis could be detected by the appearance of dark patches in the labeled substrate film, specifically associated with cells expressing MT1-MMP. This focal degradation

TABLE 1. The degradation of gelatin films by cells expressing MT1-MMP:
Effect of TIMP-1, TIMP-2 and proMMP-2

| Vector | Exogenous effector | MT1-MMP cell surface immunofluorescence | Film degradation ($\mu m^2 \pm$ S.E.) |
|---|---|---|---|
| pEE14 | – | – | 0 |
| pEE14.MT1 | – | + | 1651 ± 254 |
| pEE14.E$^{217}$-A MT1 | – | + | 0 |
| pEE14.MT1 | TIMP-1 | + | 1066 ± 131 |
| pEE14.MT1 | TIMP-2 | + | 0 |
| pEE14.MT1 | CT1399 | + | 0 |
| pEE14.MT1 | proMMP-2, 1ng | + | 3803 ± 527 |
| pEE14.MT1 | proMMP-2, 10ng | + | 5004 ± 520 |
| pEE14.MT1 | proMMP-2, 100 ng | + | 8409 ± 1048 |

NOTE: Chinese hamster ovary cells were plated on to gelatin-Texas Red films and transfected with the expression vector pEE14 alone, with the vector containing the cDNA for MT1-MMP or its inactive mutant in which Glu$^{217}$ had been modified to Ala. TIMP-1 (10 μg/well), TIMP-2 (10 μg/well), a synthetic peptide hydroxamate inhibitor CT1399 (10 μM), or proMMP-2 were added as indicated. After 16 hr cells were stained with a sheep antibody to MT1-MMP[10] observed by confocal microscopy and scored for cell surface immunofluorescence and focal degradation of the films.

could be scored using confocal microscopy and Biorad Cosmos software (TABLE 1). Tracks of film clearance were frequently observed, similar to phagokinetic tracks, illustrating proteolysis during cell migration. CHO cells transfected with an inactive mutant form of MT1-MMP, where Glu$^{217}$ had been modified to Ala (E$^{217}$A MT1-MMP), gave detectable enzyme at the surface of the cells, but no degradation of the gelatin film. When cells transfected with the cDNA for MT1-MMP were placed on the films in the presence of recombinant TIMP-2, no film degradation took place; in contrast addition of recombinant TIMP-1 had no effect. Addition of recombinant proMMP-2 to the system gave a large increase in the extent of film degradation, proportional to the amount of MMP-2 added (TABLE 1). We can conclude that MT1-MMP can effectively proteolyze the matrix adjacent to cells and that activation of proMMP-2 at the cell surface leads to markedly enhanced activity. TIMP-1 is ineffective in regulating these processes because of its inability to inhibit MT1-MMP.[11]

## ACTIVATION OF OTHER MMPS BY MT1-MMP

The work of a number of laboratories, including ours, had established that the cell-based MT1-MMP activation mechanism for proMMP-2 involved the binding of the proenzyme to an MT1-MMP/TIMP-2 complex ("receptor") on the cell surface through interaction between the C-terminal domain of proMMP-2 and the C-terminal domain of TIMP-2.[5,7,11,12] Processing of proMMP-2 to a Leu[38] intermediate form may then be initiated by an adjacent free and active MT1-MMP molecule. This initial cleavage event destabilizes the structure of the proMMP-2 propeptide domain, and autoproteolysis then proceeds in an MMP-2 concentration-dependent manner, which releases the rest of the propeptide domain and fully active MMP-2. In cell culture studies, the enzyme concentration in solution is low, and deletion of either the proMMP-2 C-terminal domain or the transmembrane domain of MT1-MMP abolishes proMMP-2 activation, emphasizing that the binding mechanism involving the MT1-MMP/TIMP-2 complex on the cell surface acts as a concentration mechanism, which is crucial for the efficiency of activation. It appears that addition of small sub equimolar amounts of TIMP-2 to cells expressing MT1-MMP can enhance proMMP-2 activation, since this increases the concentration of the MT1-MMP/ TIMP-2 receptor for proMMP-2 on the cell surface. However, at high TIMP-2 concentrations all the MT1-MMP molecules on the cell surface are complexed with TIMP-2 and, although proMMP-2 binding occurs, no active MT1-MMP remains to initiate processing. This concept has been substantiated by cell free kinetic studies of the effects of the MT1-MMP/TIMP-2 "receptor" as a mode of concentrating proMMP-2 in order to promote autoproteolysis.[13] This suggests that proMMP-2 activation is regulated by the amount of TIMP-2 secreted by MT1-MMP expressing cells as well as by the extent of MT1-MMP activation. C-terminal domain mutants of proMMP-2, TIMP-2, and MT1-MMP were used to further support the above mechanism. Soluble constructs of full-length and catalytic domain MT1-MMP were used to demonstrate that binding with TIMP-2 occurs primarily through N-terminal domain interactions of both enzyme and inhibitor, leaving the C-terminal domain of TIMP-2 free for interactions with proMMP-2. The rate of autolytic activation of proMMP-2 initiated by MT1-MMP cleavage could be potentiated by concentration of the proenzyme, for example, by binding to heparin.[13] Residues 568-631 of the proMMP-2 C-terminal domain are important in the formation of the heparin binding site, since replacement of this region with the corresponding stromelysin-1 sequence abolished binding to heparin and the potentiation of activation. The same region of MMP-2 was required for the binding of both its pro and active form to TIMP-2, whereas residues 418-474 were not required. A similar pattern was seen using cell membrane–associated MT1-MMP: residues 568-631 were necessary for both the binding and activation of MMP-2, whereas residues 418-474 were not. Neither region was required for activation in solution. The addition of TIMP-2 to HT1080 membrane preparations overexpressing MT1-MMP depleted of endogenous TIMP-2 resulted in potentiation of proMMP-2 activation. This effect was dependent upon TIMP-2 binding to MT1-MMP, rather than to an independent membrane site. Together the data supported the hypothesis that TIMP-2 forms a receptor by complex formation with MT1-MMP and that this regulates the efficient generation of functionally active MMP-2.

Besides the activation of proMMP-2 it has been shown that MT1-MMP may be responsible for the activation of proMMP-13, either directly or via MMP-2 activation[14] (FIG. 1). In both cases, processing of the prodomain of proMMP-13 occurs via a 56-kDa intermediate, yielding a final 48 kDa form. Similarly, the activation of proMMP-13 by MT1-MMP can potentiate the membrane processing of the catalytically inactive mutant of MMP-2, proE[375]-A MMP-2 (FIG. 2). As for proMMP-2 processing by MT1-MMP, proMMP-13 processing to the intermediate form is inhib-itable by TIMP-2, but not by TIMP-1 or aprotinin (FIG. 3), suggesting that the activity of MT1-MMP is critical for this stage. Further processing of the intermediate may be partially autolytic and TIMP-1 sensitive (FIG. 3). We have also found that, like proM-MP-2, cellular activation of proMMP-13 requires the presence of the C-terminal do-main. Pro$\Delta_{249-451}$ MMP-13 was not activated by MT1-MMP–containing cell membrane preparations at similar concentrations to the wild type proMMP-13 (V. Knäuper and G. Murphy, unpublished material; FIG. 4). By contrast, pro$\Delta_{249-451}$ MMP-13 and proMMP-13 activated with similar kinetics in the presence of the orga-nomercurial 4-aminophenyl mercuric acetate or at high concentrations with soluble MT1-MMP (data not shown). In the case of proMMP-2 it can be shown that appro-priate concentrations of TIMP-2 can potentiate the membrane-associated MT1-MMP processing and activation mechanism[12,13] (FIG. 5B). Using kinetic methods to ana-lyze TIMP-2–MMP-2 interactions, we deduced that the specific binding of the C-ter-minal regions of TIMP-2 and MMP-2 were important features of the membrane-associated activation.[13] However, in the case of proMMP-13, the mechanism of in-teraction with cell membranes is less clear. The C-terminal subdomain of TIMP-2 does not contribute significantly to the rate constant $k_{on}$ for complex formation be-tween enzyme and inhibitor[15] (TABLE 2). We could demonstrate that, in contrast to

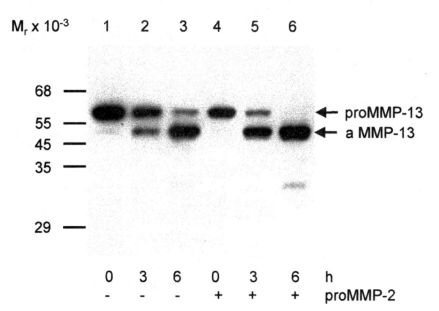

**FIGURE 1.** Activation of proMMP-13 by membranes from cells transfected with MT1-MMP and TIMP-2; potentiation by proMMP-2. Mouse myeloma cells were stably transfected with human MT1-MMP and human TIMP-2 cDNAs and the membrane fraction was prepared.[29] Human proMMP-13 was incubated with membranes at 37°C for *lane 1*, 0 hr; *lane 2*, 3 hr; *lane 3*, 6 hr; *lane 4*, 0 hr with proMMP-2; *lane 5*, 3 hr with proMMP-2; *lane 6*, 6 hr with proMMP-2. ProMMP-13 processing was monitored using SDS polyacry-lamide gel electrophoresis and immunoblotting using a specific antibody to MMP-13.

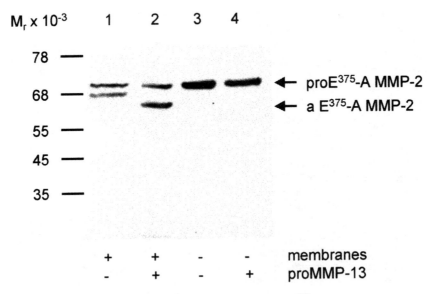

**FIGURE 2.** Effect of proE$^{375}$-A MMP-2 on the processing of proE$^{375}$-A MMP-2 by membrane-associated MT1-MMP. MT1-MMP membranes (prepared as in FIGURE 1) were incubated (inc.) at 37°C with the inactive form of proMMP-2 in which Glu$^{375}$ has been modified to Ala. *Lane 1,* 12 hr. inc.; *lane 2,* in the presence of proMMP-13; *lane 3,* omission of membranes, proE$^{375}$-A MMP-2 inc. 12 hr alone; *lane 4,* omission of membranes, proE$^{375}$-A MMP-2 and proMMP-13 12-hr inc. The incubations were processed by SDS polyacrylamide gel electrophoresis and immunoblotting with an antibody to proMMP-2.

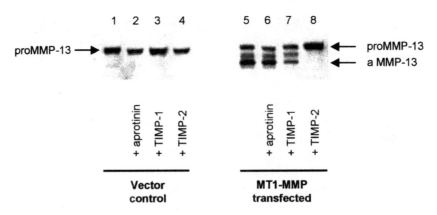

**FIGURE 3.** The action of TIMPs and aprotinin on the processing of proMMP-13 by MT1-MMP–expressing cells. HT1080 cells transfected with a vector containing MT1-MMP or no cDNA (Butler *et al.*[13]) were used to assess the effect of proteinase inhibitors on exogenous proMMP-13 processing; *lanes 1–4,* 24-hr incubations with cells containing vector alone and *lanes 5–8,* 24 hr incubations with cells expressing MT1-MMP: *lanes 1, 5* proM-MP-13; *lanes 2, 6* proMMP-13 and aprotinin; *lanes 3, 7* proMMP-13 and TIMP-1; *lanes 4, 8,* proMMP-13 and TIMP-2.TIMP-1 and aprotinin do not prevent proMMP-13 processing, but TIMP-2 does.

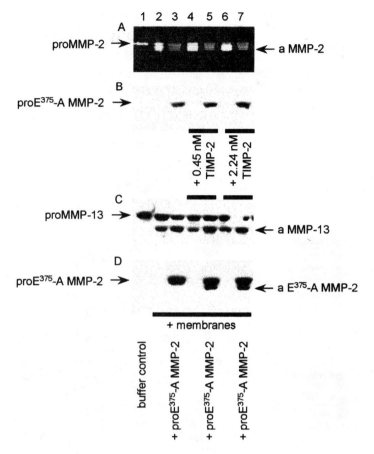

**FIGURE 6.** The effect of TIMP-2 and proE$^{375}$-A MMP-2 on the processing of proMMP-2 and proMMP-13 by membrane-associated MT1-MMP. **(A)** TIMP-2 potentiates and proE$^{375}$-A MMP-2 inhibits the MT1-MMP-mediated processing of proMMP-2. Cell membranes containing MT1-MMP and TIMP-2 were prepared as in FIGURE 1. *Lane 1,* proMMP-2 incubated alone at 37°C for 11 hr; *lane 2* proMMP-2 incubated at 37°C, 11 hr in the presence of MT1-MMP membranes; *lane 3,* as *lane 2* in the presence of a tenfold molar excess of proE$^{375}$-A MMP-2; *lane 4,* as *lane 2* in the presence of 0.45 nM TIMP-2; *lane 5,* as *lane 4* in the presence of proE$^{375}$-A MMP-2; *lane 6,* as *lane 2* in the presence of 2.24 nM TIMP-2; *lane 7,* as *lane 6* in the presence of proE$^{375}$A MMP-2. **(B)** Immunoblotting of the samples from FIGURE 5A using an antibody to MMP-2. Wild-type gelatinase is below the detection limit for the antibody. *Lanes 3, 5, and 7* show proE$^{375}$-A MMP-2 in an unprocessed form. **(C)** TIMP-2 has no effect on the MT1-MMP membrane-processing of proMMP-13, but proE$^{375}$-A MMP-2 potentiates processing. *Lane 1,* proMMP-13 incubated alone at 37°C, 11 hr; *lane 2,* proMMP-13 incubated at 37°C, 11 hr in the presence of MT1-MMP membranes; *lane 3,* as *lane 2* in the presence of a tenfold molar excess of proE$^{375}$-A MMP-2; *lane 4,* as *lane 2* in the presence of 0.45 nM TIMP-2; *lane 5,* as *lane 4* in the presence of proE$^{375}$-A MMP-2; *lane 6,* as *lane 2* in the presence of 2.24 nM TIMP-2, *lane 7,* as *lane 6* in the presence of proE$^{375}$-A MMP-2. **(D)** Immunoblot of samples from FIGURE 5C using an antibody to MMP-2. Wild-type proMMP-2 is below the detection limit but E$^{375}$-A MMP-2 is detected and shown to be processed in *lanes 5 and 7.*

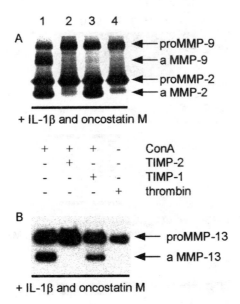

**FIGURE 7.** Effect of TIMPs and thrombin on the activation of proMMP-13, MMP-2 and MMP-9 by SW1353 cells. SW1353 cells were treated with IL1-$\beta$ (10 ng/ml), oncostatin M (50 ng/ml), and conA (50 $\mu$g/ml in DMEM/Hams F-12, 0.2% lactalbumin hydrolysate for 3 days to induce synthesis of proMMP-13 *(lanes 1–3)*. TIMP-2 (3.3 $\mu$g/ml, *lane 2*), TIMP-1 (0.42 $\mu$g/ml, *lane 3*), were added with the IL-1$\beta$ and oncostatin M. ConA was replaced with thrombin *(lane 4)*. **(A)** MMP-9 and MMP-2 were assessed by gelatin gel zymography (inverted image). **(B)** MMP-13 was assessed by SDS polyacrylamide gel electrophoresis and immunoblotting with an antibody to MMP-13.

In the presence of ConA activation of MMP-2, MMP-9 and MMP-13 was observed and TIMP-2 prevented all three processes. TIMP-1 prevented proMMP-9 activation, but not proMMP-2, nor MMP-13. Thrombin caused a modest processing of proMMP-2, but not MMP-9 or MMP-13.

**TABLE 2. TIMP-2 shows weak interactions with the C-domain of human MMP-13**

| Inhibitor | $k_{on}$ ($\times 10^{-6}$ M$^{-1}$. s$^{-1}$) | |
|---|---|---|
| | MMP-13 | $\Delta$249-451 MMP-13 |
| TIMP-2 | 1.41 | 0.81 |
| $\Delta$187-194 TIMP-2 | 1.80 | 0.69 |
| $\Delta$128-194 TIMP-2 | 0.29 | 0.32 |

NOTE: Rate constants for the inhibition of active full-length MMP-13 and $\Delta_{249-451}$ MMP-13 by full–length and truncated forms of TIMP-2 were determined using a quenched fluorescent peptide substrate assay at 25°C.[15]

This activation cascade may also operate in *in vivo* situations. For example, we have been using a method for inducing cartilage repair in lesions that do not penetrate the underlying bone[19] in order to investigate the regulatory signals and underlying mechanisms involved in cartilage repair. The method uses a staged delivery of growth factors to attract cells into partial thickness defects, and the space to be repaired is defined by filling the defect with a biodegradable matrix such as gelatin. Following formation of partial thickness defects *in vivo* in pigs, we monitored the repair process by immunofluorescence microscopy to determine which MMPs and TIMPs are expressed during the different stages of repair. At two days after wounding, macrophages identified using well-characterized cell marker antibodies were observed attached to the intact cartilage surface and to the cut cartilage matrix in the defects. These macrophages were seen to synthesize and secrete MMP-9 focally onto damaged but not onto intact cartilage surfaces. By day 8 defects were populated with increased numbers of predominantly macrophages containing MMP-9, and macrophages synthesizing MMP-9 were seen to have penetrated the necrotic zone bordering the defect, and the cartilage matrix stained throughout the necrotic area. Also by day 8 chondrocytes adjacent to and below the necrotic areas had undergone clonal expansion, and these clonal chondrocytes showed synthesis of MMP-13 in addition to increased MMP-1 and MMP-3 (R. Hembry, J. Tyler and G. Murphy, unpublished material). MMP-13 is normally restricted to hypertrophic chondrocytes[20,21] and has been shown to activate MMP-9[22]: the presence of both these enzymes on the edges of the defects may well compromise mechanical function.

## MATRIX METALLOPROTEINASE-19

One of the most recently cloned members of the MMP family is MMP-19.[23,24] Its primary sequence suggested that it is a typical secreted protein with identifiable propeptide, catalytic and hemopexin-like C-terminal domains, as found in other family members. Detailed sequence comparisons with other MMPs suggest that MMP-19 is not closely related to the collagenases or stromelysins and is the first member of a new sub-class, a hypothesis that is upheld by its chromosomal location at q14 on the long arm of chromosome 12.[24] Three unusual sequence features are an oligo-glutamic acid containing "hinge" connecting the catalytic and C-domains, an apparently free cysteine in the catalytic domain, and an extended C-terminal "tail." The cloned protein was expressed as inclusion bodies in *E. coli* and solubilized, purified, and refolded. The proenzyme had a mol. wt. of 58 kDa and could be activated with trypsin to a form that cleaved the general MMP peptide substrate Mca-Pro-Leu-Gly-Leu-Dpa-Ala-Arg-NH$_2$ with a $k_{cat}/k_M$ of $1.93 \times 10^3$ M$^{-1}$·s$^{-1}$. The proenzyme has a propensity to self-activate, which can be attributed to the fact that the propeptide "cysteine switch" motif is PRCGL̲E̲D, as compared with PRCGX̲PD in most MMPs. Mutational studies by Park *et al.*[25] have shown that modification of this Pro-residue in MMP-3 generates autoactivating forms. The N-terminal sequence of autoactivated MMP-19 was found to be [99]YLLLGRW. MMP-19 preferentially hydrolyzed a quenched fluorescent peptide "stromelysin substrate" Mca-Pro-Leu-Ala-Nva-Ala-Arg-NH$_2$ over a collagenase-favored peptide sequence.[24] MMP-19 shows little propensity for the cleavage of extracellular matrix components, although activity

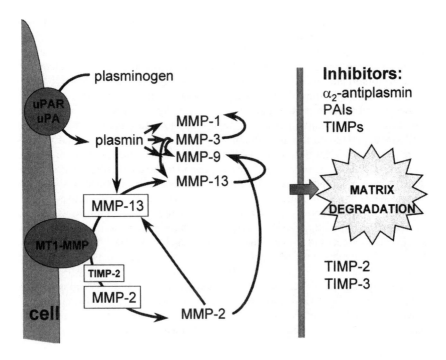

**FIGURE 8.** Cell surface–associated activation cascades for matrix metalloproteinases. The extracellular activation of most proMMPs is probably limited to the pericellular environment where cell-associated proteinase activities can function in an inhibitor-depleted environment. Key initiators of the MMP activation cascades are thought to be: MT1-MMP (and perhaps MT2-MMP and MT3-MMP), which is activated intracellularly, and urokinase-like plasminogen activator, which is bound to its cell surface receptor. The production of active or partially active MMPs allows a cascade of autolytic and heterolytic cleavages to occur, generating fully active enzymes. The efficiency of these interactions is dependent on mechanisms for the concentration of MMPs at the cell surface or by binding to matrix proteins.

against large tenascin C and gelatin (preferential cleavage of the $\alpha_2$(I)-chain) has been demonstrated. The activity versus gelatin is so limited that enzymatic activity cannot be demonstrated by gelatin zymography. Type I collagen and laminin are not cleaved (not shown). MMP-19 is able to degrade aggrecan, but cleaves it at a novel site as well as that of other MMPs (J. Stracke, A. Merryweather, and G. Murphy, unpublished material).[26] The catalytic domain of the enzyme shows slow association rate constants for binding TIMP-1 ($k_{on} = 1.22 \times 10^4$ $M^{-1}s^{-1}$) and TIMP-3 ($k_{on} = 3.73 \times 10^4$ $M^{-1}s^{-1}$). TIMP-2 binding is considerably faster, but accurate values have not yet been determined (J. Stracke, V. Knäuper and G. Murphy, unpublished material).

MMP-19 is widely distributed at the mRNA level, transcripts being significantly detectable by Northern analysis of adult human placenta, lung, pancreas, ovary, spleen, and intestine. This is unusual in that MMPs are not abundantly produced by adult cells in normal tissues. Preliminary studies of cell lines suggest that MMP-19 may be constitutively expressed, although upregulation by cytokines can occur (unpublished data). It appears to be associated with the surface of cells, an observation first made by Sedlacek *et al.*,[28] who described its localization on the surface of phytohaemagglutinin-stimulated peripheral blood leukocytes. (Interestingly, the Northern analysis of tissues by Pendas *et al.*[24] did not detect MMP-19 mRNA in leukocyte extracts.). However, MMP-19 does not have an identifiable transmembrane motif, although it has an extended C-terminal "tail" beyond the Cys[472] residue, involved in forming the typical MMP-β-propeller disulfide bond:

RPRTIDTTPSGNTTPSGTGITLDTTLSATETTFEY.

This sequence contains a putative N-glycosylation site and multiple Thr residues, which may be involved in cell surface interactions. Alternatively, the acidic hinge region between the catalytic and C-domains may confer binding properties, for example, to cell surface proteoglycans. MMP-19 has also been identified as the autoantigen RASI-1, found in the synovial membranes of a patient with rheumatoid arthritis and a circulatory autologous antibody.[27,28] By means of immunohistochemical techniques[27] the protein appears to be localized to the blood vessels of the synovium and of other normal tissues probably associated with smooth muscle cells. This enzyme is clearly an exciting addition to the MMP repertoire, meriting further work on its biochemistry and cell biology as well as an *in vivo* approach to the analysis of its function.

## ACKNOWLEDGMENTS

We thank all the colleagues who have contributed to this work and Jill Gorton for help preparing the manuscript. Our work is funded by the Arthritis and Rheumatism Campaign, the Wellcome Trust, the Nuffield Foundation and the Medical Research Council, U.K., and the European Union Biomed II Program.

## REFERENCES

1. SATO, H., T. TAKINO, Y. OKADA, J. CAO, A. SHINAGAWA, E. YAMAMOTO & M. SEIKI. 1994. A matrix metalloproteinase expressed on the surface of invasive tumour cells. Nature **370:** 61–65.

2. TAKINO, T., H. SATO, A. SHINAGAWA & M. SEIKI. 1995. Identification of the second membrane-type matrix metalloproteinase (MT-MMP-2) gene from a human placenta cDNA library—MT-MMPs form a unique membrane-type subclass in the MMP family. J. Biol. Chem. **270:** 23013–23020.

3. WILL, H. & B. HINZMANN. 1995. cDNA sequence and mRNA tissue distribution of a novel human matrix metalloproteinase with a potential transmembrane segment. Eur. J. Biochem. **231:** 602–608.

4. PUENTE, X.S., A.M. PENDAS, E. LLANO, G. VELASCO & C. LOPEZ-OTIN. 1996. Molecular cloning of a novel membrane-type matrix metalloproteinase from a human breast carcinoma. Cancer Res. **56:** 944–949.

5. LOHI, J., K. LEHTI, J. WESTERMARCK, V.M. KÄHÄRI & J. KESKI-OJA. 1996. Regulation of membrane-type matrix metalloproteinase-1 expression by growth factors and phorbol 12-myristate 13-acetate. Eur. J. Biochem. **239:** 239–247.

6. STANTON, H., J. GAVRILOVIC, J. ATKINSON, M.-P. D'ORTHO, K.M. YAMADA, L. ZARDI & G. MURPHY. 1998. The activation of proMMP-2 (gelatinase A) by HT1080 fibrosarcoma cells is promoted by culture on a fibronectin substrate and is concomitant with an increase in processing of MT1-MMP (MMP-14) to a 45 kDa form. J. Cell Sci. **111:** 2789–2798.

7. PEI, D.Q. & S.J. WEISS. 1996. Transmembrane-deletion mutants of the membrane-type matrix metalloproteinase-1 process progelatinase A and express intrinsic matrix-degrading activity. J. Biol. Chem. **271:** 9135–9140.

8. OHUCHI, E., K. IMAI, Y. FUJII, H. SATO, M. SEIKI & Y. OKADA. 1997. Membrane type 1 matrix metalloproteinase digests interstitial collagens and other extracellular matrix macromolecules. J. Biol. Chem. **272:** 2446–2451.

9. D'ORTHO, M.P., H. WILL, S. ATKINSON, G. BUTLER, A. MESSENT, J. GAVRILOVIC, B. SMITH, R. TIMPL, L. ZARDI & G. MURPHY. 1997. Membrane-type matrix metalloproteinases 1 and 2 exhibit broad-spectrum proteolytic capacities comparable to many matrix metalloproteinases. Eur. J. Biochem. **250:** 751–757.

10. D'ORTHO, M.P., H. STANTON, M. BUTLER, S.J. ATKINSON, G. MURPHY & R.M. HEMBRY. 1998. MT1-MMP on the cell surface causes focal degradation of gelatin films. FEBS Lett. **421:** 159–164.

11. WILL, H., S.J. ATKINSON, G.S. BUTLER, B. SMITH & G. MURPHY. 1996. The soluble catalytic domain of membrane type 1 matrix metalloproteinase cleaves the propeptide of progelatinase A and initiates autoproteolytic activation—Regulation by TIMP-2 and TIMP-3. J. Biol. Chem. **271:** 17119–17123.

12. STRONGIN, A.Y., I. COLLIER, G. BANNIKOV, B.L. MARMER, G.A. GRANT & G.I. GOLDBERG. 1995. Mechanism of cell surface activation of 72-kDa type IV collagenase. Isolation of the activated form of the membrane metalloprotease. J. Biol. Chem. **270:** 5331–5338.

13. BUTLER, G.S., M.J. BUTLER, S.J. ATKINSON, H. WILL, T. TAMURA, S.S. VAN WESTRUM, T. CRABBE, J. CLEMENTS, M.P. D'ORTHO & G. MURPHY. 1998. The TIMP2 membrane type 1 metalloproteinase "receptor" regulates the concentration and efficient activation of progelatinase A—A kinetic study. J. Biol. Chem. 273: 871–880.

14. KNÄUPER, V., H. WILL, C. LOPEZ-OTIN, B. SMITH, S.J. ATKINSON, H. STANTON, R.M. HEMBRY & G. MURPHY G. 1996. Cellular mechanisms for human procollagenase-3 (MMP-13) activation: evidence that MT1-MMP (MMP-14) and gelatinase A (MMP-2) are able to generate active enzyme. J. Biol. Chem. **271:** 17124–17131.

15. KNÄUPER, V., S. COWELL, B. SMITH, C. LOPEZ-OTIN, M. O'SHEA, H. MORRIS, L. ZARDI & G. MURPHY. 1997. The role of the C-terminal domain of human collagenase-3 (MMP-13) in the activation of procollagenase-3, substrate specificity, and tissue inhibitor of metalloproteinase interaction. J. Biol. Chem. **272:** 7608–7616.

16. WARD, R.V., S.J. ATKINSON, P.M. SLOCOMBE, A.J.P. DOCHERTY, J.J. REYNOLDS & G. MURPHY. 1991. Tissue inhibitor of metalloproteinases-2 inhibits the activation of 72 kDa progelatinase by fibroblast membranes. Biochim. Biophys. Acta **1079:** 242–246.

17. COCKETT, M.I., G. MURPHY, M.I. BIRCH, J.P. O'CONNELL, T. CRABBE, A.T. MILLICAN, I.R. HART & A.J.P. DOCHERTY. 1997. Matrix metalloproteinases and metastatic cancer. Biochem. Soc. Symp. **63:** 295–313.

18. COWELL, S., V. KNÄUPER, M.L. STEWART, M.P. D'ORTHO, H. STANTON, R.M. HEMBRY, C. LOPEZ-OTIN, J.J. REYNOLDS & G. MURPHY. 1998. Induction of matrix metalloproteinase activation cascades based on membrane-type 1 matrix metalloproteinase: associated activation of gelatinase A, gelatinase B and collagenase 3. Biochem. J. **331:** 453–458.

19. HUNZIKER, E.B. & L.C. ROSENBERG. 1996. Repair of partial-thickness defects in articular cartilage: Cell recruitment from the synovial membrane. J. Bone Joint Surg. **78A:** 721–733.

20. JOHANSSON, N., U. SAARIALHO-KERE, K. AIROLA, R. HERVA, L. NISSINEN, J. WESTERMARCK, E. VUORIO, J. HEINO & V.M. KÄHÄRI. 1997. Collagenase-3 (MMP-13) is expressed by hypertrophic chondrocytes, periosteal cells, and osteoblasts during human fetal bone development. Dev. Dyn. **208:** 387–397.

21. STÅHLE-BÄCKDAHI, M., B. SANDSTEDT, K. BRUCE, A. LINDAHL, M.G. JIMENEZ, J.A. VEGA & C. LOPEZ-OTIN. 1997. Collagenase-3 (MMP-13) is expressed during human fetal ossification and re-expressed in postnatal bone remodeling and in rheumatoid arthritis. Lab. Invest. **76:** 717–728.

22. KNÄUPER, V., B. SMITH, C. LOPEZ-OTIN & G. MURPHY. 1997. Activation of progelatinase B (proMMP-9) by active collagenase-3 (MMP-13). Eur. J. Biochem. **248:** 369–373.

23. COSSINS, J., T.J. DUDGEON, G. CATLIN, A.J.H. GEARING & J.M. CLEMENTS. 1996. Identification of MMP-18, a putative novel human matrix metalloproteinase. Biochem. Biophys. Res. Commun. **228:** 494–498.

24. PENDAS, A.M., V. KNÄUPER, X.S. PUENTE, E. LLANO, M.G. MATTEI, S. APTE, G. MURPHY & C. LOPEZ-OTIN. 1997. Identification and characterization of a novel human matrix metalloproteinase with unique structural characteristics, chromosomal location, and tissue distribution. J. Biol. Chem. **272:** 4281–4286.

25. PARK, A.J., L.M. MATRISIAN, A.F. KELLS, R. PEARSON, Z. YUAN & M. NAVRE. 1991. Mutational analysis of the transin (rat stromelysin) autoinhibitor region demonstrates a role for residues surrounding the "cysteine switch". J. Biol. Chem. **266:** 1584–1590.

26. HUGHES, C.E., B. CATERSON, A.J. FOSANG, P.J. ROUGHLEY & J.S. MORT. 1995. Monoclonal antibodies that specifically recognize neoepitope sequences generated by 'aggrecanase' and matrix metalloproteinase cleavage of aggrecan: application to catabolism in situ and in vitro. Biochem. J. **305:** 799–804.

27. KOLB, C., S. MAUCH, H.H. PETER, U. KRAWINKEL & R. SEDLACEK. 1997. The matrix metalloproteinase RASI-1 is expressed in synovial blood vessels of a rheumatoid arthritis patient. Immunol. Lett. **57:** 83–88.

28. SEDLACEK, R., S. MAUCH, B. KOLB, C. SCHÄTZLEIN, H. EIBEL, H.H. PETER, J. SCHMITT & U. KRAWINKEL. 1998. Matrix metalloproteinase MMP-19 (RASI 1) is expressed on the surface of activated peripheral blood mononuclear cells and is detected as an autoantigen in rheumatoid arthritis. Immunobiology **198:** 408–423.

29. ATKINSON, S.J., T. CRABBE, S. COWELL, R.V. WARD, M.J. BUTLER, H. SATO, M. SEIKI, J.J. REYNOLDS & G. MURPHY. 1995. Intermolecular autolytic cleavage can contribute to the activation of progelatinase A by cell membranes. J. Biol. Chem. **270:** 30479–30485.

# The Next Generation of MMP Inhibitors

## Design and Synthesis

BISWANATH DE,[a] MICHAEL G. NATCHUS, MENYAN CHENG,
STANISLAW PIKUL, NEIL G. ALMSTEAD, YETUNDE O. TAIWO,
CATHERINE E. SNIDER, LONGYIN CHEN, BOBBY BARNETT,
FEI GU, AND MARTIN DOWTY

*Procter & Gamble Pharmaceuticals, HCRC, 8700 Mason-Montgomery Road, Mason, Ohio 45040-8006, USA*

ABSTRACT: Since their inception during the eighties, MMP inhibitors (MMPIs) have gone through several cycles of metamorphosis. The design of early MMPIs was based on the cleavage site of peptide substrates. The second generation contained a substituted succinate scaffold (e.g., marimastat) coupled to a modified amino acid residue. The lower molecular weight analogs with multiple substitution possibilities produced a series of MMP inhibitors with varying degrees of selectivity for various MMPs. The introduction of sulfonamides in the midnineties added a new dimension to this field. The simplicity of synthesis coupled with high potency (e.g., CGS-27023A, AG-3340) produced a number of clinical candidates. This review highlights some of the key features that contributed to the discovery of this novel series of MMP inhibitors.

### INTRODUCTION

Matrix metalloproteases (MMPs) with a divalent $Zn^{2+}$ at the active site constitute an important class of endoproteinases involved in extracellular matrix remodeling. These proteases are secreted as proenzymes which are subsequently processed by other proteases to release the active forms. Under normal physiological conditions, the proteolytic activities are controlled by maintaining a balance between synthesis of the active forms and inhibition of the same by tissue inhibitors of matrix metalloproteases (TIMPs).[1–6] In pathological conditions this fine balance is tipped more towards degradation leading to pathogenesis. Consequently, there has been significant interest in the development of MMP inhibitors with the understanding that such agents will be able to control the aberrant regulation of MMP production, thus leading to amelioration of various disease states such as osteoarthritis and tumor metastasis.[7]

The design of early MMP inhibitors was based on the scissile site sequence of peptide substrates (FIG. 1).[8] The catalytic domains of MMPs identified thus far maintain a high degree of homology. Consequently, many of the early succinate-based MMP inhibitors (MMPIs) exhibited a broad-spectrum inhibition profile.

---

[a]Corresponding author.

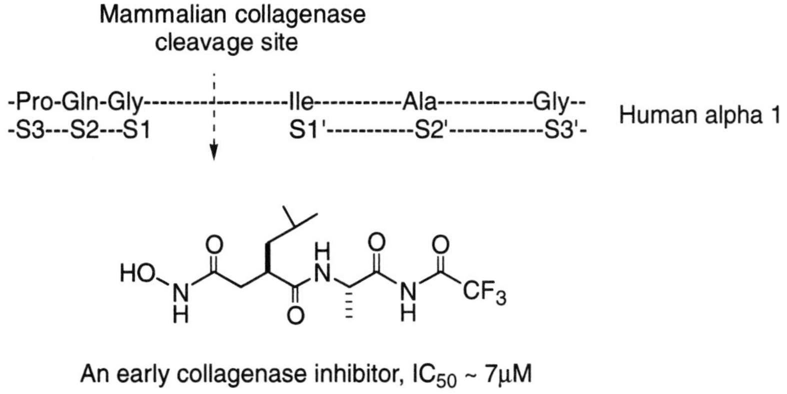
Mammalian collagenase
cleavage site
-Pro-Gln-Gly---------------------Ile-----------Ala------------Gly--
-S3---S2---S1 S1'-----------S2'------------S3'-
Human alpha 1
An early collagenase inhibitor, IC50 ~ 7μM

**FIGURE 1.** Substrate-based design of MMP inhibitors.

The availability of a vast pool of SAR data coupled with crystallography work with MMP-1 as well as MMP-3 have helped to define key binding interactions. The discovery of a deep S1′ pocket in MMP-3 as opposed to a restricted passage resulting from an Arg residue in the S1′ tunnel in MMP-1 added momentum towards the development of both MMP-1 and MMP-3 selective inhibitors.[9–11] As shown in FIGURE 2, the hydroxamic acid unit with at least 4-point attachments truly behaves like a molecular magnet, the significance of which becomes clear as one converts the same to its corresponding carboxylic acid with a concomitant 100–1000-fold loss in binding potency. The finely tuned pKa of the hydroxamic acid moiety (pKa ~ 8) helps maintain key H-bonding interactions with Glu-202 at physiological pH.[12] The amidic proton forms a stable hydrogen bond with Ala-165. The two oxygen atoms chelate the tricoordinated $Zn^{2+}$ with two nearly identical bonds (lengths ~ 1.8 to

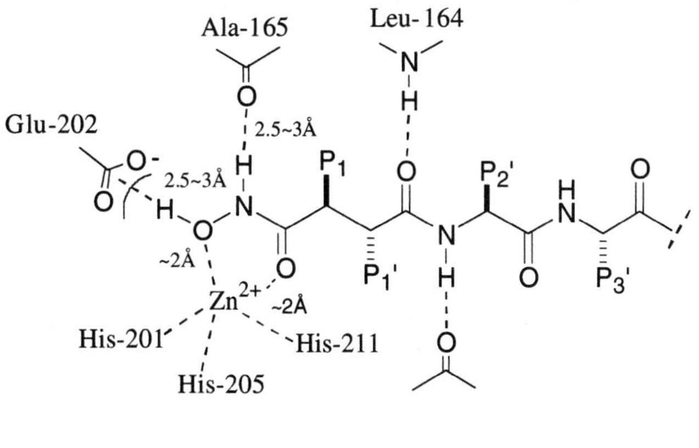
Ala-165
Leu-164
Glu-202
2.5~3Å
2.5~3Å
~2Å
~2Å
His-201
His-205
His-211

**FIGURE 2.** A general overview of key enzyme–inhibitor interactions.

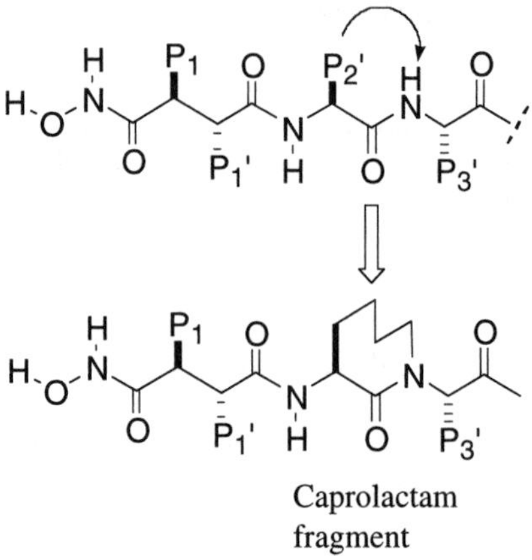

Caprolactam
fragment

**FIGURE 3.** Design of caprolactam-based MMP inhibitors.


2.2 Å). The remaining residues (P1, P1′ etc.) play a significant role in filling specific pockets, endowing the individual inhibitors with unique potency and selectivity.


## CAPROLACTAM-BASED INHIBITORS

In order to understand the significance of various substituents as well as the geometry of the active conformations, the early design strategy at P&GP was focused on incorporating cyclic residues at various parts of the inhibitor backbone. Considering the stringent binding requirements around the hydroxamic acid unit, the plan was to move away from this center and introduce a ring structure by attaching the P2′ residue to the amidic nitrogen. The first target in this project was the caprolactam-based inhibitor (FIG. 3), which could be modified by varying key residues to optimize potency, selectivity and finally efficacy in an animal model of arthritis.[13]

As described in SCHEME 1, the monosubstituted succinate intermediate **2** was synthesized via Evans oxazolidinone chemistry and subsequently coupled to the caprolactam moiety to provide the protected *t*-butyl ester **4,** which on further manipulation gave the desired hydroxamic acid **5** as a single diastereomer (see REF. 12).

For the synthesis of disubstituted succinate intermediate **11**, an alkylation followed by equilibration sequence was used. The mixture of diastereomers were carried all the way to the benzyl hydoxamate stage at which point the diastereomers were separated by column chromatography or crystallization. The 2S diastereomer **12** was finally deprotected under controlled hydrogenolysis conditions to give the desired hydroxamic acid **16**. Additional analogues, **17** and **18**, were synthesized by further elaboration of the allylic double bond in **12,** which gave **13** and **14** (see SCHEME 2).

Examination of the *in vitro* binding potencies for the various caprolactam-derived analogues indicated that the unsubstituted compound **6** did not possess high potency

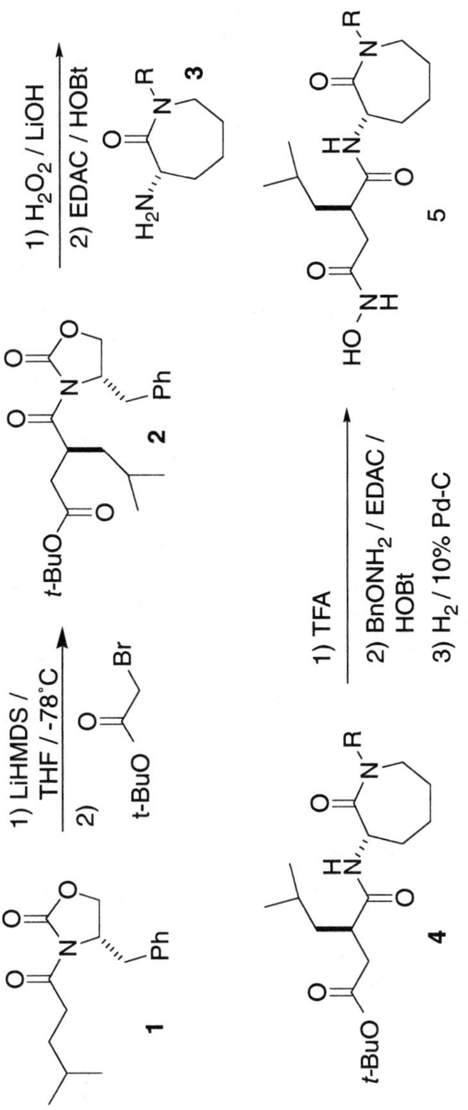

**SCHEME 1.** Synthesis of caprolactam-based MMP inhibitor.

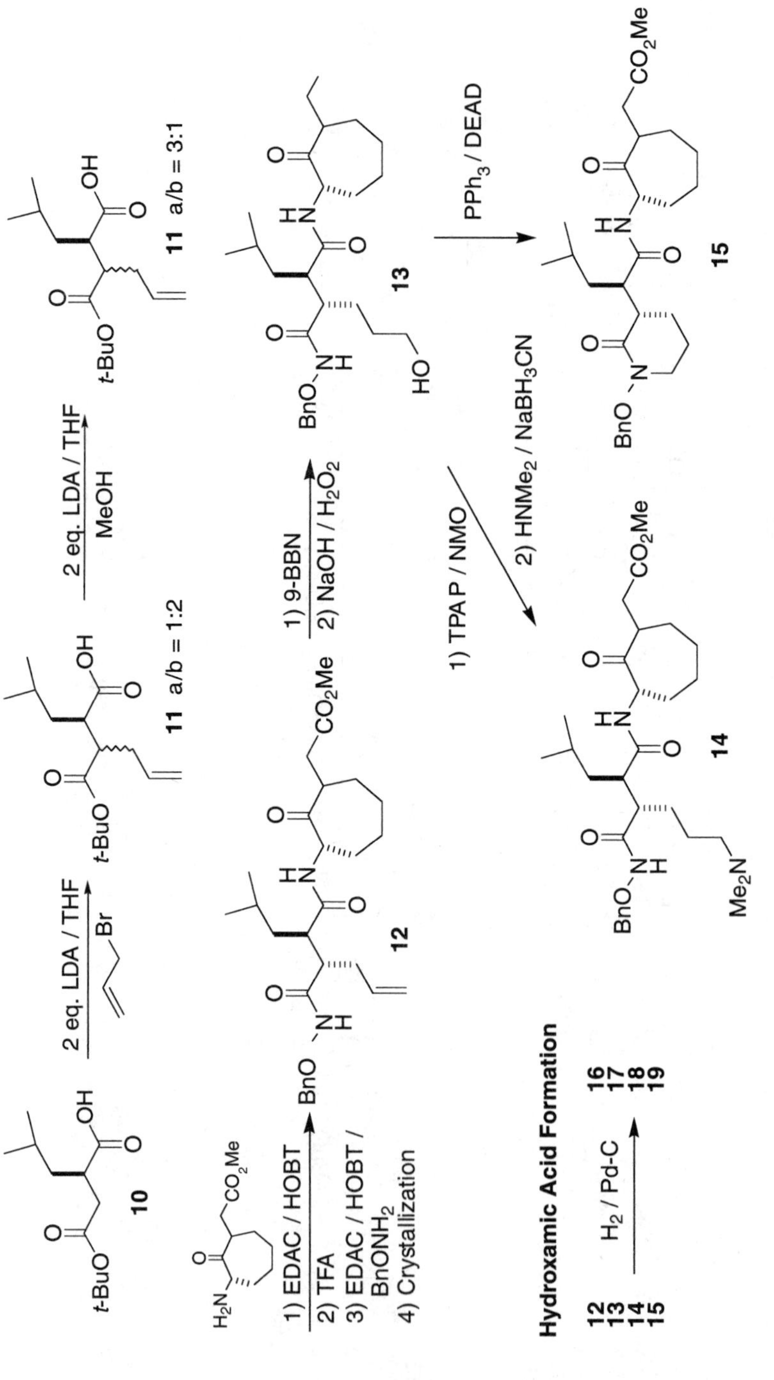

**SCHEME 2.** Synthesis of disubstituted succinate-based MMP inhibitors

**TABLE 1.** *In vitro* binding data for esters

|          |           |                                   | IC$_{50}$ (nM)$^a$ |        |
|----------|-----------|-----------------------------------|--------|--------|
| Compound | R$_1$     | R$_2$                             | MMP-1  | MMP-3  |
| 6        | *i*-Pr    | H                                 | 3,900  | 11,700 |
| 7        | *i*-Pr    | CH$_2$CO$_2$Me                    | 225    | 11,900 |
| 8        | *n*-pentyl| CH$_2$CO$_2$Me                    | 129    | 2,900  |
| 9        | *n*-heptyl| CH$_2$CO$_2$Me                    | 10     | 1,040  |

$^a$Single determination, truncated human MMP-1 and MMP-3 used for binding.

**TABLE 2.** *In vitro* binding data for ethers

|          |                 |             | IC$_{50}$ (nM)$^a$ |       |
|----------|-----------------|-------------|--------|-------|
| Compound | R$_1$           | R$_2$       | MMP-1  | MMP-3 |
| 20       | H               | *n*-heptyl  | 1,890  | 250   |
| 21       | Me              | *n*-heptyl  | 112    | 30    |
| 16       | *n*-propyl      | *i*-Pr      | 83     | 94    |
| 17       | 3-hydroxypropyl | *i*-Pr      | 72     | 102   |
| 18       | 3-dimethylamino | *i*-Pr      | 237    | 1,410 |
| 19       | Cyclized between R1 and NH | *i*-Pr | 2,300 | 5,200 |

$^a$Single determination, truncated human MMP-1 and MMP-3 used for binding.

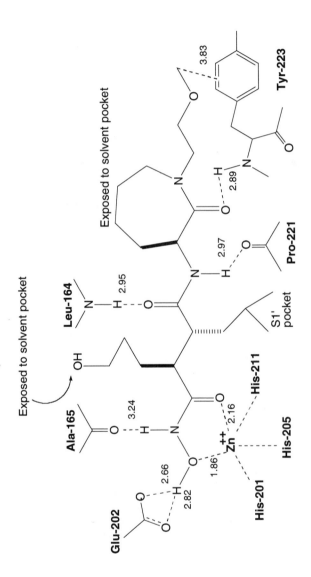

**FIGURE 4.** Key interactions of **17** with truncated human MMP-3

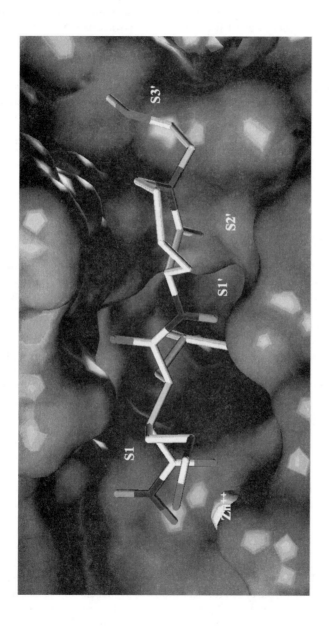

**FIGURE 4.** /continued.

against either MMP-1 or MMP-3. It was evident that some key pockets needed to be filled. The introduction of a carbomethoxy group at P3′ (see $R_2$ groups in TABLE 1) provided an early window for SAR exploration. Interestingly, elongation of the $R_1$ side chain enhanced the binding potency for MMP-1, an observation that can not be readily explained on the basis of the fact that S1′ pocket in MMP-1 is partially blocked by an Arg-residue (see details in REF. 11).

In an attempt to further explore the P1 site, the carbomethoxy group used at the P3′ site in earlier analogs was replaced by a nonhydrolyzable methoxyethyl side chain (TABLE 2). A definite trend began to appear with the first few analogues. The introduction of a small methyl group at P1 contributed not only towards overall potency, but also selectivity (compare **21** vs. **20**). In general, the inhibitors displayed enhanced potency for both enzymes with a concomitant loss of selectivity.

In order to further probe the importance of the hydroxamate N-H proton, the P1 residue was connected to this nitrogen directly providing compound **19**, which subsequently failed to exhibit any affinity for either enzyme. This finding once again confirms the importance of the Ala-164 hydrogen-bonding for the preservation of potency (FIG. 2).

On the basis of some unique properties exhibited by the caprolactam-based MMP inhibitors, an attempt was made to gain additional insight into the binding of **17** to truncated MMP-3. As shown in FIGURE 4, the X-ray structure of the co-crystal was solved and the key interactions were found to be well preserved. The hydoxamate ligand was seen to coordinate the catalytic $Zn^{2+}$ with two nearly identical bonds. The hydroxypropyl group projected much more towards solvent as opposed to being in the expected S1 pocket. The propyl residue of the caprolactam ring seemed to occupy the so-called open S2′ pocket (open to solvent). The methoxyethyl side chain filled the S3′ pocket with some significant hydrophobic interactions with Tyr-223 (aromatic ring to OMe distance ∼ 3.4–3.8 Å) as well as the isobutyl residue of Leu-164 (average distance ∼ 4 Å). Despite the introduction of a 7-membered ring into the backbone, some expected interactions (Pro-221 and Tyr-223 hydrogen bonding) were maintained. Overall, the binding mode of **17** was similar to the reported structure for many classical MMP inhibitors with succinate backbones.[9]

## PHOSPHINAMIDES

While the work on caprolactam-based MMP inhibitors was in progress, an attempt was made to identify the minimum binding requirements for MMP inhibitors in order to design smaller and, if possible, monocyclic inhibitors. It was clear from the outset of this program that the hydroxamic acid moiety was probably the most important structural element responsible for the high binding affinities observed for various MMPs. Consequently, the initial strategy was to connect P1 through P2′ residues of a typical succinate-based inhibitor (FIG. 3). The disclosure of sulfonamide-based inhibitors at this time added further support to the understanding that a potent inhibitor can be designed with fewer points of attachments.[14] Since a part of our plan was to retain the hydroxamic acid unit, two new key probes (**22** and **23**) were designed to gauge the relative importance of the Leu-164 hydrogen bonding for the preservation of binding potency.[15]

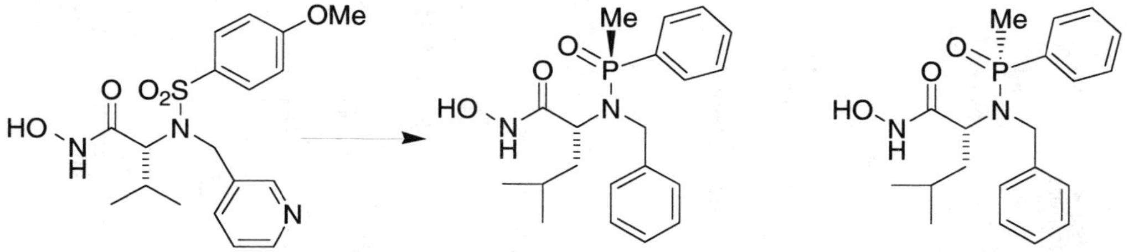

CGS-27023A

**22:** IC$_{50}$ = 20 nM for MMP-1
24 nM for MMP-3

**23:** IC$_{50}$ = 7120 nM for MMP-1
9170 nM for MMP-3

**FIGURE 5.** Design of phosphorus-based probe to investigate the role of Leu-164 H-bonding.

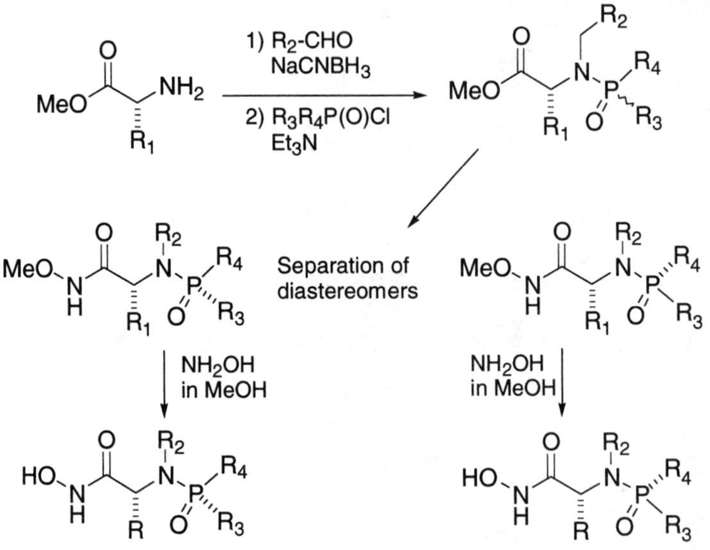

**SCHEME 3.** An overview of the synthetic strategy for phosphinamide-based MMP inhibitors.

**TABLE 3. Binding affinities for phosphinamide-based MMP inhibitors**

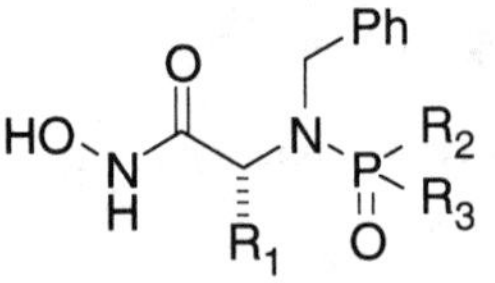

| Compound | $R_1$ | $R_2$ | $R_3$ | IC50 (nM)[a] | |
|:---:|:---:|:---:|:---:|:---:|:---:|
| | | | | MMP-1 | MMP-3 |
| 24 | Me | Me | Ph | 120 | 70 |
| 25 | Me | Et | Ph | 608 | 700 |
| 26 | Me | Ph | Ph | 6,790 | 10,300 |
| 22 | i-Bu | Me | Ph | 20 | 24 |
| 27 | i-Bu | Me | Me | 518 | 1040 |

[a]IC$_{50}$ for the more potent diastereomer.

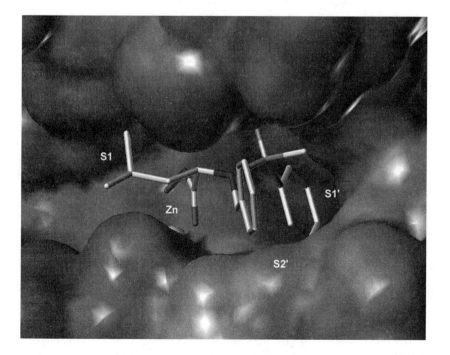

**FIGURE 6.** Key hydrogen bonding interactions of **22** to truncated MMP-3.

Most of the phosphinamide-based MMP inhibitors were synthesized following the general strategy as described in SCHEME 3 (see REF. 14 for details). The methyl ester of an amino acid was subjected to $N$-alkylation by a reductive amination sequence. The resulting secondary amine was phosphorylated using the appropriate phosphinyl chloride to afford a mixture of diastereomers, which were subsequently separated by column chromatography. The diastereomers were treated separately with methanolic hydroxyl amine to give the desired hydroxamates.

As shown in TABLE 3, a limited number of analogues were prepared to investigate the topography of the active pocket. Small substituents, like methyl and isobutyl groups, fit in well at the P1 site, whereas the environment at or around the phosphorus nucleus was quite restrictive, in terms both of size and polarity of the groups attached. The biphenylphosphinyl residue had a deleterious effect on both MMP-1 and MMP-3. The same observation was made with two methyl substituents on phosphorus. The best potency, however, was exhibited by a specific diastereomer, **22**, with a unique combination of substitution and chirality. The chirality at phosphorous was determined to be extremely important (compare **22** and **23**). These analogues clearly demonstrate that a properly oriented oxygen atom is essential to preserve the Leu-164 hydrogen bonding (FIG. 6). As shown by the crystal structure of **22** bound to truncated MMP-3, the phosphinyl oxygen clearly pointed towards the Leu-164 residue preserving an essential hydrogen bond ($\sim 2.70$ Å). Additionally, the Ala-165 residue was within reach to form a second hydrogen bond with the phosphinyl oxygen. The hydroxamate moiety chelated the catalytic $Zn^{2+}$ in the expected fashion. The benzyl substituent projected towards open solvent pocket, and the phenyl group on phosphorus partly filled the deep S1´ pocket.

Looking at the results from the previous studies it became clear that one could design small MMP inhibitors with fewer points of attachments while preserving binding potency for multiple MMPs as long as certain key interactions were preserved. This understanding provided a firm foundation for subsequent work on the design and synthesis of new MMP inhibitors.

## PIPERAZINES

With the introduction of a new generation of MMP inhibitors (see REF. 13), attempts were made to understand not only the key points of interactions, but also the geometry of the active conformation of various inhibitors. Since the P2´ residue usually projects towards the open solvent pocket, a short tether was attached between P1 and P2´ to form a small ring. In order to map the surrounding area, a heteroatom was inserted to optimize both potency and selectivity. As shown in FIGURE 7, 1,3-, 1,4- and 1,5-piperazines were chosen for investigation during the early phase of this work. This review will mostly focus on the SAR of 1,4-piperazine based MMP inhibitors.[16–18]

The design and synthesis of piperazine-based MMP inhibitors were greatly facilitated by the commercial availability of piperazine carboxylic acid in bulk quantity. The initial plan was to carry out all SAR work with a racemic mixture. As shown in SCHEME 4, the differential protection of two basic nitrogens was achieved via selective protection of the distal nitrogen with BOC-anhydride followed by sulfonation of

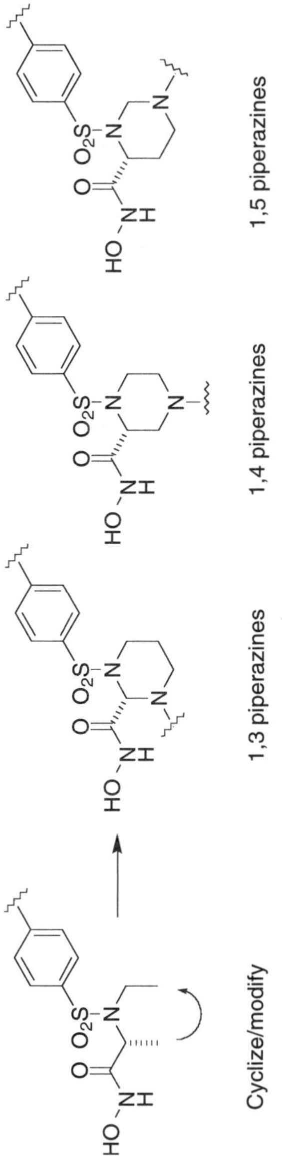

**FIGURE 7.** The design of piperazine-based MMP inhibitors.

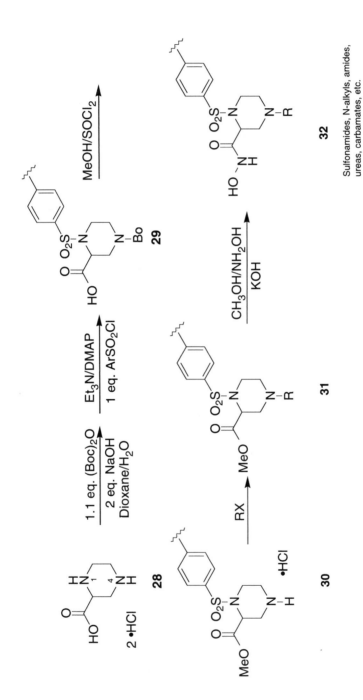

**SCHEME 4.** Synthesis of cyclic MMP inhibitors.

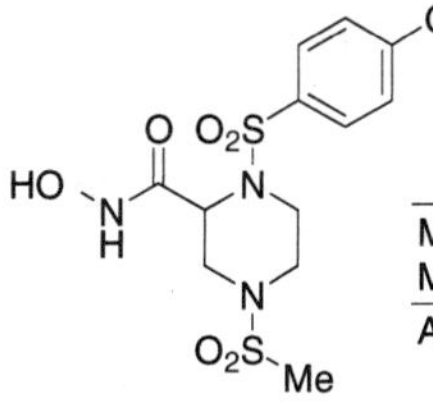

**33**
MMP-1: 8 nM
MMP-3: 3 nM
Absorption 51-62%

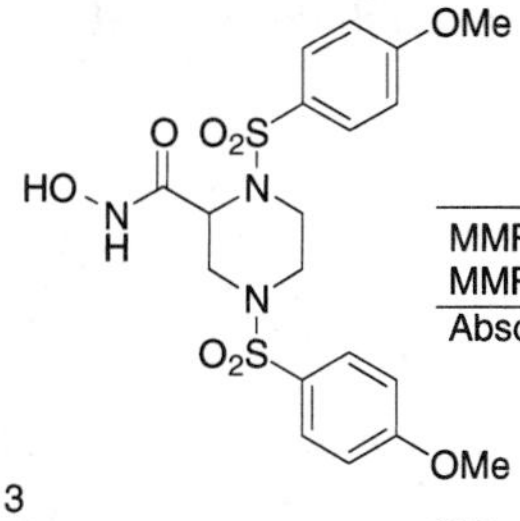

**34**
MMP-1: 76 nM
MMP-3: 25 nM
Absorption 5%

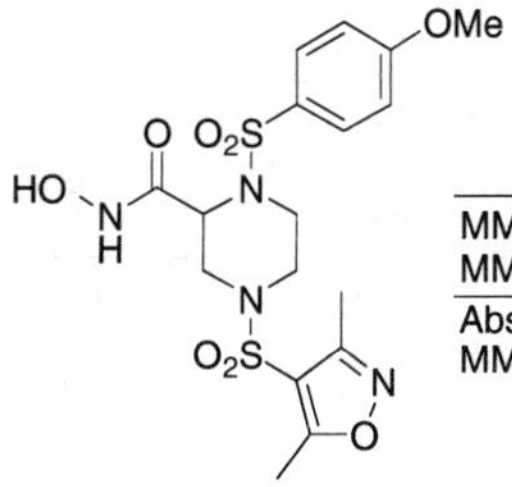

**35**
MMP-1: 53 nM
MMP-3: 3 nM
Absorption 2%
MMP-3 selection

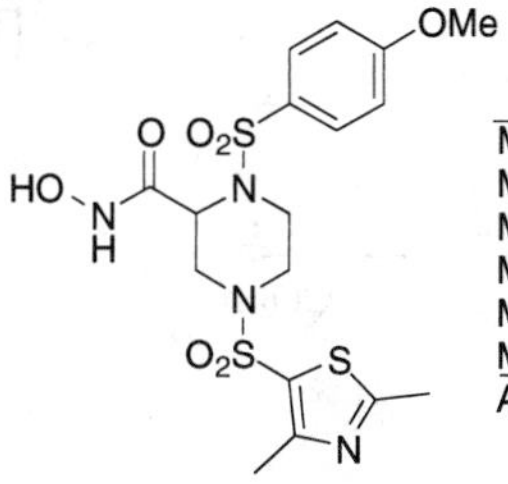

**36**
MMP-1: 30 nM
MMP-3: 2 nM
MMP-7:> 1000 nM
MMP-8: 2 nM
MMP-9: 2 nM
MMP-13: 0.8 nM
Absorption: 37%

**FIGURE 8.** Sulfonamide-derived MMP inhibitors.

the remaining basic nitrogen. The secondary amine, **30**, served as a key intermediate for a wide variety of analogues leading ultimately to the identification of **49** as the lead candidate of this series.

The first series of bis-sulfonamide–based analogues were found to be potent and generally broad-spectrum inhibitors. Although **35** showed a hint of selectivity, the size and polarity of the 4N-substituents, in general, did not have a significant effect on either potency or selectivity. In some cases, major differences were seen in the absorption profile, but it was difficult to identify the distinguishing elements which endowed the inhibitors with improved pharmacokinetic profiles. For example, in

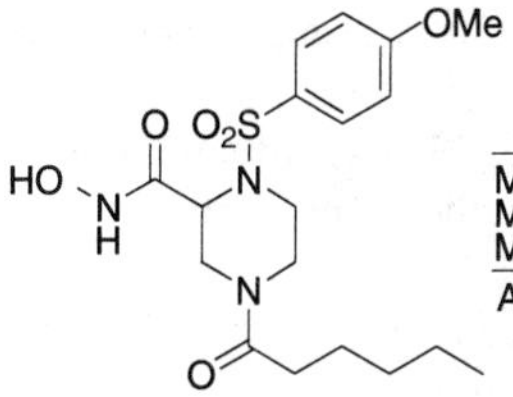

**37**
MMP-1: 115 nM
MMP-3: 39 nM
MMP-13: 2 nM
Absorption: 18%

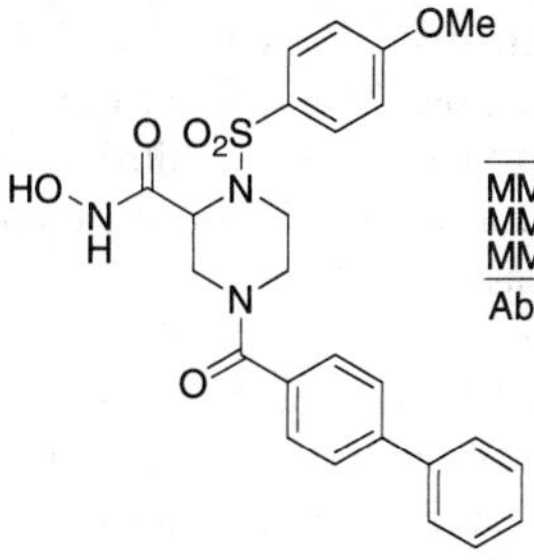

**38**
MMP-1: 101 nM
MMP-3: 15 nM
MMP-13: 3 nM
Absorption: >90%

**FIGURE 9.** Amide-derived MMP inhibitors.

**TABLE 4. Tertiary amide-derived MMP inhibitors.**

| Compound | $R_1$ | IC$_{50}$(nM) | | |
| --- | --- | --- | --- | --- |
| | | MMP-1 | MMP-3 | MMP-13 |
| **39** | COMe | 73 | 18 | nd |
| **38** | | 102 | 15 | 3 |
| **40** | | 25 | 7 | 1 |
| **41** | | 2647 | 776 | 3 |

terms of absorption profile, compound **33** with a small methyl group was comparable to **36** containing a polar aminothiazole group. At the same time both compounds **34** and **35** with comparable 4N-substituents failed to exhibit any meaningful difference.

The amides (**37** and **38**) derived from **30** were also found to be quite potent, with a respectable margin of selectivity between MMP-1 and MMP-13. All analogues, in general, exhibited a high affinity for MMP-13 (IC$_{50}$s < 10 nM) and moderate to high absorption profile. Some additional data as presented in TABLE 4 describe some unique features. For example, a small acetamide group in **39** did not have any significant effect either on selectivity or on potency. A bulkier but linear biphenyl substituent in **38** maintained the same trend. However, a small furan group present in **40** helped improve potency to some extent. The most noteworthy compound of this lot, **41**, turned out to be a poor inhibitor of both MMP-1 and MMP-3 while maintaining a high affinity for MMP-13. This difference, specifically in the absence of crystal data, cannot be easily explained.

Alkylamine-based MMP inhibitors derived from the secondary amine, **30**, exhibited unique affinity for both MMP-1 and MMP-3. As noted in TABLE 5, the secondary amine is reasonably potent for both enzymes. Adding a small methyl group led to a net loss of potency by a factor of ~5–6. A bulkier lipophilic group, like benzyl, helped restore the potency with a concomitant improvement in absorption profile. This was not entirely surprising given the fact that polar molecules oftentimes do not

TABLE 5. Tertiary amine-derived MMP inhibitors

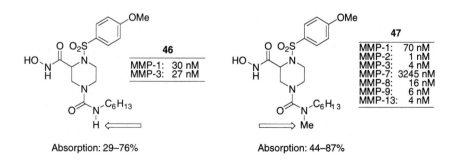

| Compound | R | IC$_{50}$(nM) | | Absorption |
|:---:|:---:|:---:|:---:|:---:|
| | | MMP-1 | MMP-3 | |
| 39 | -H | 175 | 48 | 4% |
| 40 | -Me | 1260 | 269 | nd |
| 41 | -CH$_2$Ph | 324 | 75 | 98% |
| 42 | -(CH$_2$)$_3$Ph | 1440 | 206 | nd |

get across cell membranes with great mobility. Further elongation of the chain had a deleterious effect on binding, leading to a net loss of potency for both MMP-1 and MMP-3. During the early phase of this program, it was assumed that all substituents attached to the distal nitrogen are pointing towards the solvent phase, thus displaying a poor SAR correlation. However, upon the accumulation of additional data, it became clear that this distinct pocket offered more than just access to the open solvent medium.

The availability of a large pool of isocyanates and the ease of synthesis of ureas by various methods added momentum towards the design and synthesis of a large number of urea-based MMP inhibitors. The following two examples from this series

FIGURE 10. Urea-derived MMP inhibitors.

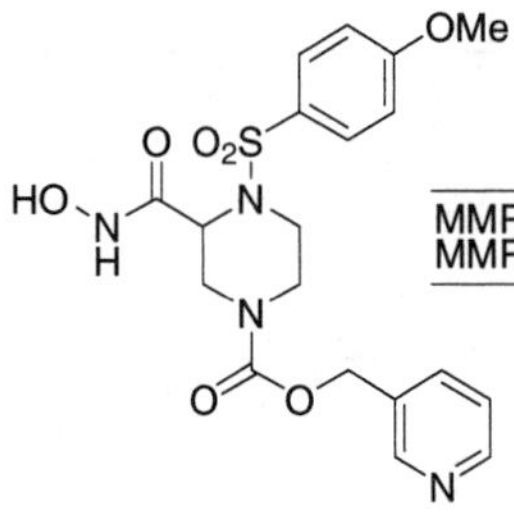


**48**

| | |
|---|---|
| MMP-1: | 17 nM |
| MMP-3: | 9 nM |

Absorption 10%

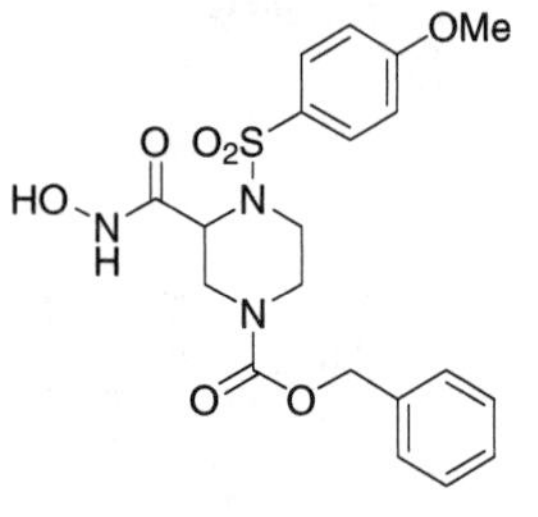


**49**

| | |
|---|---|
| MMP-1: | 24 nM |
| MMP-3: | 18 nM |
| MMP-7: | 232 nM |
| MMP-9: | 2 nM |
| MMP-13: | 1 nM |

*Absorption: 59%
*Explant $IC_{50}$ ~ 43 nM
*Orally active in rat arthritis model

**FIGURE 11.** Carbamate-derived MMP inhibitors.

highlight some of the unique features. The unsubstituted urea, **46**, was potent and displayed comparable affinity for both MMP-1 and MMP-3. Surprisingly, the addition of a small methyl group in **47** had a significant impact on selectivity while maintaining a favorable absorption profile. Once again, in the absence of any crystallographic data, it may be difficult to rationalize such a swing in selectivity.

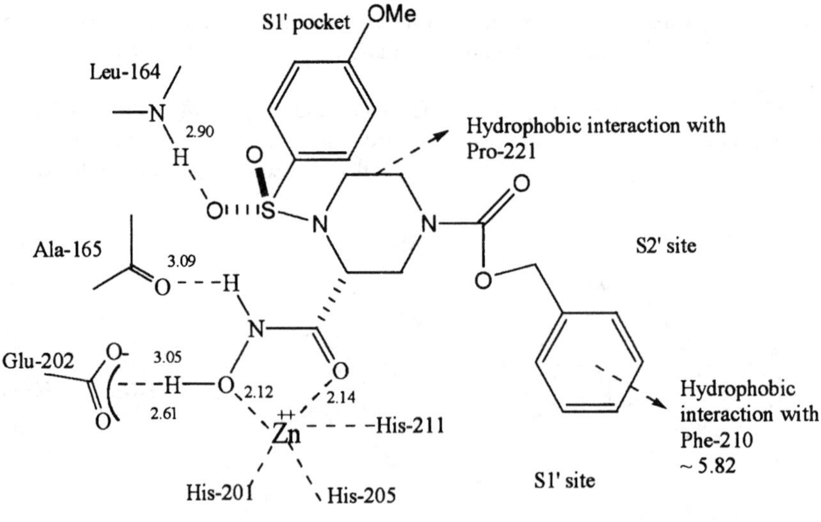


**FIGURE 12.** Data from the composite crystal structure of **49** and truncated MMP-3.

The last batch of MMP inhibitors derived from carbamates also adhered to the observed trend seen with most of the piperazine-based inhibitors. For example, both **48** and **49** were potent and displayed comparable selectivity for MMP-1 and MMP-3. However, the analogue **49**, with a relatively hydrophobic benzyloxycarbonyl group, displayed better absorption profile than **49** with a polar pyridine ring.

Overall, **49** represented a typical member of piperazine-derived MMP inhibitors and was chosen for further advancement. As the reported data indicate, the lead analogue **49** not only demonstrated efficacy in the *in vitro* cartilage explant model but was also found to be efficacious in a rat arthritis model as well.

Attempts were also made to identify various key interactions which endow these inhibitors with such high potency and, in some cases, desirable selectivity. After several attempts, it was possible to obtain a composite crystal of **49** and truncated MMP-3 by the diffusion method, in which a solution of the inhibitor was allowed to soak into a few crystals of MMP-3.

It is evident from the picture that the hydroxamic acid unit was properly poised for chelation to $Zn^{2+}$ (Zn-O lengths are 2.12 and 2.14 Å). The methoxyphenyl group went deep into the S1´ pocket, while one of the sulfonamide oxygens formed a hydrogen bond to Leu-164 (~ 2.90 Å). The piperazine ring maintained a semi-chair conformation projecting the benzyloxycarbonyl group into an intermediate position between S1 and S2´. The latter part of this observation was critical. It was initially thought that the distal N-substituents occupied the S2´ pocket, which is basically an open solvent cavity. It appears now, at least for compound **49**, that the long benzyloxy group occupies a distinct position in space. This may explain some unique fluctuations in SAR documented for various analogues.

Finally, the development history of MMP inhibitors presented in this review represents a typical trend documented throughout the industry as well as in academia thus far. The new generation of inhibitors tends to be smaller, less structurally complex, and more selective. The resurrection of carboxylic acid–based MMP inhibitors is a welcome trend since one can now make selective MMP inhibitors with optimized pharmacokinetic profiles (*vide* Bayer in Phase-II/III for tumor metastasis).[19] Considering the vast potential of MMP inhibitors for the treatment of various diseases like cancer and osteoarthritis, one can only be optimistic thinking that multiple drug candidates will eventually emerge from this field of research.

## ACKNOWLEDGMENTS

A large number of dedicated scientists at P&GP contributed towards the success of the MMP project. We wish to thank all of them for their dedication and hard work. Given below is a partial list of those scientists whose work was presented in this review: Chris Wahl, Kelly McDow-Dunham, Rimma Bradley, Lisa Williams, and Glen Mieling.

## REFERENCES

1. BIRKEDAL-HANSEN, H., W. MOORE, M.K. BODDEN, L.J. WINDSOR, B. BIRKEDAL-HANSEN, A. DECARLO, J.A. Engler. 1993. Matrix metalloproteinases: a review. Crit. Rev. Oral. Biol. Med. **4:** 197–250.

2. MATRISIAN, L.M. 1990. Metalloproteinases and their inhibitors in matrix remodeling. Trends Genet. **6:** 212–125.
3. MURPHY, G. & A.J.P DOCHERTY. 1992. The matrix metalloproteinases and their inhibitors. Am. J. Respir. Cell. Mol. Biol. **7:** 120–125.
4. DOCHERTY, A.J.P. & G. MURPHY. 1990. The tissue metalloproteinase family and the inhibitor TIMP: a study using cDNAs and recombinant proteins. Ann. Rheum. Dis. **49:** 469–479.
5. WOESSNER, F.J. 1991. Matrix metalloproteinases and their inhibitors in connective tissue remodeling. FASEB **5:** 2145–2154.
6. BECKETT, R.P., A.H. DAVIDSON, A.H. DRUMMOND, P. HUXLEY, & M. WHITTAKER. 1996. Drug Discovery Today **1:** 16–26.
7. BECKETT, R. P. & M. WHITTAKER. 1998 Exp. Opin. Ther. Patents **8:** 259–282.
8. HENDERSON, B., A.J.P. DOCHERTY, & N.R.A BEELEY. 1990. Design of inhibitors of articular cartilage destruction. Drugs of the Future **15:** 495–508.
9. WILLIAMSON, R.A. et al. 1994. Solution structure of the active domain of tissue inhibitor of metalloproteinases-2. A new member of the OB fold protein family. Biochemistry **33:** 11745–11759.
10. SPURLINO, J.C. et al. 1994. 1.56, a structure of mature truncated human fibroblast collagenase. Proteins: Structure, Function Genetics **19:** 98–109.
11. BROWNER, M.F. 1994. Matrix metalloproteases: structure-based drug discovery targets. Perspect. Drug Discovery Design **2:** 343–351.
12. BABINE, R.E. & S.L. BENDER. 1997. Molecular recognition of protein-ligand complexes: Applications to drug design. Chem. Rev. **97:** 1359–1472.
13. NATCHUS, M.G., M. CHENG, C.T. WAHL, S. PIKUL, N.G. ALMSTEAD, R.S. BRADLEY, Y.O. TAIWO, G.E. MIELING, C.M. DUNAWAY, C.E. SNIDER, J.M. MCIVER, B.L. BARNETT, S.J. MCPHAIL, M.B. ANASTASIO & B. DE. 1998. Design and synthesis of conformationally-constrained MMP inhibitors. Bioorg. Med. Chem. Lett. **16:** 2077–2080.
14. MACPHERSON, L.J. & D.T. PARKER, inventor; Ciba-Geigy AG, assignee. 1993. European patent 0 606 046 A1. Date of application: June 1.
15. PIKUL, S. et al. 1998. Design and synthesis of phosphinamide-based hydroxamic acids as inhibitors of matrix metalloproteinases, J. Med. Chem. Submitted for publication.
16. De, B., M.G. Natchus, S. Pikul, N.G. Almstead, M. Cheng, et al., inventors; P&GP, assignee. 1996. World patent 9808825-A1. Date of application: August 28.
17. PIKUL, S., K.L. MCDOW DUNHAM, N.G. ALMSTEAD, B. DE, M.G. NATCHUS, M.V. ANASTASIO, S.J. MCPHAIL, C.E. SNIDER, Y.O. TAIWO, T. RYDEL, C.M. DUNAWAY, F. GU, & G.E. MIELING. 1998. Discovery of potent, achiral matrix metalloprotease inhibitors. J. Med. Chem. **41:** 3568–3571.
18. PIKUL, S. et al., inventors; P&G, assignee. 1996. World patent 9808823-A1. Date of application: August 28.
19. KLUENDER, H. et al., inventors; Bayer, assignee. 1995. World patent 96/15096. Date of application: November 15.

# Design and Synthetic Considerations of Matrix Metalloproteinase Inhibitors

JERAULD S. SKOTNICKI,[a] ARIE ZASK, FRANCES C. NELSON,
J. DONALD ALBRIGHT, AND JEREMY I. LEVIN

*Chemical Sciences, Wyeth-Ayerst Research, Pearl River, New York 10965, USA*

**ABSTRACT:** Experimental evidence confirms that the matrix metalloproteinases (MMPs) play a fundamental role in a wide variety of pathologic conditions that involve connective tissue destruction including osteoarthritis and
rheumatoid arthritis, tumor metastasis and angiogenesis, corneal ulceration,
multiple sclerosis, periodontal disease, and atherosclerosis. Modulation of
MMP regulation is possible at several biochemical sites, but direct inhibition
of enzyme action provides a particularly attractive target for therapeutic intervention. Hypotheses concerning inhibition of specific MMP(s) with respect
to disease target and/or side-effect profile have emerged. Examples are presented of recent advances in medicinal chemistry approaches to the design of
matrix metalloproteinase inhibitors (MMPIs), approaches that address structural requirements and that influence potency, selectivity, and bioavailability.
Two important approaches to the design, synthesis, and biological evaluation
of MMPIs are highlighted: (1) the invention of alternatives to hydroxamic acid
zinc chelators and (2) the construction of nonpeptide scaffolds. One current
example in each of these two approaches from our own work is described.

## INTRODUCTION

The proteolytic action of matrix metalloproteinases (MMPs) represents critical
and necessary biochemical processes leading to connective tissue destruction with
resulting disease manifestation. These zinc-requiring enzymes are endogenously
regulated by $\alpha_2$-macroglobulin and tissue inhibitors of metalloproteinases (TIMPs),
and this regulation will define the balance between normal physiological tissue remodeling and these pathological events related to this connective tissue breakdown.
For therapeutic intervention, obstruction is possible at one or more biochemical sites
(induction, production, zymogen activation, activity) of the MMP cascade, but direct
inhibition of enzyme action by synthetic agents that bind to the catalytic site is a particularly compelling and validated objective. These target molecules are amenable to
design for optimization of chemical, biochemical, and pharmacologic properties, as
well as structural diversity. Hypotheses have been generated and tested concerning
the inhibition of specific MMP(s) with respect to disease target and/or side effect
profile, but unequivocal evidence correlating these relationships is still unavailable.

Orally active matrix metalloproteinase inhibitors (MMPIs) with improved metabolic stability and duration of action have been identified. Recent advances in me-

[a]Address for telecommunication: Phone, 914/732-3817; fax, 914/732-5561; e-mail:
skotnij@war.wyeth.com

dicinal chemistry approaches to the design of MMPIs addressing structural
requirements that influence potency, selectivity, and bioavailability have been com-
prehensively reviewed[1-20] and will not be discussed in detail. Some small-molecule
MMPIs also inhibit TNF-α converting enzyme (TACE); these results are detailed in
recent reviews.[21-25]

## DESIGN OF MATRIX METALLOPROTEINASE INHIBITORS

A productive approach for the identification of MMPIs has been the design of
peptide and peptide-like compounds which combine backbone features (P1, P1´,
P2´, P3´ regions) which would favorably interact with the enzyme subsites (S1, S1´,
S2´, S3´ pockets) and functionality capable of binding zinc (chelator) in the catalytic
site (FIG. 1). This medicinal chemistry attack is focused on the design of new chemical entities that can be used to test hypotheses and probe structure-activity issues to
yield compounds with not only optimal chemical, physico-chemical, and pharmacologic properties, but also enhanced bioavailability and pharmacokinetic attributes.
Improvements in synthetic methods and strategy, in structural techniques and information, and in pharmacological methods and tools have added precision in defining,
refining, and evaluating these requirements for potency and selectivity. On the basis
of extensive exploration of analogue structure-activity studies, high affinity for these
subsites in conjunction with effective zinc chelation has been achieved, resulting in
the generation of an opulent array of potent, competitive enzyme inhibitors. Specific
enzyme recognition and thus selectivity (e.g., MMP-3 versus MMP-1) can be attained by structural modification of the P1´ (primary) and P2´/P3´ region(s), taking
advantage of distinctions in the depth and composition of the S1´, S2´, and S3´ pockets of the individual MMPs. The nature, extent, position and orientation of these
backbone modifications have been shown to exert a substantial effect on enzyme rec-

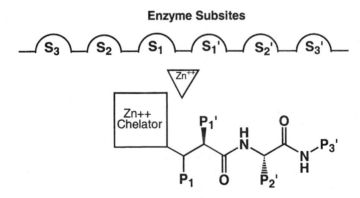

**FIGURE 1.** Design of peptide-based MMPIs.

ognition. Though the majority of these inhibitors contain the hydroxamic acid group as the zinc chelator, alternative chelators (*vide infra*) have been identified. The introduction of appropriate high-affinity (P1′ to S1′; P2′ to S2′) substituents on the peptide backbone may compensate for the resulting diminished chelating capacity and yield potent compounds.

Two areas of active MMPI research are focused on the identification of alternative chelators to the ubiquitous hydroxamic acid moiety and the design and synthesis of nonpeptide scaffolds which incorporate the hydroxamate chelating group. While addressing issues of low oral bioavailability and poor duration of action exhibited by most peptide-based compounds, both strategies provide opportunities for improved design and synthetic versatility, for augmentation of chemical and pharmacologic properties, for the generation of new biochemical probes and paradigms, and for facile entry to novel and unanticipated chemical structures. Noncomprehensive exemplification of molecules exploring these strategies are depicted in FIGS. 2 and 3. Herein is summarized an example of our work in these areas.

## ALTERNATIVE CHELATORS

As illustrated in FIGURE 2 and described in detail in the aforementioned reviews, a substantial synthesis effort has been expended in the search for non-hydroxamate zinc chelating groups. As part of a collaborative effort (Affymax and Wyeth-Ayerst Research), we have recently reported[26,27] the rationale for the design and synthesis of peptide-like mercaptoalcohols and mercaptoketones as the replacement of the hydroxamic acid group. Both the ketone and alcohol platforms are consistent with bidentate chelation. These compounds are devoid of the amide NH proximal to the sulfhydryl group, and are thus distinguished from mercapto-amides[28,29] to thereby allow examination of the importance of this hydrogen bond capability (FIG. 3).

One essential aspect[27] of this work involved the development of asymmetric syntheses for the individual succinyl mercaptoalcohol isomers, which are then converted to the corresponding ketones. In this way, we were able to examine the influence of alcohol stereochemistry on potency.

The succinyl fragment of the mercaptoalcohols and mercaptoketones is readily available in enantiomerically pure form from the manipulation of the epimeric iodolactones via a thiol displacement and lactone hydrolysis protocol, followed by a peptide coupling to incorporate the P2′ substituents. Details of the syntheses have been reported.[27] The mercaptoalcohols in which the hydroxyl and P1′ groups are *syn* are the most potent members of the series. The inhibitory potency of these analogues versus MMP-1 (TABLE 1) proved to be relatively independent of the size of the P1′ group, while MMP-3 activity was significantly improved for the longer heptyl side chain analogues. The isobutyl derivatives are therefore extremely potent and selective MMP-1 inhibitors, while the heptyl series provides a more broad-spectrum inhibition of the MMPs. Little difference was observed between the malonyl[26] and succinyl[27] series of inhibitors.

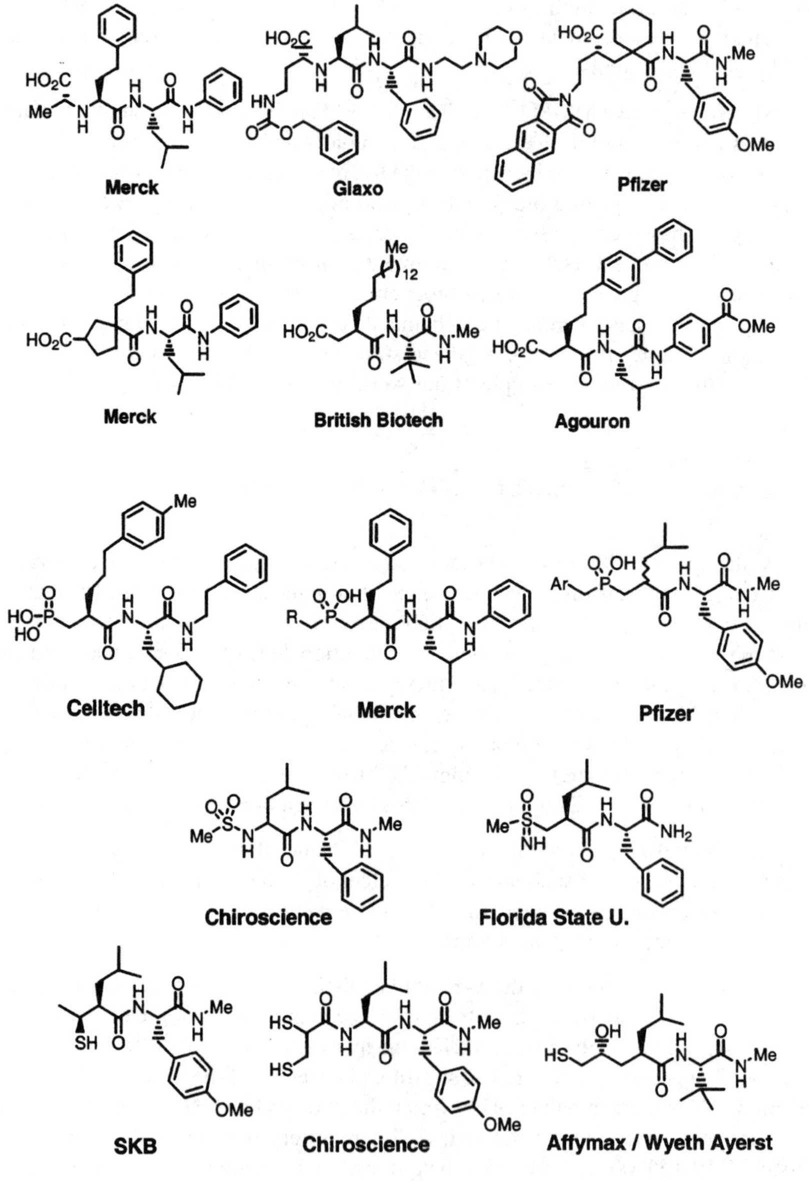

**FIGURE 2.** Representative nonhydroxamate MMPIs.

**TABLE 1.** *In vitro* activity (IC$_{50}$, nM) of mercaptoalchols and ketones

| | MMP-1 | MMP-3 |
|---|---|---|
| | 5 | 470 |
| | 11 | 480 |
| | 46 | 3700 |
| | 11 | 470 |
| | 10 | 8 |
| | 140 | 430 |
| | 5 | 9 |

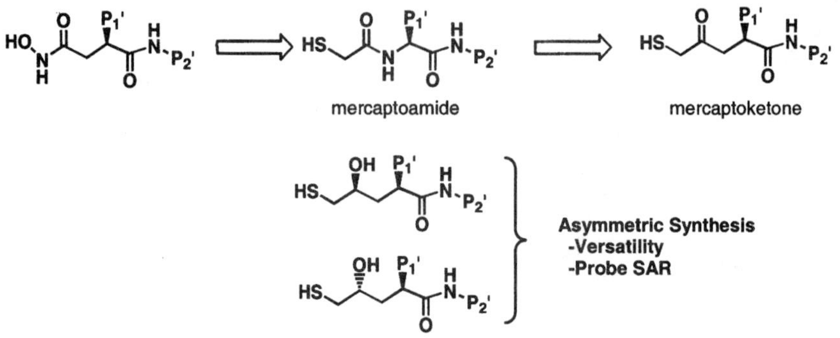

**FIGURE 3.** Design of mercaptoalcohols/mercaptoketones.

**TABLE 2.** Diazepine *in vitro* activity (IC$_{50}$, nM)

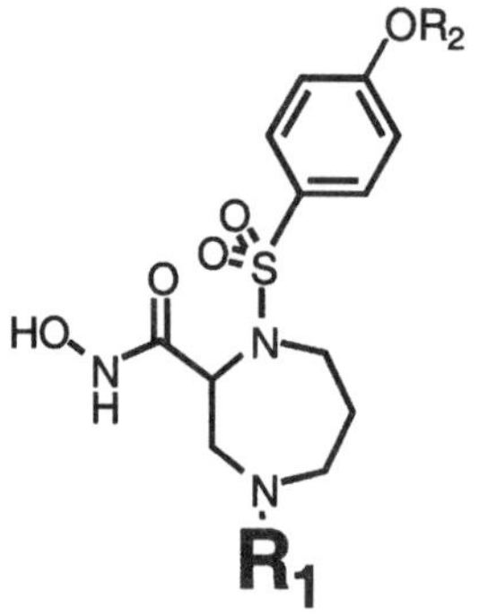

| R$_1$ | R$_2$ | MMP-1 | MMP-9 | MMP-13 |
|---|---|---|---|---|
| -H (HCl) | Me | 91 | 60 | 5.1 |
| -C(O)tBu | Me | 251 | 4.8 | 11 |
| -C(O)OtBu | Me | 618 | 9.1 | 26 |
| -C(O)CH$_2$NHtBOC | Me | 690 | 23 | 16 |
| -C(O)CH$_2$NH$_2$ (HCl) | Me | 703 | 157 | 46 |
| -C(O)Ph | Me | 45 | 2.0 | 2.2 |
| -C(O)Ph | Ph | 22 | 1.2 | 1.3 |
| -C(O)NHPh | Me | 309 | 18 | 17 |
| -CH$_2$Ph | Me | 445 | 31 | 65 |
| CGS 27023A | | 15 | 8.8 | 8.2 |

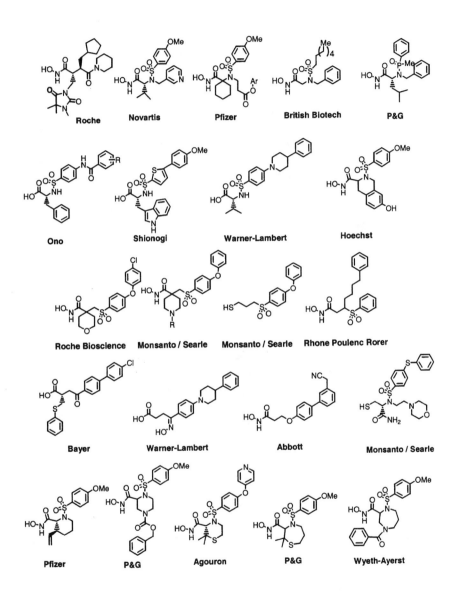

**FIGURE 4.** Representative nonpeptide MMPIs.

## NONPEPTIDE MATRIX METALLOPROTEINASE
## INHIBITOR SCAFFOLDS

With the disclosure of CGS 27023A, the lead compound of the first series of potent, orally active, nonpeptidic MMP inhibitors by Ciba-Geigy,[30,31] a flurry of activity was initiated that marked the genesis of a new era of MMPI design. The scope and extent of these efforts are enormous and have been summarized elsewhere.[1–5,8–10,12–14] These observations have resulted in a broader approach to the design of MMPIs, with an emphasis on new classes of molecules and numerous discoveries of compounds with varying selectivity profiles (FIG. 4). These new scaffolds generated new design paradigms, and this exploration has furnished several advanced candidates in short order (*vide infra*).

One aspect of our work involves the design and synthesis of diazepine sulfonamide hydroxamic acids.[32] Conceptually, these compounds can be considered to be cyclic analogues of the sulfonamide-hydroxamate MMPI scaffold. Thus, annulation of the P1 substituent of CGS 27023A via a nitrogen linker to the P2′ substituent

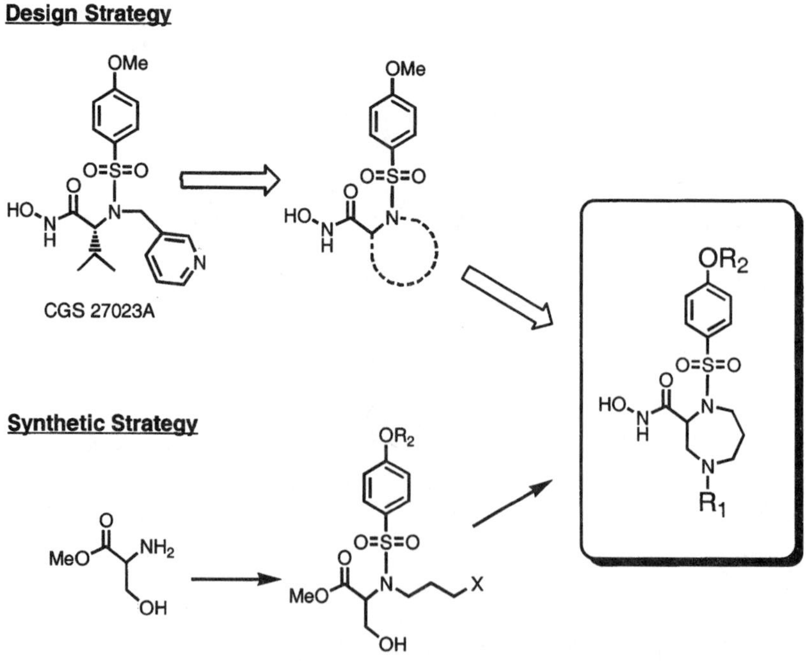

**Design Strategy**

CGS 27023A

**Synthetic Strategy**

**FIGURE 5.** Diazepine MMPIs: strategy for design and synthesis.

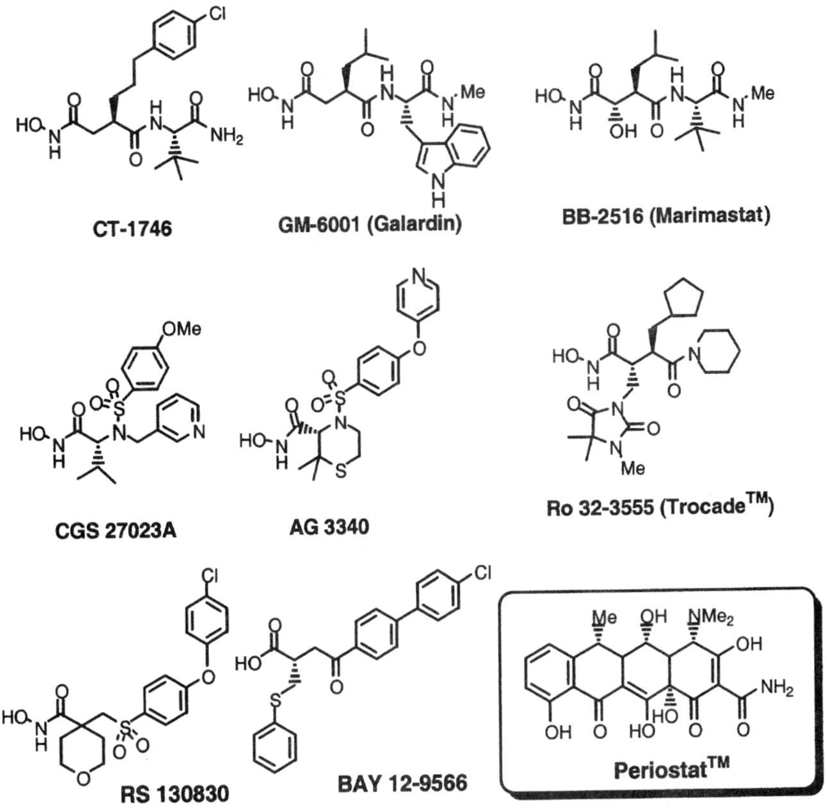

**FIGURE 6.** Representative advanced candidates for MMPI.

borne by the sulfonamide nitrogen results in the diazepine scaffold shown in FIGURE 5. The compounds are readily available starting from serine. It was hoped that the conformational restraint imposed by the use of the diazepine ring system would be advantageous for binding to the active site of the MMPs.

Several of the compounds synthesized proved to be more potent than CGS 27023A *in vitro* versus MMP-9 and MMP-13 (TABLE 2). The most potent members of the series were the diazepine benzamides, which also were equipotent to CGS 27023A in a dialysis implant bioactivity model.[33] The biaryl ether analogue was also comparable to CGS 27023A in inhibiting collagen degradation in a cartilage explant model. Alkyl amides, carbamates, and urea derivatives of the diazepine scaffold proved to be less potent *in vitro* and *in vivo*.

## SUMMARY

The design and synthesis of inhibitors of matrix metalloproteinases continues to be a prominent area of pharmaceutical research. Rapid advances in all aspects of the drug discovery process have energized these research efforts and resulted in the introduction of a number of small molecules of diverse structural make-up into clinical trials or advanced preclinical status (FIG. 6). Both peptide- and nonpeptide-based inhibitors are in clinical studies for various indications. The tetracycline Periostat™ (doxycycline hyclate)[34] is the first MMPI to obtain approval for periodontal disease.[35] Significant synthetic, modeling, and biochemical effort remains directed toward the identification of selective MMPIs, although questions remain unanswered as to the absolute necessity or desirability of inhibition of specific MMPs at the expense of others.

The search for new and improved alternatives to the hydroxamic acid zinc chelator continues. Numerous classes of surrogates have been exploited with varying success. In the present case, asymmetric syntheses of a series of succinyl mercaptoalcohols and mercaptoketones as putative alternative zinc chelators have been accomplished to furnish low nanomolar MMPIs.

The MMPI field has evolved to focus on nonpeptide inhibitors that are selective and orally active with good duration of action. Distinct and innovative scaffolds have been designed to improve our understanding of these biochemical processes and to better probe issues related to both selectivity and bioavailability. In the present case, a series of diazepine hydroxamates has been identified which furnish potent, selective, and orally active MMPIs.

## REFERENCES

1. SKOTNICKI, J.S., J.I. LEVIN, L.M. KILLAR & A. ZASK. 1999. Matrix metalloproteinase inhibitors. In Metalloproteinases as Targets for Anti-inflammatory Drugs. D. Bradshaw, J.S. Nixon & K. Bottomley, Ed. Birkhauser Verlag. Basel. In press.
2. BOTTOMLEY, K.M., W.H. JOHNSON & D.S. WALTER. 1998. Matrix metalloproteinase inhibitors in arthritis. J. Enzyme Inhib. 13: 79–101.
3. WHITTAKER, M. & P. BROWN. 1998. Recent advances in matrix metalloproteinase inhibitor research and development. Curr. Opin. Drug Disc. Develop. 1: 157–164.
4. BECKETT, R.P. & M. WHITTAKER. 1998. Matrix metalloproteinase inhibitors 1998. Exp. Opin. Ther. Patents 8: 259–282.
5. WATSON, S.A. & G. TIERNEY. 1998. Matrix metalloproteinase inhibitors. BioDrugs : 325–336.
6. ROTHENBERG, M.L., A.R. NELSON & K.R. HANDE. 1998. New drugs on the horizon: matrix metalloproteinase inhibitors. Oncologist 3: 271–274.
7. SUMMERS J.B. & S.K. DAVIDSEN. 1998. Matrix metalloproteinase inhibitors and cancer. Annu. Rep. Med. Chem. 33: 131–140.
8. LEVY, D.E. & A.M. EZRIN. 1997. Matrix metalloproteinase inhibitor drugs. Emerging Drugs 2: 205–230.
9. DENIS, L.J. & J. VERWEIJ. 1997. Matrix metalloproteinase inhibitors: present achievements and future prospects. Invest. New Drugs 15: 175–185.
10. WHITE, A.D., T.M.A. BOCAN, P.A. BOXER, J.T. PETERSON & D. SCHRIER. 1997. Emerging therapeutic advances for the development of second generation matrix metalloproteinase inhibitors. Curr. Pharm. Design 3: 45–58.
11. BROWN, P.D. 1997. Matrix metalloproteinase inhibitors in the treatment of cancer. Medical Oncology 14: 1–10.

12. ZASK, A, J.I. LEVIN, L.M. KILLAR & J.S. SKOTNICKI. 1996. Inhibition of matrix metalloproteinases: structure based design. Curr. Pharm. Design **2**: 624–661.

13. HAGMANN, W.K., M.W. LARK & J.W. BECKER. 1996. Inhibition of matrix metalloproteinases. Annu. Rep. Med. Chem. **31**: 231–240.

14. BECKETT, R.P. 1996. Recent advances in the field of matrix metalloproteinase inhibitors. Exp. Opin. Ther. Patents **6**: 1305–1315.

15. BECKETT, R.P., A.H. DAVIDSON, A.H. DRUMMOND, P. HUXLEY & M. WHITTAKER. 1996. Recent advances in matrix metalloproteinase inhibitor research. Drug Discovery Today **1**: 16–26.

16. CAWSTON, T.E. 1996. Metalloproteinase inhibitors and the prevention of connective tissue breakdown. Pharmacol. Ther. **70**: 163–182.

17. PORTER, J.R., T.A. MILLICAN & J.R. MORPHY. 1995. Recent developments in matrix metalloproteinase inhibitors. Exp. Opin. Ther. Patents **5**: 1287–1296.

18. MORPHY, J.R., T.A. MILLICAN & J.R. PORTER. 1995. Matrix metalloproteinase inhibitors: current status. Curr. Med. Chem. **2**: 743–762.

19. BEELEY, N.R.A., P.R.J. ANSELL & A.J.P. DOCHERTY. 1994. Inhibitors of matrix metalloproteinases (MMPs). Curr. Opin. Ther. Patents **4**: 7–16.

20. SCHWARTZ, M.A. & H.E. VAN WART. 1992. Synthetic inhibitors of bacterial and mammalian interstitial collagenases. Prog. Med. Chem. **29**: 271–334.

21. NELSON, F.C. & A. ZASK. 1999. The therapeutic potential of small molecule TACE inhibitors. Exp. Opin. Invest. Drugs. In press.

22. LOWE, C. 1998. Tumour necrosis factor-α Antagonists and their therapeutic applications. Exp. Opin. Ther. Patents **8**: 1309–1322.

23. SHIRE, M.G. & G.E. MULLER. 1988. TNF-α inhibitors and rheumatoid arthritis. Exp. Opin. Ther. Patents **8**: 531–544.

24. BLACK, R.A., T.A. BIRD & K.M. MOHLER. 1997. Agents that block TNF-α synthesis or activity. Annu. Rep. Med. Chem. **32**: 241–250.

25. DAVIDSON, S.K. & J.B. SUMMERS. 1995. Inhibitors of TNF-α Synthesis. Exp. Opin. Ther. Patents **5**: 1087–1100.

26. CAMPBELL, D.A., X.-Y. XIAO, D. HARRIS, S. IDA, R. MORTEZAEI, K. NGU, L. SHI, D. TEIN, Y. WANG, M. NAVRE, D.V. PATEL, M.A. SCHARR, J.F. DIJOSEPH, L.M. KILLAR, C.L. LEONE, J.I. LEVIN & J.S. SKOTNICKI. 1998. Malonyl α-mercaptoketones and α-mercaptoalcohols, a new class of matrix metalloproteinase inhibitors. Bioorg. Med. Chem. Lett. **8**: 1157–1162.

27. LEVIN J.I., J.F. DIJOSEPH, L.M. KILLAR, M.A. SHARR, J.S. SKOTNICKI, D.V. PATEL, X.-Y. XIAO, L. SHI, M. NAVRE & D.A. CAMPBELL. 1998. The asymmetric synthesis and *in vitro* characterization of succinyl mercaptoalcohol and mercaptoketone inhibitors of matrix metalloproteinases. Bioorg. Med. Chem. Lett. **8**: 1163–1168.

28. BAXTER, A.D., J. BIRD, R. BHOGAL, T. MASSIL, K.J. MINTON, J. MONTANA & D.A. OWEN. 1997. A novel synthesis of matrix metalloproteinase inhibitors for the treatment of inflammatory disorders. Bioorg. Med. Chem. Lett. **7**: 897–902.

29. BESZANT, B., J. BIRD, L.M. GASTER, G.P. HARPER, I. HUGHES, E.H. KARRAN, R.E. MARKWELL, A.J. MILES-WILLIAMS & S.A. SMITH. 1993. Synthesis of novel modified dipeptide inhibitors of human collagenase: β-mercapto carboxylic acid derivatives. J. Med. Chem. **36**: 4030–4039.

30. PARKER, D.P., L.J. MACPHERSON, R. GOLDSTEIN, M.R. JUSTICE, L.J. ZHU, M. CAPPARELLI, L.W. WHALEY, C. BOEHM, E.M. O'BYRNE, R.L. GOLDBERG & V.S. GANU. 1994. CGS 27023A: A novel, potent, and orally active matrix metalloprotease inhibitor. Presented at the Seventh International Conference of the Inflammation Research Association, White Haven, PA.

31. MacPherson, L.J., E.K. Bayhurt, M.P. Capparelli, B.J. Carroll, R. Goldstein, M.R. Justice, L. Zhu, S. Hu, R.A. Melton, L. Fryer, R.L. Goldberg, J.R. Doughty, S. Spirito, V. Blancuzzi, D. Wilson, E.M. O'Byrne, V. Ganu & D.T. Parker. 1997. Discovery of CGS 27023A, A non-peptidic, potent, and orally active stromelysin inhibitor that blocks cartilage degradation in rabbits. J. Med. Chem. **40:** 2525–2532.

32. Levin J.I., J.F. DiJoseph, L.M. Killar, A. Sung, T. Walter, M.A. Sharr, C.E. Roth, J.S. Skotnicki & J.D. Albright. 1998. The synthesis and biological activity of a novel series of diazepine MMP inhibitors. Bioorg. Med. Chem. Lett. **8:** 2657–2662.

33. DiJoseph, J.F. & M.A. Sharr. 1998. Dialysis tubing implant assay in the rat: a novel in vivo method for identifying inhibitors of matrix metalloproteinases. Drug Develop. Res. 43: 200–205.

34. Health News Daily, 10/2/98; Pink Sheet, 10/5/98; Marketletter 10/12/98.

35. Ryan, M.E., S. Ramamurthy & L.M. Golub. 1996. Matrix metalloproteinases and their inhibition in periodontal treatment. Curr. Opin. Peridontol. **3:** 85–96.

# Insights into MMP–TIMP Interactions

WOLFRAM BODE,[a,b] CARLOS FERNANDEZ-CATALAN,[a] FRANK GRAMS,[a,c]
FRANZ-XAVER GOMIS-RÜTH,[a,d] HIDEAKI NAGASE,[e] HARALD TSCHESCHE,[f]
AND KLAUS MASKOS[a]

[a]Max-Planck-Institut für Biochemie, D-82152 Martinsried, Germany

[e]Department of Biochemistry and Molecular Biology,
University of Kansas Medical Center, Kansas City, Kansas 66160, USA

[f]Universität Bielefeld, Abteilung Biochemie I, D-33615 Bielefeld, Germany

**ABSTRACT:** The proteolytic activity of the matrix metalloproteinases (MMPs) involved in extracellular matrix degradation must be precisely regulated by their endogenous protein inhibitors, the tissue inhibitors of metalloproteinases (TIMPs). Disruption of this balance can result in serious diseases such as arthritis and tumor growth and metastasis. Knowledge of the tertiary structures of the proteins involved in such processes is crucial for understanding their functional properties and to interfere with associated dysfunctions. Within the last few years, several three-dimensional structures have been determined showing the domain organization, the polypeptide fold, and the main specificity determinants of the MMPs. Complexes of the catalytic MMP domains with various synthetic inhibitors enabled the structure-based design and improvement of high-affinity ligands, which might be elaborated into drugs. Very recently, structural information also became available for some TIMP structures and MMP–TIMP complexes, and these new data elucidated important structural features that govern the enzyme–inhibitor interaction.

## INTRODUCTION

The matrix metalloproteinases (MMPs, matrixins) form a family of structurally and functionally related zinc endopeptidases. Collectively, these MMPs are capable *in vitro* and *in vivo* of degrading all kinds of extracellular matrix protein components and are thus implicated in many connective tissue remodeling processes.[1] Normally, the degenerative potential of the MMPs is mainly held in check by the endogenous tissue inhibitors of metalloproteinases (TIMPs). Disruption of this MMP–TIMP balance can result in disorders such as rheumatoid and osteoarthritis, atherosclerosis, tumor growth, metastasis, and fibrosis (for recent overviews, see, for example, Refs. 2–4). Therapeutic inhibition of MMPs is a promising approach for treatment of some of these diseases, and the MMP structures and their TIMP complexes are therefore attractive targets for rational inhibitor design (for recent literature, see Refs. 5 and 6).

[a]Address for telecommunication: Phone, ++49 89 8578 2676/8; fax, ++49 89 8578 3516; e-mail bode@biochem.mpg.de
[c]Present address: Boehringer-Mannheim, TR-CS II, D-68305 Mannheim, Germany.
[d]Present address: Departament de Biologia de Molecular I Cellular, Centre d'Investigacio i Desenvolupament C.S.I.C., Barcelona, Spain.

To date, 17 different human MMPs have been identified and/or cloned which share significant sequence homology and a common multi-domain organization; several counterparts have been found in other vertebrates and invertebrates and from plant sources,[7] together forming the MMP or matrixin subfamily A of the metalloproteinase M10 family.[8] According to their structural and functional properties, the MMP family can be subdivided into five groups: (i) the collagenases (MMPs-1, 8 and 13); (ii) the gelatinases A and B (MMPs-2 and 9); (iii) the stromelysins 1 and 2 (MMPs-3 and 10); (iv) a more heterogeneous subgroup containing matrilysin (MMP-7), enamelysin (MMP-20), the *mmp20* gene product, macrophage metalloelastase (MMP-12), and MMP-19 (together making up the "classical" MMPs); and (v) the membrane-type-like MMPs (MT-MMPs-1 to 4 and stromelysin-3, MMP-11). These MMPs share a common multi-domain structure, but are glycosylated to different extents and at different sites. According to sequence alignments,[7,9] the assembly of these domains might have been an early evolutionary event, followed by diversification.[10]

All MMPs are synthesized with an ~20 amino acid–residue signal peptide and are (probably except for the MT-MMP-like furin-processed proteinases[11–13]) secreted as latent pro-forms; these pro-proteinases consist of an ~80-residue N-terminal pro-domain followed by the ~170-residue catalytic domain (cd; see FIGURES 1 and 2), which in turn (except for matrilysin) is covalently connected through a 10- to 70-residue Pro-rich linker to an ~195-residue C-terminal haemopexin-like domain (hld). In the MT-MMPs the polypeptide chain possesses an additional 75- to 100-residue extension, which seems to constitute a connecting peptide, a transmembrane segment, and a short cytoplasmatic fragment.[12] Removal of this hld in the collagenases eliminates their characteristic capability to cleave triple-helical collagen, but does not significantly affect the hydrolytic activity toward gelatin, casein, or synthetic substrates (see Ref. 14). In both gelatinases, the catalytic domains have an additional 175 amino–acid residue insert comprising three fibronectin-related type II modules conferring gelatin and collagen binding.

The TIMP family currently includes four different members (TIMPs-1 to 4,[15–18]), which after optimal topological superposition exhibit 41 to 52% sequence identity (see FIG. 5 and Refs. 19 and 20). Besides their inhibitory role, these TIMPs seem to have other functions such as growth factor-like and anti-angiogenic activity (see, for example, Refs. 21 and 22). The TIMP cDNAs encode an ~25-residue leader peptide, followed by the 184- to 194-amino acid residue mature inhibitor (see FIGS. 4 and 5). Virtually all TIMPs form tight 1:1 complexes with MMPs, that is, except for the rather weak interaction between TIMP-1 and MT1- and 2-MMPs[12,23–25], the TIMPs do not seem to differentiate much between the various MMPs.[26] TIMPs-1 and 2 are unique in that they also bind to the pro-forms of gelatinase B and A, respectively[27]; the complex between MT1-MMP and TIMP-2 seems to act as a cell-surface-bound "receptor" for progelatinase A activation *in vivo*, using these noninhibitory interactions between TIMP-2 and progelatinase A.[24,28,29] Removal of the C-terminal third of the TIMP polypeptide chain gives rise to so-called N-terminal TIMP domains (N-TIMPs), which retain most of their reactivity towards their target MMPs.[30,31]

Only in early 1994, the first X-ray crystal structures of the catalytic domains (blocked by various synthetic inhibitors) of human fibroblast collagenase/MMP-1[32–34] and human neutrophil collagenase/MMP-8[34–36] and an NMR structure of the catalytic domain

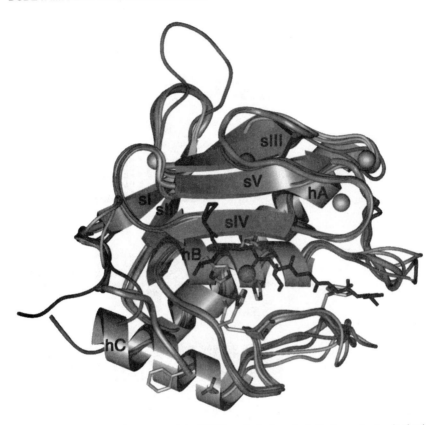

**FIGURE 1.** Ribbon structure of the MMP catalytic domain (cd) shown in standard orientation (figure made with SETOR[72]). The catalytic domain of the Phe100 form of the MMP-8 cd[36] shown as a ribbon together with a modeled heptapeptide substrate (*dark grey*) is superimposed with the cds of MMP-3, MMP-1, MMP-14, MMP-7 (*ropes*). The catalytic and the structural zinc (*center and top*) and the three calcium ions (*flanking*) are shown as *dark and light grey spheres*, and the three zinc liganding residues His218, 222, and 228, Met236 of the Met-turn, the conserved Pro238 and Tyr240 of the wall-forming segment, and the N-terminal Phe100 and the first Asp250 forming the surface-located salt bridge are given with all atoms. The chain segment forming the extra domain of both gelatinases will be inserted in the sV-hB loop (*center, right*) and presumably extends to the right side.

of stromelysin-1/MMP-3[37] became available, which were later complemented by additional cd structures of MMP-1,[38,39] matrilysin/MMP-7,[40] MMP-3,[41–45] MMP-8,[46] and MT1-MMP.[19] In 1995, the first X-ray structure of an MMP pro-form, the C-terminally truncated pro-stromelysin-1, was published,[41,44] and the first and only structure of a mature full-length MMP, namely of porcine fibroblast collagenase/MMP-1,[47] was described. At the same time, structures of the isolated hemopexin-like domains from human gelatinase A[48,49] and from collagenase-3/MMP-13[50] were also reported.[51] Of the TIMPs, a first preliminary NMR model of human N-TIMP-2 was presented in 1994,[52] which showed that the polypeptide framework of the N-terminal part of the

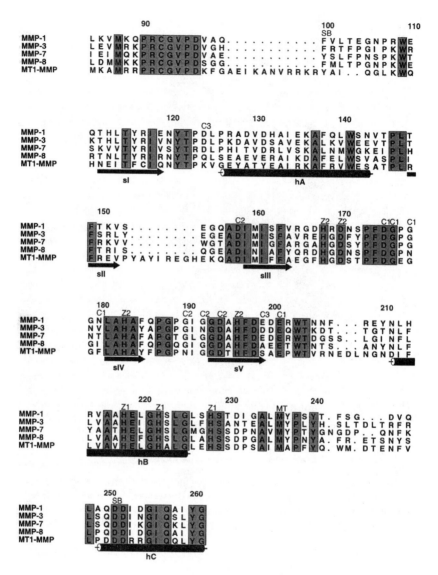

**FIGURE 2.** Alignment of the catalytic domains of MMP-1, MMP-3, MMP-7, MMP-8 and MMP-14, made according to topological equivalencies.[19] The numbering is that of MMP-1. Location and extent of α-helices and β-strands are given by *cylinders and arrows,* and *symbols* show the characteristic Met (MT) and the residues involved in main and side chain interactions with the catalytic (Z1) and the structural zinc (Z2), the first (C1), second (C2) and third calcium ion (C3), and in the surface-located salt bridge (SB) of the cds. Fully conserved residues are *enboxed.*

TIMPs resembles so called OB-fold proteins; a refined N-TIMP-2 has recently been published, describing some enhanced mobility of contacting inhibitor segments.[53,54] The first structure of a complete TIMP, human deglycosylated TIMP-1, in complex with the catalytic domain of human MMP-3, was published only in 1997,[45] followed by the X-ray structure of TIMP-2 in complex with the catalytic domain of MT1-MMP.[19]

In the following, the structures of the MMPs and TIMPs and their detailed interactions will be presented. On the basis of their topological equivalencies, for both the MMPs (FIG. 2) and the TIMPs (see FIG. 5) structure-based sequence alignments will be given. The MMP nomenclature used is based on the cDNA sequence of (human) fibroblast collagenase/MMP-1 as the reference MMP (see FIGURE. 2). For the assignment of peptide substrate residues and substrate recognition sites on the proteinase, the nomenclature of Schechter and Berger[55] will be used: P1, P2, etc., and P1′, P2′, etc., indicate the residues in N- and C-terminal direction of the scissile peptide bond of a bound peptide substrate (analogue), and S1, S2, etc., and S1′, S2′, etc., the opposite binding sites on the enzyme. TIMP residues will be given with TIMP-1/TIMP-2 numbers.

## STRUCTURES AND MECHANISMS

### *MMPs*

#### *Catalytic Domain*

The catalytic domains of the MMPs exhibit the shape of an oblate ellipsoid. In the "standard" orientation, which in this article as well as in most other MMP papers, is preferred for the display of MMPs, a small active-site cleft notched into the flat ellipsoid surface extends horizontally across the domain to bind peptide substrates from left to right (see FIGURE 1). This cleft harboring the "catalytic zinc" separates the smaller "lower subdomain" from the larger "upper subdomain."

This upper subdomain formed by the first three-quarters of the polypeptide chain (up to Gly225) consists of a five-stranded β-pleated sheet, flanked by three surface loops on its convex side and by two long regular α-helices on its concave side embracing a large hydrophobic core (FIG. 1). The polypeptide chain starts on the molecular surface of the lower subdomain, passes β-strand sI, the amphipatic α-helix hA, and β-strands sII, sIII, sIV and sV, before entering the "active-site helix" hB (for nomenclature, see FIGURE 1). In the classical MMPs, strands sII and sIII are connected by a relatively short loop bridging sI; in the MT-MMPs, however, this loop is expanded into the spur-like, solvent-exposed "MT-MMP-specific loop" of hitherto unknown function. In all MMPs, strands sIII and sIV are linked via an "S-shaped double loop," which is clamped via the "structural zinc" and the first of two to three bound calcium ions to the β-sheet. This S-loop extends into the cleft-sided "bulge" continuing in the antiparallel "edge strand" sIV; this bulge-edge segment is of prime importance for binding of peptidic substrates and inhibitors (FIG. 3a). The sIV-sV connecting loop together with the sII-sIII bridge sandwiches the second bound calcium. After strand sV, the chain passes the large open sV-hB loop before entering the active-site helix hB; this helix provides the first (218) and the second His (222), which coordinate the catalytic zinc, and the "catalytic Glu219" in between, all of them representing the N-terminal part of the "zinc binding consensus sequence" HEXXHXXGXXH (FIG. 2) characteristic of the metzincin superfamily.[56,57]

This active-site helix abruptly stops at Gly225, where the peptide chain bends down, descends (presenting the third zinc liganding histidine, His228) and runs through a wide right-handed spiral (catalytic domain's "chin"), terminating in the 1,4-tight "Met-turn" (of the strongly conserved sequence Ala234-Leu-Met236-Tyr237, with an obligatory methionine residue at turn position 3). The chain then turns back to the molecular surface to an (except in human stromelysin-3) invariant Pro238, forms with a conserved Pro238-X-Tyr240 segment (the "S1′ wall forming segment") the outer wall of the S1′ pocket, runs through another wide ("specificity") loop of slightly variable length and conformation, before it passes the C-terminal α-helix hC, which ends with the conserved Tyr260-Gly261 residue pair.

The overall structures of all MMP cds known so far are very similar, with the collagenase structures resembling one another most, and MMP-7 and the MT1-MMP structures deviating most (FIG. 1). Larger main chain differences occur (i) in the N-terminal segment up to Pro107 (depending on the length of the N-terminus and the presence of a TIMP); (ii) in the sII-sIII bridge (with the elongated and more exposed MT loop); (iii) in the sV-hB loop; and (iv) in the specificity loop (FIG. 1). In both gelatinases, the approximately seven residues between Trp203-Thr204 and Leu/Phe/Ile212 are replaced by a 183-residue insert, which probably forms a large adjacent domain consisting of three tandem copies of fibronectin type II-like modules. The specificity loop is shortest in MMP-1; those of MMPs-3, 8, and 14 (with three and two additional residues) resemble one another, while that of MMP-7 (two) deviates most (FIG. 1). Besides the catalytic zinc, all MMP catalytic domains possess another zinc ion, the structural zinc, and two (MMP-8, MT1-MMP) or three bound calcium ions (MMP-1, MMP-3, MMP-7) (FIGS. 1 and 2). The structural zinc and the first (probably most tightly bound) calcium are sandwiched between the double-S loop and the outer face of the β-sheet. The second calcium ion is sandwiched between the sIV-sV loop and sIII, while the loop immediately following sV encircles the third calcium ion, with Asp124 forming one ligand and determining the presence of this calcium site.

For the fibroblast and the neutrophil collagenases, a several-fold larger activity has been demonstrated for active enzymes starting with a highly conserved Phe100 residue compared with species truncated for one or two more residues (a phenomenon also called "superactivity"[58,59]). Position and fixation of the N-terminus of the mature classical MMPs indeed seem to depend on the presence of the N-terminal amino acid Phe/Tyr100, that is, on the accurate processing/tailoring of the MMP precursors.[35,36] In cases that it starts with (the highly conserved) Phe100 (FIG. 2), the N-terminal heptameric segment preceding the conserved Pro107-Lys/Arg-Trp109 triple is tightly packed against a hydrophobic surface groove made by the C-terminal helix hC and the descending segment centering around the third His ligand of the catalytic zinc; the N-terminal Phe100 ammonium group makes a surface-located salt bridge with the side chain carboxylate of an Asp250, which is the first residue of a strictly invariant helix hC based Asp250-Asp251 pair.[36] The side chain of the second Asp is buried in a solvent-filled protein cavity and hydrogen-bonded via the Met-turn to the first zinc liganding His218. Via this path, formation of the Phe100...Asp250 salt bridge might be signaled to the active center. In the absence of an N-terminal Phe/Tyr100, the N-terminal (hexa)peptide preceding Pro107 is disordered and might interfere with substrate binding.

*Specificity Determinants*

Bounded at the upper rim by the bulge-edge segment and the second part of the S-loop, and at the lower side by the third zinc-liganding imidazole and the S1´ wall-forming segment, the active-site cleft of all MMPs is relatively flat at the left ("non-primed") side, but descends into the molecular surface at the catalytic zinc and to the right ("primed") side, leveling-off again to the surface further to the right. In unliganded MMPs, the catalytic zinc residing in its center is coordinated by the three imidazole Nε2 atoms of the three histidines (His218, 222, and 228) and by a fixed water molecule, which simultaneously is in hydrogen bond distance to the carboxylate group of the catalytic Glu219. In case of MMP complexes with bidentate inhibitors (such as those with a hydroxamic acid function), this water is replaced by two oxygen atoms, which together with the three imidazoles coordinate the catalytic zinc in a trigonal-bipyrimidal (penta-coordinate) manner.[35] As in all other metzincins[56,57] the zinc-imidazole ensemble of the MMPs is placed above the distal ε-methyl-sulfur moiety of the strictly conserved Met236 in the Met-turn, which forms a hydrophobic base of still unclear function (FIG. 1).

Immediately to the right of the catalytic zinc the S1´ specificity pocket invaginates, which in size and shape considerably differs among the various MMPs (FIG. 3b). This pocket is mainly formed (seen in standard orientation) by (i) the initial part of the active-site helix hB ("back side"), (ii) the somewhat mobile[42] phenolic side chain of Tyr240 ("right-hand flank"), (iii) the main chain atoms of the underlying wall forming segment Pro-X-Tyr ("front side"), (iv) the flat side of the first zinc liganding His218 imidazole ("left side"), and (v) the Leu/Ile/Val235 residue of the Met-turn, which together with the Leu/Tyr/Arg214 or the Arg243 side chain (if present) form its "bottom" or line it towards the second exit opening at the lower molecular surface, respectively.

Nearly all of the synthetic inhibitors analyzed so far in MMP complexes contain a chelating group (such as a hydroxamic acid, a carboxylate, or a thiol group) for zinc ion ligation, and a peptidic or peptidomimetic moiety mimicking peptide substrate binding to the substrate recognition site. Of the synthetic inhibitors published in complex with an MMP, only the Pro-Leu-Gly-hydroxamic acid inhibitor[35,36,60] binds to the left-hand subsites (the nonprimed subsites S3 to S1) alone, antiparallel to the edge strand ("left-side inhibitor"). In the majority of synthetic inhibitors studied so far this peptidic moiety interacts in an extended manner with the primed right-hand subsites ("right-side inhibitors"), inserting between the (antiparallel) bulge-edge segment and the (parallel) S1´ wall-forming segment of the cognate MMP under formation of a three-stranded mixed β-sheet (FIG. 3a). An L-configured P1´-like side chain is perfectly arranged to extend into the hydrophobic bottleneck of the S1´ pocket.[60] This P1´–S1´ interaction is the main determinant for the affinity of inhibitors and the cleavage position of peptide substrates. Depending on the length and character of residue 214 harbored in the N-terminal part of the active-site helix hB, the size of the S1´ pocket differs considerably among the MMPs (FIG. 3b). In MMP-1 and MMP-7, the side chains of Arg214 and Tyr214, respectively, extend into

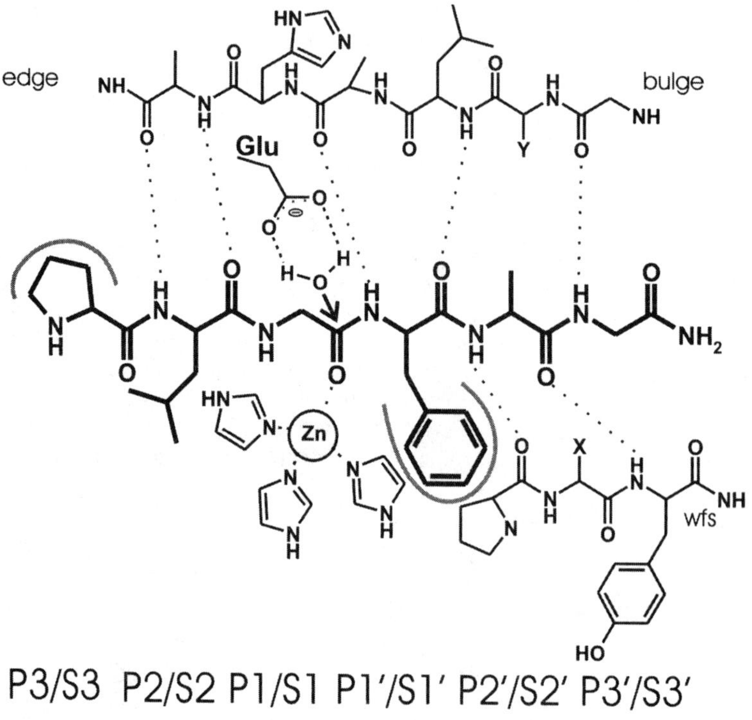

## P3/S3  P2/S2  P1/S1  P1'/S1'  P2'/S2'  P3'/S3'

**FIGURE 3a.** Peptide substrate and inhibitor interaction and specificity. Schematic drawing of the putative encounter complex between a Pro-Leu-Gly-Leu-Ala-Gly-amide hexapeptide substrate and the MMP active site.[46] The substrate polypeptide chain *(bold connections)* lies antiparallel to the bulge-edge strand *(top)* and parallel to the S1' wall-forming segment *(bottom)* forming up to five and two, respectively, inter-main-chain hydrogen bonds *(dashed lines)*.

the S1' opening, limiting it to a size and shape still compatible with the accommodation of medium-sized P1' residues, but less for very large side chains, in agreement with peptide cleavage studies on model peptides.[61–63] Some more recent MMP-1 structures show, however, that the Arg214 side chain can swing out of its normal site, thus also allowing binding of synthetic inhibitors with larger P1' side chains (M. Browner, personal communications). The smaller Leu214 residues of MMP-3 and MMP-14 (and probably also of MMPs-2 and 9) do not bar the internal S1' "pore," which extends right through the molecule to the lower surface and resembles a long solvent-filled "tube" (FIG. 3b). In spite of a small Leu214 residue, however, the S1' pocket of MMP-8 is of medium size and is closed at the bottom, due to the Arg243 side chain extending into the S1' space from the specificity loop.

Second in importance for substrate specificity seems to be the interaction of the P3 residue (in collagen cleavage sites always a Pro residue) with the mainly hydrophobic S3 pocket (FIG. 3a). The S2 site is a shallow depression extending on top of the imidazole ring of the second zinc liganding His; its polarity character might be

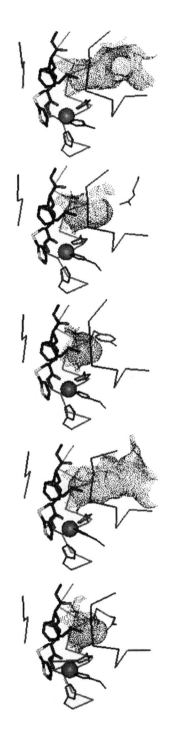

**FIGURE 3b.** Peptide substrate and inhibitor interaction and specificity. Comparison of the S1′ pockets (*dot surface*) of (*from left to right*) MMP-1, MMP-3, MMP-7, MMP-8, and MMP14. Besides the Cα plots of the bulge-edge strand (*dark-grey, top*), of the active-site helix (*light grey, center*) and the segment comprising the Met-turn and the S1′ wall-forming segment (*dark grey, bottom*), the full structure of the inhibitor lead batimastat alias BB-94 (as bound to MMP-8[60], the catalytic zinc (*sphere*), the side chains of the three zinc liganding His residues, and the side chains of residues Arg214, Tyr214 and Arg243 restricting the S1′ pockets of MMP-1, MMP-7, and MMP-8 in size are shown.

influenced by residue 227 preceding the third zinc liganding His, His228. Longer side chains of P1 residues (in collagen cleavage sites mostly a Gly) are placed in the surface groove lined by the His183 side chain of the edge strand together with the last bulge residue 180; depending on this latter side chain, P1 side chains of differing size and polarity might be preferred. P2′ side chains extend away from the surface, squeezed between the bulge rim and the side chain of the middle residue of the Pro-X239-Tyr wall-forming segment; the strength of interaction will be determined particularly by the nature of bulge residue 180 and residue 239, which in the MMPs-14, 15, and 16 and MMP-11 is an exposed Phe.[19] Further to the right side the molecular surface again has a hydrophobic/polar depression, which could accommodate P3′ side chains of differing nature.

By replacing the zinc chelating groups of such peptidic left- and right-side inhibitors by a normal peptide bond, a contiguous peptide substrate was constructed, indicating the probable binding geometry of a normal substrate-MMP encounter complex[46] (FIG. 3a). Accordingly, the peptide substrate chain is aligned in an extended manner to the continuous bulge-edge segment, under formation of an antiparallel two-stranded β-pleated sheet, which expands on the right-hand side into a three-stranded mixed parallel-antiparallel sheet due to additional alignment with the S1′ wall-forming segment. The bound peptide substrate (such as the hexapeptide shown in FIGURE 3a) forms five and two inter main-chain hydrogen bonds, respectively, to both crossing-over MMP segments. Similar to the reaction mechanism previously suggested for the more distantly related zinc endopeptidase thermolysin,[64] the MMP-catalyzed cleavage of the scissile peptide bond will probably proceed via a general-base mechanism.[46] The carbonyl group of the scissile bond (Gly-Phe in FIGURE 3a) is directed nearly toward the catalytic zinc and is strongly polarized. The zinc-bound water molecule is activated by the carboxylate/carboxylic acid of the catalytic Glu219, the pK of which might (particularly upon substrate/inhibitor binding) be shifted to higher pH values, due to packing in the protein matrix without charge-stabilizing internal hydrogen bonds (the importance of this Glu219 for proteolytic activity in MMPs has been demonstrated through replacement with Asp, Ala, and Glu residues,[65,66] resulting in mutants with lowered or extremely low catalytic activity, respectively). The activated water molecule squeezed between the Glu219 carboxylate group and the scissile peptide bond carbonyl is properly oriented to attack via its lone pair orbital this electrophilic carbonyl carbon. The tetrahedral intermediate is presumably stabilized by both the zinc and the carbonyl group of the first Ala residue of the edge strand sIV (FIG. 3a). Simultaneously, one water proton could be transferred via the Glu carboxylate (acting as a proton shuttle) to the amino group, which after breakage of the peptide bond and transfer of a second proton could leave the enzyme–substrate complex together with the N-terminal substrate fragment. Remarkably, there is no other electrophil (such as His231 in thermolysin[64] or Tyr149 in astacin[67,68]) in the catalytic zinc environment of the MMPs, which could further stabilize the carboxy anion of the presumed tetrahedral intermediate; a frequently observed water molecule suggested to be activated by the carbonyl group of Pro238 does not seem to be placed suitably for this purpose.

## *TIMPs*

### *TIMP Structure*

The TIMPs have the shape of an elongated contiguous wedge consisting of an N-terminal segment (Cys1 to residue Pro5), an all-β-structure left-hand part, an all-helical center, and a β-turn structure to the right (according to the "front view" in FIGURE 4).[45] The N- and the C-terminal halves of the polypeptide chain form two opposing subdomains. The N-terminal subdomain exhibits a so-called **OB**-fold, known for a number of **o**ligosaccharide/**o**ligonucleotide **b**inding proteins.[52] This region consists of a five-stranded β-pleated sheet of Greek-key topology rolled into a closed β-barrel of elliptical cross-section. The narrower opening of this barrel is bounded by the sB-sC loop, while its wider exit is (in contrast to other OB-fold proteins) covered by an extended segment connecting strands sC and sD, designated as "connector."[45] After leaving the barrel, the polypeptide chain passes two helices,

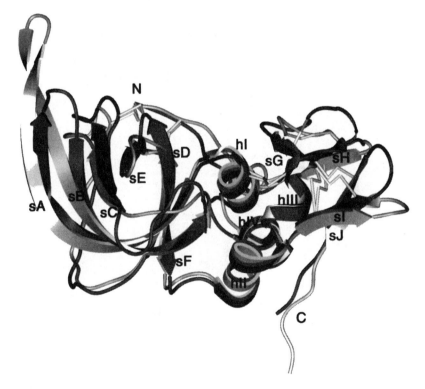

**FIGURE 4.** Tertiary structure of the TIMPs (SETOR[72]). The wedge-like TIMP-1 (*dark grey ribbon*[45]) and TIMP-2 molecules (*light grey ribbon*[19]) are superimposed upon minimizing the deviations between all equivalent α-carbon atoms.[19] This front view is related to the standard view (FIG. 1) by a 90-degree rotation about a horizontal. The polypeptide chains start with their N-terminal segment (marked with N, *center top*), then build up the N-terminal (*left*) and the C-terminal subdomain (*right*), before terminating in the flexible tails (marked with C, *bottom right*).

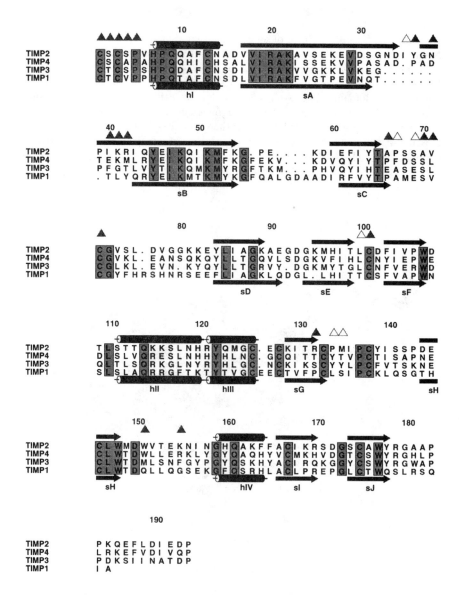

**FIGURE 5.** Alignment of TIMPs-1 to 4, made according to topological equivalencies.[19] The numbering is according to TIMP-2. Location and extent of α-helices and β-strands are given by *cylinders and arrows,* and *symbols* show residues involved in main and side chain interactions with the target proteinase. Fully conserved residues are *enboxed.*

forms a two-stranded β-sheet, runs through a wide multiple-turn loop, and terminates in a β-hairpin sheet. The last three (TIMP-1) to ten (TIMP-2) C-terminal residues do not exhibit a defined conformation and presumably form a flexible tail on the TIMP surface.[19]

The TIMP edge is formed by five separate chain segments, namely, the extended N-terminal segment Cys1-Pro5 flanked by the sA-sB loop and the sC-connector loop on the left-hand side, and by the sG-sH loop and the multiple-turn loop on the right-hand side (FIGS. 4 and 5). The N-terminal segment is tightly connected to the adjacent sC-connector and to the underlying sE-sF loop via disulfide bridges Cys1-Cys70/72 and Cys3-Cys99/101 (FIG. 4). Particularly remarkable features of TIMP-2 are the quite elongated sA-sB β-hairpin loop, which does not follow the OB-barrel curvature, but is twisted and extends away, and the much longer negatively charged flexible C-terminal tail. TIMP-4 seems to share these features with TIMP-2, while TIMP-3 exhibits a short sA-sB loop and a long but not negatively charged tail[19,20] (FIG. 4). The TIMP topologies differ considerably, in spite of 40% overall sequence identities.

*The TIMP-MMP Inhibition Mechanism*

In complexes with MMPs, the wedge-shaped TIMPs bind with their edge into the entire length of the active-site cleft of their cognate MMPs,[19,45] removing surfaces of about 1300 $\text{Å}^2$ of each molecule from contact with bulk solvent and some rigidification of the participating loops[54] upon complex formation (FIG. 6). The majority of all intermolecular contacts is made by the N-terminal segment Cys1-Pro5, the sC-connector loop and the connecting disulfide bridge; in the case of TIMP-2 (and probably also TIMP-4), participation of the sA-sB loop in intermolecular contacts is considerable. The first five TIMP residues Cys1 to Pro5 bind to the MMP active-site cleft in a substrate- or product-like manner, that is, similar as P1, P1´, P2´, P3´ and P4´ peptide substrate residues insert between the bulge and the wall-forming segments, forming five intermolecular inter-main chain hydrogen bonds (compare FIGURE 3a). The sC-connector loop, in particular residues Ser68/Ala70 and Val69/Val71, in contrast, interact in a somewhat substrate-inverse manner with the left-hand subsites S2 and S3.[19,45]

Cys1 is located directly above the catalytic zinc, with its N-terminal α-amino nitrogen and its carbonyl oxygen atoms placed directly above the catalytic zinc, coordinating it together with the three imidazole rings from the cognate MMP. The α-amino group of Cys1 approximately occupies the site of the bound "attacking" water molecule and forms a hydrogen bond to one carboxylate oxygen atom of the catalytic Glu219. In spite of this close interaction, the Glu219Asp mutation in MMP-2 does not significantly affect TIMP-1 binding.[65] The Thr/Ser side chain of the second TIMP residue extends into the S1´ pocket of the cognate MMP, similar to the side chain of a P1´ peptide substrate residue, without filling this pocket properly; the Ser2 side chain oxygen of TIMP-2 is hydrogen-bonded to the second carboxylate oxygen of the catalytic Glu219[19]; mutagenesis experiments with N-TIMP-1 have shown that replacement of Thr2 by other natural amino acids can lead to TIMP species, which are able to discriminate much more between different MMPs.[30] The Cys3, Val/Ser4, and Pro5 side chains contact subsites S2´, S3´ and S4´ in a manner expected for substrate P2´, P3´ and P4´ side chains (compare FIGURE 3a).

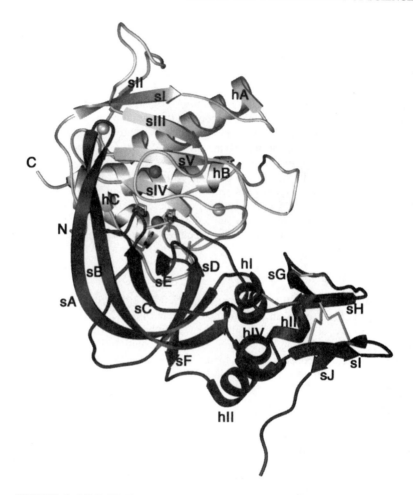

**FIGURE 6.** MMP-TIMP complex. The complex formed by the MT1-MMP catalytic domain *(top, light-colored ribbon)* and TIMP-2 *(bottom, dark-colored ribbon)*[19] is shown in front view (SETOR[72]). All disulfide bridges are given, and the zinc and calcium ions are represented as *dark grey and light grey spheres*. The wedge-shaped inhibitor binds with its edge made by six segments into the entire active-site cleft of the MMP-14, which in this view is directed towards the bottom. The N-terminal Cys1 of TIMP-2 is located on top of the catalytic zinc, and the first five N-terminal residues bind to the active site in a substrate-like manner.

It is noteworthy that a very similar "substrate-like" interaction has been found in two crystal forms of "non-inhibited" MMP-1 cds, where the opened-out N-terminal segment (Leu102-Thr-Glu-Gly105) of one molecule inserts into the S1 to S3′ sub-sites of a symmetry-related molecule,[38] utilizing identical subsite and intermolecular hydrogen bond interactions as observed in the MMP–TIMP complexes. A similar peptide–metalloproteinase interaction has, furthermore, been observed in complexes between serralysins and endogenous inhibitors produced by various *Serratia*-like bacteria; these ~100-amino acid residue inhibitors interact via a flexible N-terminal strand (of sequence Ser1-Ser2-Leu3-Arg4) with the S1 to S3′ subsites of their target serralysin, in a roughly similar manner as the N-terminal TIMP tetrapeptide does in the MMP-3-TIMP-1 complex[69]; neither the rest of the inhibitor molecules nor the other interactions bear any further resemblance with the TIMPs, however.

In the MT1-MMP-TIMP-2 complex (FIG. 6), the quite long sA-sB hairpin loop of TIMP-2 folds alongside the S-loop over the rim of the active-site cleft and reaches up to the β-sheet of its cognate MMP[19]; in spite of the relatively large overall interface between the sA-sB loop and the molecular MMP surface, most of the intermolecular contacts do not seem to be designed for optimal complementarity. The other edge loops of TIMP (namely the sE-sF loop, the sG-sH loop, and the multiple-turn loop; FIG. 4) are involved in a relatively small number of intermolecular contacts, with both C-terminal edge loops (in particular in TIMP-1) showing relatively high flexibility. Binding data indicate, however, that the interaction of the C-terminal TIMP subdomains with a cognate MMP might be of much more importance for the TIMP inhibition of the gelatinases and MMP-13,[14] probably correlating with tighter intermolecular contacts.

Superposition experiments with the structure of full-length porcine MMP-1[47] show that the C-terminal hld of full-length MMPs as positioned in that structure would be compatible with TIMP binding.[45] In such a TIMP–full-length MMP complex both domains, the TIMP and the hld, would just touch one another, in agreement with kinetic binding studies showing (except for gelatinases A and B and MMP-13; see Ref. 14) that the C-terminal MMP domains contribute relatively little to TIMP binding.[26,30] The negatively charged C-terminal tail of TIMP-2 seems to facilitate the noninhibitory binding to the progelatinase A hld mainly via electrostatic interactions. This specific TIMP-2-tail…MMP-2-hld interaction is important for formation of the MT1-MMP-TIMP-2-progelatinase complex implicated in progelatinase activation.[70,71]

## REFERENCES

1. WOESSNER, J.F., JR. 1991. Matrix metalloproteinases and their inhibitors in connective tissue remodeling. FASEB J. **5:** 2145–2155.
2. NAGASE, H., S.K. DAS, S.K. DEY, J.L. FOWLKES, W. HUANG & K. BREW. 1997. *In* Inhibitors of Metalloproteinases in Development and Disease. S.P. Hawkes, D.R. Edwards, and R. Khokha, Eds.) Harwood Academic Publishers. Lausanne, Switzerland.
3. JOHNSON, L.L., R. DYER & D.J. HUPE. 1998. Matrix metalloproteinases. Curr. Opin. Chem. Biol. **2:** 466–471.
4. YONG, V.W., C.A. KREKOSKI, P.A. FORSYTH, R. BELL & D.R. EDWARDS. 1998. Matrix metalloproteinases and diseases of the CNS. Trends Neurosci. **21:** 75–80.
5. BECKETT, R. & M. WHITTAKER. 1998. Matrix metalloproteinase inhibitors 1998. Exp. Opin. Ther. Patents **8:** 259–282.
6. BOTTOMLEY, K.M., W.H. JOHNSON & D.S. WALTER. 1998. Matrix metalloproteinase inhibitors in arthritis. J. Enz. Inhib. **13:** 79–101.
7. MASSOVA, I., L.P. KOTRA, R. FRIDMAN & S. MOBASHERY. 1998. Matrix metalloproteinases: structures, evolution, and diversification. FASEB J. **12:** 1075–1095.
8. RAWLINGS, N.D. & A.J. BARRETT. 1995. Evolutionary families of metallopeptidases. Meth. Enzymol. **248:** 183–229.

9. SANG, Q.A. & D.A. DOUGLAS. 1996. Computational sequence analysis of matrix metalloproteinases. J. Prot. Chem. **15**: 137–160.
10. MURPHY, G.J., G. MURPHY & J.J. REYNOLDS. 1991. The origin of matrix metallopro-
    teinases and their familial relationships. FEBS Lett. **289**: 4–7.
11. SATO, H., T. TAKINO, Y. OKADA, J. CAO, A. SHINAGAWA, E. YAMAMOTO & M. SEIKI.
    1994. A matrix metalloproteinase expressed on the surface of invasive tumour cells.
    Nature **370**: 61–65.
12. SATO, H., T. KINOSHITA, T. TAKINO, K. NAKAYAMA & M. SEIKI. 1996. Activation of a
    recombinant membrane type 1-matrix metalloproteinase (MT1-MMP) by furin and
    its interaction with tissue inhibitor of metalloproteinases (TIMP)-2. FEBS Lett. **393**:
    101–104.
13. PEI, D. & S.J. WEISS. 1996. Transmembrane-deletion mutants of the membrane-type
    matrix metalloproteinase-1 process progelatinase A and express intrinsic matrix-
    degrading activity. J. Biol. Chem. **271**: 9135–9140.
14. MURPHY, G. & V. KNÄUPER. 1997. Relating matrix metalloproteinase structure to
    function: why the "hemopexin" domain? Matrix Biol. **15**: 511–518.
15. DOCHERTY, A.J.P., A. LYONS, B.J. SMITH, E.M. WRIGHT, P.E. STEPHENS,
    T.J.R. HARRIS, G. MURPHY & J.J. REYNOLDS. 1985. Sequence of human tissue inhib-
    itor of metalloproteinases and its identity to erythroid potentiating activity. Nature
    **318**: 65–69.
16. STETLER-STEVENSON, W.G., H.C. KRUTZSCH & L.A. LIOTTA. 1989. Tissue inhibitor of
    metalloproteinase (TIMP-2). J. Biol. Chem. **264**: 17374–17378.
17. APTE, S.S., B.R. OLSEN & G. MURPHY. 1995. The gene structure of tissue inhibitor of
    metalloproteinases (TIMP)-3 and its inhibitory activities define the distinct TIMP
    gene family. J. Biol. Chem. **270**: 14313–14318.
18. GREENE, J., M. WANG, Y.E. LIU, L.A. RAYMOND, C. ROSEN & Y.E. SHI. 1996. Molec-
    ular cloning and characterisation of human tissue inhibitor of metalloproteinase 4. J.
    Biol. Chem. **271**: 30375–30380.
19. FERNANDEZ-CATALAN, C., W. BODE, R. HUBER, D. TURK, J.J. CALVETE, A. LICHTE,
    H. TSCHESCHE & K. MASKOS. 1998. Crystal structure of the complex formed by the
    membrane type 1-matrix metalloproteinase with the tissue inhibitor of metallopro-
    teinases-2, the soluble progelatinase A receptor. EMBO J. **17**: 5238–5248.
20. DOUGLAS, D.A., Y.E. SHI & Q.A. SANG. 1997. Computational sequence analysis of
    the tissue inhibitor of metalloproteinase family. J. Prot. Chem. **16**: 237–255.
21. GOMEZ, D.E., D.F. ALONSO, H. YOSHIJI & U.P. THORGEIRSSON. 1997. Tissue inhibi-
    tors of metalloproteinases: structure, regulation and biological functions. Eur. J. Cell
    Biol. **74**: 111–122.
22. CAWSTON, T. 1998. Matrix metalloproteinases and TIMPs: properties and implications
    for the rheumatic diseases. Mol. Med. Today **4**: 130–137.
23. BUTLER, G.S., H. WILL, S.J. ATKINSON & G. MURPHY. 1997. Membrane-type-2 matrix
    metalloproteinase can initiate the processing of progelatinase A and is regulated by
    the tissue inhibitors of metalloproeinases. Eur. J. Biochem. **244**: 653–657.
24. WILL, H., S.J. ATKINSON, G.S. BUTLER, B. SMYTH & G. MURPHY. 1996. The soluble
    catalytic domain of membrane type 1 matrix metalloproteinase cleaves the propep-
    tide of progelatinase A and initiates autocatalytic activation. J. Biol. Chem. **271**:
    17119–17123.
25. ZUCKER, S., M. DREWS, C. CONNER, H.D. FODA, Y.A. DeCLERCK, K.E. LANGLEY,
    W.F. BAHOU, A.J.P. DOCHERTY & J. CAO. 1998. Tissue inhibitor of metalloprotein-
    ase-2 (TIMP-2) binds to the catalytic domain of the cell surface receptor, membrane
    type 1-matrix metalloproteinase 1 (MT1-MMP). J. Biol. Chem. **273**: 1216–1222.
26. MURPHY, G. & F. WILLENBROCK. 1995. Tissue inhibitors of matrix metalloendopepti-
    dases. Meth. Enzymol. **248**: 496–510.

27. STRONGIN, A.Y., I.E. COLLIER, U. BANNIKOV, B.L. MARMER, G.A. GRANT & G.I. GOLDBERG. 1995. Mechanism of cell surface activation of 72-kDa type IV collagenase. J. Biol. Chem. **270:** 5331–5338.

28. STRONGIN, A.Y., B.L. MARMER, G.A. GRANT & G.I. GOLDBERG. 1993. Plasma membrane-dependent activation of the 72-kDa type IV collagenase is prevented by complex formation with TIMP-2. J. Biol. Chem. **268:** 14033–14039.

29. KINOSHITA, T., H. SATO, T. TAKINO, M. ITOH, T. AKIZAWA & M. SEIKI. 1996. Processing of a precursor of 72-kilodalton type IV collagenase/gelatinase A by a recombinant membrane-type 1 matrix metalloproteinase. Cancer Res. **56:** 2535–2538.

30. HUANG, W., Q. MENG, K. SUZUKI, H. NAGASE & K. BREW. 1997. Mutational study of the amino-terminal domain of human tissue inhibitor of metalloproteinases 1 (TIMP-1) locates an inhibitory region for matrix metalloproteinases. J. Biol. Chem. **272:** 22086–22091.

31. MURPHY, G., A. HOUBRECHTS, M.I. COCKETT, R.A. WILLIAMSON, M. O'SHEA & A.J.P. DOCHERTY. 1991. The N-terminal domain of human tissue inhibitor of metalloproteinases retains metalloproteinase inhibitory activity. Biochemistry **30:** 8097–8102.

32. LOVEJOY, B., A. CLEASBY, A.M. HASSELL, K. LONGLEY, M.A. LUTHER, D. WEIGL, G. MCGEEHAN, A.B. MCELROY, D. DREWRY, M.H. LAMBERT & S.R. JORDAN. 1994. Structure of the catalytic domain of fibroblast collagenase complexed with an inhibitor. Science **263:** 375–377.

33. BORKAKOTI, N., F.K. WINKLER, D.H. WILLIAMS, A. D'ARCY, M.J. BROADHURST, P.A. BROWN, W.H. JOHNSON & E.J. MURRAY. 1994. Structure of the catalytic domain of human fibroblast collagenase complexed with an inhibitor. Nature Struct. Biol. **1:** 106–110.

34. STAMS, T., J.C. SPURLINO, D.L. SMITH, R.C. WAHL, T.F. HO, M.W. QORONFLEH, T.M. BANKS & B. RUBIN. 1994. Structure of human neutrophil collagenase reveals large S1′ specificity pocket. Nature Str. Biol. **1:** 119–123.

35. BODE, W., P. REINEMER, R. HUBER, T. KLEINE, S. SCHNIERER & H. TSCHESCHE. 1994. The X-ray crystal structure of the catalytic domain of human neutrophil collagenase inhibited by a substrate analogue reveals the essentials for catalysis and specificity. EMBO J. **13:** 1263–1269.

36. REINEMER, P., F. GRAMS, R. HUBER, T. KLEINE, S. SCHNIERER, M. PIEPER, H. TSCHESCHE & W. BODE. 1994. Structural implications for the role of the N-terminus in the 'superactivation' of collagenases—a crystallographic study. FEBS Lett. **338:** 227–233.

37. GOOLEY, P.R., J.F. O'CONNELL, A.I. MARCY, G.C. CUCA, S.P. SALOWE, B.L. BUSH, J.D. HERMES, C.K. ESSER, W.K. HAGMANN, J.P. SPRINGER & B.A. JOHNSON. 1994. NMR structure of inhibited catalytic domain of human stromelysin-1. Nature Struct. Biol. **1:** 111–118.

38. LOVEJOY, B., A.M. HASSELL, M.A. LUTHER, D. WEIGL & S.R. JORDAN. 1994. Crystal structures of recombinant 19-kDa human fibroblast collagenase complexed to itself. Biochemistry **33:** 8207–8217.

39. SPURLINO, J.C., A.M. SMALLWOOD, D.D. CARLTON, T.M. BANKS, K.J. VAVRA, J.S. JOHNSON, E.R. COOK, J. FALVO, R.C. WAHL, T.A. PULVINO, J.J. WENDOLOSKI & D.L. SMITH. 1994. 1.56Å structure of mature truncated human fibroblast collagenase. Proteins Struct. Funct. Genet. **19:** 98–109.

40. BROWNER, M.F., W.W. SMITH & A.L. CASTELHANO. 1995. Matrilysin-inhibitor complexes: common themes among metalloproteases. Biochemistry **34:** 6602–6610.

41. BECKER, J.W., A.I. MARCY, L.L. ROKOSZ, M.G. AXEL, J.J. BURBAUM, P.M.D. FITZGERALD, P.M. CAMERON, C.K. ESSER, W.K. HAGMANN, J.D. HERMES & J.P. SPRINGER. 1995. Stromelysin-1: Three-dimensional structure of the inhibited catalytic domain and of the C-truncated proenzyme. Prot. Sci. **4:** 1966–1976.

42. DHANARAJ, V., Q.-Z. YE, L.L. JOHNSON, D.J. HUPE, D.F. ORTWINE, J.B. DUNBAR, J.R. RUBIN, A. PAVLOVSKY, C. HUMBLET & T.L. BLUNDELL. 1996. X-ray structure of a hydroxamate inhibitor complex of stromelysin catalytic domain and its comparison with members of the zinc metalloproteinase superfamily. Structure **4:** 375–386.

43. VANDOREN, S.R., A.V. KUROCHKIN, W. HU, Q.Z. YE, L.L. JOHNSON, D.J. HUPE & E.R. ZUIDERWEG. 1995. Solution structure of the catalytic domain of human stromelysin complexed with a hydrophobic inhibitor. Protein Sci. **4:** 2487–2498.

44. WETMORE, D.R. & K.D. HARDMAN. 1996. Roles of the propeptide and metal ions in the folding and stability of the catalytic domain of stromelysin (matrix metalloproteinase 3). Biochemistry **35:** 6549–6558.

45. GOMIS-RÜTH, F.X., K. MASKOS, M. BETZ, A. BERGNER, R. HUBER, K. SUZUKI, N. YOSHIDA, H. NAGASE, K. BREW, G.P. BOURENKOV, H. BARTUNIK & W. BODE. 1997. Mechanism of inhibition of the human matrix metalloproteinase stromelysin-1 by TIMP-1. Nature **389:** 77–81.

46. GRAMS, F., P. REINEMER, J.C. POWERS, T. KLEINE, M. PIEPER, H. TSCHESCHE, R. HUBER & W. BODE. 1995. X-ray structures of human neutrophil collagenase complexed with peptide hydroxamate and peptide thiol inhibitors. Implications for substrate binding and rational drug design. Eur. J. Biochem. **228:** 830–841.

47. LI, J.-Y., P. BRICK, M.C. O'HARE, T. SKARZYNSKI, L.F. LLOYD, V.A. CURRY, I.M. CLARK, H.F. BIGG, B.L. HAZLEMAN, T.E. CAWSTON & D.M. BLOW. 1995. Structure of full-length porcine synovial collagenase reveals a C-terminal domain containing a calcium-linked, four-bladed β-propeller. Structure **3:** 541–549.

48. LIBSON, A., A. GITTIS, I. COLLIER, B. MARMER, G. GOLDBERG & E.E. LATTMAN. 1995. Crystal structure of the hemopexin-like C-terminal domain of gelatinase A. Nature Struct. Biol. **2:** 938–942.

49. GOHLKE, U., F.-X. GOMIS-RÜTH, T. CRABBE, G. MURPHY, A.J.P. DOCHERTY & W. BODE. 1996. The C-terminal (haemopexin-like) domain structure of human gelatinase A (MMP2): structural implications for its function. FEBS Lett. **378:** 126–130.

50. GOMIS-RÜTH, F.X., U. GOHLKE, M. BETZ, V. KNÄUPER, G. MURPHY, C. LOPEZ-OTIN & W. BODE. 1996. The helping hand of collagenase-3 (MMP-13): 2.7Å crystal structure of its C-terminal haemopexin-like domain. J. Mol. Biol. **264:** 556–566.

51. BODE, W. (1995) A helping hand for collagenases: the haemopexin-like domain. Structure **3:** 527–530.

52. WILLIAMSON, R.A., G. MARTORELL, M.D. CARR, G. MURPHY, A.J. DOCHERTY, R.B. FREEDMAN & J. FEENEY. 1994. Solution structure of the active domain of tissue inhibitor of metalloproteinases-2. A new member of the OB fold protein family. Biochemistry **33:** 11745–11759.

53. WILLIAMSON, R.A., M.D. CARR, T.A. FRENKIEL, J. FEENEY & R.B. FREEDMAN. 1997. Mapping the binding site for matrix metalloproteinase on the N-terminal domain of the tissue inhibitor of metalloproteinases-2 by NMR chemical shift perturbation. Biochemistry **36:** 13882–13889.

54. MUSKETT, F.W., T.A. FRENKIEL, J. FEENEY, R.B. FREEDMAN, M.D. CARR & R. WILLIAMSON. 1998. High resolution structure of the N-terminal domain of tissue inhibitor of metalloproteinases-2 and characterization of its interaction site with matrix metalloproteinase-3. J. Biol. Chem. **273:** 21736–21743.

55. SCHECHTER, I. & A. BERGER. 1967. On the size of the active site in proteases. I. Papain. Biochem. Biophys. Res. Commun. **27:** 157–162.

56. BODE, W., F.-X. GOMIS-RÜTH & W. STÖCKER. 1993. Astacins, serralysins, snake venom and matrix metalloproteinases exhibit identical zinc-binding environments (HEXXHXXGXXH and Met-turn) and topologies and should be grouped into a common family, the 'metzincins.' FEBS Lett. **331:** 134–140.

57. STÖCKER, W., F. GRAMS, U. BAUMANN, P. REINEMER, F.X. GOMIS-RÜTH, D.B. MCKAY & W. BODE. 1995. The metzincins—topological and sequential relations between the astacins, adamalysins, serralysins, and matrixins (collagenases) define a superfamily of zinc-peptidases. Prot. Sci. **4:** 823–840.

58. KNÄUPER, V., G. MURPHY & H. TSCHESCHE. 1996. Activation of human neutrophil procollagenase by stromelysin 2. Eur. J. Biochem. **235:** 187–191.

59. NAGASE, H. 1997. Activation mechanisms of matrix metalloproteinases. Biol. Chem. **378:** 151–160.

60. GRAMS, F., M. CRIMMIN, L. HINNES, P. HUXLEY, M. PIEPER, H. TSCHESCHE & W. BODE. 1995. Structure determination and analysis of human neutrophil collagenase complexed with a hydroxamate inhibitor. Biochemistry **34:** 14012–14020.

61. NETZEL-ARNETT, S., G.B. FIELDS, H. BIRKEDAL-HANSEN & H.E. VAN WART. 1991. Sequence specificities of human fibroblast and neutrophil collagenase. J. Biol. Chem. **266:** 6747–6755.

62. NETZEL-ARNETT, S., Q.X. SANG, W.G.I. MOORE, M. NAVRE, H. BIRKEDAL-HANSEN & H.E. VAN WART. 1993. Comparative sequence specificities of human 72- and 92-kDa gelatinases (type IV collagenases) and PUMP (matrilysin) Biochemistry **32:** 6427–6432.

63. NIEDZWIECKI, L., J. TEAHAN, R.K. HARRISON & R.L. STEIN. 1992. Substrate specificity of the human matrix metalloproteinase stromelysin and the development of continuous fluoremetric assays. Biochemistry **31:** 12618–12623.

64. MATTHEWS, B.W. 1988. Structural basis of the action of thermolysin and related zinc peptidases. Acc. Chem. Res. **21:** 333–340.

65. CRABBE, T., S. ZUCKER, M.I. COCKETT, F. WILLENBROCK, S. TICKLE, J.P. O'CONNELL, J.M. SCOTHERN, G. MURPHY & A.J.P. DOCHERTY. 1994. Mutation of the active site glutamic acid of human gelatinase A: effects on latency, catalysis, and the binding of tissue inhibitor of metalloproteinases-1. Biochemistry **33:** 6684–6690.

66. WINDSOR, L.J., M.K. BODDEN, B. BIRKEDAL-HANSEN, J.A. ENGLER & H. BIRKEDAL-HANSEN. 1994. Mutational analysis of residues in and around the active site of human fibroblast-type collagenase. J. Biol. Chem. **269:** 26201–26207.

67. BODE, W., F.X. GOMIS-RÜTH, R. HUBER, R. ZWILLING & W. STÖCKER. 1992. Structure of astacin and implications for activation of astacins and zinc-ligation of collagenases. Nature **358:** 164–166.

68. GRAMS, F., V. DIVE, A. YIOTAKIS, I. YIALLOUROS, S. VASSILIOU, R. ZWILLING, W. BODE & W. STÖCKER. 1996. Structure of astacin with a transition-state analogue inhibitor. Nature Struct. Biol. **3:** 671–675.

69. BAUMANN, U., M. BAUER, S. LETOFFE, P. DELEPELAIRE & C. WANDERSMAN. 1995. Crystal Structure of a complex between *Serratia marcescens* metallo-protease and an inhibitor from *Erwinia chrysanthemi.* J. Mol. Biol. **248:** 653–661.

70. CAO, J., H. SATO, T. TAKINO & M. SEIKI. 1995. The C-terminal region of membrane type matrix metalloproteinase is a functional transmembrane domain required for progelatinase A activation. J. Biol. Chem. **270:** 801–805.

71. BUTLER, G.S., M.J. BUTLER, S.J. ATKINSON, H. WILL, T. TAMURA, S.S. VAN WESTRUM, T. CRABBE, J. CLEMENTS, M.P. D'ORTHO & G. MURPHY. 1998. The TIMP2 membrane type 1 metalloproteinase "receptor" regulates the concentration and efficient activation of progelatinase A. J. Biol. Chem. 273, 871-880.

72. EVANS, S.V. 1993. SETOR: hardware lighted three-dimensional solid model representations of macromolecules. J. Mol. Graph. **11:** 134–138.

# Aggrecanase

## A Target for the Design of Inhibitors of Cartilage Degradation

ELIZABETH C. ARNER,[a] MICHAEL A. PRATTA, CARL P. DECICCO,
CHU-BIAO XUE, ROBERT C. NEWTON, JAMES M. TRZASKOS,
RONALD L. MAGOLDA, AND MICKY D. TORTORELLA

*Inflammatory Diseases Research and Chemical and Physical Sciences,
DuPont Pharmaceuticals Company, Wilmington, Delaware 19880-0400, USA*

**ABSTRACT:** In arthritic diseases there is a gradual erosion of cartilage that leads to a loss of joint function. Aggrecan, which provides cartilage with its properties of compressibility and elasticity, is the first matrix component to undergo measurable loss in arthritis. This loss of aggrecan appears to be due to an increased rate of degradation, that can be attributed to proteolytic cleavage of the core protein within the interglobular domain (IGD). Two major sites of cleavage have been identified within the IGD. One, between the amino acids $Asn^{341}$-$Phe^{342}$, where the matrix metalloproteinases (MMPs) have been shown to clip; and the other, between $Glu^{373}$-$Ala^{374}$, which is attributed to a novel protease, "aggrecanase." We have generated aggrecanase in conditioned media from IL-1–stimulated bovine nasal cartilage and have used an enzymatic assay to evaluate this proteinase activity. In these studies we follow the generation of aggrecanase and MMPs in response to IL-1 in this system and examine the contribution of these enzymes in aggrecan degradation. Our data suggest that aggrecanase is a key enzyme in cartilage aggrecan degradation that represents a novel target for cartilage protection therapy in arthritis.

Degenerative joint diseases are commonly characterized by the destruction of the cartilage extracellular matrix where loss of proteoglycan (aggrecan) from cartilage is an early event that precedes the proteolytic breakdown of other extracellular matrix macromolecules.[1] As a result, the tissue loses its capacity to resist compression under load, which eventually leads to mechanical destruction of the cartilage.

At present the biochemical mechanisms involved in aggrecan catabolism are unclear. Loss of aggrecan from the tissue is thought, at least in part, to be mediated by interleukin-1 or perhaps by TNF-$\alpha$.[2,3] Explant cultures of cartilage from a variety of species and anatomical sites have been extensively used to study the effects of these cytokines on the metabolism of aggrecan. These studies have led to the observations that there are multiple sites of cleavage along the aggrecan protein core,[4,5,6] but that the major and perhaps initial site of cleavage occurs within the interglobular domain (IGD) located between the G1 and G2 globular domains of aggrecan.[7,8]

[a]Address for correspondence: Elizabeth C. Arner, Principal Research Scientist, DuPont Pharmaceuticals Company, Experimental Station E400/4239, Wilmington, Delaware 19880-0400 Phone, 302/695-7078; Fax, 302/695-7873; e-mail, Elizabeth.C.Arner@dupontpharma.com

Many matrix metalloproteinases (MMP-1, -2, -3, -7, -8, -9 and -13) have been shown to cleave aggrecan *in vitro* at the $Asn^{341}$-$Phe^{342}$ site,[8–11] and G-1 fragments with this cleavage site have been identified within articular cartilage bound to hyaluronic acid.[8] However, recent studies from a number of experimental systems have indicated that a novel proteinase termed "aggrecanase" cleaves at the $Glu^{373}$-$Ala^{374}$ bond of the IGD and plays a central role in the catabolism of aggrecan in arthritic diseases.[12–13] C-terminal fragments with the new N-terminus, ARGSV, formed by cleavage between the amino acid residues $Glu^{373}$-$Ala^{374}$ have been identified in synovial fluid of patients with osteoarthritis,[13] inflammatory joint disease,[11] and in the media from cartilage explant and chondroctye cultures stimulated with interleukin-1 or retinoic acid.[7,14–15] These data suggest that cleavage at this site plays a role in cartilage matrix degradation.

Both MMPs and aggrecanase are induced in cartilage explants stimulated with IL-1. However, there is a strong positive correlation between aggrecan degradation as monitored by GAG release and the generation of aggrecanase-cleaved fragments with the N-terminus, ARGS.[16] Studies from our laboratory also indicate that compounds that block specific cleavage at the aggrecanase site inhibit proteoglycan degradation in response to IL-1 and exhibit a correlation in potency between these effects, suggesting that aggrecanase plays an important role in mediating IL-1–induced cartilage breakdown.[16]

We have developed a method for generating soluble aggrecanase activity in media from IL-1–stimulated bovine nasal cartilage as well as an enzymatic assay for following this activity.[17,18] Using these tools, in the work reported herein, we have further characterized the role of aggrecanase and MMPs in cartilage aggrecan degradation by following the time course of generation of these enzymes in media from cartilage explant cultures stimulated with IL-1, and the relationship of their appearance with the degradation of the matrix. We also investigated the efficiency of aggrecanase versus MMP cleavage of cartilage aggrecan since the contribution of a protease to aggrecan catabolism may be influenced by a number of factors including the efficiency of substrate cleavage.

## MATERIALS AND METHODS

### Materials

Dulbecco's modified Eagle's medium (DMEM) and fetal bovine serum (FBS) were from Gibco (Grand Island, NY). The IL-1 used was a soluble, fully active recombinant human IL-1β produced as described previously.[19] The specific activity was $1 \times 10^7$ units/mg of protein, with 1 unit being defined as the amount of IL-1 that generated half-maximal activity in the thymocyte proliferation assay. Antibody BC-3, which recognizes the new N-terminus, ARGS, on aggrecan fragments produced by "aggrecanase," was produced as previously described,[20] and antibody AF-28, which recognizes the new N-terminus, FFGVG, on aggrecan fragments produced by MMP cleavage, was a generous gift from Dr. Amanda Fosang (University of Melbourne, Parkville, Australia). The monoclonal antibody, MAB2005, which recognizes a protein-related epitope associated with the binding region of chondroitin sulfate chains, was purchased from Chemicon International (Temecula, CA).

The MAB2005 antibody reacts with the protein-related epitope only after chondroitinase ABC digestion.

Chondroitinase ABC lyase (*Proteus vulgaris*) (EC 4.2.2.4), keratanase (*Pseudomonas sp.*) (EC 3.2.1.103), and keratanase II (*Bacillus sp.*) were from Seikaguku (Kogyo, Japan). The antisera to stromelysin-1 (MMP-3) was a sheep anti-human MMP-3 polyclonal antisera produced at DuPont. The hydroxamic acid MMP inhibitor, XS309, (3S-[3R*, 2-[2R*, 2-(-(R*,S*)]-hexahydro-2-[2-[2-(hydroxyamino)-1-methyl-2-oxoethyl]-4-methyl-1-oxopentyl]-N-methyl-3-pyridazinecarboxamide) was synthesized at DuPont as previously described.[21]

## Cartilage Cultures

Cartilage was obtained from bovine nasal septa, and uniform slices were prepared under sterile conditions. Cartilage slices were incubated in Dulbecco's Modified Eagle medium containing penicillin, streptomycin, amphotericin B, and neomycin (100 IU/ml, 100 μg/ml, 0.25 μg/ml and 50 μg/ml, respectively) in the absence or presence of recombinant human IL-1β at 37°C in an atmosphere of 95% air/5% $CO_2$.[22] At the end of incubation, media were removed and stored at −70°C for analysis. Proteoglycan breakdown was determined by measuring the amount of sulfated glycosaminoglycan (GAG) released from the cartilage into the media using a dimethylmethylene blue dye assay.[23]

## Enzymatic Digestion of Freeze-Thawed Cartilage

Bovine nasal cartilage slices were frozen and thawed three times to render the chondrocytes non-viable and then used as a substrate for digestion by aggrecanase or MMP-3. Freeze-thawed (F/T) cartilage slices were incubated with aggrecanase or MMP-3 for 0–48 hr. Buffer samples were assayed for sulfated glycosaminoglycans (GAG) using the dimethylmethylene blue dye assay.[24] Fragments exhibiting the neoepitope, ARGSV, generated by cleavage at the aggrecanase site were assessed by BC-3 Western blot,[21] and fragments exhibiting for the neoepitope, FFGVG, generated by cleavage at the MMP site, were assessed by AF-28 Western blot.[24]

## Aggrecanase Enzymatic Assay

IL-1–stimulated bovine nasal cartilage conditioned media (25 μl) that contained active aggrecanase was incubated with purified bovine aggrecan (500 nM) in a final volume of 100 μl of 20mM Tris/100mM NaCl/10mM $CaCl_2$ buffer at pH 7.5 at 37°C in the absence or presence of inhibitors.[17,18] At the end of the incubation, the reaction was quenched with 20mM EDTA and the aggrecanase-generated products were detected by BC-3 Western analysis as described below.

## Aggrecan Western Blot Analysis

Aggrecan products were enzymatically deglycosylated with chondroitinase ABC (0.1 units/10 μg GAG) for 2 hr at 37°C and then with keratanase (0.1 units/10 μg GAG) and keratanase II (0.002 units/10μg GAG) for 2 hr at 37°C in buffer containing 50mM sodium acetate, 0.1 M Tris/HCl, pH 6.5.[21] After digestion, the samples were precipitated with 5 volumes of acetone, reconstituted in an appropriate volume of SDS-PAGE sample buffer (0.5 M Tris-HCl, pH 6.8, 4% SDS, 0.005% bromophenol blue, 20% glycerol) (Novex, San Diego, CA), and analyzed by SDS/PAGE under nonreducing conditions on 4–12% gradient polyacrylamide gels.[25] The separated proteins were transferred to PVDF membranes and immunolocated with a 1:500 dilution of the BC-3 antibody, a 1:1000 dilution of the AF-28 antibody, or a 1:2000 dilution of the MAB2005 antibody. Subsequently, membranes were incubated with an alkaline phosphatase conjugated goat anti-mouse IgG, and aggrecan catabolites visualized by incubation with the appropriate substrate.[26] For determination of aggrecanase activity, BC-3 blots were quantitated by scanning densitometry and the integrated pixel density of the bands minus that of the media blanks or net pixel density (n.p.d.) was taken as a measure of enzymatic activity.

### MMP-3 Western Blot

TCA precipitable protein from culture media was dissolved in SDS-PAGE sample buffer containing 2.5% mercaptoethanol, heated at 95°C for 10 min and run on a 10–20% Tricine SDS-PAGE gel (Novex, San Diego, CA). Protein was transferred to nitrocellulose and probed with 1:500 dilution of sheep anti-human MMP-3 antisera in 20 mM Tris, pH 7.5, containing 0.5 M NaCl, 1% BSA and 0.05% Tween 20 for 18 hr at 4°C. The blot was then incubated with rabbit anti-sheep IgG HRP (Pierce, Rockford, IL) at a 1:1000 dilution for 1 hr at room temperature and developed in 4-chloronapthol.

### MMP-3 FELISA

MMP-3 protein levels in media were determined using a fluorescent enzyme-linked immunosorbent assay (FELISA), as described previously.[27] Briefly, standards or samples were added to a 96-well plate containing a nitrocellulose membrane forming the bottom of the wells and incubated overnight at 4°C. The wells were blocked with 1% BSA in Tris buffered saline for 1 hr at room temperature, and the membrane was then probed with a 1:500 dilution of sheep anti-human MMP-3 antisera for 1 hr. Rabbit anti-sheep IgG alkaline phosphatase was then added at a 1:5000 dilution and incubated for 1 hr at room temperature. The alkaline phosphatase fluorescent substrate, AttoPhos, was then added and incubated at room temperature for 1 hr. Reactions were quenched with EDTA and the fluorescence determined (excitation, 420 nm; emission, 550 nm) using a Perkin Elmer LS50B fluorescent reader.

### Gelatin Zymography

Culture media were diluted 1:1 with sample buffer, incubated at room temperature for 10 min, and run on a 10% SDS-PAGE gels containing 0.1% gelatin and in running buffer (0.24 M Tris, 2 M glycine, 35 mM SDS, pH 8.3) for 90 min at 125 volts. Proteins were then renatured by incubating the gels in 2.5% Triton X-100 in water for 30 minutes at room temperature and then incubated at 37°C overnight in developing buffer (50 mM Tris, 0.2 M NaCl, 5 mM $CaCl_2$, 0.02% Brij 35, pH 7.6). After incubation gels were stained in 0.25% Coomassie Brilliant Blue R-250 for 4 hr at room temperature and destained in distilled water containing 30% methanol and 10% glacial acetic acid to reveal zones of lysis within the gelatin matrix. EDTA (5mM), E64 (10 μg/ml), pepstatin (1 μg/ml), or benzamidine HCl (10mM) or XS309

(1 μM) were added to the developing buffer to identify which classes of proteinases were responsible for lysis of the gelatin. In some cases conditioned medium samples were incubated with 4-aminophenylmercuric acetate (APMA) for 3 hr at 37°C to activate latent matrix metalloproteinases before loading to the substrate gels.

### Caseinase Enzymatic Assay

Casein labeled with resorufin (43.5 μM) was incubated with protease-containing media in a final volume of 200 μl in 50 mM Tris-HCl, 5 mM CaCl$_2$ at pH 7.8 at 37°C as described previously.[28] The reactions were quenched with 20 nM EDTA and undigested substrate precipitated by adding TCA to a final concentration of 5%. Samples were filtered through a 96-well filtration plate (Millipore Co., Bedford, MA), and filtrates were collected and neutralized by addition of 2 M Tris to each well. The absorbance was then read at 575 nm, and the concentration of the resorufin-labeled peptides in the filtrate was used as a measure of proteolytic activity. Some conditioned medium samples were assayed in the presence of 2mM APMA to activate latent matrix metalloproteinases. With some samples EDTA was included to inhibit metalloproteinase activity, or XS309 was added to inhibit matrix metalloproteinase activity.

### RESULTS

Bovine nasal cartilage was incubated with 500 ng/ml IL-1 for 16–18 days with media change every 2 days through day 6 when ~90% of the cartilage aggrecan had been lost (data not shown). Incubation with IL-1 was then continued without media change to allow accumulation of protease activity in the media. To evaluate the generation of protease over time, samples of conditioned media were taken from cultures at various times during incubation with IL-1. Aggrecanase activity was detected by incubation of the media with 500nM aggrecan monomer substrate and monitoring aggrecanase-generated products by BC-3 Western blot analysis.[17,18] Media quenched with EDTA prior to incubation with the substrate served as assay blanks to assess fragments present in the media prior to enzymatic assay.

Aggrecanase activity was seen in the media by day 2 and increased with increasing time of incubation (FIG. 1). On day 2 and day 4, high background levels of BC-3-reactive fragments, generated from endogenous aggrecan during degradation of the matrix in response to IL-1, were observed in media. After day 6 of culture, when the cartilage had been depleted of aggrecan, background BC-3–reactive fragments were no longer detected.

Conditioned media were evaluated for the presence of gelatinolytic activity over the time course of aggrecanase generation using gelatin zymography (FIG. 2A). A faint band of gelatinolytic activity corresponding to proMMP-9, the 92-kDa gelatinase, was observed on day 4, and this band increased in intensity at later time points. By day 8, when the aggrecan substrate had been completely depleted from the cartilage, a band corresponding to active MMP-9 was also detected in the media and increased in intensity with increasing time of incubation. By day 10, bands of gelatinolytic activity were detected at ~66 kDa, 60 kDa, and 52 kDa and a doublet seen at ~200 kDa. By day 16–18 two additional bands could be detected between

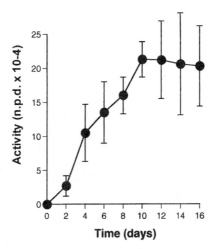

**FIGURE 1.** Time course of generation of aggrecanase activity in media from IL-1–stimulated bovine nasal cartilage. Bovine nasal cartilage slices were incubated with 500 ng/ml IL-1 for 18 days, with media being replaced every 2 days for the first 6 days of culture. Samples of media taken at various times during culture were assayed for aggrecanase activity by incubating with purified aggrecan monomer substrate for 4 hr at 37°C and evaluating fragments formed by specific cleavage at the aggrecanase site by BC-3 Western blot analysis. Media quenched with EDTA prior to incubation with substrate served as assay blanks. Cumulative aggrecanase activity as net pixel density (n.p.d.) is plotted versus days of incubation. Data represent the mean and standard deviation for two separate cultures.

100 and 200 kDa. Treatment of the day-6 media with the mercurial agent, APMA, was carried out to activate the zymogens of matrix metalloproteinases. APMA transformed the 92-kDa gelatinase into an 85-kDa species consistent with active MMP-9 (FIG. 2B). In addition, the gelatinase migrating at approximately 62 kDa was converted to the lowest mass species (~56 kDa) seen prior to APMA activation, indicating that this ~62-kDa band also represented a pro-MMP and that the lowest mass band likely represents the active form of this gelatinase. All the gelatinases detected were shown to be metalloproteases by the complete loss of gelatinolytic activity when the zymogram was incubated with EDTA (FIG. 2C). Inhibitors of the serine, cysteine, and aspartate protease classes had no effect on gelatinase activity under similar conditions. XS309, a potent, broad-spectrum inhibitor of MMPs that is ineffective against aggrecanase, also resulted in a complete loss of gelatinolytic activity (FIG. 2d), suggesting that these gelatinases are matrix metalloproteases and that they do not represent aggrecanase.

MMP-3 is induced by stimulation of chondrocytes with IL-1. Therefore, to assess the production of MMP-3 in the culture media during the course of aggrecanase generation, media samples were evaluated by MMP-3 FELISA. MMP-3 protein was induced by IL-1 and increased with time of incubation (FIG. 3A). Evaluation of media by MMP-3 Western blot (FIG. 3B) indicated that on day 2–8 only the zymogen form of MMP-3 protein was present in the culture media, but from day 10–16 after deple-

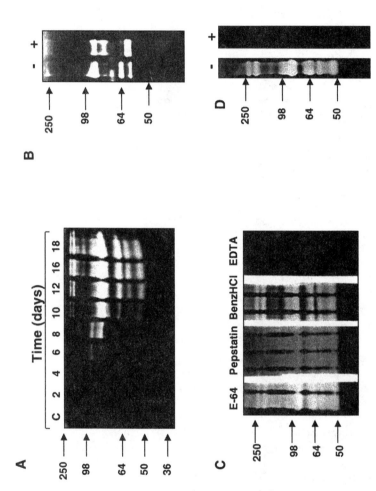

**FIGURE 2.** Time course of generation of gelatinolytic activity in media from IL-1–stimulated bovine nasal cartilage. Media from the experiment described in FIGURE 1 were analyzed for gelatinolytic activity by gelatin zymography. Equal volume of media samples taken at various times during culture were diluted 1:20 with water and evaluated for gelatinolytic activity (**A**). Undiluted medium from day-6 cultures was analyzed with (+) or without (−) treatment with APMA to activate the zymogens of matrix metalloproteases prior to zymographic analysis (**B**). Media samples from day 16 of culture were diluted 1:20 with water and analyzed with the inclusion of EDTA (5 mM), E-64 (10 μg/ml), pepstatin (1 μg/ml), or benzamidine HCl (10 mM) in the zymogram developing buffer (**C**). Media samples from day 16 of culture were diluted 1:20 with water and analyzed in the presence (+) or absence (−) of 1 μM XS309 in the developing buffer (**D**).

tion of aggrecan from the matrix, both the pro and active form of the enzyme were detected.

Conditioned media were evaluated for caseinolytic activity over the time course of aggrecanase generation (FIG. 4) using a casein-resorufin assay.[28] Casein, which is a substrate for a number of proteases, has been used to monitor activity of metallo-proteases, including that of MMP-3 and MMP-1.[29] MMPs can be present in both an active form and in a latent zymogen form that requires activation. Therefore, media were assayed without activation as a measure of active enzyme, and in the presence of APMA, which activates any latent MMPs and gives a measure of total protease activity. On day 2–6 all caseinolytic activity in the media was present in the zymogen form, as activation with APMA was required to detect activity. From day 8–16 the amount of latent activity remained relatively constant, while the active protease ac-

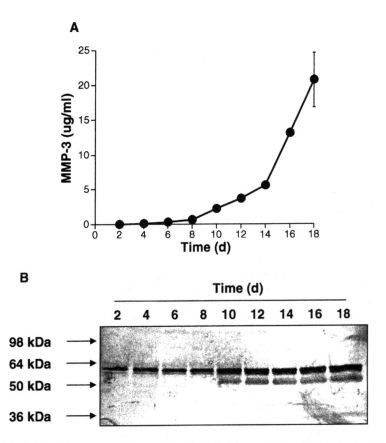

**FIGURE 3.** Time course of generation of MMP-3 in media from IL-1–stimulated bo-vine nasal cartilage. Media from cultures described in FIGURE 1 were analyzed for MMP-3 protein by MMP-3 FELISA as described in *Materials and Methods*, and MMP-3 equiva-lents per ml of media were plotted versus time of incubation (**A**). To evaluate the distribu-tion of MMP-3 between the zymogen and mature forms, these media were analyzed by MMP-3 Western blot analysis (**B**).

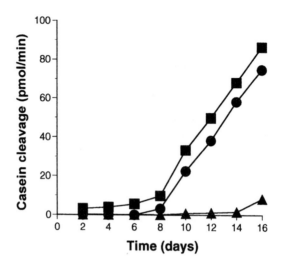

**FIGURE 4.** Time course of generation of caseinolytic activity in media from IL-1–stimulated bovine nasal cartilage. Media from cultures described in FIGURE 1 were analyzed for caseinolytic activity using a casein resorufin substrate, and activity (pmoles casein cleaved per minute) was plotted versus incubation time for samples assayed without APMA activation (●), assayed following APMA activation (■), and assayed in the presence of 10 mM EDTA following APMA activation (▲).

tivity increased over time. The majority of both the active and latent protease activity could be blocked with EDTA, indicating that this activity is primarily due to metalloproteases. Activity was also blocked by XS309, a broad potent MMP inhibitor that lacks the ability to inhibit aggrecanase (data not shown), thus indicating that this caseinolytic activity is due to MMPs but not to aggrecanase.

Although no active MMPs were detected in the culture media during the time periods when aggrecan degradation occurred, this does not rule out the presence of active enzyme within the matrix. Therefore, to determine whether MMPs, if active within the matrix, would be expected to dominate over aggrecanase in cleaving aggrecan, we examined the efficiency of aggrecanase versus MMP-3 in cleaving cartilage aggrecan. Using the native aggrecan in freeze-thawed bovine nasal cartilage as a substrate and monitoring the release of GAG from the cartilage as a measure of cleavage, we found that both MMP-3 and aggrecanase caused a time-dependent cleavage of aggrecan. However, digestion with aggrecanase resulted in more efficient cleavage.

Complete release of aggrecan from the freeze-thawed cartilage occurred by 48 hours of incubation with ~5pM aggrecanase (FIG. 5A), which approximates the concentration found in media from IL-1–stimulated live bovine nasal cartilage. Western blot analysis using the BC-3 antibody confirmed that this cleavage occurred at the $Glu^{373}$-$Ala^{374}$ site (FIG. 5B); no AF-28 reactive fragments were generated (data not shown), indicating that aggrecanase did not cleave at the MMP site.

By comparison, no increase in GAG release was detected over a 48-hour incubation with 5pM MMP-3 (FIG. 5A). At a concentration of 100 nM, comparable to the

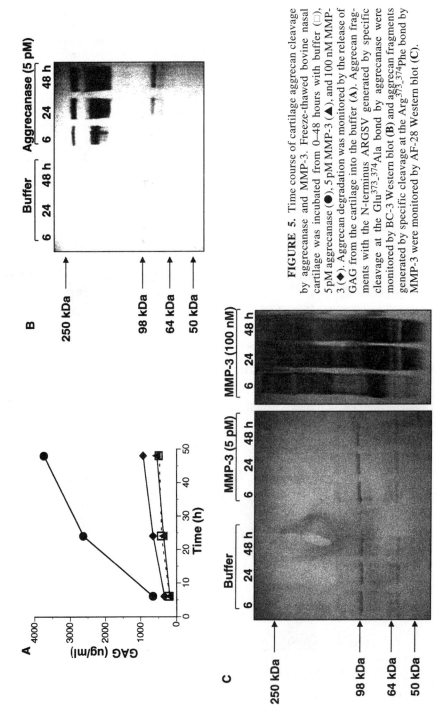

**FIGURE 5.** Time course of cartilage aggrecan cleavage by aggrecanase and MMP-3. Freeze-thawed bovine nasal cartilage was incubated from 0–48 hours with buffer (□), 5 pM aggrecanase (●), 5 pM MMP-3 (▲), and 100 nM MMP-3 (◆). Aggrecan degradation was monitored by the release of GAG from the cartilage into the buffer (**A**). Aggrecan fragments with the N-terminus ARGSV generated by specific cleavage at the Glu$^{373}$-$^{374}$Ala bond by aggrecanase were monitored by BC-3 Western blot (**B**) and aggrecan fragments generated by specific cleavage at the Arg$^{373}$-$^{374}$Phe bond by MMP-3 were monitored by AF-28 Western blot (**C**).

concentration of proMMP-3 in media from IL–stimulated cartilage, active MMP-3 caused a time-dependent cleavage of aggrecan, but at a slower rate than seen with aggrecanase. As expected, cleavage by MMP-3 resulted in the generation of AF-28-reactive fragments (FIG. 5C), indicating that cleavage occurred at the $Asn^{341}$-$Phe^{342}$ bond. Comparison of the rate of cleavage by aggrecanase and MMP-3 indicates that aggrecanase is dramatically more efficient in cleaving aggrecan.

These results were confirmed using isolated bovine aggrecan monomers and, following cleavage, using an antibody that recognizes an epitope associated with the CS binding region on deglycosylated aggrecan (FIG. 6). Disappearance of the 370 kDa band, representing intact aggrecan, occurred within 15 minutes with 5 pM aggrecanase, while a decrease in intensity of this band did not occur until 24 hours with

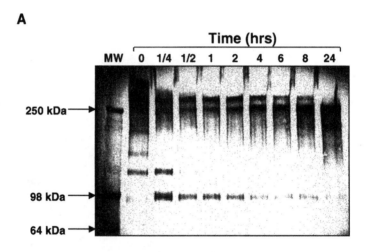

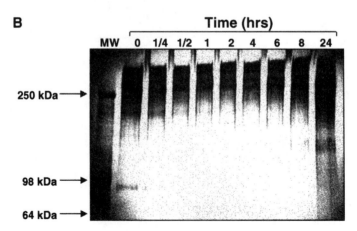

**FIGURE 6.** Time course of isolated aggrecan cleavage by aggrecanase and MMP-3. Bovine aggrecan monomers were incubated for various times with either 5 pM aggrecanase (**A**) or 10 ηM MMP-3 (**B**) and cleavage followed by CSPG Western blot.

10 nM MMP-3. These data suggest that, at the concentrations present in media from IL-1–stimulated cartilage undergoing degradation, aggrecanase would dominate in cleaving the aggrecan molecule.

## DISCUSSION

Aggrecanase activity accumulated in media during incubation of bovine nasal cartilage with IL-1. Although high background levels of BC-3-reactive fragments were present in media at early time points when active aggrecan degradation was oc-curing, activity could be detected as an increase in these levels of BC-3-reactive fragments. The formation of BC-3-reactive products from endogenous aggrecan dur-ing the course of matrix depletion (day 2 to day 6) confirms that active aggrecanase is present within the tissue during the first few days of culture. Previous studies from our laboratory demonstrated that time is necessary for the aggrecanase cleavage products to be produced in cartilage cultures in response to IL-1 and that new protein synthesis is required to generate aggrecanase activity.[16] However, aggrecanase-gen-erated fragments could be detected in cartilage cultures following 8 hr of stimulation with IL-1 when GAG release was first detected, again supporting the induction of active aggrecanase corresponding with degradation of cartilage aggrecan.

IL-1 stimulation of bovine nasal cartilage also resulted in the induction of gelat-inolytic activity in the conditioned media over time. Inhibitor studies indicated that all of the gelatinases detected were metalloproteases as their zymographic activity could be completely inhibited by the metalloprotease inhibitor, EDTA, but not by E-64, pepstatin, or benzamidine hydrochloride, which inhibit cysteine, aspartate, and serine proteases, respectively. Importantly, the gelatinase detected at times when the aggrecan matrix was undergoing degradation (day 2–6) was present as the zy-mogen form. Only at later time periods were active gelatinases detected.

XS309, a potent MMP inhibitor with $K_i$s in the nanomolar range against MMP-1, -2, -3, -8, and -9, was inactive in inhibiting aggrecanase activity at concentrations up to ~ 10 μM. However, 1 μM XS309 completely blocked all bands of gelatinolytic activity detected in aggrecanase-containing media by gelatin zymography. These data indicate that the gelatinases induced in response to IL-1 are MMPs and further suggest that aggrecanase does not cleave gelatin and that aggrecanase activity is not represented by the MMP-2, MMP-9, or other gelatinases induced in response to IL-1 stimulation. Of note, MMP inhibitors which are active against aggrecanase enzymat-ic activity were effective in blocking IL-1–induced aggrecan degradation in cartilage explant cultures.[16] but XS309 was found to be inactive (IC50 > 300uM).

The generation of caseinolytic activity was monitored using a casein-resorufin substrate.[28] MMP-1 and MMP-3 are reported to readily digest casein,[29,30] while MMP-2 and MMP-9 are much less effective.[29] The total caseinolytic activity in con-ditioned media increased over time of culture and the properties of this activity are consistent with its being due to a matrix metalloproteinase(s). From day 2 to day 6, APMA activation was required to detect activity, indicating that all of the protease was present in the zymogen form at these early time periods. Active enzyme was de-tected by day 8 and increased with time of culture. These results are consistent with our findings evaluating MMP-3 protein production by Western blot. Only the 57-kDa proMMP-3 could be detected in the media on day 2 to day 8, but after depletion of aggrecan from the matrix, both active and latent MMP-3 were detected.

Both latent and total caseinolytic activity were inhibited by inclusion of 10mM EDTA, indicating that this activity was due to a metalloproteinase, while inhibition by the broad spectrum MMP inhibitor, XS309, suggests that the activity is due to a matrix metalloproteinase(s). This inhibition of the caseinolytic activity in conditioned media by XS309, which does not inhibit aggrecanase activity, suggests that aggrecanase does not cleave casein and that the caseinolytic activity generated in response to IL-1 does not represent aggrecanase.

Although no active MMPs were detected in the culture media during aggrecan degradation, this does not rule out the presence of active enzyme within the cartilage matrix. In fact, we have been able to demonstrate the presence of active MMP-3 in cartilage extracts at 48 hr of stimulation with IL-1 (Arner and Tortorella, unpublished data). However, data showing that aggrecanase is more efficient than MMP-3 in cleaving aggrecan suggests that even in the presence of active MMPs within the matrix, aggrecanase cleavage would be favored. Using enzyme concentrations of 5 pM, which approximates the concentration of active aggrecanase present in culture media from IL-1–stimulated cartilage, aggrecanase, but not MMP-3, induced aggrecan degradation as monitored by GAG release from F/T cartilage slices. At 100 nM (the concentration of proMMP-3 detected in culture media) active MMP-3 does cleave aggrecan, but still at a much lower rate than 5 pM aggrecanase, indicating that aggrecanase is much more efficient at cleaving aggrecan and would dominate in cleaving this substrate.

The conclusions from these data are supported by work from our laboratory using neoepitope antibodies to the new N-terminus and C-terminus generated upon cleavage of aggrecan at the Glu$^{373}$-Ala$^{374}$ aggrecanase site or the Asn$^{341}$-Phe$^{342}$ MMP site to follow aggrecan catabolism in response to IL-1.[31] These studies indicated that cleavage occurred exclusively at the aggrecanase site. No induction of fragments containing the FFGVG or DIPES epitope (formed by cleavage at the MMP site) was detected within the matrix or media of cartilage explants during IL-1–stimulated proteoglycan degradation; only fragments containing the ARGSV and NITEGE epitope were induced.

## SUMMARY

We have shown that: (1) aggrecanase activity in bovine nasal cartilage culture media is induced in response to IL-1 and this induction correlates with the degradation of cartilage aggrecan; (2) MMPs induced in response to IL-1 are present in culture media as the inactive zymogen form during aggrecan degradation; (3) at concentrations produced in response to IL-1, aggrecanase is more efficient than MMP-3 in cleaving the aggrecan substrate; (4) aggrecan cleavage products possess N- or C-termini indicating that they have been formed by cleavage at the aggrecanase site; and (5) inhibitors of aggrecanase enzymatic activity are effective in blocking cartilage aggrecan degradation.[16–18] Taken together these data suggest that aggrecanase is the key enzyme for aggrecan cleavage during stimulated cartilage degradation and thus provides a target for the design of inhibitors as potential therapeutics to prevent the loss of cartilage in arthritis.

## ACKNOWLEDGMENTS

We thank Dr. Amanda Fosang, University of Melbourne, for the kind gift of the AF-28 monoclonal antibody, and Dr. Robert Copeland for MMP $K_i$s and Liana Bauerle for expert technical assistance.

## REFERENCES

1. MANKIN, H.J. & L. LIPPIELLO. 1970. Biochemical and metabolic abnormalities in articular cartilage from osteo-arthritic human hips. J. Bone Joint Surg. **52A:** 424–434.

2. MORALES, T.I. & V.C. HASCALL. 1989. Factors involved in the regulation of proteoglycan metabolism in articular cartilage. Arthritis Rheum **32**(10): 1197–1201.

3. TYLER, J.A. 1985. Chondrocyte-mediated depletion of articular cartilage proteoglycans in vitro. Biochem. J. **225:** 493–507.

4. ILIC, M.Z., C.J. HANDLEY, H.C. ROBINSON & M.T. MOK. 1992. Mechanism of catabolism of aggrecan by articular cartilage. Arch. Biochem. Biophys. **294**(1): 115–22.

5. LOULAKIS, P., A. SHRIKHANDE, G. DAVIS & C.A. MANIGLIA. 1992. N-terminal sequence of proteoglycan fragments isolated from medium of interleukin-1-treated articular-cartilage cultures. Putative site(s) of enzymic cleavage. Biochem. J. **284:** 589–593.

6. SANDY, J.D., A.H.K. PLAAS & T.J. KOOB. 1995. Pathways of aggrecan processing in joint tissues. Implications for disease mechanism and monitoring. Acta Orthop. Scand. (Suppl 266) **66:** 26–32.

7. LARK, M.W., J.T. GORDY, J.R. WEIDNER, J. AYAIA, J.H. KIMURA, H.R. WILLIAMS, R.A. MUMFORD, C.R. FLANNERY, S.S. CARISONI, M. IWATAI & J.D. SANDY. 1995. Cell-mediated catabolism of aggrecan. Evidence that cleavage at the "aggrecanase" site (Glu373-Ala374) is a primary event in proteolysis of the interglobular domain. J. Biol. Chem. **270:** 2550–2556.

8. FLANNERY, C.R., M.W. LARK & D. SANDY. 1992. Identification of a stromelysin cleavage site within the interglobular domain of human aggrecan. Evidence for proteolysis at this site in vivo in human articular cartilage. J. Biol. Chem. **267:** 1008–1014.

9. FOSANG, A.J., P.J. NEAME, K. LAST, T.E. HARDHINGHAM, G. MURPHY & J.A. HAMILTON. 1992. The interglobular domain of cartilage aggrecan is cleaved by PUMP, gelatinases, and cathepsin B. J. Biol. Chem. **267:** 19470–19474.

10. FOSANG, A.J., K. LAST, V. KNAUPER, P.J. NEAME, G. MURPHY, T.E. HARDINGHAM, H. TSCHESCHE & J.A. HAMILTON. 1993. Fibroblast and neutrophil collagenases cleave at two sites in the cartilage aggrecan interglobular domain. Biochem. J. **295:** 273–276.

11. FOSANG, A.J., K. LAST, V. KNAUPER, G. MURPHY & P.J. NEAME. 1996. Degradation of cartilage aggrecan by collagenase-3 (MMP-13). FEBS Lett 380: 17–20.

12. LOHMANDER, L.S., P.J. NEAME & J.D. SANDY. 1993. The structure of aggrecan fragments in human synovial fluid. Evidence that aggrecanase mediates cartilage degradation in inflammatory joint disease, joint injury, and osteoarthritis. Arth. Rheum. **36:** 1214–1222.

13. SANDY, J.D., C.R. FLANNERY, P.J. NEAME & L.S. LOHMANDER. 1992. The structure of aggrecan fragments in human synovial fluid. Evidence for the involvement in osteoarthritis of a novel proteinase which cleaves the Glu 373-Ala 374 bond of the interglobular domain. J. Clin. Invest. **69:** 1512–1516.

14. SANDY, J.D., R.E. BOYNTON & C.R. FLANNERY. 1991. Analysis of the catabolism of aggrecan in cartilage explants by quantitation of peptides from the three globular domains. J. Biol. Chem. 266: 8198–8205.

15. SANDY, J.D., P.J. NEAME, P.L. BOYNTON & C.R. FLANNERY. 1991. Catabolism of aggrecan in cartilage explants. Identification of a major cleavage site within the interglobular domain. J. Biol. Chem. **266:** 8683–8685.

16. ARNER, E.C., C.E. HUGHES, C.P. DECICCO, B. CATERSON, M.D. TORTORELLA. 1998. Cytokine-induced cartilage proteoglycan degradation is mediated by aggrecanase. Osteoarthritis Cart. **6:** 214–228.

17. TORTORELLA, M.D., J.M. TRZASKOS & E.C. ARNER. 1997. Identification and characterization of an assay which defines the cartilage degrading enzyme, "aggrecanase" [abstract]. Trans. Orthop. Res. Soc. **22:** 452.

18. ARNER, E.C., M.A. PRATTA, J.M. TRZASKOS, C.P. DECICCO & M.D. TORTORELLA. 1999. Generation and characterization of aggrecanase: a soluble, cartilage-derived aggrecan-degrading activity. J. Biol. Chem. **274:** 6594–6601.

19. HUANG, J.J., R.C. NEWTON, K. PEZZELLA, M. COVINGTON, T. TAMBLYN, S.J. RUTLEDGE, M. KELLEY, J. GRAY & Y. LIN. 1987. High-level expression in Escherichia coli of a soluble and fully active recombinant interleukin-1 beta. Mol. Biol. Med. **4:** 169–181.

20. HUGHES, C.E., B. CATERSON, A.J. FOSANG, P.J. ROUGHLEY & J.S. MORT. 1995. Monoclonal antibodies that specifically recognize neoepitope sequences generated by "aggrecanase" and matrix metalloproteinase cleavage of aggrecan: application to catabolism in situ and in vitro. Biochem. J. **305:** 799–804.

21. XUE, C.-B., R.J. CHERNEY, C.P. DECICCO, W.F. DEGRADO, X. HE, C.N. HODGE, I.C. JACOBSON, R.L. MAGOLDA & E.C. ARNER, inventors; DuPont Merck Pharmaceutical Company, assignee. 1997. WO 9718207 A2, May 22.

22. ARNER, E.C. & M.A. PRATTA. 1991. Modulation of interleukin-1-induced alterations in cartilage metabolism by activation of protein kinase C. Arthritis Rheum. **34:** 1006–1013.

23. FARNDALE, R.W., C.A. SAYERS & A.J. BARRETT. 1982. A direct spectrophotometric microassay for sulfated glycosaminoglycans cartilage cultures. Connect. Tissue Res. **9:** 247–248.

24. FOSANG, A.J., K. LAST, P. GARDINER, D.C. JACKSON & L. BROWN. 1995. Development of a cleavage-site-specific monoclonal antibody for detecting metalloproteinase-derived aggrecan fragments: detection of fragments in human synovial fluids. Biochem. J. **310:** 337–343.

25. LAEMMLI, U.K. 1970. Cleavage of structural proteins during the assembly of the head of bacteriophage T4. Nature **227:** 680–685.

26. HUGHES, C.E., B. CATERSON, R.J. WHITE, P.J. ROUGHLEY & J.S. MORT. 1992. Monoclonal antibodies recognizing protease-generated neoepitopes from cartilage proteoglycan degradation. Application to studies of human link protein cleavage by stromelysin. J. Biol. Chem. **267:** 16011–16014.

27. TORTORELLA, M.D. & E.C. ARNER. 1997. A fluorescent enzyme-linked immunosorbent assay (FELISA) for stromelysin using a polyclonal antisera to human stromelysin with broad cross-reactivity. Inflamm. Res. **46** (Suppl. 2): S120–S121.

28. TORTORELLA, M.D. & E.C. ARNER. 1997. A high-throughput assay for stromelysin using a casein-resorufin substrate. Inflamm. Res. **46** (Suppl. 2): S122–S123.

29. MANICORT, D.-H. & V. LEFEBVRE. 1993. An assay for matrix metalloproteinases and other proteases acting on proteoglycans, casein, or gelatin. Anal. Biochem. **215:** 171–179.
30. VAES, G., Y. EECHOUT, G. LENAERS-CLAEYS, C. FRANCOIS-GILLET & J.E. DRUETZ. 1978. The simultaneous release by bone explants in culture and the parallel activation of procollagenase and of a latent neutral proteinase that degrades cartilage proteoglycans and denatured collagen. Biochem. J. **172:** 261–274.
31. PRATTA, M.A., M.D. TORTORELLA, R.C. NEWTON, J.M. TRZASKOS, R.L. MAGOLDA, C.P. DECICCO, A.A. COLE, B.L. SCHUMACHER, K.E. KEUTTNER & E.C. ARNER. 1998. Aggrecan degradation in interleukin-1-stimulated bovine nasal cartilage explants is up-regulated exclusively by aggrecanase-mediated cleavage [abstract]. Trans. Orthop. Res. Soc. **23:** 177.

# Tissue Inhibitors of Matrix Metalloproteinases in Cancer

LAURENCE BLAVIER, PATRICK HENRIET, SUZAN IMREN, AND
YVES A. DECLERCK[a]

*Division of Hematology-Oncology,*
*Department of Pediatrics and Department of Biochemistry and Molecular Biology,*
*Childrens Hospital Los Angeles and University of Southern California,*
*Los Angeles, California 90027, USA*

ABSTRACT: Tissue inhibitors of metalloproteinases (TIMPs) play a key regulatory role in the homeostasis of the extracellular matrix (ECM) by controlling the activity of matrix metalloproteinases (MMPs). Some TIMPs have a second function as well, unrelated to their antiMMP activity, which affects cell proliferation and survival. The role of these inhibitors in cancer has been the subject of extensive investigations that have examined their biological activity in tumor growth, invasion, metastasis and angiogenesis, as well as their potential use in the diagnosis and treatment of human cancer.

## TIMPs: A FAMILY OF NATURAL INHIBITORS OF MATRIX METALLOPROTEINASES THAT CONTROL THE HOMEOSTASIS OF THE EXTRACELLULAR MATRIX

The homeostasis of the extracellular matrix (ECM) is controlled by a delicate balance between synthesis of ECM proteins, production of matrix-degrading extracellular proteases, and the presence of protease inhibitors. Among these proteases are the matrix metalloproteinases (MMPs), a family of $Zn^{2+}$-dependent endopeptidases that proteolytically degrade most of the components of the ECM. In the extracellular milieu, the activity of these proteases is tightly regulated by specific inhibitors known as tissue inhibitors of MMPs (TIMPs). Four members of the TIMP family have been so far described and their genes have been isolated and characterized in human and other mammalian species (TABLE 1). These inhibitors have many structural and functional properties in common including a primary amino-acid sequence that includes 12 disulfide bonded cysteine residues and a tertiary structure that identifies two distinct domains (N-terminal and C-terminal domains), each containing three overlapping disulfide bonds. The N-terminal domain contains a region of higher homology among the four TIMPs and is responsible for their anti-metalloproteinase activity *vis-à-vis* all members of the MMP family. This domain contains residues that interact with the $Zn^{2+}$-binding pocket of active MMPs. The three-dimensional structure of TIMP-2 has been recently resolved by X-ray crystallography to 2.1-

[a]Address for correspondence: Y.A. DeClerck, Childrens Hospital Los Angeles, MS #54, 4650 Sunset Blvd., Los Angeles, California 90027. Phone, 323/669-5648, fax, 323/664-9455, email, declerck@hsc.usc.edu.

**TABLE 1. Characteristics of TIMPs 1–4**

| Characteristics | TIMP-1 | TIMP-2 | TIMP-3 | TIMP-4 |
|---|---|---|---|---|
| MR | 28 kDa | 21 kDa | 24 kDa | 22 kDa |
| Tissue specificity (mouse) | Bone, ovary, heart | Lung, heart, muscle, brain, skin, vessels, testis, ovary, placenta | Kidney, heart, brain, lung, ovary, placenta | Heart, brain, muscle, ovary |
| Inhibition of tumor invasion | + | + | + | + |
| Growth promoting activity | + | + | n.d. | n.d. |
| Inhibition of apoptosis | + | n.d. | n.d. | n.d. |
| Inhibition of angiogenesis | + | + | + | + |

n.d. = no data.

Ångstrom resolution.[1] The N-terminal domain (residues 1 to 110) is folded into a beta-barrel, similar to the oligonucleotide/oligosaccharide binding fold found in DNA-binding proteins, and the C-terminal domain (residues 111 to 194) contains a parallel beta-hairpin plus a beta–loop-beta motif. Upon binding of TIMP-2 to an MMP, an internal rotation of 13 degrees between the two domains occurs. Sequence differences among the TIMPs seem to be preferentially located to loop regions and may be responsible for particular properties of the various TIMPs. The crystal structures of a TIMP-1/MMP-3 complex and TIMP-2/MT1-MMP and TIMP-2/MMP-3 complexes have been also reported.[2–4] These analyses have pointed to specific residue sequences in TIMPs that are involved in binding to the active domain of MMPs. In the case of TIMP-1, a first sequence involves $Cys^1$-$Thr^2$-$Cys^3$-$Ala^4$ and a second sequence is made of $Ser^{68}$ and $Val^{69}$. In the case of TIMP-2, binding sequences include residues 1 to 11, 27 to 41, 68 to 73, 87 to 90 and 97 to 104. The C-terminal domain of TIMP is more variable among the four TIMPs and binds to the hemopexin domain of several proMMPs (proMMP-9 in the case of TIMP-1 and proMMP-2 in the case of TIMP-2 and TIMP-4), allowing for the formation of TIMP-proMMP complexes that retain an antiMMP activity. In addition the C-terminal domain of TIMP-2 is also involved in its high-affinity binding to the cell surface.[5]

Despite these structural differences it is still unclear whether TIMPs play specific and different roles in controlling the remodeling of the ECM. The analysis of the temporo-spatial expression of TIMPs in adult mouse organs and during murine embryonic development has, however, provided helpful information indicating that some TIMPs may have an organ-specific role. These data are summarized in TABLE 1. For example, they suggest that TIMP-1 may play a more specific role in bone development,[6,7] TIMP-2 in organs of the reproductive system, the lung, the skin and the heart,[8] TIMP-3 in the kidney, the heart, the lung and the ovary,[9,10] and TIMP-4 in the heart and the muscle.[11] In the murine placenta, TIMP-1 is not expressed, but both TIMP-2 and TIMP-3 are abundantly expressed by spongiotrophoblasts during the second half of gestation. This latter observation suggests that TIMP-2 and TIMP-3 are involved in limiting the invasive process of the placenta.[8]

## TUMOR SUPPRESSOR ACTIVITY OF TIMPs

The importance of the degradation of the ECM during cancer progression has been well documented, and there has been an abundant literature—including work from our own laboratory—supporting the concept that TIMPs act as tumor suppressor genes because of their anti-metalloproteinase activity and their protective role on the ECM. Such evidence has been initially obtained from experiments in which TIMPs were shown to inhibit the development of metastasis in experimental metastatic animal models.[12,13] Later it has been shown that overexpression of TIMPs by cDNA transfection in malignant cells inhibits not only local invasion but also tumor growth.[14–22] More recently, investigators have demonstrated that overexpression of TIMP-1 in transgenic mice inhibits tumor development.[23–25] Experimental evidence exists that TIMPs may be involved in the earlier stages of tumor formation since TIMP-1 downregulation in nontumorigenic Swiss 3T3 fibroblasts confers on these cells a tumorigenic and metastatic phenotype.[26]

There are several mechanisms that support an anti-tumor effect of TIMPs. First, TIMPs have an anti-angiogenic activity either by a direct effect on endothelial cell proliferation or by their ability to downregulate the activity of MMPs required for endothelial cell migration and invasion. TIMP-1 and TIMP-2 have been in part purified from cartilage as anti-angiogenic factors.[27] TIMP-2 inhibits the growth of basic FGF-stimulated endothelial cells[28] and inhibits tumor vascularization *in vivo*.[29] TIMP-3 inhibits endothelial cell motility, and the proliferation of stimulated endothelial cell placed in a collagen gel or implanted in the chicken chorioallantoic membrane.[30] As a second mechanism, it has been recently recognized that MMPs can degrade extracellular proteins other than structural ECM proteins (FIG. 1). Some of these proteins form complexes with growth factors and control their bioavailability. For example, insulin-like growth factor-binding protein-3 (IGFBP-3) is degraded by several MMPs, and, as a result of its proteolytic degradation, IGF is released and biologically more active.[31] By preventing the degradation of IGFBP by MMPs, TIMPs can therefore play an important role in limiting the bioavailability of this growth factor. A third mechanism that supports an anti-tumor activity of TIMPs is based on the increasingly recognized fact that the ECM can control essential cellular functions such as growth, differentiation, and apoptosis.[32] In our laboratory we have, for example, demonstrated that TIMP-2 inhibits the growth of human melanoma cells when the cells are grown in the presence of intact fibrillar collagen.[15] This antiproliferative effect involves inhibition of cell-spreading, upregulation of the cyclin-dependent kinase inhibitor p27[KIP-1], and a block at the G1-S transition (Henriet *et al.*, in preparation). This growth-inhibitory effect is lost as MMPs produced by the melanoma cells degrade fibrillar collagen, but is maintained if the degradation of collagen is prevented in the presence of excess TIMP-2.[15] Thus by maintaining the

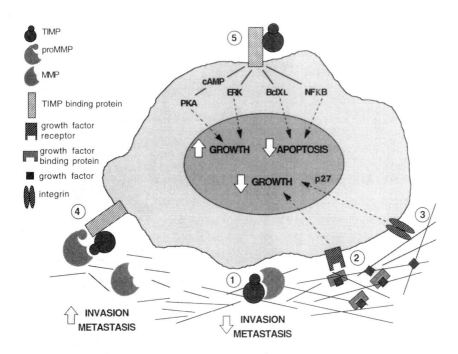

**FIGURE 1.** Paradoxical effects of TIMPs on cancer cell. The scientific literature on TIMPs has pointed to five potentially different effects on cancer cell behavior. Some (1 to 3) are dependent on the effect of TIMP on the ECM and others (4 and 5) are independent of the ECM. (1) As an inhibitor of active MMPs, TIMP protects the ECM from the proteolytic degradation and prevents invasion and metastasis. (2) By preventing the degradation of growth factor–binding proteins by MMPs, TIMP limits the bioavailability of growth factors in the ECM and inhibits cell growth. (3) Contact between cells and intact ECM (such as fibrillar collagen) upregulates p27 and blocks cell cycle progression. (4) TIMP-2 acts as an adaptor molecule enhancing proMMP-2 activation at the cell surface by MT1-MMP. (5) TIMP regulates cell proliferation and apoptosis by binding to a putative cell-associated receptor.

integrity of the ECM, TIMPs maintain an indirect control over malignant cell proliferation.

## TUMOR PROMOTER ACTIVITY OF TIMPs

There is also experimental evidence suggesting that TIMPs may function in a manner that promotes rather than suppresses tumor progression. For example, in colon cancer, higher tissue levels of TIMP-1 have been observed in association with more (rather than less) invasive stages[33] and in non-Hodgkin's lymphoma, TIMP-1 expression is a positive index of malignant progression.[34] In bladder and breast cancers, elevated levels of TIMP-2 in tumor tissues predict an unfavorable rather than a favorable outcome.[35,36] The mechanisms supporting a paradoxically positive effect of TIMPs in tumor progression are not fully understood and have been the subject of intensive investigations which have pointed to a dual function of TIMPs (FIG. 1). It has now become clear that in addition to inhibiting the activity of MMPs, certain TIMPs have a function that either enhances proteolytic degradation of the ECM or directly affects cell survival and growth. In the case of TIMP-2, it is now apparent that it plays the role of an adaptor molecule that regulates proMMP-2 activation at the cell surface.[37,38] TIMP-2 can bind to proMMP-2 via C-terminal interaction, and to the active site of MT1-MMP, a membrane-associated MMP that activates proMMP-2, by its N-terminal domain. This dual binding brings proMMP-2 close to the cell surface, where it can be activated by additional MT1-MMP molecules. It has also been reported that some TIMPs can directly affect cell growth and survival independent of their interacting with MMPs. TIMP-1 and TIMP-2 stimulate the growth of a large variety of normal and malignant cells *in vitro*,[39,40] including erythroid precursor cells,[41,42] and TIMP-1 but not TIMP-2 can suppress programmed cell death in malignant Burkitt cells.[43] The mechanisms by which TIMPs affect essential cellular functions such as proliferation and death have not been yet fully understood. It is known, however, that TIMP-1 and TIMP-2 can bind to the cell surface with high affinity ($K_d$ in the nM range),[5,39,40,44] suggesting that TIMPs may behave as ligands such as growth factors and cytokines. However, the nature of a putative cell-associated "TIMP receptor" has not yet been resolved. In the case of TIMP-2, MT1-MMP could be such a "receptor," but there has been no evidence that MT1-MMP relays intracellular signals upon TIMP-2 binding and behaves as a true receptor. The signaling pathways involved in TIMP regulation over proliferation and apoptosis have begun to be explored. It has been shown that TIMP-2 stimulates adenylate cyclase and cAMP-dependent activation of protein kinase A[44] and that TIMP-1 and TIMP-2 increase the tyrosine phosphorylation of proteins and MAP kinase activity[45] as they stimulate cell growth. In the case of apoptosis, TIMP-1 up-regulates the expression of the anti-apoptotic protein Bcl-X$_L$ but not Bcl-2 and decreases NFκB activity.[43]

The concept that an inhibitor of matrix-degrading proteases can have a stimulatory rather that an inhibitory effect on tumor progression is not entirely novel and has been previously shown for an inhibitor of serine proteases, the plasminogen activator inhibitor-1 (PAI-1). High levels of PAI-1 have been reported in more advanced stages of cancer of the breast, lung, and prostate,[46–48] and PAI-1 has been recently shown to play an important regulatory role in angiogenesis. In PAI-1-deficient mice there is no angiogenic response to tumors implanted subcutaneously, and a response can be established upon transfection of the endothelial cells with PAI-1.[49] The ability of PAI-1 to bind to the ECM protein, vitronectin, and to interfere with cell attachment to vitronectin may be an important element. It has been suggested that PAI-1 promotes the detachment and migration of tumor cells and endothelial cells, and therefore enhances the formation of distant metastasis and that it stimulates angiogenesis.[50-52]

## AN UNANSWERED QUESTION

In view of the dual and paradoxical functions of TIMP described above, an important question remains to be addressed: What are the factors that direct the activity of TIMP toward tumor suppression or tumor promotion? Several factors that are likely to play a key role can be identified, including local concentration of TIMP, cellular and pericellular distribution, association with proMMPs, and the presence of a putative "TIMP receptor." The growth-promoting activity of TIMPs has been observed with concentrations of TIMP-1 around 100 ng/ml and of TIMP-2 around 10 ng/ml,[40] but no growth stimulation has been reported at concentrations above 1 µg/ml, which are the concentrations at which TIMPs inhibit the degradation of ECM *in vitro*.[53] TIMP-1 and TIMP-2 are present in the serum of normal individuals at concentrations of 100 ng/ml and 50 ng/ml,[54] respectively, but no information is available on the concentrations of these inhibitors in tumor tissues, and it therefore remains unknown whether overexpression of TIMPs in tissues could have a stimulatory effect of cell proliferation *in vivo*. Whether TIMP is present in a free form or as a complex with proMMP is another important factor. When bound to proMMP-2, TIMP-2 is more likely to promote activation of this MMP at the cell surface and to favor ECM degradation, whereas in a free form it can interact with active MMPs. The selective presence of a "TIMP receptor" could also be a factor discriminating between cells that respond and do not respond to TIMPs; however—as previously discussed—the existence of such a receptor is still unproven.

## DIAGNOSTIC AND THERAPEUTIC VALUE
## OF TIMPs IN HUMAN CANCER

As cancer-associated proteins are discovered, the question of their use as prognostic markers to predict clinical outcome in patients has always been examined. The observation that TIMPs are expressed in tumor tissues and are present in the serum of cancer patients[54] has raised the possibility that TIMP levels could predict a (favorable) clinical outcome and a (low) risk of metastatic spread. The data generated in this regard have indicated that this is clearly not the case. Serum levels of TIMP-2 in cancer patients have been variable and of no predictive value,[54] and in tumor specimens higher rather than lower levels of TIMP-2 have been observed in more advanced stages of cancer.[35,36] Considering the complexity of the mechanisms by which TIMPs can control the activity of MMPs, it is not surprising that the single examination of TIMP levels does not predict the ability of a specific cancer to invade or metastasize. As an alternative, investigators have proposed to use the analysis of the balance between MMPs and TIMPs in tumors as a predictor of clinical outcome. For example, the analysis of the MMP-2/TIMP-2 ratio by RT-PCR revealed lower ratios in breast tumor specimens from patients without lymph node involvement and higher ratios in patients with lymph node involvement.[55] Whether such an approach could be of benefit in identifying patients at higher risk of metastasis who may benefit from more intensive therapy is presently uncertain.

Because of their well-demonstrated anti-invasive, anti-metastatic, and anti-angiogenic activities, TIMPs have been considered as therapeutic agents in multiple preclinical animal models of cancer. In a first approach, the use of recombinant (r)TIMP to prevent metastasis has been tested. Repeated intraperitoneal injections of rTIMP-1 and rTIMP-2 have been shown to inhibit lung colonization of B16 melanoma cells in mice.[12,13,56] However, the rapid *in vivo* clearance of these recombinant proteins (the TIMP-2 plasma half-life is 4 min) and the amount of inhibitor required (0.9 mg/animal) have precluded these recombinant proteins from any systemic use. As an alternate approach, it would be possible to develop modified TIMPs or TIMP mutants that would have better pharmacologic properties and efficacy and possibly less toxicity than synthetic inhibitors. For example, the combination of TIMP-2 with polyethylene glycol (pegylation) does not affect its anti-metalloproteinase activity, but substantially extends its plasma half-life (to 2.5 hours). Our knowledge of the three- dimensional structure of TIMP/MMP complexes could also be used to design synthetic peptides that would have better properties than full-length TIMPs. Whether this approach would have any therapeutic value in cancer is currently unknown. A third approach would take advantage of the two facts that TIMPs are natural inhibitors and that they exert their effect in the extracellular milieu. Following this line of thought, we have tested in our laboratory the possibility of overexpressing TIMPs in tumor cells by cDNA transfection *in vivo* using retrovirally mediated gene transfer (gene therapy). Using a Moloney leukemia–based retroviral vector, we have demonstrated that co-injection in mice of tumor cells with retroviral producer cells containing a TIMP-2 cDNA results in the overexpression of TIMP-2 in approximately 13% of the tumor cell population. Overexpression of TIMP-2 in these cells is sufficient to inhibit tumor growth by 75% and to block local invasion.[20] However to be of practical use in metastasis in human cancer, it would be necessary to deliver the vector to disseminated tumor masses *in vivo*. Currently, the majority of gene therapy protocols in cancer are based on the use of local vector delivery, and targetable vectors suitable for systemic delivery are not yet available. A last approach would be to use molecules that have the ability to upregulate the expression of TIMPs in tumor tissues. So far this possibility has not been seriously explored since only a small number of nonspecific agents have been shown to upregulate TIMPs. For example TIMP-1 expression is upregulated by phorbol and serum factors including TGFβ, IL-1 and IL-6,[57] and TIMP-2 expression is upregulated by cAMP.[58] However, because of the dual function of TIMP previously discussed, it may not be possible to predict whether these agents will inhibit or stimulate tumor growth.

## CONCLUSIONS AND FUTURE DIRECTIONS

The involvement of TIMPs in cancer has been the subject of continuous scientific interest over the last 10 years, and these investigations have pointed not only to a role of TIMPs in invasion and metastasis, but also in tumor growth, apoptosis, transformation, and angiogenesis (FIG. 2). The report that TIMPs can not only block tumor invasion and metastasis, but also inhibit the growth of primary tumors has been a fundamental observation in support of the use of synthetic MMP inhibitors in cancer clinical trials that are currently ongoing.[59,60] Experiments testing the possibility of using TIMP as a therapeutic agent in cancer have not produced significant results, however, largely because of the pharmacologic properties of TIMPs and the current

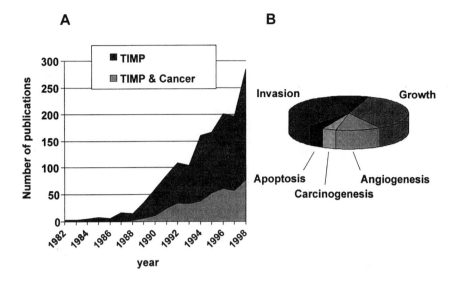

**FIGURE 2.** Scientific publications on TIMP between 1982 and 1998. **A.** Number of publications listed in Medline that contain TIMP or TIMP and cancer. **B.** Relative distribution of publications containing TIMP and cancer according to the following key words: invasion, growth, angiogenesis, carcinogenesis and apoptosis.

limitation in gene delivery. A better understanding of the dual function of TIMPs, and in particular of the signaling pathways involved in their effect on tumor growth and survival, is also needed to further determine whether TIMP could be of potential benefit in cancer treatment. Until that time, important information on the effect of inhibition of MMP activity in cancer will be obtained from synthetic inhibitors.

## ACKNOWLEDGMENTS

This work was in part supported by Grant CA42919 from the National Institutes of Health to Y.A. DeClerck.

## REFERENCES

1. TUUTTILA, A., E. MORGUNOVA, U. BERGMANN, Y. LINDQVIST, K. MASKOS, C. FERNANDEZ-CATALAN, W. BODE, K. TRYGGVASON & G. SCHNEIDER. 1998. Three-dimensional structure of human tissue inhibitor of metalloproteinases-2 at 2.1 Å resolution. J. Mol. Biol. **284:** 1133–1140.
2. MUSKETT, F.W., T.A. FRENKIEL, J. FEENEY, R.B. FREEDMAN, M.D. CARR & R.A. WILLIAMSON. 1998. High resolution structure of the N-terminal domain of tissue inhibitor of metalloproteinases-2 and characterization of its interaction site with matrix metalloproteinase-3. J. Biol. Chem. **273:** 21736–21743.

3. FERNANDEZ-CATALAN, C., W. BODE, R. HUBER, D. TURK, J.J. CALVETE, A. LICHTE, H. TSCHESCHE & K. MASKOS. 1998. Crystal structure of the complex formed by the membrane type 1-matrix metalloproteinase with the tissue inhibitor of metalloproteinases-2, the soluble progelatinase A receptor. EMBO J. **17:** 5238–5248.

4. GOMIS-RÜTH, F.X., K. MASKOS, M. BETZ, A. BERGNER, R. HUBER, K. SUZUKI, N. YOSHIDA, H. NAGASE, K. BREW, G.P. BOURENKOV, H. BARTUNIK & W. BODE. 1997. Mechanism of inhibition of the human matrix metalloproteinase stromelysin-1 by TIMP-1. Nature **389:** 77–81.

5. KO, Y.C., K.E. LANGLEY, E.A. MENDIAZ, V.P. PARKER, S.M. TAYLOR & Y.A. DECLERCK. 1997. The C-terminal domain of tissue inhibitor of metalloproteinases-2 is required for cell binding but not for antimetalloproteinase activity. Biochem. Biophys. Res. Commun. **236:** 100–105.

6. FLENNIKEN, A.M. & B.R. WILLIAMS. 1990. Developmental expression of the endogenous TIMP gene and a TIMP-lacZ fusion gene in transgenic mice. Genes Dev. **4:** 1094–1106.

7. NOMURA, S., B.L. HOGAN, A.J. WILLS, J.K. HEATH & D.R. EDWARDS. 1989. Developmental expression of tissue inhibitor of metalloproteinase (TIMP) RNA. Development **105:** 575–583.

8. BLAVIER, L. & Y.A. DECLERCK. 1997. Tissue inhibitor of metalloproteinases-2 is expressed in the interstitial matrix in adult mouse organs and during embryonic development. Mol. Biol. Cell **8:** 1513–1527.

9. APTE, S.S., K. HAYASHI, M.F. SELDIN, M.G. MATTEI, M. HAYASHI & B.R. OLSEN. 1994. Gene encoding a novel murine tissue inhibitor of metalloproteinases (TIMP), TIMP-3, is expressed in developing mouse epithelia, cartilage, and muscle, and is located on mouse chromosome 10. Dev. Dyn. **200:** 177–197.

10. LECO, K.J., R. KHOKHA, N. PAVLOFF, S.P. HAWKES & D.R. EDWARDS. 1994. Tissue inhibitor of metalloproteinases-3 (TIMP-3) is an extracellular matrix-associated protein with a distinctive pattern of expression in mouse cells and tissues. J. Biol. Chem. **269:** 9352–9360.

11. LECO, K.J., S.S. APTE, G.T. TANIGUCHI, S.P. HAWKES, R. KHOKHA, G.A. SCHULTZ & D.R. EDWARDS. 1997. Murine tissue inhibitor of metalloproteinases-4 (*Timp*-4): cDNA isolation and expression in adult mouse tissues. FEBS Lett. **401:** 213–217.

12. ALVAREZ, O.A., D.F. CARMICHAEL & Y.A. DECLERCK. 1990. Inhibition of collagenolytic activity and metastasis of tumor cells by a recombinant human tissue inhibitor of metalloproteinases. J. Natl. Cancer Inst. **82:** 589–595.

13. SCHULTZ, R.M., S. SILBERMAN, B. PERSKY, A.S. BAJKOWSKI & D.F. CARMICHAEL. 1988. Inhibition by human recombinant tissue inhibitor of metalloproteinases of human amnion invasion and lung colonization by murine B16-F10 melanoma cells. Cancer Res. **48:** 5539–5545.

14. DECLERCK, Y.A., N. PEREZ, H. SHIMADA, T.C. BOONE, K.E. LANGLEY & S.M. TAYLOR. 1992. Inhibition of invasion and metastasis in cells transfected with an inhibitor of metalloproteinases. Cancer Res. **52:** 701–708.

15. MONTGOMERY, A.M., B.M. MUELLER, R.A. REISFELD, S.M. TAYLOR & Y.A. DECLERCK. 1994. Effect of tissue inhibitor of the matrix metalloproteinases-2 expression on the growth and spontaneous metastasis of a human melanoma cell line. Cancer Res. **54:** 5467–5473.

16. KHOKHA, R., M.J. ZIMMER, S.M. WILSON & A.F. CHAMBERS. 1992. Up-regulation of TIMP-1 expression in B16-F10 melanoma cells suppresses their metastatic ability in chick embryo. Clin. Exp. Metastasis **10:** 365–370.

17. KHOKHA, R., M.J. ZIMMER, C.H. GRAHAM, P.K. LALA & P. WATERHOUSE. 1992. Suppression of invasion by inducible expression of tissue inhibitor of metalloproteinase-1 (TIMP-1) in B16-F10 melanoma cells. J. Natl. Cancer Inst. **84:** 1017–1022.

18. KOOP, S., R. KHOKHA, E.E. SCHMIDT, I.C. MACDONALD, V.L. MORRIS, A.F. CHAMBERS & A.C. GROOM. 1994. Overexpression of metalloproteinase inhibitor in B16F10 cells does not affect extravasation but reduces tumor growth. Cancer Res. **54:** 4791–4797.

19. KHOKHA, R. 1994. Suppression of the tumorigenic and metastatic abilities of murine B16-F10 melanoma cells in vivo by the overexpression of the tissue inhibitor of the metalloproteinases-1. J. Natl. Cancer Inst. **86:** 299–304.

20. IMREN, S., D.B. KOHN, H. SHIMADA, L. BLAVIER & Y.A. DECLERCK. 1996. Overexpression of tissue inhibitor of metalloproteinases-2 *in vivo* by retroviral mediated gene transfer inhibits tumor growth and invasion. Cancer Res **56:** 2891–2895.

21. WANG, M.S., Y.L.E. LIU, J. GREENE, S.J. SHENG, A. FUCHS, E.M. ROSEN & Y.E. SHI. 1997. Inhibition of tumor growth and metastasis of human breast cancer cells transfected with tissue inhibitor of metalloproteinase 4. Oncogene **14:** 2767–2774.

22. ANAND-APTE, B., L. BAO, R. SMITH, K. IWATA, B.R. OLSEN, B. ZETTER & S.S. APTE. 1996. A review of tissue inhibitor of metalloproteinases-3 (TIMP-3) and experimental analysis of its effect on primary tumor growth. Biochem. Cell Biol. **74:** 853–862.

23. MARTIN, D.C., U. RUTHER, O.H. SANCHEZ SWEATMAN, F.W. ORR & R. KHOKHA. 1996. Inhibition of SV40 T antigen-induced hepatocellular carcinoma in TIMP-1 transgenic mice. Oncogene **13:** 569–576.

24. KRÜGER, A., J.E. FATA & R. KHOKHA. 1997. Altered tumor growth and metastasis of a T-cell lymphoma in *timp-1* transgenic mice. Blood **90:** 1993–2000.

25. KRÜGER, A., O.H. SANCHEZ-SWEATMAN, D.C. MARTIN, J.E. FATA, A.T. HO, F.W. ORR, U. RÜTHER & R. KHOKHA. 1998. Host TIMP-1 overexpression confers resistance to experimental brain metastasis of a fibrosarcoma cell line. Oncogene **16:** 2419–2423.

26. KHOKHA, R., P. WATERHOUSE, S. YAGEL, P.K. LALA, C.M. OVERALL, G. NORTON & D.T. DENHARDT. 1989. Antisense RNA-induced reduction in murine TIMP levels confers oncogenicity on Swiss 3T3 cells. Science **243:** 947–950.

27. MOSES, M.A., J. SUDHALTER & R. LANGER. 1990. Identification of an inhibitor of neovascularization from cartilage. Science **248:** 1408–1410.

28. MURPHY, A.N., E.J. UNSWORTH & W.G. STETLER-STEVENSON. 1993. Tissue inhibitor of metalloproteinases-2 inhibits bFGF-induced human microvascular endothelial cell proliferation. J. Cell Physiol. **157:** 351–358.

29. VALENTE, P., G. FASSINA, A. MELCHIORI, L. MASIELLO, M. CILLI, A. VACCA, M. ONISTO, L. SANTI, W.G. STETLER-STEVENSON & A. ALBINI. 1998. TIMP-2 overexpression reduces invasion and angiogenesis and protects B16F10 melanoma cells from apoptosis. Int. J. Cancer **75:** 246–253.

30. ANAND-APTE, B., M.S. PEPPER, E. VOEST, R. MONTESANO, B. OLSEN, G. MURPHY, S.S. APTE & B. ZETTER. 1997. Inhibition of angiogenesis by tissue inhibitor of metalloproteinase-3. Invest. Ophthalmol. Vis. Sci. **38:** 817–823.

31. FOWLKES, J.L., J.J. ENGHILD, K. SUZUKI & H. NAGASE. 1994. Matrix metalloproteinases degrade insulin-like growth factor-binding protein-3 in dermal fibroblast cultures. J. Biol. Chem. **269:** 25742–25746.

32. JULIANO, R. 1996. Cooperation between soluble factors and integrin-mediated cell anchorage in the control of cell growth and differentiation. Bioessays **18:** 911–917.

33. LU, X.Q., M. LEVY, I.B. WEINSTEIN & R.M. SANTELLA. 1991. Immunological quantitation of levels of tissue inhibitor of metalloproteinase-1 in human colon cancer. Cancer Res. **51:** 6231–6235.

34. KOSSAKOWSKA, A.E., S.J. URBANSKI, S.A. HUCHCROFT & D.R. EDWARDS. 1992. Relationship between the clinical aggressiveness of large cell immunoblastic lymphomas and expression of 92 kDa gelatinase (type IV collagenase) and tissue inhibitor of metalloproteinases-1 (TIMP-1) RNAs. Oncol. Res. **4:** 233–240.

35. Visscher, D.W., M. Hoyhtyea, S.K. Ottosen, C.M. Liang, F.H. Sarkar, J.D. Crissman & R. Fridman. 1994. Enhanced expression of tissue inhibitor of metalloproteinase-2 (TIMP-2) in the stroma of breast carcinomas correlates with tumor recurrence. Int. J. Cancer 59: 339–344.

36. Grignon, D.J., W. Sakr, M. Toth, V. Ravery, J. Angulo, F. Shamsa, J.E. Pontes, J.C. Crissman & R. Fridman. 1996. High levels of tissue inhibitor of metalloproteinase-2 (TIMP-2) expression are associated with poor outcome in invasive bladder cancer. Cancer Res. 56: 1654–1659.

37. Shofuda, K., K. Moriyama, A. Nishihashi, S. Higashi, H. Mizushima, H. Yasumitsu, K. Miki, H. Sato, M. Seiki & K. Miyazaki. 1998. Role of tissue inhibitor of metalloproteinases-2 (TIMP-2) in regulation of pro-gelatinase A activation catalyzed by membrane-type matrix metalloproteinase-1 (MT1-MRP) in human cancer cells. J. Biochem. (Tokyo) 124: 462–470.

38. Butler, G.S., M.J. Butler, S.J. Atkinson, H. Will, T. Tamura, S.S. Van Westrum, T. Crabbe, J. Clements, M.P. D'Ortho & G. Murphy. 1998. The TIMP2 membrane type 1 metalloproteinase "receptor" regulates the concentration and efficient activation of progelatinase A - A kinetic study. J. Biol. Chem. 273: 871–880.

39. Hayakawa, T., K. Yamashita, K. Tanzawa, E. Uchijima & K. Iwata. 1992. Growth-promoting activity of tissue inhibitor of metalloproteinases-1 (TIMP-1) for a wide range of cells. A possible new growth factor in serum. FEBS Lett. 298: 29–32.

40. Hayakawa, T., K. Yamashita, E. Ohuchi & A. Shinagawa. 1994. Cell growth-promoting activity of tissue inhibitor of metalloproteinases-2 (TIMP-2). J. Cell Sci. 107: 2373–2379.

41. Stetler-Stevenson, W.G., N. Bersch & D.W. Golde. 1992. Tissue inhibitor of metalloproteinase-2 (TIMP-2) has erythroid-potentiating activity. FEBS Lett. 296: 231–234.

42. Avalos, B.R., S.E. Kaufman, M. Tomonaga, R.E. Williams, D.W. Golde & J.C. Gasson. 1988. K562 cells produce and respond to human erythroid-potentiating activity. Blood 71: 1720–1725.

43. Guedez, L., W.G. Stetler-Stevenson, L. Wolff, J. Wang, P. Fukushima, A. Mansoor & M. Stetler-Stevenson. 1998. In vitro suppression of programmed cell death of B cells by tissue inhibitor of metalloproteinases-1. J. Clin. Invest. 102: 2002–2010.

44. Corcoran, M.L. & W.G. Stetler-Stevenson. 1995. Tissue inhibitor of metalloproteinase-2 stimulates fibroblast proliferation via a cAMP-dependent mechanism. J. Biol. Chem. 270: 13453–13459.

45. Yamashita, K., M. Suzuki, H. Iwata, T. Koike, M. Hamaguchi, A. Shinagawa, T. Noguchi & T. Hayakawa. 1996. Tyrosine phosphorylation is crucial for growth signaling by tissue inhibitors of metalloproteinases (TIMP-1 and TIMP-2). FEBS Lett. 396: 103–107.

46. Foekens, J.A., M.P. Look, H.A. Peters, W.L. van Putten, H. Portengen & J.G. Klijn. 1995. Urokinase-type plasminogen activator and its inhibitor PAI-1: predictors of poor response to tamoxifen therapy in recurrent breast cancer. J. Natl. Cancer Inst. 87: 751–756.

47. Foekens, J.A., M. Schmitt, W.L. Van Putten, H.A. Peters, M.D. Kramer, F. Janicke & J.G. Klijn. 1994. Plasminogen activator inhibitor-1 and prognosis in primary breast cancer. J. Clin. Oncol. 12: 1648–1658.

48. Andreasen, P.A., L. Kjoller, L. Christensen & M.J. Duffy. 1997. The urokinase-type plasminogen activator system in cancer metastasis: A review. Int. J. Cancer 72: 1–22.

49. BAJOU, K., A. NOEL, R.D. GERARD, V. MASSON, N. BRUNNER, C. HOLST-HANSEN, M. SKOBE, N.E. FUSENIG, P. CARMELIET, D. COLLEN & J.M. FOIDART. 1998. Absence of host plasminogen activator inhibitor 1 prevents cancer invasion and vascularization. Nat. Med. **4:** 923–928.

50. CHAPMAN, H.A. 1997. Plasminogen activators, integrins, and the coordinated regulation of cell adhesion and migration. Curr. Opin. Cell Biol. **9:** 714–724.

51. WALTZ, D.A., L.R. NATKIN, R.M. FUJITA, Y. WEI & H.A. CHAPMAN. 1997. Plasmin and plasminogen activator inhibitor type 1 promote cellular motility by regulating the interaction between the urokinase receptor and vitronectin. J. Clin. Invest. **100:** 58–67.

52. DENG, G., S.A. CURRIDEN, S. WANG, S. ROSENBERG & D.J. LOSKUTOFF. 1996. Is plasminogen activator inhibitor-1 the molecular switch that governs urokinase receptor-mediated cell adhesion and release? J. Cell Biol. **134:** 1563–1571.

53. DECLERCK, Y.A., T.D. YEAN, D. CHAN, H. SHIMADA & K.E. LANGLEY. 1991. Inhibition of tumor invasion of smooth muscle cell layers by recombinant human metalloproteinase inhibitor. Cancer Res. **51:** 2151–2157.

54. FUJIMOTO, N., J. ZHANG, K. IWATA, T. SHINYA, Y. OKADA & T. HAYAKAWA. 1993. A one-step sandwich enzyme immunoassay for tissue inhibitor of metalloproteinases-2 using monoclonal antibodies. Clin. Chim. Acta **220:** 31–45.

55. ONISTO, M., M.P. RICCIO, P. SCANNAPIECO, C. CAENAZZO, L. GRIGGIO, M. SPINA, W.G. STETLER-STEVENSON & S. GARBISA. 1995. Gelatinase A/TIMP-2 imbalance in lymph-node-positive breast carcinomas, as measured by RT-PCR. Int. J. Cancer. **63:** 621-626.

56. DECLERCK, Y.A., O.A. ALVAREZ, H. SHIMADA, S.M. TAYLOR & K.E. LANGLEY. 1994. Tissue inhibitors of metalloproteinases: role in tumor progression. Contrib. Nephrol. **107:** 108–115.

57. LECO, K.J., L.J. HAYDEN, R.R. SHARMA, H. ROCHELEAU, A.H. GREENBERG & D.R. EDWARDS. 1992. Differential regulation of TIMP-1 and TIMP-2 mRNA expression in normal and Ha-ras-transformed murine fibroblasts. Gene **117:** 209–217.

58. TANAKA, K., Y. IWAMOTO, Y. ITO, T. ISHIBASHI, Y. NAKABEPPU, M. SEKIGUCHI & Y. SUGIOKA. 1995. Cyclic AMP-regulated synthesis of the tissue inhibitors of metalloproteinases suppresses the invasive potential of the human fibrosarcoma cell line HT1080. Cancer Res. **55:** 2927–2935.

59. RASMUSSEN, H.S. & P.P. MCCANN. 1997. Matrix metalloproteinase inhibition as a novel anticancer strategy: A review with special focus on batimastat and marimastat. Pharmacol. Ther. **75:** 69–75.

60. WOJTOWICZ-PRAGA, S., J. TORRI, M. JOHNSON, V. STEEN, J. MARSHALL, E. NESS, R. DICKSON, M. SALE, H.S. RASMUSSEN, T.A. CHIODO & M.J. HAWKINS. 1998. Phase I trial of marimastat, a novel matrix metalloproteinase inhibitor, administered orally to patients with advanced lung cancer. J. Clin. Oncol. **16:** 2150–2156.

# The Regulation of MMPs and TIMPs in Cartilage Turnover

TIM CAWSTON,[a] CARON BILLINGTON, CATHERINE CLEAVER,
SARAH ELLIOTT, WANG HUI, PAUL KOSHY, BILL SHINGLETON,
AND ANDREW ROWAN

*Department of Rheumatology, The Medical School,*
*University of Newcastle upon Tyne, Newcastle upon Tyne NE2 4HH, England*

ABSTRACT: The treatment of cartilage with mediators initiates the breakdown of proteoglycan followed by collagen. This is accompanied by the modulation of different proteinases and inhibitors that include members of the
MMP family and TIMPs. We have evidence that a chondrocyte membrane-
associated metalloproteinase cleaves aggrecan. This activity is rapidly induced
after stimulation with IL-1 and OSM and is not inhibited by TIMPs-1 and -2
but is inhibited by synthetic MMP inhibitors. This same combination of cytokines also upregulates the collagenases with the subsequent release of collagen
fragments, and there is a close correlation between the amount of collagen released and collagenase activity produced. Collagen release can be prevented
after treatment with specific inhibitors of MAP kinases, inhibitors of MMP
transcription, synthetic metalloproteinase inhibitors, TIMPs and treatment of
cartilage with agents that upregulate TIMPs. The results from bovine cartilage
culture models show that collagen release occurs when TIMP levels are low,
collagenases are upregulated and then subsequently activated.

The matrix metalloproteinases (MMPs) are a unique family of enzymes that in concert can degrade all the components of the extracellular matrix.[1] These potent enzymes are made in a proenzyme form, and activation occurs after leaving the cell.
Extracellular activity is also controlled by specific inhibitors called tissue inhibitor
of metalloproteinases (TIMPs),[2] which bind to the active forms of the enzyme, forming 1:1 complexes, and block activity. The inflammatory cytokine interleukin-1
(IL-1) can induce resorption of cartilage *in vitro*[3] and *in vivo* when injected into rabbit joints[4]; it is also found in joint fluids from patients with rheumatoid arthritis.[5] IL-
1 is known to stimulate the production of the MMPs, collagenase-1, and
stromelysin-1 from human synovial cells and chondrocytes.[6] Raised levels of these
enzymes have been localized at both the mRNA and protein level within diseased
cartilage from different rheumatic diseases.[7] Consequently, the MMPs are a valid
therapeutic target for inhibition in the rheumatic diseases.[8]

A number of studies have shown that IL-1-stimulated cartilage model systems
can be used to test the effectiveness of proteinase inhibitors. The release of glycosaminoglycan (GAG) fragments can be prevented by the addition of low molecular weight synthetic inhibitors,[9] but not by the addition of TIMPs.[10] However, while

[a]Address for telecommunication: Phone, 0191 222 5363; fax, 0191 222 5455 e-mail,
T.E.Cawston@ncl.ac.uk

the release of GAG fragments from cartilage appears to be rapid in response to IL-1,[4,11] it is also quickly resynthesized by chondrocytes.[12] In contrast, the release of collagen from cartilage is much slower, is less reproducible, and appears to be irreversible as resynthesis is difficult to achieve.[13]

In this study we have examined the release of both collagen and proteoglycan fragments from resorbing cartilage in response to a combination of IL-1 and oncostatin M (OSM) and we have studied different agents that block this release by the downregulation of MMPs.

## MATERIALS AND METHODS

Chemicals were obtained from the following suppliers: IL-1$\alpha$ was a generous gift from GlaxoWellcome, Stevenage, UK. OSM and interleukin-4 (IL-4) was obtained from R&D Systems, Abingdon, UK. All other chemicals/biochemicals were analytical grade reagents obtained from Merck Ltd, Poole, UK or have been previously described.[10]

### Cartilage Degradation Assay

Control culture medium was Dulbecco's modification of Eagle's medium (DMEM) containing 25 mM HEPES supplemented with sodium bicarbonate (0.5 g/l), glutamine (2 mM), streptomycin (100 $\mu$g/ml), penicillin (100 U/ml), and amphotericin (2.5 $\mu$g/ml). Cartilage slices were dissected from bovine nasal septum cartilage and discs (2 × 2 mm) cut and washed twice in HBSS. Three discs were incubated at 37°C in control medium (600 $\mu$l) in a 24-well plate for 24 hours for stabilization. Control medium (600 $\mu$l) with or without cytokines was added and the plate incubated at 37°C for 7 days. The supernates were harvested and replaced with fresh medium containing identical test reagents to day 0. The experiment was continued for a further 7 days and day-7 and -14 supernates were stored at −20°C until assay. For experiments investigating the prevention of collagen degradation then IL-4 and transforming growth factor-$\beta$ (TGF-$\beta$) were added to the culture medium at day 0 and day 7 at the concentrations shown. For human cartilage assays articular cartilage was dissected from joint replacement samples, and three pieces of cartilage were incubated and treated as described for bovine cartilage. In order to determine the total GAG and OHPro content of the cartilage fragments, the remaining cartilage was digested with papain (2.5 $\mu$g/ml) in 0.1 M phosphate buffer, pH 6.5, containing 5 mM EDTA and 5 mM cysteine hydrochloride, incubating at 65°C until digestion was complete (16 hours).

### Proteoglycan Degradation

Media samples and papain digests were assayed for sulfated GAG (as a measure of proteoglycan release) using a modification of the 1,9-dimethylmethylene blue dye binding assay.[14] Sample or standard (40 $\mu$l) was mixed with dye reagent (250 $\mu$l), prepared as described in the well of a microtiter plate, and the absorbance at 525 nm determined immediately. Chondroitin sulfate from shark fin (5–40 $\mu$g/ml) was used as a standard. The complex formed with 1,9-dimethylmethylene blue decreases absorbance at 525 nm.

### Collagen Degradation

Hydroxyproline release was assayed (as a measure of collagen degradation) using a microtiter plate modification of the method of Bergman and Loxley.[15] Chloramine T (7% [w/v]) was diluted 1:4 in acetate-citrate buffer (57 g sodium acetate, 37.5 g tri-sodium citrate, 5.5 g citrate acid, 385 ml propan-2-ol /liter water). *P*-dimethylaminobenzaldehyde (DAB; 20 g in 30 ml 60% perchloric acid) was diluted 1:3 in propan-2-ol. Specimens were hydrolyzed in 6 M HCl for 20 hours at 105°C and the acid removed on a centrifugal evaporator. The residue was dissolved in water and 40 µl sample or standard (hydroxyproline; 5–30 µg/ml) added to microtiter plate together with chloramine T reagent (25 µl) and then DAB reagent (150 µl) after 4 min. The plate was covered and heated to 60°C for 35 min and the absorbance at 560 nm was determined.

### Immortalized Chondrocytes

T/C28a4 chondrocytes[16] were grown to 70% confluence in DMEM +10% fetal calf serum (FCS). Medium was replaced with DMEM supplemented with 1% acid-treated FCS (ATFCS) and incubated for 24 hours. Medium was replaced with DMEM + ATFCS containing the following agents: IL-1 (1 ng/ml), OSM (10 ng/ml), retinoic acid ($10^{-6}$ M), IL-4 (10 ng/ml), or TGF-β (2 ng/ml). Cells were incubated with the cytokines, alone or in combination, for 24 hours. Medium was harvested and stored at −20°C until assayed.

### Enzyme and Inhibitor Assays

For experiments using human cells, TIMP-1 and collagenase (MMP-1) were assayed by specific ELISAs.[17] For experiments using bovine cartilage, collagenase activity was determined by the diffuse fibril assay[10] using $^3$H-acetylated collagen. Inhibitory activity was assayed by the addition of samples to a known amount of active collagenase in the diffuse fibril assay. Aggrecanase activity was assayed as described.[18]

### Preparation of Aggrecanase

Bovine nasal chondrocytes were prepared and harvested at passage 2 after stimulation with IL-1 and OSM and a membrane fraction prepared.[18] The plasma membrane fraction and a sample of the cytosolic fraction was incubated with purified bovine aggrecan overnight at 37°C.[18] SDS final sample buffer was added and incubated at 100°C for 3 minutes and samples were separated by SDS-PAGE. After Western blotting onto nitrocellulose the degraded aggrecan fragments were detected using the antibody T767 as described.[18]

## RESULTS

### *Aggrecan Degradation in Human Cartilage*

Human cartilage was treated with IL-1 (5 ng/ml), OSM (50 ng/ml), and a mixture of IL-1+OSM (5 ng/ml and 50 ng/ml) in qaudruplicate wells. Medium was removed on day 7 and the amount of GAG measured. The % release of total GAG is shown after treatment with media alone, IL-1, OSM, and IL-1+OSM. A marked increase in proteoglycan release is seen in response to IL-1 and OSM (FIG. 1).

### *Chondrocyte Membrane Aggrecanase*

When chondrocyte membranes were prepared from cells treated for 24 hours with IL-1+OSM, a large amount of associated aggrecanase activity was detected on the membranes (FIG. 2). Little aggrecanase activity could be demonstrated in the cytosolic fraction.

### *Collagen Release and the Stimulation of MMP-1 in Human Articular Cartilage*

Assay of human articular cartilage culture medium (day 7, 14, and 21) for hydroxyproline showed a similar increase in the release of collagen into the medium in response to IL-1+OSM with a cumulative total of 16.03 ± 7.39% of the total collagen (data not shown). Assay of the medium for MMP-1 and TIMP-1 demonstrates that the cumulative levels of MMP-1 released into the medium are increased by IL-1 alone but a large synergistic increase in the amount of MMP-1 protein was seen in

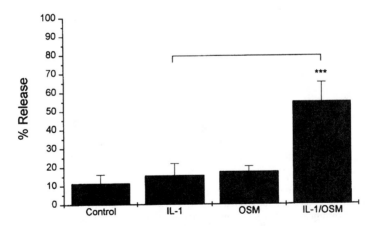

**FIGURE 1.** The effect of IL-1 in combination with OSM on the release of GAG fragments from human articular cartilage. Human articular cartilage was cultured in quadruplicate in 600 μL of medium alone, medium + 5 ng/ml IL-1, medium + 50 ng/ml OSM and medium + 5 ng/ml IL-1 + 50 ng/ml OSM. Medium was removed on day 7 and replaced, removed at day 14 and replaced, and then harvested at day 21. The remaining cartilage was digested with papain and the levels of GAG released on day 7 were measured and expressed as a % of the total. Results are expressed as mean ± SD. *** = $p < 0.001$.

**West Blot**                           1              2        **AbT767**

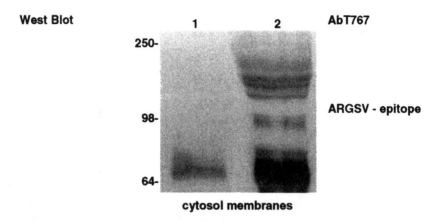

**cytosol membranes**

**FIGURE 2.** Detection of aggrecanase on bovine chondrocyte membranes after stimulation with IL-1+OSM. Confluent bovine chondrocytes cultured with IL-1+OSM for 24 hours were harvested and lysed. Cell nuclei were removed by centrifugation and the supernatant spun at 33,000$g$ for 1.5 hr. Fractions of the supernatant and the resuspended membrane pellet were assayed for aggrecanase activity. High levels of activity were associated with the membrane pellet.

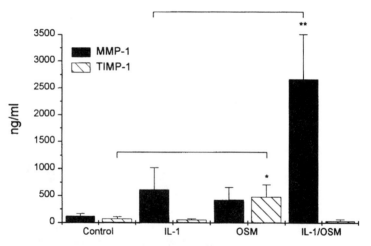

**FIGURE 3.** The cumulative levels of MMP-1 and TIMP-1 released into the medium from human articular cartilage treated with IL-1+OSM. Human articular cartilage was cultured in quadruplicate in 600 μL of medium alone, medium + 5 ng/ml IL-1, medium + 50 ng/ml OSM, and medium + 5 ng/ml IL-1 + 50 ng/ml OSM. Medium was removed on day 7 and replaced, removed at day 14 and replaced, and then harvested at day 21. The levels of MMP-1 and TIMP-1 were measured by ELISA at each time point and the total amount of each measured. IL-1 alone stimulated the release of MMP-1, and a large, synergistic increase in MMP-1 levels was seen in combination with OSM. Although OSM alone increased TIMP-1 levels, this effect was reversed when OSM was combined with IL-1. The results are expressed as a cumulative total mean ± SD released by day 21. ** = $p < 0.01$; * = $p < 0.05$.

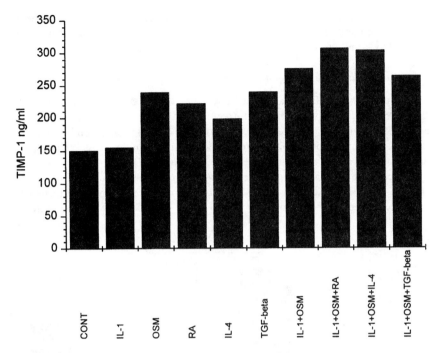

**FIGURE 4.** Upregulation of TIMP-1 levels by growth factors and retinoic acid. T/C28a4 chondrocytes were treated with medium alone, medium + IL-1 (1 ng/ml), OSM (10 ng/ml), retinoic acid ($10^{-6}$M), IL-4 (10 ng/ml), TGF$\beta$ (2ng/ml), IL-1+OSM, IL-1+OSM+retinoic acid, IL-1+OSM+IL-4, IL-1+OSM+TGF$\beta$. The medium was harvested at 24 hours and the levels of TIMP-1 protein measured by ELISA. OSM, retinoic acid, IL-4, and TGF-$\beta$ all increased levels of TIMP-1, and some further increase in TIMP-1 levels were seen when these agents were combined. Results from a representative experiment are shown.

the wells treated with IL-1+OSM (FIG. 3). OSM alone increased TIMP-1 production, but TIMP-1 levels were markedly reduced when IL-1+OSM was present (FIG. 4). IL-1+OSM markedly increased aggrecanase activity on chondrocyte membranes and proteoglycan release from cartilage. Collagen release was also increased with raised levels of MMP-1 and lowered levels of TIMP-1 in the medium at day 14.

### Control of TIMP-1 Production in Chondrocytes by Growth Factors and Retinoic Acid

T/C28a4 chondrocytes[16] were treated with a variety of growth factors and the levels of TIMP-1 measured by ELISA[17] after 24 hours. IL-1 (1 ng/ml) had no effect on TIMP-1 levels, whilst OSM (10 ng/ml), TGF-$\beta$ (2 ng/ml), retinoic acid ($10^{-6}$M) and IL-4 (10 ng/ml) stimulated the production of TIMP-1. Some increase in levels were seen when these agents were combined. It was decided to test the effectiveness of two of these agents, IL-4 and TGF-β, on reversing the release of collagen fragments from bovine nasal cartilage to determine whether the upregulation of TIMP-1 was able to block the action of the collagenases and the subsequent release of collagen.

### Control of Cartilage Collagen Breakdown by Control of TIMP-1 and MMP-1

Bovine nasal cartilage was cultured as described in *Materials and Methods*. IL-1+OSM released >90% collagen by day 14 (FIG. 5). This release was inhibited in a dose-dependent way by increasing concentrations of IL-4 from 2 ng/ml to 50 ng/ml (FIG. 5). Total inhibition of collagen release was seen at 50 ng/ml IL-4. When the levels of collagenolytic activity were measured, low collagen release was accompanied by low collagenolytic activity. Little effect was seen on the levels of TIMPs (data not shown). Collagen release was also inhibited in a dose-dependent way by increasing concentrations of TGF-β from 1 ng/ml to 50 ng/ml. TGF-β at >10 ng/ml completely

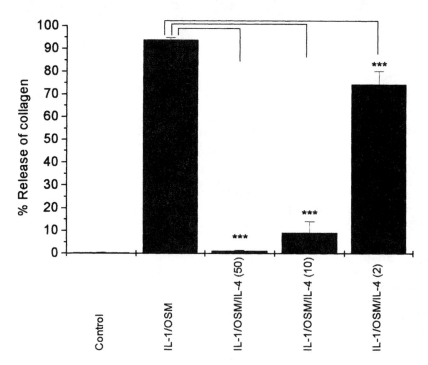

**FIGURE 5.** The effect of IL-4 on the release of collagen fragments from bovine nasal cartilage treated with IL-1+OSM. Bovine nasal cartilage was cultured in quadruplicate in 600 μL of medium alone, medium + 1 ng/ml IL-1 + 10 ng/ml OSM, medium + IL-1+OSM + IL-4 at 2, 10, and 50 ng/ml. Medium was removed on day 7 and replaced and removed at day 14, and the remaining cartilage digested with papain. The levels of OH-proline as a measure of collagen release were measured in day-7 and -14 medium and the papain digests. The % of the total collagen released with each treatment is shown. IL-4 significantly inhibited the release of collagen by IL-1+OSM. Results are expressed as mean ± SD. *** = $p < 0.001$.

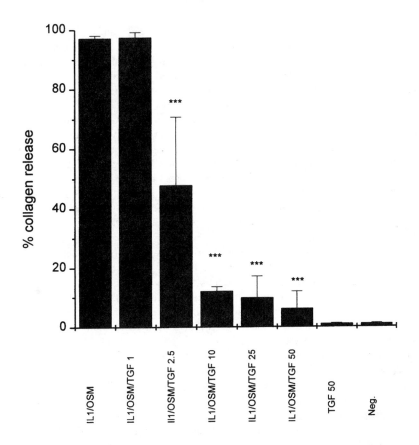

**FIGURE 6.** The effect of TGF-β on the release of collagen fragments from bovine nasal cartilage treated with IL-1+OSM. Bovine nasal cartilage was cultured in quadruplicate in 600 μL of medium alone, medium+IL-1+OSM, and medium+IL-1+OSM+TGF-β at 1, 2.5, 10, 25, and 50 ng/ml. Medium was removed on day 7 and replaced and removed at day 14 and the remaining cartilage digested with papain. The levels of OH-proline as a measure of collagen release were measured in day-7 and -14 medium and the papain digests. The % of the total collagen released with each treatment is shown in day-14 medium samples. TGF-β significantly inhibited the release of collagen induced by IL-1+OSM at concentrations greater than 2.5 ng/ml. Results are expressed as mean ± SD. *** = $p < 0.001$.

blocked the release of cartilage collagen fragments (FIG. 6) and collagenolytic activity in the culture medium. Little effect was observed on TIMP activity (data not shown). These studies indicate that upregulation of collagenases is required with subsequent activation before collagen destruction can occur. Reducing the total activity of the collagenases is sufficient to block cartilage collagen resorption.

## DISCUSSION

Previous studies investigating the release of proteoglycan in response to IL-1 alone showed variable effects[19] and only 30–40% of these samples responded. In this study not only did the inclusion of OSM with IL-1 initiate a more rapid release, but also 98% of the samples tested to date responded by releasing proteoglycan. The release of collagen, which is rarely released in response to IL-1 alone, was less dramatic. Only 50% of the samples responded, and in very few did the level of release exceed 20% of the total collagen. Although this effect is greater than with IL-1, alone the reasons for the variable response are not known. Certainly the samples of human cartilage obtained at joint operations are from an elderly population and the number of cells in these cartilage samples are relatively low. Other growth factors could be needed to elicit the responses which are not present in some samples of human cartilage. Alternatively, inhibitory molecules, which block the response, could be present in unresponsive samples. Cartilage is not the only human tissue to respond to IL-1+OSM, as human tendon also releases collagen fragments in response to this combination of cytokines.[20]

Analysis of the proteoglycan fragments from patients' cartilage *in vivo* has shown that the proteoglycan is specifically cleaved between the G1 and G2 domains of human aggrecan to reveal a new N-terminus that has the sequence ARGSV.[21] This cleavage is effected by an enzyme, putatively named aggrecanase, but it is not yet characterized. We used antibodies that specifically recognize this new epitope generated by aggrecan cleavage and showed that an enzyme present on chondrocyte membranes specifically cleaved aggrecan at this point. IL-1+ OSM treatment markedly upregulated this activity.

This study also shows that to degrade collagen the collagenases need to be upregulated. Although TIMPs are upregulated early in the response of cartilage to growth factors, this effect is not sustained. In contrast, the upregulation of the collagenases is sustained; and once these enzymes are activated, they exceed the local concentration of TIMPs, and collagen destruction then proceeds. IL-4 and TGF-β are able to downregulate the collagenases, which could explain the protection of cartilage collagen seen with these growth factors and cytokines in this study.

## ACKNOWLEDGMENTS

We thank the Arthritis Research Campaign for their support of this work and Chiroscience Ltd., Cambridge for supporting the work on aggrecanase.

## REFERENCES

1. WOESSNER, J.F., JR. 1991. Matrix metalloproteinases and their inhibitors in connective tissue remodeling. FASEB J. **5:** 2145–2154.
2. CAWSTON, T.E., W.A. GALLOWAY, E. MERCER, G. MURPHY & J.J. REYNOLDS. 1981. Purification of a rabbit bone inhibitor of collagenase. Biochem. J. **195:** 159–165.
3. SAKLATVALA, J., L.M.C. PILSWORTH, S.J. SARSFIELD, J. GAVRILOVIC & J.K. HEATH. 1984. Pig catabolin is a form of interleukin-1. Biochem. J. **224:** 461–466.

4. PETTIPHER, E.R., G.A. HIGGS & B. HENDERSON. 1986. Interleukin-1 induces leukocyte infiltration and cartilage proteoglycan degradation in the synovial joint. Proc. Natl. Acad. Sci. USA **85:** 8749–8753.

5. WOOD, D.D., E.J. IHRIE, C.A. DINARELLO & P.L. COHEN. 1983. Isolation of an interleukin-1-like factor from human joint effusions. Arthritis Rheum. **26:** 975–983.

6. BUNNING, R.A.D., H.J. RICHARDSON, A. CRAWFORD, H. SKJODT, D. HUGHES, D.B. EVANS, M. GOWEN, P.R.M. DOBSON, B.L. BROWN & R.G.G. RUSSELL. 1986. The effect of IL-1 on connective tissue metabolism and its relevance to arthritis. Agents Actions (Suppl.) **18:** 131–152.

7. BRINCKERHOFF, C.E. 1991. Joint destruction in arthritis: Metalloproteinases in the spotlight. Arthritis Rheum. **34:** 1073–1075.

8. CAWSTON, T.E. & A. ROWAN. 1998. Prevention of cartilage breakdown by matrix metalloproteinase inhibition - a realistic therapeutic target. Brit. J. Rheumatol. **37:** 353–356.

9. NIXON, J.S., K.M.K. BOTTOMLEY, M.J. BROADHURST, P.A. BROWN, W.H. JOHNSON, G. LAWTON, J. MARLEY, A.D. SEDGWICK & S.E. WILKINSON. 1991. Potent collagenase inhibitors prevent interleukin-1-induced cartilage degradation in vitro. Int. J. Tiss. Reac. **13:** 237–243.

10. ANDREWS, H.J., T.A. PLUMPTON, G.P. HARPER & T.E. CAWSTON. 1992. A synthetic peptide metalloproteinase inhibitor, but not TIMP, prevents the breakdown of proteoglycan within articular cartilage *in vitro*. Agents Actions **37:** 147–154.

11. DINGLE, J.T., D.P. PAGE THOMAS, B. KING & D.R. BARD. 1987. In vivo studies of articular tissue damage mediated by catabolin/interleukin-1. Ann. Rheum. Dis. **46:** 527–533.

12. THOMAS, D.P.P., B. KING, T. STEPHENS & J.T. DINGLE. 1991. In vivo studies of cartilage regeneration after damage induced by catabolin/interleukin-1. Ann. Rheum. Dis. **50:** 75–80.

13. FELL, H.B. & R.W. JUBB. 1980. The breakdown of collagen by chondrocytes. J. Pathol. **130:** 159–167.

14. FARNDALE, R.W., D.J. BUTTLE & A.J. BARRETT. 1986. Improved quantitation and discrimination of sulphated glycosaminoglycans by use of dimethylmethylene blue. Biochem. Biophys. Acta **883:** 173–177.

15. BERGMAN, I. & R. LOXLEY. 1963. Two improved and simplified methods for the spectrophotometric determination of hydroxyproline. Anal. Biochem. **35:** 1961–1965.

16. GOLDRING, M.B., J.R. BIRKDALE, L-F. SUEN, R. YAMIN, S. MIZUNO, J. GLOWACKI, J.L. ARBISER & J.F. APPERLEY. 1994. Interleukin 1β modified gene expression in immortalized human chondrocytes. J. Clin. Invest. **94:** 2307–2316.

17. BIGG, H.F. & CAWSTON, T.E. 1996. Effect of retinoic acid in combination with PDGF or TGFβ on TIMP-1 and MMP-1 secretion from human skin and synovial fibroblasts. J.Cell.Physiol. **166:** 84–93.

18. BILLINGTON, C.J., I.M. CLARK & T.E. CAWSTON. 1998. An aggrecan degrading activity associated with chondrocyte membranes. Biochem. J. **336:** 207–212.

19. ISMAIEL, S., R.M. ATKINS, M.F. PEARCE, P.A. DIEPPE & C.J. ELSON. 1992. Susceptibility of normal and arthritic cartilage to degradative stimuli. Brit. J. Rheumatol. **31:** 369–373

20. CAWSTON, T.E., V.A. CURRY, C.A. SUMMERS, I.M. CLARK, G.P. RILEY, P.F. LIFE, J.R. SPAULL, M.B. GOLDRING, P.J.T. KOSHY, A.D. ROWAN & W.D. SHINGLETON. 1998. The role of oncostatin M in animal and human connective tissue turnover and its localization within the rheumatoid joint. Arthritis Rheum. **41:** 1760–1771.

21. SANDY, J.D., C.R. FLANNERY, P.J. NEAME & L.S. LOHMANDER. 1992. The structure of aggrecan fragments in human synovial fluid. Evidence for the involvement in osteoarthritis of a novel proteinase which cleaves the Glu 373-Ala 374 bond of the interglobular domain. J. Clin. Invest. **89:** 1512–1516.

# Scientific Basis of a Matrix Metalloproteinase-8 Specific Chair-side Test for Monitoring Periodontal and Peri-implant Health and Disease

TIMO SORSA,[a,b] PÄIVI MÄNTYLÄ,[a] HANNE RÖNKÄ,[c] PEKKA KALLIO,[a] GUN-BRITT KALLIS,[c] CHRISTINA LUNDQVIST,[c] DENIS F. KINANE,[d] TUULA SALO,[e] LORNE M. GOLUB,[f] OLLI TERONEN,[g] AND SARI TIKANOJA[c]

[a]Department of Periodontology, Institute of Dentistry, University of Helsinki, Helsinki, Finland

[c]Medix Biochemica Oy Ab, Kauniainen, Finland

[d]Department of Periodontology, Glasgow University School of Dentistry, Glasgow, Scotland

[e]Departments of Oral Surgery and Pathology, University of Oulu, Oulu, Finland

[f]Department of Oral Biology and Pathology, SUNY at Stony Brook, Stony Brook, NY 11794, USA

[g]Departments of Oral and Maxillofacial Surgery, University of Helsinki and University Central Hospital, Helsinki, Finland

ABSTRACT: Matrix metalloproteinases (MMPs), especially collagenase-2 (MMP-8), are key mediators of irreversible tissue destruction associated with periodontitis and peri-implantitis. MMP-8 is known to exist in elevated amounts and in active form in the gingival crevicular fluid (GCF) and peri-implant sulcular fluid (PISF) from progressing periodontitis and peri-implantitis lesions and sites, respectively. (Sorsa *et al.* Ann. N.Y. Acad. Sci. 737: 112–131 [1994]; Teronen *et al.* J. Dent. Res. 76: 1529–1537 [1997]). We have developed monoclonal antibodies to MMP-8 (Hanemaaijer *et al.* J. Biol. Chem. 272: 31504–31509 [1997]) that can be used in a chair-side dipstick test to monitor the course and treatment of periodontitis and peri-implantitis. Monoclonal and polyclonal antibody tests for MMP-8 coincided with the classical functional collagenase activity test from GCF and PISF (Sorsa *et al.* J. Periodont. Res. 22: 386–393 [1988]) in periodontal and peri-implant health and disease. In future a chair-side functional and/or immunological MMP-test can be useful to diagnose and monitor periodontal and peri-implant disease and health.

[b]Address for correspondence: Dr. Timo Sorsa, Department of Periodontology, Institute of Dentistry, Univeristy of Helsinki, P.O. Box 41 (Mannerheimintie 172), FIN-00014, Helsinki, Finland. Phone, -358-9-19127 294; fax, -358-9-19127 519; e-mail: tsorsa@hammas.helsinki.fi

## INTRODUCTION

Periodontal diseases are very common bacterially induced inflammatory conditions which lead to the destruction of the soft and hard tooth-supporting tissues, that is, the gingiva, periodontal ligament, root cementum, and alveolar bone.[1] During the inflammatory process of periodontitis, tooth-supporting tissues undergo irreversible destruction. It is generally accepted that although periodontal disease is initiated by subgingival microflora (or the metabolic products of these organisms), the molecular mediators of irreversible periodontal tissue destruction are primarily generated through the host response to these organisms.[1,2] Prevalence, extent, severity and rate of progression of periodontitis are influenced by age, genetics, systemic health, environment, and even lifestyle.[1] At one time it was believed that periodontitis, once established, progressed continuously and inevitably, in a straight-line relationship with age ultimately causing loss of teeth. However, detailed measurements of loss of attachment site-specifically over time, that is, true longitudinal observations, gave results that seemed to conflict with the concept of continuous disease progression. Thus periodontal destruction is not considered to progress continuously but rather in an episodic manner, with site-specific "bursts" of destructive activity and periods of quiescence, and possibly of repair. There is great individual variation in the pattern of destruction, and this also varies within individuals over time.[1] Overall, the long-term chronic and short-term episodic nature of periodontitis, the wide range of severity of disease affecting different teeth and sites within subjects, and the requirement for repeated treatment over a long period of time indicate that diagnostic information extending beyond the traditional diagnosis of periodontal diseases might be very useful. Long-term treatment and prevention of periodontal diseases should be founded on diagnostic tests based on etiopathogenic factors rather than just "clinical experience," and in this regard biochemically based diagnostic kits may prove to be crucial.

## BRIEF SUMMARY OF DIAGNOSIS OF
## PERIODONTAL DISEASE AND PERI-IMPLANTITIS

The conventional diagnostic tools for periodontitis and peri-implantitis are probes, tooth and implant mobility analysis, and radiographic examination. These can provide information about the extent of periodontal and peri-implant tissue loss/destruction, but cannot at present predict the risk of progression of periodontitis and peri-implantitis.[1]

Bacteriological investigation has been used to identify periodontal pockets and peri-implantitis lesions harboring potentially pathogenic microbial species. The most modern methods can even be used chair-side. However, the problem with such microbiological investigations is that they may only indicate pockets that have undergone periodontal and peri-implant destruction, but cannot detect sites at which the disease is currently active. Attempts to relate microbiological findings to the clinical course of periodontitis and peri-implantitis have encountered complications mainly because of disagreement about which species (if any) are the principal and putative periodontopathogens.[1,3]

Various host-derived biochemical markers have been suggested in relation to diagnosis of periodontal disease as well as peri-implantitis.[2,4–19] They include gingival crevicular fluid (GCF), peri-implant sulcular fluid (PISF), salivary and mouthrinse analysis of inflammation associated molecules and tissue-destruction markers such as cytokines, various enzymes and other proinflammatory mediators originating from supporting periodontal and peri-implant tissues and inflammatory cells. Aspartate amino transferase (AST) and lactate dehydrogenase (LDH) are soluble enzymes normally confined to the cell cytoplasm but released by dead or dying cells.[1] Since cell death is considered to be an essential feature of periodontal disease, it has been suggested that these enzymes must be released during tissue destruction.[1] LDH levels have been found to correlate with probing depth, and to be related at least to some extent to periodontitis disease activity. However, the correlations were less marked than those for neutrophil enzymes (such as β-glucuronidase and elastase).[1,2,4,15] Serum and cerebrospinal-fluid AST levels have been used for years in medicine as biochemical indicators of cell death and necrosis.[1] Overall, the major problems with assay systems of these kind are changing background values and low specificities to reflect actual progression of periodontitis. GCF AST assay can be undertaken using the commercially available PocketWatch™ Periodontal Tissue Monitor System.[1,4] In addition, host tissue breakdown products, such as collagen telopeptides and osteocalcin in GCF, have been found to reflect the progression of periodontitis correlating to GCF host enzymes (i.e., MMPs) evidently causing the breakdown.[10,11,17]

Other GCF and PISF enzyme assay methods include measurements of overall protease activities (both host-cell and bacteria-derived proteases), for example, by Periocheck® and PerioScan™.[18,19] In respect to host enzymes, gelatinase and collagenase activities in GCF and peri-implant sulcular fluid (PISF) have been determined by various functional means on the basis of their catalytic properties.[2,4,6,8,9,16] So far, tests introduced commercially have been associated with a lack of specificity in relation to both enzymes and detection of periodontal disease activity, mostly because of similar activities between both bacterial and nonspecific host-cell enzymes, and to some serum-derived enzymes.[1,3,4] In fact, recent studies have addressed and evaluated the cellular origins and enzyme forms and species of certain neutral proteinases involved in the progression of periodontitis. The matrix metalloproteinases (collagenase-2 [MMP-8], gelatinase B [MMP-9]) and serine proteinase (elastase) have been studied for their individual ability to assess accurately the prevalence of active disease(s).[2,4,15,16] Although important knowledge has been derived from these studies on the etiology and pathology of periodontal disease, the enzyme assays so far developed for the individual neutral proteinase enzymes are not applicable to routine dental office use. The analysis time is long, the cost of testing is expensive, and the chair-side complexity is high.

Chen et al.[5] recently conducted a study in which they compared the initial diagnostic potential of GCF elastase, cathepsin B, MMP-8, and $\alpha_2$-macroglobulin to reflect the clinical efficiency of conventional periodontal treatment comprising scaling and root planing. Chen et al.[5] showed that mean patient clinical parameters and GCF enzyme and inhibitor total decreased significantly after treatment, and for GCF concentrations especially MMP-8 showed a significant decrease after treatment associated with improvement of clinical parameters. Overall, MMP-8 activities and levels in GCF appear to reflect progression of periodontitis and successful periodontal treatment.[4–7,9–12]

## MATRIX METALLOPROTEINASE-8 IN PERIODONTAL DISEASE

An increased body of evidence indicates that irreversible connective-tissue degradation in periodontitis results from an imbalance between matrix metalloproteinases (MMPs) and the specific tissue inhibitors of matrix metalloproteinases (TIMPs). Pathologically elevated excessive collagenase activity in inflamed periodontal tissues results in degradation of the connective tissue.[2] There have been reports that collagenase activity of human cell origin in gingival tissues and GCF in periodontitis patients and in peri-implant sulcular fluid (PISF) of peri-implantitis patients was higher than in healthy subjects, and that this activity correlated with disease activity.[4–7] During progression of periodontitis marked elevations in levels of the active collagenase (originally when compared to other MMPs this was shown to be matrix metalloproteinase-8[6,12] ([MMP-8], see below) were noted in GCF at the time of detection of connective-tissue attachment loss.[9] On the other hand, recent data reported by Havemouse-Poulsen *et al.*[20] indicated that collagen-degrading activity (even cytokine-inducible), apparently representing mostly fibroblast-type collagenolytic MMPs (MMP-1 and -2) from various periodontitis patients' fibroblastic cell lines, was significantly lower than that of fibroblastic cells from periodontally healthy control subjects. Nonetheless, direct extracts of untreated inflamed gingival tissue specimens taken from periodontitis patients in comparison to healthy non-inflamed gingival tissue specimens have been shown to contain pathologically elevated levels of collagenase-2 (MMP-8) in catalytically competent active form.[21,22] It is thus concluded that determination of GCF and PISF collagenase activities (i.e., MMP-8) could be effective in diagnosing progressive periodontitis and peri-implantitis lesions.[5,6,8,9,12]

Recent results of immunochemical analyzes making use of antibodies specific for neutrophil collagenase-2 (MMP-8), fibroblast-type collagenase-1 (MMP-1), and collagenase-3 (MMP-13)[8,9] are consistent with results of several previous enzymological studies that showed the major interstitial collagenase in adult periodontitis GCF and peri-implantitis PISF to be MMP-8 (90–95% of collagenase activity).[7–12] MMP-8 is catalytically very efficient in degrading gingival collagen in comparison to MMP-1.[2,20] The problem with the collagenase activity assays is that although they make use of recognized "fingerprint-characteristics" of GCF or PISF collagenases (MMP-1, -8 and -13) in relation to natural or synthetic proteinase-substrates, they do not differentiate between genetically distinct collagenases clearly enough.[2,4,20] GCF and PISF collagenases can at present be identified most accurately by specific immunochemical assays.[5–13] MMP-8 would seem to be the best marker of periodontal disease and peri-implantitis, especially if determined immunochemically.[5,8,10–13]

MMP-8 is mainly a polymorphonuclear leukocyte-(PMN-)specific matrix metallo-proteinase stored in specific granules of PMNs.[2,4,9,11–14] MMP-8 activity and release are regulated by factors such as cytokines (tumor necrosis factor -$\alpha$ or interleukin-1$\beta$ [IL-1$\beta$]) and various bacterial virulence factors that affect MMP-8 release by PMN degranulation following PMN infiltration to the site of inflammation.[2]

These factors also induce *de novo* synthesis of MMP-8 by certain other non-PMN-lineage cells in the oral cavity such as gingival and periodontal ligament fibro-blasts.[12,13,22] During periodontitis, MMP-8 released from PMNs in a latent, inactive proform becomes activated by the independent and/or combined actions of host- and microbial-derived proteases and reactive oxygen species (ROS) produced by trig-gered PMNs.[14] The irreversible periodontal soft- and hard-tissue destruction usually associated with deepening of the periodontal pockets differentiates periodontitis from preceding gingivitis, and is reflected in diseased GCF by high levels of activat-ed MMP-8.[2,4,6,9,10,11] In gingivitis, there can be slightly elevated MMP-8 levels in GCF, but the MMP-8 will be in the latent, inactive proform.[6,9,12] During active phas-es of periodontitis, MMP-8 levels in GCF are significantly elevated, and the MMP-8 will be almost completely converted to the active form.[4–12] Our results and those of others suggest that the best marker of active periodontal disease and peri-implantitis is likely to be activated MMP-8.[4–12,21,22]

## EFFECT OF PERIODONTAL TREATMENT ON GCF AND GINGIVAL TISSUE MMP-8 LEVELS

Conventional periodontal treatment, involving scaling and root planing, and re-cently described adjunctive drug-treatment to inhibit and downregulate gingival and GCF MMP-8 have been shown to reduce the pathologically excessive collagenase activities in periodontitis GCF to values close to those of periodontally healthy GCF.[2,4,5–12,23]

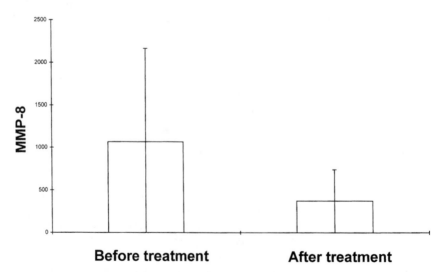

**FIGURE 1.** Effect of conventional periodontal treatment comprising of scaling and root planing on gingival crevicular fluid (GCF) MMP-8 levels in adult patients with peri-odontitis; the number of diseased periodontitis (>4 mm–deep lesions) sites is 31. The reduc-tion in GCF MMP-8 levels (ng/ml), expressed as means ± S.D., reflected the improvement in clinical parameters of periodontitis lesions.

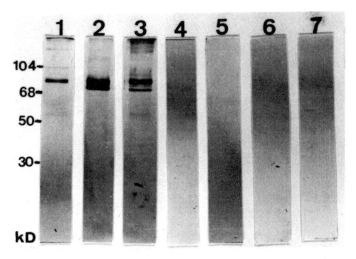

**FIGURE 2.** Western blot analysis of matrix metalloproteinase-8 or collagenase-2 in gingival crevicular fluid (GCF) from deep (>6 mm) periodontitis lesions ($n = 2$) before and after conventional periodontal treatment comprising of scaling and root planing. *Lane 1*: Human neutrophil MMP-8 (500 ng); observe 75-kDa proMMP-8; *lanes 2 and 3*: GCF samples (5 μg protein) from >6-mm periodontitis lesions; observe in addition to 75-kDa proMMP-8 significant amounts of about 5–10-kDa lower molecular weight active MMP-8 forms in untreated periodontitis GCF; *lanes 4 and 5*: same GCF samples as in *lanes 2 and 3* after periodontal treatment (scaling and root planing); *lane 6 and 7*: GCF samples (5 μg) protein from periodontally healthy (<3 mm deep) sites. *Lanes 1–7* stained with specific anti-MMP-8 antibody according to the method of Golub *et al.*[10] and Sorsa *et al.*[12] Corresponding immunoblot data from GCF and PISF has been observed, also for other MMPs such as MMP-9 and MMP-13 (see Sorsa *et al.*,[12] Golub *et al.*,[10] and Teronen *et al.*[8] Molecular weight of standard proteins are indicated on the *left*.

Medication of adult periodontitis (AP) patients and rats in endotoxin (LPS)-induced well-established tissue-destructive periodontitis animal model with anti-collagenase drugs (such as synthetic MMP inhibitors) and downregulators (such as doxycycline, chemically modified non-antimicrobial tetracycline-derivates [CMTs], bisphosphonates and their combinations) have been shown to result in reduced MMP-8 (and other MMPs) activities and levels associated with clinically beneficial effects.[8–12,23,24] We have recently developed monoclonal antibodies (mAbs) against PMN MMP-8 that detect both PMN- and mesenchymal-cell-derived MMP-8 (mMMP-8).[13] We have also developed an immunofluorescence assay utilizing the mAbs against MMP-8 (MMP-8 IFMA).[13] Using this MMP-8 IFMA, the collagenase activity assay,[6] and immunoblot analysis[12] employing polyclonal antibodies specific for PMN and mMMP-8,[6–13] we analyzed GCF samples from AP patients on 2-month and 4-month low-dose doxycycline (LDD) regimens.[10,11] Data from the 2-month LDD study showed that inhibition of MMP-8 resulted in significant reduction of GCF collagenase levels and that this coincided well with clinical improvements in AP patients' periodontal status, resulting in a gain in periodontal attachment lev-

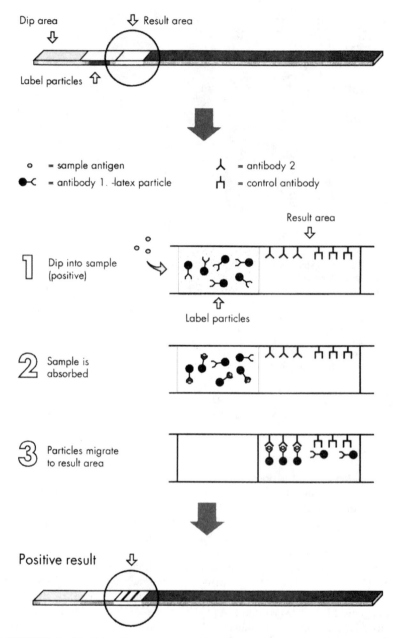

**FIGURE 3.** Principle of the immunochromatographic dipstick test for MMP-8 in gingival crevicular fluid (GCF) and peri-implant sulcular fluid (PISF). The immunoassay can conveniently be adapted alone and/or in any combinations for other MMPs, MMP-related/associated molecules (TIMPs, NGAL, etc.), their split variants, proinflammatory mediators/cytokines, growth factors and enzymes (elastase, cathepsins, tryptase, trypsins, etc.), as well as microbes, viruses, fungi, other parasites and their enzymes and other (virulence) factors.

els.[10,11] In the 4-month LDD study, reductions in immunologically determined MMP-8 levels coincided with downregulation of the cytokine IL-1β.[11] IL-1β can induce neutrophil degranulation and mesenchymal-cell-derived MMP-8 synthesis, release, and activation.[2,13]

When MMP-8 IFMA, immunoblotting, and collagenase activity determinations were undertaken to monitor the effects of conventional periodontal treatment on AP patients, the results revealed that proper scaling and root planing significantly reduced the GCF MMP-8 levels at periodontitis sites, clearly reflecting the clinical efficiency of successful conventional periodontal therapy (FIGS. 1 and 2).

One prerequisite for the development of a chair-side MMP-8 dipstick test for periodontitis was the availability of monoclonal antibodies to MMP-8.[13] Other considerations included the ability to manufacture immunochromatographic tests. The chair-side assay briefly described: GCF is collected from gingival crevices by filter paper strips, from which the samples are extracted with specimen extraction solution. The test is based on immunochromatography.[25] It involves two monoclonal antibodies to human MMP-8. One is bound to blue latex particles to act as the detecting label. The other antibody is immobilized on a carrier membrane to catch label particles and indicate a positive result. When the dip area of the dipstick is placed in an extracted sample, the dipstick absorbs liquid, which starts to flow up the dipstick. If the sample contains MMP-8 it binds to the antibody attached to the latex particles. The particles are carried by the liquid flow and, if MMP-8 is bound to them, they bind to the catching antibody. A blue line (positive line) will appear in the result area if the concentration of MMP-8 in the sample exceeds the cut-off value for the test. A second blue line confirms correct performance of the test (FIG. 3).

Once the assay has been shown to have a significant relationship to periodontitis by (i) differentiating periodontal health and disease and (ii) by successfully reflecting the effect of various beneficial periodontitis treatments,[26] longitudinal studies are to be conducted to find out whether there is an association between MMP-8 and disease progression. This also typically includes another critical point of periodontal monitoring, which occurs during the maintenance phase of therapy in order to observe whether the initial therapeutic result has had long-term stability. The biggest problem of longitudinal evaluation of a new periodontal test is the lack of a generally accepted "gold standard" for disease activity.[1] In longitudinal studies, a measurable clinical attachment loss is usually accepted as the criterion for progression of periodontitis and peri-implantitis. These longitudinal studies are now in progress with the developed chair-side MMP-8-assay. Regarding the future development of MMP assays for diagnosing *different stages* in the progression of periodontitis, recent studies have shown that gingival epithelial cells and macrophages, in addition to bone cells, can express a third type of collagenase (collagenase-3 or MMP-13).[10,11,27] Cell death in periodontal lesions associated to course of periodontitis may be reflected in GCF as elevated levels of released intracellular split variants of these as well as other MMPs.[28] Moreover MMP-13 has been detected in elevated levels in GCF of periodontitis patients and has been found to be reduced, to a *greater* extent than MMP-8, in the fluid of the periodontal lesion (i.e., pocket) when the patients were adminstered a long-term (2-month) regimen of an MMP-inhibitory drug (i.e., sub-antimicrobial doxycycline).[10,11] Thus, it is tempting to speculate that, in the future: (1) a chair-side MMP-8 diagnostic assay could be used to identify the "acute" phase of periodontitis; (2) an MMP-13 chair-side assay might serve as a "marker" of the more "chronic" stage of the disease (reflecting, in part, pathologically excessive collagenase activity generated by crevicular epithelial cells,[27] promoting their migration along the root surface,[27] and by bone cells,[10,11] mediating alveolar bone loss, both cells types cooperating to produce the periodontal pocket); and (3) both chair-side tests could be used as combination MMP-8 and MMP-13 (or other MMPs) dipstick in the same periodontal pocket to allow the clinician more accurately to predict during the examination of the patient the rate of disease progress and the response to therapy.

## SUMMARY

Matrix metalloproteinases (MMPs), especially collagenase-2 (MMP-8), are key mediators of tissue destruction associated with periodontitis and peri-implantitis. MMP-8 is known to exist in elevated amounts and in active form in the gingival crevicular fluid (GCF) and peri-implant sulcular fluid (PISF) of progressing periodontitis and peri-implantitis sites, respectively. We have used MMP-8 antibodies to monitor the course and treatment of periodontitis and peri-implantitis. Monoclonal and polyclonal antibody tests for MMP-8 coincided with the classical functional collagenase activity test from GCF and PISF in periodontal and peri-implant health and disease. In future a (chair-side) immunological or functional oral fluid MMP test, or any measurement of other inflammatory molecules and mediators can be useful to diagnose, follow, monitor, and predict periodontal and peri-implant diseases and health.

## ACKNOWLEDGMENTS

This research was supported by grants from Technology Development Center (Tekes) of Finland, the Finnish Dental Society, and the EVO Clinical Research Grant (TKILO 19).

## REFERENCES

1. MANSON, J.D. & B.M. ELEY. 1996. Outlines of Periodontics, 3rd ed. :1–285, Wright Publisher, Elsevier. Oxford.
2. RYAN, M.E., N.S. RAMAMURTHY & L.M. GOLUB. 1996. Matrix metalloproteinases and their inhibitors in periodontal treatment. Curr. Op. Periodontol., **3:** 85–96.
3. HAFFAJEE, A.D. & S.S. SOCRANSKY. 1994. Microbiological etiological agents of destructive periodontal diseases. Periodontol. 2000 **5:** 78–111.
4. MCCULLOCH, C.A.G. 1994. Host enzymes in gingival crevicular fluid as diagnostic indicators of periodontitis. J. Clin. Periodontol. **21:** 497–506.
5. CHEN, H.-Y., S.W. COX, B.M. ELEY, P. MÄNTYLÄ, H. RÖNKÄ & T. SORSA. 1999. Matrix metalloproteinase-8, elastase, cathepsin B, α2-macroglobulin levels in gingival crevicular fluid from chronic adult periodontitis patients. Submitted for publication.
6. SORSA, T., V.-J. UITTO, K. SUOMALAINEN, M. VAUHKONEN & S. LINDY. 1988. Comparison of interstitial collagenases from human gingiva, sulcular fluid and polymorphonuclear leukocytes. J. Periodont. Res. **23:** 386–393.

7. INGMAN, T., T. TERVAHARTIALA, Y. DING, H. TSCHESCHE, D.F. KINANE, A. HAERIAN, Y.T. KONTTINEN & T. SORSA. 1996. Matrix metalloproteinases and their inhibitors in gingival crevicular fluid and saliva of periodontitis patients. J. Clin. Periodontol. **23:** 1127–1132.

8. TERONEN, O., Y.T. KONTTINEN, C. LINDQUIST, T. SALO, T. INGMAN, A. LAUHIO, Y. DING, S. SANTAVIRTA & T. SORSA. 1997. Human neutrophil collagenase MMP-8 in peri-implant sulcus fluid and its inhibition by clodronate. J. Dent. Res. **76:** 1477–1485.

9. LEE, W., S. AITKEN, J. SODEK & C.A.G. MCCULLOCH. 1995. Evidence for a direct relationship between neutrophil collagenase activity and periodontal tissue destruction in vivo: role of active enzyme in human periodontitis. J. Periodontal Res. **30:** 23–30.

10. GOLUB, L.M., H.M. LEE, R.A. GREENWALD, M.E. RYAN, T. SORSA, T. SALO & W.V. GIANNOBILE. 1997. A matrix metalloproteinase inhibitor reduces bone-type collagen degradation fragments and specific collagenases in gingival crevicular fluid during adult periodontitis. Infl. Res. **46:** 310–319.

11. GOLUB, L.M., H.M. LEE, M.E. RYAN, W.V. GIANNOBILE, J. PAYNE & T. SORSA. 1998. Tetracyclines and their analogs inhibit connective tissue breakdown by multiple nonantimicrobial mechanisms. Adv. Dent. Res. **12:** 12–26.

12. SORSA, T., Y. DING, T. SALO, A. LAUHIO, O. TERONEN, T. INGMAN, H. OHTANI, N. ANDOH, S. TAKIHA & Y.T. KONTTINEN. 1994. Effects of Tetracyclines on neutrophil, gingival and salivary collagenases. Ann. N.Y. Acad. Sci. **732:** 112–131.

13. HANEMAAIJER, R., T. SORSA, Y.T. KONTTINEN, Y. DING, H. RÖNKÄ, M. SUTINEN, H. VISSER, V.M.W. VAN HINSBERG, T. HELAAKOSKI, T. KAINULAINEN, H. TSCHESCHE & T. SALO. 1997. Matrix metalloproteinase-8 is expressed in human rheumatoid fibroblasts and endothelial cells. Regulation by TNF-α and doxycycline. J. Biol. Chem. **272:** 31504–31509.

14. SORSA, T., T. INGMAN, K. SUOMALAINEN, M. HAAPASALO, Y.T. KONTTINEN, O. LINDY, H. SAARI & V.-J. UITTO. 1992. Identification of proteases from potent periodontopathogenic bacteria as activators of latent human neutrophil- and fibroblast-type collagenases. Infect. Immun. **60:** 4491–4495.

15. PALCANIS, K.G., I.K. LARJAVA, B.R. WELLS, K.A. SUGGS, J.R. LANDIS, D.E. CHADWICK & M.K. JEFFCOAT. 1992. Elastase as an indicator of periodontal disease progression. J. Periodontol. **63:** 237–242.

16. TENG, Y.T., J. SODEK & C.A.G. MCCULLOCH. 1992. Gingival crevicular fluid gelatinase and its relationship to periodontal disease in human subjects. J. Periodont. Res. **27:** 544–552.

17. GIANNOBILE, W.V., S.E. LYNCH, R.G. DENMARK, D.W. PAQUETTE, J.P. FIORELLINI & R.C. WILLIAMS. 1995. Crevicular fluid osteocalcin and pyridoline cross-linked carboxyterminal telopeptide of type 1 collagen (ICTP) as markers of rapid bone turnover in periodontitis. A pilot study in beagle dogs. J. Clin. Periodontol. **22:** 903–910.

18. BOWERS, J.E. & R.T. ZAHRADNIK. 1989. Evaluation of a chairside gingival protease tes for use in periodontal diagnosis. J. Clin. Dent. **1:** 106–109.

19. AMALFITANO, J., A.B. DE FILIPPO, W.A. BRETZ & W.J. LOESCHE. 1993. The effects of incubation length and temperature on the specifity and sensitivity of the BANA (*N*-benzoyl-DL-arginine-naphthylamide) test. J. Periodontol. **64:** 848–852.

20. HAVEMOUSE-POULSEN, A., P. HOLMSTRUP, K. STOLTZE & H. BIRKEDAL-HANSEN. 1998. Dissolution of type 1 collagen fibrils by gingival fibroblasts isolated from patients of various periodontitis categories. J. Periodont. Res. **33:** 280–291.

21. GOLUB, L.M., T. SORSA, H.M. LEE, S. CIANCIO, D. SORBI, N.S. RAMAMURTHY, B. GRUBER, T. SALO & Y.T. KONTTINEN. 1995. Doxycycline inhibits neutrophil (PMN)-type matrix metalloproteinases in human adult periodontitis gingiva. J. Clin. Periodontol. **22:** 100–109.

22. SORSA, T., P.L. LUKINMAA, U. WESTERLUND, T. INGMAN, Y. DING, H. TSCHESCHE, Y.T. KONTTINEN, T. HELAAKOSKI & T. SALO. 1996. The expression, activation and chemotherapeutic inhibition of matrix metalloproteinase-8 in inflammation. *In* Biological Mechanisms of Tooth Movement and Craniofacial Adaptation. Z. Davidovitch & L. Norton, Eds. 317–323. Harvard Society for Orthodontic Advancement. EBSCO Media. AL.

23. SORSA, T., L.M. GOLUB, H.M. LEE, M.E. RYAN, G.-B. KALLIS, C. LUNDQVIST, H. RÖNKÄ, T. SALO, P. MÄNTYLÄ & S. TIKANOJA. 1998. The anti-collagenolytic effect of low-dose doxycycline (LDD) regimen in adult human periodontitis can be monitored by the immunological test for neutrophil collagenase (MMP-8). J. Dent. Res. **77:** 647.

24. LLAVANERAS, A., P. HEIKKILÄ, T. SORSA, O. TERONEN, T. SALO, M.E. RYAN, L.M. GOLUB & N.S. RAMAMURTHY. 1999. CMT-8/Clodronate combination synergistically inhibits bone loss and matrix metalloproteinases in LPS-induced periodontitis. This volume.

25. SUTHERLAND, R.M. & B. SIMPSON. 1990. Advances in simple immunoassays for decentralized testing. Adv. Clin. Chem., **28:** 93–108.

26. MOHD, M., S. SAID, L. SANDER, M. RÖNKÄ, T. SORSA & D.F. KINANE. 1999. GCF levels of MMP-3 and MMP-8 following placement of bioresorbable membranes. J. Clin. Periodontol. In press.

27. TERVAHARTIALA, T., E. PIRILÄ, A. CEPONIS, T. SALO, Y.T. KONTTINEN, P. MAISI & T. SORSA. 1999. Expression of matrix metalloproteinases-13 and -8 in human periodontitis gingiva. J. Dent. Res. **78:** 187.

28. HU, S.-I., M. KLEIN, M. CAROZZA, J. REDISKE, J. PEPPARD & J.-S. QI. 1999. Indentification of a splice variant of neutrophil collagenase (MMP-8). FEBS Lett. **443:** 93–108.

# MMP-9 Activity in Urine from Patients with Various Tumors, as Measured by a Novel MMP Activity Assay Using Modified Urokinase as a Substrate

R. HANEMAAIJER,[a,b] C.F.M. SIER,[c] H. VISSER,[a] L. SCHOLTE,[a] N. VAN LENT,[a] K. TOET,[a] K. HOEKMAN,[d] AND J.H. VERHEIJEN[a]

[a]Gaubius Laboratory, TNO-PG, 2301 CE Leiden, the Netherlands

[c]Department of Molecular Genetics, San Raffaele Scientific Institute, Milan, Italy

[d]Department of Oncology, University Hospital VU, Amsterdam, the Netherlands

ABSTRACT: Matrix metalloproteinases (MMPs) play an important role in many pathologic processes, but their activities are difficult to determine since no simple specific and/or chromogenic substrate exists. We have developed a novel MMP activity assay using a modified urokinase as a substrate. Protein engineering enabled the plasmin activation site in this urokinase to be substituted by a specific activation site recognized by MMPs. In this way the MMP activity can be monitored via urokinase activity as measured by a simple chromogenic assay. The assay was made specific for MMP-9 by a capture step using MMP-9-specific antibodies that do not interfere with MMP-activity. This assay monitors the amount of active enzyme as well as the latent, but potentially active proform. Using this assay the levels of MMP-9 were investigated in urine from patients with various kinds of carcinoma. High levels of both active and latent MMP-9 were detected in urine from patients with carcinoma of the bladder, whereas hardly any activity was observed in urine from healthy controls. MMP-9 in urine was present in its intact form. Surprisingly, MMP-9 was also increased in the urine of patients with nonurogenital carcinoma. Therefore, measurement of urinary MMP-9 activity levels may be a convenient diagnostic tool for various types of carcinoma.

## INTRODUCTION

Proteolytic enzymes play an important role in the invasive behavior of tumors. At various stages of tumor development local proteolysis is required—for example, in tumor growth and tumor angiogenesis, invasion, and metastasis. Local proteolysis and matrix remodeling are regulated by a delicate balance between the production, activation, and inhibition of proteolytic enzymes, resulting in guarded spatial and temporal activity. The matrix metalloproteinases (MMPs) form a family of at least fourteen enzymes participating in matrix remodeling. MMPs are the only proteases that are collectively able to degrade all of the components of the extracellular matrix,

[b]Address for correspondence: R. Hanemaaijer, Ph.D., Gaubius Laboratory TNO-PG, P.O. Box 2215, 2301 CE Leiden, the Netherlands. Phone, +31-71-5181446; fax, +31-71-5181904; e-mail, R.Hanemaaijer@PG.TNO.NL

such as fibrillar and nonfibrillar collagens, fibronectin, laminin, elastin, and basement membrane glycoproteins.[1] Numerous *in vitro* studies using cell lines have shown an overexpression of MMPs to be associated with invasive phenotype (for a review see Ref. 2), but high MMP amounts were also detected *in vivo* in tumor tissue compared to normal tissue.[3,4] In many tumors an association was observed between MMP expression and metastatic potential or bad survival prognosis,[5–13] underlining the significance of these proteinases for the development of malignancy of the tumors.[4,11] Enhanced levels of MMPs have also been found in blood from cancer patients,[14–16] but determinations in blood are difficult because of the sensitivity of MMP-ELISAs for variations in sampling.[17] The elevated concentrations in the circulation might be a reflection of the presence of these enzymes in the actual cancer tissue. Recently, enhanced MMP-2 and MMP-9 activity has also been detected in urine from carcinoma patients.[18,19] These urinary MMPs might also originate from the tumor and hence be indicative of its proteolytic capacity. To investigate a possible clinical use of urinary MMP analysis, we determined MMP-9 activity levels in urine from patients with various kinds of carcinoma. A recently developed method for the detection of MMP-9 activity in human body fluids was used.[20,21] This assay measures the amount of active enzyme as well as the total amount, including the inactive proform, which might also be relevant with respect to carcinogenesis.

## MATERIALS AND METHODS

### Materials

The urokinase-specific substrate pyro-Glu-Gly-Arg-pNA (S-2444) was obtained from Chromogenix (Mölndal, Sweden). MMP-9 was obtained from Fuji (Japan). APMA (*p*-aminophenylmercuric acetate) was purchased from Sigma. All molecular biology reagents were from Gibco Life Sciences.

Urine samples were collected before surgery and urine specimens from sex- and age-matched healthy volunteers (age range 34–80 years) were used as controls. All samples were snap-frozen and kept at −80°C. Before measurement the urine specimens were quickly thawed at 37°C and centrifuged at 5000×*g* for 5 minutes.

### MMP-9 Activity Assay

MMP-9 activity was measured as described previously.[21] In brief, MMP-9 was captured from biological fluids using MMP-9-specific monoclonal antibody–coated 96-well plates. The wells were washed three times with PBS-T (phosphate buffered saline solution containing 0.05% [v/v] Tween-20) and incubated with 125-µl assay buffer (50 mM Tris-HCl, pH 7.6, 150 mM NaCl, 5 mM CaCl$_2$, 1 µM ZnCl$_2$ and 0.01% [v/v] Brij-35), to which 15 µl (50 µg/ml) modified pro-urokinase (UKcol) and 10 µl (6 mM stock) chromogenic substrate S-2444 was added. Color development was recorded by measurement of A$_{405}$ using a Titertek Multiskan 8-channel photometer.[20] For measurement of total activity (already active plus latent MMP-9) in biological fluid, the immobilized MMP-9 was incubated with assay buffer containing 0.5 mM APMA for 2 hr, after which UKcol and S-2444 were added and activity was recorded.

## Miscellaneous Methods

Gelatin zymography was carried out as described previously.[22,23] Modified urokinase was expressed and purified as described by Verheijen *et al.*[20] Assay of uPA antigen was performed using ELISA as described by Koolwijk *et al.*[24]

## Statistics

The data are expressed as units of MMP-9 activity, in which a unit is defined as $1000*\Delta A_{405}/hr^2$. Group results are given as median values. The Mann-Whitney U test was used to compare differences between groups, considering $p < 0.05$ significant.

## RESULTS AND DISCUSSION

Only few assays have been developed to measure MMP antigen or activity. Gelatin zymography is a generally used assay in which gelatinases can be measured in a semiquantitative way.[25] A few ELISAs have now been developed which detect only one or few of the various forms in which MMP-9 can be present (latent, complex, active). The development of MMP-9 activity measurements has been hampered by the fact that no chromogenic substrates exist for MMPs. Some activity assays using fluorescent substrates were developed, but these were not specific for the various MMPs and special laboratory equipment was needed.[26,27] Recently we have developed a new MMP activity assay, using modified urokinase as a substrate,[20] by which overall MMP activity can be measured as can activity of specific MMPs, and which in addition can separately measure already active enzyme and the latent form, which is measured after activation *in vitro*.[21] The assay has been made specific for MMP-9 by introduction of a capture step using an MMP-9-specific monoclonal antibody.[21] An additional advantage of the capture step is that no interference takes place with disturbing factors in the biological fluid, which may hamper the analysis of MMP activity (e.g., the observed high background signal in urine if fluorescent peptides are used).

Using the MMP-9 activity assay, we measured MMP-9 activity in urine from patients with bladder carcinoma as well as healthy controls. In the urine of patients with bladder carcinoma both already active and latent MMP-9 was detectable (FIG. 1, *left*). In age-matched healthy controls no or only low levels of already active MMP-9 or latent MMP-9 could be detected (FIG. 1, *left*). To test whether the MMP-9 activity represented intact MMP-9 or a MMP-9 degradation product, gelatin zymography was performed. FIGURE 2 shows that intact 92-kDa MMP-9 is present in the urine from bladder carcinoma patients.

The presence of MMP-9 activity in these urine samples is in accordance with previous data of Davies *et al.*[5] and Moses *et al.*,[19] who showed the presence of MMP-9 antigen in urine by zymography. The MMP-9 in the urine may originate directly from the bladder carcinoma since urine may be regarded as the conditioned medium of the bladder. Here we show that already active MMP-9 is present, indicating that the balance between active MMP and its inhibitor TIMP is in favor of the MMP in urine. This suggests that the presence of MMP-9 in urine is not a general leakage

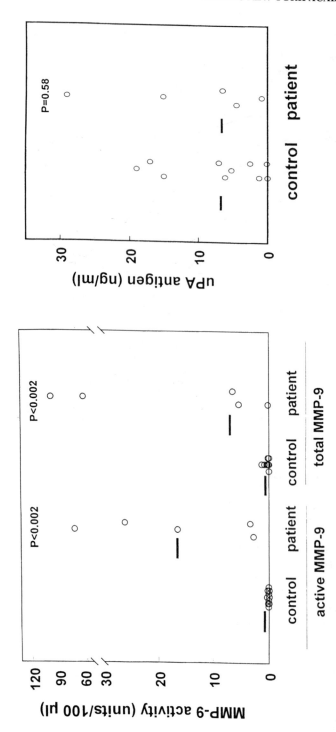

**FIGURE 1.** MMP-9 activity and uPA antigen in urine from bladder carcinoma patients and healthy controls. *Left:* Urine from bladder carcinoma patients (patient) and healthy controls (control) was used for MMP-9 activity measurements as described in the Materials and Methods section. Active MMP-9 represents MMP-9 activity that was directly measured following capture on the plate. Total MMP-9 activity represents the bound MMP-9 fraction which was activated by incubation with 0.5 mM APMA for 2 hr at 37°C before activity was determined. MMP-9 activity is expressed as units, in which a unit is $1000*\Delta A_{405}/hr^2$. The *open circles* represent the individual controls ($n = 10$) and patients ($n = 5$). The *short lines* represent the median values. *Right:* uPA antigen was measured in the same urine samples using an ELISA as described in the Materials and Methods section. The *open circles* represent the individual numbers; the *short lines* the median values.

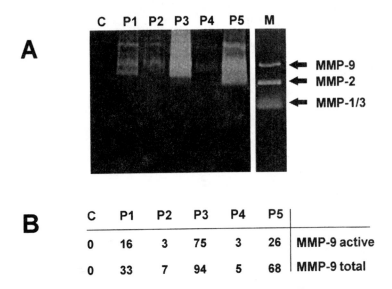

**FIGURE 2.** Gelatin zymogram of urine samples from bladder carcinoma patients. Urine from 5 different patients with bladder carcinoma was diluted 1:1 in PBS and applied for gelatin zymography (**A**). The corresponding MMP-9 activity, as measured by the MMP-9 activity assay, is given for comparison (**B**). C represents a healthy control; P represents the different patients.

from the bladder, because it was shown that in general, but also in bladder cancer, in addition to MMP-9, TIMP-1 and TIMP-2 are often increased in the invasive phenotype.[28,29] To test whether the MMP-9 in the urine represented a general activation or leakage in the urogenital tract, we also measured urokinase plasminogen activator (uPA). This protease is synthesized by the renal epithelium and therefore is directly secreted in the urine. No significant difference ($p = 0.58$) between uPA levels in urine specimens from patients and healthy control was observed (FIG. 1, *right*).

We also tested urine from patients with other tumors, some of which were outside the genitourinary tract. Surprisingly, in patients with breast, lung, esophageal, and ovarian cancer, levels of total MMP-9 activity and, in most cases, of already active MMP-9 as well were significantly increased in urine as compared to the levels in healthy controls (FIG. 3). Obviously, the increase of MMP-9 activity was not restricted to urine from patients with bladder carcinoma. It is not known what the source of MMP-9 activity is in these urine samples. Some reports describe the presence of proteases like cathepsin B in urine from patients with breast tumor and from controls,[30] pepsinogen,[31] or fragments of plasminogen (of which one fragment, the so-called angiostatin, was shown to inhibit angiogenesis). This fragment was only observed in the urine of tumor-bearing patients, and is thought to be induced by the primary tumor itself.[32,33]

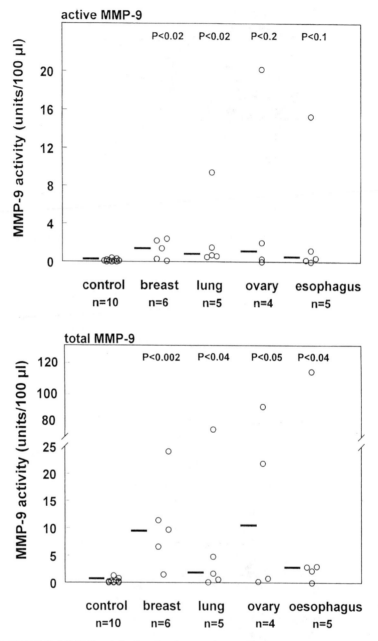

**FIGURE 3.** MMP-9 activity in urine from patients with various kinds (nonbladder) of carcinoma. MMP-9 activity in urine from patients with breast, lung, ovarian, and esophagal cancer was measured as described in the legend of FIGURE 1. *Top:* Active MMP-9 activity. *Bottom:* Total MMP-9 activity. The *open circles* represent the individual controls ($n = 10$) and patients ($n = 5$). The *short lines* represent the median values.

In conclusion, measurement of MMP-9 activity by a chromogenic activity assay is well suited for the analysis of MMP-9 activity in urine from cancer patients. Future research will show whether the levels of MMP-9 activity in urine can be used as a diagnostic tool for the degree of cancer and the effectiveness of therapy.

## REFERENCES

1. STETLER-STEVENSON, W.G., L.A. LIOTTA & D.E. KLEINER. 1993. Extracellular matrix 6: role of matrix metalloproteinases in tumor invasion and metastasis. FASEB J. **7:** 1434–1441.
2. RAY, J.M. & W.G. STETLER-STEVENSON. 1994. The role of matrix metalloproteases and their inhibitors in tumour invasion, metastasis and angiogenesis. Eur. Resp. J. **7:** 2062–2072.
3. DUFFY, M.J. & K. MCCARTHY. 1998. Matrix metalloproteinases in cancer: prognostic markers and targets for therapy. Int. J. Oncol. **12:** 1343–1348.
4. SIER, C.F.M., F.J.G.M. KUBBEN, S. GANESH, M.M. HEERDING, G. GRIFFIOEN, R. HANEMAAIJER, J.H.J.M. VAN KRIEKEN, C.B.H.W. LAMERS & H.W. VERSPAGET. 1996. Tissue levels of matrix metalloproteinases MMP-2 and MMP-9 are related to the overall survival of patients with gastric carcinoma. Br. J. Cancer **74:** 413–417.
5. DAVIES, B., J. WAXMAN, H. WASAN, P. ABEL, G. WILLIAMS, T. KRAUSZ, D. NEAL, D. THOMAS, A. HANBY & F. BALKWILL. 1993. Levels of matrix metalloproteases in bladder cancer correlate with tumor grade and invasion. Cancer Res. **53:** 5365–5369.
6. KAWAMATA, H., S. KAMEYAMA, K. KAWAI, Y. TANAKA, L. NAN, D.H. BARCH, W.G. STETLER-STEVENSON & R. OYASU. 1995. Marked acceleration of the metastatic phenotype of a rat bladder carcinoma cell line by the expression of human gelatinase A. Int. J. Cancer **63:** 568–575.
7. GOHJI, K., N. FUJIMOTO, T. KOMIYAMA, A. FUJII, J. OHKAWA, S. KAMIDONO & M. NAKAJIMA. 1996. Elevation of serum levels of matrix metalloproteinase-2 and -3 as new predictors of recurrence in patients with urothelial carcinoma. Cancer **78:** 2379–2387.
8. GOHJI, K., N. FUJIMOTO, I. HARA, A. FUJII, A. GOTOH, H. OKADA, S. ARAKAWA, S. KITAZAWA, H. MIYAKE, S. KAMIDONO & M. NAKAJIMA. 1998. Serum matrix metalloproteinase-2 and its density in men with prostate cancer as a new predictor of disease extension. Int. J. Cancer **79:** 96–101.
9. BEREND, K.R., A.P. TOTH, J.M. HARRELSON, L.J. LAYFIELD, L.A. HEY & S.P. SCULLY. 1998. Association between ratio of matrix metalloproteinase-1 to tissue inhibitor of metalloproteinase-1 and local recurrence, metastasis, and survival in human chondrosarcoma. J. Bone Joint Surg. Am. **80:** 11–17.
10. MUNCK-WIKLAND, E., K. HESELMEYER, J. LINDHOLM, R. KUYLENSTIERNA, G. AUER & G. ENGEL. 1998. Stromelysin-3 mRNA expression in dysplasias and invasive epithelial cancer of the larynx. Int. J. Oncol. **12:** 859–864.
11. MURRAY, G.I., M.E. DUNCAN, P. O'NEIL, W.T. MELVIN & J.E. FOTHERGILL. 1996. Matrix metalloproteinase-1 is associated with poor prognosis in colorectal cancer. Nat. Med. **2:** 461–462.
12. ALLGAYER, H., R. BABIC, B.C. BEYER, K.U. GRUTZNER, A. TARABICHI, F.W. SCHILD-BERG & M.M. HEISS. 1998. Prognostic relevance of MMP-2 (72-kD collagenase IV) in gastric cancer. Oncology **55:** 152–160.
13. VAISANEN, A., H. TUOMINEN, M. KALLIOINEN & T. TURPEENNIEMI-HUJANEN. 1996. Matrix metalloproteinase-2 (72 kD type IV collagenase) expression occurs in the early stage of human melanocytic tumor progression and may have prognostic value. J. Pathol. **180:** 283–289.
14. ZUCKER, S., R.M. LYSIK, M.H. ZARRABI & U. MOLL. 1993. Mr 92.000 type IV collagenase is increased in plasma of patients with colon cancer and breast cancer. Cancer Res. **53:** 140–146.

15. BAKER, T., S. TICKLE, H. WASAN, A. DOCHERTY, D. ISENBERG & J. WAXMAN. 1994 Serum metalloproteinases and their inhibitors: markers for malignant potential. Br. J. Cancer **70:** 506–512.

16. TORII, A., Y. KODERA, K. UESAKA, T. HIRAI, K. YASUI, T. MORIMOTO, Y. YAMAMURA, T. KATO, T. HAYAKAWA, N. FUJIMOTO & T. KITO. 1997. Plasma concentration of matrix metalloproteinase 9 in gastric cancer. Br. J. Surg. **84:** 133–136.

17. JUNG, K., C. LAUBE, M. LEIN, R. LICHTINGHAGEN, H. TSCHESCHE, D. SCHNORR & S.A. LOENIG. 1998. Kind of sample as preanalytical determinant of matrix metalloproteinases 2 and 9 and tissue inhibitor of metalloproteinase 2 in blood. Clin. Chem. **44:** 1060–1062.

18. MARGULIES, I.M., M. HOYHTYA, C. EVANS, M.L. STRACKE, L.A. LIOTTA & W.G. STETLER-STEVENSON. 1992. Urinary type IV collagenase: elevated levels are associated with bladder transitional carcinoma. Cancer Epidemiol. Biomark. Prev. **1:** 467–474.

19. MOSES, M.A., D. WIEDERSCHAIN, K.R. LOUGHLIN, D. ZURAKOWSKI, C.C. LAMB & M.R. FREEMAN. 1998 Increased incidence of matrix metalloproteinases in urine of cancer patients. Cancer Res. **58:** 1395–1399.

20. VERHEIJEN, J.H., N.M.E. NIEUWENBROEK, B. BEEKMAN, R. HANEMAAIJER, H.W. VERSPAGET, H.K. RONDAY & A.H.F. BAKKER. 1997. Modified proenzymes as artificial substrates for proteolytic enzymes: colorimetric assay of bacterial collagenase and matrix metalloproteinase activity using modified pro-urokinase. Biochem. J. **323:** 603–609.

21. HANEMAAIJER, R., H. VISSER, Y. KONTTINEN, P. KOOLWIJK & J.H. VERHEIJEN. 1998. A novel and simple immunocapture assay for determination of gelatinase-B (MMP-9) activities in biological fluids: saliva from patients with Sjögren's syndrome contain increased latent and active gelatinase-B levels. Matrix Biol. **17:** 657–665.

22. HANEMAAIJER, R., P. KOOLWIJK, L. LE CLERCQ, W.J.A. DE VREE & V.W.M. VAN HINSBERGH. 1993. Regulation of matrix metalloproteinase (MMP) regulation in human vein and microvascular endothelial cells. Biochem. J. **296:** 803–809.

23. HANEMAAIJER, R., T. SORSA, Y.T. KONTTINEN, Y. DING, M. KYLMÄNIEMI, H. VISSER, V.W.M. VAN HINSBERGH, T. HELAAKOSKI, T. KAINULAINEN, H. RÖNKÄ, H. TSCHESCHE & T. SALO. 1997. Matrix metalloproteinase-8 is expressed in rheumatoid synovial fibroblasts and endothelial cells: regulation by TNFα and doxycycline. J. Biol. Chem. **272:** 31504–31509.

24. KOOLWIJK, P., M.G.M. VAN ERCK, W.J.A. DE VREE, M.A. VERMEER, H.A. WEICH, R. HANEMAAIJER & V.W.M. VAN HINSBERGH. 1996. Cooperative effect of TNFα, bFGF and VEGF on the formation of tubular structures of human microvascular endothelial cells in a fibrin matrix: role of urokinase activity. J. Cell Biol. **132:** 1177–1188.

25. KLEINER, D.E. & W.G. STETLER-STEVENSON. 1994. Quantitative zymography: detection of picogram quantities of gelatinases. Anal. Biochem. **218:** 325–329.

26. NAGASE, H. & G. FIELDS. 1996. Human matrix metalloproteinase specificity studies using collagen sequence-based synthetic peptides. Biopolymers **40:** 399–341.

27. BEEKMAN, B., J. DRIJFHOUT, W. BLOEMHOFF, H. RONDAY, P. TAK & J. KOPPELE. 1996. Convenient fluorometric assay for matrix metalloproteinase activity and its application in biological media. FEBS Lett. **390:** 221–225.

28. KHOKHA, R. 1994. Suppression of the tumorigenic and metastatic abilities of murine B16-F10 melanoma cells in vivo by the overexpression of the tissue inhibitor of the metalloproteinases-1. J. Natl. Cancer Inst. **86:** 299–304.

29. NARUO, S., H. KANAYAMA, H. TAKIGAWA, S. KAGAWA, K. YAMASHITA & T. HAYAKAWA. 1994. Serum levels of a tissue inhibitor of metalloproteinases-1 (TIMP-1) in bladder cancer patients. Int. J. Urol. **1:** 228–231.
30. DENGLER, R., T. LAH, D. GABRIJELCIC, V. TURK, H. FRITZ & B. EMMERICH. 1991. Detection of cathepsin-B in tumor cytosol and urine of breast cancer patients. Biom. Biochim. Acta **50:** 555–560.
31. AOKI, T., T. TAKASAKI, J. MORIKAWA, T. YANO & H. WATABE. 1994. Electrophoretic analysis of a gastric cancer-associated acid proteinase using a highly sensitive detection system. Biol. Pharm. Bull. **17:** 1358–1363.
32. O'REILLY, M.S., L. HOLMGREN, C. CHEN & J. FOLKMAN. 1996. Angiostatin induces and sustains dormancy of human primary tumors in mice. Nature Med. **2:** 689–692.
33. O'REILLY, M.S., L. HOLMGREN, Y. SHING, C. CHEN, R.A. ROSENTHAL, M. MOSES, W.S. LANE, Y. CAO, E.H. SAGE & J. FOLKMAN. 1994. Angiostatin: a novel angiogenesis inhibitor that mediates the suppression of metastases by a Lewis lung carcinoma. Cell **79:** 315–328.

# Fluorogenic MMP Activity Assay for Plasma Including MMPs Complexed to $\alpha_2$-Macroglobulin

B. BEEKMAN,[a] J.W. DRIJFHOUT,[b] H.K. RONDAY,[c,d] AND J.M. TeKOPPELE[a,e]

[a]Gaubius Laboratory, TNO Prevention and Health, Leiden, the Netherlands

[b]Immunohaematology and Bloodbank, Leiden University Medical Center, Leiden, the Netherlands

[c]Department of Rheumatology, Leyenburg Hospital, The Hague, the Netherlands

[d]Department of Rheumatology, Leiden University Medical Center, Leiden, the Netherlands

ABSTRACT: Elevated MMP activities are implicated in tissue degradation in, e.g., arthritis and cancer. The present study was designed to measure MMP enzyme activity in plasma. Free active MMP is unlikely to be present in plasma: upon entering the circulation, active MMP is expected to be captured by the proteinase inhibitor $\alpha_2$-macroglobulin ($\alpha$2M). Reconstituted MMP-13/$\alpha$2M complex was unable to degrade collagen (MW 300,000) in contrast to the low-molecular-weight fluorogenic substrate (MW < 1500). Limited access of high-MW substrates to the active site of MMPs captured by $\alpha$2M presents the most likely explanation. Consistently, the high-MW inhibitor TIMP (MW ~28,000) was unable to inhibit MMP/$\alpha$2M enzyme activity, whereas the low-MW inhibitor BB94 (MW ~500) effectively suppressed enzyme activity. By using fluorogenic substrates with Dabcyl/Fluorescein as quencher/fluorophore combin-ation, sensitive MMP-activity assays in plasma were achieved. Spiking of active MMP-13 and MMP-13/$\alpha$2M complex, and inhibitor studies with TIMP-1 and BB94, indicated that active MMPs are efficiently captured by $\alpha$2M in plasma. MMP activity was even detected in control plasma, and was significantly increased in plasma from rheumatoid arthritis patients.

In various pathologic conditions in which MMPs are implicated, such as arthritis and cancer, elevated MMP levels in plasma or serum have been observed.[1–11] Mainly, antibody-based methods were used that are able to detect active MMP, inactive proMMP and TIMP-inhibited MMP forms. Free active MMP is unlikely to be present in serum or plasma: upon entering the circulation, all active MMP will likely be captured by the proteinase inhibitor $\alpha_2$-macroglobulin ($\alpha_2$M).[12] The MMP/$\alpha_2$M complex in blood may well represent a marker for local excess of free active enzyme over inhibitor, and thus for matrix degradation. Hence, a local excess of active MMP in a pathologic tissue will be cleared by secretion into the circulation, followed by neutralization by $\alpha_2$M. This interesting MMP/$\alpha_2$M complex is usually not detected

[e]Address for correspondence: Dr. J.M TeKoppele, Gaubius Laboratory, TNO Prevention and Health, P.O. Box 2215, 2301 CE Leiden, the Netherlands. Phone, +31 71 5181384; fax, +31 71 5181904; e-mail, JM.TeKoppele@pg.tno.nl

by means of antibody-based techniques since the epitopes on the enzyme are shielded by the high molecular weight inhibitor. Therefore, an activity assay for the MMP/$\alpha_2$M complex was developed that gives the unique opportunity to investigate whether MMP activity in plasma or serum can serve as a prognostic or diagnostic marker of tissue degradation.

Upon complexation by $\alpha_2$M, the captured MMP may still be able to convert small peptide substrates (MW < 1,500) that diffuse into the complex. Therefore, fluorogenic substrates were chosen to monitor MMP activity in plasma. This type of substrate consists of a short amino acid sequence, recognized by MMPs, to which a fluorophore is attached. The fluorescence is intramolecularly quenched by an absorbing moiety (quencher) also coupled to the peptide. Upon cleavage, quenching is lost, and an increase in fluorescence is measured proportional to the amount of hydrolyzed substrate.[13,14]

Recently, we have designed fluorogenic substrates using EDANS/Dabcyl as a fluorophore/quencher combination, and we have successfully applied these to monitor MMP activity in synovial fluid.[13,14] However, these substrates will not provide the sensitivity needed for plasma assays, as the level of MMP activity in plasma is expected to be approximately 200-fold lower than in synovial fluid.[15] The main reason for the low sensitivity is related to the suboptimal fluorescence characteristics of EDANS (low absorption and fluorescence quantum yield). Therefore, in the present study, EDANS was replaced by fluorescein. The novel highly quenched substrates were efficiently hydrolyzed by MMPs, and were successfully applied in monitoring MMP activity in plasma.

## MATERIAL AND METHODS

### *Substrates*

Dabcyl-Gaba-Pro-Gln-Gly-Leu-Cys-Ala-Lys-NH$_2$ and Dabcyl-Gaba-Arg-Pro-Lys-Pro-Val-Glu-Nva-Ala-Arg-Cys-Gly-NH$_2$ were synthesized as described before[16]; fluorescein was coupled via the cystein's thiol function using iodacetamide fluorescein, giving TNO211-F and TNO003-F, respectively. Catalytic efficiencies for all MMP-1, -2, -3, -8, -9 and -13 were determined at 25°C.[14]

### *Formation of MMP-13/$\alpha_2$M Complex*

MMP-13 was incubated with a concentration range of $\alpha_2$M to form MMP-13/$\alpha_2$M complexes (ranging from ratios of 0.01 to 100) at 30°C for two hours in 50 mM Tris (pH 7.5), 5 mM CaCl$_2$, 150 mM NaCl, 1 $\mu$M ZnCl$_2$, 0.01% Brij-35, 0.02% NaN$_3$ (buffer A). Thereafter, the complexes were mixed with fluorogenic substrate TNO211-F (2.5 $\mu$M, see below), or type II collagen (100 $\mu$g/ml) in buffer A. Cleavage of the collagen into TC$_A$ and TC$_B$ fragments after overnight incubation at 30°C was visualized by non-reducing SDS-PAGE (10% polyacrylamide gel).

### *Fluorogenic Substrate Conversion Incubations*

Fluorescence, resulting from enzyme-mediated conversion of the fluorogenic substrates, was monitored in a CytofluorII fluorimeter (Perseptive Biosystems), using black, round-bottom 96-well plates (Dynatech). Fluorescence was measured using standard 485-nm (excitation) and 530-nm (emission) bandpass filters. The initial increase in fluorescence was plotted against the incubation time; the slope (rate of increase in fluorescence) was calculated by linear regression.

Incubations in plasma (final condition: 5-fold dilution) were performed at 25°C in the presence of the non-MMP inhibitor cocktail Complete™, EDTA-free (Boehringer Mannheim) at a substrate concentration of 2.5 µM.

## RESULTS

### *Substrate Characteristics: Conversion by MMP/$\alpha_2$M Complex*

The catalytic efficiencies of TNO211-F and TNO003-F towards MMPs were determined (FIG. 1). TNO211-F is hydrolyzed at reasonable rates by all MMPs tested,

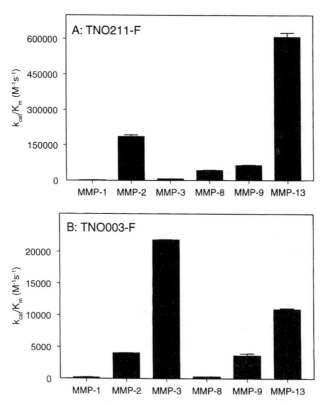

**FIGURE 1.** Catalytic efficiencies ($k_{cat}/K_m$) of fluorogenic substrates TNO211-F and TNO003-F towards MMPs. $k_{cat}/K_m$ values (in $M^{-1}s^{-1}$) were determined at 25°C as described before.[14] TNO211-F (Dabcyl-Gaba-Pro-Gln-Gly ♦ Leu-Cys(Fluorescein)-Ala-Lys-NH$_2$, ♦ = cleavage site) appeared to be a general MMP substrate, with preferential cleavage by MMP-13. TNO003-F (Dabcyl-Gaba-Arg-Pro-Lys-Pro-Val-Glu ♦ Nva-Ala-Arg-Cys(Fluorescein)-Gly-NH$_2$) was preferentially cleaved by MMP-3.

A. Substrate: collagen type II

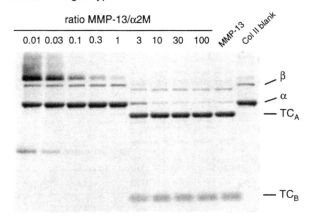

B. Substrate TNO211-F

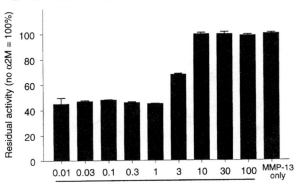

ratio MMP-13/α2M

**FIGURE 2.** High and low molecular weight substrate conversion by MMP-13/$\alpha_2$M complexes. MMP-13 and $\alpha_2$M, and MMP-13 only, were pre-incubated at 30°C for 2 hours using ratios varying from 1:100 to 100:1 (concentration of MMP-13 is constant). Subsequently, either collagen type II (MW ~ 300,000) or the low molecular weight substrate TNO211-F (MW ~ 1,440) was added. Substrate conversion by the MMP-13/$\alpha_2$M complexes and free MMP-13 was monitored by SDS-PAGE (for collagen type II, panel **A**) or by quantification of fluorescence-increase (for TNO211-F, panel **B**). (**A**) Free MMP-13 degraded collagen type II almost completely into the characteristic TC$_A$ and TC$_B$ fragments in the overnight incubation. When an excess of MMP-13 over $\alpha_2$M was present (i.e., ratios 3:1 through 100:1), the collagen was hydrolyzed similarly. In contrast, from ratios 1:1 down to 1:100 (excess of $\alpha_2$M), the activity of MMP-13 towards collagen was completely blocked, indicating full inhibition by $\alpha_2$M. (**B**). TNO211-F was cleaved by free MMP-13 and by α2M/MMP-13 complex (even when an excess of α2M over MMP-13 was present). MMP-13/$\alpha_2$M showed approximately 43% of the activity against TNO211-F compared to that of free MMP-13. Altogether, this experiment demonstrated that low molecular weight substrates like TNO211-F are hydrolyzed by MMP/α2M complex, whereas high molecular weight collagen does not reach the active site of the MMP/α2M complex.

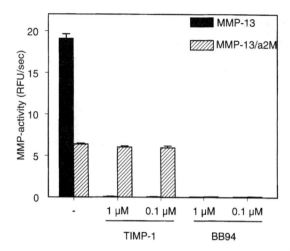

**FIGURE 3.** Inhibition of MMP-13/$\alpha_2$M activity by high and low molecular weight inhibitors. MMP-13 or MMP-13/$\alpha_2$M (1:100) was incubated with TNO211-F in the presence of the high molecular weight inhibitor TIMP-1 (MW ~ 28,000) or the low molecular weight inhibitor BB94 (MW ~ 500). Free MMP-13 (0.5 nM) was fully blocked by both TIMP-1 and BB94. In contrast, the MMP-13/$\alpha_2$M complex (containing 0.5 nM MMP-13) was not inhibited by TIMP-1, even at a 2000-fold molar excess of inhibitor. BB94, however, did fully block the MMP-13/$\alpha_2$M complex. These data corroborate the finding that only small inhibitors and substrates can enter the MMP-13/$\alpha_2$M complex.

with preferential cleavage by MMP-13, whereas TNO003-F is cleaved most efficiently by MMP-3.

Complexes of MMP-13 with $\alpha_2$-macroglobulin ($\alpha_2$M) also converted the small fluorogenic substrates. Conversion of TNO211-F by the MMP-13/$\alpha_2$M complex was approximately 43% of that for free MMP-13 (FIG. 2). Thus, even after complexation with $\alpha_2$M, MMP-13 degrades small substrates diffusing into the complex. In contrast, the hydrolysis of collagen type II was completely blocked when an equimolar amount or an excess of $\alpha_2$M relative to MMP-13 was present (FIG. 2). Thus large substrates like collagen type II cannot reach the active site of the enzyme, indicating the effect of steric hindrance by $\alpha_2$M.

The limited access of the MMP/$\alpha_2$M complex does not only hold for large substrates, like collagen type II, but also for high molecular weight inhibitors like TIMP. Whereas small inhibitors like the hydroxamate BB94 efficiently blocked the activity of the MMP-13/$\alpha_2$M complex, TIMP-1 did not show any effect, even at a 2000-fold excess of TIMP-1 (FIG. 3).

## MMP ACTIVITY IN PLASMA

For measurements in plasma, our hypothesis was that all the active MMP will be captured by $\alpha_2$M, as the concentration of the latter is as high as 4 $\mu$M.[17] To investigate this, MMP-13 (10–500 pM), or its complex with $\alpha_2$M (inhibitor present in a

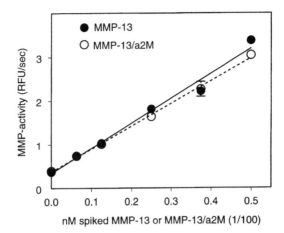

**FIGURE 4.** Activity of spiked MMP-13 and MMP-13/$\alpha_2$M in plasma. A concentration range (0–0.5 nM) of MMP-13 and MMP-13/$\alpha_2$M (1:100) was spiked to plasma, and the resulting activity was monitored using substrate TNO211-F. A linear curve was obtained for both MMP-13 and its complex with $\alpha_2$M. Remarkably, no difference in activity was obtained whether MMP-13 or MMP-13/$\alpha_2$M was spiked: a lower substrate turnover by MMP-13/$\alpha_2$M was expected (see FIGS. 2 and 3). Seemingly, upon spiking to plasma, MMP-13 is rapidly captured by endogenous $\alpha_2$M, which is present at ~4 μM.

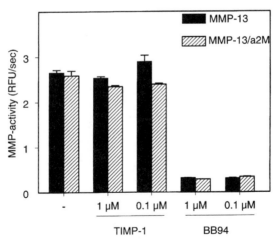

**FIGURE 5.** Inhibition of plasma spiked with MMP-13 and MMP-13/$\alpha_2$M activity by high and low molecular weight inhibitors. Inhibition of plasma spiked with MMP-13 (0.5 nM) and MMP-13/$\alpha_2$M (1:100, 0.5 nM MMP-13) was observed only with the low molecular weight inhibitor BB94. No inhibition was found with TIMP-1, even when plasma was spiked with free MMP-13. This result corroborates the finding that MMP-13 is directly converted into its $\alpha_2$M complex upon addition to plasma (shown in FIGURE 2), and thus cannot be inhibited by TIMP-1.

100-fold excess relative to MMP-13), was added to human plasma, and the resulting activity was measured with substrate TNO211-F. The recovered activity was proportional to the spiked concentration of enzyme (FIG. 4). Spiking of MMP-13/$\alpha_2$M showed identical substrate conversion rates compared to spiked MMP-13. This is in contrast with the data obtained in buffer, where $\alpha_2$M-MMP-13 exhibited approximately 43% of the activity of free MMP-13 against the TNO211-F (FIGS. 2 and 3). This indicates that, upon spiking in plasma, all the MMP-13 is rapidly converted into its $\alpha_2$M complex. Inhibition experiments of the spiked MMP-13 and MMP-13/$\alpha_2$M by TIMP-1 and BB94 corroborated this finding: TIMP-1 could not inhibit the spiked activity, whereas low molecular weight inhibitor BB94 completely blocked the spiked activity of MMP-13 (FIG. 5).

From these spiking and inhibitor studies in plasma, it appeared that the MMP activity measured is likely to originate from MMP/$\alpha_2$M complexes only, indicating that only small substrates can be used for activity measurements in such a bodily fluid. In plasma from patients with rheumatoid arthritis, increased levels of MMPs are observed especially for MMP-3.[2,4,6,18] In a pilot study the MMP activity in plasma from seven patients was monitored with substrate TNO003-F, which is preferentially converted by MMP-3. In all plasmas the measured activity could be inhibited by the MMP-inhibitor BB94 ($p < 0.05$, see FIG. 6A). However, not all measured substrate

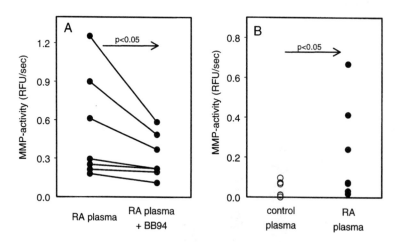

**FIGURE 6.** MMP-activity in plasma from control and rheumatoid arthritis patients. Heparinized plasma from seven randomly selected rheumatoid arthritis patients was incubated with fluorogenic substrate TNO003-F (2.5 µM), in either the absence or presence of 10 µM BB94, for 1 hour, and the initial increase in fluorescence (RFU/sec) was quantified. In all RA plasma samples, the RFU/sec activity was inhibited by the MMP-inhibitor BB94, indicating the presence of MMP activity. Substrate conversion was not totally inhibited by BB94, indicating nonspecific turnover of the substrate. Therefore, the activity inhibited by BB94 was regarded as MMP activity. MMP-activity was detected in the plasma of control subjects and in the plasma from RA patients ($n = 7$ and $n = 7$, respectively). A significantly higher MMP activity was found in RA plasma ($p < 0.05$) compared to control plasma. All incubations were performed in the presence of Complete™, EDTA-Free, a non-MMP inhibitor cocktail.

conversion was inhibited by BB94, indicating nonspecific turnover (despite the presence of the non-MMP inhibitor cocktail "Complete™, EDTA-free"). Therefore, only the activity in plasma that could be inhibited by 10 μM BB94 was regarded as MMP activity. In that respect, the MMP activity found in plasma from control subjects was significant lower than that in the RA plasmas ($p < 0.05$; FIG. 6B).

## CONCLUSION

For the first time, we have shown that MMP activity can be measured in plasma using fluorogenic substrates. The measurable MMP activity in plasma is derived from active MMPs captured by $\alpha_2 M$, indicating that only small peptides able to penetrate the $\alpha_2 M$–MMP complex are suitable substrates. In this respect, the designed fluorogenic substrates, utilizing Dabcyl/Fluorescein as quencher/fluorophore combination, serve as highly sensitive and suitable tools to measure MMP activity in plasma.

MMP activity was even detected in control plasma, and was significantly increased in plasma from rheumatoid arthritis patients (using an MMP-3 selective substrate). A longitudinal study in which clinical parameters are also collected will provide information whether MMP activity in plasma can be used as a prognostic marker for joint destruction in patients with arthritis. In addition, the potential role as a marker in other types of disease, such as cancer will be investigated.

## ACKNOWLEDGMENTS

This work was supported by a grant from Pfizer Central Research, Groton, CT, USA.

## REFERENCES

1. GRUBER, B.L., D. SORBI, D.L. FRENCH, M.J. MARCHESE, G.J. NUOVO, R.R. KEW & L.A. ARBEIT. 1996. Clin. Immunol. Immunopathol. **78:** 161–171.
2. YOSHIHARA, Y., K. OBATA, N. FUJIMOTO, K. YAMASHITA, T. HAYAKAWA & M. SHIMMEI. 1995. Arthritis Rheum. **38:** 969–975.
3. MANICOURT, D.-H., N. FUJIMOTO, K. OBATA & E.J.-M.A. THONAR. 1994. Arthritis Rheum. **37:** 1774–1783.
4. MANICOURT, D.-H., N. FUJIMOTO, K. OBATA & E.J.-M.A. THONAR. 1995. Arthritis Rheum. **38:** 1031–1039.
5. SASAKI, S., H. IWATA, N. ISHIGURO, K. OBATA & T. MIURA. 1994. Clin. Rheumatol. **13:** 228–233.
6. ZUCKER, S., R.M. LYSIK, M.H. ZARRABI, R.A. GREENWALD, B. GRUBER, S.P. TICKLE, T.S. BAKER & A.J.P. DOCHERTY. 1994. J. Rheumatol. **21:** 2329–2333.
7. HAYASAKA, A., N. SUZUKI, N. FUJIMOTO, S. IWAMA, E. FUKUYAMA, Y. KANDA & H. SAISHO. 1996. Hepatology **24:** 1058–1062.
8. BRENNAN, FM., K.A. BROWNE, P.A. GREEN, J.M. JASPAR, R.N. MAINI & M. FELDMANN. 1997. Br. J. Rheumatol. **36:** 643–650.
9. TORII, A., Y. KODERA, K. UESAKA, K. YASUI, T. MORIMOTO, Y. YAMAMURA, T. KATO, T. HAYAKAWA, N. FUJIMOTO & T. KITO. 1997. Br. J. Surg. **84:** 133–136.

10. ENDO, K., Y. MAEHARA, H. BABA, M. YAMAMOTO, S. TOMISAKI, A. WATANABE, Y. KAKEJI & K. SUGIMACHI. 1997. Anticancer Res. **17:** 2253–2258.
11. JUNG, K., L. NOWAK, M. LEIN, D. SCHNORR & S.A. LOENING. 1997. Int. J. Cancer **74:** 220–223.
12. BARRETT, A.J. 1981. Methods Enzymol. **80:** 737–754.
13. BEEKMAN, B., J.W. DRIJFHOUT, W. BLOEMHOFF, H.K. RONDAY, P.P. TAK & J.M. TEKOPPELE. 1996. FEBS Lett. **290:** 221–225.
14. BEEKMAN, B., B. VAN EL, J.W. DRIJFHOUT, H.K. RONDAY & J.M. TEKOPPELE. 1997. FEBS Lett. **418:** 305–309.
15. TAYLOR, D.J., N.T. CHEUNG & P.T. DAWES. 1994. Ann. Rheum. Dis. **53:** 768–772.
16. DRIJFHOUT, J.W., J. NAGEL, B. BEEKMAN, J.M. TEKOPPELE & W. BLOEMHOFF. 1996. *In* Peptides: Chemistry, Structure and Biology, P.T.P. Kaumaya & R.S. Hodges, Eds.: 129–131, Mayflower Scientic Ltd., England.
17. SOTTRUP-JENSEN, L. 1989. J. Biol. Chem. **264:** 11539–11542.
18. TAYLOR, D.J., N.T. CHEUNG & P.T. DAWES. 1994. Ann. Rheum. Dis. **53:** 768–772.

# MMP Inhibition in Abdominal Aortic Aneurysms

## Rationale for a Prospective Randomized Clinical Trial

ROBERT W. THOMPSON,[a,b] AND B. TIMOTHY BAXTER[c]

[a]Departments of Surgery, Radiology, and Cell Biology and Physiology,
Washington University School of Medicine, St. Louis, Missouri 63110, USA

[c]Departments of Surgery, Cell Biology, and Anatomy,
University of Nebraska Medical Center, Omaha, Nebraska 68198, USA

ABSTRACT: Abdominal aortic aneurysms (AAAs) represent a chronic degenerative condition associated with a life-threatening risk of rupture. The evolution of AAAs is thought to involve the progressive degradation of aortic wall elastin and collagen, and increased local production of several matrix metalloproteinases (MMPs) has been implicated in this process. We have previously shown that tetracycline derivatives and other MMP inhibitors suppress aneurysm development in experimental animal models of AAA. Doxycycline also reduces the expression of MMP-2 and MMP-9 by human vascular wall cell types and by AAA tissue explants *in vitro*. To determine whether this strategy might have a role in the clinical management of small AAA, we examined the effect of doxycycline on aortic wall MMP expression *in vivo*. Patients were treated with doxycycline (100 mg p.o. bid) for 7 days prior to elective AAA repair, and aneurysm tissues were obtained at the time of surgery ($n = 5$). Tissues obtained from an equal number of untreated patients with AAA were used for comparison. By reverse transcription-polymerase chain reaction and Southern blot analysis, MMP-2 and MMP-9 were both found to be abundantly expressed in the aneurysm wall. Preoperative treatment with doxycycline was associated with a 3-fold reduction in aortic wall expression of MMP-2 and a 4-fold reduction in MMP-9 ($p < 0.05$ compared to untreated AAA). These preliminary results suggest that even short-term treatment with doxycycline can suppress MMP expression within human AAA tissues. Given its pleiotropic effects as an MMP inhibitor, doxycycline may be particularly effective in suppressing aortic wall connective tissue degradation. While it remains to be determined whether MMP inhibition will have a clinically significant impact on aneurysm expansion, it is expected that this question can be resolved by a properly designed prospective randomized clinical trial.

[b]Address for correspondence: Robert W. Thompson, M.D., Section of Vascular Surgery, Washington University School of Medicine, 9901 Wohl Hospital, 4960 Children's Place, St. Louis, Missouri 63110. Phone, 314/362-7410; fax, 314/747-3548; e-mail: thompsonr@msnotes.wustl.edu.

## INTRODUCTION

### *Clinical Significance of Abdominal Aortic Aneurysms*

Abdominal aortic aneurysms (AAAs) represent a chronic degenerative condition associated with aging and atherosclerosis.[1,2] AAAs are characterized by segmental weakening and dilatation of the aortic wall, and they carry a life-threatening risk of rupture. While the immediate risk associated with a small asymptomatic AAA is quite low, the natural history of these lesions is one of gradual expansion over a period of years.[3–11] If undetected and unrepaired, a substantial number of AAAs will eventually spontaneously rupture.

More than 45,000 patients presently undergo elective surgical repair of AAAs each year in the United States and another 15,000 patients die unexpectedly from ruptured aneurysms.[12] There is accumulating evidence that the incidence of AAA is increasing worldwide despite a general decline in other forms of atherosclerotic cardiovascular disease,[12,13] with recent estimates suggesting that aneurysms affect up to 6–9% of the U.S. population over 65 years of age.[14,15] Current census figures indicate that about 12.8% of the U.S. population is over the age of 65, a figure projected to increase dramatically over the next two decades.[16] By extrapolation to the entire U.S. population, it can be estimated that the number of patients affected by aneurysm disease may be as high as 1.5 million at present and up to 2.7 million by the year 2025.

The only available treatment for AAA is surgical replacement of the diseased aorta, an operative procedure reserved for those at significant risk of rupture. The size of the aneurysm has proven to be the best predictor of rupture; because this risk begins to rise exponentially for aneurysms greater than 5.5 cm in diameter, elective repair is currently recommended for most patients with an AAA larger than 5 cm.[12] The operative mortality risk for elective AAA repair is less than 4% in most centers,

**TABLE 1. Estimated rates of expansion for small AAA[a]**

| Author/reference | Number of patients | Measured aneurysm growth rates (mm per month) | | |
|---|---|---|---|---|
| | | 3.0–3.9 cm | 4.0–4.9 cm | All AAA |
| Bernstein et al.[3] | 49 | 0.26 | 0.51 | 0.39 |
| Kremer et al.[4] | 35 | 0.19 | 0.18 | 0.17 |
| Bernstein & Chan[5] | 110 | 0.39 | 0.36 | 0.37 |
| Cronenwett et al.[6] | 67 | 0.79 | 0.45 | 0.62 |
| Delin et al.[7] | 35 | 0.54 | 0.47 | 0.51 |
| Sterpetti et al.[8] | 125 | 0.25 | 0.56 | 0.41 |
| Nevitt et al.[9] | 103 | 0.26 | 0.46 | 0.36 |
| Limet et al.[10] | 114 | 0.53 | 0.69 | 0.61 |
| Totals/Mean | 638 | 0.40 | 0.46 | 0.43 |
| | | 0.40 cm/yr | 0.55 cm/yr | 0.52 cm/yr |

[a]Data abstracted from K.A. Wilson et al.[11]

and the morbidity of treatment continues to decline, particularly with the application of minimally invasive methods.[12] However, it is notable that all current approaches to the management of aneurysm disease are predicated on early detection of asymptomatic AAAs and, unfortunately, most aneurysms are only identified by serendipity. While ultrasound screening programs would effectively identify nearly all patients with AAA, it has also been shown that the vast majority of these aneurysms are less than 4.5 cm diameter.[17] In the absence of alternative forms of treatment, current management of patients with a small AAA follows a policy of "watchful waiting": clinical observation, with interval measurements of aneurysm size, until elective surgical repair is indicated.[18–20] Follow-up studies have demonstrated that the average rate of aneurysm expansion is about 0.5 cm per year (TABLE 1), indicating that a large number of patients will eventually require repair, or succumb to rupture of the aneurysm during their remaining lifetime. It is therefore evident that any therapeutic strategy proven to reduce the rate of expansion of an aneurysm would be a welcome advance.

## Pathophysiology of Aneurysmal Degeneration

Although the cause of aneurysmal degeneration is still unknown, it is widely recognized that AAAs are closely associated with chronic (transmural) inflammation and the destruction of connective tissue proteins within the outer aortic wall.[1, 2, 21–23] The development of aneurysmal dilatation is primarily attributed to the depletion of medial and adventitial elastin, whereas rupture of the aneurysm is generally thought to involve the additional degradation of adventitial collagen.[24–31] Because the biochemical events occurring within the aneurysm wall are superimposed upon the constant tensile stress associated with arterial blood pressure, hemodynamic forces also contribute to aneurysmal degeneration, as well as the risk of rupture itself. Indeed, it is likely that the pathophysiology of aneurysm disease operates through a gradual imbalance between factors acting to weaken the aortic wall and a compensatory "wound healing" response acting to resist tensile wall stress (FIGURE 1).

While numerous connective tissue proteinases have been described in human AAA tissues, the greatest emphasis has been placed on members of the matrix metalloproteinase (MMP) family.[32–47] Previous studies have focused on 92-kD gelatinase/type IV collagenase (MMP-9; gelatinase B), the most prominent elastolytic enzyme secreted by AAA tissues in organ culture and in vivo.[37–39] MMP-9 is generally an inducible product of mononuclear phagocytes, the cell type to which it is most often localized in human AAA tissues.[38] 72-kD gelatinase/type IV collagenase (MMP-2; gelatinase A) is another elastolytic MMP found in human AAA, where it is expressed by vascular smooth muscle cells, fibroblasts and endothelium, most often in areas adjacent to mononuclear phagocytes.[40,41,46] Recent evidence indicates that soluble factors elaborated by aneurysm-infiltrating macrophages can induce the expression of MMP-2 by fibroblasts and vascular smooth muscle cells.[46,48] Additional MMPs expressed in human aneurysm tissues include macrophage metalloelastase (MMP-12), an enzyme which localizes to macrophages and residual elastin fiber fragments within the degenerating aortic wall.[47] The production of elastolytic MMPs within human AAA tissue has suggested that they might serve as therapeutic targets in this disease; thus, MMP inhibition has been proposed as a novel

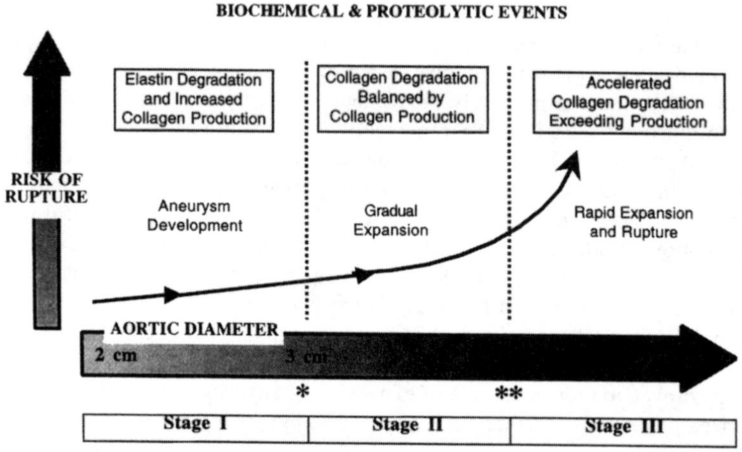

**BIOCHEMICAL & PROTEOLYTIC EVENTS**

**FIGURE 1.** Relationships between the biochemical events occurring within aneurysm tissue and the clinical progression of aneurysm disease. Aneurysms are depicted in this scheme to evolve through three clinical stages: Stage I, early aneurysm development; Stage II, gradual aneurysm expansion; and Stage III, rapid aneurysm expansion and rupture. Aneurysms are clinically detectable at the end of Stage I (*), while the threshold for elective surgical repair generally occurs between Stages II and III (**). Emerging pharmacotherapeutic approaches to aneurysm disease are principally targeted toward reducing either hemodynamic wall stress or the activity of matrix-degrading metalloproteinases thought to be involved in the gradual expansion and accelerated progression of established aortic aneurysms.

strategy to suppress ongoing connective tissue degradation, and thereby aneurysm expansion, in patients with small asymptomatic AAA.[1,2]

## MMP INHIBITION IN EXPERIMENTAL AAA

### Elastase-Induced AAA in Rats

Animal models have become increasingly useful to examine the potential of MMP inhibitors to suppress aneurysm development and growth *in vivo*. The majority of these studies have employed an elastase-induced rat model of AAA, in which an isolated segment of the infrarenal aorta is cannulated and briefly perfused with porcine pancreatic elastase.[49] Under the proper experimental conditions, there is only minimal structural damage to the medial elastic lamellae immediately after elastase perfusion and the aorta only dilates about 30–50%. Aortic wall structure and diameter also remain stable for several days after elastase perfusion, yet the damaged aorta begins to progressively expand thereafter, enlarging to aneurysmal proportions within 7 to 14 days. Importantly, the delayed onset of aortic dilatation is temporally and spatially associated with aortic wall infiltration by mononuclear phagocytes, increased local expression of elastolytic metalloproteinases (including MMP-2 and

MMP-9), and pronounced destruction of the medial elastic lamellae.[49–63] The aortic wall response to elastase perfusion therefore recapitulates many of the morphologic and biochemical events thought to occur in human AAA, encouraging the use of this model for pathophysiologic and pharmacologic investigations.

Glucocorticoids and leukockyte-depleting (anti-CD18) antibodies, both of which reduce the aortic wall inflammatory response, effectively suppress elastase-induced aneurysmal dilatation.[55,56] Not surprisingly, treatment with these agents is associated with structural preservation of the elastic lamellae, an effect consistent with diminished inflammatory cell production of MMPs within the aortic wall. Nonsteroidal anti-inflammatory drugs (NSAIDs), such as indomethacin, also prevent the development of elastase-induced AAA.[57] Although NSAIDs do not act to diminish inflammatory cell infiltration of the elastase-injured aorta, indomethacin is associated with a reduction in MMP expression by mononuclear phagocytes and preservation of medial elastin. Recent evidence indicates that these effects are mediated by inhibition of the inducible (COX-2) isoform of cyclooxygenase and, consequently, a reduction in aortic wall $PGE_2$ production.[63] Evidence that prostaglandin-dependent pathways are involved in the regulation of at least some MMPs supports this mechanism.[64]

Specific pharmacological inhibition of MMPs has also been examined in the process of elastase-induced AAA. Petrinec *et al.* first demonstrated the feasibility of this strategy in experiments using doxycycline, a synthetic tetracycline antibiotic with direct MMP-inhibiting properties.[58] Doxycycline prevents the development of aneurysmal dilatation with an efficacy comparable to those of previously used agents; although it has no discernible effect on the aortic wall inflammatory response, the elastic lamellae are well preserved in doxycycline-treated animals. At doses of 60 mg/kg/day by subcutaneous injection, doxycycline appears to suppress aortic wall production of MMP-9 and the activation of MMP-2. However, effective inhibition of aneurysmal dilatation can also be achieved at lower doses, with half-maximal effects at ~ 6 mg/kg/day, where no suppression of aortic wall MMP production is observed.[61] Recent studies indicate that non-antibiotic chemically modified tetracyclines (CMTs) are as effective as doxycycline in the elastase-induced rat model, and that inhibition of aneurysmal degeneration can also be achieved with hydroxamate-based MMP inhibitors, such as BB-94 and RS 132908.[60–62] The cumulative results of these studies support the notion that MMP inhibition is a feasible strategy by which to suppress aneurysmal degeneration *in vivo*.

### Tetracyclines: Pleiotropic Mechanisms of MMP Inhibition

Since the initial discovery that tetracyclines have anti-metalloproteinase properties, much experimental work has focused on the hypothesis that these compounds act in a direct inhibitory fashion on MMP activity.[65–68] While this was initially thought to involve interactions between tetracyclines and the active site zinc required for MMP activity, it is also possible that tetracyclines act at other molecular sites. One of these targets may include the nonactive site calcium ion present in MMPs, which might subsequently influence tertiary structure, proteinase activity, and enzyme stability in the extracellular space.[67] It is also recognized that tetracyclines act in a relatively nonspecific fashion with regard to different MMPs, and that they inhibit disintegrin metalloproteinases and membrane-type MMPs as well as more traditional MMP family members.[69] Through a variety of chemical modifications in the tetracycline nucleus, it has been possible to demonstrate that the structural elements responsible for MMP inhibition are independent of those conferring antibiotic activity.[66,68] This has led to the development of CMTs as a new class of tetracycline-based MMP inhibitors devoid of antibiotic effects.

Despite considerable progress, there are several unresolved issues with regard to the basic mechanism(s) of action of tetracyclines in limiting MMP-mediated tissue destruction. For example, tetracyclines are relatively weak inhibitors of MMP activity *in vitro*, displaying inhibitory constants in the micromolar range in most assay systems.[67] This compares unfavorably with peptide hydroxamates and other compounds specifically designed to interact with the MMP active site, which often exhibit MMP inhibition at low nanomolar concentration.[70] Nonetheless, tetracyclines are frequently as effective as other MMP inhibitors *in vivo* and their beneficial influence on connective tissue destruction can often be achieved at remarkably low dose schedules.[68,71] One explanation for this paradox may be better pharmacokinetics and tissue absorption compared to those of other currently available MMP inhibitors. Importantly, a second possibility arises from recent observations indicating that tetracyclines can also suppress the activation of inducible nitric oxide synthase (iNOS) in macrophages and other cell types by post-transcriptional mechanisms.[72–74] Because nitric oxide serves as a potent inflammatory mediator, and because oxidative nitration may specifically mediate extracellular activation of proMMPs,[75] tetracycline-induced suppression of iNOS may amplify any direct inhibitory effects on MMP activities *in vivo*.

Finally, emerging studies demonstrate that tetracyclines may also act to reduce steady-state levels of mRNA for at least some MMPs. This has been demonstrated most prominently for MMP-2 in cultured human skin keratinocytes,[76] for MMP-8 in rheumatoid synovial fibroblasts and endothelial cells,[77] and for interleukin-1β-induced transcription of MMP-3,[78] but it remains to be determined how specific this effect is for different MMPs and for different cell types. At present, it appears likely that the combination of molecular and cellular pathways potentially affected by tetracyclines offers a particularly effective means by which to influence an overall reduction in MMP-mediated tissue destruction. Coupled with their clinical availability, low cost and well-recognized safety profile, the use of tetracyclines appears to represent an ideal starting point for clinical studies on MMP-inhibition in a variety of conditions.

## PRELIMINARY STUDIES IN PATIENTS

On the basis of favorable experimental results *in vitro* and in elastase-induced models of AAA, doxycycline might be an attractive pharmacological agent by which to test the efficacy of MMP-inhibiting strategies in patients.[79] It is not yet known, however, whether tetracycline derivatives can inhibit MMP activity in the complex local environment of human aortic aneurysms or if this might translate into effective suppression of connective tissue destruction *in vivo*. It has been difficult to reliably measure the effects of doxycycline on MMP activity in soluble protein extracts from patients with AAA, indicating that alternative means to demonstrate *in vivo* effects of MMP inhibitors will be needed to test these strategies. As a first step toward addressing these concerns, we examined whether clinically relevant concentrations of doxycycline can inhibit MMP production in human vascular wall cell types and in AAA tissue explants *in vitro*. Secondly, we sought to determine whether short-term administration of doxycycline can achieve satisfactory inhibition of MMP expression in human aneurysm tissue *in vivo*. As described below, these initial investigations were based on detecting diminished tissue expression of MMP-2 and MMP-9 mRNA as a representative measure of therapeutic effect.

## EXPERIMENTAL METHODS

### MMP Expression in Cultured Vascular Wall Cell Types and AAA Tissue

Vascular SMC and adventitial fibroblasts were obtained from normal human aortic tissue by explant techniques and cultured in Dulbecco's modified essential medium (DMEM) containing 10% fetal calf serum. Subconfluent cells in early passage were examined under basal culture conditions and following stimulation with macrophage/lymphocyte-conditioned medium, as previously described.[46] Cells were exposed to doxycycline (20 µg/ml) for 24 hours, and RNA was isolated from cell lysates. The expression of MMP-2 and MMP-9 was measured by Northern blot analysis, using ethidium bromide staining of the cRNA to demonstrate equivalent loading of total RNA. To determine doxycycline concentrations that might be achieved during chronic therapy *in vivo*, five normal adult volunteers were given 200 mg doxycycline by mouth. Plasma doxycycline levels were measured by HPLC 2 hours after drug ingestion.

### Aneurysm Patients, Drug Treatment, and Tissue Samples

Patients scheduled to undergo elective surgical repair of an infrarenal AAA were recruited according to a protocol approved by the Washington University Human Research Studies Committee. Each patient was provided with doxycycline hyclate capsules (Danbury Pharmacal, Inc., Danbury, CT, a subsidiary of Schein Pharmaceutical, Inc., Florham Park, NJ). Patients were instructed to take 100 mg doxycycline twice daily with meals, beginning 7 days prior to surgery, and continuing up to the morning of the scheduled operation. An untreated control group consisted of an equal number of patients undergoing elective AAA repair during the same time period. For each group of patients, full-thickness aortic wall specimens were obtained from the anterolateral aspect of the aneurysm wall during the routine conduct of AAA repair. All tissue specimens were immersed in liquid nitrogen and maintained at −70°C prior to nucleic acid extraction.

### Analysis of MMP Expression in AAA Tissues

Aortic tissue samples were pulverized under liquid nitrogen and total RNA was isolated by guanidinium isothiocyanate-phenol-chloroform extraction. All samples were normalized to the same amount of total RNA for analysis by reverse transcription-polymerase chain reaction (RT-PCR) and each sample was analyzed in duplicate, along with controls for genomic DNA (absence of reverse transcriptase) and for nonspecific DNA contamination (absence of RNA template). First-strand cDNA synthesis was performed in a total reaction volume of 20 μl using 0.1 μg of total RNA, 20 U RNase inhibitor, 2.5 μM random hexamers, 1 mM dNTPs and 50 U murine leukemia virus reverse transcriptase, as provided in the GeneAmp RNA PCR kit from Perkin-Elmer (Norwalk, CT). Samples were incubated at 42°C for 15 min and the reaction was terminated by heating to 99°C for 5 minutes. Reverse transcription products served as the template for PCR amplification, using primer pairs specific for human MMP-2, MMP-9, and beta-actin. PCR amplifications were performed in a 100-μl reaction volume with 10 mM Tris-HCl buffer containing 50 mM KCl, 2 mM MgCl$_2$, 100 pmol (each) forward and reverse complement primers, and 2.5 units of AmpliTaq DNA polymerase. Reactions included 4 min at 95°C for denaturation and 30 cycles of 1 min at 95°C, 1 min at 55°C, and 1 min at 72°C; samples were then incubated for 7 min at 72°C for final extension prior to holding at 4°C. A 30-μl aliquot of each sample was resolved by 1.2% agarose electrophoresis in the presence of 5 ng/ml ethidium bromide, and DNA was visualized under ultraviolet light to detect the presence of PCR amplification products at the anticipated size (385 bp for MMP-2, 371 bp for MMP-9, and 193 bp for beta-actin).

RT-PCR products were transferred to Hybond N+ nylon membranes (Amersham) by standard Southern techniques. The specificity of each product was confirmed by hybridization with cDNA oligonucleotide probes for MMP-2, MMP-9 or beta-actin, using an ECL 3′ oligolabeling and detection system (Amersham). After pre-hybridization, membranes were incubated at 42°C for 90 min with 10 ng/ml labeled cDNA probe. Membranes were washed under stringent conditions, incubated with horseradish peroxidase-conjugated anti-fluorescein and ECL detection reagents, and then exposed to radiographic film. The abundance of each product was measured by densitometry and recorded as the beta-actin-normalized ratio for each sample. The mean densitometric ratios for MMP-2 and MMP-9 were determined for all samples of aortic tissue obtained from untreated AAA patients and doxycycline-treated AAA patients, and compared using the Student's *t*-test.

## RESULTS

### *Effect of Doxycycline on MMP Expression in Cultured Aortic Wall Cell Types*

As shown in the zymogram in Figure 2A, doxycycline inhibited the production of MMP-2 in vascular SMC and adventitial fibroblasts. This inhibition corresponded with a reduction in the steady-state levels of MMP-2 mRNA. Equivalent loading is demonstrated by ethidium bromide staining of total RNA. As illustrated by Figure 2B, doxycycline reduced vascular SMC MMP-2 levels slightly at 5 μg/ml and by 3-fold at a concentration of 20 μg/ml. More recent work has shown a reduction in MMP-2 protein levels in AAA tissue incubated in 5 μg/ml doxycycline. This concentration may be particularly relevant since the plasma concentration of doxycycline in normal volunteers 2 hours after ingestion of 200 mg was 5.7 ± 0.7 μg/ml.

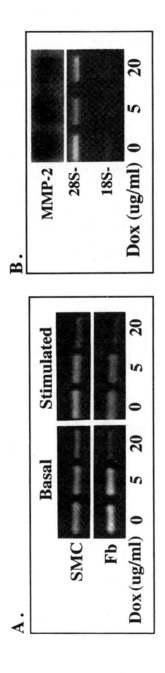

**FIGURE 2.** Effect of doxycycline on MMP-2 expression *in vitro*. (**A**) Cultured human vascular smooth muscle cells (SMC) and adventitial fibroblasts (Fb) were exposed to varying concentrations of doxycycline (Dox), either under basal conditions or following stimulation by macrophage/lymphocyte-conditioned medium. By gelatin zymography, Dox appeared to reduce the amount of MMP-2 produced by each cell type in a concentration-dependent fashion. (**B**) By Northern blot analysis, Dox also reduced the amount of MMP-2 mRNA in SMC lysates (28S and 18S ribosomal RNAs are shown as a loading control).

## AAA Patients and Tolerance of Drug Treatment

The untreated control patients and the doxycycline-treated patients were similar with respect to mean age and all had aneurysms of relatively large size (TABLE 2). No doxycycline-treated patients reported any significant side effects during preoperative drug therapy.

## Effect of Doxycycline on Aortic Wall Expression of MMPs

FIGURE 3 illustrates the results of RT-PCR analysis for MMP-2 and MMP-9 in human AAA tissues. By Southern hybridization, amplification products corresponding to MMP-2 and MMP-9 were detected in nearly all AAA tissue samples, and each of the untreated patients exhibited abundant expression of both MMPs in the aneurysm wall (FIG. 3A, lanes 1 to 5). In contrast, AAA tissue from patients treated with doxycycline exhibited diminished amounts of both MMP-2 and MMP-9 (FIG. 3A, lanes 6 to 10). MMP-2 was only slightly detectable in two of the five treated patients (lanes 7 and 9), and MMP-9 was undetectable in three (lanes 7, 8 and 9). There was no discernible correlation between the reduction in MMP expression and patient age or AAA size. As determined by semi-quantitative analysis (FIG. 3B), doxycycline treatment was associated with a 3-fold reduction in MMP-2 ($1.14 \pm 0.41$ treated vs. $3.48 \pm 0.42$ untreated; $p < 0.05$), and a 4-fold reduction in MMP-9 ($0.73 \pm 0.47$ treated vs. $3.13 \pm 0.77$ untreated; $p < 0.05$).

**TABLE 2. Clinical characteristics of patients with AAA**

| | | | *Untreated control patients* | | |
|---|---|---|---|---|---|
| Patient No. | Gender | Age (yr) | AAA Size (cm) | | |
| 52 | F | 63 | 9.5 | | |
| 55 | F | 73 | 7.5 | | |
| 307 | M | 83 | 8.0 | | |
| 377 | M | 67 | 5.0 | | |
| 406 | F | 64 | 6.5 | | |
| Mean ± SE | | $70.0 \pm 3.7$ | $7.3 \pm 0.8$ | | |
| | | | *Doxycycline-treated patients* | | |
| Patient No. | Gender | Age (yr) | AAA Size (cm) | Wt (kg) | Dose (mg/kg/d) |
| 421 | M | 61 | 4.9 | 99.7 | 2.01 |
| 546 | M | 82 | 6.7 | 63.0 | 3.17 |
| 557 | M | 69 | 7.0 | 78.4 | 2.55 |
| 564 | M | 74 | 7.0 | 88.4 | 2.26 |
| 568 | M | 68 | 4.0 | 53.5 | 3.74 |
| Mean ± SE | | $70.8 \pm 3.5$ | $5.9 \pm 0.6$ | $76.6 \pm 8.4$ | $2.75 \pm 0.31$ |

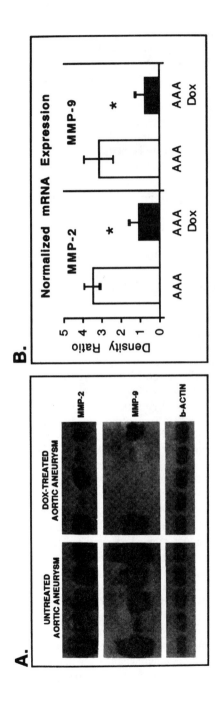

**FIGURE 3.** Effect of doxycycline on aortic wall expression of MMP-2 and MMP-9 *in vivo*. Tissues were obtained from untreated patients with AAA and patients who had been treated with doxycycline (Dox) for 7 days prior to surgery. (**A**) Southern blots of RT-PCR products specific for MMP-2, MMP-9 and b-actin. (**B**) Relative density ratios for MMP-2 and MMP-9 in AAA tissues, normalized to b-actin (mean ± SE, $n = 5$ patients in each group). AAA, untreated patients with aortic aneurysms; AAA Dox, doxycycline-treated patients with aortic aneurysms ($*p < 0.05$, AAA vs. AAA Dox, Student's *t*-test).

Doxycycline is recognized to have pleiotropic effects on MMPs *in vivo*, including direct effects on MMP activity, inhibition of inflammatory signaling pathways, and suppression of MMP production. The results of these preliminary studies suggest that doxycycline can effectively reduce the steady-state expression of MMP-2 and MMP-9 mRNA transcripts, both in cultured vascular wall cell types and within human AAA tissues. It will be important to clarify the molecular mechanisms by which doxycycline exerts these effects in cultured cells (i.e., transcriptional regulation *versus* diminished mRNA stability), and these clinical observations will need to be extended to a larger number of patients. The effects of doxycycline on MMP expression will also need to be correlated with plasma and tissue drug levels achieved during different dose schedules. Nonetheless, these data suggest for the first time that even short-term administration of doxycycline can exert potentially important effects on MMPs within human aneurysm tissue, and that analysis of MMP mRNA expression may provide a measurable endpoint to quantify these effects in patients.

## PROSPECTS FOR A RANDOMIZED CLINICAL TRIAL

Therapeutic strategies proven to limit the growth of aortic aneurysms have the potential to dramatically influence the overall clinical approach taken to this disease. Most importantly, they might result in a shift in clinical management to include routine population screening and medical management (pharmacotherapy) for small asymptomatic AAAs.[79,80] This approach would be clinically favorable to the extent that it (i) facilitated identification and close clinical follow-up of patients with small AAA, and (ii) enhanced the early detection of unsuspected AAAs already large enough to require operation. Each of these outcomes would help to eliminate unnecessary deaths from ruptured AAAs. This shift in clinical management would also be economically feasible to the extent that it might limit the overall number of patients that ultimately require surgical repair during their remaining lifetime, either under elective conditions or, more importantly, as emergencies. On the basis of experimental evidence and the preliminary studies discussed here, doxycycline is emerging as an attractive agent by which to begin testing these possibilities in the clinical setting. It is therefore useful to consider how the effectiveness of doxycycline or other MMP inhibitors might be evaluated in a randomized clinical trial in patients with AAA.

FIGURE 4 illustrates how the effectiveness of a pharmacotherapeutic agent for small AAAs might be demonstrated in clinical trial and how these effects might produce an overall reduction in the number of patients for whom aneurysm repair is ultimately required. For example, based on current evidence, the average rate of aneurysm expansion is expected to be about 0.5 cm per year.[3–11] A small AAA measuring 3.0 cm diameter is thereby predicted to require surgical repair, upon reaching a size of 5.0 cm, approximately 48 months after its initial detection (FIG. 4A). Given a resolution of about 0.5 cm diameter to detect an increase in aneurysm size by ultrasound or CT scanning, patients receiving a therapy that had complete (100%) effectiveness would be radiologically distinguishable from untreated controls as early as 12 months (FIG. 4A, point A). Even those who received a therapy with only 50% effectiveness would be radiologically distinguishable from untreated controls by 24

## A. Expansion Rate

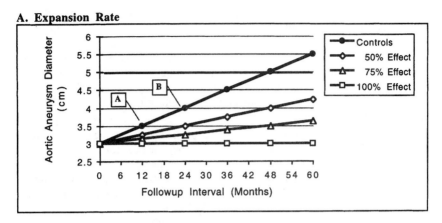

## B. Eventual Need for Operation

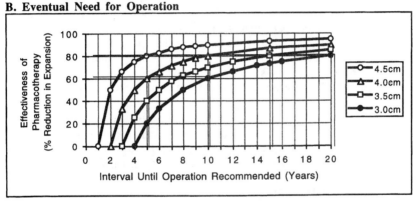

**FIGURE 4.** Predicted effect of pharmacotherapy on patients with small AAAs. (**A**) The potential effects on aneurysmal expansion rate were calculated for pharmacologic interventions with varying degrees of effectiveness based on a 3.0-cm-diameter AAA with an average expansion rate of 0.5 cm per year. Abdominal imaging studies are currently capable of discriminating a difference in aortic diameter of 0.5 cm. (**B**) The potential effects on the eventual need for operation were calculated for pharmacologic interventions of varying degrees of effectiveness and AAAs of different sizes based on the assumption that operation is recommended for all AAAs greater than 5.0 cm in diameter.

months (FIG. 4A, point B). Thus, given a sufficient number of patients, a clinical trial of therapy for small AAAs can be expected to yield interpretable results within 2 years for any pharmacological effect that reduces the rate of aneurysm expansion by 50% or more.

FIGURE 4B illustrates the potential long-term impact of an effective pharmacotherapeutic approach to AAA. As pharmacologically induced differences in the rate of aneurysm expansion become magnified over time, the need for surgical repair can also be used as a measurable endpoint of therapeutic effectiveness. Thus, a 3.0-cm-diameter AAA in an untreated control patient is expected to reach 5.0 cm in diameter by 48 months, whereas an aneurysm patient receiving a therapy with 100% effectiveness would not progress to require repair at any interval. Importantly, aneurysms in patients receiving a therapy with only 50% effectiveness would still not reach a diameter of 5.0 cm for 8 years (96 months), and aneurysms in patients receiving a therapy with a 75% effectiveness would not reach a diameter of 5.0 cm for 16 years (192 months). In each case, patients given a therapy that is only partially effective would still experience a substantial prolongation of the interval between initial detection and the potential need for repair.

A second consideration revealed by this analysis is that any therapy resulting in a 60% reduction in expansion rate will likely prolong the need for surgery for at least 5 years in all patients with an AAA of 4.0 cm in diameter or less in initial size (FIG. 4B). Similarly, any therapy that is 80% effective will prolong the need for surgery for at least 10 years in all patients with an AAA less than 4.0 cm in diameter. Finally, a therapy capable of reducing the rate of aneurysmal expansion by 90% would essentially eliminate the eventual need for operation for all patients with aneurysms less than 4.0 cm in diameter by prolonging the expected interval of expansion to 20 years or more. Although pharmacotherapy would not be intended for the treatment of patients with initial aortic aneurysm diameters of 5.0 cm or greater, it is also notable that a therapy with 95% effectiveness could postpone the need for operation for about 4 years in this group. In patients who might be at particularly high risk for operation, this interval might well be sufficient to prevent the need for aneurysm repair altogether.

Current data on life expectancy indicate that a substantial number of patients with small AAAs will succumb to other disease processes during the intervals described by this analysis. Thus, for a patient population anticipated to have an average age of 65 years at the time of initial AAA detection, these estimates translate into a highly significant reduction in the overall need for aneurysm repair in the general population. Most importantly, it is clear that significant benefits would be achieved even by pharmacotherapeutic approaches that have only partial effects (50% or more) on the rate of aneurysm expansion. These considerations illustrate how the overall management of AAA might be favorably influenced by coupling new pharmacologic treatment options with the development of population-based ultrasound screening programs for this disease.

Using doxycycline as a representative MMP inhibitor, design of a randomized, prospective, multicenter clinical trial in patients with small AAAs is currently under way. Because doxycycline exerts a broad range of effects, and because the clinical implications of aneurysm disease are substantial, such studies must be carefully designed. It is notable that MMPs are regulated at many different levels and that the potential effects of doxycycline on endogenous tissue inhibitors (TIMPs), *in vivo* activators of MMPs, or compensatory synthesis of matrix proteins within AAA tissue are still unknown. Moreover, aneurysm growth is often sporadic and the potential duration of therapy that might be necessary is only speculative at present. Thus, it is important to emphasize that the use of doxycycline for inhibition of aneurysmal disease cannot be recommended outside of carefully conducted clinical trials.

The proposed trial will compare the effect of two different doses of doxycycline with placebo, using the rate of aneurysm expansion as the principal endpoint. Secondary endpoints will include the ultimate extent of aneurysm expansion, the need for surgical repair, the risk of aneurysm rupture, as well as compliance and tolerance to pharmacologic therapy. A number of biological markers of aneurysm disease will also be assessed, including plasma levels of MMP-9, connective tissue degradation products, and proinflammatory cytokines to determine their utility in predicting a satisfactory response to therapy.[81–83] It is anticipated that a total of 350 patients will be recruited over the first 2 years of the study, with follow-up examinations planned at 6-month intervals up to 4 years. Successful completion of this effort will provide the first real insight into the clinical feasibility of MMP inhibition as a strategy for AAA disease.

## ACKNOWLEDGMENTS

This work was supported by Grants HL53409 (BTB) and HL56701 (RWT) from the National Heart, Lung, and Blood Institute, the Veterans Administration Merit Review (BTB), and the Pacific Vascular Research Foundation (BTB and RWT). Members of the Section of Vascular Surgery at Washington University helped to recruit patients for the clinical studies, and valuable technical assistance was provided by Drs. Yuri Perdisky, Yoshifumi Itoh, Dongli Mao, and John A. Curci.

## REFERENCES

1. GRANGE, J.J., V. DAVIS & B.T. BAXTER. 1997. Pathogenesis of abdominal aortic aneurysm: update and look toward the future. Cardiovasc. Surg. **5:** 256–265.
2. THOMPSON, R.W. 1996. Basic science of abdominal aortic aneurysms: Emerging therapeutic strategies for an unresolved clinical problem. Curr. Opin. Cardiol. **11:** 504–518.
3. BERNSTEIN, E.F., R.B. DILLEY, L.E. GOLDBERGER, B.B. GOSINK & G.R. LEOPOLD. 1976. Growth rates of small abdominal aortic aneurysms. Surgery **80:** 765–773.
4. KREMER, H., B. WEIGOLD, W. DOBRINSKI, M.A. SCHREIBER & N. ZOLLNER. 1984. Sonographische verlaufsbeobachtungen von bauchaortenaneurysmen. Klin. Wochenschr. **62:** 1120–1125.
5. BERNSTEIN, E.F. & E.L. CHAN. 1984. Abdominal aortic aneurysm in high-risk patients. Ann. Surg. **200:** 255–263.
6. CRONENWETT, J.L., T.F. MURPHY & G.B. ZELENOCK. 1985. Actuarial analysis of variables associated with the rupture of small abdominal aortic aneurysms. Surgery **98:** 472–483.
7. DELIN, A., H. OHLSEN & J. SWEDENBORG. 1985. Growth rate of abdominal aortic aneurysms as measured by computed tomography. Br. J. Surg. **72:** 530–532.
8. STERPETTI, A.V., R.V. SCHULTZ, R.J. FELDHAUS, S.E. CHENG & D.J. PEETZ. 1987. Factors influencing enlargement rate of small abdominal aortic aneurysms. J. Surg. Res. **43:** 211–219.
9. NEVITT, M.P., D.J. BALLARD & J.W. HALLETT. 1989. Prognosis of abdominal aortic aneurysms. N. Engl. J. Med. **321:** 1009–1014.
10. LIMET, R., N. SAKALIHASSAN & A. ALBERT. 1991. Determination of the expansion rate and incidence of rupture of abdominal aortic aneurysms. J. Vasc. Surg. **14:** 540–548.
11. WILSON, K.A., K.R. WOODBURN, C.V. RUCKLEY & F.G.R. FOWKES. 1997. Expansion rates of abdominal aortic aneurysms: Current limitations in evaluation. Eur. J. Vasc. Endovasc. Surg. **13:** 521–526.
12. HOLLIER, L.H., L.M. TAYLOR & J. OCHSNER. 1992. Recommended indications for operative treatment of abdominal aortic aneurysms. J. Vasc. Surg. **15:** 1046–1056.

13. COLE, C.W., G.B. HILL, J. LINDSAY, W.P. MICKELSON, C. MILLS & D.T. WIGLE. 1994. Proceedings of the workshop on the control of abdominal aortic aneurysm. Chronic Dis. Canada Suppl. **15:** S1–S64.

14. ALCORN, H.G., S.K. WOLFSON, K. SUTTON-TYRRELL, L.H. KULLER & D. O'LEARY. 1996. Risk factors for abdominal aortic aneurysms in older adults enrolled in the Cardiovascular Health Study. Arterioscler. Thromb. Vasc. Biol. **16:** 963–970.

15. LEDERLE, F.A., G.R. JOHNSON, S.E. WILSON, E.P. CHUTE, F.N. LITTOOY, D. BANDYK, W.C. KRUPSKI, G.W. BARONE, C.W. ACHER & D.J. BALLARD (for the Aneurysm Detection and Management [ADAM] Veterans Affairs Cooperative Study Group). 1997. Prevalence and associations of abdominal aortic aneurysm detected through screening. Ann. Intern. Med. **126:** 441–449.

16. U.S. BUREAU OF THE CENSUS. 1996. Current Population Reports: P25–1130.

17. COLLIN, J., L. ARAUJO & J. WALTON. 1990. A community detection program for abdominal aortic aneurysm. Angiology **41:** 53–58.

18. KATZ, D.A., B. LITTENBERG & J.L. CRONENWETT. 1992. Management of small abdominal aortic aneurysms: early surgery vs. watchful waiting. JAMA **268:** 2678–2686.

19. KATZ, D.A. & J.L. CRONENWETT. 1994. The cost-effectiveness of early surgery versus watchful waiting in the management of small abdominal aortic aneurysms. J. Vasc. Surg. **19:** 980–991.

20. SMALL ANEURYSM TRIAL PARTICIPANTS. 1995. The U.K. Small aneurysm trial: design, methods and progress. Eur. J. Vasc. Endovasc. Surg. **9:** 42–48.

21. SHAH, P.K. 1997. Inflammation, metalloproteinases, and increased proteolysis: an emerging paradigm in aortic aneurysm. Circulation **96:** 2115–2117.

22. PEARCE, W.H. & A.E. KOCH. 1996. Cellular components and features of immune response in abdominal aortic aneurysms. Ann. N. Y. Acad. Sci. **800:** 175–185.

23. BROPHY, C.M., J.M. REILLY, G.J.W. SMITH & M.D. TILSON. 1991. The role of inflammation in nonspecific abdominal aortic aneurysm disease. Ann. Vasc. Surg. **5:** 229–233.

24. ZATINA, M.A., C.K. ZARINS, B.L. GEWERTZ & S. GLAGOV. 1984. Role of medial lamellar architecture in the pathogenesis of aneurysms. J. Vasc. Surg. **1:** 442–448.

25. DOBRIN, P.B., W.H. BAKER & W.C. GLEY. 1984. Elastolytic and collagenolytic studies of arteries: Implications for the mechanical properties of aneurysms. Arch. Surg. **119:** 405–409.

26. CAMPA, J.S., R.M. GREENHALGH & J.T. POWELL. 1987. Elastin degradation in abdominal aortic aneurysms. Atherosclerosis **65:** 13–21.

27. BAXTER, B.T., G.S. MCGEE, V.P. SHIVELY, I.A. DRUMMOND, S.N. DIXIT, M. YAMAUCHI & W.H. PEARCE. 1992. Elastin content, cross-links, and mRNA in normal and aneurysmal human aorta. J. Vasc. Surg. **16:** 192–200.

28. WHITE, J.V., K. HAAS, S. PHILLIPS & A.J. COMEROTA. 1993. Adventitial elastolysis is a primary event in aneurysm formation. J. Vasc. Surg. **17:** 371–381.

29. CHANG, M.H. & M.R. ROACH. 1994. The composition and mechanical properties of abdominal aortic aneurysms. J. Vasc. Surg. **20:** 6–13.

30. DOBRIN, P.B. & R. MRKVICKA. 1994. Failure of elastin or collagen as possible critical connective tissue alterations underlying aneurysmal dilatation. Cardiovasc. Surg. **2:** 484–488.

31. WHITE, J.V. & S.D. SCOVELL. 1999. Etiology of abdominal aortic aneurysms: the structural basis for aneurysm formation. *In* Diagnosis and Treatment of Aortic and Peripheral Arterial Aneurysms. K.D. Calligaro, M.J. Dougherty & L.H. Hollier, Eds.: 3–12. W. B. Saunders, Philadelphia, PA.

32. VINE, N. & J.T. POWELL. 1991. Metalloproteinases in degenerative aortic disease. Clin. Sci. **81:** 233–239.

33. HERRON, G.S., E. UNEMORI, M. WONG, J.H. RAPP, M.H. HIBBS & R.J. STONEY. 1991. Connective tissue proteinases and inhibitors in abdominal aortic aneurysms: Involvement of the vasa vasorum in the pathogenesis of aortic aneurysms. Arterioscler. Thromb. **11**: 1667–1677.

34. REILLY, J.M., C.M. BROPHY & M.D. TILSON. 1992. Characterization of an elastase from aneurysmal aorta which degrades intact aortic elastin. Ann. Vasc. Surg. **6**: 499–502.

35. IRIZARRY, E., K.M. NEWMAN, R.H. GANDHI, G.B. NACKMAN, V. HALPERN, S. WISHNER, J.V. SCHOLES & M.D. TILSON. 1993. Demonstration of interstitial collagenase in abdominal aortic aneurysm disease. J. Surg. Res. **54**: 571–574.

36. NEWMAN, K.M., A.M. MALON, R.D. SHIN, J.V. SCHOLES, W.G. RAMEY & M.D. TILSON. 1994. Matrix metalloproteinases in abdominal aortic aneurysm: Characterization, purification, and their possible sources. Conn. Tiss. Res. **30**: 265–276.

37. NEWMAN, K.M., Y. OGATA, A.M. MALON, E. IRIZARRY, R.H. GANDHI, H. NAGASE & M.D. TILSON. Identification of matrix metalloproteinases 3 (stromelysin-1) and 9 (gelatinase B) in abdominal aortic aneurysm. Arterioscler. Thromb. **14**: 1315–1320.

38. THOMPSON, R.W., D.R. HOLMES, R.A. MERTENS, S. LIAO, M.D. BOTNEY, R.P. MECHAM, H.G. WELGUS & W.C. PARKS. 1995. Production and localization of 92-kD gelatinase in abdominal aortic aneurysms: an elastolytic metalloproteinase expressed by aneurysm-infiltrating macrophages. J. Clin. Invest. **96**: 318–326.

39. MCMILLAN, W.D., B.K. PATTERSON, R.R. KEEN, V.P. SHIVELY, M. CIPOLLONE & W.H. PEARCE. 1995. In situ localization and quantification of mRNA for 92-kD type IV collagenase and its inhibitor in aneurysmal, occlusive, and normal aorta. Arterioscler. Thromb. Vasc. Biol. **15**: 1139–1144.

40. FREESTONE, T., R.J. TURNER, A. COADY, D.J. HIGMAN, R.M. GREENHALGH & J.T. POWELL. 1995. Inflammation and matrix metalloproteinases in the enlarging abdominal aortic aneurysm. Arterioscler. Thromb. Vasc. Biol. **15**: 1145–1151.

41. MCMILLAN, W.D., B.K. PATTERSON, R.R. KEEN & W.H. PEARCE. 1995. In situ localization and quantification of seventy-two-kilodalton type IV collagenase in aneurysmal, occlusive, and normal aorta. J. Vasc. Surg. **22**: 295–305.

42. PATEL, M.I., J. MELROSE, P. GHOSH & M. APPLEBERG. 1996. Increased synthesis of matrix metalloproteinases by aortic smooth muscle cells is implicated in the etiopathogenesis of abdominal aortic aneurysms. J. Vasc. Surg. **24**: 82–92.

43. THOMPSON, R.W. & W.C. PARKS. 1996. Role of matrix metalloproteinases in abdominal aortic aneurysms. Ann. N.Y. Acad. Sci. **800**: 157–174.

44. TAMARINA, N.A., W.D. MCMILLAN, V.P. SHIVELY & W.H. PEARCE. 1997. Expression of matrix metalloproteinases and their inhibitors in aneurysms and normal aorta. Surgery **122**: 264–271.

45. ELMORE, J.R., B.F. KEISTER, D.P. FRANKLIN, J.R. YOUKEY & D.J. CAREY. 1998. Expression of matrix metalloproteinases and TIMPs in human abdominal aortic aneurysms. Ann. Vasc. Surg. **12**: 221–228.

46. DAVIS, V., R. PERSIDSKAIA, L. BACA-REGEN, Y. ITOH, H. NAGASE, Y. PERSIDSKY, A. GHORPADE & B.T. BAXTER. 1998. Matrix metalloproteinase-2 production and its binding to the matrix are increased in abdominal aortic aneurysms. Arterioscler. Thromb. Vasc. Biol. **18**: 1625–1633.

47. CURCI, J.A., S. LIAO, M.D. HUFFMAN, S.D. SHAPIRO & R.W. THOMPSON. 1998. Expression and localization of macrophage elastase (matrix metalloproteinase-12) in abdominal aortic aneurysms. J. Clin. Invest. **102**: 1900–1910.

48. LEE, E., A.J. GRODZINSKY, P. LIBBY, S.K. CLINTON, M.W. LARK & R.T. LEE. 1995. Human vascular smooth muscle cell-monocyte interactions and metalloproteinase secretion in culture. Arterioscler. Thromb. **15**: 2284–2289.

49. ANIDJAR, S., J.L. SALZMANN, D. GENTRIC, P. LAGNEAU, J.P. CAMILLERI & J.B. MICHEL. 1990. Elastase induced experimental aneurysms in rats. Circulation **82:** 973–981.

50. ANIDJAR, S., P.B. DOBRIN, M. EICHORST, G.P. GRAHAM & G. CHEJFEC. 1992. Correlation of inflammatory infiltrate with the enlargement of experimental aortic aneurysm. J. Vasc. Surg. **16:** 139–147.

51. GADOWSKI, G.R., M.A. RICCI, E.D. HENDLEY & D.B. PILCHER. 1993. Hypertension accelerates the growth of experimental aortic aneurysms. J. Surg. Res. **54:** 431–436.

52. GADOWSKI, G.R., D.B. PILCHER & M.A. RICCI. 1994. Abdominal aortic aneurysm expansion rate: effect of size and beta-adrenergic blockade. J. Vasc. Surg. **19:** 727–731.

53. HALPERN, V.J., G.B. NACKMAN, R.H. GANDHI, E. IRIZARRY, J.V. SCHOLES, W.G. RAMEY & M.D. TILSON. 1994. The elastase infusion model of experimental aortic aneurysms: synchrony of induction of endogenous proteinases with matrix destruction and inflammatory cell response. J. Vasc. Surg. **20:** 51–60.

54. THOMPSON, R.W., S. LIAO, D. PETRINEC, D.R. HOLMES, J.M. REILLY, H.G. WELGUS & W.C. PARKS. 1995. Sequential expression of metallogelatinases during elastase-induced aneurysmal degeneration of the rat aorta: correlations with aortic dilatation and the destruction of medial elastin (abstract). FASEB Journal **9:** A967.

55. RICCI, M.A., G. STRINDBERG, J.M. SLAIBY, R. GUIBORD, L.J. BERGERSEN, P. NICHOLS, E.D. HENDLEY & D.B. PILCHER. 1996. Anti-CD 18 monoclonal antibody slows experimental aortic aneurysm expansion. J. Vasc. Surg. **23:** 301–307.

56. DOBRIN, P.B., N. BAUMGARTNER, S. ANIDJAR, G. CHEJFEC & R. MRKVICKA. 1996. Inflammatory aspects of experimental aneurysms: effect of methylprednisolone and cyclosporin. Ann. N. Y. Acad. Sci. **800:** 74–88.

57. HOLMES, D.R., D. PETRINEC, W. WESTER, R.W. THOMPSON & J.M. REILLY. 1996. Indomethacin prevents elastase-induced abdominal aortic aneurysms in the rat. J. Surg. Res. **63:** 305–309.

58. PETRINEC, D., S. LIAO, D.R. HOLMES, J.M. REILLY, W.C. PARKS & R.W. THOMPSON. 1996. Doxycycline inhibition of aneurysmal degeneration in an elastase-induced rat model of abdominal aortic aneurysm: preservation of aortic elastin associated with suppressed production of 92-kD gelatinase. J. Vasc. Surg. **23:** 336–346.

59. HINGORANI, A., E. ASCHER, M. SCHEINMAN, W. YORKOVICH, P. DEPIPPO, C.T. LADOULIS & S. SALLES-CUNHA. 1998. The effect of tumor necrosis factor binding protein and interleukin-1 receptor antagonist on the development of abdominal aortic aneurysms in a rat model. J. Vasc. Surg. **28:** 522–526.

60. BIGATEL, D.A., J.R. ELMORE, D.J. CAREY, G. CIZMECI-SMITH, D.P. FRANKLIN & J.R. YOUKEY. 1999. The matrix metalloproteinase inhibitor BB-94 limits expansion of experimental abdominal aortic aneurysms. J. Vasc. Surg. **29:** 130–138.

61. CURCI, J.A., D. PETRINEC, S. LIAO, L.M. GOLUB & R.W. THOMPSON. 1998. Pharmacologic suppression of experimental abdominal aortic aneurysms: a comparison of doxycycline and four chemically-modified tetracyclines. J. Vasc. Surg. **28:** 1082–1093.

62. MOORE, G., S. LIAO, J.A. CURCI, B.C. STARCHER, R.L. MARTIN, R.T. HENDRICKS, J.J. CHEN & R.W. THOMPSON. 1999. Suppression of experimental abdominal aortic aneurysms by systemic treatment with a hydroxamate-based matrix metalloproteinase inhibitor (RS 132908). J. Vasc. Surg. **29:** 522–532.

63. MIRALLES, M., W.N. WESTER, G.A. SICARD, R.W. THOMPSON & J.M. REILLY. 1998. Indomethacin inhibits expansion of experimental aortic aneurysms via inhibition of the COX2 isoform of cyclooxygenase. J. Vasc. Surg. In press.

64. PENTLAND, A.P., S.D. SHAPIRO & H.G. WELGUS. 1995. Agonist-induced expression of tissue inhibitor of metalloproteinases and metalloproteinases by human macrophages is regulated by endogenous prostaglandin E2 synthesis. J. Invest. Dermatol. **104:** 52–57.

65. GOLUB, L.M., H.M. LEE, G. LEHRER, A. NEMIROFF, T.F. MCNAMARA, R. KAPLAN & N.S. RAMAMURTHY. 1983. Minocycline reduces gingival collagenolytic activity during diabetes: Preliminary observations and a proposed new mechanism of action. J. Periodont. Res. **18:** 516–526.

66. GOLUB, L.M., T.F. MCNAMARA, G.D'ANGELO, R.A. GREENWALD & N.S. RAMAMURTHY. 1987. A non-antibacterial chemically-modified tetracycline inhibits mammalian collagenase activity. J. Dent. Res. **66:** 1310–1314.

67. GREENWALD, R.A., L.M. GOLUB, N.S. RAMAMURTHY, M. CHOWDHURY, S.A. MOAK & T. SORSA. 1998. In vitro sensitivity of the three mammalian collagenases to tetracycline inhibition: Relationship to bone and cartilage degradation. Bone **22:** 33–38.

68. RYAN, M.E., N.S. RAMAMURTHY & L.M. GOLUB. 1996. Matrix metalloproteinases and their inhibition in periodontal treatment. Curr. Opin. Periodontol. **3:** 85–96.

69. GEARING, A.J., P. BECKETT, M. CHRISTODOULOU, M. CHURCHILL, J.M. CLEMENTS, M. CRIMMIN, A.H. DAVIDSON, A.H. DRUMMOND, W.A. GALLOWAY & R. GILBERT. 1995. Matrix metalloproteinases and processing of pro-TNF-alpha. J. Leuk. Biol. **57:** 774–777.

70. BROWN, P.D. 1998. Synthetic inhibitors of matrix metalloproteinases. *In* Matrix Metalloproteinases. W. C. Parks & R. P. Mecham, Eds. :243–261. Academic Press. San Diego, CA.

71. CROUT, R.J., H.M. LEE, K. SCHROEDER, H. CROUT, N.S. RAMAMURTHY, M. WIENER & L.M. GOLUB. 1996. The "cyclic" regimen of low-dose doxycycline for adult periodontitis: a preliminary study. J. Periodontol. **67:** 506–514.

72. AMIN, A.R., M.G. ATTUR, G.D. THAKKER, P.D. PATEL, P.R. VYAS, R.N. PATEL, I.R. PATEL & S.B. ABRAMSON. 1996. A novel mechanism of action of tetracyclines: effects on nitric oxide synthases. Proc. Natl. Acad. Sci. USA **93:** 14014–14019.

73. AMIN, A.R., R.N. PATEL, G.D. THAKKER, C.J. LOWENSTEIN, M.G. ATTUR & S.B. ABRAMSON. 1997. Post-transcriptional regulation of inducible nitric oxide synthase mRNA in murine macrophages by doxycycline and chemically modified tetracyclines. FEBS Lett. **410:** 259–264.

74. D'AGOSTINO, P., F. ARCOLEO, C. BARBERA, G. DI BELLA, M. LA ROSA, G. MISIANO, S. MILANO, M. BRAI, G. CAMMARATA, S. FEO & E. CILLARI. 1998. Tetracycline inhibits the nitric oxide synthase activity induced by endotoxin in cultured murine macrophages. Eur. J .Pharmacol. **346:** 283–290.

75. RAJAGOPALAN, S., X.P. MENG, S. RAMASAMY, D.G. HARRISON & Z.S. GALIS. 1996. Reactive oxygen species produced by macrophage-derived foam cells regulate the activity of vascular matrix metalloproteinases in vitro: implications for atherosclerotic plaque stability. J. Clin. Invest. **98:** 2572–2579.

76. UITTO, V.J., J.D. FIRTH, L. NIP & L.M. GOLUB. 1994. Doxycycline and chemically modified tetracyclines inhibit gelatinase A (MMP-2) gene expression in human skin keratinocytes. Ann. N. Y. Acad. Sci. **732:** 140–151.

77. HANEMAAIJER, R., T. SORSA, Y.T. KONTTINEN, Y. DING, M. SUTINEN, H. VISSER, V.W. VAN HINSBERGH, T. HELAAKOSKI, T. KAINULAINEN, H. RONKA, H. TSCHESCHE & T. SALO. 1997. Matrix metalloproteinase-8 is expressed in rheumatoid synovial fibroblasts and endothelial cells. Regulation by tumor necrosis factor-alpha and doxycycline. J. Biol. Chem. **272:** 31504–31509.

78. JONAT, C., F.Z. CHUNG & V.M. BARAGI. 1996. Transcriptional downregulation of stromelysin by tetracycline. J. Cell Biochem. **60:** 341–347.

79. THOMPSON, R.W., S. LIAO & J.A. CURCI. 1998. Therapeutic potential of tetracycline derivatives to suppress the growth of abdominal aortic aneurysms. Adv. Dent. Res. **12:** 159–165.
80. HAK, E., R. BALM, B.C. EIKELBOOM, G.J.M. AKKERSDIJK & Y. VAN DER GRAAF. 1996. Abdominal aortic aneurysm screening: an epidemiological point of view. Eur. J. Vasc. Endovasc. Surg. **11:** 270–278.
81. MCMILLAN, W.D. & W.H. PEARCE. 1999. Increased plasma levels of metalloproteinase-9 are associated with abdominal aortic aneurysms. J. Vasc. Surg. **29:** 122–127.
82. JUVONEN, J., H.M. SURCEL, J. SATTA, A.M. TEPPO, A. BLOIGU, H. SYRJALA, J. AIRAKSINEN, M. LEINONEN, P. SAIKKU & T. JUVONEN. 1997. Elevated circulating levels of inflammatory cytokines in patients with abdominal aortic aneurysm. Arterioscler. Thromb. Vasc. Biol. **17:** 2843–2847.
83. SATTA, J., T. JUVONEN, K. HAUKIPURO, M. JUVONEN & M.I. KAIRALUOMA. 1995. Increased turnover of collagen in abdominal aortic aneurysms, demonstrated by measuring the concentration of the aminoterminal propeptide of type III procollagen in peripheral and aortal blood samples. J. Vasc. Surg. **22:** 155–160.

# Effect of Matrix Metalloproteinase Inhibition on Progression of Atherosclerosis and Aneurysm in LDL Receptor–Deficient Mice Overexpressing MMP-3, MMP-12, and MMP-13 and on Restenosis in Rats after Balloon Injury

MARGARET FORNEY PRESCOTT,[a] WILBUR K. SAWYER,
JEAN VON LINDEN-REED, MICHAEL JEUNE, MARY CHOU,
SHARI L. CAPLAN, AND ARCO Y. JENG

*Metabolic and Cardiovascular Research Department, Novartis Institute for Biomedical Research, Summit, New Jersey 07901, USA*

ABSTRACT: The broad-spectrum MMP inhibitor CGS 27023A was tested to determine its potential as a therapy for atherosclerosis, aneurysm, and restenosis. LDL receptor–deficient (LDLr -/-) mice fed a high-fat, cholic acid–enriched diet for 16 weeks developed advanced aortic atherosclerosis with destruction of elastic lamina and ectasia in the media underlying complex plaques. Lesion formation correlated with a 4.6- to 21.7-fold increase in MMP-3, -12, and -13 expression. Treatment with CGS 27023A (p.o., b.i.d. at 50 mg/kg) had no effect on the extent of aortic atherosclerosis ($36 \pm 4\%$ versus $30 \pm 2\%$ in controls), but both aortic medial elastin destruction and ectasia grade were significantly reduced (38% and 36%, respectively, $p < 0.05$). In the rat ballooned-carotid-artery model, CGS 27023A (12.5 mg/kg/day via osmotic minipump) reduced smooth muscle cell migration at 4 days by 83% ($p < 0.001$). Intimal lesions were reduced by 85% at 7 days ($p < 0.001$), but intimal smooth muscle proliferation was unaffected, and inhibitory efficacy was lost with time. At 12 days, intimal lesion reduction was less potent (52%, $p < 0.01$). At 3 and 6 weeks, reductions of 11% and 4%, respectively, were not significant. This demonstrates that it is essential to include late time points when the ballooned-carotid-artery model is employed to ensure that lesion size does not "catch up" when a compound solely inhibits smooth muscle cell migration. In summary, MMP inhibitor therapy delayed but did not prevent intimal lesions, thereby demonstrating little promise to prevent restenosis. In contrast, MMP inhibitor therapy may prove useful to retard progression of aneurysm.

[a]Address for correspondence: Dr. Margaret F. Prescott, Novartis Institute for Biomedical Research, 556 Morris Avenue, LSB 2143, Summit, New Jersey 07901. Phone, 908/277-7506; fax, 908/277-4756; e-mail, margaret.prescott@pharma.novartis.com

179

## INTRODUCTION

Matrix metalloproteinases (MMPs) have been shown to be involved in many cellular processes thought to play a role in cardiovascular disease, including smooth muscle cell migration and extracellular matrix degradation.[1-3] The MMPs expressed in human atherosclerotic plaques or at sites of aneurysm include collagenases-1 and -3 (MMP-1 and -13), stromelysin-1 (MMP-3), gelatinase B (MMP-9), and metalloelastase (MMP-12).[4-7] Overexpression of MMP-2, -3 and -9 as well as several membrane-type MMPs has been demonstrated in preclinical models of restenosis.[8-11] In this study, we tested the broad-spectrum MMP inhibitor, CGS 27023A, in animal models of atherosclerosis, aneurysm, and balloon injury in an attempt to determine whether MMP inhibitors have therapeutic potential in these cardiovascular indications.

## METHODS

### Matrix Metalloproteinase Inhibitor

CGS 27023A is an orally active hydroxamic acid MMP inhibitor with potent activity against a variety of MMPs including the collagenases MMP-1 and -13 ($IC_{50}$ = 25 and 6 nM), stromelysin-1 or MMP-3 ($IC_{50}$ = 16 nM), the gelatinases MMP-2 and -9 ($IC_{50}$ = 10 and 8 nM), and metalloelastase or MMP-12 ($IC_{50}$ = 12 nM).[12] Inhibitory effect is achieved via interaction with the active-site zinc on the activated matrix metalloproteinase.[12] CGS 27023A has previously been shown to be chondroprotective and to inhibit osteoarthritis in both the rabbit menisectomy and the guinea pig models of spontaneous osteoarthritis.[13]

### LDL Receptor–Deficient Mouse Model of Atherosclerosis and Aneurysm

Male low-density lipoprotein (LDL) receptor–deficient mice (LDLr -/-) (B6129FI/LDLR; Jackson Laboratories, Bar Harbor, ME) were placed on a high-fat, high-cholesterol diet containing cholic acid (18% butter fat, 1% cholesterol, 0.5% cholic acid, 20% casein; DYETS, Bethlehem, PA) at 4–6 weeks of age. An additional group of LDLr -/- animals were fed normal chow in order to determine mRNA expression of MMPs in non-lesioned vessels ($n = 12$). High-fat-diet–fed animals were orally gavaged twice daily either with CGS 27023A (50 mg/kg, b.i.d.) ($n = 18$) or vehicle ($n = 18$). After 16 weeks of the respective diets, animals were anesthetized with pentobarbital, bled via cardiac puncture, and perfused *in situ* for 2 minutes with physiological saline solution at 90 mm Hg pressure.

For those animals allocated for mRNA analysis, the entire aorta from the atrial sinus to ileal bifurcation was immediately removed, cleaned of adventitial fat and connective tissue, and snap frozen ($n = 6$). The cDNA probes for MMP-3, -9, -12, and -13 were obtained by PCR amplification as described by Jeng *et al.*[14] Northern blots containing 10 μg of total aortic mRNA (2 pools per group, $n = 3$ aortae each pool) were used for hybridization with the four cDNAs.

Those animals allocated for histologic and morphologic study were perfusion-fixed an additional 5 minutes with 10% formalin immediately following *in situ* perfusion with saline solution, as described above. Immersion fixation in formalin was continued *ex vivo* for a minimum of 24 hours. Entire aortae were then carefully dissected from the animals, cleaned of adventitial fat and connective tissue, opened longitudinally, pinned, and photographed for image analysis of gross atherosclerotic lesion area. Gross lesion and total aortic area were determined using a computerized image analysis system (Empire Imaging, Milford, NJ) with Image ProPlus software (Media Cybernetics, Silver Spring, MD). Atherosclerotic lesion area was expressed as a percent of total aortic area.

Once photographed, aortae were cross-sectioned into 5-mm-long segments, processed, embedded in paraffin, and semi-thin-sectioned for histologic and morphologic evaluation of the lesion and aneurysm formation. Aneurysm formation was determined in two aortic arch segments per animal using four Verhoff-stained cross-sections per segment. Only those segments exhibiting intimal atherosclerotic lesion were included. Medial elastolysis and ectasia (dilatation into the adventitia) were quantitated using a modification of the method described by Carmeliet *et al.*[15] Elastolytic sites within any of the layers of medial elastic lamella were counted (including the internal and external elastic laminae) and the ectasia (dilatation and destruction) stage was determined using the system illustrated and described in FIGURE 1.

Plasma total cholesterol was measured using an enzymatic, colorimetric assay (SIGMA, St. Louis, MO).

### Rat Carotid Injury Model

The left common carotid artery of male Sprague Dawley rats (Taconic Farms) weighing 350–500 g was subjected to balloon-catheter injury ($n = 12$–18 per group). CGS 27023A was dissolved in 50% DMSO and administered at 12 mg/kg/day via Alzet 2ML2 or 2ML4 minipumps implanted subcutaneously. Compound or 50% DMSO vehicle was administered continuously from 5 days prior to balloon injury until sacrifice at 4, 7, 12, 21, and 42 days post injury. One-half hour prior to sacrifice, Evans Blue dye was administered i.v. to allow visualization of non-endothelialized areas. Carotid arteries were perfusion-fixed under pressure (100 mm Hg) and dissected from the animal and two samples from the central blue region of each carotid artery were processed for histologic study. Four cross sections of each left carotid artery were measured for intimal and medial area, with extent of intimal lesion expressed as the intimal/medial ratio. The four cross-sections selected for lesion measurement represented the largest lesions observed in each animal. Smooth muscle cell migration was determined by counting the number of smooth muscle cells that had migrated to the intima by day 4. Smooth muscle cell proliferation was determined at 7 days after balloon injury using PCNA immunostaining. The number of cells undergoing proliferation was quantitated and expressed as a percent of the total number of intimal cells.

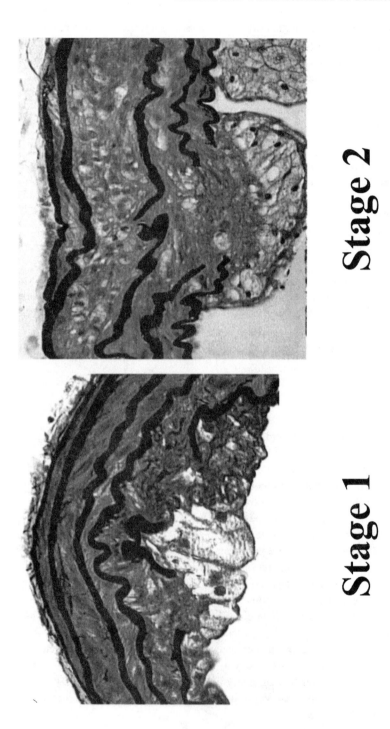

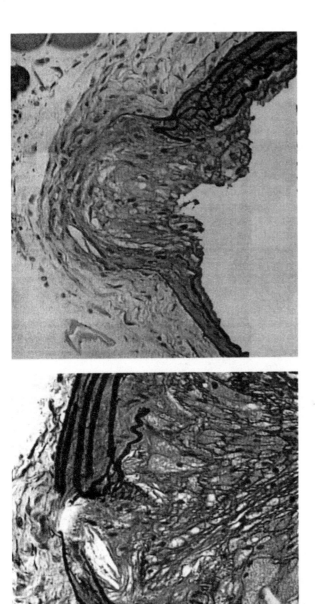

**FIGURE 1.** The method used to determine the stage of ectasia of the media underlying intimal atherosclerotic plaques (modified after Carmeliet *et al.*[15]). Stage 0: no elastolysis despite intimal lesion. Stage 1: elastolysis of only the internal elastic lamina (IEL). Stage 2: elastolysis of the IEL plus one or more elastic layers within the tunica media. Stage 3: elastolysis of the IEL, all elastic layers within the tunica media, as well as the external elastic lamina (EEL). Stage 4: elastolysis of all elastic layers and bulging into the adventitia.

## RESULTS

### *LDL Receptor–Deficient Mouse Model of Atherosclerosis and Aneurysm*

Development of advanced atherosclerotic lesions overlying areas of medial elastin destruction was observed in the aortae of untreated LDLr -/- mice after 16 weeks on a high-fat diet (FIG. 2). Complex atherosclerotic lesions contained lipid-filled foam cells, cholesterol clefts, and fibrous caps. Elastolysis and dilatation of the media were observed solely within the media underlying complex plaques in which cholesterol clefts were usually prominent. In contrast, LDLr -/- mice fed a normal diet for 16 weeks did not develop intimal atherosclerotic lesions, and no elastin degradation or dilatation was noted within the media.

The mRNA expression of MMP-3 (1.8 kb), MMP-12 (1.8 kb), and MMP-13 (2.9 kb) was markedly elevated in atherosclerotic aortae from LDLr -/- mice fed the high-fat diet. In contrast, message for these MMPs was either not present or was very low in aortae from LDL -/- mice on the normal diet that did not develop atherosclerosis (FIG. 3). Analysis by densitometry showed that the mRNA levels of MMP-3, -12, and -13 were increased by 4.6-, 21.7-, and 11.2-fold, respectively, in LDLr -/- mice fed the high-fat diet compared with those fed the normal diet (FIG. 4). No significant mRNA expression of MMP-9 was detected in LDLr -/- mice fed either the high-fat or normal diets (data not shown). No significant effect on the mRNA levels of MMP-3, -12, or -13 was observed following treatment with CGS 27023A.

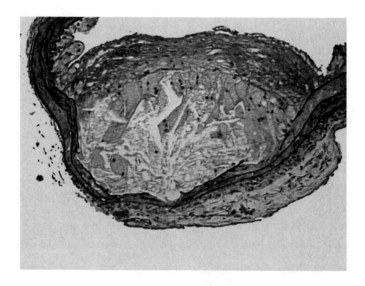

**FIGURE 2.** Abdominal aorta of an LDLr -/- mouse fed high-fat diet for 16 weeks. An advanced atherosclerotic lesion containing lipid-rich foam cells, cholesterol clefts, and a pronounced fibrous cap can be seen invading the underlying medial layer. Elastolysis of the internal elastic lamina and elastic lamella within the tunica media can be seen. Medial elastolysis was often observed underlying plaques containing extensive cholesterol cleft formation. Stage = 2.

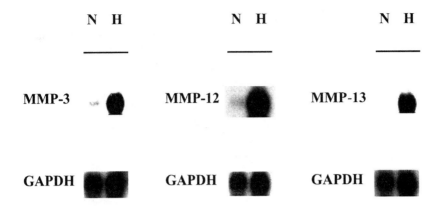

**FIGURE 3.** Northern blot analysis of MMP-3, -12, and -13 mRNA expression. Comparison was between aortae of LDLr -/- rats that had been fed a normal (N) or a high-fat diet (H) for 16 weeks. Aortic atherosclerosis and medial elastolysis developed in mice fed the high-fat diet, but not in those fed the normal diet.

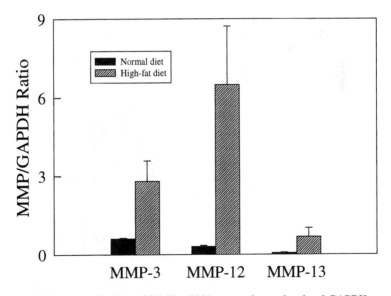

**FIGURE 4.** Normalization of MMP mRNA expression to levels of GAPDH expression. Comparison of levels in aorta from normal-diet-fed animals with those fed a high-fat diet demonstrated a 4.6-, 21.7-, and 11.2-fold increase in MMP-3, -12, and -13, respectively. No expression of MMP-9 was observed in aortae from either normal-diet or high-fat-diet-fed LDLr -/- mice (data not shown).

**TABLE 1.** Effect of CGS 27023A on plasma cholesterol, extent of aortic athero-sclerosis, medial elastic lamina cuts, and ectasia grade in LDLr -/- mice following 16 weeks of high-fat diet

| Group | Total cholesterol | Aortic atherosclerosis | Cuts in elastic lamina | Ectasia stage |
|-------|-------------------|------------------------|------------------------|---------------|
| Vehicle | 2458 ± 98 | 30% ± 2 | 4.1 ± 0.4 | 1.4 ± 0.1 |
| CGS 27023A | 2485 ± 157 | 36% ± 4 | 2.5 ± 0.6* | 0.9 ± 0.2* |

NOTE: CGS 27023A administered orally, b.i.d. at 50 mg/kg for 16 weeks concomitant with a high-fat diet. Values expressed as mean ± SEM.
   $*p < 0.05$.

CGS 27023A treatment had no effect on plasma cholesterol levels or on the extent of aortic atherosclerosis after 16 weeks of high-fat diet (TABLE 1). Gross lesion extent was similar in CGS 27023A-treated and vehicle-treated aortae (36% ± 4% versus 30% ± 2%, $p > 0.2$). In contrast, elastolysis and ectasia were significantly reduced by daily CGS 27023A treatment (TABLE 1). Elastolysis of one or more elastic layers within the media (the internal and/or external laminae, and/or the inner layers of elastic lamellae) was reduced by 38% ($p < 0.05$). Using the grading system in FIGURE 1 we showed the stage of medial ectasia (dilatation and destruction) to be significantly reduced by CGS 27023A treatment (36%, $p < 0.05$).

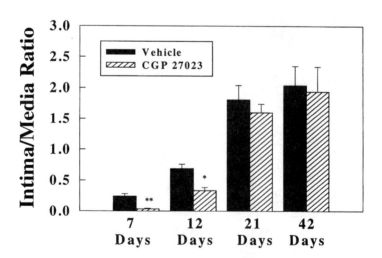

**FIGURE 5.** Intimal lesion size, expressed as intimal area/medial area, at 7, 12, 21, and 42 days after balloon injury to the carotid artery. The MMP inhibitor CGS 27023A reduced lesion size 85% at 7 days (**$p < 0.001$) and 52% at 12 days (*$p < 0.01$). No statistically significant effect on lesion size was observed at 21 or 42 days (3 or 6 weeks).

## Balloon Injury Studies

Treatment with CGS 27023A potently inhibited smooth muscle cell migration into the intima by 83% at 4 days after balloon injury ($p < 0.01$). Although intimal lesion size was significantly reduced at both 7 and 12 days after injury (FIG. 5), the effect was more pronounced at 7 days than at 12 days (reductions of 85% and 52%, respectively). PCNA analysis at 7 days after injury demonstrated similar rates of proliferation of intimal smooth muscle cells in CGS 27023A and vehicle-treated animals ($30 \pm 5\%$ versus $28 \pm 6\%$, respectively). By 3 weeks after ballooning, intimal lesion size in CGS 27023A-treated carotid arteries was reduced by only 11%, which was not statistically significant compared to vehicle-related animals ($p = 0.47$). At 6 weeks after ballooning, lesion size was similar in treated and vehicle control carotid arteries (FIG. 5).

## DISCUSSION

The aim of this study was to use the broad-spectrum MMP inhibitor CGS 27023A as a tool to determine the potential for MMP inhibitor therapy to prevent atherosclerosis, aneurysm, and restenosis. Treatment with the MMP inhibitor reduced both medial elastolyis and ectasia, suggesting that MMP inhibitor therapy might limit progression of aortic aneurysms. In contrast, despite the ability to potently reduce smooth muscle migration, MMP inhibitor therapy delayed, but did not prevent, restenotic lesion formation and had no effect on the development of atherosclerosis.

This study is the first to demonstrate that in LDLr -/- mice fed a high-fat diet, development of complex atherosclerosis and medial elastolysis (aneurysm) is accompanied by overexpression of MMP-3, -12, and -13. Similarly, in a companion article in this volume,[14] we demonstrated that development of advanced atherosclerosis and aneurysm in fat-fed ApoE-deficient mice was also accompanied by overexpression of MMP-3, -12, and -13. Our data are consistent with previous reports of MMP overexpression in human cardiovascular disease. MMP-3 and -13 have been shown to be expressed within advanced human atheroma, but not in normal vessels.[16,17]. Likewise, MMP-12 has recently been shown to be overexpressed within human abdominal aneurysms.[7] Our failure to demonstrate overexpression of MMP-9 in either fat-fed LDLr -/- or ApoE -/- aortae[14] was unanticipated since MMP-9 overexpression in human aneurysm has been widely reported.[4-6] Lack of demonstration of MMP-9 expression in the present and companion[14] studies does not appear to be due to lack of efficacy of our cDNA probe, since MMP-9 overexpression was demonstrated in mouse lungs in a TNFα-induced injury model using the same probe (Jeng and Prescott, unpublished material).

Although the MMP inhibitor CGS 27023A did not prevent development of advanced aortic atherosclerosis in the fat-fed LDLr -/- mice, treatment was found to reduce destruction of the elastic layers within the media and to retard subsequent ectasia (dilatation). These data are in agreement with those of Carmeliet *et al.*[15] who found that the extent of atherosclerosis was similar in fat-fed ApoE -/- mice and in fat-fed ApoE -/- mice crossed with uPA -/- mice (ApoE -/- × uPA -/-). They also demonstrated that uPA deficiency protected against medial elastolysis and ectasia.[15] The authors suggested that plasmin is a significant activator of pro-MMPs *in vivo* and hypothesized that reduced elastolysis observed in the ApoE -/- × PA -/- mice was due to reduced MMP activity.[15] Our demonstration of reduced elastolysis in LDLr -/- mice treated with CGS 27023A is also consistent with a recent report by Bigatel *et al.*[18] in which the broad-spectrum MMP inhibitor batimistat (BB-94) reduced development of elastase-induced abdominal aneurysm in rats by approximately 28%. The rather modest reduction in the extent of aneurysm following batimistat treatment,[18] as well as the 38% reduction we currently report with CGS 27023A, may be due to suboptimal pharmacokinetics. Alternatively, the presence of other elastin-degrading enzymes not inhibited by MMP inhibitor therapy could also account for the limited efficacy. Indeed, several cysteine and serine metalloproteases have been reported at sites of human atherosclerosis and aneurysm.[19–22]

We[23] and others[24,25] have previously demonstrated that MMP inhibitor therapy reduces smooth muscle cell migration after injury, but does not inhibit neointimal smooth muscle cell proliferation. Smooth muscle cell migration is the primary mechanism for formation of early lesions after balloon injury, whereas lesion expansion at later time points is due to neointimal smooth muscle cell proliferation and deposition of extracellular matrix.[26,27] Since CGS 27023A solely inhibited smooth muscle cell migration, it is logical that the intimal lesion size in treated rats would, with time, "catch up" to the lesion size observed in vehicle-treated rats. Our clear demonstration that efficacy was lost over time shows that when the rat carotid injury model is used to test potential anti-restenotic compounds, long term (>3 week) time points should be included to ensure that no "catch up" occurs.

Our finding that intimal lesion size in MMP inhibitor treated and untreated rats was similar at both 3 and 6 weeks after injury suggests that MMP inhibitor therapy did not result in excessive deposition of extracellular matrix. Since Tyagi *et al.*[28] had demonstrated that collagenase levels were reduced and TIMP-1 (an endogenous MMP inhibitor) levels were elevated in late human restenotic lesions, we had been concerned that MMP inhibitor therapy might exacerbate late lesion formation. Our study demonstrated, however, that as late as 6 weeks after injury, exacerbation was not observed. An analogous finding was recently reported using a dorsal skin incision model in which the MMP inhibitor GM6001 significantly increased wound strength, but did not result in hyperplastic scar formation.[29]

Similar to our findings with late lesions after balloon injury, the ability of CGS 2702A to potentially inhibit smooth muscle migration did not result in less extensive atherosclerosis. After 16 weeks of high-fat diet, LDLr -/- plaques contained lipid-filled macrophage foam cells, cholesterol clefts, and smooth muscle cell-rich fibrous caps of varying thickness. It is possible that MMP inhibitor therapy may result in increased extracellular matrix accumulation of the plaque fibrous cap and thus result in plaque "stabilization." A recent study in rabbits demonstrated that lipid lowering reduced MMP activity and increased collagen content of atheroma.[30] The composition of atherosclerotic lesions in fat-fed LDLr -/- mice after CGS 27023A treatment is currently being analyzed.

## ACKNOWLEDGMENTS

The inventor of CGS 27023A was David T. Parker, Novartis (formerly Ciba-Geigy) Pharmaceuticals.

## REFERENCES

1. PAULY, R.R., A. PASSANTI, C. BILATO, R. MONTICONE & L. CHENG *et al.* 1994. Migration of cultured vascular smooth muscle cells through a basement membrane barrier requires type IV collagenase activity and is inhibited by cellular differentiation. Circ. Res. **75**: 41–54.
2. KENAGY, R.D., S. VERGEL, E. MATTSSON, M. BENDECK, M.A. REIDY & A.W. CLOWES. 1996. The role of plasminogen, plasminogen activators, and matrix metalloproteinases in primate arterial smooth muscle cell migration. Arterioscler. Thromb. Vasc. Biol. **16**: 1373–1382.
3. WERB, Z., R.M. HEMBRY, G. MURPHY & J. AGGELLER. 1986. Commitment to expression of the metalloendopeptidases, collagenase and stromelysin: relationship of inducing events to changes in cytoskeletal architecture. Cell Biol. **102**: 697–702.
4. NEWMAN, K.M., J. JEAN-CLAUDE, H. LI, J.V. SCHOLES, Y. OGATA, H. NAGASE & M.D. TILSON. 1994. Cellular localization of matrix metalloproteinases in the abdominal aortic aneurysm wall. J. Vasc. Surg. **20**: 814–820.
5. DOLLERY, C.M., J.R. McEWAN & A.M. HENNEY. 1995. Matrix metalloproteinases and cardiovascular disease. Circ. Res. **77**: 863–868.
6. TAMARINA, N.A., W.D. MCMILLAN, V.P. SHIVELY & W.H. PEARCH. 1997. Expression of matrix metalloproteinases and their inhibitors in aneurysms and normal aorta. Surgery **122**: 264–271.
7. CURCI, J.A., S. LIAQ, A.D. HUFFMAN, S.D. SHAPIRO & R.W. THOMPSON. 1998. Expression and localization of macrophage elastase (matrix metalloproteinase-12) in abdominal aortic aneurysms. J. Clin. Invest. **102**: 1900–1910.
8. ZEMPO, N., R.D. KENAGY, Y.P. AU, M. BENDECK, M.M. CLOWES, M.A. REIDY & A.W. CLOWES. 1994. Matrix metalloproteinases of vascular wall cells are increased in balloon-injured rat carotid artery. J. Vasc. Surg. **20**: 209–217.
9. WEBB, K.E., A.M. HENNEY, S. ANGLIN, S.E. HUMPHRIES & J.R. McEWAN. 1997. Expression of matrix metalloproteinases and their inhibitor TIMP-1 in the rat carotid artery after balloon injury. Arterioscler. Thromb. Vasc. Biol. **17**: 1837–1844.
10. SHOFUDA, K., Y. NAGASHIMA, K. KAWAHARA, H. YASUMITSU, K. MIKI & K. MIYAZAKI. 1998. Elevated expression of membrane-type 1 and 3 matrix metalloproteinases in rat vascular smooth muscle cells activated by arterial injury. **78**: 915–923.
11. JENKINS, G.M., M.T. CROW, C. BILATO, Y. GLUZBAND, W.S. RYU, Z. LI, W. STETLER-STEVENSON, C. NATLER, J.P. FROEHLICH, E.G. LAKATTA & L. CHENG. 1998. Increased expression of membrane-type matrix metalloproteinase and preferential localization of matrix metalloproteinase-2 to the neointima of balloon-injured rat carotid arteries. Circulation **97**: 82–90.
12. MACPHERSON, L.J., E.K. BAYBURG, M.P. CAPARELLI, B.J. CARROLL, R. GOLDSTEIN, M.R. JUSTICE, L. ZHU, S. HU, R.A. MELTON, L. FRYER, R.L. GOLDBERG, J.R. DOUGHTY, S. SPIRITO, V. BLANCUZZI, D. WILSON, E.M. O'BYRNE, V. GANU & D.T. PARKER. 1997. Discovery of CGS 27023A, a non-peptidic, potent, and orally active stromelysin inhibitor that blocks cartilage degradation in rabbits. J. Med. Chem. **40**: 2525–2532.
13. O'BYRNE, E.O., V. BLANCUZZI, H. SINGH, L.J. MACPHERSON, D.T. PARKER & E.D. ROBERTS. 1999. Chrondroprotective activity of a matrix metalloproteinase inhibitor, CGS 27023A in animal models of osteoarthritis. *In* Advances in Osteoarthritis. S. Tanaka & C. Hamanishi, Eds.: 63–173. Springer-Verlag. Tokyo.

14. JENG, A.Y., CHOU, W.K. SAWYER, S.L. CAPLAN, J. VON LINDEN-REED, M. JEUNE & M.F. PRESCOTT. 1999. Enhanced expression of matrix metalloproteinase-3, -12, and -13 mRNAs in the aorta of apolipoprotein E–deficient mice with advanced atherosclerosis. Ann. N.Y. Acad. Sci. This volume.

15. CARMELIET, P., L. MOONS, R. LIJNEN, M. BAES, V. LEMAITRE, P. TIPPING, A. DREW, Y. EECKHOUT, S. SHAPIRO, F. LUPU & D. COLLEN. 1997. Urokinase-generated plasmin activates matrix metalloproteinases during aneurysm formation. Nature Genet. **17:** 439–444.

16. SCHONBECK, U., F. MACH, G.K. SUKKOVA, E. ATKINSON, E. LEVESQUE, M. HERMAN, P. GRABER, P. BASSET & P. LIBBY. 1999. Expression of stromelysin-3 in atherosclerotic lesions: regulation via CD40-CD40 ligand signaling in vitro and in vivo. J. Exp. Med. **189:** 843–853.

17. SUKHOVA, G., U. SCHOENBECK, E. RABKIN, F.J. SCHOEN, A.R. POOLE, R.C. BILLINGHURST & P. LIBBY. 1998. Colocalization of interstitial collagenses MMP-1 & MMP-13 with sites of cleaved collagen indicates their role in plaque rupture. Circulation **17:** 1–48.

18. BIGATEL, D.A., J.R. ELMORE, D.J. CAREY, G. CIZMECI-SMITH, D.P. FRANKLIN & J.R. YOUKEY. 1999. The matrix metalloproteinase inhibitor BB-94 limits expansion of experimental abdominal aortic aneurysms. J. Vasc. Surg. **29:** 130–138.

19. RAO S.K., K.V. REDDY & J.R. COHEN. 1996. Role of serine proteases in aneurysm development. Ann. N.Y. Acad. Sci. **800:** 131–137.

20. GACKO, M. & L. CHYCAEWSKI. 1997. Activity and localization of cathepsin B, D and G in aortic aneurysm. Int. Surg. **82:** 398–402.

21. GACKO, M. & S. GLOWINSKI. 1998. Cathepsin D and cathepsin L activities in aortic aneurysm wall and parietal thrombus. Clin. Chem. Lab. Med. **36:** 449–452.

22. BUCKMASTER, M.J., J.A. CURCI, P.R. MURRAY, S. LIAO, B.T. ALLEN, G.A. SICARD & R.W. THOMPSON. 1999. Source of elastin-degrading enzymes in mycotic aortic aneurysms: bacteria or host inflammatory response? Cardiovasc. Surg. **7:** 16–26.

23. PRESCOTT, M.F., W.K. SAWYER & J. VON LINDEN-REED. 1995. Matrix metalloproteinase inhibitor-induced reduction of smooth muscle migration does not inhibit late lesion formation following ballooning. FASEB J. **9:** A855.

24. ZEMPO, N. N. KOYAMA, R.D. KENAGY, H.J. LEA & W. CLOWES. 1996. Regulation of vascular smooth muscle cell migration and proliferation in vitro and in injured rat arteries by a synthetic matrix metalloproteinase inhibitor. Arterioscler. Thromb. Vasc. Biol. **16:** 28–32.

25. BENDECK, M.P., C. IRVIN & M.A. REIDY. 1996. Inhibition of matrix metalloproteinase activity inhibits smooth muscle cell migration but not neointimal thickening after arterial injury. Circ. Res. **78:** 38–43.

26. CLOWES, A.W., M.A. REIDY & M.M. CLOWES. 1983. Mechanisms of stenosis after arterial injury. Lab. Invest. **49:** 208–215.

27. CLOWES, A.W. & S.M. SCHWARTZ. 1985. Significance of quiescent smooth muscle migration in the injured rat carotid artery. Circ. Res. **56:** 139–145.

28. TYAGI, S.C., L. MEYER, R.A. SCHMALTZ, H.K. REDDY & D.J. VOELKER. 1995. Proteinases and restenosis in the human coronary artery: extracellular matrix production exceeds the expression of proteolytic activity. Atherosclerosis **116:** 43–57.

29. WITTE, M.B., F.J. THORNTON, T. KIYAMA, D.T. EFRON, G.S. SCHULTZ, L. MOLDAWER & A. BARBUL. 1998. Metalloproteinase inhibitors and wound healing: a novel enhancer of wound strength. Surgery **124:** 464–470.

30. AIKAWA, M., E. RABKIN, Y. OKADA, S.J. VOGLIC, S.K. CLINTON, C.E. BRINCKERHOFF, G.K. SUKHOVA & P. LIBBY. 1998. Lipid lowering by diet reduces metalloproteinase activity and increases collagen content of rabbit atheroma: a potential mechanism of lesion stabilization. Circulation **97:** 2433–2444.

# Treatment of Osteoporosis with MMP Inhibitors

S. WILLIAMS, J. BARNES, A. WAKISAKA, H. OGASA, AND C.T. LIANG[a]

*Gerontology Research Center, National Institute on Aging,
National Institutes of Health, Baltimore, Maryland 21224, USA*

ABSTRACT: In the current study, we examined the effects of minocycline on the osteopenia of ovariectomized (OVX) aged rats using the marrow ablation model. This injury induces rapid bone formation followed by bone resorption in the marrow cavity. Old female rats were randomly divided into five groups: sham, OVX, OVX+minocycline (5–15 mg/day, orally), OVX+17β-estradiol (25 µg/day, subcutaneously), and OVX+both agents. Rats were OVX, treated with minocycline and/or estrogen, followed by marrow ablation. Bone samples were collected 16 days post–marrow ablation. X-ray radiography of bones operated on showed that treatment of OVX old rats with minocycline increased bone mass in diaphyseal region. Diaphyseal bone mineral density (BMD) was measured by DEXA scan. Diaphyseal BMD of OVX rats was increased 17–25% by treatment with 5–15 mg of minocycline or 17β-estradiol. The effects of minocycline and estrogen treatments on the expression of osteoblast and osteoclast markers were also examined. Northern and dot blot analysis of RNA samples showed that treatment of OVX aged rats with minocycline increased the expression of type I collagen (COL I) (49%) and decreased that of interleukin-6 (IL-6) (31%). In contrast, estrogen treatment decreased the expression of interleukin-6 (IL-6) (39%), carbonic anhydrase II (CA II) (36%), and osteopontin (OP) (37%). Neither minocycline nor 17β-estradiol had an effect on the expression of osteocalcin (OC) and alkaline phosphatase (AP). To elucidate the mechanism by which minocycline prevented the loss of bone in OVX aged rats, we examined the colony-formation potential of bone marrow stromal cells in *ex vivo* cultures. Minocycline stimulated the colony-forming efficiency of marrow stromal cells derived from old animals. We have therefore concluded that the modest increase in BMD noted in OVX aged rats, in response to minocycline treatment, may be due to a change in bone remodeling that favors bone formation; and the anabolic effect of minocycline is likely due to its effect on the expression of COL I and/or the metabolism of osteoprogenitor cells.

## INTRODUCTION

The generalized loss of bone, the development of osteoporosis, and the subsequent occurrence of fractures all increase with age. While BMD declines in both men and women with age, many women start with lower bone mineral density and show an accelerated loss at menopause due to a decline in estrogen production. Current concepts suggest that estrogen deficiency leads to an increase in bone resorption,

[a]Address for correspondence: C. Tony Liang, Gerontology Research Center, National Institute on Aging National Institutes of Health, 4940 Easterns Avenue, Baltimore, Maryland 21224. Phone, 410/558-8468; fax, 410/558-8317; e-mail, liangt@grc.nia.nih.go

probably secondary to an increase in osteoclast number and collagenase activity stimulated by an increase in interleukin levels.[1–3] The causes of the decline of BMD with age are probably complex and include a reduction in bone formation activity.[4,5]

Previously, we tested a new therapy using minocycline to treat osteoporosis. The general concept for this approach is based on the inhibitory effect of minocycline on collagenase.[6] In a published report, we showed that minocycline can effectively prevent the loss of BMD and trabecular bone in OVX old rats.[7] Interestingly, in addition to its effect on inhibiting bone resorption, minocycline is capable of stimulating bone formation in these animals. The stimulatory effect of tetracyclines on bone formation has also been reported in other experimental models.[8]

In the marrow ablation model, the removal of bone marrow from the femur induces a rapid and massive cycle of bone formation and resorption.[9,10] Following marrow ablation, there is a rapid proliferation of mesenchymal cells, followed by the deposition of bone beginning at day 5[9,10]; the induction of collagenase on day 9[11] is followed by the rapid resorption of the newly deposited bone and the re-establishment of the bone marrow. Aged rats show significant deficits in the induction of genes involved in bone formation as well as the amount of bone deposited[11] owing, in part, to a deficiency of osteoprogenitor cells.[12,13]

In the current study, we tested the effect of minocycline treatment on the expression of osteoblast markers and correlated that to changes in BMD in femurs of OVX aged rats following bone marrow ablation. We showed that minocycline treatment stimulated the expression of specific genes related to bone formation and increased BMD in operated-upon femurs. Minocycline also stimulated the colony formation efficiency of cultured marrow stromal cells which are enriched with osteoprogenitor cells.

## MATERIALS AND METHODS

### *Animals*

Female Wistar rats, 22 months old, were supplied by the Animal Resource Facility at the Gerontology Research Center, National Institute on Aging. Rats were maintained *ad libitum* on National Institutes of Health rat chow, which contained 1.2% calcium and 0.95% phosphorus. They were housed in facilities allowing 12-hr light and 12-hr dark cycles. The animal protocol used in this study was reviewed and approved by the Animal Care and Use Committee at the Gerontology Research Center, National Institute on Aging.

### *Ovariectomy and Treatment with Estrogen and Minocycline*

Female rats were anesthetized with sodium pentobarbital intraperitoneally (i.p.) at a dose of 10 mg/kg. A transverse incision was made inferior to the rib cage on the dorsolateral body wall. The uterine tubes were exteriorized and clamped and the ovaries were excised. Treatment with minocycline (5–15 mg/d) and/or 17β-estradiol (24 μg/d) began on day 1 post ovariectomy and continued throughout the entire experiment. Bone marrow ablation was carried out at day 14 post ovariectomy.

### Bone Marrow Ablation

Bone marrow ablation was performed as described previously.[11] In brief: after anesthetizing each animal, an initial transverse incision at the anterior surface of the flexed knee was extended longitudinally along the medial surface of the joint. After exposing the intercondylar surface, a 1.6-mm burr hole was drilled into the marrow cavity. Marrow was aspirated by vacuum suction through Teflon tubing. The cavity was then flushed with sterile normal saline solution. The skin was closed with wound clips. Rats were sacrificed at day 16 following ablation. One leg from each rat was collected and stored in liquid nitrogen and or at $-75°$ for RNA preparation. The other leg was preserved in 10% buffered formalin and used for bone mineral density (BMD) scanning and X-ray radiography.

### X-Ray Radiography

Femurs from the OVX control and OVX+minocycline (10 mg/day) groups as described above were placed side by side and X-ray radiography was performed using a Siemens Mammomat-2 designed for use in mammography.

### BMD Measurements

Bone mineral density of the femurs was assessed by dual-energy X-Ray absorptiometry (DEXA) scanning. Scans were performed blinded, by a single examiner, and repeated once. Each scan was analyzed by the appendicular mode of the small animal program from LUNAR. BMD levels were compared using the unpaired *t*-test.

### RNA Preparation and Analysis

Bones were removed from liquid nitrogen and pulverized with a mortar and pestle that were prechilled on dry ice. Total RNA was extracted from the tissue using the procedure as described.[14] Total RNA was fractionated on a 1% agarose gel (FMC Co., Rockland, ME) and transferred to Genescreen Plus membrane (New England Nuclear, Boston, MA). Northern and dot blot analyses were carried out as described previously.[15]

### Bone Marrow Stromal Cells

Bone marrow stromal cells were collected from femurs and cultured as described.[16] Confluent primary cells were detached, pooled, and used for the first passage. Effect of minocycline on colony forming efficiency was determined.

## RESULTS

### X-Ray Radiographs

Radiographs of femurs from old OVX control animals and those OVX but treated with minocycline (10 mg/d) are shown in FIGURE 1. These samples were collected at day 16 after marrow ablation. The femoral samples from the OVX animals treated with minocycline showed increased density in diaphysis as well as other regions

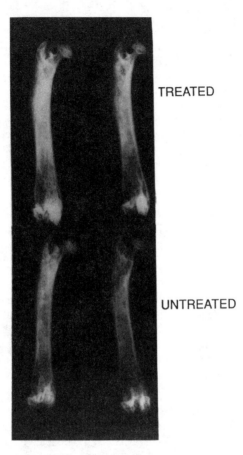

**FIGURE 1.** X-ray radiographs of femurs from OVX old rats treated with minocycline. Femurs were collected from OVX old rats at day 16 after marrow ablation. Two representative samples from the control and minocycline-treated groups are shown.

compared to those from the OVX controls. Bone histology was also examined at the mid section of femurs. Samples from OVX animals treated with minocycline showed more cancellous bone than did OVX control animals (data not shown).

### *Bone Density Measurements*

The effects of minocycline and 17β-estradiol on diaphyseal BMD in femurs of old OVX rats at day 16 after marrow ablation are shown in FIGURE 2. Minocycline treatment, at doses of 5–15 mg/d, increased BMD 17–25% ($p < 0.01$). As expected, the BMD levels of the 17β-estradiol treated rats were also significantly higher than in the OVX controls, ($p < 0.01$). However, combined treatment with minocycline (10 mg/d) and 17β-estradiol did not increase BMD above the levels seen with either agent alone.

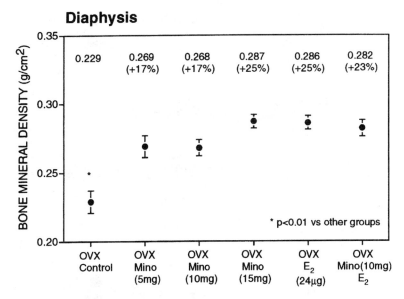

**FIGURE 2.** Effect of minocycline and estrogen treatment on BMD in femurs of old OVX rats. Diaphyseal BMD was determined by DEXA. Data shown are mean ± S.E. for 8–10 femurs from each groups.

### *Gene Expression*

Expression of genes associated with osteoblasts and osteoclasts was examined in femoral samples collected at day 16 after marrow ablation. As shown in TABLE 1, the expression of type I collagen (COL I) mRNA was increased with minocycline treatment of OVX rats ($p < 0.05$), while estrogen treatment did not alter COL I ex-

**TABLE 1. Effect of minocycline treatment on the expression of bone-related genes in aged female rats at day 16 after marrow ablation**

| | mRNA level (arbitrary unit) | | | | | |
|---|---|---|---|---|---|---|
| | COL I | OP | AP | OC | IL-6 | CAII |
| OVX control | 1.00 | 1.00 | 1.00 | 1.00 | 1.00 | 1.00 |
| OVX+minocycline | 1.49±0.08* | 0.92±0.05 | 0.92±0.05 | 1.10±0.05 | 0.69±0.04* | 0.90±0.05 |
| OVX+E$_2$ | 1.01±0.09 | 0.63±0.04* | 1.00±0.04 | 1.04±0.05 | 0.61±0.05* | 0.64±0.04* |
| OVX+minocycline+E$_2$ | 0.95±0.10 | 0.70±0.05* | 0.97±0.04 | 1.04±0.03 | 0.76±0.11* | 0.91±0.09 |

NOTE: OVX aged rats were treated with minocycline (10 mg/d) or estrogen 1 day after ovariectomy. Bone marrow ablation was performed at day 14 postovariectomy, and bone samples were collected at day 16 after marrow ablation. Femoral RNA samples were prepared, and the levels of mRNAs were determined by dot blot hybridization. The level of each mRNA was normalized with β-actin. Data shown are mean ± SE of 8–10 RNA samples from each group.
* $p < 0.05$ vs. OVX control, unpaired *t*-test.

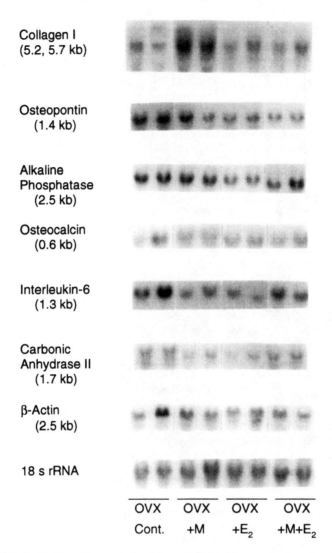

**FIGURE 3.** Northern blot analysis of femoral RNA. Two randomly selected samples from each groups are shown.

pression in OVX animals. The expression of interleukin-6 (IL-6) mRNA was signif-icantly decreased both by minocycline and estrogen treatment as compared to OVX animals. Interestingly, estrogen, but not minocycline, decreased the expression of the osteopontin (OP). The expression of carbonic anhydrase II (CA II) mRNA, an osteoclast-enriched marker, was significantly decreased with estrogen treatment. The expression of alkaline phosphatase (AP) and osteocalcin (OC) was also examined and no apparent effect by minocycline and estrogen treatment was observed. A typical Northern analysis of RNA samples was also shown (FIG. 3). In a separate experiment, femoral RNA samples were also prepared from rats 6 days after marrow ablation. Effects of minocycline and estrogen on the expression of COL I, OP, and IL-6 were similar to that observed above (data not shown).

### Colony-Forming Efficiency

To examine the mechanism by which minocycline treatment increases bone formation activity, we examined the effect of minocycline on colony formation of bone marrow stromal cells derived from old animals. Minocycline at 0.3–3 µg/ml stimulated the colony-forming efficiency in a dose-dependent manner (FIG. 4). The maximal stimulation was obtained at 3 µg/ml.

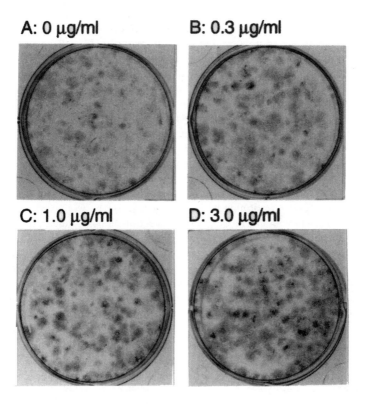

**FIGURE 4.** Colony-forming efficiency of marrow stromal cells. Effect of minocycline on colony-forming efficiency was examined in the first passage of bone marrow stromal cells derived from old rats.

## DISCUSSION

We were interested in testing the effects of minocycline, a known inhibitor of collagenase, on the density of bone from old animals subjected to both ovariectomy and marrow ablation. These models were employed since ovariectomy in the old animal mimics the accelerated loss of bone at menopause, while marrow ablation activates a cycle of bone formation followed by resorption[9,10]; and thus should increase the ease of assessing bone-enhancing agents. Previous work done in this laboratory has shown that the response to marrow ablation is related to a defined sequence in the expression of genes implicated in bone formation and resorption.[11] Of special interest is the fact that collagenase expression immediately precedes the onset of bone resorption in this system. This suggests that collagenase may play a role in resorption. This study documents the changes in the femurs of OVX aged rats subjected to bone marrow ablation injury and in some cases to treatment with minocycline and/or estrogen.

In preliminary experiments, we showed that minocycline treatment can increase the bone density in femurs collected from OVX aged rats. The effect of minocycline on BMD was then analyzed and compared to that of estrogen treatment. Ovariectomized animals receiving minocycline showed greater BMD in diaphysis than that observed in the untreated OVX animals. Indeed, diaphyseal BMD in the minocycline-treated animals was comparable to that of 17β-estradiol-treated group, particularly at the higher dose of 15 mg/day. Minocycline at 15 mg/day or 17β-estradiol at 24 μg/day not only prevented the loss in BMD associated with OVX, but also surpassed that observed in the sham controls (data not shown).

The effects of minocycline and estrogen on the expression of osteoblast and osteoclast markers were also examined. 17β-estradiol is known to suppress cytokine production, which reduces the formation of osteoclasts and the expression of collagenase,[1–3] while minocycline is a known inhibitor of collagenase[6] and collagenase gene expression is turned on at day 9 after bone marrow ablation and is coincident with the onset of bone resorption in this model.[11] However, while various observations are consistent with this mode of estrogen action on bone, minocycline appears to increase bone density primarily by some other means. In the current study, a different spectrum of effects on gene expression in bone was noted with estrogen as compared to minocycline. While both reduced the expression of IL-6, estrogen but not minocycline suppressed the expression of CA II, a marker for osteoclasts, and OP, an extracellular matrix protein that is also highly expressed in osteoclasts[17] and participates in osteoclast function and bone resorption.[18,19] Minocycline, but not estrogen, increased the expression of COL I, which is the most abundant extracellular protein in bone. These results are consistent with the proposed action of 17β-estradiol in suppressing osteoclast differentiation and thereby bone resorption. However, the effects observed in the minocycline-treated animals are not consistent solely with an inhibition of resorption as postulated,[20] but rather with a stimulation of bone formation.

To elucidate the mechanism by which minocycline stimulates bone formation, we examined the colony-formation efficiency of bone marrow stromal cells. It is well established that marrow stromal cells contain a small population of cells that can adhere to the plastic surface of the culture dish, express alkaline phosphatase, differentiate to osteoblasts, and form bone tissue when these cells are implanted to syngeneic host.[21] Because these cells have potential to form bone, we call them osteoprogenitor cells. The finding that minocycline stimulated the colony formation of marrow stromal cells derived from old animals suggests that minocycline may increase bone formation activity by upregulating the population of osteoprogenitor cells. Our finding that oral administration of minocycline can stimulate the femoral expression of COL I, the most abundant matrix protein in bone, is consistent with this hypothesis.

## REFERENCES

1. PACIFICI, R. *et al.* 1991. Effect of surgical menopause and estrogen replacement on cytokine release from human blood mononuclear cells. Proc. Natl. Acad. Sci. USA **88:** 5134–5138.
2. MUNDY, G.R. 1993. Role of cytokines in bone resorption. J. Cell. Biochem. **53:** 296–300.
3. MIYAURA, C. *et al.* 1995. Endogeneous bone-resorbing factors in estrogen deficiency: cooperative effect of IL-1 and IL-6. J. Bone Min. Res. **10:** 1365–1373.
4. IRVING, J.T. *et al.* 1981. Ectopic bone formation and aging. Clin. Orthop. **154:** 249–253.
5. NISHIMOTO, S.K. *et al.* 1985. The effect of aging on bone formation in rat: biochemical and histological evidence for decreased bone formation capacity. Calcif. Tissue Int. **37:** 617–624.
6. GOLUB, L.M. *et al.* 1991. Tetracyclines inhibit connective tissue breakdown: new therapeutic implications for an old family of drugs. Crit. Rev. Oral Biol. Med. **2:** 297–321.
7. WILLIAMS, S. *et al.* 1996. Minocycline prevents the decrease in bone mineral density and trabecular bone in ovariectomized aged rats. Bone **19:** 637–644.
8. BAIN, S. *et al.* 1997. Tetracycine prevents cancellous bone loss and maintains near-normal rate of bone formation in streptozotocin diabetic rats. Bone **21:** 147–153.
9. AMSEL, S. *et al.* 1969. The significance of intramedullary cancellous bone formation in repair of bone marrow tissue. Anat. Rec. **164:** 103–112.
10. PATT, H.M. *et al.* 1975. Bone marrow regeneration after local injury: a review. Exp. Hemat. **3:** 135–148.
11. LIANG, C.T. *et al.* 1992. Impaired bone activity in aged rats: alterations at the cellular and molecular levels. Bone **13:** 435–441.
12. TANAKA, H. *et al.* 1996. Mitogenic activity but not phenotype expression of rat osteoprogenitor cells in response to IGF-I is impaired in aged rats. Mech. Aging Dev. **92:** 1–10.
13. QUARTO, R. *et al.* 1995. Bone progenitor cells deficits and the age-associated decline in bone repair capacity. Calcif. Tissue Int. **56:** 123–129.
14. TANAKA, H. *et al.* 1994. In vivo and in vitro effects of insulin-like growth factor-I (IGF-I) on femoral mRNA expression in old rats. Bone **15:** 647–653.
15. TANAKA, H. *et al.* 1996. Effect of age on the expression of insulin-like growth factor-I, interleukin-6 and transforming growth factor-β mRNAs in rat femurs following marrow ablation. Bone **18:** 473–478.
16. TANAKA, H. *et al.* 1995. Effect of platelet-derived growth factor on DNA synthesis and gene expression in bone marrow stromal cells derived from adult and old rats. J. Cell. Physiol. **164:** 367–375.
17. MERRY, K. *et al.* 1993. Expression of osteopontin mRNA by osteoblasts and osteoclasts in modeling adult human bone. J. Cell. Sci. **104:** 1013–1020.
18. REINHOLT, F.P. *et al.* 1990. Osteopontin-a possible anchor of osteoclasts to bone. Proc. Natl. Acad. Sci. USA **83:** 4473–4475.

19. ARAI, N. *et al.* 1993. Osteopontin expression during bone resorption: an in situ hybridization study of induced ectopic bone in the rat. Bone Minerals **22:** 129–145.
20. RIFKIN, B.R. *et al.* 1994. Modulation of bone resorption by tetracyclines. Ann. N.Y. Acad. Sci. **723:** 165–179.
21. OWEN, M.E. *et al.* 1987. Clonal analysis in vitro of osteogenic differentiation of marrow stromal CFU-F. J. Cell. Sci. **87:** 731–738.

# Clinical Trials of a Stromelysin Inhibitor

## Osteoarthritis, Matrix Metalloproteinase Inhibition, Cartilage Loss, Surrogate Markers, and Clinical Implications

RICHARD L. LEFF[a]

*Bayer Corporation, 400 Morgan Lane, West Haven, Connecticut 06516, USA*

Articular cartilage is composed of an abundant extracellular matrix that is rich in collagen and sulfated proteoglycans. The containment of proteoglycans within the collagen network provides cartilage with the compressibility and elasticity necessary to protect and cushion the subchondral bone. During the development of osteoarthritis, the physical characteristics of the cartilage matrix become disrupted and a loss of collagen and proteoglycan from cartilage occurs, which is the hallmark of the disease.[1]

A family of structurally related matrix-degenerating proteinases, the matrix metalloproteinases (MMPs), has been implicated in the pathophysiology of osteoarthritis and rheumatoid arthritis, as well as in other arthritides.[2–5] A variety of cell types, including chondrocytes and synoviocytes, secrete these matrix proteinases *in vivo* and *in vitro,* and disease is associated with an increase in the concentrations of MMPs in plasma and synovial fluid. MMP-3 (stromelysin-1) and other cartilage-degrading enzymes have varying and overlapping activities in animal and human cartilage. Inhibition of the activity of these degradative enzymes may halt or slow the progression of osteoarthritis, and ameliorate the course of rheumatoid arthritis. However, no therapy specifically designed to inhibit MMPs—and thus prevent cartilage loss—is yet proven for the treatment of osteoarthritis.

Steps toward developing an MMP inhibitor for the potential treatment of osteoarthritis, as discussed in this article, are:

(1) selecting a molecular target;

(2) performing preclinical and early clinical trials;

(3) identifying short-term markers of cartilage loss;

(4) identifying long-term measures of cartilage loss; and

(5) establishing clinical implications of slowing cartilage loss.

## STROMELYSIN AS A MOLECULAR TARGET

Stromelysin-1 (MMP-3) was selected as a key molecular target for inhibition because it degrades collagen and aggrecan and because it activates other MMPs.[5] Col-

---

[a]Address for telecommunication: Phone, 203/812-5140; fax, 203/812-5306.

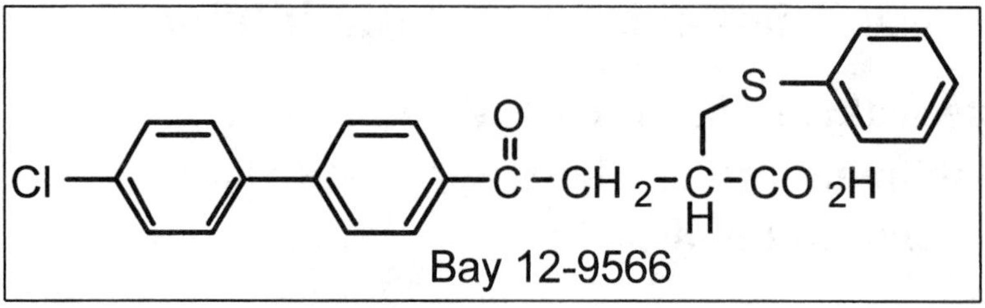

**FIGURE 1.** Non-peptidic chemical structure of BAY 12-9566.

lagenase-1 (MMP-1) inhibition was negatively selected against, because of its widespread expression and potential for adverse effects.

BAY 12-9566, a stromelysin inhibitor, was discovered at Bayer Corporation. BAY 12-9566 has a unique, non-peptidic, biphenyl structure (FIG. 1), and inhibits MMP-3 activity *in vivo* and *in vitro*. BAY 12-9566 also inhibits other MMPs such as MMP-2, -8, -9, and -13. Hence, BAY 12-9566 is being developed as a possible treatment to halt the progression of osteoarthritis. In preclinical experiments, BAY 12-9566 at micromolar concentrations inhibited cartilage plug degradation. Nanomolar concentrations of BAY 12-9566 inhibited release of cartilage breakdown products from equine cartilage explants.[6]

## PRECLINICAL ANIMAL STUDIES AND EARLY CLINICAL TRIALS

BAY 12-9566 has been tested in animal models of osteoarthritis, including guinea pig and canine models of traumatic osteoarthritis.[7] At daily doses of 1–15 mg/kg, BAY 12-9566 was effective at reducing cartilage loss by 50–60% compared with controls, as measured by cartilage histopathology. Animal models of osteoarthritis have demonstrated efficacy by allowing researchers to take advantage of pathologic examinations that are not feasible in human studies. Therefore, human studies must use surrogate markers that represent cartilage loss over long periods of time.

To date, BAY 12-9566 has been tested in studies with 300 osteoarthritis patients for up to 3 months. BAY 12-9566 is administered in a convenient, once-daily dose, does not need to be taken with food, and appears to be well tolerated. It penetrates into synovial fluid at approximately 60% of serum level, most likely because of protein binding. BAY 12-9566 has been detected in human cartilage in studies in patients undergoing elective joint replacement.

Examination for safety remains mainly empirical, but the lack of severe musculoskeletal findings thus far with this compound is promising. No musculoskeletal adverse effects have been observed to date in human trials. Such adverse effects are not expected in future trials, given the absence of such events in past studies with BAY 12-9566. Moreover, musculoskeletal abnormalities have not been observed in long-term animal models. With regard to safety, only mild to moderate, reversible, asymp-

D 97-004 Platelets

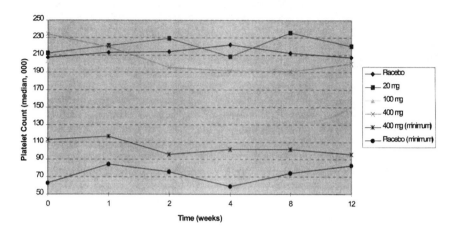

**FIGURE 2.** BAY 12-9566 in osteoarthritis patients produced a mild reduction in platelet counts at the highest dose. Shown are the median and lowest platelet counts in a 3-month study with osteoarthritis patients (D97-004) over time given placebo or BAY 12-9566 at doses of 20 mg, 100 mg, or 400 mg taken once daily. The normal platelet range is 130,000–400,000 per cubic millimeter. A mild decrease is apparent by 2 weeks for only those patients receiving the 400-mg dose.

tomatic laboratory abnormalities have been reported to date. For example, in human studies, mild, reversible lowering of platelet counts was observed with no associated bleeding (FIG. 2).

## SHORT-TERM TESTING OF BAY 12-9566 IN OSTEOARTHRITIS

The goal of early clinical development of BAY 12-9566 in osteoarthritis is to test an MMP inhibitor in humans to determine whether there is sufficient evidence of efficacy over relatively short periods of time to warrant long-term human testing. "Short-term" osteoarthritis trials generally refer to trials of less than 1 year.[8] The long-term goal is to find a treatment for the loss of cartilage in osteoarthritis.

Human osteoarthritis is the most common disabling disease, and can affect any diarthrodial joint: hands, knees, hips, spine, etc. Fortunately, for most people, cartilage loss is slowly progressive. Knee and hip cartilage can be measured by the joint space width on standard radiography. Joint space width in healthy adult knees and hips is typically 4–6 mm, which includes two surfaces of hyaline cartilage. Cartilage loss (as measured by joint space narrowing) in large joints in osteoarthritic patients is slow, on average 0.2 mm per year.[9] Cartilage loss alone poorly correlates with clinical disease, however, and clinical worsening is often difficult to document until the joint is replaced.

An ideal cartilage-preserving treatment would safely slow the progression of disease and be convenient to administer. Whereas safety can be assessed empirically, efficacy should be determined by *a priori* variables that assess whether a compound is effective and at what dose. Testing any MMP inhibitor for the treatment of human osteoarthritis should focus on cartilage preservation, particularly:

    (1)  biochemical markers of cartilage metabolism;
    (2)  other indirect markers of cartilage loss (synovial fluid volume as observed by magnetic resonance imaging [MRI], ultrasound [US], and c-reactive protein [CRP]);
    (3)  physical measures of cartilage (X-ray, MRI, US); and
    (4)  symptoms.

Variables 1 and 2 listed above are postulated to change more quickly with an MMP inhibitor than are the subsequent variables; therefore these first two variables are reasonable surrogate endpoints for short-term trials in osteoarthritic patients.

There are several unknowns for any of these four variables, including:

    (1)  length of time necessary for occurrence of a change in the variable;
    (2)  heterogeneity of different joints;
    (3)  heterogeneity of different cartilage surfaces within a joint;
    (4)  difficulty with sampling and registering identical areas of cartilage;
    (5)  variation of fluxes among the compartments of the body (cartilage, synovial fluid, blood, and urine);[10]
    (6)  underlying heterogeneity of osteoarthritis in humans.[1]

Although the clinical relevance of change detected in any of these surrogate endpoints for short-term testing in osteoarthritic patients is unknown, the following may be useful in determining what dose of an MMP inhibitor may be effective in human osteoarthritis:

    (1)  *Ex vivo* assays (ideally with human cartilage) for a decrease in cartilage degradation.
    (2)  Decrease in markers of cartilage degradation relative to markers of synthesis in cartilage, synovial fluid, blood, and urine.
    (3)  Changes in water content/quality and cartilage volume as revealed by MRI.
    (4)  Changes in synovial fluid volume as revealed by MRI or US.
    (5)  Subtle changes in the physicochemical properties of cartilage as revealed by high-resolution US, electromechanical flow, or indentation resistance.

Surrogate endpoints are not validated, because to date no agent has been demonstrated to preserve cartilage in humans. Such surrogate endpoints could include markers of cartilage metabolism (soluble and *in situ*). There are, however, inherent difficulties in measuring such markers. The serum concentration of soluble markers can be altered by a number of processes, such as changes in clearance from the synovial fluid, while *in situ* markers in cartilage are difficult to obtain and may reflect selection bias.

Early clinical testing of MMP inhibitors in humans therefore should focus on: (1) surrogate endpoints, as mentioned above, and (2) clinical effects resulting from altering cartilage metabolism and subsequently symptoms. Tetracycline-related compounds are not reported to have a large clinical benefit to osteoarthritis patients, so large symptomatic changes are unlikely to occur if such compounds are administered as osteoarthritis therapy. Clinical effects could be demonstrated if the loss of cartilage is inhibited, resulting in a smaller amount of breakdown products being released. Beneficial changes from MMP inhibition that could result in clinical improvement in short-term studies could reflect less cartilage breakdown, leading to:

(1) decreased release of cartilage breakdown products into synovial fluid;
(2) lower synovial fluid volume;
(3) less synovial inflammation;
(4) decreased stretching of capsule;
(5) alteration in ligaments of stretch receptors or insertion sites;
(6) subtle changes in bone (subchondral or periosteal).

## GOAL OF LONG-TERM TESTING OF BAY 12-9566 IN OSTEOARTHRITIS

"Long-term" refers to trials in osteoarthritis patients of greater than 1 year (typically 2–3 years or more) that demonstrate whether an MMP inhibitor has benefit by altering validated, longitudinal measures of cartilage loss.[11] The means of such measures of cartilage loss are listed in TABLE 1, along with some important advantages and disadvantages.

Of the methods listed in the table, X-rays represent the generally accepted approach, while MRI offers great promise because of its ability to examine all the structures of a joint.

**TABLE 1. Advantages and disadvantages of methods to measure cartilage**

| Cartilage Measure | Advantages | Disadvantages |
|---|---|---|
| X-ray | Reproducibility and longitudinal changes known; widely used; generally accepted by regulatory agencies | Indirectly examines cartilage by imaging bone; difficult to discern slow changes over time; limited clinical correlations |
| MRI | Examines entire joint, including underlying bone | Reproducibility and change over time unknown; hardware and software often change |
| Arthroscopy | Directly views cartilage | Invasive; difficult to standardize reading; unable to image below surface |
| Ultrasonography | Images cartilage, including below the surface | Can be invasive (intraarticular); difficult to register location |
| Biochemical markers | Directly analyzes content of cartilage or released fragments | Flux among compartments varies; relevance to cartilage structure preservation not determined |

## MEASURES OF CARTILAGE LOSS AND SYMPTOM
## PROGRESSION IN LONG-TERM TESTING

Selection of the target patient population and the structural and clinical endpoints are essential for long-term clinical testing in humans. Improved radiographic measures and clinical parameters that are correlated over time will be important for the development of disease/structure-modifying drugs. Current tools for measuring cartilage loss (structural measures) are crude, poorly correlated with clinical outcome, and not validated in humans. There is not a high correlation between indirect measures of cartilage loss, such as biochemical markers, and symptoms.

Improved tools to measure cartilage loss will become validated as newly designed drugs undergo testing in humans. Standardized X-rays that are precisely performed with patients in clinically relevant positions[9] and MRI[12] are likely to be the best measures of structural changes in osteoarthritis in coming years. Although cartilage may vary among and within synovial joints, the basic underlying degradative process in osteoarthritis is likely to be similarly altered by MMP inhibition in these joints.

## SUMMARY

Long-term trials with BAY 12-9566, a stromelysin inhibitor, have been initiated in osteoarthritis. Validation of the long-term, clinical relevance of early markers has to be tested in these and other trials. Detection of the slowing of cartilage loss (as gauged by measures of joint space narrowing and by other techniques, such as MRI) remains to be proven, but now may be possible in intervention trials.

## ACKNOWLEDGMENTS

I would like to acknowledge Charles Maniglia, Harold Kluender, and all the Bayer researchers in West Haven, Connecticut, who discovered a new class of potentially useful matrix metalloproteinase inhibitors that may increase the understanding of a wide variety of diseases and, it is hoped, result in novel treatments. Special thanks are also extended to the many Bayer development employees, investigators, consultants, and patients who participated and will participate in clinical trials to determine the role of BAY 12-9566 in human osteoarthritis.

## REFERENCES

1. DEQUEKER, J. & L. VAN DE PUTTE. 1994. Disorders of bone, cartilage, and connective tissue. *In* Rheumatology. J.H. Klippel & P.A. Dieppe, Eds. Mosby-Year Book Europe Ltd. London.
2. WOESSNER, J.F., JR. & Z. GUANJA-SMITH. 1991. Role of metalloproteinases in human osteoaritis. J. Rheumatol. Suppl. 27: 99–101.
3. BIRKEDAL-HANSEN, H., W.G. MOORE, M.K. BODDEN, L.J. WINDSOR, B. BIRKEDAL-HANSEN, A. DECARLO & J.A. ENGLER. 1993. Matrix metalloproteinases: a review. Crit. Rev. Oral Biol. Med. 4(2): 197–250.

4. LOHMANDER, L.S., L.A. HOERRNER & M.W. LARK. 1993. Metalloproteinases, tissue inhibitor, and proteoglycan fragments in knee synovial fluid in human osteoarthritis. Arth. Rheum. **36**(2): 181–189.

5. NAGASE, H. 1997. Activation of matrix metalloproteinases. Biol. Chem. **378:** 151–160.

6. BILLINGHURST R.C., K. O'BRIEN, A.R. POOLE & C.W. MCILWRAITH. Inhibition of articular cartilage degradation in culture by a novel non-peptidic matrix metalloproteinase inhibitor. This volume.

7. CHAU, T., G. JOLLY, J.M. PLYM et al. 1998. Inhibition of articular cartilage degradation in dog and guinea pig models of osteoarthritis by the stromelysin inhibitor, BAY 12-9566. Arth. Rheum. **41**(9S): S300.

8. 1996. Design and conduct of clinical trials in patients with osteoarthritis: Recommendations from a task force of the Osteoarthritis Research Society. Special Report. Osteoarth. Cartil. **4:** 217–243.

9. BUCKLAND-WRIGHT, J.C. 1997. Current status of imaging procedures in the diagnosis, prognosis and monitoring of osteoarthritis. Baillieres Clin. Rheumatol. **11**(4): 727–748.

10. MYERS, S.I., B.L. O'CONNOR & K.D. BRANDT. 1996. Accelerated clearance of albumin from the osteoarthritis knee: implications for interpretation of concentrations of "cartilage markers" in synovial fluid. J. Rheumatol. **23**(10): 1744–1747.

11. OMERACT III. 1997. Outcome measures in arthritis clinical trials. Cairns, Australia, April 16–19, 1996. Proceedings. J. Rheumatol. **24**(4): 763–802.

12. LOEUILLE, D., P. OLIVIER, D. MAINARD, P. GILLET, P. NETTER & A. BLUM. 1998. Review: magnetic resonance imaging of normal and osteoarthritic cartilage. Arth. Rheum. **41**(6): 963–975.

# Experimental Models to Identify Antimetastatic Drugs: Are We There Yet?

## A Position Paper

STANLEY ZUCKER[a]

*Veterans Administration Medical Cancer, Northport, New York 11768, USA*

*State University of New York at Stony Brook, Stony Brook, New York 11794, USA*

## IDENTIFICATION OF ANTICANCER AGENTS

The linchpin to the development of new types of anticancer treatment is the establishment of experimental models that accurately mimic events of human pathophysiology. Let us recall that more than 20 years elapsed before scientists came to appreciate the limitations of experimental models used for standard chemotherapy drug screening. Many effective chemotherapeutic agents in use today languished for decades on pharmacologists' shelves because these drugs showed limited antiproliferative activity in the standard L-1210 mouse leukemia model used to identify potentially active agents.

Consider the huge amount of time, money, and, most importantly, talent that has been invested in clinical testing of antiproliferative, cytocidal agents in the treatment of solid forms of cancer. On the basis of results of animal models, an entire industry developed around the principle that killing rapidly growing cancer cells is the key to curing cancer only to realize that this is not the case with the most common organ sites of human cancer (lung, prostate, breast, and GI tract). As we subsequently learned, rapid cell proliferation is not a consistent finding with most types of solid cancer. Recently we have learned that defects in the apoptosis pathway in cancer cells may be a more important factor in controlling tumor growth. The limited effectiveness of antiproliferative drugs has been a stimulus for the search for other approaches to treat solid cancer.

## BASIC CONCEPTS OF CANCER METASTASIS DERIVED FROM EXPERIMENTAL MODELS

It is generally agreed that the dissemination of cancer cells to distant sites in the body (metastasis) is best evaluated in an intact animal. Small animals, primarily mice and rats, were employed initially because of cost, convenience, and availability. Extrapolation from rodent studies has been invaluable in the development of current theories and concepts of cancer metastasis in humans. Drawing upon these experi-

---

[a]Address for correspondence: Stanley Zucker, M.D. (Mail Code #151), VA Medical Center, Northport, New York 11768, USA. Phone, 516/261-4400, ext. 2861; fax, 516/544-5317; e-mail, zucker.stanley@northport.va.gov

ments, we can describe the following steps/stages of the metastatic process: (1) cancer cell adhesion; (2) motility; (3) local invasion; (4) penetration of cancer cells into blood vessels and lymphatics; (5) arrest of cancer cells in distant organs; (6) reverse cancer cell penetration of blood vessels; and (7) cancer cell proliferation in distant organs. Experiments performed using specialized *in vivo* models such as the chick chorioallantoic membrane have provided results that support the principles identified in rodent experiments. Data derived from *in vitro* experiments using cultivated cancer cells have generally been consistent with the basic tenets derived from animal experiments.

This understanding of the pathophysiology of cancer invasion and metastasis forms the basis for the development of drugs designed to interfere with these processes. Basing their work on the principle that local cancer invasion and distant metastasis require cancer cells to release proteolytic enzymes that digest surrounding connective tissues, the pharmaceutical industry has recently invested hundreds of millions of dollars in trying to develop safe and effective protease inhibitors as treatment for metastasis. The focus has been primarily on matrix metalloproteinases (MMPs), which function extracellularly at neutral pH or at the plasma membrane. Industrial interest in inhibition of the plasminogen activator system as a modality of cancer treatment has lagged behind because of the uncertainty of the role of plasmin in human cancer.

## MMP INHIBITORS IN CANCER TREATMENT

Let us look more closely at the development of our concepts of the role of MMPs in cancer. Although collagenase-1 (the first MMP described in cancer) was initially identified as an important factor in tadpole metamorphosis in the 1960s, the function of collagenase and other MMPs in cancer invasion did not became a central focus of scientific investigation until the mid 1980s. The identification of high levels of MMPs in immortalized cancer cell lines led to the conclusion that these enzymes were required for cancer invasion and metastasis; numerous experiments in animals have been published in support of this concept. Soon thereafter it was shown that tissue inhibitors of metalloproteinases (TIMPs) purified from natural biologic sources were able to reverse the degradative effects of MMPs in cancer. A natural outcome of this research was the development and testing of drugs that effectively inhibit MMPs. For the most part, hydroxamic acid–based inhibitors of MMPs have had a fine record in decreasing cancer metastasis and invasion in experimental models. Four of these MMP inhibitors (Marimastat, AG-3340, CSG27023A, and Bay 12-9566) and a modified tetracycline with anti-MMP properties (COL-3) are currently being evaluated as treatment for patients with advanced cancer with the hope of controlling progression of the disease. Preliminary clinical reports in limited numbers of patients with cancer have described a decrease in tumor marker measurements (CA-125, PSA) in patient serum as evidence of antitumor effect. Uncertainty about the correlation between decrease in these tumor marker measurements and control of tumor growth, however, has dampened enthusiasm for these surrogate measurements.

## CANCER CELLS ARE NOT THE SOURCE
## OF MOST MMPS IN A TUMOR

Shortly after the pharmaceutical industry's foray into developing anti-MMP therapy, studies of human pathological material reported that most of the MMPs and plasminogen activators in cancer tissues were produced by peritumoral fibroblasts rather than by the cancer cells themselves. These data fit with an earlier observation that cancer cells produce factors (*E*xtracellular *M*atrix *M*etallo*PR*oteinase *IN*ducer-EMMPRIN) that stimulate surrounding fibroblasts to actually produce MMPs. The fibroblasts in a tumor can be considered the drones of the cancer cell, making enzymes that destroy host tissues rather than serving the function of repairing damaged tissue (the natural function of fibroblasts). This new information concerning the cellular origin of MMPs in tumors may be of crucial importance to the outcome of using synthetic MMP inhibitors to treat cancer. Endothelial cell MMPs are also important in tumor angiogenesis during the early stages of blood vessel ingrowth into an expanding tumor.

## INJECTING CANCER CELLS INTO MICE: A PREDICTIVE MODEL
## FOR ANTIMETASTASIC DRUG DEVELOPMENT IN HUMANS?

How much confidence can we have in experimental animal models that are currently being used to screen and identify antimetastatic/antiangiogenic agents? I propose that we are forging ahead too quickly to develop antimetastatic treatments without sufficient understanding of the differences between the disease process in humans and experimental animals. Obvious examples of major differences between experimental animal and spontaneous human cancer include: (1) more rapid cell proliferation of primary and metastatic tumors in animals; (2) less local cancer invasion in animal tumors (resulting in pseudo capsule formation) and larger tumors growing out through the skin; (3) production of MMPs by experimental cancer cells rather than the stromal cell production observed in human cancer; (4) experimental cancer cell lines selected by *in vitro* growth characteristics versus spontaneous *in vivo* selection in human cancers; (5) higher frequency and more bizarre chromosomal abnormalities in cancer cell lines; (6) tumor angiogenesis less prominent in experimental cancers as exemplified by more central tumor necrosis; (7) differences between cytokine control of MMP synthesis in human and animal models. These differences bring into question which observations made with transplantable cancers in rodents will be directly applicable to cancers in humans.

## RECOMMENDATIONS FOR DEVELOPING NEW
## EXPERIMENTAL METASTASIC MODELS

I propose that we desperately need innovative approaches to develop new experimental models to explore various aspects of cancer dissemination. Although great strides have been made and a breakthrough may be soon at hand, it would be naive to think that we have an in-depth understanding of this scientific area.

More-natural animal models of cancer (carcinogen-induced cancer, transgenic mice, hereditary tumor models, etc.) are worthy of exploration. Comparative studies in humans and rodents examining MMPs involved in tumor angiogenesis are needed. *In vivo* and *in vitro* systems that explore the interaction of cancer cells and stromal cells are required to fully appreciate the human pathophysiology. A better understanding of the role of cytokines in MMP regulation in human cancer is needed.

Another practical problem is that the testing of drugs designed to halt cancer progression is considerably different from the testing of cytotoxic drugs. The end-point of a cytotoxic drug effect can be evaluated in a much shorter period. Furthermore, whereas hydroxamic acid–derived inhibitors of MMPs display their anticancer effects within weeks of instituting treatment in experimental animals, the same effect in humans may not be evident for many months or even years.

I recognize that it would be folly to recommend that we slow down new drug development until the validity of experimental metastasic models for humans can be documented because the impact of cancer in modern society is too great to delay the process. But I do recommend that we reexamine the principles and model systems upon which we will base our future scientific investments.

## CLINICAL DRUG TRIALS IN HUMANS

Another problem that we are encountering in the clinical testing of antimetastatic agents relates to the clinical stage of cancer in patients selected for initial drug trials. Based on principles developed for screening antiproliferative cancer drugs, new antimetastatic/antiangiogenic drugs are being tested initially on patients who have widely metastatic cancer. It is quite conceivable that these effective new drugs may not be identified in this patient population and that these drugs may be discarded inappropriately. Ideally, drugs that are designed to interfere with metastasis should be tested in patients at risk for developing disseminated cancer, rather than in patients with already demonstrable metastases. This experimental design will be more difficult in diseases such as breast and colon cancer, where the cytotoxic efficacy of chemotherapy has been proven in the adjuvant setting. Clinicians experienced in cytotoxic drug trials may be reluctant to add antimetastatic drugs to a continually evolving combination of cytotoxic drugs. In other circumstances, such as gastric cancer and prostate cancer, where adjuvant chemotherapy is ineffective, early use of new categories of anticancer drugs may be more readily received.

The magnitude and extensive cost of adjuvant studies using antimetastatic drugs will have an important influence on future drug trials. Patience and diligence in selection of drugs for specific cancer types and stages cannot be overemphasized. Competition for patient recruitment to different types of protocols will become an ever-increasing reality. All of these factors further emphasize the importance of developing new experimental models for screening of the noncytotoxic cancer drugs of the future.

# Measurement of Matrix Metalloproteinases and Tissue Inhibitors of Metalloproteinases in Blood and Tissues

## Clinical and Experimental Applications

STANLEY ZUCKER,[a,b] MICHELLE HYMOWITZ,[a] CATHLEEN CONNER,[a] HOSEIN M. ZARRABI,[a] ADAM N. HUREWITZ,[b] LYNN MATRISIAN,[c] DOUGLAS BOYD,[d] GARTH NICOLSON,[e] AND STEVE MONTANA[a]

[a]*Departments of Medicine and Research, Veterans Administration Medical Center, Northport, New York 11768, USA*

[b]*State University of New York, Stony Brook, New York 11794, USA*

[c]*Vanderbilt University, Nashville, Tennessee 37232, USA*

[d]*University of Texas MD Anderson Cancer Center, Houston, Texas 77030, USA*

[e]*Institute for Molecular Medicine, Huntington Beach, California 92649, USA*

ABSTRACT: The balance between production and activation of MMPs and their inhibition by TIMPs is a crucial aspect of cancer invasion and metastasis. On the basis of the concept that MMPs synthesized in tissues seep into the bloodstream, we have examined MMP levels in the plasma of patients with cancer. In colorectal, breast, prostate, and bladder cancer, most patients with aggressive disease have increased plasma levels of gelatinase B. In patients with advanced colorectal cancer, high levels of either gelatinase B or TIMP complex were associated with shortened survival. We propose that these assays may be clinically useful in characterizing metastatic potential in selected kinds of cancer. In rheumatoid arthritis and systemic lupus erythematosus (SLE), serum and plasma levels of stromelysin-1 were ~ 3–5-fold increased. Fluctuating serum stromelysin-1 levels in SLE did not correspond with change in disease activity. In SLE, stromelysin-1 may be a component of the chronic tissue repair process rather than being responsible for inciting tissue damage. On the basis of these observations, we conclude that measurement of plasma/serum MMP and TIMP levels may provide important data for selecting and following patients considered for treatment with drugs that interfere with MMP activity.

## ROLE OF MMPs, TIMPs, AND EMMPRIN IN CANCER INVASION AND METASTASIS

Production and activation of matrix metalloproteinases (MMPs) in tumors is an important aspect of cancer invasion and metastasis. Tissue inhibitors of metallopro-

[b]Address for correspondence: Stanley Zucker, M.D. (Mail Code 151), VA Medical Center, Northport, New York 11768, USA. Phone, 516/261-4400, ext. 2861; fax, 516/544-5317; email: zucker.stanley@northport.va.gov

teinases (TIMPs) counteract the proteolytic activity of MMPs. A delicate balance between activation and inhibition of MMPs (gelatinases, collagenases, stromelysins, matrilysin) at the invasive edge of a cancer controls the degradation of extracellular matrix (ECM) and subsequent dissemination of the cancer. A correlation between high levels of MMPs in cancer tissues and aggressive behavior of experimental tumors has been repeatedly demonstrated. Paradoxically, TIMP levels are often increased in cancer tissues, but this compensatory increase presumably is inadequate to counteract ECM degradation.[1,2] MMPs are also involved in many aspects of normal tissue development and repair involving cell motility, release of growth factors bound in tissues, and remodeling of the ECM.

A basic limitation in many studies performed to date is the uncertainty of whether MMPs are involved in the initiation of tissue damage in a disease process or are involved in the repair mechanism. Likewise, it is unclear whether TIMPs are produced in response to increased MMP production or whether they are independently controlled. In this regard, we are only beginning to understand the relationships between MMP and TIMP production by epithelial, stromal, and endothelial cells in disease-targeted tissue. At present, it is appreciated that the concept of MMPs versus TIMPs as the controlling factors in cancer invasion and metastasis is an oversimplification of a much more complicated series of events.

Although experimental studies initially suggested that cancer cells themselves produce the matrix metalloproteinases required for degradation of the extracellular matrix, more recent studies using *in situ* hybridization techniques have indicated that most MMPs are produced by stromal cells within human tumors (breast, gastrointestinal, lung, prostate), rather than by the cancer cells themselves. Gelatinase A (GLA),[3–7] collagenase-1,[8] stromelysin-3,[9–11] collagenase-3, and membrane type 1-MMP (MT1-MMP)[9,11] have been identified in stromal fibroblasts in distinct patterns, especially in proximity to invading cancer cells.[12] Gelatinase B (GLB) has been localized to inflammatory cells (macrophages and neutrophils).[4,5,13,14] Stromelysin-1 and stromelysin-2 were infrequently identified in human tumors. Matrilysin represents the major exception since this MMP is produced by carcinoma cells, rather than by stromal cells.[15,16]

Membrane type 1-MMP (MT1-MMP) is recognized to be an important physiologic activator of progelatinase A in benign and malignant conditions.[17–19] The physiologic activation mechanism for other secreted MMPs remains to be determined. By means of an immunohistochemical technique, MT1-MMP protein has been reported to be primarily localized in and on gastric carcinoma cells along with gelatinase A.[20] The mechanism of transport of MT1-MMP to the cancer cell surface after synthesis in stromal cells remains to be elucidated. The functional activity of MT1-MMP after release from the cell surface is also uncertain.

TIMP-1, TIMP-2, TIMP-3, and TIMP-4 form physiologically irreversible complexes with all types of activated MMPs in the amino terminal portion of the enzymes. In addition, TIMP-1 forms a specific complex with latent gelatinase B, and TIMP-2 forms a specific complex with latent gelatinase A in the carboxy terminal regions of the respective enzymes. These unique latent complexes lead to stabilization of the enzyme activation mechanism. Many types of cancer cells and peritumoral mesenchymal cells secrete increased concentrations of both TIMPs and gelatinases leading to the formation of complexes of MMPs and TIMPs extracellularly.[3,21] These complexes subsequently leach into the blood stream, where they can be identified by immunoassays.[22]

An explanation for the role of stromal cells in cancer has come from the seminal work of Biswas, *et al.*, who demonstrated that carcinoma cells produce a stimulatory factor, *E*xtracellular *M*atrix *M*etallo*P*roteinase *In*ducer (EMMPRIN, originally designated Tumor Collagenase Stimulatory Factor, TCSF), which induces stromal fibroblasts to produce MMPs.[23, 24] In recent *in situ* hybridization studies of breast and lung cancer, EMMPRIN was identified in cancer cells, while adjacent stromal fibroblasts produce gelatinase A and MT1-MMP. EMMPRIN has also been identified in normal epithelial cells (breast ductal cells, keratinocytes), suggesting similarities between control of MMP synthesis in normal and malignant tissues.

Another important aspect of MMP and TIMP production in cancer is the question of whether these proteins are produced early or late in the disease. Stromelysin-3 and matrilysin are present early in tumor development (colon adenomas and breast carcinoma *in situ*), at a time when tumors are not known to be invasive or metastatic; this observation suggests that stromelysin-3 and matrilysin may not be involved in metastasis.

## TISSUE MMPs AND TIMPs AS PROGNOSTIC MARKERS IN CANCER: CORRELATION WITH STAGE

Increased tumor tissue levels of gelatinase B, gelatinase A, collagenase-1, and MT1-MMP, identified by either mRNA measurement, bioassay, or immunohistochemistry, have been found to correlate with advanced cancer stage in gastrointestinal, breast, prostate, and bladder cancer.[10,14,25,26] Zeng *et al.* demonstrated by Northern blot analysis that an increased ratio of tumor/normal mucosa gelatinase B mRNA correlated significantly with the status of distant metastasis and Dukes' clinical staging for colorectal cancer.[27,28] The presence of activated gelatinase A and gelatinase B in human cancer tissue extracts, as detected by substrate zymography, has also been proposed as a cancer marker in aggressive breast and colon adenocarcinomas.[29,30] While Liabakk *et al.* likewise reported increased levels of gelatinase B and gelatinase A, as well as activated gelatinase A in colorectal cancer tissues, no correlation between gelatinase levels and survival was noted.[31] In a study of tumor specimens obtained from patients with various stages of colon cancer, Murnane *et al.* noted that activated gelatinase A appeared early and persisted, whereas gelatinase B became more prominent during the advanced stages of colon cancer. Cathepsins B and L were elevated in early-stage cancer, but then decreased. These authors suggested that gelatinase A is involved in the initiation and maintenance of malignancy and gelatinase B is required for distant spread.[32] Nielsen *et al.* emphasized that neutrophils and macrophages infiltrating colon[33] and breast cancer tissue are the major source of gelatinase B, with macrophage turnover of gelatinase B being rapid. Interaction of T cells with macrophages via the gp39-CD40 counter receptors[34] is one potential mechanism for induction of macrophage gelatinase B. Increased levels of TIMP-1 and TIMP-2 mRNA and protein have also been identified in malignant stromal tissues.[11,15,27,35] A correlation between TIMP-1 levels and Dukes' stage of colorectal cancer has been proposed.[15,35]

Using specific monoclonal antibodies in immunoassays, Stearns reported that higher levels of latent and activated gelatinase A in prostate tissue extracts correlated with more aggressive (higher Gleason score) and metastatic cancer.[36] Increased gelatinase B secretion and a high ratio of gelatinase B to gelatinase A in short-term tissue culture was observed with prostate cancer as opposed to benign prostatic hypertrophy.[37] Likewise, using *in situ* hybridization, high tissue expression of gelatinase B and gelatinase A and low expression of TIMP-1 and TIMP-2 were independent predictors of poor outcome in prostate cancer.[6] Using reverse transcriptase PCR to measure gene expression in tissue obtained at surgery, Kanayama *et al.* reported that gelatinase A, TIMP-2, and MT1-MMP are useful prognostic indicators in patients with bladder cancer. Patients with high expression of any of these three genes had a worse prognosis than those with low expression after radical cystectomy.[38]

In view of the fact that other types of proteinases (i.e., plasminogen activator, cathepsin B, cathepsin D) and proteinase inhibitors (i.e., plasminogen activator inhibitor, calpain) have also been implicated in cancer metastasis, it is not surprising that the correlation between concentration of a single proteinase in tissues or blood and clinical outcome is often disputed.

## MEASUREMENT OF MMPs AND TIMPs IN BODY FLUIDS

Increased serum levels of the commonly used tumor markers (CEA, PSA, CA-125, HCG) provide useful information about the body tumor burden; these tests however provide little information that is predictive of the biologic behavior of a particular form of cancer. On the basis of detection of high levels of MMPs in human cancer tissues and the identification of MMPs in human plasma, we proposed that tissue MMPs may leach into the blood stream in increased amounts in patients with biologically aggressive cancer and thereby provide unique markers that might be useful for predicting metastasis. To explore this hypothesis, sandwich-type enzyme-linked immunosorbent assays (ELISA) have been developed to measure the concentration of GLB, GLB:TIMP-1 complexes, GLA, GLA:TIMP-2 complexes, and stromelysin-1 in the plasma of patients with cancer.

Numerous investigators have demonstrated that MMPs and TIMPs are readily measured in plasma and serum by ELISAs which employ specific polyclonal or monoclonal antibodies. Mean levels of gelatinase A,[39] TIMP-1,[40] TIMP-2,[41] stromelysin-1,[40] gelatinase B, collagenase-1,[42] and matrilysin[43] in normal plasma/serum range from ~500 ng/ml to ~10 ng/ml (order of appearance reflects their relative concentration). It should be remembered that the measurement of gelatinase B in serum is unreliable, since it reflects, in large part, *in vitro* MMP release occurring during degranulation of blood neutrophils;[44,45] plasma measurements are not confounded by this artifact. One of the limitations of ELISAs is that the results achieved using different antibodies and protein standards can vary, which complicates comparison of results between various reports.[40,42] Another confounding factor affecting clinical MMP measurements is that MMPs are capable of binding to connective tissue matrix[46]; hence increased local secretion of MMPs in disease may not necessarily be translated into increased plasma levels. The degradation and excretion pathways for MMPs and TIMPs in the body have not been examined to date. Thus, we can only assume that high blood levels of these proteins reflect increased production rather than diminished excretion.

In development of a prototype sandwich ELISA, Zucker et al. demonstrated that plasma gelatinase A was significantly increased (>50%) in the second half of pregnancy as compared to early pregnancy or the nonpregnant state.[39] In contrast to expectations, plasma gelatinase A levels were not significantly increased in patients with advanced gastrointestinal cancer, breast cancer, gynecologic cancer, lung cancer, and lymphoma-leukemia as compared to levels in normal individuals. On the basis of in vitro studies, it was proposed that endothelial cells and other normal cells make a sizable contribution to plasma levels of gelatinase A in the intact animal, and may thereby obfuscate detection of increased levels of enzyme originating from solid tumors.[47] Fujimoto et al. likewise reported that serum gelatinase A levels were not increased in stomach cancer and pancreatic cancer, but were increased (~30%) in hepatocellular carcinoma, hyperthyroidism, and biliary cirrhosis.[48] Although Gabrisa et al.[49] reported increased serum gelatinase A in patients with metastatic lung cancer as compared to localized lung cancer, the differences were small.

In a study of patients with urothelial cancer, Gohji et al. demonstrated that the ratio of gelatinase A to TIMP-2 in serum was significantly higher in patients with recurrent cancer. Disease-free survival in patients with high gelatinase A:TIMP-2 ratios was poor as compared to patients with lower ratios; this ratio was shown to be an independent prognostic indicator of urinary tract cancer recurrence.[50] High serum stromelysin-1 and gelatinase A levels were also shown to be predictors for recurrent cancer in patients with advanced bladder cancer after complete tumor resection.[51] In contrast to TIMP-2, serum levels of TIMP-1 were reported by Naruo et al. to be increased in patients with bladder cancer; increased TIMP-1 levels correlated with progression of cancer as demonstrated by increased invasion and metastasis.[52] Increased serum gelatinase A levels have been demonstrated in patients with prostate cancer; a correlation with clinical course in patients with bone metastasis was demonstrated.[53]

Gelatinase B measurements in blood have provided more encouraging results in cancer than have been achieved with the gelatinase A assay. Using a pair of monoclonal antibodies to gelatinase B in a sandwich ELISA, Zucker et al. demonstrated that gelatinase B was significantly increased in the plasma of patients with breast cancer and gastrointestinal tract cancer (metastatic and nonmetastatic) as compared to normal subjects.[45] Likewise, latent gelatinase B:TIMP-1 complexes (TIMP complexes) were significantly increased in the plasma of patients with gastrointestinal tract cancer and female genitourinary tract cancer as compared to control subjects. Of interest, some patients had increased plasma levels of gelatinase B:TIMP-1 complexes and normal gelatinase B levels, and vice versa. When results from both the gelatinase B and the gelatinase B:TIMP-1 complex ELISAs were combined, increased levels of either gelatinase B, TIMP complexes, or both were found in 36% of patients with gastrointestinal cancer and 65% of patients with genitourinary tract cancer. Most importantly, clinical follow-up (40 months) of patients with stage IV (metastatic) gastrointestinal cancer indicated that the length of survival of patients with increased plasma levels of gelatinase B or TIMP complexes was significantly shorter than that of patients with normal plasma levels (4 months vs. 20 months, respectively).[22] On the basis of these studies, Zucker *et al.* proposed that the assay of gelatinase B and TIMP complexes in plasma may be clinically useful in predicting survival in certain subsets of cancer patients. An inverse relationship was identified between plasma gelatinase B and TIMP complexes versus CEA, suggesting that production of gelatinase B (presumably by tumor macrophages) occurs primarily in colorectal cancers with low CEA production. It remains to be determined whether gelatinase B and TIMP complexes will be a marker for metastasis or prognosis in patients with earlier stage colorectal cancer.[45] A practical limitation to these tests in cancer patients undergoing treatment is that plasma gelatinase B levels have been reported to decrease in parallel with the drop in blood granulocytes in patients treated with chemotherapy.[44] An unsatisfying aspect of these studies is that we do not understand the connection between the inflammatory cell infiltrates which are responsible for the production of high levels of gelatinase B and TIMP-1 and the poor prognosis in this subset of patients with colorectal cancer.

In these initial studies, even though significant differences exist between groups, the bimodal distribution of plasma MMPs and TIMP complexes in the normal population limits the practical value of these assays for the individual patient since there is considerable overlap in test results between the cancer patient population and the healthy population.[39,44,45]

Employing a more sensitive revised ELISA consisting of a polyclonal capture antibody and a monoclonal detecting antibody to gelatinase B, we have recently demonstrated that gelatinase B is increased in the plasma of more than 50% of patients with breast cancer, prostate cancer, and bladder cancer, but not gynecologic cancers (ovary, cervix, uterus, vagina). In treated patients with prostate cancer, no correlation was noted between the plasma concentration of gelatinase B and prostate-specific antigen (PSA), indicating that MMPs do not correlate with body tumor burden. Using this ELISA, we have examined the effect of therapy on fluctuations of plasma gelatinase B in patients with breast cancer. Both chemotherapy and hormonal therapy resulted in considerable fluctuations in plasma gelatinase B levels: in some instances there appeared to be a correlation with clinical progression/regression in disease activity, but not uniformly (FIG. 1). Elevated plasma levels of gelatinase B have also been reported in patients with hepatocellular carcinoma.[54]

In a comprehensive study of serum MMPs and TIMPs, Baker *et al.* reported increased levels of TIMP-1 and collagenase-1 and lower levels of TIMP-2 in patients with prostate cancer as compared to control subjects.[55] Patients with metastatic cancer had significantly higher levels of collagenase-1 than did control subjects. In agreement with other studies,[56] serum stromelysin-1 was not increased in patients with cancer. Baker *et al.* emphasized that it is uncertain whether collagenase-1 and TIMP-1 are tumor-derived or result from tumor stromal cell expression in response to tumor cell growth.[55] Jung *et al.* confirmed the finding of increased plasma levels of TIMP-1 in prostate cancer, but also reported increased levels of stromelysin-1 in prostate cancer patients with metastases.[57] On the basis of our studies in rheumatoid disease,[56,58] we suspect that the minimal increase in stromelysin-1 in prostate cancer reflects an inflammatory type of response rather than a cancer effect.

We have recently developed an ELISA for measurement of matrilysin (MMP-7) using a rabbit polyclonal antibody to recombinant antigen as the capture antibody

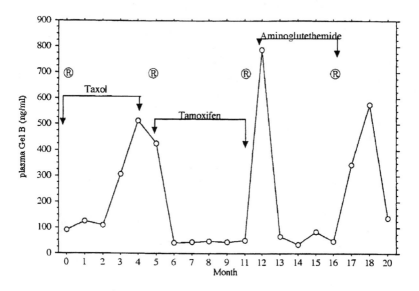

**FIGURE 1.** Line chart showing the fluctuation of plasma gelatinase B (measured by ELISA) in a 57-year-old woman with breast cancer who was undergoing chemotherapy and hormonal therapy. This patient had invasive ductal carcinoma of the breast with a large axillary lymph node metastasis. A mastectomy had been performed one year earlier and was followed by adjuvant chemotherapy. The cancer recurred (®) (listed as month zero), and the patient was treated with Taxol for 4 months. Skull metastases were then noted, and tamoxifen was initiated. New metastatic lesions were noted at 12 months, and aminoglutethemide was started. Disease progression at 16 months was followed again by a rise in plasma gelatinase B. The patient died at 20 months after operation.

and a rat monoclonal antibody as the detecting antibody. In preliminary studies, it became obvious that a bimodal distribution of matrilysin values was noted within a control population of men and women; healthy people had either plasma matrilysin levels <5 ng/ml or >100 ng/ml. The comparable bimodal distribution of matrilysin levels was also noted in macrophage-conditioned media collected from healthy volunteers. Measurement of matrilysin in cancer patients demonstrated increased plasma levels only in patients with bladder cancer, but not in breast, GI, lung, gynecologic, or prostate cancer (FIG. 2). These data are contrary to expectation since high levels of matrilysin have been identified in pathologic tissues and cell lines derived from patients with GI, breast, and prostate cancer.[59]

Of interest, gelatin zymography has recently been used to examine the frequency of detection of MMPs in the urine of patients with cancer. Gelatinase A and gelatinase B were demonstrated in the urine of most patients with active breast, bladder, and prostate cancer, but infrequently in patients with cancer and no evidence of disease or in healthy control subjects. An unclassified high molecular weight species of gelatinase (125–150 kDa) was also demonstrated in the urine of cancer patients.[60]

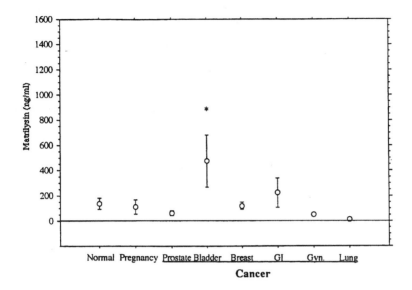

**FIGURE 2.** Point chart showing the mean ± standard error of plasma matrilysin in normal individuals, pregnant women, and patients with prostate, bladder, breast, GI, gynecologic, and lung cancers. Plasma matrilysin was measured by ELISA. Bladder cancer was the only group in which plasma levels were significantly above those of patients in the normal group (as indicated by the *asterisk*).

## STROMELYSIN-1 LEVELS IN BLOOD OF PATIENTS WITH INFLAMMATORY DISEASE

Numerous reports described increased levels of serum/plasma stromelysin-1 levels in patients with rheumatoid arthritis and to a lesser degree in patients with osteoarthritis and gout. Approximately 80% of patients with rheumatoid arthritis had 3-fold increased blood levels of stromelysin-1; higher serum levels of stromelysin-1 correlated with certain parameters of disease activity.[55,56,61–63] Synovial fluid stromelysin-1 levels are several hundred-fold higher than serum levels, which is consistent with local stromelysin-1 production in the joint and subsequent leaching into the blood stream. In arthritis, an excellent correlation between paired serum and synovial fluid levels of stromelysin-1 suggests that serum levels accurately reflect intra-articular events.[61] Serum collagenase-1 is increased to a lesser degree than is stromelysin-1 in patients with arthritis.

Elevated serum stromelysin-1 levels in inflammatory bowel disease[64] and graft-versus-host disease suggests that mesenchymal cell production of stromelysin-1 is under the control of numerous cytokines (i.e., TNF, IL-1), thereby limiting the usefulness of stromelysin-1 measurements in differential diagnosis of inflammatory diseases.

Systemic lupus erythematosus, the prototype autoimmune disease, is another disease in which we might anticipate enhanced MMP production. In a recent study by Zucker *et al.*, serum MMP levels and SLE Disease Activity Indices were measured in a large number of patients with SLE and in healthy controls. Results indicated a 3–5-fold increase in serum stromelysin-1 levels in patients with SLE as compared to healthy subjects. Contrary to expectations, serial measurements of stromelysin-1 in individual patients with SLE did not correlate with fluctuation in disease activity scores. Collagenase-1, gelatinase A, and TIMP-1 levels were not significantly increased in the serum of patients with SLE as compared to control subjects.[58] These data suggested that stromelysin-1 may not be primarily involved in the initial tissue damage in SLE, but rather participates in a later aspect of inflammation involving tissue repair. If this theory is correct, the use of MMP-inhibiting drugs may prove to be detrimental in diseases in which the MMP repair mechanism is beneficial to the organism. In these instances, MMP inhibitory drugs may be useful early in disease, when the destructive aspects of MMPs are prominent, but not during the later repair stages of disease. The importance of clinical trials to clarify these issues cannot be underestimated.

## SERUM MMPs AND TIMPs IN LIVER DISEASE

In contrast to cancer and arthritis, many disease processes are associated with increased extracellular matrix deposition rather than degradation. The hallmark of tissue injury in response to chronic excess alcohol consumption is increased hepatic fibrosis (cirrhosis). Collagen accumulation reflects not only enhanced synthesis, but also results from a failure of collagen degradation to keep pace with collagen production. Several groups have demonstrated increased serum and liver tissue levels of TIMP-1 in patients with alcoholic cirrhosis associated with liver fibrosis.[65] Serum TIMP-1 levels correlated more closely with the histologic degree of liver fibrosis than with hepatic inflammation or necrosis. Collagenase-1, gelatinase A, gelatinase B, and stromelysin-1 are also produced by the liver in response to injury. It is well established that collagenase-1 activity is decreased as the fibrosis progresses in cirrhotic livers.[66] A recent study has demonstrated that patients with hemochromatosis and hepatic fibrosis have increased ratios of TIMP-1 compared to either collagenase-1, gelatinase A, or stromelysin-1.[67] In patients with chronic hepatitis C treated with interferon, responders had an increase in serum collagenase-1 and a decrease in serum TIMP-1; nonresponders had the reverse effect, suggesting that interferon may exert a beneficial effect on hepatic fibrosis.[68]

## SERUM MMPs AND TIMPs IN VASCULAR DISEASE

There is considerable current interest in the role of MMPs in cardiovascular disease, especially in regard to the pathogenesis of dissecting aortic aneurysms and rupture of atherosclerotic plaques.[69] Of interest, Hirohata *et al.* have described a drop in serum collagenase-1 and TIMP-1 persisting for several days after acute myocardial infarction and then a return to normal levels. A correlation with cardiac function was noted suggesting the involvement of collagenase in the healing process during cardiac remodeling.[70]

## PLASMA MMPs AND TIMPs IN PREGNANCY

It is well established that MMP function is important in many aspects of fetal implantation, pregnancy, and delivery. In a recent study, Tu *et al.* demonstrated that plasma gelatinase B levels are more than 15-fold increased from week 19 to 36 of pregnancy. Plasma gelatinase B levels during spontaneous labor increased an additional 3-fold regardless of gestational age.[71] The tissue origin of gelatinase B in pregnancy is presumed to be the choriodecidual membranes. In contrast, Kolben *et al.*[72] reported that plasma concentrations of gelatinase B rise between the 10th and 40th week of pregnancy. A role for IL-8 in increasing the release of gelatinase B and collagenase-2, resulting in cervical ripening during labor, has been proposed.[73] A similar pattern of gestational plasma levels is seen for collagenase-1, except that the prenatal levels do not differ from those of nongravid controls, but become dramatically elevated at term labor.[74] In uncomplicated pregnancies, serum TIMP-1 levels were low during pregnancy until the 37th week, when the levels rose to that of the nonpregnant state. Serum TIMP-1 levels rose during labor and the postpartum period. The contrast between changes of MMPs and TIMP during pregnancy serve to re-emphasize the balance/imbalance occurring during various physiologic states.[75]

## MEASUREMENT OF MMP AND TIMP IN
## BRONCHOALVEOLAR LAVAGE FLUID (BAL)

Patients with serious lung disease frequently are subjected to bronchoscopy for diagnostic or therapeutic purposes. We have examined BAL samples from 30 patients with infectious, inflammatory, and malignant lung disease to determine whether the concentration of MMPs and TIMPs reflect the underlying disease process. Gelatin zymography and ELISAs were performed. The results indicated that: (1) gelatinase B is the dominant MMP in BAL ($605 \pm 159$ pM); (2) the mean concentrations of gelatinase A ($24 \pm 1$ pM), collagenase-1 ($93 \pm 28$ pM), and stromelysin-1 ($80 \pm 7$ pM) were considerably lower; (3) the concentration of TIMP-2 ($724 \pm 81$ pM) exceeded that of TIMP-1 ($284 \pm 156$ pM); (4) only modest differences were noted between levels of MMPs and TIMPs in these patients; and (5) patients with asthma had significantly higher levels of gelatinase B ($2454 \pm 1454$ pM) and TIMP-2 than any of the other groups, including lung cancer (FIG. 3). By contrast to these BAL results, the normal plasma levels of gelatinase A and gelatinase B are ~7000 pM and 100 pM, respectively. This reversal of the relative concentration of these two MMPs in BAL as contrasted with normal plasma suggests that the gelatinase B in BAL is not simply a transudate of plasma, but rather is produced locally in the lung, primarily by macrophages lining the bronchi and alveoli. Furthermore, these data confirm that local MMP synthesis is often increased in nonmalignant diseases equivalent to malignancy. This has an impact on the interpretation of plasma MMP measurements in differential diagnosis.

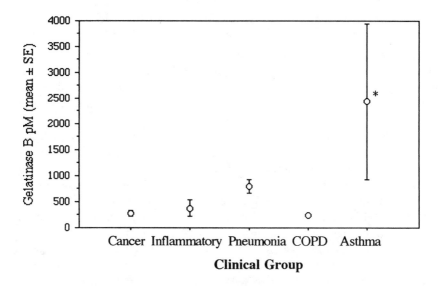

**FIGURE 3A.** Gelatinase B concentrations (pM) from BAL samples in each of five clinical groups with means ± SEM. Gelatinase B concentrations in patients with lung cancer and COPD are more tightly grouped around the mean. Patients with asthma had significantly higher gelatinase B than did each of the other groups.

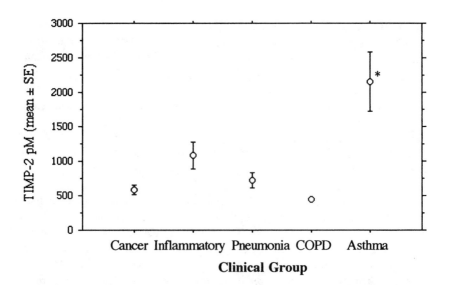

**FIGURE 3B.** Concentrations of TIMP-2 (pM) in BAL samples from five clinical groups with means ± SEM also shown. Significant differences ($p < 0.05$) are shown (*). Asthmatic patients had the most elevated levels.

## PROSPECTS FOR THE FUTURE

In conclusion, we propose that the assay of metalloproteinases and their complexes with TIMPs in the plasma of patients with cancer may be clinically useful in characterizing the metastatic potential of certain types of cancer. These measurements will also be useful in nonmalignant diseases characterized by excess tissue degradation. From a theoretical point of view, the detection of activated gelatinases or complexes of activated gelatinase with the noncorresponding TIMP (gelatinase A: TIMP-1 complex or gelatinase B:TIMP-2 complex) may provide a better plasma test for monitoring metastasis in patients with cancer than the assay of latent enzymes and latent MMP:TIMP complexes since activation of MMPs is ultimately required for matrix degradation. Improvement in the sensitivity of these "complex" ELISAs will be required before their full potential can be recognized, since the concentration of activated gelatinase complexes in plasma appears to make up less than 1% of the level of the latent enzymes.

## ACKNOWLEDGMENTS

This research was supported by Grant No. DAMD 17-95-1 from the Unites States Army Medical Research and Development Command, a VA Merit Review Grant, and a Carol Baldwin Cancer Grant (SUNY at Stony Brook).

## REFERENCES

1. STETLER-STEVENSON, W.G. 1995. Proteinase A activation during tumor cell invasion. Invasion Metastasis **14:** 259–268.
2. BIRKEDAL-HANSEN, H., W.G.I. MOORE, M.K. BODDEN *et al.* 1993. Matrix metalloproteinases: a review. Crit. Rev. Bio. Med. **42:** 197–250.
3. POULSOM, R., M. PIGNATELLI, W.G. STETLER-STEVENSON *et al.* 1992. Stromal expression of 72 kda type IV collagenase (MMP-2) and TIMP-2 mRNAs in colorectal neoplasia. Am. J. Pathol. **141:** 389–394.
4. EMMERT-BUCK, M.R., M.J. ROTH, Z. ZHUANG *et al.* 1994. Increased gelatinase A (MMP-2) and cathepsin B activity in invasive tumor regions of human colon cancer samples. Am. J. Pathol. **145:** 1285–1290.
5. PYKE, C., E. RALFKIAER, K. TRYGGVASON *et al.* 1993. Messenger RNA for two types of type IV collagenases is located in stromal cells in human colon cancer. Am. J. Pathol. **142:** 359–365.
6. WOOD, M., K. FUDGE, J.L. MOHLER *et al.* 1997. In situ hybridization studies of metalloproteinases 2 and 9 and TIMP-1 and TIMP-2 in human prostate cancer. Clin. Exp. Metastasis **15:** 246–258.
7. POLETTE, M., C. GILLES, B. NAWROCKI *et al.* 1997. TCSF expression and localization in human lung and breast cancer. J. Histochem. Cytochem. **45:** 703–709.
8. HEWITT, R. E., I.H. LEACH, D.G. POWE *et al.* 1991. Distribution of collagenase and tissue inhibitors of metalloproteinases (TIMP) in colorectal tumours. Int. J. Cancer **49:** 666–672.
9. OKADA, A., J.-B. BELLOCQ, N. ROUYER *et al.* 1995. Membrane-type matirx metalloproteinase (MT-MMP) gene is expressed in stromal cells of human colon, breast, and head and neck carcinomas. Proc. Natl. Acad. Sci. USA **92:** 2730–2734.

10. PORTE, H., E. CHASTRE, S. PREVOT et al. 1995. Neoplastic progression of human colorectal cancer is associated with overexpression of stromelysin-3 and BM-40/SPARC genes. Int. J. Cancer **64:** 70–75.
11. URBANSKI, S.J., D.R. EDWARDS, N. HERSHFIELD et al. 1993. Expression pattern of metalloproteinases and their inhibitors changes with progression of human sporatic colorectal neoplasia. Diagnostic Mol. Pathol. **2:** 81–89.
12. HEPPNER, K.J., L.M. MATRISIAN, R.A. JENSEN et al. 1996. Expression of most matrix metalloproteinase family members in breast cancer represents a tumor-induced host response. Amer. J. Pathol. **149:** 273-282.
13. ZENG, Z.S., Y. HUANG, A.M. COHEN et al. 1996. Prediction of colorectal cancer relapse and survival via tissue RNA levels of matrix metalloproteinase-9. J. Clin. Oncol. In press.
14. JEZIORSKA, M., N.Y. HABOUBI, P.F. SCHOFIELD et al. 1994. Distribution of gelatinase B (MMP-9) and type IV collagen in colorectal carcinoma. Int. J. Colorect. Dis. **9:** 141–148.
15. NEWELL, K.J., J.P. WITTY, W.H. RODGERS et al. 1994. Expression and localization of matrix-degrading metalloproteinases during colorectal tumorigenesis. Molec. Carcinogenesis **10:** 199–206.
16. YOSHIMOTO, M., F. ITOH, H. YAMAMOTO et al. 1993. Expression of MMP-7 (PUMP-1) mRNA in human colorectal cancers. Int. J. Cancer **54:** 614–618.
17. SATO, H., T. TAKINO, Y. OKADA et al. 1994. A matrix metalloproteinase expressed on the surface of invasive tumor cells. Nature **370:** 61–65.
18. STRONGIN, A.Y., I. COLLIER, G. BANNICOV et al. 1995. Mechanism of cell surface activation of 72-kDa type IV collagenase. Isolation of the activated form of the membrane metalloproteinase. J. Biol. Chem. **270:** 5331–5338.
19. ZUCKER, S., C. CONNER, B.I. DiMASSIMO et al. 1995. Thrombin induces the activation of progelatinase A in vascular endothelial cells: Physiologic regulation of angiogenesis. J. Biol. Chem. **270:** 23730–23738.
20. NOMURA, H., H. SATO, M. SEIKI et al. 1995. Expression of membrane-type matrix metalloproteinase in human gastric carcinomas. Cancer Res. **55:** 3263–3266.
21. BASSET, P., J.P. BELLOCQ, C. WOLF et al. 1990. A novel metalloproteinase gene specifically expressed in stromal cells of breast carcinomas. Nature **348:** 699–704.
22. ZUCKER, S., R.M. LYSIK, B.I. DiMASSIMO et al. 1995. Plasma assay of gelatinase B:tissue inhibitor of metalloproteinase (TIMP) complexes in cancer. Cancer **76:** 700–708.
23. KATAOKA, H., R. DeCASTRO, S. ZUCKER et al. 1993. The tumor cell-derived collagenase stimulating factor, TCSP, increases expression of interstitial collagenase, stromelysin and 72 kDa gelatinase. Cancer Res. **53:** 3155–3158.
24. GUO, H., S. ZUCKER, M. GORDON et al. 1997. Stimulation of metalloproteinase production by recombinant EMMPRIN from transfected CHO cells. J. Biol. Chem. **272:** 24–27.
25. ALLGAYER, H., R. BABIC, B.C. M. BEYER et al. 1998. Prognostic relevance of MMP-2 (72-kD collagenase IV) in gastric cancer. Oncology **55:** 152–160.
26. MURRAY, G.I., M.E. DUNCAN, P. O'NEIL et al. 1996. Matrix metalloproteinase-1 is associated with poor prognosis in colorectal cancer. Nature Med. **2:** 461–462.
27. ZENG, Z. S. & J. G. GUILLEM. 1995. Distinct pattern of matrix metalloproteinase 9 and tissue inhibitor of metalloproteinase 1 mRNA expression in human colorectal cancer and liver metastases. Br. J. Cancer **72:** 575–582.
28. ZENG, Z. S., Y. HUANG, A.M. COHEN et al. 1996. Prediction of colorectal cancer relapse and survival via tissue RNA levels of matrix metalloproteinase-9. J. Clin. Oncol. **14:** 3133–3140.
29. BROWN, P.D., R. E.BLOXIDGE, E. ANDERSON et al. 1993. Expression of activated gelatinase in human invasive breast cancer. Clin. Exp. Metastasis **11:** 183–189.
30. DAVIES, B., D.W. MILES, L.C. HAPERFIELD et al. 1993. Activity of type IV collagenases in benign and malignant breast disease. Br. J. Cancer **67:** 1126–1131.
31. LIABAKK, N.-B., E. TALBOT, R.A. SMITH et al. 1996. Matrix metalloproteinase 2 (MMP-2) and matrix metalloproteinase 9 (MMP-9) type IV collagenases in colorectal cancer. Cancer Res. **56:** 190–196.

32. MURNANE, M.J., S. SHUJA, E. DEL RE *et al.* 1997. Characterizing human colorectal carcinomas by proteolytic profile in vivo **11:** 209–216.
33. NIELSEN, B.S., S. TIMSHEL, L. KJEDSEN *et al.* 1996. 92 kDa type IV collagenase (MMP-9) is expressed in neutrophils and macrophages but not in malignant epithelial cells in human colon cancer. Int. J. Cancer **65:** 57–62.
34. MALIK, N., B.W. GREENFIELD, A.F. WAHL, *et al.* 1996. Activtion of human monocytes through CD40 induces matrix metalloproteinases. J. Immunol. **156:** 3952–3960.
35. LU, X., M. LEVY, I.B. WEINSTEIN *et al.* 1991. Immunologic quantification of levels of tissue inhibitor of metalloproteinase-1 in human colon cancer. Cancer Res. **51:** 6231–6235.
36. STEARNS, M. & M.E. STEARNS. 1996. Evidence for increased activated metalloproteinase 2 (MMP-2a) expression associated with human prostate cancer progression. Oncology Res. **8:** 69–75.
37. FESTUCCIA, C., M. BOLOGNA, C. VICENTINI *et al.* 1996. Increased matrix metalloproteinase-9 secretion in short-term cultures of prostatic tumor cells. Int. J. Cancer **69:** 386–393.
38. KANAYAMA, H.-O., K.-Y. YOKOTO, Y. KUROKAWA *et al.* 1998. Prognostic value of matrix metalloproteinase-2 and tissue inhibitor of metalloproteinase-2 expression in bladder cancer. Cancer **82:** 1359–1366.
39. ZUCKER, S., R.M. LYSIK, M. GURFINKEL *et al.* 1992. Immunoassay of type IV collagenase/gelatinase (MMP-2) in human plasma. J. Immunol. Meth. **148:** 189–198.
40. COOKSLEY, S., J.P. HIPKISS, S.P. TICKLE *et al.* 1990. Immunoassays for the detection of human collagenase, stromelysin, tissue inhibitor of metalloproteinases (TIMP) and enzyme-inhibitor complexes. Matrix **10:** 285–291.
41. FUJIMOT, N., J. ZHANG, K. IWATA *et al.* 1993. A one step sandwich immunoassay for tissue inhibitor of metalloproteinases-2 using monoclonal antibodies. Clin. Chim. Acta **220:** 31–45.
42. CLARK, I.M., L. POWELL, J.K. WRIGHT *et al.* 1992. Monoclonal antibodies against human fibroblast collagenase and the design of an enzyme-linked immunosorbent assay to measure total collagenase. Matrix **12:** 475–480.
43. OHUCHI, E., I. AZUMANO, S. YOSHIDA *et al.* 1996. A one-step enzyme immunoassay for human matirx metalloproteinase 7 (matrilysin) using monoclonal antibodies. Clin. Chim. Acta **244:** 181–198.
44. KJELDSEN, L., O.W. BJERRUN, D. HOVGAARD *et al.* 1992. Human neutrophil gelatinase: A marker for circulating neutrophils. Purification and quantification by enzyme linked immunosorbent assay. Eur. J. Haematol. **49:** 180–191.
45. ZUCKER, S., R.M. LYSIK, M.H. ZARRABI *et al.* 1993. Mr 92,000 type IV collagenase is increased in plasma of patients with colon cancer and breast cancer. Cancer Res. **53:** 140–146.
46. MOSCATELLI, D. & D.B. RIFKIN. 1988. Membrane and matrix localization of proteinases: a common theme in tumor invasion and angiogenesis. Biochim. Biophys. Acta **948:** 67-84.
47. ZUCKER, S., R.M. LYSIK, M.H. ZARRABI *et al.* 1992. Type IV collagenase/gelatinase (MMP2) is not increased in plasma of patients with cancer. Cancer Epidemiol. Biomark. and Prevent. **1:** 475–479.
48. FUJIMOTO, N., N. MOURI, K. IWATA *et al.* 1993. A one-step sandwich enzyme immunoassay for human matrix metalloproteinase 2 (72-kDa gelatinase/type IV collagenase) using monoclonal antibodies. Clin. Chim. Acta **221:** 91–103.
49. GARBISA, S., G. SCAGLIOTTI, L. MASIERO *et al.* 1992. Correlation of serum metalloproteinase levels with lung cancer metastasis and response to therapy. Cancer Res. **52:** 4548–4549.
50. GOHJI, K., N. FUJIMOTO, A. FUJII *et al.* 1996. Prognostic significance of curculating matrix metalloproteinase-2 to tissue inhibitor of metalloproteinases-2 ratio in recurrence of urothelial cancer after complete resection. Cancer Res. **56:** 3196–3198.
51. GOHJI, K., N. FUJIMOTO, T. KOMIYAMA *et al.* 1996. Elevation of serum levels of matrix metalloproteinase-2 and -3 as new predictors of recurrence in patients with urothelial carcinoma. Cancer **78:** 2379–2387.

52. NARUO, S., H.-O. KANAYAMA, H. TAKIGAWA et al. 1994. Serum levels of tissue inhibitor of metalloproteinases-1 (TIMP-1) in bladder cancer patients. Int. J. Cancer **1:** 228–231.
53. GOHJI, K., N. FUJIMOTO, I. HARA et al. 1998. Serum matrix metalloproteinase-2 and its density in men with prostate cancer as a new predictor of disease extension. Int. J. Cancer **79:** 96–101.
54. HAYASAKA, A., N. SUZUKI, N. FUJIMOTO et al. 1996. Elevated plasma levels of matrix metalloproteinase-9 (92-kd type IV collagenase/gelatinase B) in hepatocellular carcinoma. Hepatology **24:** 1058–1062.
55. BAKER, T., S. TICKLE, H. WASAN et al. 1994. Serum metalloproteinases and their inhibitors: markers for malignant potential. Br. J. Cancer **70:** 506–512.
56. ZUCKER, S., R.M. LYSIK, M.H. ZARRABI et al. 1994. Elevated plasma stromelysin levels in arthritis. J. Rheumatol. **21:** 2329–2333.
57. JUNG, K., L. NOWAK, M. LEIN et al. 1997. Matrix metalloproteinases 1 and 3, tissue inhibitor of metalloproteinase-1/tissue inhibitor in plasma of patients with prostate cancer. Int. J. Cancer **74:** 220–223.
58. ZUCKER, S., N. MIAN, M. DREWS et al. 1998. Increased serum stromelysin-1 in systemic lupus erythematosus: lack of correlation with disease activity. J. Rheumatol. **26:** 78–80.
59. WILSON, C.L. & L.M. MATRISIAN. 1996. Matrilysin: an epithelial matrix metallopro.teinase with potentially novel functions. Int. J. Biochem. Cell. Biol. **28:** 123–136.
60. MOSES, M.A., D. WIEDERSCHAIN, K.R. LOUGHLIN et al. 1998. Incresed incidence of matrix metalloproteinases in urine of cancer patients. Cancer Res. **58:** 1395–1399.
61. TAYLOR, D.T., N.T. CHEUNG & P.T. DAWES. 1994. Increased serum pro-MMP-3 in inflammatory arthritides: a potential indicator of synovial inflammatory monokine activity. Ann. Rheum. Dis. **21:** 768–772.
62. MANICOURT, D.-H., N. FUJIMOTO, K. OBATA et al. 1995. Levels of circulating collagenase, stromelysin-1, and tissue inhibitor of matrix metalloproteinases 1 in patients with rheumatoid arthritis. Relation to serum levels of keratin sulfate and systemic parameters of inflammation. Arth. Rheum. **38:** 1031–1039.
63. SASAKI, S., H. IWATA, N. ISHIGURO et al. 1994. Detection of stromelysin in synovial fluid and serum from patients with rheumatoid arthritis and osteoarthritis. Clin. Rheumatol. **13:** 228–233.
64. BAILEY, C.J., R.M. HEMBRY, A. ALEXANDER et al. 1994. Distribution of the matrix metalloproteinases stromelysin, gelatinase A and B, and collagenase in Chron's disease and normal intestines. J. Clin. Pathol. **47:** 113–116.
65. LIEBER, C.S. 1994. Alcohol and the liver: 1994 update. Gastroenterology **106:** 1085–1105.
66. ARTHUR, M.J.P. 1995. Collagenase and liver fibrosis. J. Hepatology **22** (Suppl. 2): 43–48.
67. GEORGE, D.K., G.A. RAMM, L.W. POWELL et al. 1998. Evidence for altered hepatic matrix degradation in genetic haemochromatosis. Gut **42:** 715–720.
68. ARAI, M., M. NIIOKA, K. MARUYAMA et al. 1996. Changes in serum levels of metalloproteinases and their inhibitors by treatment of chronic hepatitis C with interferon. Digest. Dis. Sci. **41:** 995–1000.
69. DOLLERY, C.M., J.R. MCEWAN & A.M. HENNEY. 1995. Matrix metalloproteinases and cardiovascular disease. Circ. Res. **77:** 863–868.
70. HIROHATA, S., S. KUSACHI, M. MURAKAMI et al. 1997. Time dependent alterations of serum matrix metalloproteinase-1 and metalloproteinase-1 tissue inhibitor after successful reperfusion of acute myocardial infaction. Heart **78:** 278–284.
71. TU, F.F., R.L. GOLDENBERG, M.B. DUBARD, T. TAMURA et al. 1998. Plasma matrix metalloproteinase-9 (MMP-9) levels as predictors of spontaneous preterm birth. Obstet. Gynecol. **92:** 446–449.
72. KOLBEN, M., A. LOPENS, J. BLASER et al. 1996. Proteases and their inhibitors are indicative of gestational age. Eur. J. Obstet. Gynecol. **68:** 59–65.

73. OSMERS, R.G.W., B.C. ADELMANN-GRILL, W. RATH *et al.* 1995. Biochemical events in cervical ripening dilatation during pregnancy and parturition. J. Obstet. Gynaecol. **21:** 185–194.
74. RAJABI, M., D.D. DEAN & J.F. WOESSNER. 1987. High levels of serum collagenase in premature labor: A potential biochemical marker. Obstet. Gynecol. **69:** 6166–6170.
75. CLARK, I.M., J.J. MORRISON, G.A. HACKETT *et al.* 1994. Tissue inhibitors of metalloproteinases: Serum levels during pregnancy and labor, term and preterm. Obstet. Gynecol. **83:** 532–537.

# Preclinical and Clinical Studies of MMP Inhibitors in Cancer

ALAN H. DRUMMOND,[a] PAUL BECKETT, PETER D. BROWN,
ELISABETH A. BONE, ALAN H. DAVIDSON, W. ALAN GALLOWAY,
ANDY J.H. GEARING, PHIL HUXLEY, DAVID LABER, MATTHEW McCOURT,
MARK WHITTAKER, L. MICHAEL WOOD, AND ANNETTE WRIGHT

*British Biotech Pharmaceuticals Limited, Watlington Road,
Oxford, OX4 5LY, United Kingdom*

ABSTRACT: The role of matrix metalloproteinases in tumor angiogenesis and growth is now well recognized for models of both human and animal cancer. Clinical studies currently under way with the prototype matrix metalloproteinase inhibitor, marimastat, will establish whether inhibitors of these enzymes are of benefit in the treatment of different types of human cancer. On chronic therapy in humans, marimastat induces a reversible tendinitis that can also be detected in certain animal species. This paper compares the ability of broad-spectrum and various types of selective matrix metalloproteinase inhibitors to induce tendinitis and to exhibit anticancer effects in an animal cancer model. Under conditions in which both systemic exposure and inhibitor potency are controlled, selective inhibitors are less pro-tendinitic, but are weaker anticancer agents than broad-spectrum agents such as marimastat. The clinical relevance of these findings is discussed.

## INTRODUCTION

Despite enormous attention and research over the last 30–40 years, cancer remains one of the major causes of death worldwide. While there is little doubt that the advent of newer and more powerful cytotoxic agents, such as paclitaxel, have led to an improvement in the treatment of the disease, to this day only a few types of cancer can be regarded as truly curable. Death rates from cancer remain stubbornly high in spite of our vastly improved understanding of the causes and factors leading to the progression of the disease. In short, the role of drug therapy, as with surgery, radiation and other treatments, remains much more limited than might have been hoped for after so many years of concentrated effort.

With only a few exceptions, pharmacological approaches to the treatment of cancer have relied, until now, on the doubtful ability of cytotoxic agents to distinguish between normal and cancerous cells. Their use at doses which are the maximum that can be tolerated by cancer patients has led to some success, but there remains an urgent need for novel therapies that target characteristics that distinguish tumors other than that their constituent cells multiply more frequently than most normal cells.

[a]Address for telecommunication: Phone, (44) 1865 781132; fax, (44) 1865 780804; e-mail, drummond@britbio.co.uk

Such considerations have led pharmaceutical companies and academic groups to other anticancer approaches that focus on immunological differences between normal and tumor cells, the hormonal dependence of certain tumors, or the overriding need for rapidly growing tumors to gain access to a robust blood supply in the process known as tumor angiogenesis.

There is now considerable evidence that one class of metalloenzyme, the matrix metalloproteinase (MMP) or matrixin family, plays a key role in tumor angiogenesis as well as in other features of malignancy, namely, local growth/invasion and metastasis.[1] In each case, there is a necessity for the degradation of the stromal matrix during the process and, either directly or indirectly, the tumor is able to achieve this via MMP action.

The purpose of this review is to examine the potential, limitations, and status of MMP inhibition in the treatment of cancer.

## MATRIX METALLOPROTEINASES

The MMPs are a growing family of $Zn^{++}$-dependent enzymes whose original members were described on the basis of their ability to degrade extracellular matrix substrates such as the collagens and proteoglycans. Most of the members of the current, enlarged family would still qualify on this basis, although some of the more recently identified enzymes remain to be fully investigated in this respect. Subdivisions of MMPs can be made on the basis of their preference for certain substrates. For example, there are three "collagenases," all of which share the rare ability to degrade triple-helical collagen at neutral $pH^{2-4}$ (fibroblast collagenase [MMP-1], neutrophil collagenase [MMP-8] and collagenase-3 [MMP-13]), and two "gelatinases", which degrade basement membrane collagens[5,6] (72kDa gelatinase [MMP-2] and 92kDa gelatinase [MMP-9]). Indeed, from a pharmacological perspective, one of the key problems in optimizing the dosing of MMP inhibitors is the fact that multiple enzymes from the MMP family can act on individual matrix protein substrates (see below).

A second issue is that, in many cases, the preferential substrates for MMP members are not known. Moreover, because some enzymes, such as stromelysin-3, do not appear to act on matrix substrates at all, but rather on serpins (serine proteinase inhibitors),[6] it is likely that the consequences of inhibiting MMPs will not be limited to protection of matrix proteins.

## MEMBRANE PROTEIN "SHEDDASES"

In the early 1990s, workers at British Biotech discovered that the production of soluble tumor necrosis factor (TNF-$\alpha$) was dependent on the action of a matrix metalloproteinase-like enzyme.[7] Some MMP inhibitors were found to be powerful inhibitors of TNF-$\alpha$ release in all cells, tissues, and species examined both *in vitro* and *in vivo*.[8]

A range of other membrane proteins were known at the time to be hydrolyzed by uncharacterized proteinases that, at least in some cases, were $Zn^{++}$-dependent and

1,10-phenanthroline-inhibitable.[9] Further studies demonstrated that the release or shedding of transforming growth factor (TGF-$\alpha$), macrophage colony-stimulating factor (M-CSF), epidermal growth factor (EGF), and stem cell factor (SCF, c-kit ligand) were also powerfully inhibited by the same inhibitors with similar structure–activity relationships (Ref. 10 and L.A. Needham, unpublished material). In contrast to results with conventional MMP enzymes, the effects of some of the natural MMP inhibitors (TIMPs)[11] on these processes were not marked, suggesting that even although recombinant proTNF-$\alpha$ could be hydrolyzed to soluble TNF-$\alpha$ by recombinant MMPs, these enzymes may not be the natural "sheddases." Workers at Immunex and Glaxo Wellcome have reported subsequently that the enzyme responsible for the hydrolysis of proTNF-$\alpha$, TNF-$\alpha$ convertase or TACE, is a member of the reprolysin family of $Zn^{++}$-metalloproteinases, the closest known analogues of MMPs.[12,13]

It remains unclear whether multiple reprolysin family members take part in the cleavage of the wide range of membrane proteins now known to be blocked by MMP inhibitors, or whether TACE itself is a particularly promiscuous enzyme. The key issue raised by these discoveries is that certain MMP inhibitors do more than inhibit the breakdown of matrix proteins *in vivo*. The functional consequence of these actions remains largely unknown at present.

## MATRIX METALLOPROTEINASE AND CANCER

There is now a very substantial body of data indicating a role for a number of different MMPs in the local invasion, metastasis, and angiogenesis that are associated with malignant tumors.[14] These data derive from studies in a range of human cancer types and utilize a number of different experimental approaches—immunological, biochemical, and genetic technologies. Because of the complexity of MMP expression and activation and the consequent presence in tissue samples of enzyme–inhibitor complexes, in addition to proenzyme and active enzyme, most of the techniques merely reveal that there is increased expression of MMPs in cancer versus normal tissue rather than the presence of active MMPs.

Even with this proviso, it is evident that the overexpression of MMPs is correlated with the malignant phenotype. In some cases, a correlation has been reported between expression of an MMP and either prognosis or tumor staging,[15,16] but this is not a universal finding, perhaps because our ability to detect the levels of active MMP reliably in a tissue or extract is limited. Nevertheless, it is rare not to find overexpression of a number of MMP enzymes in human tumor tissue. Key questions remain, however. Is any one of the sixteen or so known MMPs routinely associated with cancerous tissue? Do individual human cancer types exhibit characteristic overexpression of groups of MMPs? The answers to these questions are under active study currently, but on first pass there is little to suggest that either question can be answered in the affirmative.

Where analysis of multiple MMPs have been conducted in human cancer, it is routine to find expression of groups of MMPs, rather than one individual enzyme subtype. Human colorectal cancer, for example, has been reported to be associated with the overexpression of at least six separate MMP—matrilysin (MMP-7), fibroblast collagenase (MMP-1), 72- and 92-kDa gelatinases (MMPs-2 and 9), and stromelysins-1 and -3 (MMP-3 and MMP-11).[17] Whether all contribute to the underlying pathology remains unclear at present, and studies with selective MMP inhibitors may be the best way to clarify this.

## MATRIX METALLOPROTEINASE INHIBITORS

The original drug design concepts for MMP inhibitors have been reviewed recently and are the subject of other articles in this volume.[18,19] MMP inhibitor design has evolved over the last 10–15 years and now relies on structure-based design approaches rather than knowledge of the matrix protein sequences that are hydrolyzed by MMPs. This advance has increased the diversity of compounds that retain good MMP inhibitor activity. Nevertheless, the nature of the $Zn^{++}$-binding group employed in these inhibitors remains, as before, largely restricted to hydroxamic acids, carboxylic acids, and, to a lesser extent, thiols. The most important inhibitors are based on hydroxamic acid. Batimastat (BB-94) was the first MMP inhibitor to be widely studied in animal models of cancer.[20] It is a potent broad-spectrum and particularly insoluble MMP inhibitor that, upon i.p. administration to rodents, leads to sustained blood level of 50–100 ng/ml throughout the day.

In retrospect, it was its favorable pharmacokinetic profile in rodents following i.p. injection that generated our considerable interest in MMPs and cancer. It is effective in preventing metastasis in animal models in cancer, as well as in slowing the local invasion and angiogenesis associated with tumor growth.[20] More soluble inhibitors are also effective in animal models of cancer when given i.p. or, with appropriate compounds, p.o. However, when plasma exposure is controlled by continuous infusion, doses of around 10-fold lower can be used to the same effect, and it is likely that the anticancer effects of certain MMP inhibitors have been underestimated historically due to a poor pharmacokinetic profile. Drugs such as marimastat, also a hydroxamate and an analogue of batimastat, are an example of this; only more recent infusion experiments with marimastat confirm the strong anticancer efficacy seen readily with batimastat following i.p. administration. (Data not shown.)

Like batimastat, marimastat is a broad-spectrum MMP inhibitor which has limited ability to block sheddase enzyme.[21] However, it is orally bioavailable in all species so far examined and, because of this, has been in clinical development since 1994. Although other orally bioavailable compounds have followed marimastat into the clinic (AG 3340, Ro 32-3555, CGS 27023A and Bay 12-9566), it is with marimastat that the largest database on the effects and side effects of MMP inhibitors in human cancer patients exists.

The clinical development program for marimastat owes little to previous work with cytotoxic anticancer agents. Firstly, because of its benign toxicological profile, it was possible to conduct normal phase I studies on healthy volunteers. These confirmed the dose-proportional plasma exposure seen in animals and indicated a terminal elimination half-life of 7–10 hours.

Marimastat was well tolerated and the only significant finding was a serial rise in ALT in one subject receiving 200 mg twice daily (20 times the dose used routinely in phase III studies).[22] This led to phase II studies in patients that, for the most part, were unconventional in classical oncology terms.

A series of studies focused on cancer types in which clinical progression is associated with escalation in circulating cancer antigen levels and in which cytoreductive therapy leads to a fall in antigen levels. The intent was to examine the rate of increase in cancer antigen levels in the month before and in the month following the start of marimastat treatment. A decrease in the rate of rise in cancer antigens in the month in which marimastat was administered might indicate a biological effect of the compound and, hence, facilitate the choice of dose for the pivotal phase III studies to follow. The six cancer antigen studies in four cancer types (prostate, ovarian, colorectal and pancreatic cancer) suggested that a dose of 10 mg twice daily was the most appropriate for future studies.[23]

The other important outcome of these studies was the identification of the clinically relevant side effect associated with marimastat treatment.[23] This was an incidence of musculoskeletal pain and inflammation seen particularly in chronic treatment with high doses of marimastat (>25 mg twice daily). This "tendinitis" was reversible within 2 weeks of stopping marimastat treatment, but did not respond well to either nonsteroidal anti-inflammatory drugs or low doses of prednisone (10–20 mg/day).

These human studies echoed the effects seen with marimastat in 6-month toxicology studies in the marmoset, another species in which the drug is both well absorbed orally and has a reasonably long half-life. (Data not shown.)

In contrast, there was no indication of this in the rat, where it is difficult to maintain significant blood levels of marimastat throughout the day following oral dosing (data not shown).

The tendinitic effects of marimastat in human cancer patients occur only slowly at doses below 25 mg twice a day—after 5 months of treatment, the incidence is around 30%.[23] Nevertheless, despite these side effects, which are modest for anticancer drugs, it was important to understand their genesis and whether they could be avoided with other MMP inhibitors.

The marmoset is not an ideal species for this work, both because it is a primate and because it requires prolonged dosing to exhibit clinical or histological signs of the tendinitis. However, further work in this species demonstrated that the pro-tendinitic effects were likely to be a mechanism-based effect: the inactive enantiomer of marimastat at equivalent doses did not elicit signs of tendinitis in 6-month marmoset studies.

A more practical model system was necessary to facilitate research into the causes of the side effect and this was provided when studies in the rat showed that controlled exposure to marimastat by infusing the drug over 14 days resulted in tendinitis which was clinically and histologically similar to that seen in the marmoset (Wood *et al.*, manuscript in preparation). As with the marmoset, this was shown to be mechanism-based (the enantiomer of marimastat again proving inactive). A study was therefore begun to examine whether MMP inhibitors that were more limited in the range of MMP enzymes that were inhibited (narrow-spectrum inhibitors) or were even more broad-spectrum MMP inhibitors than marimastat itself (compounds that in addition blocked sheddases) could avoid causing tendinitis, whilst retaining marimastat-like anticancer efficacy.

There has been considerable debate on the relative merits of broad- and narrow-spectrum inhibitors as both anticancer and pro-tendinitic agents. It has always been

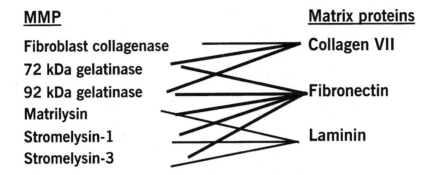

**FIGURE 1.** Elevated expression of MMP enzymes in human colorectal cancer: implications for matrix degradation.

our belief that the expression of multiple MMP types in human cancer and their overlapping substrate preferences (FIG. 1) militate against narrow-spectrum inhibitors as active anti-tumor drugs.

Nevertheless, some studies have demonstrated activity in cancer models with narrower-spectrum inhibitors than marimastat (e.g., CT-1746).[24] We therefore set out to compare the relative merits of a range of MMP inhibitors in the rat tendinitis model and in a mouse B16 melanoma model, in which local growth of the tumor was measured in the flank of the animal. In both models, exposure to MMP inhibitors was controlled by constant drug infusion to around 250 ng/ml (tendinitis, rat) or around 120 ng/ml (cancer, mouse). Four compounds were selected which had different MMP inhibitory profiles: (a) a broad-spectrum MMP inhibitor comparable in activity and potency to marimastat; (b) a broad-spectrum inhibitor which, in addition, was a powerful sheddase inhibitor; (c) a so-called "shallow-pocket" (collagenase-selective) inhibitor; and (d) a "deep-pocket" (gelatinase-selective) inhibitor that, in essence, is a fibroblast collagenase and matrilysin knockout compound. In all four cases, the innate inhibitory potency against the enzymes that were sensitive to the drugs were comparable. Thus, by the combination of controlled and sustained plasma exposure and the choice of equipotent inhibitors that differed only in regard to the spectrum of MMPs that could be inhibited, we hoped to gauge accurately the relative propensity of the four classes of MMP inhibitor to both elicit tendinitis and exhibit anticancer activity.

The results of this work are summarized in TABLE 1. Like marimastat, the broad-spectrum MMP inhibitor exhibited anticancer efficacy but caused tendinitis. At similar levels of MMP inhibitor exposure (see above), neither collagenase-selective nor gelatinase-selective inhibitors caused tendinitis. However, neither showed effective anticancer action at these levels of exposure. In contrast, a broad-spectrum MMP inhibitor with additional ability to block sheddases such as TACE, retained anticancer efficacy without inducing the clinical signs of tendinitis.

Studies of a similar nature are under way in other animal models of cancer in an attempt to confirm or deny the generality of these findings. However, they suggest

**TABLE 1. A comparison of broad-spectrum and narrow-spectrum MMP inhibitors as anticancer and pro-tendinitic agents**

| Type of inhibitor | Activity in | |
| --- | --- | --- |
| | Tendinitis model | Cancer model |
| (A) Broad-spectrum MMP inhibitor | Yes | Yes |
| (B) Broad-spectrum MMP inhibitor with anti-sheddase action | No | Yes |
| (C) Collagenase-selective MMP inhibitor | No | No |
| (D) Gelatinase-selective MMP inhibitor | No | No |

NOTE: The potency of these MMP inhibitors against MMP-1, MMP-2, MMP-9, and MMP-3 were (A) 4, 8, 70, and 15 nM, respectively; (B) 10, 80, 70, and 30 nM, respectively; (C) 7, 200, 300, and 1000 nM, respectively; and (D) 200, 2, 10, and 40 nM, respectively. Blood levels in these animals were sustained throughout the duration of the study at 150–300 ng/ml (tendinitis) and 50–190 ng/ml (cancer).

(a) that narrow spectrum MMP inhibitors are inherently weaker anticancer agents than are broad-spectrum inhibitors such as marimastat, and (b) that some product of sheddase action contributes to the fibroplasia or inflammatory events that are associated with tendinitis.

## ACKNOWLEDGMENTS

The authors wish to thank all the members of the British Biotech MMPI team for their help over the last decade.

## REFERENCES

1. BROWN, P.D. 1997. Matrix metalloproteinase inhibitors in the treatment of cancer. Med. Oncol. **14:** 1–10.
2. WILHELM, S.M., A.Z. EISEN, M. TETER et al. 1986. Human fibroblast collagenase: glycosylation and tissue specific levels of enzyme synthesis. Proc. Natl. Acad. Sci. USA **83:** 3756–3760
3. HASTY, K.A., T.F. POURMOTABBED, G.I. GOLDBERG et al. 1990. Human neutrophil collagenase: a distinct gene product with homology to other matrix metalloproteinases. J. Biol. Chem. **265:** 11421–11424.
4. FREIJE, J.M., I. BIEZ-ITZA, M. BALBIN et al. 1994. Molecular cloning and expression of collagenase-3, a novel human matrix metalloproteinase produced by breast carcinomas. J. Biol. Chem. **269:** 16766–16773.
5. COLLIER, I.E., S.M. WILHELM, A.Z. EISEN et al. 1988. H-ras oncogene-transformed human bronchial epithelial cells (TBE-1) secrete a single metalloproteinase capable of degrading basement membrane collagen. J. Biol. Chem. **263:** 6579–6587.
6. PEI, D., G. MAJMUDAR & S.J. WEISS. 1994. Hydrolytic inactivation of a breast carcinoma cell-derived serpin by human stromelysin-3. J. Biol. Chem. **269:** 25849–25855.
7. CRIMMIN, M.J., W.A. GALLOWAY & A.J.H. GEARING, inventors; British Biotech, assignee. 1993. U.S. patent 5,691,382. Date of application: November 12th.

8.  GEARING, A.J.H., P. BECKETT, M. CHRISTODOULOU *et al.* 1994. Processing of tumour necrosis factor–alpha precursor by metalloproteinases. Nature **370:** 555–557.

9.  EHLERS, M.R.W. & J.F. RIORDAN. 1991. Membrane proteins with soluble counterparts: role of proteolysis in the release of transmembrane proteins. Biochemistry **30:** 10065–10074.

10. ARRIBAS, J., L. COODLY, P. VOLLMER *et al.* 1996. Diverse cell surface protein ectodomains are shed by a system sensitive to metalloprotease inhibitors. J. Biol. Chem. **27:** 11376–11382.

11. BLACK, R.A., F.H. DURIE, C. OTTEN-EVANS *et al.* 1996. Relaxed specificity of matrix metalloproteinases and TIMP insensitivity of tumor necrosis factor α production suggest the major TNFα-converting enzyme is not an MMP. Biochem. Biophys. Res. Commun. **225:** 400–405.

12. BLACK, R.A., C.T. RAUCH, C.J. KOZLOSKY *et al.* 1997. A metalloproteinase disintegrin that releases tumour necrosis factor α from cells. Nature **385:** 729–732.

13. MOSS, M.L. S-L.C. JIN, M.E. MILLA *et al.* 1997. Cloning of a disintegrin metalloproteinase that processes TNFα. Nature **385:** 733–736.

14. STETLER-STEVENSON, W.G., S. AZNOVOORIAN & L.A. LIOTTA. 1993. Tumor cell interactions with the extracellular matrix during invasion and metastasis. Ann. Rev. Cell. Biol. **9:** 541–573.

15. MURRAY, G.I., M.E. DUNCAN, P. O'NEIL *et al.* 1996. Matrix metalloproteinase-1 is associated with poor prognosis in colorectal cancer. Nature Med. **2:** 461–462.

16. VAN DER STAPPEN, J.W.J., T. HENDRICKS & T. WOBBES. 1990. Correlation between collagenolytic activity and grade of histological differentiation in colorectal tumours. Int. J. Cancer **45:** 1071–1078.

17. DUNCAN, M.E., G.I. MURRAY, P. O'NEIL, W.T. MELVIN & J.E. FOTHERGILL. 1996. Expression of matrix metalloproteinase in colorectal cancer. Biochem. Soc. Trans. **24:** 3295.

18. BECKETT, P., A.H. DAVIDSON, A.H. DRUMMOND *et al.* 1996. Recent advances in matrix metalloproteinase inhibitor research. Drug. Dev. Today **1:** 16–26

19. WHITTAKER, M. & P.D. BROWN. 1998. Recent advances in matrix metalloproteinase inhibitor research and development. Curr. Opin. Drug Disc. Dev. **1:** 157–164.

20. NGO, J., A. GRAUL & J. CASTANER. 1996. Batimastat. Drugs Future **21:** 1215–1220.

21. DRUMMOND, A.H., P. BECKETT, E.A. BONE *et al.* 1995. BB-2516: an orally bioavailable matrix metalloproteinase inhibitor with efficacy in animal cancer models. Proc. Am. Assoc. Cancer Res. **36:** 100.

22. MILLAR, A.W., P.D. BROWN, J. MOORE *et al.* 1998. Results of single and repeat dose studies of the oral matrix metalloproteinase inhibitor marimastat in healthy volunteers. Br. Clin. Pharmacol. **45:** 21–26.

23. NEMUNAITIS, J., C. POOLE, J. PRIMROSE *et al.* 1998. Combined analysis of studies of the effects of the matrix metalloproteinase inhibitor marimastat on serum tumor markers in advanced cancer: selection of a biologically active and tolerable dose for longer term studies. Clin. Cancer Res. **4:** 1101–1111.

24. AN, Z., X. WANG, N. WILLMOTT *et al.* 1997. Conversion of a highly malignant colon cancer from an aggressive to a controlled disease by oral administration of a metalloproteinase inhibitor. Clin. Exp. Metastasis. **15:** 184–195.

# Broad Antitumor and Antiangiogenic Activities of AG3340, a Potent and Selective MMP Inhibitor Undergoing Advanced Oncology Clinical Trials

D.R. SHALINSKY,[a,b] J. BREKKEN,[a] H. ZOU,[a] C.D. McDERMOTT,[a] P. FORSYTH,[c] D. EDWARDS,[c] S. MARGOSIAK,[d] S. BENDER,[e] G. TRUITT,[f] A. WOOD,[f] N.M. VARKI,[g] AND K. APPELT[h]

Departments of [a]Pharmacology, [d]Biochemistry, [e]Chemistry, and [h]Ophthalmology Research, Agouron Pharmaceuticals, San Diego, California 92121, USA

[c]Tom Baker Cancer Centre, University of Calgary, Calgary, Alberta, Canada

[f]Department of Oncology, Hoffmann-La Roche Inc., Nutley, New Jersey, USA

[g]Cancer Center, University of California, San Diego, La Jolla, California, USA

ABSTRACT: We studied AG3340, a potent metalloproteinase (MMP) inhibitor with pM affinities for inhibiting gelatinases (MMP-2 and -9), MT-MMP-1 (MMP-14), and collagenase-3 (MMP-13) in many tumor models. AG3340 produced dose-dependent pharmacokinetics and was well tolerated after intraperitoneal (i.p.) and oral dosing in mice. Across human tumor models, AG3340 produced profound tumor growth delays when dosing began early or late after tumor implantation, although all established tumor types did not respond to AG3340. A dose–response relationship was explored in three models: COLO-320DM colon, MV522 lung, and MDA-MB-435 breast. Dose-dependent inhibitions of tumor growth (over 12.5–200 mg/kg given twice daily, b.i.d.) were observed in the colon and lung models; and in a third (breast), maximal inhibitions were produced by the lowest dose of AG3340 (50 mg/kg, b.i.d.) that was tested. In another model, AG3340 (100 mg/kg, once daily, i.p.) markedly inhibited U87 glioma growth and increased animal survival. AG3340 also inhibited tumor growth and increased the survival of nude mice bearing androgen-independent PC-3 prostatic tumors. In a sixth model, KKLS gastric, AG3340 did not inhibit tumor growth but potentiated the efficacy of Taxol. Importantly, AG3340 markedly decreased tumor angiogenesis (as assessed by CD-31 staining) and cell proliferation (as assessed by bromodeoxyuridine incorporation), and increased tumor necrosis and apoptosis (as assessed by hematoxylin

[b]Address for correspondence: Agouron Pharmaceuticals, Inc., 4245 Sorrento Valley Blvd., San Diego, California 92121. Phone, 619/622-3006; fax, 619/622-5999; e-mail, shalinsky@agouron.com

Abbreviations: b.i.d., twice daily; q.i.d., four times daily; BrdU, bromodeoxyuridine; FCS, fetal calf serum; H & E, hematoxylin and eosin; i.p., intraperitoneal; $K_i$, 50% inhibitory concentration; MMP, matrix metalloproteinase; MMP-1, collagenase-1; MMP-2, gelatinase A; MMP-7, matrilysin; MMP-9, gelatinase B; MMP-3, stromelysin; MMP-13, collagenase-3; MMP-14, MT-MMP-1; LC-MS/MS, liquid chromatography in tandem mass spectroscopy; MTD, maximum tolerated dose; NSCLC, non-small cell lung cancer; $T_{1/2\beta}$, terminal elimination half-life; TIMP, tissue inhibitor of metalloproteinase.

and eosin and TUNEL staining). These effects were model dependent, but angiogenesis was commonly inhibited. AG3340 had a superior therapeutic index to the cytotoxic agents, carboplatin and Taxol, in the MV522 lung cancer model. In combination, AG3340 enhanced the efficacy of these cytotoxic agents without altering drug tolerance. Additionally, AG3340 decreased the number of murine melanoma (B16-F10) lesions arising in the lung in an intravenous metastasis model when given in combination with carboplatin or Taxol. These studies directly support the use of AG3340 in front-line combination chemotherapy in ongoing clinical trials in patients with advanced malignancies of the lung and prostate.

## INTRODUCTION

Lung and prostate cancers are among a group of malignancies that undergo extensive invasion and metastasis, and respond poorly to chemotherapies.[1,2] New treatments are needed to improve the outlook for patients with advanced malignant diseases such as lung cancer. Many new agents, including cytotoxic agents and biologic response modifiers, are under development in an attempt to circumvent the resistance that often develops to chemotherapy.[3–6] New therapies that have progressed to the clinic have produced higher response rates and survival times compared to established treatment regimens. However, more effective treatments are still needed.

Inhibition of matrix metalloproteinase (MMP) activity represents a promising noncytotoxic approach towards improving the therapy of aggressive, metastatic disease.[7–12] MMPs are the principal secreted proteases required for degradation of extracellular matrix. Degradation of this matrix, especially the basement membrane, is essential for tumor invasion, metastasis, growth, and angiogenesis.[12,13] At least four classes of MMPs—the gelatinases, collagenases, stromelysins, and membrane-type MMPs—have been identified. Overexpression of MMP-2 (gelatinase A), its specific activator, MMP-14 (MT-MMP-1), and MMP-9 (gelatinase B) is particularly problematic, because these enzymes degrade basement membranes, leading to metastasis and tumor-associated angiogenesis.[12,14–18] Imbalance of expression and activation of MMPs and their natural tissue inhibitors, TIMPs, is a hallmark of aggressive and invasive behavior in many forms of human cancer.[19–25]

The concept that MMPs promote tumor progression is well established. Several MMP inhibitors inhibit metastasis, cellular proliferation, and angiogenesis in tumor models. Hua and Muschel have reported that inhibition of MMP-9 transcription in a rat sarcoma tumor model eliminated metastasis *in vivo*.[26] It has also been reported that angiogenesis is reduced in MMP-2-deficient mice,[16] supporting a direct role for MMP-2 in promoting angiogenesis. TIMPs[27–30] and exogeneous inhibitors of MMPs[31–33] slow tumor growth and inhibit metastasis in animal tumor models, supporting the contention that MMPs promote malignancies. An increasing body of literature suggests that MMPs (MMP-1, MMP-2, MMP-3, MMP-7) are also upregulated under conditions of enhanced carcinogenesis.[16,34–36]

In the search for more effective therapies against solid tumors, we employed protein structure-based drug design to synthesize AG3340, a selective, nonpeptidic inhibitor of MMPs,[37] that potently inhibits MMP-2 and -9 (the gelatinases), MMP-3, MMP-13 and MMP-14 (MT-MMP-1), but spares MMP-1 and MMP-7. The gelatinases and MT-MMP-1 are likely the most relevant targets that should be suppressed for oncology indications.[7–12] AG3340 is the most potent of the MMP inhibitors under development, possessing $K_i$ values of 30–300 pM for inhibiting these critical MMPs, as compared to other inhibitors that block MMP activities with nM to µM affinities. Most other inhibitors also lack the selectivity for these MMPs, with the exception of BAY-12-9566,[38] which also selectively inhibits MMP-2 and MMP-14.

Santos *et al.* initially reported that AG3340 has favorable oral pharmacokinetics in rats and antimetastatic activity against Lewis lung carcinoma *in vivo*.[39] Subsequently, we have investigated the antitumor activity and pharmacokinetics of AG3340 in a number of human and murine tumor models.[40–46] We report here that AG3340 has activity against many human tumor models, both as a single agent and in combination chemotherapy, in immunodeficient mice, and against murine B16-F10 tumors in a model of intravenous metastasis. Along with excellent pharmacokinetics and tolerance, these data support ongoing clinical trials of AG3340 in frontline combination chemotherapy in the treatment of advanced malignancies of the lung and prostate.

## MATERIALS AND METHODS

### *Reagents*

AG3340, 3(S)-2,2-dimethyl-4-[4-pyridin-4-yloxy)-benzenesulfonyl]-thimorpholine-3-carboxylic acid hydroxyamide ($M_r$ 423.5) was synthesized at Agouron Pharmaceuticals, Inc. as previously described.[37] AG3340 was generally solubilized in water (pH 2.3) for oral administration. The solution was filtered under sterile conditions, stored in a refrigerator, and made fresh every 2 weeks. This vehicle was used in all human tumor studies except for the MDA-MB-435 model, in which AG3340 was suspended in 0.5% carboxymethylcellulose : 0.1% Pluronic F68 (Sigma Chemical Corp., St. Louis, MO) in water. Additionally, AG3340 was prepared as a suspension in the same vehicle for i.p. administration in SCID mice bearing U87 tumors. However, in C57BL/6 mice, AG3340 was administered i.p. as an acidic solution. BrdU was purchased from Calbiochem (La Jolla, CA) and stored desiccated at 0–5°C. The antibody to BrdU (RPN202) was purchased from Amersham Inc. (Arlington Heights, IL). Biotinylated goat anti-mouse IgG and 3,3-diaminobenzidine were purchased from Ventana Medical Systems (Tucson, AZ). The antibody to CD-31 and goat anti-rat antibody were purchased from Pharmingen (La Jolla, CA). The Fluorescein In Situ Cell Death Detection Kit to assess apoptosis was purchased from Boehringer Mannheim (Germany).

### *Cell Lines*

Human colon COLO-320DM,[47] and androgen-independent PC-3,[48] and U87 glioma[49] cells were obtained from the American Tissue Culture Collection (Rockville, MD). Human NSCLC MV522[50] tumor cells were kindly provided by Dr. Michael Kelner (UC, San Diego). Human MDA-MB-435 breast[51] tumor cells were kindly provided by Dr. Patricia Steeg (National Cancer Institute) and human KKLS gastric carcinoma cells[29] were kindly provided by Dr. Yutaka Takahashi of the Cancer Research Institute at Kanazawa University (Kanazawa, Japan). COLO-

320DM, MV522, PC-3 cells, and KKLS cells were cultured in RPMI 1640 medium supplemented with 10% FCS and L-glutamine (Mediatech, Inc., Herndon, VA). MDA-MB-435 cells were cultured in Leibovitz's L-15 medium supplemented with 10% FCS, 10 µg/ml insulin, 5 µg/ml cortisol, and 16 µg/ml glutathione. U87 cells were grown in DMEM media containing 12% FCS and L-glutamine. Murine B16-F10 cells were cultured in DMEM medium supplemented with high glucose and 10% FCS. All cells were cultured at 37°C in 95% air:$CO_2$.

## *Animals*

Female athymic Balb/c *nu/nu* nude mice (age 6–8 weeks) were obtained from Bantin & Kingman Universal Limited (Fremont, CA) or from Charles River Laboratories (Wilmington, MA). Male SCID-NOD mice were purchased from the Cross Cancer Institute (Edmonton, Canada) and were housed in the vivarium at the University of Calgary. Immunodeficient animals were housed under sterilized conditions in class II hoods, in accordance with procedures approved by the Institutional Animal Care and Use Committee at Agouron Pharmaceuticals, Inc. C57BL/6 mice were purchased from Bantin & Kingman Universal Limited and were housed in microisolator cages separately from immunodeficient rodents. Rodents in the vivarium at Agouron Pharmaceuticals, Inc. were healthy as confirmed by a sentinel program utilizing nude heterozygote mice housed in isolation cages in animal housing rooms. Mice were generally kept in quarantine for 1 week after arrival and had access to sterilized food and water *ad libitum.* General health was assessed daily.

## *Tumor Biology: Human Xenograft Studies*

### *Studies Conducted at Agouron Pharmaceuticals, Inc.*

Human colon, prostate, lung, gastric and murine melanoma studies were conducted in the Laboratory Animal Resource Center at Agouron Pharmaceuticals, Inc. Studies using COLO-320DM cells were initiated by harvesting serially passaged tumors ($\cong$ 500–1000 mm$^3$) from donor athymic mice and preparing 1- to 2-mm pieces for implantation into naive mice. Tumor cells or pieces were implanted bilaterally (2 sites/mouse). Studies using other human cell lines were initiated by harvesting exponentially growing cells from cell culture and preparing suspensions for s.c. implantation. Mice were randomized, ear-punched for identification, and housed in groups of 3/cage after tumor implantation; each study consisted of control and AG3340-treated groups containing 10–12 animals/group. Tumors were generally allowed to establish for 5 days prior to beginning dosing with AG3340 or vehicle (sterile water, pH 2.3). AG3340 was administered orally using sterile 20 g × 1.5 in. intragastric feeding needles (Popper and Sons, Inc., New Hyde Park, NY). Animals were dosed 7 days/week, b.i.d., at approximately 9 am and 4 pm. Therefore, the total daily dose for a group given 100 mg/kg AG3340, b.i.d., was 200 mg/kg.

Tumor growth was assessed by calculating volumes after measuring the length and width of subcutaneous tumors with electronic calipers. Volumes were calculated using the formula $\frac{1}{2}(length)(width)^2$. Additionally, effects of dosing regimens, vehicles and AG3340 on body weights and general health of mice were assessed throughout experiments.

*Studies Conducted at the University of Calgary*

Human U87 glioma studies were conducted under approval by the Institutional Animal Care and Use Committee. Studies were initiated by implanting $5 \times 10^6$ cells/site in control and AG3340-treated groups containing 5–8 animals/group. Tumors were allowed to establish for 2 to 4 weeks before i.p. dosing with vehicle or AG3340 began. Animals were dosed once daily, 5 days/wk (M–F); on Saturday, a double dose was given, and no dose was given on Sunday. Tumor areas were measured according to the area formula *(length)* × *(width)*.

*Studies Conducted at Hoffmann–La Roche, Inc.*

Human MDA-MB-435 breast cancer studies were conducted under approval by the Institutional Animal Care and Use Committee at Hoffmann–La Roche, Inc. (Nutley, NJ). Studies were initiated by implanting $1.5 \times 10^6$ cells in the mammary fat pad of athymic mice in control and AG3340-treated groups ($n = 10$/group). Tumors were allowed to establish for > 2 weeks before oral b.i.d. dosing with vehicle or AG3340 began on a regimen of 7 days dosing/week. Tumor volumes were measured using the formula $\frac{1}{2}$ *(length)(width)*$^2$.

### *Tumor Biology: Murine Intravenously Induced Metastasis Study*

B16-F10 cells in log-phase were detached from culture with 0.25% trypsin-EDTA, rinsed with media containing FCS, centrifuged, and resuspended in media without serum. Cells were then placed on ice. Experiments were conducted by randomizing control and AG3340-treated animals into 12 animals/group. $10^5$ B16-F10 cells were implanted in 0.2-ml tumor into the tail-vein of C57BL/6 mice using 1-ml sterile syringes fitted with 30-g sterile needles. Cellular viabilities exceeded 90% and viabilities were maintained over the 1–2 hr period required to implant tumor cells into mice.

### *Tumor Collection: Histology*

Tumors were collected after about 48 to 62 days of study as indicated in the Results section. One hour prior to collection, mice were dosed i.p. with 100 mg/kg BrdU (10 ml/kg). After mice were euthanized by cervical dislocation, tumors were surgically removed using sterile instruments, fixed in 10% buffered formalin, and embedded in paraffin. Five-μm sections of tumors were cut using a Leica RM2025 microtome, embedded in paraffin using a Leica Tissue Embedding Console System (Leica EG1160), fixed on slides, deparaffinized, and stained with H & E. Tumor morphologic patterns were analyzed in a masked fashion by a medical histopathologist (N. Varki) with extensive experience in studying preclinical tumor biology models.

### *Tumor BrdU Staining*

To assess cellular proliferation in tumors *in vivo,* animals were dosed with BrdU[52,53] prior to sacrifice. Five-μm tumor sections were fixed on positively charged slides and deparaffinized. The sections were heated for 20 min at pH 5.7 to induce epitope retrieval, and exposed to a monoclonal antibody to BrdU (1:50) for

25 min at ambient temperature followed by biotinylated goat anti-mouse IgG. Peroxidase-conjugated streptavidin was then added, and a brown color for positive staining developed upon addition of 3,3-diaminobenzidine. Sections were then counterstained with hematoxylin. The distribution and intensity of BrdU staining was assessed by quantifying the number of positive staining cells in regions of tumors containing viable carcinoma cells. At least 4 randomly chosen 200 or 400X fields/tumor were assessed in separate experiments ($n \geq 3$ tumors/group/experiment).

### Tumor Anti-CD-31 Staining

Angiogenesis was assessed in tumors by CD-31 staining.[54,55] Frozen sections were cut and allowed to air-dry. Sections were fixed in acetone, blocked, and washed in 10% goat serum with 1% BSA, and exposed to the primary rat anti-mouse antibody to CD-31 for an hour followed by biotinylated anti-rat IgG2a. Alkaline phosphatase-conjugated streptavidin was then added. After adding the alkaline phosphatase substrate, tissues were counterstained with Nuclear Fast Red. Rabbit anti-human-Factor VIII was used as a positive control. CD-31 staining was quantified by counting blood vessels in at least 5 randomly-chosen representative 200X fields/tumor from a minimum of 4 tumors/group.

### Tumor Apoptosis Staining

To assess apoptosis[56] frozen sections of PC-3 and colon tumors were stained with the Fluorescein In Situ Cell Death Detection Kit. Cells undergoing apoptosis were stained green under fluorescent light. The extent of apoptosis was assessed in a masked fashion (N. Varki) by randomly estimating the percentage of apoptotic cells per tumor in a minimum of 4 tumors/group. In the case of PC-3 prostatic tumors, 4–8 tumors/group were analyzed.

### Plasma Concentrations of AG3340 after Oral Dosing

Control mice and mice pretreated for 4 weeks with AG3340 were given a single oral dose of AG3340. In naive mice, blood was collected at time zero and 0.5, 1, 2, 4, 8 and 12 hr after dosing. In pretreated mice, blood was collected on the 30th day of study, 17 hr after the previous day's dose, and at the same time points as for naive mice. Mice were anesthetized with metafane, blood was collected by cardiac puncture, and plasma was obtained by centrifugation (10 min at 3,000 RPM (Sorvall RT7, Newtown, CT). Plasma samples were stored at −70°C until analyzed by liquid chromatography in tandem mass spectroscopy (LC MS/MS). Pharmacokinetic values of AG3340 were calculated using WinNonlin Professional Network Version 1.5 (Science Consulting Inc., Lexington KY) with noncompartmental analysis for extravascular administration (model 200; trapezoidal rule).

### Extraction and LC MS/MS Analysis of AG3340

AG3340 in mouse plasma was quantified using an LC-MS/MS method. Plasma samples were spiked with an internal standard (AG3340 d8) and extracted with methyl *tert*-butyl ether. Extracts were evaporated to dryness and reconstituted with

100 μl of a mixture containing 1% formic acid:acetonitrile (85:15). Twenty-five μl of reconstituted sample containing AG3340 was injected into the LC-MS/MS system. Chromatography was performed on a reverse-phase HPLC column (Zorbax SB C18 column [2.1 mm × 50 mm] at a flow rate of 0.2 ml/min) using a Hewlett Packard 1090 Chem Station equipped with an autoinjector. A mobile phase of 90% $H_2O$:acetic acid (99.9:0.1 v/v; solvent A) and 10% acetonitrile:acetic acid (99.9:0.1 v/v; solvent B) was used to equilibrate the HPLC column. The analysis was run on a gradient from 90:10 A:B to 40:60 A:B over 4.5 min. The retention time for both AG3340 and the internal standard was approximately 3.5 min. Mass detection was achieved on a Micromass Quattro II mass spectrometer using positive ion electrospray at m/z 424 → 128.9 and dwell time of 0.3 sec. Peak area ratios of AG3340/internal standard *vs.* concentration were fitted to a weighted ($1/y^2$) linear regression curve. All curve data and calculated concentrations of AG3340 were obtained using MassLynx (version 2.2) software. The dynamic range of quantification was 0.1–500 ng/ml and acceptable quality control values were ≤ 15%.

### *Isolation of MMPs for Enzymatic Inhibition Assays*

The catalytic domain of human collagenase-1, the catalytic and fibronectin-like portion of human progelatinase A (pro MMP-2), a C-terminally truncated prostromelysin-1, human matrilysin, the propeptide, and catalytic domain of human collagenase-3 were expressed as fusion proteins with ubiquitin in *E. coli.*[57] After purification of the fusion protein, the fibroblast collagenase-1 catalytic domain was released either by treatment with purified, active stromelysin-1 (1:50 w/w ratio), which generated nearly 100% N-terminal Phe1, or by autoprocessing the concentrated collagenase-1 fusion. The other MMPs were activated autocatalytically or with *para* amino phenyl acetate followed by final purification with chelate chromatography.

### In vitro *Enzymatic Assay Procedure*

Assays were performed in 50 mM Tricine, pH 7.5, 200 mM sodium chloride, 10 mM calcium chloride, 0.5 mM zinc acetate containing 2% dimethyl sulfoxide. Stock solutions of inhibitors and substrate were prepared in 100% DMSO at 20 mM and 6 mM, respectively.

The assay method was based on the hydrolysis of MCA-Pro-Leu-Gly-Leu-DPA-Ala-Arg-$NH_2$ (American Peptide Co.) at 37°C.[58] The fluorescence changes were monitored with a Perkin-Elmer LS-50B fluorimeter using an excitation wavelength of 328 nm and an emission wavelength of 393 nm. Substrate (10 μM) and inhibitor were diluted into the assay from a solution in 100% DMSO, and controls substituted an equal volume of DMSO so that the final DMSO concentration from inhibitor and substrate dilution in all assays was 2%. The concentration of enzyme in the assay ranged from 60 pM for gelatinase A to 850 pM for stromelysin and is a function of the enzyme's respective $k_{cat}/K_m$ for the methyl coumarin peptide substrate. Proper determination of steady-state rates of substrate cleavage required assay lengths of 60 min to allow for complete equilibration of the enzyme–inhibitor complex.

## Tight-Binding Kinetic Assays and Analysis

AG3340 binds quite tightly to MMPs and in some cases the concentration of inhibitor evaluated was similar to or less than the concentration of enzyme in the assay. During these instances tight-binding kinetic analysis was needed to determine the correct $K_i$ (50% inhibitory concentration). The classic equation of Morrison[59] was employed using nonlinear curve-fitting techniques to calculate and correct the free inhibitor concentration for the amount of inhibitor that was bound to enzyme during the experiment. In addition, the $K_m$ for the MCA substrate is greater than the solubility limit of 200 μM so that at the 10 μM concentration used, the assumption that $[S] \ll K_m$ is valid and $K_{i,app} = K_i$.

## Data Analyses

Data were plotted as mean ± SEM unless otherwise indicated. Statistical analyses of data were performed using unpaired Student's *t*-test with one- or two-tailed comparison. Differences of $p < 0.05$ were considered to be significantly different from control.

## RESULTS

### Potent Selective Inhibition of MMPs

The structure of AG3340 and $K_i$ values for inhibition of MMPs are shown in FIGURE 1 and TABLE 1, respectively. AG3340 potently and selectively inhibits MMP-2, MMP-9[i], MMP-3, MMP-13, and MMP-14,[i] with $K_i$ values of 30–300 pM. In comparison, AG3340 inhibited the enzymatic activities of MMP-1 and MMP-7 166- and 1080-fold less potently, respectively, than MMP-2. The high potency of AG3340 as an MMP inhibitor arises from the following key inhibitor–protein interactions: (1) bidentate coordination of the hydroxamate function to the catalytic zinc; (2) a key hydrogen bond from the backbone NH of Ala-166; and (3) the near optimal space filling of the S1′ specificity pocket by the pyridyloxyphenyl moiety.[37]

### Growth Inhibition by AG3340 in Human Tumor Models

AG3340 was tested in a variety of human xenograft models with administration orally, b.i.d., or i.p., once daily. In all models, AG3340 was well tolerated. AG3340 did not alter body weights or the health of treated animals compared to vehicle-treated controls in most models. However, AG3340 increased animal survival and also attenuated ill health associated with tumor burden in some models.

### COLO-320DM Colon Tumors

Studies were initiated by serially implanting tumor pieces. Oral, b.i.d. dosing with AG3340 began 5 days later. Control tumors achieved volumes of 1674 ± 285 mm³ (mean ± SEM) after 50 days (FIG. 2A). AG3340 dose-dependently delayed tu-

[i]Robert Martin, Roche Bioscience, Inc., personal communication.

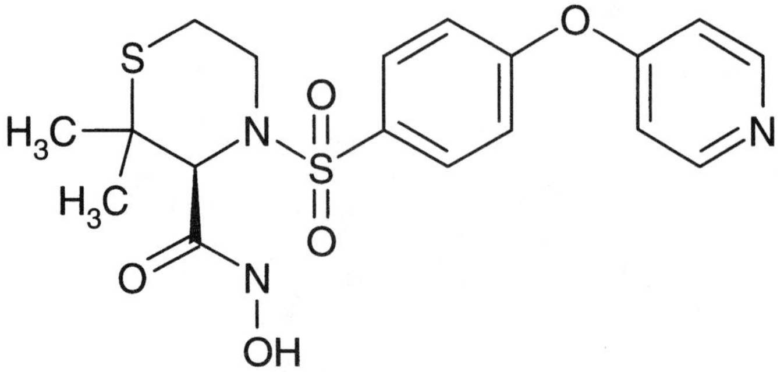

**FIGURE 1.** Structure of AG3340.

mor growth over the range of 12.5 to 100 mg/kg per dose, given b.i.d.[40] One hundred mg/kg AG3340 per dose (b.i.d.) decreased tumor volumes by $79 \pm 9\%$ in three experiments after $\approx 45$ days of study. As shown, AG3340 produced profound growth delays of $\approx 10$, 20, and $>> 30$ days after b.i.d. treatment with 12.5, 25, and 100 mg/kg AG3340 per dose, respectively.[j]

### Effects of AG3340 on Tumor Angiogenesis, Proliferation, and Apoptosis in Colon Tumors

AG3340 markedly decreased tumor angiogenesis as assessed by CD-31 staining (FIG. 5C and D) and cellular proliferation as assessed by BrdU incorporation (FIG. 7, TABLE 2) in COLO-320DM tumors *in vivo* when dosing began 5 days after tumor implantation. These inhibitions were on the order of 50 and $\geq 90\%$, respectively. The inhibition of BrdU incorporation was maintained for at least 2 weeks when dosing was stopped after 48 days of continuous 7 day/wk treatment,[k] indicating that AG3340 had long-lasting effects on the suppression of cellular proliferation. However, AG3340 did not inhibit the growth of established COLO-320DM tumors nor inhibit BrdU incorporation when treatment began 22 days after tumor implantation (data not shown), suggesting that AG3340 interfered with the early establishment of these tumors. Additionally, AG3340 increased apoptosis by 2.9-fold ($15.7 \pm 11.7\%$ [mean $\pm$ SD] vs. $5 \pm 0\%$ in controls; $n = 5$–7 tumors/group) in the region of tumors containing viable carcinoma cells ($p < 0.04$, one-tailed $t$-test).

### MV522 NSCLC Lung Tumors

Control tumors became palpable 5 days after implantation, at which time oral dosing began on a b.i.d. regimen (FIG. 2B). AG3340 decreased tumor growth dose-dependently.[42] After 45 days,[l] tumor volumes in mice treated with 50 and 200 mg/

[j]Tumor growth delay was calculated on the basis of tumors achieving volumes of 1,000 mm$^3$.
[k]D.R. Shalinsky, unpublished data.
[l]Day 45 was chosen arbitrarily. Growth stasis was evident by the 24th day of study.

**TABLE 1.  Inhibition of MMP activity by AG3340**

| Enzyme | $K_i$ (nM)[a] | Number of experiments |
|---|---|---|
| MMP-2 (gelatinase A) | 0.05 ± 0.02 | 4 |
| MMP-13 (collagenase-3) | 0.03 ± 0.01 | 2 |
| MMP-3 (stromeylsin-1) | 0.3 ± 0.1 | 3 |
| MMP-1 (collagenase-1) | 8.3 ± 0.8 | 1 |
| MMP-7 (matrilysin) | 54 ± 7 | 1 |

[a]Values = mean ± SD from experiments performed in triplicate. Values for MMP-9 and -14 of 0.26 and 0.33 nM were determined as referenced in Footnote *i*.

**TABLE 2.  Effects of AG3340 on tumor cell proliferation as assessed by BrdU incorporation**

| Tumor type | Time of tumor collection (days) | Number of BrdU-positive cells per 200 X field | | % Decrease | No. of experiments | *p* value |
|---|---|---|---|---|---|---|
| | | Control (vehicle) | AG3340 (100 mg/kg per dose[a]) | | | |
| COLO-320DM colon | 48 | 236.5 ± 37.5[b] | 10 ± 2 | 96 | 2 | < 0.01 |
| MV522 lung | 39 | 104.9 ± 32.5 | 92.2 ± 12.7[a] | 12 | 1 | > 0.1 |
| U87 glioma | 55 | 131.3 ± 31.8 | 45.0 ± 44.8 | 66 | 1 | 0.003 |
| MDA-MB-435 breast | 56 | 47.3 ± 9.6 | 38.9 ± 7.3 | 18 | 1 | 0.1 |
| PC-3 prostate | 62 | n.q.[c] | n.q. | — | — | — |

[a]The dose was 100 mg/kg per dose b.i.d. for COLO-320DM, MV522, MDA-MB-435, and PC-3-bearing animals, except for MV522 tumor-bearing animals, in which AG3340 was given at 200 mg/kg per dose b.i.d. U87 tumor-bearing animals were given AG3340 only once daily.

[b]Values are mean ± SD from three to eight separate tumors, examining three to five separate 200 or 400X fields/tumor.

[c]n.q., not quantified.

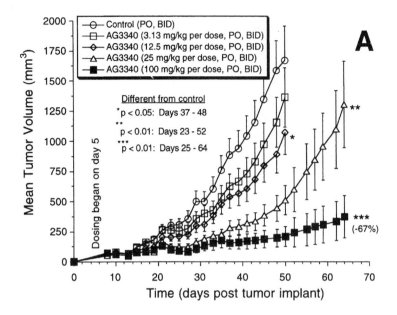

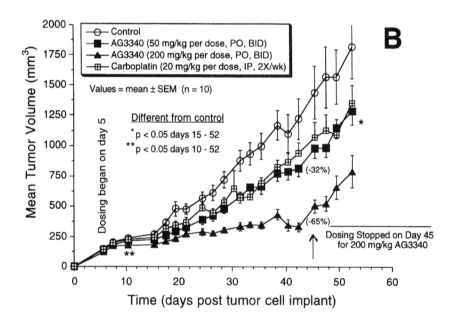

**FIGURE 2.** Inhibition of the growth of human COLO-32ODM colon (**A**), MV522 lung (**B**), U87 glioma (**C**), and MDA-MB-435 breast (**D**) tumors by AG3340. (Panels **C** and **D** are on facing page.) Dosing began 5 days after tumor implantation in colon and lung cancer tumors, and after 30 and 17 days in glioma and breast tumors, respectively.

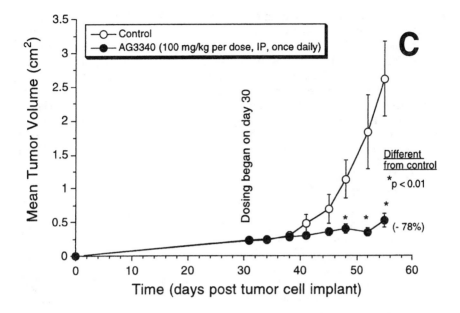

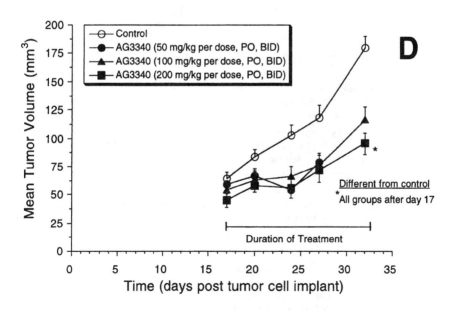

**FIGURE 2/continued.** AG3340 produced significant tumor growth delays in each model. Values are mean ± SEM of 10–12 mice/point, except for glioma tumors, $n = 4$–8. Statistically significant decreases are indicated in figures. **B** and **C** were reprinted by permission from *Clinical Cancer Research*.

kg AG3340, b.i.d., were significantly decreased by $31.8 \pm 2.7$ and $65.1 \pm 7.1\%$, respectively, of control tumor volumes ($p < 0.05$). Growth inhibitions were nearly complete as shown by a flattened growth curve over days 24–42 by 200 mg/kg AG3340 given b.i.d. The high dose of AG3340 did not surpass its maximum tolerated dose (MTD). In comparison, the cytotoxic agent, carboplatin, did not inhibit the growth of these tumors by more than 29% at its MTD (FIG. 2B), confirming the chemoresistance that has been reported for this model.[50] Interestingly, tumors regrew when dosing with AG3340 was stopped, suggesting that AG3340 produced cytostatic effects on tumor growth. AG3340 markedly and dose-dependently decreased angiogenesis and increased tumor necrosis in this model (FIGS. 5A and B and FIG. 6).

## U87 Glioma Tumors

This model was employed to test the effects of AG3340 on established tumors. Tumors were allowed to reach a relatively large size of $\approx 0.25$ cm$^2$ before treatment began. SCID mice bearing U87 tumors were treated once daily with 100 mg/kg AG3340 given i.p. Control tumors ultimately reached a size of $\approx 2.5$ cm$^2$. AG3340 significantly decreased tumor size by 78% ($p < 0.01$; FIG. 2C) before tumors began to regrow.[43] Mice treated with AG3340 survived on average for 101 days before becoming moribund as compared to 61 days for vehicle-treated controls, demonstrating an $\approx 1.7$-fold increase in survival (data not shown).[43] AG3340 markedly decreased tumor cellular proliferation as assessed by BrdU incorporation (TABLE 2) and invasion as assessed by histological analyses of tumors.[43]

## MDA-MD-435 Breast Tumors

This model was also used to test the effects of AG3340 on established tumors. Oral dosing with AG3340, b.i.d., was initiated when tumors reached a volume of $64 \pm 6$ mm$^3$ (FIG. 2D). AG3340 inhibited tumor growth similarly by $\approx 50\%$ over the range of 50 to 200 mg/kg per dose ($p \leq 0.02$), indicating that tumor growth inhibition was maximum at the low dose of AG3340. A dose–response relationship may have been developing over the course of study, but this experiment was not conducted long enough to determine this. AG3340 significantly decreased MDA-MB-435 tumor growth by a similar magnitude in three of four separate studies. In the fourth, the magnitude of inhibition was similar, but inhibitions did not reach statistical significance. In that study, tumor BrdU incorporation into tumor cells was not different between control and AG3340-treated animals (TABLE 2).

## PC-3 Prostate Cancer Tumors

Prostate tumors grew slowly compared to other tumor models. AG3340, administered orally, b.i.d., at 100 mg/kg per dose, was begun 5 days after tumor implantation. A representative study is shown in FIGURE 3. Small growth inhibitions reached statistical significance after 22 days ($p < 0.02$) of study.[49] The average inhibition of tumor growth was $35.9 \pm 8.9\%$ after 55 days of study ($p < 0.05$) ($n = 2$). Tumors ultimately re-grew, demonstrating that AG3340 produced long-lasting growth delays,

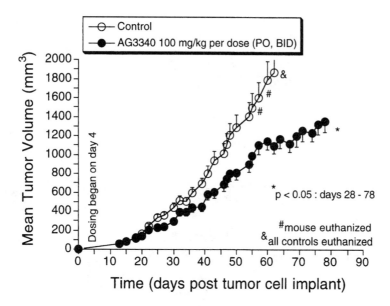

**FIGURE 3.** Tumor growth delay produced by AG3340 in male nude mice bearing human PC-3 prostatic tumors. Values are mean ± SEM of 12 mice/point with two tumors/mouse. All control animals became moribund and were euthanized on day 62; AG3340-treated mice survived more than 80 days (tumor data only shown through day 78).

but did not permanently suppress growth in this model. Importantly, AG3340 attenuated body weight losses by 50% and ill health associated with PC-3 tumor burden, leading to a significant increase in survival.[44]

### Induction of Necrosis and Apoptosis in PC-3 Tumors

AG3340 markedly increased necrosis in PC-3 tumors (FIGURE 4A vs. 4B), confounding the assessment of tumor growth inhibition. Because physical caliper measurements included both viable and nonviable areas of tumors, actual growth inhibition was underestimated. After 62 days, undifferentiated control tumors were composed primarily of viable cells (FIG. 4A). FIGURE 4C shows a higher-magnification section of the control tumor shown in FIGURE 4A. Small regions around the tumor periphery contained both nonviable and viable cells (FIG. 4C), but viable cells predominated. In contrast, AG3340 dramatically increased the necrotic space in tumors, producing a sharp border between viable and nonviable cells. This border is clearly delineated at lower and higher magnifications in H & E-stained tumors (FIGS. 4B and D, respectively). On the basis of masked analyses, there was a 2-fold increase in the percentage of necrosis in tumors, indicating that inhibitions produced by AG3340 were actually ≈ 70% in this model. To investigate further, apoptosis in tumors was assessed by TUNEL assay (FIGS. 4E and F). AG3340 increased the percentage of apoptosis in the viable region of carcinoma cells by

2.4-fold, from $10.6 \pm 4.1\%$ (mean $\pm$ SEM) to $25.0 \pm 6.6\%$ ($p < 0.04$, one-tailed $t$-test; $n = 4$–8 tumors/group). Thus, AG3340 markedly increased PC-3 tumor cell death, and the relatively modest tumor growth inhibitions (FIG. 3) determined by physical caliper measurements did not reflect the prominent growth inhibitions that were revealed by histological analyses.[44]

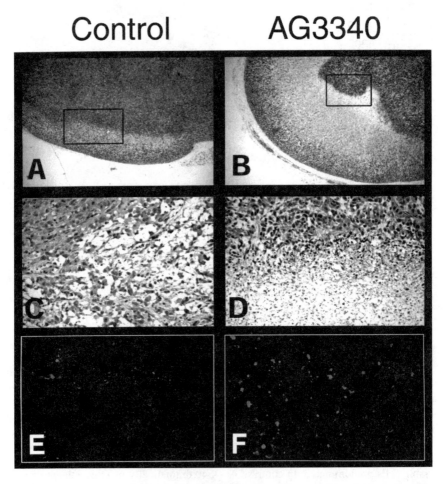

**FIGURE 4.** Morphologic pattern and apoptotic (TUNEL) staining of PC-3 tumors in the presence and absence of AG3340 treatment. Tumors were collected on day 62 of the study shown in FIGURE 3. Control tumors were stained with H & E (**A**, $\times$ 40). The *box* in **A** is shown at higher magnification in panel **C** ($\times$200). AG3340 (**B** and **E**) markedly increased tumor necrosis compared to controls. A sharp border between viable (*upper*) and necrotic (*lower*) cells is shown in **B**, and in a *boxed region* of **B** at higher magnification (**D**). TUNEL staining was increased 2.4-fold in the viable regions of tumors by AG3340 (**F**) as compared to controls (**E**), as described in the Results section.

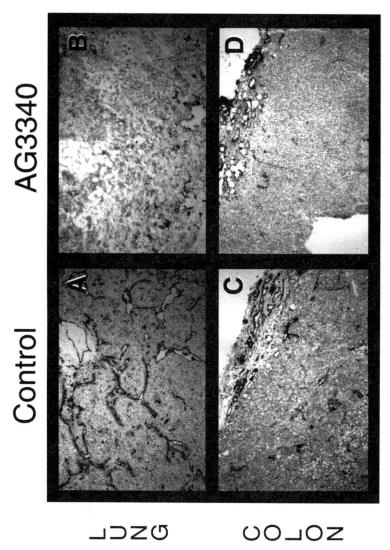

**FIGURE 5.** AG3340 inhibited angiogenesis as assessed by CD-31 staining in NSCLC MV522 lung and COLO-320DM colon tumors. Representative tumors are shown. Control lung (**A**) and colon (**C**) tumors had high staining. AG3340 (100 mg/kg per dose, b.i.d. [colon, **D**]; 200 mg/kg per dose, b.i.d. [lung, **B**]) decreased CD-31 staining in both tumor types. AG3340 also markedly increased necrosis in lung tumors (**B**). Lung and colon tumors were collected on days 48 and 39 of study, respectively (×200). **A** and **B** were reprinted by permission from *Clinical Cancer Research.*

## *Tumor Angiogenesis as Assessed by CD-31 Staining*

To assess angiogenesis, tumors were stained with an antibody to CD-31, an endothelial cell marker that is expressed on new vessels.[54] Representative CD-31 staining from human non-small cell lung and colon cancers in vehicle and AG3340-treated tumors is shown in FIGURE 5. Control lung and colon tumors had high staining (lung [FIG. 5A] > colon [FIG. 5C]), indicating that they were vascularized. After 52 days of study, $97.5 \pm 24.5$ (mean $\pm$ SD) vessels per 200X field were counted in control lung tumors. AG3340, administered orally, b.i.d., decreased angiogenesis by 45 and 77% after dosing with 50 and 200 mg/kg per dose, respectively (FIG. 6).[42] The dramatic effect of AG3340 on decreasing angiogenesis was associated with a large increase in tumor necrosis at a 200 mg/kg dose of AG3340 (FIG. 5B). Decreases in angiogenesis were associated with increases in tumor necrosis, in conjunction with inhibition of tumor growth (FIG. 2B).

In colon tumors, AG3340 decreased CD-31 staining by 46% at 100 mg/kg per dose (FIG. 5D); treated tumors had $22.2 \pm 8.2$ vessels per 20X field vs. $41.2 \pm 14.8$ vessels per field for the controls ($n = 4$ tumors/group from five randomly chosen fields/tumor; four control and two AG3340-treated tumors were scored on day 52, and two more AG3340-treated-tumors were scored on day 64; all from separate animals).[40]

In comparison, PC-3 tumors had $58 \pm 29$ CD-31-positive vessels per 200X field, and AG3340 decreased CD-31 staining by 51.7% ($p < 0.05$, one-tailed $t$-test; $n = 3$–5 tumors/group). In contrast, U87 tumors had negligible CD-31 staining in control

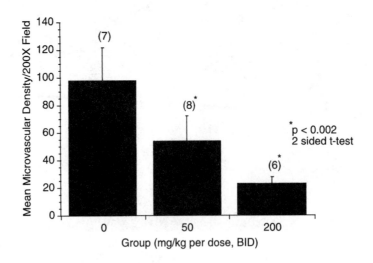

**FIGURE 6.** AG3340 dose-dependently inhibited angiogenesis in NSCLC MV522 tumors. Tumors were collected on day 39 of study. After dosing with 50 and 200 mg/kg AG33340 per dose, given b.i.d., angiogenesis was decreased by 45 and 77%, respectively ($p < 0.002$). Values are mean $\pm$ SEM from number of tumors indicated in parentheses. This figure reprinted by permission from the *Clinical Cancer Research.*

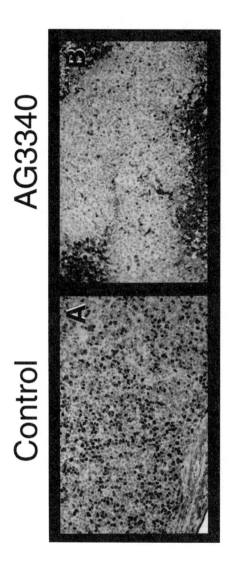

**FIGURE 7.** AG3340 decreased cellular proliferation of COLO-320DM colon tumors as assessed by BrdU staining of viable carcinoma cells. A representative control (**A**) and AG3340 (100 mg/kg, b.i.d.)-treated tumor (**B**) are shown. Nonspecific staining of necrotic areas is also shown in **B**. These results reflect control and AG3340-treated tumors from the experiment shown in FIGURE 2A and TABLE 2. Tumors were collected on day 52 (magnification × 200).

tumors and AG3340 did not alter this staining (data not shown). Angiogenesis was not assessed in KKLS tumors. In summary, AG3340 decreased angiogenesis in three of four tumor models in which CD-31 staining was sufficient for quantification.

### Tumor Cell Proliferation as Assessed by BrdU Staining

The proliferative index of tumors was determined by staining dividing cells with an antibody to BrdU. Representative colon tumor cells had extensive, uniform incorporation of BrdU into their nuclei as shown in FIGURE 7A. High BrdU staining was consistent with the rapid growth rates of these tumors (FIG. 2A). AG3340 markedly decreased BrdU staining of the viable carcinoma cells by $\geq$ 90% (FIG. 7F, TABLE 2).[40] Dose-dependent inhibitions were observed at doses $\geq$ 25 mg/kg on a b.i.d. regimen (data not shown) in conjunction with inhibition of tumor growth (FIG. 2A).

BrdU staining was also assessed in lung, glioma, breast and prostate cancer tumors (TABLE 2). AG3340 also markedly decreased the proliferative index in glioma tumors but did not affect BrdU staining in breast or prostate tumors. However, BrdU staining was quantified in only one of four breast cancer studies. In that study, decreases in tumor growth did not achieve statistical significance in contrast to the other three studies. Thus, a lack of effect in the latter study may not have been unexpected. Additionally, control BrdU staining could not be reliably quantified in the slow-growing PC-3 tumor model because incorporation was very low. Potential changes in BrdU staining were therefore not quantified in PC-3 tumors. In summary, treatment with AG3340 resulted in marked quiescence of human colon and glioma tumor cells *in vivo*.

### Plasma Concentrations of AG3340 after Oral b.i.d. Dosing

AG3340 was rapidly detected in the plasma of nude mice after oral dosing (FIG. 8).[41] Peak plasma concentrations ($C_{max}$) were observed 30 min after dosing; the terminal elimination phase occurred after 2 hr and AG3340 had a terminal elimination half-life ($T_{1/2\beta}$) of 1–2 hr in mice. Trough plasma concentrations approximating 1 ng/ml (2.36 nM) were detected after 12 hr. Thirty minutes after administration of a 100-mg/kg dose in naive mice, a $C_{max}$ of 2081 ng/ml (5 μM) was detected. Plasma concentrations were also determined in AG3340-pretreated mice by giving an additional dose of AG3340 on day 30, 17 hours after the preceding day's dose (after 29 days of daily, 100-mg/kg, b.i.d. dosing). On day 30, the trough concentration was 1 ng/ml (2.36 nM) AG3340 prior to administration of the single dose. Similar plasma concentration vs. time curves were produced in naive mice and in AG3340-pretreated mice, (FIG. 8; TABLE 3).

As shown in TABLE 3, plasma concentrations of AG3340 achieved in nude mice were dose-dependent after oral dosing. In pretreated mice given a single dose of AG330 at 25, 50, and 100 mg/kg AG3340 per dose (p.o., b.i.d.), $C_{max}$ values of 462, 729 and 2081 ng/ml, respectively, were achieved, associated with respective acute $AUC_{0-24hr}$ values of 664, 1559 and 5393 ng*hr/ml.[40] When converted to molarity, peak plasma concentrations of 1100, 1735, and 4900 nM were detected after a single dose of 25, 50 and 100 mg/kg AG3340, respectively.

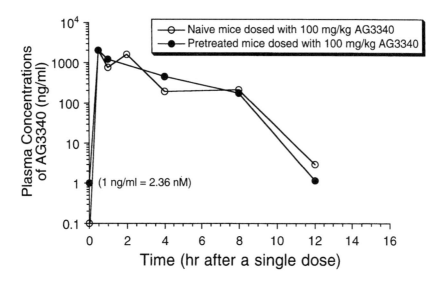

**FIGURE 8.** Plasma concentrations of AG3340 after oral dosing in naive and pretreated mice. Mice were pretreated for 29 days with 100 mg/kg per dose on a b.i.d. regimen of AG3340 prior to receiving a single dose on day 30. The "time zero" on the curve was collected 17 hr after the second b.i.d. dose on the preceding day. Values are mean of 3 mice/point. SD values were generally less than 35% (TABLE 3).

Plasma concentrations were also monitored in C57BL/6 mice given AG3340 i.p. All pharmacokinetic parameters were similar except for the $C_{max}$ and AUC values; the $C_{max}$ value after i.p. dosing was 6–10-fold greater than that produced by oral dosing, with corresponding increases in exposure (data not shown). Otherwise, the plasma vs. time curves were superimposable with the curves generated after oral dosing.

### Efficacy of AG3340 Associated with Maintaining Minimum Effective Plasma Concentrations

Because trough concentrations of ≈ 1 ng/ml were apparently associated with antitumor efficacy on the b.i.d. regimen (FIG. 8), we tested whether efficacy could be produced by maintaining plasma concentrations equal or greater to 1 ng/ml.

By dosing animals four times each day (q.i.d.) with 6.25 mg/kg AG3340, minimum plasma concentrations between 1 and 80 ng/ml were maintained in nude mice (FIG. 9A); concentrations between 1 and 20 ng/ml were produced over most of the day.[41] A total daily dose of 25 mg/kg was compared to a daily dose of 200 mg/kg, given at 100 mg/kg per dose, b.i.d. This b.i.d. dose has produced the highest efficacy in the COLO-320DM tumor model (FIG. 2A). Peak plasma concentrations and total daily exposures were reduced by 90 to 95% on the q.i.d. regimen as compared to the b.i.d. regimen (FIG. 8, TABLE 3). Despite a 12-fold reduction in dose, AG3340 inhibited tumor growth equivalently to that produced by the b.i.d. regimen (FIG. 9B), dem-

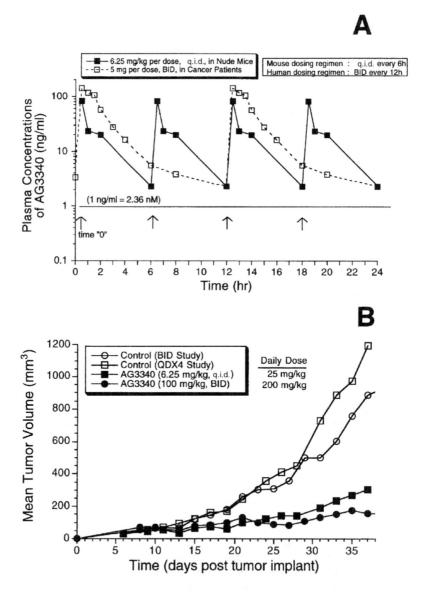

**FIGURE 9. (A)** comparison of plasma concentrations produced by AG3340 after oral dosing in nude mice and in cancer patients. Mice were given 6.25 mg/kg AG3340 four times/day (q.i.d. × 4), for a daily dose of 25 mg/kg. Plasma concentrations ranged between 1 and 80 ng/ml. In cancer patients given 5 mg AG3340, b.i.d., similar plasma concentrations were achieved. Values are mean of 3–6 specimens/point. **(B)** comparison of growth inhibition by AG3340 in the human COLO-320DM tumor model. A total daily dose of 25 mg/kg, given in four fractionated doses, produced the same growth inhibition as when AG3340 was given b.i.d. at 100 mg/kg per dose (also shown in FIG. 2A). Values are mean of 12 mice/group. This figure is reprinted by permission from *Investigational New Drugs*.

**TABLE 3. Pharmacokinetics of AG3340 in nude mice after oral administration**

| Dose of AG3340 (mg/kg) | $C_{max}$[a] (ng/ml) | $AUC_{last}$ after a single dose[b] (ng*h/ml) | Trough plasma concentrations (ng/ml) |
|---|---|---|---|
| 6.25 | 287 ± 175 | 398 ± 182 | time zero: not done<br>12 hr: < 1<br>8 hr: < 1<br>4 hr: 26 ± 9 |
| 25 | 462 ± 495 | 664 ± 248 | time zero: 4.4 ± 1.1[c]<br>12 hr: n.m.<br>8 hr: 8.5 ± 7.4 |
| 50 | 729 ± 240 | 1559 ± 302 | time zero: 3.2 ± 2.3<br>12 hr: 1.2 ± 0.2<br>8 hr: 28 ± 24 |
| 100 | 2081 ± 1675 | 5393 ± 1299 | time zero: 1.0 ± 0.2<br>12 hr: 1.2 ± 0.4<br>8 hr: 174 ± 106 |
| 100[d] | 1988 ± 642 | 5439 ± 337 | time zero: 0<br>12 hr: 3.0 ± 1.8<br>8 hr: 217 ± 105 |

NOTE: The study was conducted for 30 days on a regimen of oral b.i.d. dosing. Plasma concentrations were determined on day 30. Pretreated mice were dosed on day 30 with the same dose of AG3340 as pretreated. Portions of the table were reprinted by permission of *Investigational New Drugs*.

[a]$C_{max}$ coincided with with 30-min timepoint collected after administration of a single dose. Values are mean ± SD from 3 animals per group. The limit of detection for the assay of AG3340 was 0.1 ng/ml.

[b]AUC for the 6.25 mg/kg dose was determined over 6 hr. AUC for other doses was determined over 12 hr.

[c]The "zero" time point was collected on day 30 seventeen hours after the last dose on day 29.

[d]Naive mice were dosed acutely with 100 mg/kg AG3340. Naive mice had no detectable concentration of AG3340 in plasma at time zero.

onstrating that the efficacy was not dependent upon the total daily dose, exposure ($AUC_{0-24hr}$) or $C_{max}$. Rather, efficacy was associated with maintaining minimum effective plasma concentrations of AG3340 (≥ 1 ng/ml) in this tumor model.

For comparison, the plasma concentration vs. time curve for a 5-mg dose of AG3340 administered clinically in cancer patients is also shown in FIGURE 9A, demonstrating that plasma concentrations associated with preclinical efficacy in mice are observed in patients. A b.i.d. regimen in humans maintains these concentrations because of the slower clearance of AG3340 in humans as compared to mice, which require more frequent dosing to attain the same trough concentrations (FIG. 9A). Ongoing clinical trials will tell whether these concentrations result in a positive therapeutic index for AG3340 in cancer patients.

### Combination Therapy with AG3340 and Carboplatin or Taxol

The antitumor efficacy and tolerance of AG3340 as a single agent suggested that it would have utility in combination regimens. AG3340 was subsequently tested in combination with carboplatin or Taxol in various tumor models.

### MV522 Lung Cancer

In this NSCLC model, carboplatin did not inhibit tumor growth by more than 29% at its MTD (FIG. 2B), demonstrating an inferior therapeutic index relative to AG3340.[42] The MTD for carboplatin was used in combination studies, and AG3340 was used at a suboptimal dose (50 mg/kg, b.i.d.), because higher doses almost completely inhibited tumor growth (FIG. 2B). In combination, AG3340 and carboplatin produced inhibitions that were additive on the basis of physical caliper measurements (FIG. 10). However, inhibitions were deemed to have been greater than additive when tumor histologic patterns were taken into account. This is because the combination produced significant increases in tumor necrosis and decreases in angiogenesis relative to the single-agent treatment (data not shown, Ref. 42).

AG3340 was also tested in combination with Taxol in the same model. Taxol produced a very steep dose–response curve associated with marked toxicity (data not shown). To avoid toxicity, a dose of ≈ 50% of the MTD was used. This dose did not inhibit tumor growth. In combination with a suboptimal dose of AG3340, Taxol potentiated the efficacy of AG3340, resulting in enhanced tumor growth inhibition

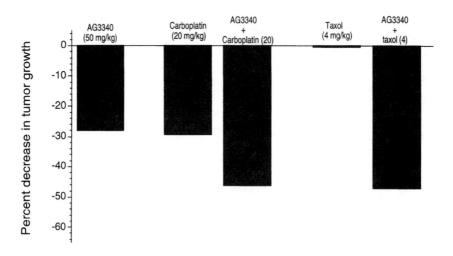

**FIGURE 10.** Enhanced inhibition of the growth of NSCLC MV522 lung tumors by AG3340 and carboplatin or Taxol. Values are the mean percent decrease in tumor growth from 10–12 mice/bar. AG3340 was administered orally at 50 mg/kg per dose, b.i.d. Carboplatin was administered i.p. twice weekly at 20 mg/kg per dose. Taxol was administered three times weekly at 3.75 mg/kg. AG330 and carboplatin additively inhibited tumor growth, whereas Taxol potentiated the efficacy of AG3340.

(FIG. 10).[42] In this NSCLC model, AG3340 was well tolerated compared to either Taxol or carboplatin, and was well tolerated in combination with either agent (data not shown).

### Combination Therapy with AG3340 and Taxol in a Human Gastric Model That Is Refractory to AG3340

Combination treatment with Taxol was also studied in a human gastric KKLS tumor model that is insensitive to AG3340 (FIG. 11). AG3340 was given orally, b.i.d. at 100 mg/kg per dose, and Taxol was given i.p, three times weekly at 3.75 mg/kg per dose. The dose of Taxol chosen was suboptimal because Taxol fully inhibited tumor growth at higher doses (data not shown). AG3340 potentiated the growth inhibition produced by Taxol, significantly increasing the difference between the growth inhibition produced by Taxol alone and that produced by Taxol in combination with AG3340 (FIG. 11). Tolerance to the combination was not different from either agent alone (data not shown). These data demonstrated that a dose of AG3340 which was ineffective when administered alone could potentiate the activity of Taxol in a tumor model that is refractory to AG3340.

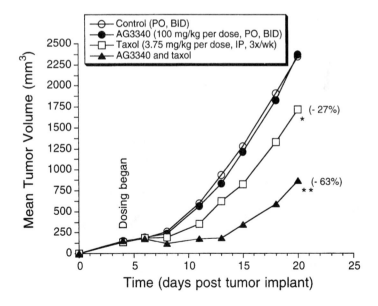

**FIGURE 11.** AG3340 potentiated the efficacy of Taxol in human KKLS gastric tumors. Values are mean of 12 animals, each bearing 2 tumors/mouse. Taxol significantly decreased tumor growth by 27%. The combination of AG3340 and Taxol decreased growth further. * $p < 0.05$ vs. control; ** $p < 0.05$ vs. Taxol.

### Antimetastatic Activity of AG3340 in an Intravenous Metastasis Model

To further explore the effects of AG3340 in combination therapy, an intravenous metastasis model was employed using B16-F10 tumor cells.[45] In this model, tumors arise in the lung after systemic extravasation following intravenous tail vein injection of tumor cells. AG3340 was given orally, b.i.d., beginning 1 day after tumor implantation. Carboplatin was administered i.p., two times weekly, or Taxol was administered i.p. three times weekly, beginning at the same time. In this model, AG3340 modestly decreased the number of lung lesions by 33%, but decreases did not reach statistical significance ($p > 0.1$). In contrast, i.p. administration of AG3340 in this model produces statistically significant decreases in lesion number.[45] Carboplatin was inactive at a dose approximating its MTD, but clearly potentiated the activity of AG3340, resulting in a 72% decrease in lesion number relative to control or monotherapies ($p < 0.02$; FIG. 12A). In comparison, Taxol was active in this model. At doses of 7.5 and 15 mg/kg (approximate MTD), lesion number was decreased by 47 and 77%, respectively ($p < 0.05$ for 15 mg/kg dose only). Thus, Taxol did not produce a statistically significant decrease in lesion number at 7.5 mg/kg, but in combination with AG3340, these agents produced a 68% decrease in lesion number relative to controls and monotherapies ($p < 0.05$; FIG. 12B). As in the human tumor models, AG3340, alone and in combination, was well tolerated. In summary, AG3340 produced superior antitumor efficacy when used in combination with the cytotoxic agents, carboplatin or Taxol, in a number of tumor models.

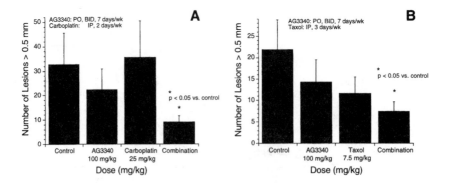

**FIGURE 12.** Enhanced inhibition of the number of B16-F10 lesions developing in the lung by AG3340 and carboplatin or AG3340 and Taxol. Values are mean ± SEM of 10–12 mice/point. AG3340 was given orally beginning 1 day after tumor cell implantation. Carboplatin or Taxol were given two or three times per week, respectively. The combination decreased the number of lung lesions to a greater extent than monotherapy ($p < 0.05$, one-sided $t$-test).

## DISCUSSION

### *Overview*

In the search for more effective chemotherapy agents, we have designed a novel MMP inhibitor, AG3340,[37] to target a key subclass of extracellular metalloproteinases, including gelatinases and MT-MMP-1 (MMP-14), that are involved in tumor invasion, metastasis, and angiogenesis.[7–12] We report here that AG3340 has significant antitumor efficacy against many human tumors *in vivo*, associated with inhibition of angiogenesis, cellular proliferation, and altered apoptosis. Furthermore, promising antitumor activities were also observed when AG3340 was administered in combination with cytotoxic agents in both AG3340-sensitive and refractory tumor models.

### *Potency and Selectivity of AG3340*

The pM potency and selectivity of AG3340 towards inhibition of MMP-2, MMP-9 and MMP-14 (TABLE 1, footnote *b*) distinguishes AG3340 from other MMP inhibitors that antagonize MMPs with nM affinities.[11,33,38] On the basis of poor affinities for inhibition of MMP-1 and MMP-7, our data suggest that these MMPs played negligible roles in the anticancer efficacies produced by AG3340 in the tumor models that were studied. Conversely, the MMPs that are potently inhibited by AG3340 may play prominent roles in promoting tumor growth.

Gelatinases represent proteases that play crucial roles in tumor progression associated with enhanced invasion, metastasis, and angiogenesis *in vitro* and *in vivo*.[7,14,16–18,26,28–30] This contention is supported by data from other MMP inhibitors that inhibit gelatinases nonselectively[31,32] and selectively.[33,38] We have hypothesized that selective inhibition of key MMPs, such as the gelatinases (MMP-2 and MMP-9) and MMP-14, relative to MMP-1, will produce a superior MMP inhibitor for oncology indications. AG3340 was designed with this inhibitory profile in mind in an attempt to decrease the adverse effects (i.e., joint toxicities, arthralgias) that have been associated with the use of broad spectrum MMP inhibitors. The Bayer Corporation has created BAY 12-9566 with a similar goal of inhibiting key MMPs and sparing inhibition of MMP-1.[38] This approach is supported by a recent report in which ubiquitous expression of MMP-1 was illustrated in normal joints.[60]

### *Growth-Inhibitory Effects of AG3340*

AG3340 was studied most often in human tumors implanted subcutaneously into nude mice. Inasmuch as these tumors generally do not metastasize from the site of implantation, effects of AG3340 were limited to a primary tumor site in these models. Physical caliper measurements showed marked growth inhibitions of 65–79%, in three of six models (lung, colon, glioma). Inhibition of growth was coupled with significant tumor growth delays. The large magnitude of these inhibitions with AG3340 compares favorably with that of other MMP inhibitors, which generally have only modest effects on primary tumor growth in xenograft models.[61,62] AG3340 also enhanced animal survival in the glioma model. In two other models (prostate and breast cancer), AG3340 moderately inhibited tumor growth by 40–50% (based on physical caliper measurements). However, growth inhibitions in the PC-3 prostatic model were largely underestimated due to significant increases in tumor necrosis that were included in the tumor measurements. After accounting for necrosis, it was determined that AG3340 had also markedly inhibited the growth of PC-3 prostatic tumors by $\approx 70\%$ ($p < 0.05$). These decreases were associated with significant tumor growth delays, attenuated ill health associated with tumor burden, and importantly, increased animal survival.[44] Thus, AG3340 clearly increased animal survival in two separate subcutaneously implanted tumor models. A survival advantage was also suggested by the long tumor growth delays produced in the MV522 tumor model. In the breast cancer model, histologic studies of the tumor were not performed so the histological basis of growth inhibition could not be assessed. In one model (KKLS gastric), AG3340 did not inhibit tumor growth. Thus, most tumors were moderately to markedly sensitive to the growth inhibitory effects of AG3340.

As a single agent, AG3340 generally inhibited tumor growth in a cytostatic manner. Profound tumor growth delays were produced by AG3340 in human colon, glioma, and prostatic tumors. In the colon and lung models, AG3340 dosing was stopped after 40 to 50 days and tumors began to regrow shortly thereafter, demonstrating that continued dosing of AG3340 was necessary to maintain tumor growth inhibition. In the colon model, increased tumor volumes were artifactual. Colon tumors filled with hemorrhagic blood and fluid after cessation of dosing.[m] However, as indicated in the Results section, BrdU incorporation in the viable carcinoma cells remained inhibited 2 weeks after dosing was stopped. Nevertheless, these data demonstrate that monotherapy with AG3340 did not permanently suppress human tumor growth, suggesting that AG3340 may have better clinical utility in combination chemotherapy regimens.

### Inhibition of Angiogenesis by AG3340

Growth inhibition produced by AG3340 across tumor models led us to examine angiogenic, proliferative, and apoptotic status in tumors. Angiogenesis was assessed by CD-31 staining of tumor-associated blood vessels in three of the four human tumor models (colon, lung, prostate) in which staining was sufficient for quantitation. These models were all sensitive to the growth-inhibitory effects of AG3340.

AG3340 markedly decreased angiogenesis in the NSCLC lung model. Inhibition was dose dependent and ranged from 45 and 77% after dosing with 50 and 200 mg/kg AG3340, given b.i.d., respectively.[42] In the colon and prostatic models, only one dose (100 mg/kg, b.i.d.) was tested; AG3340 decreased angiogenesis by approximately 50% across these models. Additionally, AG3340 decreased angiogenesis, by $\approx 46\%$, as assessed by Factor VIII staining, in a human skin graft model in SCID mice in which PC-3 cells are allowed to invade acellular human dermis in the graft.[n] AG3340 also decreased angiogenesis dose-dependently by up to 77% after i.p. administration of 1.6 to 16 mg/kg, once daily, in an ophthalmic model of neovascularization in newborn mice.[63] However, AG3340 did not appear to be a major mechanism by which AG3340 inhibited U87 glioma tumor growth, but these tumors had such low CD-31 staining in controls that it was difficult to assess potential

[m]D.R. Shalinsky, unpublished data.
[n]J. O'Leary and M. Shuman, University of California, San Francisco, personal communication.

changes. These data demonstrate that AG3340 inhibits angiogenesis across a variety of human tumor and murine ophthalmology models.

## Inhibition of Tumor Cell Proliferation by AG3340

AG3340 also profoundly inhibited other cellular markers commonly associated with neoplastic growth. For example, AG3340 markedly inhibited cellular proliferation in tumors *in vivo* as shown by inhibition of BrdU incorporation (TABLE 2). Marked inhibition of proliferation was demonstrated in the human colon and glioma models, whereas small effects were observed in the lung and breast cancer models. BrdU incorporation was too low in the prostate model to be reliably quantified. However, control MV522 lung tumors had reasonable basal incorporation of BrdU and changes could have been quantified had they occurred. AG3340 significantly inhibited MDA-MB-435 growth in three of four studies, indicating that there was some variability in response to AG3340 in the breast cancer model. Unfortunately, BrdU staining was assessed only in the fourth study, calling into question whether cellular proliferation may have been decreased in the other studies. Thus, AG3340 markedly decreased tumor cell proliferation in two of four models, demonstrating that tumor cells became quiescent in a model-dependent manner after extended treatment with AG3340 *in vivo*.

## Induction of Apoptosis by AG3340

Apoptosis was examined in two models: human prostate and colon. AG3340 increased TUNEL staining by 2.4-fold in the prostatic model, indicating that AG3340 treatment influenced the rate of tumor cell death as well as tumor cell proliferation. In the prostatic model, apoptosis was associated with tumor necrosis. In comparison, tumor necrosis was increased in colon tumors, but to a lesser extent than in the prostatic tumors. The smaller increase may have been due to a greater basal level of necrosis in the colon as compared to the prostate tumors. Additionally, AG3340 profoundly increased necrosis in NSCLC lung tumors (FIG. 4B). Investigation into an induction of apoptosis has not been conducted in these tumors. These data demonstrate convincingly that some tumors undergo extensive loss of viability associated with increased apoptosis after AG3340 treatment.

Tumor cellular quiescence or death may have resulted from direct effects of AG3340 on tumor cells[64] or indirectly from a decrease in blood supply.[5] It is unlikely that nonspecific cytotoxicity decreased tumor growth because plasma concentrations that were achieved after oral dosing (FIG. 8) did not affect tumor cell proliferation *in vitro*.[40] Based on the breadth of the antiangiogenic spectrum of activity of AG3340 *in vivo,* the antitumor efficacy of AG3340 likely results, at least in part, from an inhibition of tumor blood supply. However, because marked growth inhibitions produced by AG3340 were apparently not associated with a high degree of antiangiogenesis in human glioma tumors, and because inhibition of MMP-2 activity by genetic manipulation can induce cellular quiescence *in vitro* (in the absence of the extracellular matrices that exist *in vivo*,[64]), AG3340 may have inhibited tumor growth by other MMP-related antiproliferative mechanisms as well.

## Plasma Concentrations of AG3340 after Oral Dosing:
### Concentrations Associated with Antitumor Efficacy

Oral administration of AG3340 resulted in dose-dependent increases in plasma concentrations (TABLE 3) associated with dose-dependent decreases in tumor growth (FIGS. 2A and B). In view of the plasma concentrations achieved after oral dosing and the $K_i$ values for inhibition of MMP activity (TABLE 1), it is reasonable to expect that gelatinases would have been inhibited by doses of AG3340 $\geq$ 25 mg/kg per dose (oral, b.i.d.). As shown in FIGURE 2, AG3340 significantly delayed tumor growth at doses $\geq$ 25 mg/kg, suggesting that sufficient AG3340 was delivered to tumors to inhibit their growth. Because trough levels of $\approx$ 1 ng/ml (2.36 nM) AG3340 were achieved across effective doses, further studies were conducted to test whether efficacy was associated with maintaining trough concentrations in plasma.

Minimum plasma concentrations were maintained by dosing four times per day; the antitumor efficacy of AG3340 was determined to be associated with maintaining minimum effective plasma concentrations above 1 ng/ml (FIG. 9A). Efficacy was dissociated from total daily dose, exposure, and $C_{max}$ values in the COLO-320DM colon tumor model. Importantly, AG3340 plasma concentrations and $AUC_{0-24hr}$ values associated with preclinical efficacy (FIG. 9A) are readily achievable with b.i.d., oral dosing in healthy human volunteers[65] and in cancer patients,[41] suggesting that efficacious doses can be delivered to humans. In summary, growth inhibitory effects of AG3340 were associated with maintaining minimum effective plasma concentrations in nude mice. Ongoing clinical trials in patients with advanced malignancies will answer the question of whether these preclinical data are predictive of clinical response for AG3340.

## Rationale for Using AG3340 in Combination Chemotherapy Regimens

Clearly, the efficacy of AG3340 as a single agent across tumor models was encouraging. The lack of a permanent suppression of tumor growth across models supported a case for use of AG3340 in combination chemotherapy. A combination approach was also supported by the observation that AG3340 inhibited the growth of COLO-320DM tumors only when administered early after tumor implantation. Kerbel, Folkman, Teicher, and Brem have advocated combining antiangiogenic with cytotoxic agents to improve cancer chemotherapy.[5,6,66] Ample data collected by them, others, and ourselves in preclinical tumor models support this contention.[42,45,66–70] We agree with this approach because, as important as MMPs are for promoting aberrant cellular growth, they are not the only contributors to neoplastic progression;[71,72] inhibition of this pathway alone would therefore not be expected to fully inhibit tumor growth. Thus, AG3340 was studied in combination therapy with the goal of enhancing antitumor efficacy.

## Promising Tumor Growth Inhibition with AG3340
### in Combination with Carboplatin or Taxol

In chemoresistant NSCLC MV522 tumors, the combination of AG3340 and carboplatin or Taxol markedly enhanced tumor growth inhibition compared to results achieved by their respective monotherapies. In the case of Taxol, a strong potentiation of the efficacy of AG3340 was observed. In comparison, carboplatin had weak activity on its own that was significantly increased by co-treatment with AG3340. The magnitude of inhibition in combination was additive, as determined by physical caliper measurements, but was actually greater when tumor necrosis and inhibition of angiogenesis were taken into account.[42] Importantly, the antiangiogenic effects of AG3340 were markedly increased by the combination of AG3340 and carboplatin, suggesting that the antiangiogenic activity of AG3340 was a major mechanism of increased inhibition.

In the B16-F10 murine melanoma model, similar growth-inhibition data were generated, demonstrating enhanced efficacy by the combination of AG3340 and Taxol or carboplatin. Additionally, AG3340 potentiated the growth-inhibitory effects of taxol in KKLS gastric tumors that are insensitive to AG3340. Thus, combination treatment with AG3340 and cytotoxic agents was promising in models possessing varied sensitivities to either AG3340 or to cytotoxic agents. These results were obtained with positive therapeutic indices, indicating that AG3340 and these cytotoxic agents were well tolerated when administered together.

## SUMMARY

This report demonstrates marked antitumor efficacy of AG3340, as a single agent and in combination therapy, against many human and murine tumor types. AG3340 was well tolerated after oral and i.p. dosing, and in combination with cytotoxic agents in immunodeficient mice. Growth inhibition was associated with inhibition of tumor-associated angiogenesis and cell proliferation, and with increased tumor necrosis and apoptosis *in vivo*. Thus, this potent and selective MMP inhibitor produced a remarkable range of anticancer activities, providing a strong rationale for the clinical testing of AG3340 in oncology.

Importantly, plasma concentrations associated with preclinical antitumor efficacy are achieved in cancer patients after oral, b.i.d. dosing, providing an opportunity to directly relate our preclinical MMP research data into clinical studies. Additionally, the clinical doses being tested have been designed to selectively inhibit key MMPs, such as the gelatinases and MT-MMP-1, in cancer patients. In conclusion, these data support ongoing clinical studies of AG3340 in front-line combination chemotherapy targeted against advanced malignant lesions of the lung and prostate.

## ACKNOWLEDGMENTS

We thank Stanley Robinson, David Gonzalez, Rich Daniels, and Cecile Bowser-Robinson (Agouron Pharmaceuticals), Jhong Qiad Shi (University of Calgary), and Jeannene Butler (Hoffmann-La Roche, Inc.) for expert assistance in the conduct of animal studies. We also thank Stan Kolis (Hoffmann-La Roche, Inc.) for conduct of LC-MS/MS studies, and Dr. Scott Zook for synthesis of AG3340, Eleanor Dagostino and James Register for conducting enzyme assays, and Linda Musick (Agouron Pharmaceuticals, Inc.) for protein purification. Additionally, we thank Patricia Dill and Larry Eck (US Laboratories, Irvine, CA) for conducting BrdU staining, and Mary Ann Lawrence and Jing Xue Yu (UC, San Diego) for histology and immunohistochemistry support. Emory Emrich, John Gale, and Muizz Hasham (Agouron Pharmaceuticals, Inc.) are also gratefully acknowledged for graphic arts support.

# REFERENCES

1. RUDDON, W.R. 1955. Characteristics of human cancer. *In* Cancer Biology, 3rd ed.: 3–18. Oxford University Press. New York.
2. TRICHOPOULOS, D., L. LIPWORTH, E. PETRIDOU & H-O. ADAMI. 1997. Epidemiology of cancer. *In* Cancer: Principles and Practice of Oncology, 5th ed. V.T. DeVita, S. Hellman & S.A. Rosenberg, Eds. J.B. Lippincott. Philadelphia.
3. BUNN, JR., P.A. & K. KELLY. 1998. New chemotherapeutic agents prolong survival and improve quality of life in non-small cell lung cancer: a review of the literature and future directions. Clin. Cancer Res. **5:** 1087–1100.
4. SHEPHERD, F.A. 1997. Alternatives to chemotherapy and radiotherapy as adjuvant treatment for lung cancer. Lung Cancer **17** (Suppl. 1): 121–136.
5. FOLKMAN, J. 1971. Tumor angiogenesis: therapeutic implications. N. Engl. J. Med. **285:** 1182–1186.
6. KERBEL, R.S. 1991. Inhibition of tumor angiogenesis as a strategy to circumvent acquired resistance to anti-cancer agents. BioEssays **13:** 31–36.
7. LEVY, D.E. & A.M. EZRIN. 1997. Matrix metalloproteinase inhibitor drugs. *In* Emerging Drugs: The Prospective for Improved Medicines. Ashley Publications Ltd.
8. WOJTOWICZ-PRAGA, S.M., R.B. DICKSON & M. HAWKINS. 1997. Matrix metalloproteinase inhibitors. Invest. New Drugs **15:** 61–75.
9. MORPHY, J.R., T.A. MILLICAN & J.R. PORTER. 1995. Matrix metalloproteinase inhibitors: current status. Cur. Med. Chem. **2:** 743–762.
10. RAY, J.M. & W.G. STETLER-STEVENSON. 1996. Matrix metalloproteinases and malignant disease: recent developments. Exp. Opin. Invest. Drugs **5:** 323–335.
11. BROWN, P.D. 1997. Matrix metalloproteinase inhibitors in the treatment of cancer. Med. Oncol. **14:** 1–10.
12. LIOTTA, L.A., K. TRYGGVASON, W.G.I. MOOR & M.K. BODDEN. 1980. Metastatic potential correlates with enzymatic degradation of basement membrane collagen. Nature **284:** 67–68.
13. FIDLER, I. 1997. Molecular biology of cancer: invasion and metastasis *In* Cancer: Principles and Practice of Oncology, 5th ed. V.T. DeVita, S. Hellman, & S.A. Rosenberg, Eds. J.B. Lippincott. Philadelphia.
14. BROOKS, P.C., S. STROMBLAD, L.C. SANDERS, T.L. VON SCHALSCH, R.T. AIMES, W.G. STETLER-STEVENSON, J.P. QUIGLEY & D.A. CHERESH. 1996. Localization of matrix metalloproteinase MMP-2 on the surface of invasive cells by interaction with integrin $\alpha V\beta 3$. Cell **85:** 683–693.
15. VARNER, J. & D. CHERESH. 1997. Tumor angiogenesis and the role of vascular cell integrin $\alpha v\beta 3$. *In* Importance Advances in Oncology. V. DeVita, S. Hellman & S.A. Rosenberg, Eds. Lippincott-Raven. Philadelphia.
16. ITOH, T., M. TANIOKA, H. YOSHIDA, T. YOHSIOKA, H. NISHIMOTO & S. ITOHARA. 1998. Reduced angiogenesis and tumor progression in gelatinase-A deficient mice. Cancer Res. **58:** 1048–1051.
17. SATO, H., T. TAKINO, Y. OKADA, J. CAO, A. SHINAGAWA, E. YAMAMOTO & M. SEIKI. 1994. A matrix metalloproteinase expressed on the surface of invasive tumor cells. Nature **370:** 61–65.
18. STRONGIN, A.Y., I. COLLIER, G. BANNIKOV, B.L. MARMER, G.A. GRANT & G.I. GOLDBERG. 1995. Mechanism of cell surface activation of 72-kDA type IV collagenase. J. Biol. Chem. **270:** 5331–5338.
19. NAGASE, H. 1998. Stromelysins 1 and 2. *In* Matrix Metalloproteinases. W.C. Parks & R.P. Mecham, Eds.: 43–68. Academic Press. San Diego, CA.

20. EDWARDS, D.R., P.P. BEAUDRY, T.D. LAING, V. KOWAL, K.J. LECO & M.S. LIM. 1996. The role of tissue inhibitors of metalloproteinases in tissue remodeling and cell growth. Int. J. Obesity **20:** S9–S15.

21. NAKAGAWA, T., T. KUBOTA, M. KABUTO, K. SATO, H. KAWANO, T. HAYAKAWA & T. OKADA. 1994. Production of matrix metalloproteinases and tissue inhibitor of metalloproteinase-1 by human brain tumors. J. Neurosurg. **81:** 69–77.

22. BROWN, P.D., R.E. BLOXIDGE, N.S.A. STUART, K.C. GATTER & J. CARMICHAEL. 1993. Association between expression of activated 72-kilodalton gelatinase and tumor spread in non-small cell lung carcinoma. J. Natl. Cancer Inst. **85:** 574–578.

23. LIABAKK, N-B., I. TALBOT, R.A. SMITH, K. WILKINSON & F. BALKWILL. 1996. Matrix metalloproteinase 2 (MMP-2) and matrix metalloprotease 9 (MMP-9) Type IV collagenases in colorectal cancer. Cancer Res. **56:** 190–196.

24. HAMDY F.C., E.J. FADLON, D. COTTAM, J. LAWRY, W. THURRELL, P.B. SILCOCKS, J.B. ANDERSON, J.L. WILLIAMS & R.C. REES. 1994. Matrix metalloproteinase 9 expression in primary human prostatic adenocarcinoma and benign prostatic hyperplasia. Br. J. Cancer **69:** 177–182.

25. SIER, C.F.M., F.J.G.M. KUBBEN, S. GANESH, M.M. HEERDING, G. GRIFFIOEN, R. HANEMAAIJER, J.H. VAN KRIEKEN, C.B.H.W. LAMERS & H.W. VERSPAGET. 1996. Tissue levels of matrix metalloproteinases MMP-2 and MMP-9 are related to the overall survival of patients with gastric carcinoma. Br. J. Cancer **74:** 413–417.

26. HUA, J. & R.J. MUSCHEL. 1996. Inhibition of matrix metalloproteinase 9 expression by a ribozyme blocks metastasis in a rat sarcoma model system. Cancer Res. **56:** 5279–5284.

27. DECLERCK, Y.A., N. PEREZ, H. SHIMADA, T.C. BOONE, K.E. LANGLEY & S.M. TAYLOR. 1992. Inhibition of invasion and metastasis in cells transfected with an inhibitor of metalloproteinases. Cancer Res. **52:** 701–708.

28. KOOP, S., R. KHOKA, E.E. SCHMIDT, I.C. MACDONALD, V.L. MORRIS, A.F. CHAMBERS & A.C. GROOM. 1994. Overexpression of metalloproteinase inhibitor in B16-f10 cells does not affect extravasation but reduces tumor growth. Cancer Res. **54:** 4791–4797.

29. WATANABE, M., Y. TAKAHASHI, T. OHTA, M. MAI, T. SASAKI & M. SEIKI. 1996. Inhibition of metastasis in human gastric cancer cells transfected with tissue inhibitor of metalloproteinase I gene in nude mice. Cancer **15:** 1676–1680.

30. DECLERCK, Y.A., N. PEREZ, H. SHIMADA, T.C. BOONE, K.E. LANGLEY & S.M. TAYLOR. 1992. Inhibition of invasion and metastasis in cells transfected with an inhibitor of metalloproteinases. Cancer Res. **52:** 701-708.

31. WANG, X., X. FU, P.D. BROWN, M.J. CRIMMIN & R.M. HOFFMAN. 1994. Matrix metalloproteinase inhibitor BB-94 (batimastat) inhibits human colon tumor growth and spread in a patient-like orthotopic model in nude mice. Cancer Res. **54:** 4726–4728.

32. CHIRIVI, R.G., A. GAROFALO, M.J. CRIMMIN, L.J. BAWDEN, A. STOPPACCIARO, P.D. BROWN & R. GIAVAZZI. 1994. Inhibition of the metastatic spread and growth of B16-BL6 murine melanoma by a synthetic matrix metalloproteinase inhibitor. Int. J. Cancer **58:** 460–464.

33. AN, Z., X. WANG, N. WILLMOTT, S.K. CHANDER, S. TICKLE, A.J. DOCHERTY, A. MOUNTAIN, A.T. MILLICAN, R. MORPHY, J.R. PORTER, R.O. EPEMOLU, T. KUBOTA, A.R. MOOSSA & R.M. HOFFMAN. 1997. Conversion of highly malignant colon cancer from an aggressive to a controlled disease by oral administration of a metalloproteinase inhibitor. Clin. Exp. Metastasis **15:** 184–195.

34. CRAWFORD, H.C. & L.M. MATRISIAN. 1996. Mechanisms controlling the transcription of matrix metalloproteinase genes in normal and neoplastic cells. Enyme Protein **49:** 20–37.

35. HIMELSTEIN, B.P., R. CANETE-SOLER, E.J. BERNHARD, D.W. DILKS & R. MUSCHEL. 1994. Metalloproteinases in tumor progression: the contribution of MMP-9. Invasion Metastasis **14:** 246–258.

36. WILSON, C.L., K.J. HEPPNER, P.A. LABOSKY, B.L. HOGAN & L.M. MATRISIAN. 1997. Intestinal tumorigenesis is suppressed in mice lacking the metalloproteinase matrilysin. Proc. Natl. Acad. Sci. **94:** 1402–1407.

37. BENDER, S.L. 1997. Structure-based design of MMP inhibitors: discovery and development of AG3340. Presented at the 214th National Meeting of the American Chemical Society, Las Vegas, Nevada.

38. HIBNER, B., A. CAD, C. FLYNN, A.M. CASAZZA, G. TARABOLETTI, M. RIEPPI & R. GIAVAZZI. 1998. BAY 12-9566, a novel, biphenyl matrix metalloproteinase inhibitor, demonstrates anti-invasive and anti-angiogenic properties. Proc. Am. Assoc. Cancer Res. **39:** 2063.

39. SANTOS, O., C.D. MCDERMOTT, R.G. DANIELS & K. APPELT. 1997. Rodent pharmacokinetic and anti-tumor efficacy studies with a series of synthetic inhibitors of matrix metalloproteinases. Clin. Exp. Metastasis **15:** 499–508.

40. SHALINSKY, D.R., J. BREKKEN, N.V. VARKI, S.R. ROBINSON, R. DANIELS, S. KOLIS, S. BANSAL, H. ZOU, S. BENDER, S. ZOOK, S. MARGOSIAK, A.W. WOOD, S. WEBBER & K. APPELT. 1998. Marked inhibition of the proliferation of human adenocarcinoma colon tumors *in vivo* by orally-administered AG3340, a novel matrix metalloproteinase inhibitor. Proc. Am. Assoc. Cancer Res. **39:** 2059.

41. SHALINSKY, D.R., J. BREKKEN, H. ZOU, S. KOLIS, A. WOOD, S. WEBBER & K. APPELT. 1999. Antitumor efficacy of AG3340 associated with maintenance of minimum effective plasma concentrations and not total daily dose, exposure or peak plasma concentrations. Invest. New Drugs. In press.

42. SHALINSKY, D.R., J. BREKKEN, H. ZOU, L. BLOOM, C. MCDERMOTT, N.M. VARKI & K. APPELT. 1999. Marked antiangiogenic and antitumor efficacy of AG3340 in chemoresistant human NSCLC tumors: single agent and combination chemotherapy studies. Clin. Cancer Res. In press.

43. PRICE, A., N.B. REWCASTLE, V.W. YONG, Q.S. ZHONG. P. BRASHER, M.G. MORRIS, D. SPENCER, D. EDWARDS, K. APPELT & P. FORSYTH. 1999. Marked inhibition of tumor growth in a malignant glioma tumor model by the novel, synthetic matrix metalloproteinase (MMP) inhibitor, AG3340. Clin. Cancer Res. **5:** 845–854.

44. SHALINSKY, D.R., J. BREKKEN, H. ZOU, S. BENDER, S. ZOOK, K. APPELT, S. WEBBER & N.V. VARKI. 1998. Increased apoptosis in human androgen-independent prostatic PC-3 tumors following oral administration of a novel matrix metalloproteinase (MMP) inhibitor, AG3340, in male nude mice. Proc. Am. Assoc. Cancer Res. **39:** 4400.

45. NERI, A., B. GOGGIN, S. KOLIS, J. BREKKEN, N. KHELEMSKAYA, L. GABRIEL, S.R. ROBINSON, S. WEBBER, A.W. WOOD, K. APPELT & D.R. SHALINSKY. 1998. Pharmacokinetics and efficacy of a novel matrix metalloproteinase inhibitor, AG3340, in single agent and combination therapy against B16-F10 melanoma tumors developing in lung after IV-tail implantation in C57BL/6 mice. Proc. Am. Assoc. Cancer Res. **39:** 2060.

46. JOHNSTON, M.R., J.B.M. MULLEN, M. PAGURA, K.A. APPELT & D.R. SHALINSKY. 1998. AG3340, a novel matrix metalloproteinase (MMP) inhibitor, inhibits the growth of human large cell lung cancer tumors orthotopically implanted into the lung of athymic nude rats. Proc. Am. Assoc. Cancer Res. **39:** A2060.

47. QUINN, L.A., G.E. MOORE, R.T. MORGAN & L.K. WOODS. 1979. Cell lines from human colon carcinoma with unusual cell products, double minutes, and homogeneously staining regions. Cancer Res. **39:** 4914–4924.

48. KAIGHN, M.E., K.S. NARAYAN, Y. OHNUKI, J.F. LECHNER & L.W. JONES. 1979. Establishment and characterization of a human prostatic carcinoma cell line (PC-3). Invest. Urol. **17:** 16–23.

49. PONTEN, J. & E.H. MACINTYRE. 1968. Long term culture of normal and neoplastic human glioma. Acta Pathol. Microbiol. Scand. **74:** 465–486.

50. KELNER, M.J., T.C. MCMORRIS, L. ESTES, R. STARR, K. SAMPSON, N. VARKI & R. TAETLE. 1995. Nonresponsiveness of the metastatic human lung carcinoma MV522 xenograft to conventional anticancer agents. Anticancer Res. **15:** 867–871.

51. BRINKLEY, B.R., P.T. BEALL, L.J. WIBLE, M.L. MACE, D.S. TURNER & R.M. CAILLEAU. 1980. Variations in cell form and cytoskeleton in human breast carcinoma cells in vitro. Cancer Res. **40:** 3118–3129.

52. COHEN, M.B., F.M. WALDMAN, P.R. CARROLL, R. KERSCHMANN, K. CHEW & B.H. MAYHALL. 1993. Comparison of five histopathologic methods to assess cellular proliferation in transitional cell carcinoma of urinary bladder. Hum. Pathol. **24:** 772–778.

53. LIMAS, C., A. BIGLER, R. BLAIR, P. BERNHART & P. REDDY. 1993. Proliferative activity of urothelial neoplasms: comparison of BrdU incorporation, Ki 67 expression, and nucleolar organiser regions. Clin. Pathol. **46:** 159–165.

54. FAVALORO, E.J., N. MORAITIS, K. BRADSTOCK & J. KOUTTS. 1990. Co-expression of haemopoietic antigens on vascular endothelial cells: a detailed phenotypic analysis. Br. J. Haematol. **74:** 385–394.

55. WEIDNER, N. & J. FOLKMAN. 1997. Tumoral vascularity as a prognostic factor in cancer. *In* Importance Advances in Oncology, V. DeVita, S. Hellman & S.A. Rosenberg, Eds. Lippincott-Raven. Philadelphia.

56. WIJSMAN, J.H., R.R. JONKER, R. KEIZER, C.J.H. VAN DE VELDES, C.J. CORNELISSE & J.H.V. DIERENDONCK. 1993. A new method to detect apoptosis in paraffin sections: in situ end-labeling of fragmented DNA. J. Histochem. Cytochem. **41:** 7–12.

57. GEHRING, M.R., B. CONDON, S.A. MARGOSIAK & C.C. KAN. 1995. Characterization of the Phe-81 and Val-82 human fibroblast collagenase catalytic domain purified from Escherichia coli. J. Biol. Chem. **270:** 22507.

58. KNIGHT, C.G., F. WILLENBROCK & G. MURPHY. 1992. A novel coumarin-based peptide for sensitive continuous assays of the matrix metalloproteinases. FEBS Lett. **296:** 263–266.

59. MORRISON, J.F. 1969. Kinetics of the reversible inhibition of enyzme-catalzyzed reactions by tight binding inhibitors. Biochim. Biophys. Acta **185:** 269–286.

60. YOCUM, S., L. LOPRESTI-MORROW, L. REEVES & P. MITCHELL. 1999. MMP-13 and MMP-1 expression in tissues of normal articular joints. This volume.

61. CHAMBERS, A.F., S. WYLIE, I.C. MACDONALD, H. VARGHESE, E.E. SCHMIDT, V.L. MORRIS & A.C. GROOM. 1998. The matrix metalloproteinase inhibitor batimastat inhibits angiogenesis in liver metastases of B16F1 melanoma cells. Proc. Am. Assoc. Cancer Res. **39:** 566.

62. FLYNN, C., C. BULL, D. EBERWEIN, C. MATHERNE & B. HIBNER. 1998. Anti-metastatic activity of BAY-12-9566 in a human colon carcinoma HCT116 orthotopic model. Proc. Am. Assoc. Cancer Res. **39:** 2057.

63. RIVERO, M.E., C.R. GARCIA, C.D. MCDERMOTT, D-U. BARTSCH, G. BERGERON-LYNN, K. ZHANG, K. APPELT & W.R. FREEMAN. 1998. Intraocular properties of AG3340, a selective matrix metalloprotease inhibitor with antiangiogenic activity. Invest. Ophthalmol. Visual Sci. **39:** S585.

64. TURCK, J., A.S. POLLOCK, L.K. LEE, H.P. MARTI & D.H. LOVETT. 1996. Matrix metalloprotease 2 (gelatinase A) regulates glomerular mesangial cell proliferation and differentiation. J. Biol. Chem. **271:** 15074–15083.

65. COLLIER, M.A., G.J. YUEN, S.K. BANSAL, S. KOLIS, T.G. CHEW, K. APPELT & N.J. CLENDININ. 1997. A Phase I study of the matrix metalloproteinase inhibitor (MMP) inhibitor, AG3340 given in single doses to healthy volunteers. Proc. Am. Assoc. Cancer Res. **38:** A1491.
66. TEICHER, B.A., S.A. HOLDEN, G. ARA, E.A. SOTOMAYOR, Z.D. HUANG, Y.N. CHEN & H. BREM. 1994. Potentiation of cytotoxic cancer therapies by TNP-470 alone and with other anti-angiogenic agents. Int. J. Cancer **57:** 920–925.
67. ANDERSON, I.C., M.A. SHIPP, A.J.P. DOCHERTY & B.A. TEICHER. 1996. Combination therapy including a gelatinase inhibitor and cytotoxic agent reduces local invasion and metastatis of murine Lewis lung carcinoma. Cancer Res. **56:** 715–720.
68. MAUCERI, H.J., N.N. HANNA, M.A. BECKETT, D.H. GORSKI, M.J. STABA, K.A. STELLATO, K. BIGELOW, R. HEIMANN, S. GATELY, M. DHANABAL, G.A. SOFF, V.P. SUKHATME, D.W. KUFE & R.R. WEICHSELBAUM. 1998. Combined effects of angiostatin and ionizing radiation in antitumour therapy. Nature **394:** 287–291.
69. GIAVAZZI, R., A. GAROFALO, C. FERRI, V. LUCCHINI, E.A. BONE, S. CHIARI, P.D. BROWN, M.I. NICOLETTI & G. TARABOLETTI. 1998. Batimastat, a synthetic inhibitor of matrix metalloproteinases, potentiates the antitumor activity of Cisplatin in ovarian carcinoma xenografts. Clin. Cancer Res. **4:** 985–992.
70. JOHNSTON, M.R., J.M. MULLEN, M. PAGURA, J. BREKKEN, H. ZOU & D.R. SHALINSKY. 1999. AG3340 and carboplatin increase survival in an orthotopic nude rat model of primary and metastatic human lung cancer. Proc. Am. Assoc. Cancer Res. **40:** A1046.
71. PERKINS, A.S. & D.F. STERN. 1997. Molecular biology of cancer: oncogenes. *In* Cancer: Principles and Practice of Oncology, 5th ed. V.T. DeVita, S. Hellman & S.A. Rosenberg, Eds. J.B. Lippincott. Philadelphia.
72. CHRISTEN, R.D., S. ISONISHI, J.A. JONES, A.P. JEKUNEN, D.K. HOM, R. KRONING, D.P. GATELY, F.B. THIEBAUT, G. LOS & S.B. HOWELL. 1994. Signaling and drug sensitivity. Cancer Metastasis Rev. **13:** 175–189.

# MMP Inhibition in Prostate Cancer

BAL L. LOKESHWAR[a]

*Department of Urology, University of Miami School of Medicine, Miami, Florida 33101, USA*

**ABSTRACT:** Matrix metalloproteinases (MMPs) play a significant role during the development and metastasis of prostate cancer (CaP). CaP cells secrete high levels of MMPs and low levels of endogenous MMP inhibitors (TIMPs), thus creating an excess balance of MMPs. Established CaP cell lines that express high levels of MMPs frequently metastasize to the bone and the lungs. Drugs such as Taxol and alendronate that reduce cell motility and calcium metabolism reduce bony metastasis of xenografted CaP tumors. We tested several synthetic, nontoxic inhibitors of MMPs that can be administered orally, including doxycycline (DC) and chemically modified tetracyclines (CMTs) on CaP cells *in vitro* and on a rat CaP model *in vivo*. Among several anti-MMP agents tested, CMT-3 (6-deoxy, 6-demethyl,4-de-dimethylamino tetracycline) showed highest activity against CaP cell invasion and cell proliferation. Micromolar concentration of CMT-3 and DC inhibited both the secretion and activity of MMPs by CaP cells. When tested for *in vivo* efficacy in the Dunning rat CaP model by daily oral gavage, CMT-3 and DC both reduced the lung metastases (> 50%). CMT-3, but not DC, inhibited tumor incidence (55 ± 9%) and also reduced the tumor growth rate (27 ± 9.3%). More significantly, the drugs showed minimum systemic toxicity. Ongoing studies indicate that CMT-3 may inhibit the skeletal metastases of CaP cells and delay the onset of paraplegia due to lumbar metastases. These preclinical studies provide the basis for clinical trials of CMT-3 for the treatment of metastatic disease.

## INTRODUCTION

Carcinoma of the prostate (CaP) is a major malignant disease in developed countries. In the United States alone an estimated 186,500 Americans are expected to be diagnosed with CaP in 1998, and about 39,200 fatalities are expected.[1] Approximately 50% of prostate cancer patients have extra-prostatic disease at the time of diagnosis.[2] Furthermore, the disease relapses in a majority of the patients despite treatment of the primary tumor.[3,4] Surgery and radiation are the two common treatment modalities for patients with localized (stage A and B) or locally extensive disease (stage C).[5] Patients with inoperable conditions due to age are treated with "hormonal" therapy or radiation.[6]

The most common hormonal therapy for prostate cancer is either neo- or adjuvant androgen ablation.[7] Most prostate tumors originate from the glandular epithelial cells of the peripheral region of the prostate. The glandular epithelium of the prostate is dependent on androgens, the common male steroid hormones—testosterone and dihydrotestosterone—for survival and proliferation.[8] Depriving the prostate tumor

[a]Address for correspondence: Department of Urology (M-800), University of Miami School of Medicine, P.O. Box 016960 Miami, Florida 33101; email,blokeshw@mednet.med.miami.edu

cells of androgens by castration or by suppressing androgen production in the go-
nads with pituitary gonadotropin (LH-RH) analogue (e.g., Lupron, TAP Pharmaceuticals, Chicago, IL), inhibitor of steroid 5α-reductase (e.g., Proscar, Merck
Pharmaceuticals, Rahway, NJ), or treatment with anti-androgens (e.g., Flutamide)
almost always results in temporary tumor remission.[9] However, deprivation of androgens to prostate (androgen ablation) almost always leads to the onset of a more
aggressive, metastatic, hormone-refractory incurable phase of the disease.[10]

Since most instances of prostate cancer initially respond well to androgen ablation (which is more selective than cytoreductive chemotherapy), chemotherapy is
not currently a first line of therapy during any stage of this disease. Several unique
features of CaP contribute to this. For example, very low growth fraction in the primary tumor, lower than most proliferating tissues of the body, precludes treatment
of early or localized disease with cytoreductive chemotherapy.[11] Drug-induced morbidity, or the co-morbidity of other ailments, or simply advanced age of the patient
preclude aggressive chemotherapy of patients with advanced prostate cancer. Nevertheless, after the failure of "hormone"-related treatments, the disease is usually treated palliatively with radiation or cytoreductive chemotherapy, as its metastatic
growth leads to acute bone pain, spinal compression, and often paraplegia.[12]

Targeting the metastatic growth of prostate cancer with site-directed, nontoxic
drugs that disable tumor cells from establishing metastatic colonies is an alternative
therapeutic option. Drugs that inhibit metastatic process, but not necessarily cell proliferation, are not likely to discriminate between androgen-dependent and independent prostate tumor cells. Nontoxic inhibitors of MMPs that inhibit tumor cell
invasion, angiogenesis, and adhesion-dependent interactions of tumor cells with the
host tissue are drugs with such potential. Although a number of laboratories are engaged in developing such therapeutic agents for prostate cancer, currently few drugs
exist in clinical use that achieve desired efficacy. In the next few pages I will summarize what is known about the role of matrix metalloproteinases in prostate cancer.
I will also summarize current efforts to develop therapeutic avenues using MMP inhibition as a tool.

### Basic Mechanism of Cancer Metastasis

The cellular basis of cancer metastasis is explained by the three-step model proposed by Liotta *et al.*[13]: (1) adhesion of tumor cells to basement membrane; (2) local
proteolysis that leads to the invasion of cancer cells into stroma; and (3) tumor cell
proliferation. Thus, according to this model, the process of invasion begins by adhesive interactions between tumor cells and the extracellular matrix (ECM), leading to
the proteolysis of the basement membrane. This is followed by the migration of tumor cells through stroma, invading the capillary wall to enter blood circulation. After
entering the circulation the tumor cells are embolized throughout the circulatory system and migrate to distant organs by hemodynamic principles. In this process, the
tumor cells adhere to the capillary wall and begin to extravasate in a sequence the
reverse of the invasion process. Tumor cells then interact with the stromal components of the new organ, which results in either the elimination of tumor cells or their
colonization owing to stimulation of cell proliferation and angiogenesis. Taken together, the entire metastatic process is known to be extremely inefficient, where less
than 0.01% of tumor cells that enter circulation are capable of establishing a meta-

static colony.[14] Chambers *et al.*[15] have recently challenged the earlier concept that all steps in metastasis are equally inefficient. They showed, by direct observation of intravenously injected tumor cells, that early steps in metastasis, such as hemodynamic destruction of circulating tumor cells and extravasation, may contribute less to metastatic inefficiency than the interaction of tumor cells with the surrounding environment. Interaction with the host environment is likely to dictate either the proliferation or death of tumor cells at the metastatic site.

## Mechanism of Prostate Cancer Metastasis

At the organ level, extraprostatic spread of prostate tumors, especially to lumbar bones, is explained by two mechanisms. The first mechanism is based on the principles of hemodynamics, where the prostate cancer cells enter the circulation to reach lumbar vertebra (bone metastasis) under increased intra-abdominal pressure (Bateson's hypothesis[16]). Studies have shown a direct correlation between the size of the primary prostate tumor and the incidence of capsular invasion and distant metastasis.[17] In addition, experimentally induced bone metastasis that approaches the frequency found in prostate cancer patients can be achieved by increasing the intra-abdominal venous pressure at the time of tumor cell injection. Shevrin *et al.*[18] and later Geldof and Rao[19] showed that tail vein injection of prostate tumor cells with simultaneous vena cava clamping causes skeletal and lumbar metastases at unusually high frequency.

The second mechanism of extraprostatic spread of tumors is explained by Paget's "seed and soil" hypothesis.[20] This hypothesis states that tumor metastasis is dependent on both the tumor cells (the "seed") and the microenvironment of the organ (the "soil"). Many studies support the "soil and seed" hypothesis as a mechanism of prostate cancer metastasis. For example, although several highly aggressive CaP cell lines have been established, few are spontaneously metastatic to bone.[21] One requirement that determines organ-specific metastasis may be the tumor cells' potential to establish a bi-directional communication ("inductive interaction") with the marrow cells.[22] Very likely, the inductive interactions result in the secretion of both tumor cell– and host cell–derived factors that support tumor growth and angiogenesis. These factors include diffusible growth factors that support osteoblast growth, capillary bed formation, and inducers of matrix-degrading enzymes (e.g, MMPs, urokinase-like plasminogen activator [uPA], etc.)[23,24] that are involved in bone matrix remodeling.

## Matrix-Degrading Enzymes and Tumor Cell Invasion

The dissolution of basement membrane components such as type IV collagen, laminin, fibronectin, and proteoglycans is a critical step in the multistep cascade that leads to metastasis.[25] Tumor cells dissolve these components using a variety of matrix-degrading enzymes, including aspartyl, cysteine, serine, and matrix metalloproteinases.[26] Both stromal and epithelial cells of the prostate secrete aspartyl proteases (e.g., cathepsin D[27]), serine proteases (e.g., uPA, PSA[28]), and metalloproteinases (MMPs[29]).

Several studies have shown an association of increased production of MMPs (MMP-2, -3, -7 and -9) with malignant progression of prostate cancer.[28] For example, both our work and that of others has shown, by analyzing the primary cultures of human prostate tumor tissues, that epithelial cell cultures of malignant prostate secrete high levels of MMP-2 and MMP-9 and low levels of their inhibitors (TIMP-1 and TIMP-2).[30,31] Stearns et al.[32] have reported that, among these, MMP-2 expression is associated with Gleason sum 7 or higher. (Gleason sum is the current pathology index to evaluate malignant status of prostate tumors.[33]) Wood et al.[34] have reported that levels of both MMP-2 and MMP-9 are low in normal prostate and organ-confined tumors with Gleason sum 5 or lower, whereas they were highly expressed in high Gleason sum (8–10) tissues. A particularly significant observation about the role of MMP in CaP progression is that of increased expression of *activated* MMP-2.[32] The expression of activated MMP-2 in high-grade tumor tissues suggests that in addition to higher levels of MMP-2, MMP-2 activators such as uPA, membrane type-MMP (MMP-14), and matrilysin may also be associated with malignant progression and metastasis.[35]

### Association between MMP Expression and Metastatic Potential

Probably, the strongest evidence as yet for the role of MMPs in prostate cancer metastasis has come from studies on animal models. Stephenson et al.[39] showed that only those sublines of prostate cancer cells that produce high levels of MMPs are capable of distant metastasis when tumors are generated by intraprostatic tumor cell injection. Powell et al.[37] reported significant induction of metastatic activity associated with high-level expression of matrilysin. Studies have also been reported that androgen ablation, which frequently results in more aggressive and metastatic cancer, can lead to increased expression of MMPs. Two independent studies on rats have shown that castration leads to involution of ventral prostate with a significant increase in MMP-2, matrilysin, and uPA.[29] These studies, if confirmed in patients, raise the spectre of inadvertent promotion of metastatic potential of residual tumor cells in patients undergoing androgen-ablation therapy.

### Role of TIMPs in Prostate Cancer

The tissue inhibitors of metalloproteinases (TIMP-1 and TIMP-2) have been recognized as the "balancing" factors in MMP-induced invasion and metastasis.[38] Several studies using tumor tissue specimens and primary explant cultures have shown decreased TIMP-1 expression (mRNA) and secretion (protein) in malignant prostate cancer. For example, stimulation of a highly metastatic prostate cancer cell line (PC-3ML) with interleukin-10 (IL-10) results in upregulation of TIMP-1 and downregulation of MMP-2 and MMP-9, with concomitant decrease in liver and bone metastases.[39] Jung et al.[40] reported, however, increased levels of TIMP-1 in the plasma of patients with metastatic prostate cancer compared to the those of patients with organ-confined cancer or those with benign prostatic hyperplasia. One should be cautious, though, in comparing the significance of the levels of MMPs and TIMPs in circulation with cancer aggressiveness rather than the corresponding tissue levels of TIMPs and MMPs in active metastatic sites.

Owing to the large body of evidence suggesting the association of MMP activity and promotion of metastasis, inhibition of MMP activity by either natural or synthetic inhibitors, with relatively mild systemic toxicity, may be an important avenue to treat hormone-refractory prostate cancer.

## *Inhibitors of Matrix Degrading Enzymes as Cancer Therapeutics*

DeClerck and others[41] have shown that excess production of TIMP-1 or TIMP-2 in invasive tumor cells can reduce their metastatic potential.[42,43] The strategy of using the superinduction of TIMPs to control metastasis, although exciting, is not very practical at present, because TIMPs are large glycoproteins (23 kDa and 21 kDa) and require a significant portion of the molecule for biological activity.[44] To overcome this limitation, smaller synthetic MMP inhibitors have been developed (e.g., BB-94 [batimastat] and BB2516 [marimastat], British Biotech. Inc., Oxford, England). The hydroxamate group in these compounds reversibly chelates the zinc atom at active site of MMPs, thereby potently inhibiting MMPs.[45] Batimastat and marimastat are forerunners in the area of MMP inhibitors used for clinical application, especially for treating metastatic tumors.[46] Preclinical studies using several animal models and human xenograft models of a variety of solid tumors have established the efficacy and possible clinical application of these compounds.[47,48] Initially batimastat and recently marimastat have been introduced in clinical trials on patients with advanced, treatment-refractory aggressive cancer. Results of the orally administrable marimastat are encouraging, with only a few dose-limiting cases of toxicity reported in clinical trials.[49,50]

## *Preclinical Studies of Anti-MMP Agents in Prostate Cancer*

Although extensive studies have been conducted on the potential clinical use of MMP inhibitors in other kinds of carcinoma, only a handful of such studies have been reported on prostate cancer. Among the limited number of studies reported so far, notable ones are by Stearns and Wang,[51] who showed that paclitaxel (Taxol, Bristol-Myers, Raritan, NJ), a potent cytoreductive chemotherapeutic drug, inhibits both MMP secretion and synthesis in PC-3 ML cells. Taxol reduced metastases of PC-3 ML cells to bones, liver, and lungs in SCID mice. Although Taxol is not an inhibitor of MMP-2 enzyme activity, it interfered with the synthesis and secretion of MMP-2, but not TIMPs. In another study,[52] using the same tumor model, PC-3 ML, these investigators showed that alendronate (a bisphosphonate used for treating osteoporosis) together with Taxol blocks the establishment, growth, and metastasis of PC-3 ML tumors. Treatment with alendronate alone, however, increased soft-tissue metastases, and only partially blocked bone metastasis.

## *Clinical Studies*

So far one clinical study has reported using MMP inhibitor against prostate cancer,[53] a dose-finding study using marimastat.[54] Eighty-eight patients with aggressive stage C and D prostate cancer with poor prognosis were given marimastat orally for 4 weeks. The therapeutic response, as measured by decrease in the rate of rise of serum prostate-specific antigen (PSA) levels from 53% to 29%, was considered "encouraging." The response was dose-dependent: at a higher dose more than 50% of the patients showed a significant fall in the rate of increase in PSA. A decline in serum PSA is currently the most common surrogate marker for evaluating treatment efficacy for prostate tumors.

In the remainder of this article I will present evidence to demonstrate the efficacy of a class of nontoxic MMP inhibitors on a prostate cancer model with potential clinical application.

## TETRACYCLINES AND THEIR NON-ANTIMICROBIAL ANALOGUES ARE POTENT INHIBITORS OF COLLAGENASE

Golub and his colleagues discovered that the common antibiotic, tetracycline, is a potent inhibitor of gingival fibroblast-derived collagenase.[55] In subsequent studies, the gelatinolytic, cytolytic, and anti-angiogenic activities of tetracyclines were discovered.[56,57] This group also showed that chemical modifications of tetracycline, such as removal of the dimethyl amino group from the carbon 4 of the A ring, resulted in the loss of antimicrobial activity, but did not abolish anti-collagenolytic activity.[58] Several of the chemically modified tetracyclines (CMTs) available at present, are non-antimicrobial and inhibit the activities of collagenases and MMPs.[59] We have tested a number of CMTs for their potential as anti-metastatic agents and found that one CMT, 6-dedimethyl, 6-deoxy, 4-dedimethylamino tetracycline (CMT-3), is superior to the others.[60] The potential advantages of CMTs over conventional tetracycline with antimicrobial activity include long-term systemic administration without gastrointestinal toxicity, higher plasma accumulation, and a longer plasma clearance time and therefore lower drug dose requirement.[61]

### Investigation of Antitumor and Antimetastatic Activities of CMTs and Doxycycline(DC)

In Vitro *Studies*

Our objectives for testing these drugs against CaP cells *in vitro* were two-fold: (1) to investigate whether the CMTs or DC inhibit basement membrane invasive activity of CaP cells at physiologically achievable serum concentration levels (e.g., 5 μg/ml), and (2) to test whether the CMTs and DC are cytotoxic to tumor cells, as some studies have shown earlier.[57]

The ability of CMTs and DC to inhibit invasive activity of CaP cells *in vitro* was tested by plating cells on the top chamber of the Boyden Chemotactic chambers (Transwell plates, Costar-Corning Corporation, Cambridge, MA). Before plating the cells, the porous (12-μ) filter of the top chamber was coated with a soluble preparation of basement membrane, Matrigel (Collaborative Research/B-D Systems, Bedford MA). The bottom chamber contained human fibroblast culture-conditioned medium as a chemoattractant. Others have shown that cells invade the bottom chamber by dissolving Matrigel and migrating through the pores in the filter.[62] Various CMTs or DC were added to both top and bottom chambers at the time of plating cells. Invasive activity was quantified by estimating the fraction of invaded cells on the underside of the filter and in the bottom chamber. As we reported before, we find CMT-3 to be the most effective inhibitor of Matrigel invasion by CaP cells among the various CMTs tested. Using two CaP cell lines, PC-3 ML and DU 145, we found that CMT-3 was the most potent inhibitor of invasive activity and CMT-6 the least. Interestingly, although DC was minimally effective as an inhibitor of invasion of human CaP cells, it significantly inhibited the invasive activity of the Dunning rat MAT LyLu prostate cancer cells (FIG. 1). In subsequent studies we determined the drug concentration at which Matrigel invasion was inhibited by 50% ($IC_{50}$). CMT-3 had the lowest $IC_{50}$ among the seven CMTs tested ($IC_{50} = 1.5$ µg/ml). The $IC_{50}$ for DC was 7.9 µg/ml in DU 145 cells and 5.3 µg/ml for MAT LyLu cells.

We also investigated whether the ability of CMTs and DC to inhibit invasive activity against CaP cells is associated with their ability to inhibit MMP activity. We reported previously that both CMT-3 and DC inhibit not only the activity of MMPs secreted by tumor cells, but also the synthesis of MMPs by human CaP cell lines and MAT LyLu cells.[63] Interestingly, while CMT-3 and DC strongly inhibited the synthesis of MMP-2, the synthesis of MMP-9 was weakly inhibited by DC in human CaP cells. Furthermore, while CMT-3 inhibited MMP synthesis in a dose-dependent manner, the synthesis of TIMP-1 and TIMP-2 were weakly inhibited (Lokeshwar, Selzer, Zhu *et al.*, submitted for publication).

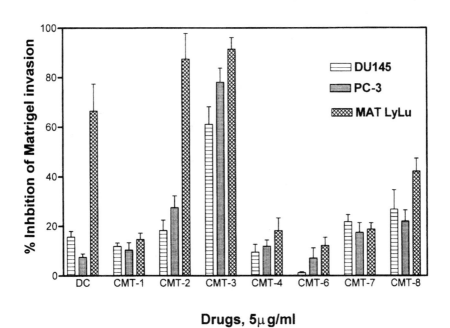

**Drugs, 5µ g/ml**

**FIGURE 1.** Inhibition of invasive potential of tumor cells by DC and CMTs. Invasion of tumor cells through the Matrigel-coated filters was assayed following 48 hours of exposure to 5 µg/ml of each drug. Only the drug diluent (0.1% dimethyl sulfoxide) was added to control wells. Percentage of cells that invaded in the control (0.1% DMSO) wells varied from $12.5 \pm 6.4\%$ for DU 145 cells to $17 \pm 4.2$ for MAT LyLu cells. 0.1% DMSO had negligible effect on invasion. Results presented are from three independent experiments.

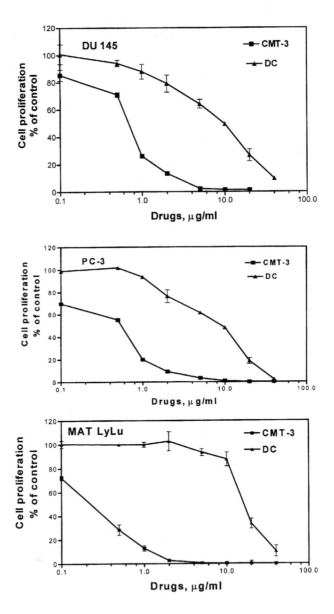

**FIGURE 2.** Effect of DC and CMT-3 on proliferation of prostate tumor cell lines. Tumor cells were incubated with various concentrations of DC or CMT-3 for 48 hours in complete culture medium. Cell proliferation activity, defined as synthesis of [$^3$H]-thymidine-labeled DNA, was assayed by 2-hour pulse-labeling the cells with [$^3$H]-thymidine as described in the text. Data presented are for three CaP cell lines. Similar results were obtained for other cell lines. *Vertical bars* represent mean ± SEM from four independent determinations.

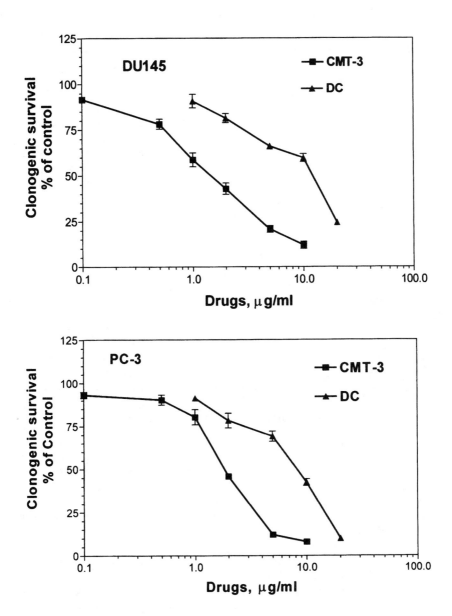

**FIGURE 3.** Clonogenic survival of CaP cells exposed to CMT-3 or DC. Cells were exposed to CMT-3 or DC for 24 hours and plated out for clonogenic survival. Colonies of cells that survived the drug treatment were counted after 7–10 days. Note that the 50% clonogenic inhibition dose is similar to that obtained using the DNA synthesis inhibition assay as described in FIGURE 2.

## Inhibition of Cell Proliferation by CMTs and DC

There are several reports in literature regarding cytotoxic and cytostatic actions of tetracycline, although there are none at present on CMTs.[64,65] We report here the effects of CMT-3 and DC on cell proliferation. Antiproliferative effects of CMT-3 and DC were assayed quantitatively using the [$^3$H]thymidine assay as described before.[67] Cells cultured in 48-well clusters were incubated with a range of concentrations of DC or CMT-3 for 48 hours. Cell proliferation activity was determined by pulse labeling the cells with [$^3$H]thymidine for 2 hours. [$^3$H]thymidine incorporation into proliferating cell DNA was stopped by adding cold 10% trichloroacetic acid to the culture wells. [$^3$H]thymidine-labeled DNA was extracted, and incorporated radioactivity was measured in a liquid scintillation counter. The results are presented in FIG. 2, which shows that the inhibition of cell proliferation by CMT-3 or DC was dose dependent. CMT-3 was significantly more potent than DC in inhibiting cell proliferation. The concentration of the drug that decreased cell proliferation by 50% (IC$_{50}$) was at least 5-fold lower for CMT-3 than that of DC.

We next examined the cytotoxic effects of DC and CMT-3 on clonogenic cell survival in two CaP cell lines, PC-3 ML and DU 145. Cells were incubated with various concentrations of drugs for 24 hours. Drugs were then washed off from the culture plates, fresh culture medium was added, and cell culture continued for 10 days in the absence of the drugs. Resulting cell colonies were fixed and stained with 0.1% crystal violet. Colonies of cells containing $\geq$ 50 cells were enumerated. As shown in FIGURE 3, the clonogenic cell survival was significantly inhibited by CMT-3 at concentrations $\geq$ 1 µg/ml, when exposed for 24 hours. The 50% clonogenic inhibition doses for CMT-3 and DC for both CaP cell lines were $1.4 \pm 0.2$ µg/ml and $10.8 \pm 0.3$ µg/ml, respectively. Thus, CMT-3 was about 10-fold more potent inhibiting clonogenic survival of two prostate cancer cell lines tested than was DC.

## Inhibition of Dunning Tumor Growth and Metastasis

Encouraged by the activity of CMT-3 and DC on cancer cells *in vitro,* we investigated whether these drugs inhibit tumor growth and metastasis *in vivo.* We chose an androgen-unresponsive spontaneously metastatic prostate tumor model, the Dunning MAT LyLu rat tumor.[68] This tumor grows rapidly upon subcutaneous injection of as low as $5 \times 10^4$ cells in Copenhagen rats. Palpable tumor growth appears in 7 to 9 days post tumor implant, with a medium tumor growth rate of 1.7 days in untreated animals. Spontaneous metastasis to lymph nodes and lungs are observed from 12 days after tumor cell injection.

The tumor-bearing rats were treated for 21 days with a daily oral gavage of CMT-3 or DC suspended in 2% carboxymethyl cellulose (2% CMC) at the dose of 20 or 40 mg/kg. Groups of seven to ten rats were dosed either 7 days before or immediately after tumor cell injection ($1 \times 10^5$ and $1 \times 10^6$ cells/site/animal, respectively). The rats in the control group were gavaged with 2% CMC. The results of two experiments where the tumors were generated by injecting $1 \times 10^6$ cells/injected site or $1 \times 10^5$ cells/injected site are summarized in TABLE 1: when a high tumor cell inoculum ($1 \times 10^6$ cells/injected site) was used to generate tumor, tumor incidence, latency, or growth rate of tumors was not affected by oral administration of CMT-3 or DC. However, the tumor latency, defined as the duration between tumor cell injec-

**TABLE 1. Growth and metastasis of MAT LyLu tumor in rats[a]**

| Experiment | Treatment group[b] | Tumor growth[c] (days to 3-cc tumor [mean ± sd]) | Tumor growth[d] (days to 10-cc tumor [median]) | Number of tumor foci in lungs (% of control) |
|---|---|---|---|---|
| (1) $1 \times 10^6$ cells/site, s.c. (10 per group) | Control | 13.2 ± 2.4 | 17 | 59.5 ± 13.9 (100) |
| | DC (40mg/kg) | 15 ± 1.3 | 17 | 43.6 ± 18.8 (73.2)* |
| | CMT-3 (40 mg/kg) | 16 ± 1.6 | 19 | 28.9 ± 15.4 (48.5)* |
| (2) $1 \times 10^5$ cells/site, s.c. (7 per group) | Control | 15.9 ± 2.0 | 21 | 77 ± 12.5 (100) |
| | DC (pre-dose, 7 days) | 16.7 ± 1.9 | 23 | 38.3 ± 12.1 (49.7)* |
| | CMT-3 (pre-dose, 7 days) | 20.2 ± 3.5 | 23 | 31.8 ± 4.7 (41.2)* |

[a] Adult male Copenhagen rats, 90–100 days old and weighing approximately 250 g, were injected s.c. with MAT LyLu cells grown in culture. Tumor growth was monitored by palpating the site of cell injection and by using vernier calipers as described.[60]

[b] Drugs were orally administered by daily gavage, starting from the day of tumor cell injection (experiment No. 1), or a pre-dose was given 7 days before tumor cell injection (experiment No. 2) to a total of 21 days. Animals in the control group were gavaged with vehicle (2% carboxymethyl cellulose).

[c] Tumor growth rate was determined as described[60] by the slope of log-linear regression of individual tumor volume measurements for each animal and then time to reach tumor growth to 3 cc was determined by interpolation of the regression line.

[d] Median time to a growth of 10-cc tumor was when >50% of the animals per group had tumors ≥10 cc.

*Values are significantly different than those for control ($p < 0.05$; *t*-test).

tion and appearance of a palpable tumor mass, significantly increased if the tumor cell inoculum was lowered from $1 \times 10^6$ cells/site to $1 \times 10^5$ cells/site. Tumor growth rate, defined as the time duration to reach a 3 cubic cm tumor, was comparable between the low and high inoculum groups in untreated animals. Tumor growth was significantly affected, however, by CMT-3 gavage in the experiments when reduced tumor cell inoculum was used. The tumor growth was significantly slowed down in CMT-3-gavaged rats, but not to an appreciable extent in the DC-treated group. Tumor growth rate was 20.2 ± 3.5 days in rats gavaged with CMT-3 (40 mg/kg, daily for 21 days) versus 16.7 ± 1.9 days (DC, 40 mg/kg) or 15.9 ± 2.0 days in the control group. In addition, in two independent studies, we observed a regression or disappearance of palpable tumor in CMT-3-treated groups, but not in the control or DC-treated groups. Tumor regression was observed in about 40% (3/7 and 4/10). Furthermore, we also observed a significant reduction in tumor incidence (55 ± 9%) in rats which were pre-dosed with CMT-3 (40 mg/kg), but not in those treated with DC. Animals with no tumor incidence lived tumor-free for up to a year, at which time they were euthanized. At necropsy we observed a scar at the site of tumor cell injection, but no live tumor cells.

Regardless of the inhibition of local tumor growth, spontaneous metastasis to the lungs was reduced significantly in groups of rats treated with DC or CMT-3. We observed, in four separate experiments, that the reductions in the number of metastatic foci were $33 \pm 12.3\%$ (DC) to $51 \pm 7.4\%$ (CMT-3).

A remarkable observation during the course of these experiments was the lack of adverse reaction (systemic toxicity) of the drugs on the animals. As a measure of toxicity, we monitored the body weights of the animals during all the experiments. The animals were weighed before injecting the tumor cells, during gavage, and until the animals were euthanized because of a large subcutaneous tumor. As shown in FIGURE 4, the all three groups treated gained weight. Interestingly, animals treated with DC or CMT-3 gained a mean weight by $8 \pm 3.4\%$ during the 3 to 4 weeks of the measurement period. These weight gains were comparable to those of the naive animals. We did not observe any other adverse effects of the two drugs during the course of this study (it spanned three years and 177 animals). Only six animals, never more than one animal per treatment group, died prematurely. The animals had no visible signs of distress or common signs of chemotherapy-induced lethargy, alopecia, or gastrointestinal abnormalities, indicating the very safe nature of the drugs.

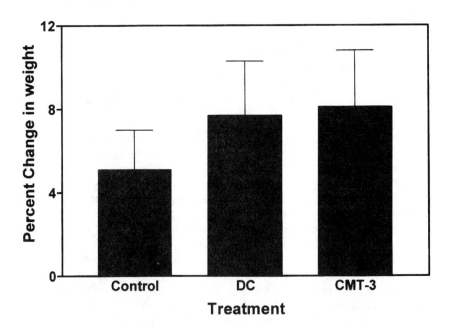

**FIGURE 4.** Change in body weight of rats treated with oral gavage of CMT-3, DC or the drug vehicle. Pooled measurements for each group before after the treatment were averaged. Data represent mean changes in weight after the treatment, just before the animals were euthanized. Data presented are from a single experiment with 10 animals per group, and similar results were observed in other experiments.

## Effect of CMT-3 on Induced Skeletal Metastasis in the MAT LyLu Model

Lumbar and vertebral bones are the common sites of prostate cancer metastases.[69] Degradation of bone matrix collagen is an essential step in skeletal invasion and skeletal remodeling during metastatic growth. We reasoned that inhibitors of MMPs should be able to reduce or eliminate bony metastasis and thus prolong the survival of rats bearing tumors that are metastatic to bone. Furthermore, in view of the classical finding that tetracyclines accumulate in bone tissues,[70] these agents may have an added advantage in inhibiting skeletal metastasis. To facilitate MAT LyLu tumor metastasis to lumbar and skeletal bones we followed the procedure of Geldof and Rao,[16] injecting MAT LyLu tumor cells into tail vein while momentarily occluding the caval vein with surgical clamps in the inferior vena cava of anesthetized rats. This procedure results in hematogenous spread of tumor cells into lumbar regions via lumbar venous plexus. Once the cells enter the lumbar region, they colonize lumbar and vertebral bones, which eventually leads to hind-limb paralysis in affected animals. The details of the experiment are described in a separate article, by Selzer *et al.* in this volume. In brief: four groups of six animals were injected with MAT LyLu tumor cells and were gavaged with CMT-3 (40 mg/kg), starting 7 or 2 days before tumor cell injection or 1 day after tumor cell injection. Animals in the control group were gavaged with 2% carboxymethyl cellulose (vehicle). The animals in all groups injected with MAT LyLu cells developed lung metastasis before they were moribund. In the group treated with vehicle alone, 5/6 (83%) of the animals had also developed hind-limb paralysis. The hind-limb paralysis might have resulted from spinal chord compression or have been due to the growth of tumors in the femur. Indeed, we were able to recover tumor cells from the cultures of marrow plugs obtained from femurs of paralyzed animals. However, in the group gavaged with CMT-3 starting 7 days pre-dose and continued until the animals were euthanized, only 1/6 (17%) developed hind-limb paralysis, and in another group treated with CMT-3, starting 2 days before tumor cell injections, none of the animals developed paralysis. In the group treated with CMT-3 one day after tumor cell injection, only 2/6 (33%) developed paralysis (see FIGURE 1 of the article of by Selzer *et al.* in this volume) and the survival in groups treated with CMT-3 increased significantly. Thus, in summary, treatment with CMT-3 results both in increase in survival and in decrease in skeletal and soft tissue metastasis. Further studies are under way in our laboratory to determine a noninvasive method of monitoring skeletal metastasis using this tumor model.

## Investigation of the Mechanism of Action for CMT-3 Induced Cytotoxicity

Since CMT-3 strongly inhibited cell proliferation *in vitro*, tumor growth, metastasis and tumor incidence *in vivo*, without much systemic toxicity, we investigated its mechanism of selective cytotoxic action. We found that the cytotoxicity of CMT-3 was limited to actively proliferating cells. For example, the $IC_{50}$ (>20 μg/ml) of CMT-3 on quiescent cells, such as serum-starved fibroblasts or confluent cultures of nontransformed cells, was at least 5- to 10-fold more than that found for continuous prostate cancer cell lines. Since many of the anti-tumor drugs which inhibit cell proliferation induce programmed cell death (PCD), we investigated whether CMT-3 also induces PCD. In these experiments, MAT LyLu cells were treated with various concentrations (1–20 μg/ml) of either CMT-3 or DC. The culture-conditioned media

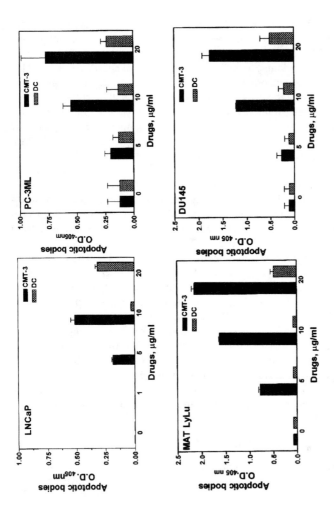

**FIGURE 5.** CMT-3 and DC induced programmed cell death (PCD) in prostate cancer cells. Cells were exposed to DC or CMT-3 for 48 hours. Culture supernatants were assayed for the presence of apoptotic bodies (free nucleosomes) using the Behringer-Mannheim cell death ELISA-Plus kit. The kit provided both positive and negative controls for the assay for comparing inter-assay consistency and specificity. The absorbance at 405 nm (OD) was directly proportional to the amount of free nucleosomes. Note that LNCaP and PC-3 cells were less apoptotic to CMT-3 or DC than were DU145 and MAT LyLu CaP cells. Induction of PCD was also independently confirmed using other parameters of cell death such as caspase-3 activation and cell surface translocation of phosphatidyl serine in drug-exposed cells (data not shown). Results presented are from at least three independent determinations.

from the drug-treated cells were assayed for PCD. Cells undergoing PCD release soluble nucleosomes into the culture medium. These nucleasomes contain H1 histone which is detected by means of an indirect-ELISA (Cell Death Detection ELISA Plus kit, BM Biochemicals, Indianapolis, IN). As we have reported before,[60] induction of PCD by CMT-3 exposure was both dose and time dependent. CMT-3 was able to induce PCD in all the seven permanent prostate cancer cell lines tested. As reported before,[60] a brief exposure ($\leq 4$ hr) to CMT-3 (10 µg/ml) led to nuclear fragmentation in > 80% of the cells. Of interest, DC induced PCD at 5–10-fold higher concentrations ($\geq 20$ µg/ml) than that of CMT-3 (FIG. 5) in our assays. The ability to induce PCD is a desirable property of an orally administrable anticancer drug. Rapid and irreversible induction of PCD should make the drug more effective, even if the peak concentration of the drug is achieved only briefly. Thus, our results and those of others[71] clearly demonstrate that CMT-3 could be an effective therapeutic drug for metastatic prostate cancer.

## SUMMARY

Results obtained by us and others support the hypothesis that MMP inhibition could be an effective approach to reduce tumor growth and metastases. Prophylactic administration of some anti-MMP agents (e.g., DC or CMT-3) may delay clinical manifestation of prostate cancer metastatic to bone. Results obtained using CMT-3 support the concept that MMP inhibition, combined with cytotoxic properties, but little systemic toxicity, could be an effective therapy for advanced cancer. We demonstrated a unique property of CMTs, an induction of apoptosis with cell-type specificity.

## ACKNOWLEDGMENTS

This work is funded in part by the National Institute of Health Grant R29 CA 61038, Department of the U.S. Army Prostate Cancer Research Program Ideas Development Grant No. DAMD 17-98-272, and Austin L. Weeks Endowment to the Department of Urology.

I am grateful to Marie G. Selzer, Bao-qian Zhu, Heather L. Houston-Clark, and Eva Escatel for generating most of the data reported here. I am also indebted to my mentor, Professor Norman L. Block, for many years of financial support, advice, and encouragement.

## REFERENCES

1. LANDIS, S.H., T. MURRAY, S. BOLDEN & P.A. WINGO. 1998. Cancer statistics, 1998. CA, Cancer J. Clin. **48(1):** 6–29.
2. KIRBY, K.S. 1996. Recent advances in the medical management of prostate cancer. Br. J. Clin. Pract. **50:** 88–93.
3. ISMAIL, M. & L.G. GOMELLA. 1997. Current treatment of advanced prostate cancer. Tech Urol. **3(1):** 16–24.
4. GITES, R.F. 1991. Carcinoma of the prostate. N. Eng. J. Med. **324:** 236–245.

5. DROLLER, M.J. 1997.Medical approaches in the management of prostatic disease. Br. J. Urol. **79**(Suppl 2): 42–52.
6. KIRBY, R.S., T.J. CHRISTMAS & M.K. BRAWER. 1996. Prostate Cancer. Mosby, London.
7. SCHRODER, F.H. 1993. Endocrine therapy for prostate cancer. Br. J. Urol. **71**: 633-640.
8. CUNHA G.R., A.A. DONJACOUR, P.S. COOKE *et al.* 1987. The endocrinology and developmental biology of the prostate. Endocr. Rev. **8**: 338–362.
9. AQUILINA J.W., J.I. LIPSKY & D.G. BOSTWICK. 1997. Androgen deprivation as a strategy for prostate cancer chemoprevention. J. Natl. Cancer Inst. **89**: 689–96.
10. NEWLING, D.W. 1996.The management of hormone refractory prostate cancer. Eur. Urol. **29**(Suppl. 2): 69–74.
11. RAGHAVAN, D., B. KOCZWARA & M. JAVLE. 1997. Evolving strategies of cytotoxic chemotherapy for advanced prostate cancer. Eur. J. Cancer **33**(4): 566–574.
12. LOGOTHETIS, C.J. 1993. Management of androgen-independent prostate carcinoma. *In* Prostate Diseases. H. Lepor & R.K. Lawson, Eds. W.B. Saunders. Philadelphia.
13. LIOTTA, L.A. & W.G. STETTLER-STEVENSON. 1993. Principles of molecular cell biology of cancer: cancer metastasis. *In* Cancer: Principles and Practice of Oncology, 4th ed., V.T. De Vita, S. Hellman & S.A. Rosenberg, Eds.: 134–149. Lippincott. Philadelphia.
14. FIDLER, I.J. 1989. Origin and biology of cancer metastasis. Cytometry **10**: 673–680.
15. CHAMBERS A.F., I.C. MACDONALD, E.E. SCHMIDT *et al.* 1995. Steps in tumor metastasis: new concepts from intra vital video-microscopy. Cancer Met. Rev. **14**: 279–301.
16. BATESON, O.V. 1942. The role of the vertebral veins in metastatic process. Ann. Int. Med. **16**: 38–45.
17. MCNEAL J.E. 1993. Prostatic carcinomas in relation to cancer origin and evaluation to clinical cancer. Cancer **71**: 984–991.
18. SHEVRIN, D.H., K.I. GORNY & S.C. KUKREJA. 1989. Patterns of metastasis by the human prostate cancer cell line PC-3 in athymic nude mice. Prostate **15**: 187–94.
19. GELDOF, A. & B.R. RAO. 1990. Prostatic tumor (R3327) skeletal metastasis. Prostate **16**: 279–290.
20. PAGET, S. 1989. The distribution of secondary growth in cancer of the breast. (reprinted) Cancer Metast. Rev. **1**: 571–573.
21. REMBRINK, K., J.C. ROMIJN, T.H. VAN DER KWAST *et al.* 1997. Orthotopic implantation of human prostate cancer cell lines: a clinically relevant animal model for metastatic prostate cancer. Prostate **31**(3): 168–174.
22. GLEAVE, M.E., J.T. HSIEH, A.C. VON ESCHENBACH & L.W.K. CHUNG. 1992. Prostate and bone fibroblasts induce human prostate cancer growth in vivo: Implications for bidirectional stromal-epithelial interaction in prostate carcinoma growth and metastasis. J. Urol. **147**: 1151–1159.
23. GLEAVE, M., J.T. HSIEH, C. GAO *et al.* 1991. Acceleration of human prostate cancer growth in vivo by factors produced by prostate and bone fibroblasts. Cancer Res. **51**: 3753–3761.
24. HOOSEIN, N.M., D.D. BOYD, W.J. HOLAS *et al.* 1991. Correlation of levels of urokinase and its receptor with the invasiveness of human prostatic carcinoma cell lines. Cancer Commun. **3**: 255-264.
25. MIGNANTI, P. & D.B. RIFKIN. 1993. Biology and biochemistry of proteinases in tumor invasion. Physiol. Rev. **73**: 161–195.
26. LIOTTA, L.A. & W.G. STETTLER-STEVENSON. 1993. Principles of molecular cell biology of cancer: cancer metastasis. *In* Cancer: Principles and Practice of Oncology, 4th ed., V.T. De Vita, S. Hellman & S.A. Rosenberg, Eds.: 134–149. Lippincott. Philadelphia.

27. NUNN, S.E., D.M. PEEHL & P. COHEN. 1997. Acid-activated insulin-like growth factor binding protein protease activity of Cathepsin D in normal and malignant prostate epithelial cells and seminal plasma. J. Cell Physiol. **171:** 196–204.

28. WEBBER, M.M., A. WAGHARY, D. BELLO & J.S. RHIM. 1996. Mini review: protease and invasion in human prostate epithelial cell lines: implications in prostate cancer prevention and intervention. Radiat. Oncol. Invest. **3:** 358–362.

29. STEARNS, M.E. & M. WANG. 1993. Type IV collagenase (Mr 72,000) expression in human prostate: benign and malignant tissue. Cancer Res. **53:** 878–883.

30. Lokeshwar, B.L., M.G. Selzer, N.L. Block & Z. Gunja-Smith. 1993. Secretion of matrix metalloproteinase and their inhibitors (TIMPs) by human prostate in explant cultures: Reduced tissue inhibitor of metalloproteinase secretion by malignant tissues. Cancer Res. **53:** 4493–4498.

31. FESTUCCIA, C., M. BOLOGNA, C. VINCENTINI *et al.* 1996. Increased matrix metalloproteinase-9 secretion in short-term tissue cultures of prostatic tumor cells. Int. J. Cancer. **69:** 386–393.

32. STEARNS M. & M.E. STEARNS. 1996. Evidence for increased activated metalloproteinase 2 (MMP-2a) expression associated with human prostate cancer progression. Oncol. Res. **8:** 69–75.

33. GLEASON, D.F. Classification of prostate carcinoma. 1966. Cancer Chemother. Rep. **50:** 125–131.

34. WOOD M., K. FUDGE, J.L. MOHLER *et al.* 1997. In situ hybridization studies of metalloproteinases 2 and 9 and TIMP-1 and TIMP-2 expression in human prostate cancer. Clin. Exp. Metastasis. **15:** 246–258.

35. POWELL, W.C., J.D. KNOX, M. NAVRE *et al.* 1993. Expression of the metalloproteinase matrilysin in DU-145 cells increase their invasive potential in severe combined immunodeficient mice. Cancer Res. **53:** 417–422.

36. STEPHENSON, R.A., C.P.N. DINNEY, K. GOHJI *et al.* 1992. Metastatic model for human prostate cancer using orthotopic implantation in nude mice. J. Natl. Cancer Inst. **84:** 951–957.

37. POWELL, W.C., F.E. DOMANN, JR., J.M. MITCHEN *et al.* 1996. Matrilysin expression in the involuting rat ventral prostate. Prostate **29:** 159–168.

38. LIOTTA L.A., P.S. STEEG & W.G. STETTLER-STEVENSON. 1991. Cancer metastasis and angiogenesis: an imbalance of positive and negative regulation. Cell **64:** 327–336.

39. STEARNS, M.E., K. FUDGE, F. GARCIA & M. WANG. 1997. IL-10 inhibition of human prostate PC-3ML cell metastases in SCID mice: IL-10 stimulation of TIMP-1 and inhibition of MMP-2/MMP-9 expression. Invasion Metast. **17:** 62–74.

40. JUNG, K., L. NOWAK, M. LIEN *et al.* 1997. Matrix metalloproteinases 1 and 3 tissue inhibitor of metalloproteinase-1 and the complex of metalloproteinase/tissue inhibitor in plasma of patients with prostate cancer. Int. J. Cancer. **74:** 220–223.

41. DECLERCK, Y.A., N. PEREZ, H. SHIMADA *et al.* 1992. Inhibition of invasion and metastasis in cells transfected with an inhibitor of metalloproteinase. Canc. Res. **52:** 701–708

42. KHOKHA, R., M.J. ZIMMER, C.H. GRAHAM *et al.* 1992, Suppression of invasion by inducible expression of tissue inhibitor of metalloproteinase-1 (TIMP-1) in B16-F10 melanoma cells. J. Natl. Cancer Inst. **84:** 1017–1022.

43. DECLERCK, Y.A., S. IMREN, A.M. MONTGOMERY *et al.* 1997. Proteases and proteases inhibitors in tumor progression. Adv. Exp. Med. Biol. **425:** 89–97.

44. BODDEN, M.K, L.J. WINDSOR, N.C.M. CATERINA *et al.* 1994. Analysis of timp-1/fib-cl complex. Ann. N.Y. Acad. Sci. **732:** 84–95.

45. BROWN, P. 1994. Clinical trials of a low molecular weight matrix metalloproteinase inhibitor in cancer. Ann. NY Acad. Sci. **732:** 217–221.

46. BEATTIE, G.J., H.A. YOUNG & J.F. SMITH. 1994. Phase I study of intra-peritoneal met-alloproteinase inhibitor BB-94 in patients with malignant ascites. *In* Abstracts of the 8th NCI-EORTC Symposium on New Drug Development, Amsterdam, March 1994.

47. DAVIES, B., P.D. BROWN, N. EAST *et al.* 1993. A synthetic matrix metalloproteinase inhibitor decreases tumor burden and prolongs survival of mice bearing human ova-rian carcinoma xenografts. Canc. Res. **53:** 2087–2091.

48. SLEDGE, G.W., JR., M. QULALI, R. GOULET *et al.* 1995. Effect of matrix metalloprotie-nase inhibitor batimastat on breast cancer regrowth and metastasis in athymic mice. J. Natl. Cancer Inst. **87:** 1546–1550.

49. MILLAR A. & P. BROWN. 1996. 360 patient meta-analysis of studies of marimastat: a novel matrix metalloproteinase inhibitor. Ann. Oncol. **7**(suppl. 5): 123.

50. POOLE, C., M. ADAMS, V. BARLEY *et al.* 1996. A dose finding study of marimastat, an oral matrix metalloprotienase inhibitor, in patients with advanced ovarian cancer. Ann. Oncol. **7**(suppl. 5): **68**.

51. STEARNS, M.E. & M. WANG. 1992. Taxol block processes essential for prostate tumor cell (PC-3ML) invasion and metastases. Cancer Res. **52:** 3776–3781.

52. STEARNS, M.E. & M. WANG. 1996. Effects of alendronate and taxol on PCML cell bone metastases in SCID mice. Invasion Metast. **16:** 116–131.

53. WOJTOWICZ-PRAGA, S.M., R.B. DICKSON & M.J. HAWKINS. 1997. Matrix metallopro-teinase inhibitors. Invest. New Drugs **14:** 62–75.

54. RASMUSSEN, H.K. & P.P. MCCANN. 1997. Matrix metalloproteinase inhibition as a novel anticancer strategy: a review with special focus on batimastat and marimastat. Pharmacol. Ther. **75:** 69–75.

55. GOLUB, L.M., N.S. RAMAMURTHY & T.F. MCNAMARA. 1991. Tetracyclines inhibit connective tissue breakdown: new therapeutic implications for an old family of drugs. Crit. Rev. Oral. Biol. Med. **2:** 297–322.

56. SIPOS, E.P., R.J. TAMARGO, J.D. WEINGERT *et al.* 1994. Inhibition of tumor angiogen-sis. Ann. N.Y. Acad. Sci. **732:** 263–272.

57. KROON, A.M., B.H.J. DONTJE, M. HOLTROP *et al.* 1984. The mitochondrial genetic system as a target for chemotherapy: Tetracycline as cytostatics. Canc. Lett. **25:** 33–80.

58. GOLUB, L.M., K. SOUMMALAINEN & T. SORSA. 1992. Host modulation with tetracy-clines and their chemically modified analogues. Curr. Opi. Dent. **2:** 80–90.

59. GOLUB, L.M., N.S. RAMAMURTHY, T.F. MCNAMARA *et al.* 1991. Tetracyclines inhibit connective tissue breakdown: new therapeutic implications for an old family of drugs. Crit. Rev. Oral Biol. Med. **2:** 297–322.

60. LOKESHWAR, B.L., M.G. SELZER, H.L. HOUSTON-CLARK *et al.* 1998. Potential applica-tion of a chemically modified non-antimicrobial tetracycline (CMT-3) against meta-static prostate cancer. Adv. Dental Res. **12:** 97–102.

61. YU, Z.M., K. LEUNG, N.S. RAMAMURTHY *et al.* 1992. HPLC determination of a chem-ically modified tetracycline: biological implications. Biochem. Med. Metab. Biol. **47:** 10–20.

62. Albini, A., Y. Iwamoto, H.K. Kleinman *et al.* A rapid in vitro assay for qauntitating the invasive potential of tumor cells. 1987. Canc. Res. **47:** 3239–3245.

63. LOKESHWAR, B.L., M.G. SELZER, B-Q. ZHU *et al.* Inhibition of tumor growth and metastasis by oral administration of a non-antimicrobial tetracycline analogue (CMT-3), and doxycycline in a metastatic prostate cancer model. Clin. Cancer Res. Submitted for publication.

64. VAN DEN BOGERT, C., B.H.J. DONTJE, M. HOLTROP *et al.* 1986. Arrest of the prolifer-ation of renal and prostate carcinomas of human origin by inhibition of mitochon-drial protein synthesis. Cancer Res. **46:** 3283–3289.

65. FIFE, R.S., G.W. SLEDGE, JR., B.J. ROTH & C. PROCTOR. 1998. Effects of doxycycline on human prostate cancer cells *in vitro*. Cancer Lett. **127:** 37–41.
66. ZUCKER, S., R.M. LYSIK, N.S. RAMAMURTHY *et al.* 1985. Diversity of melanoma plasma membrane proteinases: Inhibition of collagenolytic and cytolytic activities by minocycline. J. Natl. Cancer Inst. **75:** 517–525.
67. LOKESHWAR, B.A., S.M. FERRELL & N.L. BLOCK. 1995. Enhancement of radiation response of prostatic carcinoma by taxol: therapeutic potential for late-stage malignancy. Anticancer Res. **15:** 93–98.
68. ISAACS J.T., W.B. ISAACS, W.F. FEITZ & J. SCHERES. 1986. Establishment and characterization of seven Dunning rat prostatic cancer cell lines and their use in developing methods for predicting metastatic abilities of prostatic cancers. Prostate **9:** 261–281.
69. JACOBS, S.C. 1983. Spread of prostate cancer to bone. Urology **21:** 331–344.
70. SANDE, M.A. & G.L. MANDEL. 1990. Antimicrobial agents: tetracycline, chloramphenicaol, erythromycin and miscellaneous antibacterial agents. *In* Goodman and Gilman's The Pharmacological Basis of Therapeutics. A.G. Gilman, T.W. Rall, A.S. Nies *et al.* Eds.: 117-125. Macmillan Publishing Co. New York.
71. SEFTER, R.E.B., E.A. SEFTER, J.E. DELARCO *et al.* 1998. Chemically modified tetracyclines inhibit human melanoma cell invasion and metastasis. Clin. Exp. Metastasis. **16:** 217–225.

# A Chemically Modified Nonantimicrobial Tetracycline (CMT-8) Inhibits Gingival Matrix Metalloproteinases, Periodontal Breakdown, and Extra-Oral Bone Loss in Ovariectomized Rats

LORNE M. GOLUB,[a] NUNGAVARUM S. RAMAMURTHY,[a] ANALEYDA LLAVANERAS,[b] MARIA E. RYAN,[a] HSI MING LEE,[a] Y. LIU,[a] STEPHEN BAIN,[c] AND TIMO SORSA[d]

[a]Department of Oral Biology and Pathology, School of Dental Medicine, SUNY at Stony Brook, Stony Brook, New York 11794, USA

[b]Department of Pharmacology, Dental School, University of Central Venezuela, Caracas, Venezuela

[c]SkeleTech, Inc., Kirkland, Washington, USA

[d]Department of Periodontology, Institute of Dentistry, University of Helsinki, Helsinki, Finland

ABSTRACT: Estrogen deficiency in the postmenopausal (PM) female is the major cause of osteoporosis and may contribute to increased periodontal disease, including alveolar bone loss, seen in these women. In the current study, an animal model of PM osteoporosis, the OVX adult female rat, was studied to determine: (i) the relationship between periodontal breakdown and skeletal bone loss, and (ii) the effect of CMT-8 on gingival collagenase and bone loss. OVX rats were daily gavaged with CMT-8 (1, 2, or 5 mg/rat) for 28 or 90 days; non-OVX rats and those gavaged with vehicle alone served as controls. Elevated collagenase activity, assessed using [$^3$H-methyl] collagen as substrate in the presence or absence of APMA, was seen in the gingiva of the OVX rats, and CMT-8 therapy suppressed this effect. Western blot revealed a similar pattern for MMP-8 and MMP-13 concentrations. The changes in the gingival collagenase activity paralleled changes in periodontal bone loss, which, in turn, reflected trabecular bone density changes. Preliminary studies on PM humans administered sub-antimicrobial tetracycline as a matrix metalloproteinase inhibitor are under way.

## INTRODUCTION

Estrogen deficiency in postmenopausal women is the major cause of osteoporosis in humans. The primary clinical problem is a deficiency of bone mass resulting in pathologic fractures typically of the wrist, vertebrae, and hip.[1] The ovariectomized (OVX) rat is a widely used model of the high-turnover bone loss that occurs in postmenopausal humans[2] and is characterized by increases in the rates of *both* bone formation and bone resorption, the latter enhanced to a greater extent.[3] Ohyori *et al.*[4]

recently reviewed the two animal models most often studied—the "mature" rat model (3–5 months old), used for many morphometric and metabolic studies on OVX-induced osteoporosis, and the "aged" rat model (6–12 months old), which has the advantage of a relatively stable skeleton with time.

Regarding potential new therapies, recent studies have shown that both antimicrobial tetracyclines (TCs), such as minocycline and doxycycline, and TC analogues which have been chemically modified (CMTs) to eliminate their antimicrobial activity (e.g., removal of the dimethylamino group from carbon-4 of the "A" ring[5]), can reduce the severity of osteoporosis in the OVX rat,[4,6,7] and can increase the biomechanical strength of bone, at least in localized lesions.[8] Earlier preliminary studies found that other types of osteoporosis (e.g., diabetes-induced in the rat and estrogen-deficiency in the human) could also respond beneficially to TC therapy.[9,10] The initial rationale for using TC compounds for osteoporosis arose from the discovery, first reported in 1983–84, that these drugs can inhibit mammalian collagenases, and connective tissue breakdown *including bone resorption,* by a nonantimicrobial mechanism.[11–13] Since then, a series of CMTs have been developed which inhibit by *multiple mechanisms* several matrix metalloproteinases (MMPs), including the three collagenases (MMP-1, MMP-8, MMP-13), two gelatinases (MMP-2, MMP-9), as well as macrophage metalloelastase (MMP-12) and membrane-type MMP (MMP-14) (see Golub *et al.*[14,15] and Ryan *et al.*[16] for reviews). These inhibitory mechanisms include (but are not limited to) the ability: to block the activity of mature MMPs[11,12,17,18,45]; to prevent the conversion of pro-MMPs into their active forms[19,20]; and to downregulate the expression of (at least) some MMPs.[21–23] Regarding mechanisms of bone destruction, MMP-2 and MMP-9 activity (assessed by gelatin zymography) was recently found to be increased in the trabeculae of long bones of aged (9-month-old) osteoporotic rats.[24] Additional MMPs, including MMP-1, MMP-13, MMP-12, and MMP-14, have also been implicated in bone resorption.[25–27]

A number of studies on the MMP-inhibitory properties of TCs and CMTs have focused on the efficacy of these drugs for reducing soft tissue and bone destruction in human and animal periodontal disease.[28–32] However, only in the past decade has a relationship between postmenopausal osteoporosis and increased periodontal disease been established (see below); this relationship provides another model to explain the ability of certain *systemic* diseases or factors (diabetes mellitus, estrogen deficiency, smoking) to enhance *local* periodontal bone loss, the latter induced by the inflammatory response in the periodontium to oral microbial factors. A widely studied example of this systemic disease/local disease interaction is the increased periodontal (including alveolar bone) breakdown and osteoporosis which are both observed as complications of long-term diabetes mellitus.[15,16,33] Another example is seen in postmenopausal estrogen-deficient women, who have been reported to exhibit significantly increased periodontal (soft tissue) attachment loss, reductions in alveolar bone density, and loss of alveolar bone height, compared to appropriate control subjects including postmenopausal women on hormone replacement therapy.[34–36] Elevated levels of bone-resorption-inducing and proinflammatory cytokines (IL-1β, IL-6, and IL-8) in the gingival crevicular fluid (GCF) of the periodontal pockets of estrogen-deficient postmenopausal women provides one explanation for the increased periodontitis in these patients.[35–38] Of extreme interest, a recent large-scale epidemiologic study of 42,171 postmenopausal women carried out by Grodstein *et al.*[39] concluded that "the risk of tooth loss was lower in women who currently used hormones." Estrogen-deficiency, promoting loss of alveolar bone density and increased periodontitis, was considered an important factor contributing to the increased tooth loss in postmenopausal women who did not use hormone-replacement therapy.

In the current study, both "mature" and "aged" OVX rats (see above for rationale) were administered a CMT, CMT-8 (the nonantimicrobial analogue of doxycycline), which was recently found to be more effective than other TCs and CMTs as an inhibitor of both bone resorption in organ culture and MMP-13 (collagenase-3) activity *in vitro*; this collagenase is believed to mediate, in part, the destruction of the type I collagen matrix during bone resorption.[25] The effect of this drug on periodontal breakdown (assessed by measuring gingival collagenase activity and alveolar bone loss) was compared to its effect on skeletal bone loss in these animal models of osteoporosis.

## MATERIALS AND METHODS

Details of this animal model and the procedures described below, including the preparation and partial purification of gingival extracts, and measurement of bone loss and activities of MMPs, have been described by us previously[4,7,9,25,28] and, therefore, are only briefly discussed.

### *Animal Model of Estrogen-Deficiency Osteoporosis; Treatment with CMT-8 in Vivo*

Two types of widely used rat models of OVX-induced osteoporosis were set up.[4] In the first experiment, twenty-four 3-month old "mature" growing female rats (body weight = 285 g ± 18; purchased from Ziv Miller Farms, Inc., Zelienople, PA) were distributed into four experimental groups ($n = 6$ rats/group) as follows: Two days after ovariectomizing 12 rats, half of them began to receive, by oral gavage, CMT-8 (5 mg/rat) suspended in 2% carboxymethyl cellulose (CMC) once/day for 28 days. The remaining six OVX rats were gavaged daily with the vehicle alone. The other 12 rats (control, non-OVX group) were "sham-operated" and half of these were gavaged daily with either CMT-8 or with vehicle (2% CMC) alone.

In the second experiment, thirty-five 6-month old "aged" rats (mean body weight = 312 g ± 35) were ovariectomized or sham-operated. This age was selected because growth of bones in length and width is greatly diminished at this time,[40] making these older rats a widely used model of postmenopausal osteoporosis in humans.[4] On day = 0, five of these rats were euthanized to provide tissues at baseline for analysis. The remaining 30 rats were then distributed into the following experimental groups ($n = 6$ rats/group): 18 of the rats were ovariectomized and 12 were sham-operated. Two days after OVX, all of the rats in this group were treated by oral gavage with either 1 mg CMT-8/day, 2 mg CMT-8/day, or with vehicle alone (1 ml 2% CMC) over a 90-day time period. The sham-operated controls were gavaged daily with either 2 mg CMT-8 or with the vehicle alone.

At the end of each experimental protocol, the rats were anesthetized with a solution of ketamine and xylazine, and an intracardiac blood sample was drawn 2–4 hours after the last drug treatment to determine the serum concentration of CMT-8 in each treated rat by HPLC.[41] The rats were then euthanized and the gingiva around the teeth from the maxillary jaws (about 20 mg wet weight/rat) were dissected and pooled by group; insufficient gingival tissue was available for individual analysis. The gingival tissues were then stored at −80°C until analyzed, the maxillary jaws were removed and defleshed to quantitate alveolar (periodontal) bone loss, and the tibias were removed (the long bones were studied only in the "aged" rat model; data on the "mature" OVX rats were reported by us elsewhere[4]) to assess trabecular bone mineral density (see below).

### Assessment of Alveolar Bone Loss and Tibial Bone Density

In a manner similar to that described by us previously,[28,29] each maxillary jaw was defleshed by boiling (in additional experiments, a half-maxilla was processed for standard histologic study) and was additionally cleaned free of remaining soft tissue by soaking in 0.2N NaOH (22°C). The jaws were then stained with Lofeller's methylene blue to demarcate the cemento-enamel junction (CEJ) on each tooth, and alveolar bone loss was measured morphometrically, at 17 different sites in each half-maxilla (see Chang *et al.*[29] for details), using NIH software and a MacIntosh computer. Bone loss was defined as the distance (each unit = 0.1 mm) from the CEJ to the crest of the alveolar bone at each site.

For each tibia, the bones were dissected free of soft tissue and then stored in 70% ethanol (22°C). Bone density was measured using peripheral quantitative computer tomography (pQCT; Norland/Stratec XCTrm). The proximal metaphysis and midshaft of each tibia were scanned using a threshold for delineation of external boundary, and an area "peel" to subdivide each bone into a cortical/subcortical region and a cancellous subregion. The bone mineral density (BMD; $g/cm^3$) and areal properties of each subregion were then determined by system software.

As described by us previously,[42] dynamic histomorphometric studies using fluorochrome labels injected into the different groups of rats at several time periods and fluorescence microscopy of undecalcified sections were carried out to assess the effect of OVX and CMT-8 on both rate of bone formation and of resorption in trabecular bone of the tibiae; these data are reported elsewhere.

### Preparation of Gingival Extracts: Assessment of Activity, Type and Concentration of Collagenase and Gelatinase

As described previously,[28] the pooled gingival tissues for each group of rats were weighed, minced (all procedures at 4°C), and extracted with Tris-NaCl-$CaCl_2$ buffer (pH 7.6; 100 mg wet weight gingival tissue/5 ml buffer) containing 5 M urea. After centrifugation, the supernatant was dialyzed exhaustively against the Tris/NaCl/$CaCl_2$ buffer, and the extract was partially purified by precipitation with ammonium sulfate added to 60% saturation. Aliquots of each gingival extract were measured for protein using the Bio-Rad protein assay kit (Bio-Rad, Richmond, CA) prior to measurement of MMPs (see below).

Collagenase activity was measured in the gingival extracts using lathyritic rat skin type I collagen, labeled *in vitro* with [$^3$H] formaldehyde, as substrate.[5,11] Incubations were carried out at 22°C in the presence or absence of aminophenylmercuric acetate (APMA) added in a final concentration of 1.2 mM. The 3/4 and 1/4 collagen degradation fragments ($\alpha^A$ and $\alpha^B$) generated by the gingival extracts after incubation with the [$^3$H-methyl] collagen substrate, and the undegraded collagen components ($\alpha_1(l)$ and $\alpha_2$), were identified by a combination of SDS-PAGE and fluorography. Collagenase activity was calculated (i.e., by conversion of intact collagen $\alpha$ components to form $\alpha_1 A$ and $\alpha_2 A$) after the fluorograms were scanned with an LKB Ultroscan XL laser densitometer.

Gelatinase activity was measured using a modification of the method of McCroskery *et al.*[43] In brief, [$^3$H-methyl] type I collagen was denatured by heating at 60°C for 20 min to serve as the gelatin substrate. Aliquots of the gingival extract were preincubated with APMA and soybean trypsin inhibitor was added at a final concentration of 1.4 mM and 200 µg/mL, respectively, for 1 hr at 22°C. The [$^3$H-methyl] gelatin was then added to the incubation mixture, which was incubated for an additional 4 hr at 37°C. The reaction was terminated by addition of nonradioactive gelatin (20 mg/mL) and trichloroacetic acid at a final concentration of 45%; the mixture was cooled to 4°C and centrifuged ($13,000 \times g$); and the release of [$^3$H] labeled gelatin degradation products in the supernatant was counted in a liquid scintillation counter.

In addition, the gelatinase activity was characterized by zymography in 10% SDS polyacrylamide gels co-polymerized with 1.0 mg/mL denatured type I collagen (gelatin)[28] using Novex precast gels (Novax, Inc., San Diego, CA.)

To further characterize some of the genetically distinct types of MMPs in the different groups of rats, Western blot analysis was carried out on aliquots of the partially purified extracts of gingival tissue using specific polyclonal and monoclonal antisera.[44,45,48] In brief, freeze-dried samples of gingival extract (10–20 µg protein) were dissolved in 100 µl of 10 mM Tris-HCl (pH 7.8), run on 7.5% SDS-PAGE, and transferred to nitrocellulose filter paper. Nonspecific binding was eliminated by incubating (90 min at 37°C) with phosphate-buffered saline solution supplemented with 5% nonfat dry milk. The blots were incubated either with rabbit polyclonal or mouse monoclonal antibodies against human MMPs diluted 1 : 500 – 1 : 1000, or with nonimmune control serum diluted l:100, for l–6 hr at 20°C. After washing, the blots were incubated with biotinylated anti-rabbit or anti-mouse immunoglobulins (1:500) for l hr at 50°C and then incubated with anti-rabbit or anti-mouse antibody-alkaline phosphatase conjugate, and the color was developed by standard technique.[45] The Western blots were analyzed densitometrically (Model GS-700 Imaging Densitometer, Bio-Rad) using the Molecular Analyst® program (Image analysis system version l, 4; Bio-Rad), and the concentration of the different MMPs (including MMP-8, MMP-13, MMP-2 and MMP-9) was calculated from a standard curve constructed using recombinant MMPs.

## RESULTS

In both rat models of osteoporosis, the untreated OVX rats gained 71.7% and 66.0% *more* weight, after the 28-day (which involved "mature" rats) and 90-day ("aged" rats) protocols, respectively, than the non-OVX control groups (TABLE 1). These results are consistent with the increased depot fat observed by other investigators in estrogen-deficient rats.[46] Of interest, the OVX rats that were treated with 1, 2, or 5 mg/day CMT-8 showed 28–63% *less* body weight gain (and, presumably, less fat) than the untreated OVX rats; it should be noted, however, that treating the non-OVX control rats with CMT-8 had little or no effect on body weight gain. CMT-8 concentrations were measured, by HPLC, in the serum of the different groups of rats treated by oral gavage during both the 28-day and the 90-day time periods. In the 90-day ("aged" rat) protocol, the non-OVX control rats, given 2 mg CMT-8/day, showed a serum concentration of 5.5 μg/ml ± 1.4 (TABLE 1). The OVX rats treated with CMT-8 doses of 1 mg/day and 2 mg/day showed serum concentrations of 2.4 ± 0.9 and 9.9 ± 2.7 μg/ml, respectively. These data demonstrated that increasing the oral dose did increase the serum concentration of CMT-8. These data also demonstrated that the OVX rats exhibited an 80% greater serum concentration of the drug, at the same oral dose (2 mg/day), than did the non-OVX control rats (note: both the non-OVX control rats and the OVX rats received similar doses of CMT-8 relative to body weight, approximately 5 mg/kg). In contrast, the HPLC data for the "mature" rats treated with CMT-8 (but at a higher oral dose, i.e., 5 mg/day) showed that although the serum concentration was higher for these rats treated with this dose (both the non-OVX controls and the OVX rats received about 15 mg/kg body weight), there was no difference between these two groups; both groups showed a serum concentration of about 14 μg/ml. Presumably the serum concentra-

**TABLE 1. Body weight gain and serum CMT-8 concentration in "mature" and "aged" normal (C) and ovariectomized (OVX) female rats[a]**

| Experimental groups | "Mature" rats[b] Body weight gain (g/28 days) | Serum (μg/ml) | "Aged" rats[c] Body weight gain (g/90 days) | Serum (μg/ml) |
|---|---|---|---|---|
| NON-OVX (C) | 46 ± 2 | 0 | 97 ± 4 | 0 |
| C + CMT-8 (d or e) | 49 ± 2[d] | 14.2 ± 1.2[d] | 89 ± 3[e] | 5.5 ± 1.4[e] |
| OVX | 79 ± 2 | 0 | 161 ± 6 | 0 |
| OVX + 1 mg CMT-8/day | – | – | 103 ± 3 | 2.4 ± 0.9 |
| OVX + 2 mg CMT-8/day | – | – | 60 ± 2 | 9.9 ± 2.7 |
| O X + 5 mg CMT-8/day | 57 ± 2 | 13.6 ± 1.8 | – | – |

[a]Each value represents the mean of 6 rats/group ± SEM.
[b]Initial body weight (g) = 285 ± 18.
[c]Initial body weight (g) = 312 ± 35.
[d]C+5 mg CMT-8/day.
[e]C+2 mg CMT-8/day.

tion in the rats treated with the highest oral dose in this study reached a maximum steady state level.

## Bone Loss in Tibia Induced by OVX: Effect of CMT-8 Therapy

Bone mineral density (BMD; $g/cm^3$) and trabecular bone density ($mg/cm^3$) in the proximal tibia of the "aged" (6-month-old) OVX rats after 90-day treatment with vehicle alone (no drug) were reduced by 13.4% (data not shown) and 43.8% (FIG. 1c), respectively ($p < 0.0l$, for both measurements) compared to the non-OVX controls (bone loss was not assessed in the "mature" 3-month old rats in this experiment; see below for previous experiments using this rat model). These deficiencies in BMD and trabecular bone density were prevented by treating the OVX rats with 2 mg/day CMT-8 but not by 1 mg/day of this drug (FIG. 1c). A similar pattern was seen when trabecular bone volume (TBV) was assessed morphometrically with time (data not shown). The non-OVX control rats treated with vehicle alone or with 2 mg/day CMT-8 lost about 5% TBV during the 90-day treatment phase of the protocol. In contrast, the OVX rats treated with vehicle alone or with 1 mg/day CMT-8 lost about 21–23% TBV. However, when the dose of CMT-8 was increased to 2 mg/day, the loss of TBV in these "aged" OVX rats was reduced by about 45% ($p < 0.05$). A similar pattern of change was seen using "mature" (3-month-old) OVX rats treated for prolonged time periods with CMT-8 and was reported previously.[4]

## Alveolar Bone Loss Induced by OVX: Effect of CMT-8 Therapy

Like the data on bone loss in the tibia described above, periodontal (alveolar) bone loss was only assessed, in the current experiments, in the "aged" rat model. As shown in FIGURE 1b, alveolar bone loss, assessed by computer-assisted morphometric analysis of the distance from the CEJ on the tooth to the crest of the adjacent alveolar bone, was increased by 41.8% ($p < 0.05$) in the untreated OVX rat compared to the alveolar bone loss seen in the normal controls. Treating the OVX rats over a 90-day time period with either 1 mg/day or 2 mg/day CMT-8 reduced this pathologically excessive alveolar bone loss by 17.7% and 33.5%, respectively, a dose–response effect. However, only the beneficial effect of the higher CMT-8 dose was statistically significant ($p < 0.05$). It should also be noted (i) that this dose of CMT-8 (2 mg/day) completely "normalized" both the loss of alveolar bone (FIG. 1b) and the reduction of trabecular bone density of the tibia (FIG. 1c) induced by long-term es-

FIGURE 1. Gingival collagenase activity (a), loss of alveolar bone height (b), and trabecular bone density in tibias (c) in OVX rats: effect of CMT-8 therapy. Six-month-old female rats were sham-operated (N groups) or ovariectomized (OVX groups) and then treated by daily oral gavage with CMT-8 (1 or 2 mg per day) for 90 days. The gingiva were dissected from each rat, pooled by group, and extracted; the extracts were partially purified by ammonium sulfate precipitation; and the aliquots were incubated with [$^3$H-methyl] collagen (22°C) with APMA (1.2 mM). The collagen components and collagenase digestion fragments were assessed by SDS-PAGE and fluorography. Alveolar bone loss was assessed morphometrically in the defleshed maxillary jaws, and trabecular bone loss was measured histomorphometrically, in each rat. Each value (b and c) represents the mean ± SEM of 6 rats/group.

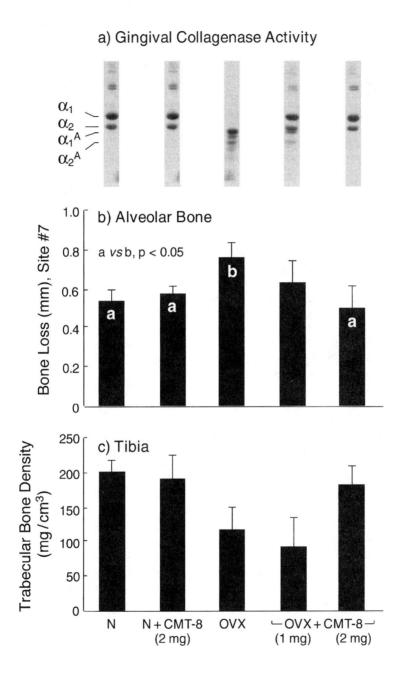

a) Gingival Collagenase Activity

$\alpha_1$
$\alpha_2$
$\alpha_1{}^A$
$\alpha_2{}^A$

b) Alveolar Bone

Bone Loss (mm), Site #7

a *vs* b, p < 0.05

c) Tibia

Trabecular Bone Density ($mg/cm^3$)

N    N + CMT-8    OVX    ⌐OVX + CMT-8⌐
     (2 mg)              (1 mg)   (2 mg)

trogen deficiency, and (ii) that although only site #7 data (i.e., bone loss between the first and second molars in the maxilla) are shown in FIGURE 1b, similar changes due to both OVX and CMT-8 therapy were seen at the additional 16 sites in the maxilla analyzed morphometrically (data not shown). These data clearly indicate that the changes in systemic bone loss (assessed by measuring trabecular bone density and trabecular bone volume) due to OVX, and its treatment with a nonantimicrobial TC analogue, paralleled the *local* changes in bone loss seen in the periodontal tissues— the latter traditionally associated only with microbially induced inflammation of these oral tissues.[16]

## Effect of OVX and CMT-8 Therapy on Gingival (Soft Tissue) Collagenases and Gelatinases

### "Mature" Rat Model

*(a) Functional assays:* Collagenase activity, assessed using [$^3$H-methyl] collagen as substrate and a combination of SDS-PAGE/fluorography, was measured in both the "mature" (3-month-old; FIG. 2) and "aged" (6-month-old; FIG. 1a) OVX rat models. The gingival extracts from the 3-month-old control (non-OVX) rats did not exhibit collagenase activity *in vitro* in the presence or absence of APMA (FIG. 2). However, ovariectomizing the "mature" rats increased the collagenase activity in these periodontal soft tissues, and this MMP appeared to exist primarily in the active form, rather than the latent pro-form of this enzyme, since its elevated activity appeared to be unaffected by the addition of APMA to the incubation mixture (FIG. 2). Treating the OVX rats with 5mg/day CMT-8 over a 28-day time period appeared to completely inhibit this excess collagenase activity regardless of whether the gingival extracts were incubated with or without APMA (FIG. 2).

Gelatinase activity assessed by zymography showed a similar pattern of change (FIG. 3). Small amounts of 68 kDa gelatinase were seen in the gingival extracts from the normal control "mature" rats, and ovariectomizing the rats appeared to dramatically increase the levels of slightly higher molecular weight forms of this MMP. Treating the OVX rats with CMT-8 for 28 days reduced this excess gelatinase activity to essentially control non-OVX levels. In a preliminary study, extract of trabecular bone in tibias from the different groups of rats (prepared as described by Mansell *et al.*[24] ) also showed increases in gelatinase activity (assessed by zymography) due to ovariectomy and reductions due to CMT-8 therapy (data not shown). The molecular identification of the types of gelatinases extracted from the gingiva of the different groups of rats is discussed below.

*(b) Western blot analyses:* The data in TABLE 2 show the relative proportions and concentrations of two different collagenases (MMP-8 and MMP-13; MMP-1 is not produced by rodent tissues[25]) in the gingival tissues of the different groups of rats, assessed by densitometric analysis of Western blots. Consistent with the data on functional collagenase activity (see FIGURES 2 and 1a) discussed above, only low levels of MMP-8 and MMP-13 were detected in the extracts of gingiva obtained from the control non-OVX rats, with MMP-13 comprising 73–83% of the total collagenase in these tissues (TABLE 2). However, a marked increase in the concentration of both of these collagenases could be seen in the gingiva of the untreated OVX rats

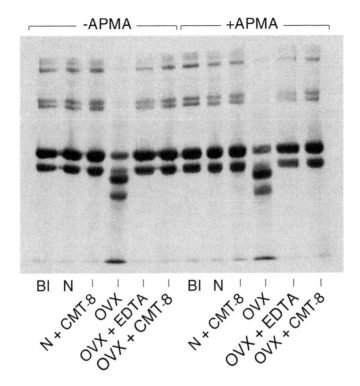

**FIGURE 2.** Collagenase activity in partially purified extract of gingiva from 3-month-old female rats after 28 days of daily oral gavage with CMT-8 or vehicle alone. Gingival extracts were incubated (22°C) with [$^3$H-methyl] collagen in the presence or absence of APMA (1.2 mM), and the collagen components ($\alpha$) and degradation fragments ($\alpha_1 A$ and $\alpha_2 A$) were assessed by a combination of SDS-PAGE and fluorography.

although MMP-8 protein expression was increased more than MMP-13; in these soft tissues, MMP-8 and MMP-13 now appeared to be present in near-equal levels (TABLE 2). Treatment of the OVX rats with CMT-8 appeared to suppress the pathologically excessive MMP-8 and MMP-13 levels in the gingival tissues to essentially those levels seen in the normal rats. Furthermore, although not shown, the Western blots indicated that these collagenases were present in the gingival extracts as different molecular weight forms including active 50 kDa, pro-forms at 65 kDa, and higher molecular weight forms (75 and 85 kDa), which may represent active forms complexed by TIMPs. Studies are now under way to more clearly identify the different molecular forms of these MMPs in the normal and OVX rats. Consistent with the data obtained by gelatin zymography (FIG. 3), MMP-9 was not detected in the gingival extracts of the different groups of rats as determined by Western blot analysis and the increased gelatinase activity in the OVX rats and its reduction by CMT-8 therapy appeared to reflect changes in MMP-2 (data not shown).

**TABLE 2. Collagenases in extract of gingiva from normal (control) and ovariecto-mized (OVX) mature female rats: effect of CMT-8 administration**

| Experimental groups | Collagenase concentration (ng/µg protein)[a] | | MMP-13 as a % of total collagenase[b] |
|---|---|---|---|
| | MMP-8 | MMP-13 | |
| Control (non-OVX) | 0.75 | 3.77 | 83.4 |
| Control + CMT-8 | 0.55 | 1.47 | 72.8 |
| OVX (no drug) | 8.17 | 6.42 | 44.0 |
| OVX + CMT-8 | 0.63 | 0.80 | 53.9 |

[a]Each value obtained from a pool of 6 rats/group.
[b]MMP-1 not found in rat tissues (current study; in agreement with the literature[28–30]); rat MMP-13 is the homologue of human MMP-1.

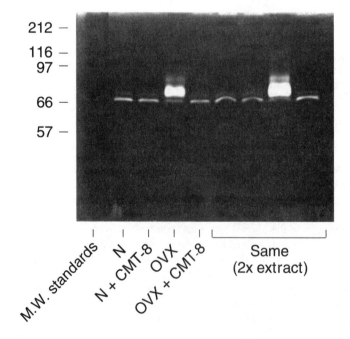

**FIGURE 3.** Gelatinase activity, assessed by enzymography, in partially purified extracts of gingiva from sham-operated (N group) and OVX rats: effect of CMT-8 therapy.

*"Aged" Rat Model*

The data in FIGURE 1, generated from the "aged" rat model of OVX, suggest a relationship between the collagenase activity in the gingival tissue of the different groups of rats, and the loss of underlying alveolar bone. Consistent with the data described above for the "mature" rat model, no collagenase activity in the gingival extracts could be seen in the normal control rats, and a dramatic increase in gingival collagenase activity was seen in the untreated OVX group. When the OVX rats were treated with two different doses of CMT-8 over a 90-day time period, the 1 mg/day dose partially inhibited, and the 2 mg/day completely inhibited this excess collagenase activity—a dose–response effect *in vivo*. As shown in FIGURES 1a and b, the changes in gingival collagenase activity, due to both OVX and treatment with CMT-8, appeared to parallel the changes in loss of underlying alveolar bone.

## DISCUSSION

Periodontal disease, the major cause of tooth loss in the adult population, is characterized by the destruction of the collagen fibers and other matrix constituents of the gingiva, periodontal ligament, and alveolar bone. Although the oral microflora, primarily the anaerobic gram-negative microorganisms that accumulate in the gingival crevice, initiate this inflammatory disease in the gingival tissues, the connective tissue breakdown that occurs in this, and subsequently in the deeper periodontal tissues, is mediated by excessive levels of activated MMPs (and other neutral proteinases such as PMN leukocyte elastase), produced by host cells (e.g., neutrophils, macrophages, fibroblasts, osteoblasts, and osteoclasts), overcoming the endogenous proteinase-inhibitor (e.g., TIMPs, $\alpha_2$-macroglobulin) shield.[47] The MMPs that have been associated with the diseased periodontal tissues include the following:

(i) MMP-8 and MMP-9 are dominant in the GCF of the periodontal pocket, although MMP-13 has recently been found in small amounts in this inflammatory exudate.[48–50]

(ii) Both types of gelatinase (MMP-2, and MMP-9), stromelysins (MMP-3 and MMP-10), and the two collagenases (MMP-1 and MMP-8) have been detected in diseased gingival tissues,[47,48,51] although Uitto *et al.*[51] recently reported that a third collagenase, MMP-13, was more widespread throughout this soft tissue, particularly expressed by epithelial cells, than previously recognized.

(iii) Alveolar bone, like other parts of the skeletal system, is presumably resorbed, in part, by MMPs (MMP-1, MMP-13, MMP-2, MMP-9, MMP-12 and MMP-14) produced by osteoblasts and osteoclasts.[24–27]

Treatment of periodontal disease has traditionally focused on the reduction of the bacterial "load" in the periodontal pocket by mechanical debridement and can also involve the use of topical or systemic antibiotics as an adjunct. A novel approach currently gaining interest is the use of "host-modulating therapy," which attempts to supplement the traditional antimicrobial treatment strategies with drugs that (1) inhibit the production of inflammatory mediators, such as the prostaglandins, or (2) block the production and activity of host-generated, tissue-destructive proteinases such as the MMPs.[57] Bisphosphonates, which act by blocking osteoclast-mediated bone resorption, are a third category of "host-modulating drugs" that are currently being explored as adjuncts in the treatment of periodontal disease; these compounds were also recently reported to exhibit MMP-inhibitory properties.[52] With regard to the second therapeutic strategy, Golub et al.[11] discovered that tetracyclines (TCs) can inhibit collagenase and other MMPs by nonantimicrobial mechanisms and subsequently developed a series of chemically modified TC analogues (called CMTs1-10) that lost their antimicrobial activity but retained (and even showed enhanced) MMP-inhibitory properties.[14–16,25] These TCs and CMTs are now recognized to have multiple mechanisms of action, including the inhibition of already-active MMPs, the inhibition of activation of latent pro-MMPs, the downregulation of MMP expression, and the protection of endogenous MMP and serine-proteinase inhibitors from proteolytic or oxidative inactivation. Golub et al.[15] recently reviewed these pleiotropic mechanisms, and the numerous diseases that might be beneficially affected by TCs and CMTs such as periodontitis, various types of arthritis, sterile corneal ulcers, aortic aneurysms, and cancer invasion and metastasis. Of the ten CMTs tested to date, one of these, CMT-5 (the pyrazole analogue), which lacks the $Ca^{2+}$ and $Zn^{2+}$ binding site at carbon-11 and carbon-12 of the TC molecule, does not inhibit MMP activity in vitro. Of those CMTs that do inhibit MMPs and also inhibit bone resorption in tissue culture, CMT-3 and CMT-8 were found to be the most effective: their $IC_{50}$ as inhibitors of bone resorption was found to be < 2 and 0.1–0.5 μg/ml, respectively, and were recently reported to be the most effective of these TC analogues as inhibitors of MMP-13 activity in vitro with an $IC_{50}$ of 0.5 μM and 0.2 μM, respectively.[25,53,63] As a result of these and other studies, one of these potent CMTs, CMT-3, has been extensively tested, and found to be effective, as an inhibitor of cancer cell gelatinase activity in vitro, cancer cell invasiveness in culture, and cancer metastasis in rats in vivo.[17,54] The other newly developed CMT, CMT-8, because of its potent efficacy as an inhibitor of both (i) bone resorption and cartilage destruction in culture, and (ii) bone/cartilage-type collagenase (MMP-13) activity in vitro,[25,53] is being tested as a potential therapeutic agent in animal models of bone loss (see below) as well as in other diseases such as diabetes.[16]

Referring again to periodontal disease, although microbially induced and mediated by excessive activity of host-generated MMPs and other tissue-destructive proteinases, it is now recognized that this local oral disease can be exacerbated by certain systemic factors. In this regard, estrogen-deficient postmenopausal women exhibit (i) increased severity of periodontitis characterized by the loss of alveolar bone height and density around the teeth (as well as systemic bone loss—osteoporosis) and soft tissue changes including excessive migration of the epithelium lining the gingival crevice resulting in the clinical sequelae of increased pocket depth and attachment loss, and (ii) increased tooth loss.[34–39] Mechanisms explaining this increased periodontal disease include the detection of elevated levels of cytokines (IL-1, IL-6, and IL-8) in the GCF of the periodontal pockets in these postmenopausal subjects; increased expression of these and other cytokines has been associated with systemic osteoporosis due to estrogen deficiency.[55,56]

In the current study, ovariectomizing female rats produced evidence of increased periodontal breakdown consistent with previous studies on estrogen-deficiency in rats[58] and in postmenopausal women.[34–39] However, this is the first report to our knowledge that an animal model of postmenopausal osteoporosis (or for that matter, humans with estrogen-deficiency) exhibits increased MMP (collagenase and gelatinase) activity in the gingiva or other periodontal tissues (alterations in estrogen and progesterone levels were previously shown to modulate collagenase activity and collagen breakdown in other soft tissues, for example, during involution of the postpartum uterus[71,72]). This alteration appeared to reflect an increase of mostly active, rather than latent pro-forms of collagenase since the production of the 3/4 α (αA) collagenase-digestion fragments by the gingival extracts from the OVX rats were seen *in vitro* in the presence or absence of APMA as a pro-MMP activator. In our initial characterization of this increased collagenase activity in the OVX rats, Western blot analysis indicated that both MMP-8 and MMP-13 were increased in the gingival tissues. This increase in collagenase protein expression and activation in the OVX rat gingiva may reflect, in part, elevated cytokine production during estrogen deficiency (see above). In this regard, Chubinskaya *et al.*[59] reported that IL-1β upregulated the expression of human MMP-8 by mesenchymal cells, and Uitto *et al.*[51] found that human MMP-13 expressed by periodontal epithelial cells in culture was increased by TNFα. However, in the current study on rat MMPs, OVX appeared to upregulate the expression of MMP-8 much more than MMP-13. Since MMP-13 in the rat is homologous with human MMP-1, perhaps this MMP (collagenase-3) is less readily induced in the rat than MMP-8 or MMP-13 in the human.

Gelatinase activity was also increased in the gingiva of the OVX rat, and analysis by zymography and Western blot indicated that MMP-2, rather than MMP-9, was upregulated. Preliminary studies in our lab on dermal wound healing in OVX rats detected increased expression of MMP-2, in both epithelial cells and fibroblasts, determined by *in situ* hybridization,[63] and the same cell types could be responsible for the increased gelatinase activity in the gingival tissues. Increases in gelatinase have also been seen in trabecular bone from OVX rats, although in this tissue both MMP-2 and MMP-9 may be affected by estrogen deficiency.[24,60] Increased cytokine levels have already been addressed as a mechanism of increased periodontal breakdown (including alveolar bone loss) and osteoporosis in postmenopausal women and in the OVX animal model. Additional mechanisms may include (but are not limited to) altered levels of other cell regulators such as nitric oxide, arachidonic acid metabolites such as $PGE_2$, and advanced glycation end-products.[15,42,61,64]

Of extreme interest, the oral administration of CMT-8 (the nonantimicrobial analogue of doxycycline) was found to "normalize" the pathologic changes in the periodontal tissues (gingiva and alveolar bone were studied) and long bones (tibia) of the OVX rat. CMT-8 administration was found:

(i) to reduce the pathologically elevated expression and activity of two collagenase (MMP-8 and MMP-13) and a gelatinase (MMP-2) in the gingival tissues; gelatinase activity (assessed by zymography) also appeared to be reduced by this treatment in trabecular bone of tibia from the OVX rats (Ramamurthy *et al.*, unpublished results), and

(ii) to "normalize" the loss of both alveolar bone height and tibial bone density in this animal model of postmenopausal osteoporosis.

Defining the therapeutic mechanisms involved remains to be determined, especially in view of the multiple nonantimicrobial pharmacologic actions of CMTs,[15] which include effects of these drugs on MMP expression, activation, and activity, as well as effects on production of cytokines, nitric oxide, and arachidonic acid metabolites (e.g., $PGE_2$). In this regard, it is interesting to note:

(i) that of all the CMTs and TCs studied to date, CMT-8 was the most potent inhibitor of MMP-13 activity *in vitro*, with an $IC_{50}$ < 0.2 µg/ml (mid-nanomolar) using either type 1 collagen or gelatin as substrates, and that this relative efficacy of the different CMTs/TCs was found to be correlated with the potency of these compounds as inhibitors of bone resorption in tissue culture;[25]

(ii) that CMT-8 is the most readily absorbed of the 10 different CMTs *in vivo* after oral gavage, producing peak ($C_{max}$) serum concentrations (e.g., 6–18 µg/ml, depending on oral dose) higher than those produced by the other CMTs[41] *and well in excess of* the concentrations needed to directly inhibit collagenase and gelatinase *in vitro* (see above); and

(iii) that although MMP-13 in the rat gingiva was not increased as much as MMP-8 (see above), it was reduced by CMT-8 therapy as much as MMP-8; this may, again, reflect the pleiotropic mechanisms of action of CMTs, that is, these compounds may act as inhibitors of MMP expression (counteracting the increased cytokine levels following OVX) as well as inhibiting the activity and activation of these proteinases extracellularly.

With regard to the MMP-inhibitory action of CMT-8, recent preliminary data (Ramamurthy, Bookbinder and Golub, unpublished work) showed that the collagenase activity (assessed by measuring the conversion of α collagen components to $α^A$ degradation fragments, using SDS-PAGE/fluorography) in extracts of gingiva from untreated OVX rats, although only partially inhibited *in vitro* by 0.2 µg/ml of CMT-8, was *completely* inhibited at final concentrations of 2 µg/ml and 20 µg/ml, levels of CMT-8 readily achieved *in vivo* in the current and in other[41] studies. These data are also consistent with previous studies[25] showing that the activity of MMP-8 and MMP-13 (which are both elevated in the gingiva of the OVX compared to normal rats) is readily inhibited *in vitro* by CMT-8.

In addition to the advantages of CMT-8 over the other CMTs and TC antibiotics (e.g., doxycycline) already discussed, (i.e., its superior *in vivo* pharmacokinetics and *in vitro* inhibitory activity against MMPs and bone resorption), this nonantimicrobial analogue of doxycycline may also be a more potent inhibitor of cytokine production by bone cells. In this regard, Kirkwood *et al.*[62] reported that CMT-8 was more effective than several TC analogues as an inhibitor of mRNA and protein expression of IL-6 in IL-1β-stimulated MC3T3-El osteoblastic cells in culture. One therapeutic cascade (in addition to others, such as direct inhibition of MMPs, already discussed) suggested by recent studies[65] is that CMT-8, by suppressing IL-6 production, could inhibit cyclooxygenase (and $PGE_2$ formation) in osteoblasts thus suppressing (a) proteinase expression in these cells and in osteoclasts and (b) osteoclast-mediated bone resorption.

Another therapeutic pathway may involve the ability of CMT-8 (and other TC analogues) to inhibit inducible nitric oxide (NO) synthase and NO production[66,67] since NO appears to be involved in osteoclast-mediated bone resorption[61,69] and periodontal disease.[64] Finally, these pleiotropic therapeutic actions of CMTs may prevent local (periodontitis) and systemic (osteoporosis) bone loss by *increasing* osteoblastic bone formation (and collagen production in soft tissues[70] as well) in addition to inhibiting bone resorption and collagenolysis.[14–16,33]

## CLINICAL IMPLICATIONS FOR LOCAL (PERIODONTITIS) AND SYSTEMIC (OSTEOPOROSIS) BONE LOSS IN HUMANS

Consistent with the animal studies described in the current report, Scarpellini *et al.*[10] found that the administration of a 3-month regimen of a *regular antimicrobial dose of doxycycline* to postmenopausal humans produced shifts in serum and urine markers of bone resorption (e.g., hydroxyprolinuria) and bone formation (e.g., serum alkaline phosphatase and osteocalcin) consistent with a partial "normalization" of the high-turnover bone loss seen in these osteoporotic women. As described by Ciancio and Ashley,[32] *sub-antimicrobial (low-dose) regimens of doxycycline* (SDD; previously shown to suppress MMP activity in the GCF and gingival tissues of periodontitis patients[30,50,57,68]), administered to humans in long-term double-blind studies, significantly reduced the severity of periodontal disease, including alveolar bone loss. On the basis of these findings, Payne, Numikkoski and Golub (unpublished results) recently completed a small *preliminary* double-blind study on postmenopausal women, diagnosed as osteoporotics, who were *not* on estrogen replacement therapy. They found that this SDD regimen, administered over a 1-year time period, appeared to: (i) reduce the severity of periodontal disease assessed clinically on the basis of soft tissue (gingival) changes and (ii) reduce the loss of alveolar bone height and density assessed by subtraction radiography and computer-assisted densitometric image analysis, respectively. More extensive studies are now being designed to determine whether SDD can reduce local periodontal breakdown (including reduced alveolar bone loss) in addition to possibly bestowing systemic beneficial effects on the course of osteoporosis. In support of this strategy, SDD was just approved by the FDA as the first MMP-inhibitor. A future goal of this research is to use the more potent CMTs (e.g., CMT-8 and CMT-3) as inhibitors of MMPs and bone destruction in other human diseases such as osteoporosis and cancer.

## ACKNOWLEDGMENTS

This research was supported by Grants R37DE-03987 and K16DE-00275 from the National Institute of Dental Research (NIH) and by grants from CollaGenex Pharmaceuticals, Inc. (Newtown, PA), the Finnish Dental Society, and the Academy of Finland.

## REFERENCES

1. SEAMAN, E., C. TSALAMANDIS, S. BOSS & G. PEARCE. 1995. Present and future osteoporosis therapy. Bone **17** (Suppl.): 235–295.
2. LEPOLA, V.T., K. KIPPO, R. HANNUNIEMI, L. LAUREN, T. VIRTAMO, T. OSTERMAN, P. JALOVAARA, R. SELLMAN & H.K. VAANANEN. 1996. Bisphosphonates clodronate and etidronate in the prevention of ovariectomy-induced osteopenia in growing rats. J. Bone Min. Res. **11:** 1508–1517.

3. OHTA, H., T. IKEDA, T. MASUZAWA, K. MAKITAK, Y. SUDA & S. NOZAWA. 1993. Differences in axial bone mineral density, serum levels of sex steroids, and bone metabolism between postmenopausal and age and body size matched premenopausal subjects. Bone **14:** 111–116.

4. OHYORI, N., T. SASAKI, K. DEBARA, N.S. RAMAMURTHY & L.M. GOLUB. 1998. Long-term therapy with a new chemically-modified tetracycline (CMT-8) inhibits bone loss in femurs of ovariectomized rats. Adv. Dent. Res. In press.

5. GOLUB, L.M., T.F. MCNAMARA, G. D'ANGELO, R.A. GREENWALD & N.S. RAMAMURTHY. 1987. A non-antibacterial chemically-modified tetracycline inhibits mammalian collagenase activity. J. Dent. Res. **66:** 1310–1314.

6. WILLIAMS, S., A. WAKISAKA, Q.Q. ZENG, J. BARNES, G. MARTIN, W.J. WECHTER & C.T. LIANG. 1996. Minocycline prevents the decrease in bone mineral density and trabecular bone in ovariectomized aged rats. Bone **19:** 637–644.

7. AOYAGI, M., T. SASAKI, N.S. RAMAMURTHY & L.M. GOLUB. 1996. Tetracycline/flurbiprofen combination therapy modulates bone remodeling in ovariectomized rats: preliminary observations. Bone **19:** 629–635.

8. ZERNICKE, R.F., G.R. WOHL, R.A. GREENWALD, S.A. MOAK, W. LENG & L.M. GOLUB. 1997. Administration of systemic matrix metalloproteinase inhibitors maintains bone mechanical integrity in adjuvant arthritis. J. Rheumatol. **24:** 1324–1331.

9. GOLUB, L.M., N.S. RAMAMURTHY, H. KANEKO, T. SASAKI, B. RIFKIN & T.F. MCNAMARA. 1990. Tetracycline administration prevents diabetes-induced osteopenia in the rat: initial observations. Res. Commun. Chem. Pathol. Pharmacol. **68:** 24–40.

10. SCARPELLINI, F., L. SCARPELLINI, S. ANDREASSI & E.V. COSMI. 1994. Doxycycline may inhibit postmenopausal bone damage: preliminary observations. Ann. N.Y. Acad. Sci. **732:** 493–494.

11. GOLUB, L.M., H.M. LEE, A. NEMIROFF, T.F. MCNAMARA, R. KAPLAN & N.S. RAMAMURTHY. 1983. Minocycline reduces gingival collagenolytic activity during diabetes: preliminary observations and a proposed new mechanism of action. J. Periodont. Res. **18:** 516–526.

12. GOLUB, L.M., N.S. RAMAMURTHY, T.F. MCNAMARA, B.C. GOMES, M.S. WOLFF, A. CASINO, A. KAPOOR, J. ZAMBON, S.G. CIANCIO & H. PERRY. 1984. Tetracyclines inhibit tissue collagenase activity: a new mechanism in the treatment of periodontal disease. J. Periodont. Res. **19:** 651–655.

13. GOMES, B.C., L.M. GOLUB & N.S. RAMAMURTHY. 1984. Tetracyclines inhibit parathyroid hormone–induced bone resorption in organ culture. Experientia **40:** 1273–1275.

14. GOLUB, L.M., N.S. RAMAMURTHY, T.F. MCNAMARA, R.A. GREENWALD & B.R. RIFKIN. 1991. Tetracyclines inhibit connective tissue breakdown: new therapeutic implications for an old family of drugs. Crit. Revs. Oral Biol. Med. **2:** 297–322.

15. GOLUB, L.M., H.M. LEE, M.E. RYAN, W.V. GIANNOBILE, J. PAYNE & T. SORSA. 1998. Tetracyclines inhibit connective tissue breakdown by multiple non-antibacterial mechanisms. Adv. Dent. Res. **12:** in press.

16. RYAN, M.E., N.S. RAMAMURTHY & L.M. GOLUB. 1996. Matrix metalloproteinases and their inhibition in periodontal treatment. Curr. Opin. Periodontol. **3:** 85–96.

17. LOKESHWAR, B.L., M.G. SELTZER, S.M. DUDAK, L.N. BLOCH & L.M. GOLUB. Inhibition of tumor growth and metastasis by oral administration of a non-antimicrobial tetracycline analog (CMT-3), and doxycycline in a metastic prostate cancer model. Clin. Cancer Res. Submitted for publication.

18. LEE, H.M., C. CAO, S. ZUCKER, T. SORSA & L.M. GOLUB. 1998. CMT-3, a modified non-antimicrobial tetracycline, inhibits MTl-MMP mediated gelatinolysis and pro-MMP-2 activation: relevance to cancer. J. Dent. Res. **77** (Spec. Issue B): abstr. # 926.

19. SORSA, T., N.S. RAMAMURTHY, A.T. VERNILLO, X. ZHANG, Y.T. KONTTINEN, B.R. RIFKIN & L.M. GOLUB. 1998. Functional sites of chemically modified tetracyclines: inhibition of the oxidative activation of human neutrophil and chicken osteoclast pro-matrix metalloproteinases. J. Rheumatol. **25:** 975–982.

20. RAMAMURTHY, N.S., A.T. VERNILLO, R.A. GREENWALD, H.M. LEE, T. SORSA, L.M. GOLUB & B.R. RIFKIN. 1993. Reactive oxygen species activate and tetracyclines inhibit rat osteoblast collagenase. J. Bone Min. Res. **8:** 1347–1253.

21. UITTO, V.J., J.D. FIRTH, L. NIP & L.M. GOLUB. 1994. Doxycycline and chemically modified tetracyclines inhibit gelatinase A (MMP-2) gene expression in human skin keratinocytes. Ann. N.Y. Acad. Sci. **732:** 140–151.

22. JONAT, C., F.Z. CHUNG & V.M. BARAGI. 1996. Transcriptional down regulation of stromelysin by tetracycline. J. Cell Biochem. **60:** 341–347.

23. COODLY-GUSDON, L., A. MCGUIRE, H. POTVIN, J.A. MCCUTCHEON, A. VERNILLO & B.R. RIFKIN. 1997. Chemically-Modified Tetracyclines affect rat osteoblastic collagenase mRNA expression. J. Dent. Res. 75 (Spec. Issue): abstr. #1637.

24. MANSELL, J.P., J.F. TARLTON & A.J. BAILEY. 1997. Expression of gelatinases within the trabecular bone compartment of ovariectomized and parathyroidectomized adult female rats. Bone **20:** 533–538.

25. GREENWALD, R.A., L.M. GOLUB, N.S. RAMAMURTHY, M. CHOWDHURY, S.A. MOAK & T. SORSA. 1998. In vitro sensitivity of the three mammalian collagenases to tetracycline inhibition: relationship to bone and cartilage degradation. Bone **22:** 33–38.

26. SATO, T., M.C. OVEJERO, P. HOU, A.M. HEEGAARD, M. KUMEGAWA, N.T. FOGED & J.M. DELAISSE. 1997. Identification of the membrane-type matrix. metalloproteinase MT1-MMP in osteoclasts. J. Cell Sci. **110:** 589–596.

27. HOU, P., M.C. OVEJERO, T. SATO, M. KUMEGAWA, N.T. FOGED & J.M. DELAISSE. 1997. MMP-12, a proteinase that is indispensable for macrophage invasion, is highly expressed in osteoclasts. J. Bone Min. Res. **12** (Suppl. 1), S417: abst. # S259.

28. GOLUB, L.M., R.T. EVANS, T.F. MCNAMARA, H.M. LEE & N.S. RAMAMURTHY. 1994. A non-antimicrobial tetracycline inhibits gingival matrix metallo-proteinases and bone loss in *Porphyromonas gingivalis*-induced periodontitis in rats. Ann. N.Y. Acad. Sci. **732:** 96–111.

29. CHANG, K.M., N.S. RAMAMURTHY, T.F. MCNAMARA, R.T. EVANS, B. KLAUSEN, P.A. MURRAY & L.M. GOLUB. 1994. Tetracyclines inhibit *Porphyromonas gingivalis*-induced alveolar bone loss in rats by a non-antibacterial mechanism. J. Periodont. Res. **29:** 242–249.

30. CROUT, R.J., H.M. LEE, K. SCHROEDER, H. CROUT, N.S. RAMAMURTHY, M. WIENER & L.M. GOLUB. 1996. The "cyclic" regimen of low-dose doxycycline for adult periodontitis: a preliminary study. J. Periodontol. **67:** 506–514.

31. GARRETT, S., D. ADAMS, C. BANDT, B. BEISWANGER, G.C. BOGLE, J. CATON, K. DONLY, C. DRISKO, W.W. HALLMAN *et al.* 1997. Two multicenter clinical trials of subgingival doxycycline in the treatment of periodontis. J. Dent. Res. **75** (Spec. Issue): abstr. # 1113.

32. CIANCIO, S. & R. ASHLEY. 1998. Safety and efficacy of sub-antimicrobial dose doxycycline therapy in patients with adult periodontitis. Advan. Dent. Res. **12:** 27–31.

33. VERNILLO, A.T., N.S. RAMAMURTHY, L.M. GOLUB & B.R. RIFKIN. 1994. The nonantimicrobial properties of tetracycline for the treatment of periodontal disease. Curr. Opin. Periodont. **2:** 111–118.

34. WOWERN, N.V., J. KLAUSEN & G. KOLLERUP. 1994. Osteoporosis: a risk factor in periodontal disease. J. Periodontol. **65:** 1134–1138.

35. PAYNE, J.B., N.R. ZACKS, R.A. REINHARDT, P.V. NUMMIKOSKI & K. PATIL. 1997. The association between estrogen status and alveolar bone density changes in postmenopausal women with a history of periodontitis. J. Periodontol. **68:** 25–31.

36. LOZA, J.C., L.C. CARPIO & R. DZIAK. 1996. Osteoporosis and its relationship to oral bone loss. Curr. Opin. Periodont. **3:** 27–33.

37. STRECKFUS, C.F., R.B. JOHNSON, T. NICK, B. TSAO & M. TUCCI. 1997. Comparison of Alveolar bone loss, alveolar bone density, salivary and gingival crevicular fluid interleukin-6 concentrations in healthy premenopausal and postmenopausal women on estrogen therapy. J. Gerontol. A Biol. Sci. Sci. Med. **52:** M343–M351.

38. PAYNE, J.B., R.A. REINHARDT, M.P. MASADA, L.M. DUBOIS & A.C. ALLISON. 1993. Gingival crevicular fluid IL-8: correlation with local IL-1β levels and patient estrogen status. J. Periodont Res. **28:** 451–453.

39. GRODSTEIN, F., G.A. COLDITZ & M.J. STAMPFER. 1996. Post-menopausal hormone use and tooth loss: a prospective study. J. Am. Dent. Assoc. **127:** 370–377.

40. SONTAG, W. 1994. Age-dependent morphometric change in the lumbar vertebrae of male and female rats: comparison with the femur. Bone **15:** 593–601.

41. LIU, Y., N.S. RAMAMURTHY, J. MARECEK, T. MCNAMARA, B.R. RIFKIN, O. KABOURIDOU & L.M. GOLUB. 1997. Chemically-modified tetracyclines (CMTs) exhibit different pharmacokinetics in vivo. J. Dent. Res., 77 (Spec. Issue A): abstract #1165.

42. BAIN, S., N.S. RAMAMURTHY, T. IMPEDUGLIA, S. SCOLMAN, L.M. GOLUB & C. RUBIN. 1997. Tetracycline prevents cancellous bone loss and maintains near-normal rates of bone formation in streptozotocin diabetic rats. Bone **21:** 147–153.

43. MCCROSKERY, P.A., J.F. RICHARDS & E. HARRIS, JR. 1975. Purification and characterization of a collagenase extracted from rabbit tumors. Biochem. J., **152:** 131–142.

44. RYAN, M.E., N.S. RAMAMURTHY, E. GOTTESMAN, R.T. EVANS, T. SORSA & L.M. GOLUB. 1997. *P. gingivalis*-induced rat model of periodontitis: active/inactive phases. J. Dent. Res. **76** (Spec. Issue): IADR abstract #2213.

45. SORSA, T., Y. DING, T. SAALO, A. LAUHIO, O. TERONEN, T. INGMAN, H. OHTORI, N. ANDOH, S. TAKIHA & Y.T. KONTTINEN. 1994. Effects of tetracyclines on neutrophil, gingival, and salivary collagenases. Ann. N.Y. Acad. Sci. **732:** 112–131.

46. GOULDING, A. & E. GOLD. 1989. A new way to induce estrogen-deficiency osteopaenia in the rat: comparison of the effects of surgical ovariectomy and administration of the LHRH against buserelin on bone resorption and composition. J. Endocrinol. **121:** 293–298.

47. BIRKEDAL-HANSEN, H. 1993. Role of cytokines and inflammatory mediators in tissue destruction. J. Periodont. Res. **28:** 500–510.

48. INGMAN, T., T. TERVAHARTIALA, Y. DING, H. TSCHESCHE, A. HAERIAN, D. KINANE, Y.T. KONTTINEN & T. SORSA. 1996. Matrix metalloproteinases and their inhibitors in gingival crevicular fluid and saliva of periodontitis patients. J. Clin. Periodontol. **23:** 1127–1132.

49. GOLUB, L.M., T. SORSA, H.M. LEE, S. CIANCIO, O. SORBI, N.S. RAMAMURTHY, B. GRUBER, T. SALO & Y.T. KONTTIMEN. 1995. Doxycycline inhibits neutrophil (PMN)-type matrix metalloproteinases in human adult periodontitis gingiva. J. Clin. Periodontol. **22:** 100–109.

50. GOLUB, L.M., H.M. LEE, R.A. GREENWALD, M.E. RYAN, T. SORSA & W.V. GIANNOBILE. 1997. A matrix metalloproteinase inhibitor reduces bone type collagen degradation fragments and specific collagenases in gingival crevicular fluid during adult periodontitis. Inflamm. Res. **46:** 310–319.

51. UITTO, V.J., K. AIROLA, M. VAALAMO, N. JOHANNSON, E.E. PUTNINS, J.D. FIRTH, J. SALONEN, C. LOPEZ-OTIN, U. SAARIALHO-KERE & V.M. KAHARI. 1998. Collagenase (matrix metalloproteinase-13) expression is induced in oral mucosal epithelium during chronic inflammation. Am. J. Pathol. **152:** 1489–1499.

52. TERONEN, O., Y.T. KONTTINEN, C. LINDQUIST, T. SALO, T. INGMAN, A. LAUHIO, Y. DING, S. SANTIVIRTA, H. VALLEALA & T. SORSA. 1997. Inhibition of matrix metalloproteinase-1 by dichloromethylene bisphosphonate (clodronate). Calcif. Tissue Int. **61:** 59–61.

53. RIFKIN, B.R., A.T. VERNILLO, L.M. GOLUB & N.S. RAMAMURTHY. 1994. Modulation of bone resorption by tetracyclines. Ann. N.Y. Acad. Sci. **732:** 165–180.

54. SEFTOR, R.E.B., E.A. SEFTOR, J.E. DE LARCO, D.E. KLEINER, J. LEFERSON, W.G. STETLER-STEVENSON, T.F. MCNAMARA, L.M. GOLUB & M.J.C. HENDRIX. 1998. Chemically-modified tetracyclines inhibit human melanoma cell invasion and metastasis. Clin. Exp. Metastasis **16:** 217–225.

55. PACIFICI, R. 1996. Estrogen, cytokines, and pathogenesis of postmenopausal osteoporosis. J. Bone Min. Res. **11:** 1043–1052.

56. MANOLOGAS, S.C. 1995. Role of cytokines in bone resorption. Bone **17** (Suppl.): 638–678.

57. GOLUB, L.M., M. WOLFF, S. ROBERTS, H. LEE, M. LEUNG & G.S. PAYONK. 1994. Treating periodontal diseases by blocking tissue-destructive enzymes. J. Am. Dental Assoc. **125:** 163–169.

58. GILLES, T., D.L. CANES, M. DALLAS, S. HOLT & L. BONEWALD. 1995. Oral bone loss increased in ovariectomized rats. J. Bone Min. Res. **10** (Suppl. 1): abst. # T354, pS443.

59. CHUBINSKAYA, S., K. HUCH, K. MIKECZ, G. SZABO, K.A. HASTY, K.E. KUETTNER *et al.* 1996. Chondrocyte matrix metalloproteinase-8: up-regulation of neutrophil collagenase by interleukin-1$\beta$ in human cartilage from knee and ankle joints. Lab. Invest. **74:** 232–240.

60. ZHAO, H., G. CAI, J. DU, Z. XIA, L. WONG & T. ZHU. 1997. Expression of matrix metalloproteinase-9 mRNA in osteoporotic bone tissues. J. Tongji Med. Univ. **17:** 28–31.

61. DAMOULIS, P.D. & P.V. HAUSCHKA. 1994. Cytokines induce nitric oxide production in mouse osteoblasts. Biochem. Biophys. Res. Commun. **201:** 924–931.

62. KIRKWOOD, K.L., L.M. GOLUB, Y. LIU & P.G. BRADFORD. Non-antimicrobial and antimicrobial tetracyclines inhibit IL-6 expression in murine osteoblasts: potential molecular mechanisms for the treatment of metabolic bone diseases. J. Bone Min. Res. Submitted for publication.

63. RAMAMURTHY, N.S., L.M. GOLUB, A.J. GWINNETT, T. SALO, Y. DING & T. SORSA. 1998. In vivo and in vitro inhibition of matrix metalloproteinases including MMP-13 by several chemically modified tetracyclines (CMTs). *In* Biological Mechanisms of Tooth Eruption, Reabsorption and Replacement by Implants. Z. Davidovitch & J. Mah, Eds.: 271–277. Harvard Soc. Adv. Orthodont. Boston.

64. LOHINAI, Z., P. BENEDEK, E. FEHER, A. GYORFI, L. ROSIVALL, A. FAZEKAS, A.L. SALZMAN & C. SZABO. 1998. Protective effects of mercaptoethylguanidine, a selective inhibitor of inducible nitric oxide synthase, in ligature-induced periodontitis in the rat. Br. J. Pharmacol. **123:** 353–360.

65. TAI, H., C. MIYAURA, C.C. PILBEAM, T. TAMURA, Y. OHSUGI, Y. KOSHIHARA, N. KOBUDORA, H. KAWAGUDRI, L.G. RAISZ & T. SUDA. 1997. Transcriptional induction of cyclooxygenase-2 in osteoblasts is involved in interleukin-6 induced osteoclast formation. Endocrinology **138:** 2372–2379.

66. AMIN, A.R., R.N. PATEL, G.D. THAKKER, C.J. LOWERSTEIN, M.G. ATTUV & S.B. ABRAMSON. 1997. Post-transcriptional regulation of inducible nitric oxide synthase mRNA in murine macrophages by doxycycline and chemically-modified tetracyclines. FEBS Lett. **410:** 259–264.

67. TRACHTMAN, H., S. FUTTENWEIT, R. GREENWALD, S. MOAK, P. SINGLAD, N. FRANK & A.R. AMIN. 1996. Chemically modified tetracyclines inhibit inducible nitric oxide synthase expression and nitric oxide production in cultured rat mesangial cells. Biochem. Biophys. Res. Commun. **229:** 243–248.
68. GOLUB, L.M., S. CIANCIO, N.S. RAMAMURTHY, M. LEUNG & T.F. McNAMARA. 1990. Low-dose doxycycline therapy: effect on gingival and crevicular fluid collagenase activity in humans. J. Periodont. Res. **25:** 321–330.
69. HUKKANEN, I., F.J. HUGHES, L.K.D. BUTTERY, S.S. GROSS, T.J. EVANS, S. SEDDON, A. RIVEROS-MORENO, I. MACINTYRE & J.M. POLAK. 1995. Cytokine-stimulated expression of inducible nitric oxide synthase by mouse, rat, and human osteoblast-like cells and its functional role in osteoblast metabolic activity. Endocrinology **136:** 5445–5453.
70. CRAIG, R.G., Z. YU, L. XU, R. BARA, N.S. RAMAMURTHY, J. BOLAND, M. SCHNEIR & L.M. GOLUB. 1998. A chemically modified tetracycline inhibits streptozotocin - induced diabetic depression of skin collagen synthesis and steady-state type 1 procollagen mRNA. Biochim. Biophys. Acta **1402:** 250–260.
71. WOESSNER, J.F., JR. 1979. Total, latent and active collagenase during the course of post-partum involution of the rat uterus: effect of oestradiol. Biochem. J. **180:** 95–102.
72. MARBAIX, E., I. KOKORINE, P. HENREIT, J. DONNEZ, P.J. COURTOY & Y. EECKHOUT. 1995. The expression of interstitial collagenase in human endometrium is controlled by progesterone and by oestradiol and is related to menstruation. Biochem. J. **305:** 1027–1030.

# MMP-Mediated Events in Diabetes

MARIA EMANUEL RYAN,[a,b] NUNGAVARUM S. RAMAMURTHY,[a]
TIMO SORSA,[c] AND LORNE M. GOLUB[a]

[a]School of Dental Medicine, Department of Oral Biology and Pathology, South Campus, State University of New York at Stony Brook, Stony Brook, New York 11794-8702, USA

[c]University of Helsinki, Department of Periodontology, Helsinki, Finland

ABSTRACT: Both Type I and Type II diabetes mellitus (DM) have been associated with unusually aggressive periodontitis. Accordingly, rat models of both types of DM were used to study (i) mechanisms mediating this systemic/local interaction and (ii) new pharmacologic approaches involving a series of chemically modified tetracyclines (CMTs) that have lost their antimicrobial but retained their host-modulating (*e.g.*, MMP-inhibitory) properties. *In vitro* experiments on tissues from Type I DM rats demonstrated that several of these CMTs were better matrix metalloproteinase (MMP) inhibitors than was antibacterial doxycycline (doxy), except for CMT-5, which, unlike the other MMP inhibitors, was found not to react with zinc. Data from *in vivo* studies on the same rat model generally supported the relative efficacy of these compounds: the CMTs and doxy were found to inhibit MMP activity, enzyme expression, and alveolar bone loss. To examine other long-term complications such as nephropathy and retinopathy, a Type II (ZDF) model of DM was studied. Treatment of these DM rats with CMT-8 produced a 37% ($p < 0.05$), 93% ($p < 0.001$), and 50% ($p < 0.01$) reduction in the incidence of cataract development, proteinuria, and tooth loss, respectively; whereas the doxy-treated ZDF rats showed little or no effect on these parameters. CMT treatment decreased mortality of the Type II ZDF diabetic animals, clearly indicating that CMTs, but not commercially available antibiotic tetracyclines (TCs), may have therapeutic applications for the long-term management of diabetes.

## INTRODUCTION

Eight major complications have been associated with diabetes to date, and these include: psychosocial problems, acute glycemic complications, adverse outcomes of pregnancy, eye disease, kidney disease, cardiovascular disease, neuropathy and wound healing problems.[1] Loe, as director of the National Institute of Dental Research, concluded[2] that the research findings of multiple epidemiologic studies demonstrated that both types of diabetes, IDDM (Type I) and NIDDM (Type II), are predictors of periodontal disease and suggested that aggressive periodontitis be recognized as an additional complication of diabetes mellitus.

It is clear that diabetes is a complex disease characterized by a number of variables that can influence the development of complications, including periodontitis. Although the exact mechanisms of action are not yet known, poor metabolic control as well as extended duration of the diabetic state are the risk factors for periodontitis

[b]Address for telecommunication: Phone, 516/632-9529; fax, 516/632-9705; e-mail, mryan1@hotmail.com

and altered host function. Of course it is most likely a combination of many factors that ultimately leads to the increased prevalence and severity of periodontitis in diabetic patients. Vascular abnormalities or degenerative vascular changes previously seen in other tissues and/or organs of the diabetic[3] also occur in the gingival tissues.[4,5] Aspects of the host response such as collagen metabolism[6,7] and leukocyte function[8,9] have also been implicated. According to this approach, an increased susceptibility to infection by periodontopathogenic microorganisms owing to suppressed neutrophil function, in addition to a reduced healing capacity associated, in part, with altered collagen metabolism, could result in the greater incidence and severity of periodontal destruction in diabetics. Elevated levels of $PGE_2$ have been detected in the blood of Type I diabetic patients as compared to nondiabetic controls,[10] which appears to be associated with a hyper-responsive monocytic $PGE_2$ trait systemically.[11] Another consequence of hyperglycemia is the alteration of circulating and immobilized proteins. When proteins such as collagen, or lipids, are exposed to aldose sugars they undergo nonenzymatic glycation[12] and oxidation,[13] resulting in the irreversible formation of advanced glycation endproducts (AGEs). These glucose-derived crosslinks contribute to reduced collagen solubility and turnover rate in diabetic animals and humans.[14] AGEs act on target cells via their recognition of cell surface polypeptide receptors. The best characterized binding site for AGEs is a member of the immunoglobulin superfamily now called the Receptor for AGE or RAGE.[15] AGEs can interact with RAGEs on cells, such as macrophages, stimulating the production of MMPs, adhesion molecules (e.g., vascular cell adhesion molecule-1 [VCAM-1][16]), cytokines (e.g., TNF, IL-1[17] and IL-6[18]) as well as other mediators. These factors may act individually in an additive fashion or synergistically to contribute to periodontal disease. The question remains as to which of these factors should be the primary focus for the best treatment of periodontitis, as well as other complications, in this compromised patient population.

## MATERIALS AND METHODS:

### Studies on Animal Models of Type I Diabetes

Virus-free, 3-month-old, adult male Sprague-Dawley rats weighing 300–350g (Charles River Labs, Wilmington, MA) were distributed into nine experimental groups, each consisting of 5 to 7 animals. Following general anesthesia by halothane (Halocarbon Lab, Inc., Hackensack, NJ) inhalation, diabetes was induced by i.v. administration of streptozotocin (70 mg/kg body weight) to all rats, except for the 7 rats in the nondiabetic control group, after 12 hours of fasting.[19] Diabetic status was confirmed weekly by using a glucose enzymatic test strip (Tes-Tape, Eli Lily, Inc., Indianapolis, IN), which revealed >2% glucose in the urine of the STZ-injected animals. Each group of diabetic rats was given 5 mg of the following treatments (15 mg/kg) daily by oral gavage for a period of 3 weeks: (1) CMT-1 (4-dedimethylamino tetracycline); (2) CMT-3 (6-demethyl 6-deoxy 4-dedimethylamino tetracycline); (3) CMT-4 (7-chloro 4-dedimethylamino tetracycline); (4) CMT-5 (pyrazole); (5) CMT-7 (12α-deoxy, 4-dedimethylamino tetracycline); (6) CMT-8 (6α deoxy-5-hydroxy-4-dedimethylamino tetracycline); (7) doxycycline; and (8) vehicle alone (2% carboxymethylcellulose (CMC); Sigma, St. Louis, MO).

## Studies on Animal Models of Type II Diabetes

One hundred twenty 2-month-old, ZDF-Gmi/fa male rats weighing ~350 g and 40 nondiabetic ZDF-Gmi lean control (LC) male rats weighing ~280 g (Genetic Models Inc., Indianapolis, IN) were distributed randomly into the treatment (tx) groups shown in TABLE 1.

**TABLE 1. Treatment groups in Type II diabetes**

| Time course | Treatment groups | | | | |
|---|---|---|---|---|---|
| | Untreated | Doxy-Tx | CMT-3 Tx | CMT-8 Tx | LCs |
| 5 mos old (3 mos tx) | 6 | 6 | 6 | 6 | 8 |
| 7 mos old (5 mos tx) | 6 | 6 | 6 | 6 | 8 |
| 9 mos old (7 mos tx) | 8 | 8 | 8 | 8 | 8 |
| Supplemental | 8 | 8 | 8 | 8 | 8 |

NOTE: Baseline core-treatment at 2 months.

The rats were fed Purina Formulab 5008 Rodent Chow, as suggested by the breeder. Diabetic status was confirmed weekly by using a glucose enzymatic test strip (Tes-Tape, Eli Lily, Inc., Indianapolis, IN), which revealed >2% glucose in the urine of the diabetic animals. The diabetic rats in each group were given 5 mg of either CMT-3, CMT-8, doxycycline, or vehicle (2% CMC), alone daily by oral gavage until the day of sacrifice or death by natural causes. Lean controls were gavaged daily with 2% CMC. The animals were weighed weekly and at sacrifice. Mortality was monitored over the course of the experiment. Previous studies[20] demonstrated that untreated ZDF-Gmi/fa male rats can survive for 10–12 months. However, there is no evidence in the literature that daily gavage for extended periods of time had ever been performed on Type II diabetic ZDF rats.

## Skin and Gingiva Analyses for MMPs

### Extraction of Tissue and Measurement of MMP Activity

Following sacrifice, the entire shaved skin (except for that over the limbs) was dissected from each rat, the adherent subcutaneous tissue was removed, and the skins were weighed. The skins were finely minced at 4°C. The gingival tissues from both maxillary and mandibular arches of each rat were excised and pooled per experimental group ($n = 5$–7 rats per group). The pooling of gingival tissues for each group was necessary because individual rats did not yield sufficient gingiva for enzyme analyses. Then, either 100 mg of minced skin for each rat or 50 mg pooled gingiva per group was extracted, and collagenase in the extracts was partially purified as previously described.[21] In brief, the samples were homogenized (all procedures at 4°C) with a glass grinder (Kontes Glass Co., Vineland, NJ) attached to a T-Line Lab stirrer (Model 106 Taboys Engineering Corp., Emerson, NJ) in 10 mM Tris-HCl buffer (pH 7.5) containing 0.4 M sucrose and 5 mM $CaCl_2$. The tissues were extracted at a constant w/v (i.e., 100 mg/6 ml buffer) ratio, and the homogenates were then centrifuged (15,000 rpm, 1 hr). The pellets were washed, extracted overnight with 50 mM

Tris-HCl (pH 7.6) containing 5M urea, 0.2 M NaCl and 5 mM $CaCl_2$, then centrifuged (15,000 rpm, 1 hr). The supernatants were collected, then dialyzed exhaustively against Tris (50 mM)/NaCl (0.2 M)/$CaCl_2$ (5 mM) buffer (pH 7.8). $(NH_4)_2SO_4$ was added to the dialysate to produce a 60% saturation and allowed to stand overnight, and the collagenase-containing precipitates were collected by centrifugation at 15,000 rpm for 90 min. The pellets were then dissolved in the Tris/NaCl/$CaCl_2$ buffer (pH 7.8) containing 0.05% Brij and exhaustively dialyzed against the same buffer. Protein content of the extracts was determined by the Coomassie Blue protein assay kit (Bio-Rad, Melville, NY).

Collagenase activity in the skin and gingival extracts from each rat was determined by a standard collagenolysis assay, which involved incubating 70 μl of the partially purified extracts with 10 μl of radiolabeled lathyritic rat skin Type I ($^3$H-methyl)-collagen at 27°C for 18 hr as described previously.[22] The reaction was terminated by the addition of 1,10 phenanthroline, carrier collagen, and 1,4 dioxane, and the undegraded collagen was precipitated by centrifugation at 20°C so that the degradation products in the supernatant could be assessed by liquid scintillation spectrophotometry. In additional assays, the radiolabeled collagen components ($\alpha_1$ and $\alpha_2$ chains) and degradation products ($\alpha^A$ and $\alpha^B$) were separated by SDS-PAGE, visualized by fluorography, and quantitated by laser densitometric scanning of the fluorograms (LKB ultrascan) after overnight incubation at 22°C, as described by Yu et al.[23] Gelatinase activity was determined by incubating 70 μl of the partially purified extracts with Type I ($^3$H-methyl)-gelatin (thermally denatured collagen) at 37°C for 4 hr using a modification of the method described by McCroskery et al.[24]; trichloracetic acid (TCA) and cold gelatin were added after the incubation to precipitate the undigested gelatin. Activation of latent MMPs, to measure total enzyme activity in the partially purified extracts, was accomplished by adding the organomercurial agent, 4-aminophenylmercuric acetate (APMA), to the incubation mixtures at a final concentration of 1 mM. To assay the different types of gelatinases in each skin or gingival extract, enzymography was performed following the method of Heussen and Dowdle.[25]

Western blot analyses of MMPs-2, -3, -8 and -9 were performed on aliquots of the partially purified gingival extracts. 100 μl of the pooled gingival extracts were lyophilized with a Speed-Vac (Savant Instruments, Inc., Holbrook, NY) and stored at −80°C. Each sample was then thawed and analyzed for MMP levels by Western blot analysis using a modification of techniques described previously.[26] In brief, the lyophilized gingival extracts were treated with Laemmli's buffer (pH 7.0) containing 5 mM dithiothreitol and heated for 5 min at 100°C. High- and low-range pre-stained SDS-PAGE standard proteins were used as molecular weight markers. The samples were electrophoresed on 7.5% SDS-polyacrylamide gels and then electrophoretically transferred to nitrocellulose membranes. Nonspecific binding was blocked by incubation with phosphate-buffered saline (PBS) containing 5% nonfat dry milk (90 mins, 37°C). The membranes were then incubated with polyclonal antibodies specific for human MMPs-2, -3, -8 and -9 diluted 1:500 (or with nonimmune control serum diluted 1:100) for 1–6 hr at 20°C. After repeated washings, the membranes were incubated with biotinylated anti-mouse immunoglobulins (1 hr, 20°C) and further incubated with anti-mouse antibody-alkaline phosphatase conjugate, and the color was developed by standard technique.

## Morphometric Analyses of Alveolar Bone Loss

After sacrifice, the animals were decapitated and the jaws defleshed, and alveolar bone loss was assessed morphometrically.[27] In brief, maxillary and mandibular jaws were removed from each rat and the gingival tissues excised and pooled per group as described above. The jaws were then boiled in water for 5 min and then defleshed by dissection. The jaws were allowed to dry and large particles were brushed off and were then soaked in 0.2 N NaOH at room temperature for 5 minutes to remove the remaining soft tissue debris; they were then rinsed with water and allowed to dry. Finally, each tooth was stained with methylene blue in Loeffler's buffer to highlight the position of the cemento-enamel junction (CEJ). The distance between the CEJ and the alveolar bone crest was measured under a dissecting microscope with a Mitutoyo Digimatic Caliper (accurate to 0.01 mm).

## Urinalysis for Protein

The evening before sacrifice the animals were housed in metabolic cages so that 18-hour urine samples could be collected after overnight fasting. The level of protein in these urine samples was monitored using Uristix 4 (Bayer Corporation, Elkhart, IN). This test is based on the protein-error-of-indicators principle. At a constant pH, the development of any green color is due to the presence of protein. Colors range from yellow for "negative" through yellow-green and green to green-blue for increasingly "positive" reactions. A shift to alkaline pH may cause false positives in this protein test. Normally no protein is detectable in the urine, although a minute amount may be excreted by the normal kidney after vigorous exercise. A color matching any block greater than Trace indicates significant proteinuria. The sensitivity for protein using this test is 15–30 mg/dL albumin. The reagent is more sensitive to albumin than globulins, hemoglobin, Bence-Jones protein, and mucoprotein; a negative result does not rule out the presence of these other proteins.

*Statistical analysis* was performed in the following manner. The standard error of the mean (SEM) was calculated from the standard deviation. The statistical significance between the groups were determined by analysis of variance (ANOVA) or by using a paired Student's *t*-test. These analyses were carried out using SigmaStat statistical software (Jandel Scientific Software, San Rafael, CA).

## RESULTS

As expected, after inducing Type I diabetes with STZ or with genetically induced Type II diabetes, the rats were severely hyperglycemic. The blood glucose concentrations were significantly higher ($p < 0.001$) than the normal control levels, with no difference between the doxycycline and CMT-treated as well as untreated diabetics.[28]

## MMP Analyses

Elevated gelatinase and collagenase activity in the skins of the Type I diabetic animals was normalized in all treatment groups ($p < 0.001$), with the exception of CMT-5, as determined by lysis assays, for both gelatinase at 37°C (FIG. 1) and col-

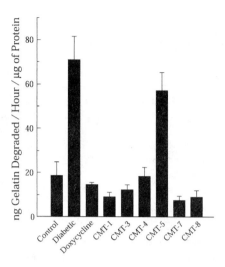

**FIGURE 1.** Effect of diabetes and CMT therapy on gelatinase activity in partially purified extract of skin. Each *bar* represents the mean ± SEM of 5–7 rats per group.

lagenase at 27°C (not shown). These data reflect total enzyme activity (i.e., the
proMMPs were activated by incubation with APMA); for collagenase, almost all of
the enzyme in the skins appeared to be in the active form with no differences between ± APMA (data not shown), whereas elevations in latent levels of gelatinase
(based on ± APMA) were evident in the diabetic animals which were decreased to
non-diabetic control or normal levels by treatment with the tetracycline analogues
(data not shown). Intracellular modulatory pathways, which could explain the reduced expression of MMPs after treatment of the diabetics with CMTs, include effects of these drugs on a variety of cytokines, inducible nitric oxide synthase (iNOS),
cyclooxygenase, and advanced glycation endproducts (AGEs) formation.[28–30] Similar patterns of change were detected when different molecular types of gelatinase
were assessed by gelatin enzymography. It should be noted that most of the elevated
gelatinase activity during diabetes was in the 92-kDa form. There were only slight
increases in the 72-kDa form of gelatinase above constitutive levels, as can be seen
in the untreated diabetic rats (FIG. 2; note that CMT-5 again was the only tetracycline
analogue to have lost its anti-MMP efficacy). In these enzymograms one can see both
latent 92-kDa and (presumably) active 88-kDa gelatinase in the diabetic rat skin extracts. Treatment of the diabetic rats with CMTs-1, -3, -4, -7 and -8 effectively eliminated the 92-kDa form of gelatinase from their skin extracts (FIG. 2). A similar
pattern of change was seen when collagenase activity was examined by SDS-PAGE
fluorography (22°C) in the presence of the organomercurial APMA. Diabetes increased collagenase activity in the extracts of skin (FIG. 3). Gelatinase is responsible
for the subsequent breakdown of the $\alpha^A$ collagen breakdown products to even smaller secondary fragments which are also seen on the fluorograms.[31] All of the compounds tested by oral gavage *in vivo,* with the exception of CMT-5, reduced the

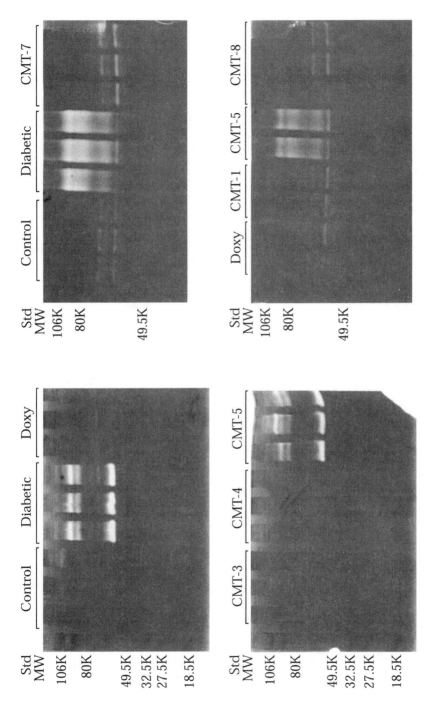

**FIGURE 2.** Enzymography (using 1 mg/ml denatured type I collagenase substrate) of diabetic rat skin gelatinase: effect of CMT therapy. Each *lane* represents partially purified extract of skin from 1 rat per group.

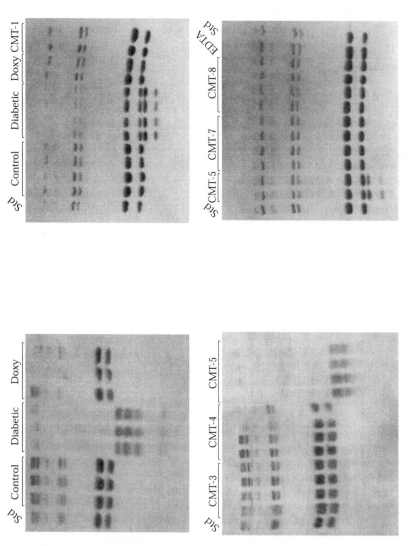

**FIGURE 3.** SDS/PAGE fluorography of diabetic rat skin collagenase: effect of CMT therapy.

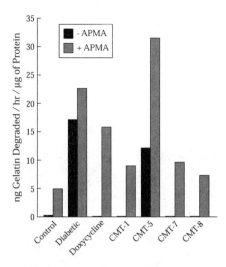

**FIGURE 4.** Effect of *in vivo* CMT administration on gelatinase activity in diabetic rat gingiva. Each value represents the mean duplicate analysis of a pool of gingival tissue obtained from 5–7 rats per group. Gelatinase activity was measured in the presence or absence of APMA added to the incubation mixture in a final concentration of 1 mM.

collagenase levels to normal control levels. The extracts of the untreated and CMT-5 treated animals mediated the breakdown and loss of the α collagen components and the formation of $α^A$ collagen fragments characteristically produced by collagenase.

Since the destruction of collagenous tissue is a major pathogenic feature of periodontal disease, the effects of treatment with the various CMTs on the matrix metalloproteinases responsible for destruction of the supporting structures of the tooth were examined. As described above for skin extracts from the Type I diabetic animals, a four-fold (or greater) elevation in matrix metalloproteinase activity was noted in the gingival extracts of the untreated diabetic animals (FIGS. 4 and 5). The elevated gelatinase and collagenase activity in these gingival extracts was significantly reduced or normalized in all treatment groups, with the exception of CMT-5, as determined by lysis assays using radiolabeled substrates, for both gelatinase at 37°C (FIG. 4) and collagenase at 27°C (data not shown). The data represented in FIGURE 4 for gelatinase demonstrate that almost all of this MMP(s) in the gingival tissues in the untreated diabetic was in the active form, whereas no active gelatinase, only latent, was seen in the controls (active gelatinase was demonstrated in the absence of APMA, and latent MMP was estimated as the difference between the total MMP (+APMA) and the active form (−APMA)). These observations were consistent with those in the zymograms showing (FIG. 5) mostly 88-kDa, rather than 92-kDa gelatinase in the untreated diabetic rats. Active gelatinase in the gingiva was completely reduced to undetectable normal levels in all treatment groups, with the exception of CMT-5, which showed only a slight reduction (nonsignificant), as measured in ng gelatin degraded/hour/µg of protein (FIG. 4). Mostly latent levels of

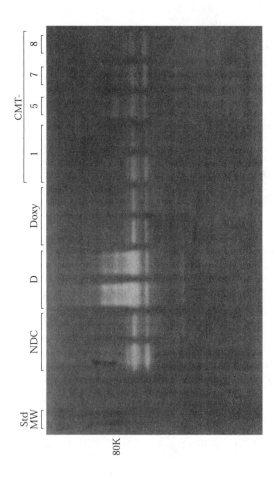

**FIGURE 5.** Gelatin zymography of partially purified extract of gingiva from nondiabetic control (NDC), untreated diabetic (D), and doxy- and CMT-treated diabetic rats.

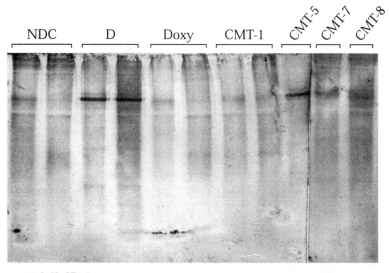

anti-MMP-8

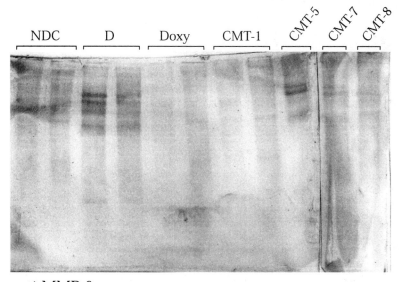

anti-MMP-9

**FIGURE 6.** Western blot analysis of MMP-8 (*top*) and MMP-9 (*bottom*) in extracts of gingiva from nondiabetic control (NDC), untreated diabetic (D), and doxy- and CMT-treated diabetic rats.

gelatinase (not active gelatinase) were seen in the doxy, CMT-1, -7 and -8 and non-diabetic control groups. Of interest, a higher percentage of latent gelatinase was seen in the CMT-5-treated diabetics than in the untreated diabetics, suggesting that although active MMP was not inhibited, the conversion of latent into active gelatinase may be reduced by CMT-5.

Western blot analyses of MMPs-2 and -3 (not shown) in the gingival extracts of the Type I diabetics showed either no detectable levels or no differences in the levels of these proteases between nondiabetic controls and the diabetics. However, elevations in protein levels of collagenase-2 (MMP-8) and 92-kDa gelatinase (MMP-9) (FIG. 6) were seen in the diabetic rats, which were normalized in all treatment groups with the exception of CMT-5 and to a lesser extent CMT-7; these data were consistent, at least in part, with the functional enzyme analyses described above. Therefore CMT-5, in contrast to earlier hypotheses that this compound only loses its ability to inhibit enzyme activity, also appeared to have lost its ability to downregulate MMP protein expression.

### Alveolar Bone Loss

FIGURE 7 shows representative examples of defleshed maxillary quadrants from the palatal view. It is evident that the Type I diabetic animals often (~45% of rats) developed severe alveolar bone loss compared to the normal nondiabetic control animals (10% incidence of severe lesions). This bone loss was particularly evident at the interproximal sites, where scooped-out intrabony defects could be seen. Administration of doxycycline to the diabetic animals appeared to prevent this destruction of the alveolar bone (FIG. 7). The dose of doxycycline administered to these animals is antibacterial, suggesting that the effect might have been due (at least in part) to the antibiotic effects of the drug. However, the non-antimicrobial CMTs, in this case CMT-8 (FIG. 7), were found to be just as effective at preventing bone destruction in the diabetic animals.

### Proteinuria

Owing to hydronephrosis (i.e., a condition that is indigenous to Type II diabetic ZDF rats and their lean control littermates, personal communication with Dr. Kevin McCarthy, LSU), the rats in the lean control group exhibited significant proteinuria (200 mg/18 hour). However, the severity of proteinuria was increased even further, by 81% ($p < 0.001$), in the ZDF rats compared to the lean control rats. A significant 74% ($p < 0.001$) reduction in proteinuria was seen in the CMT-3- and -8-treated animals (compared to vehicle-treated diabetics) after 3 months of treatment (FIG. 8; the doxy-treated ZDF rats showed a much lesser reduction in proteinuria (21% decrease) and the effect of this tetracycline was not significant. Owing to hydronephropathy in this animal model system, even the lean controls exhibited proteinuria, which was higher than that seen in the CMT-3- and -8-treated Type II diabetic animals. Of interest, treatment of the genetically induced Type II diabetic rats with the CMTs brought the levels of proteinuria down to those that have been observed in nondiabetic normal Sprague-Dawley rats. Consistent with the observations using the Type II model of diabetes (FIG. 8), CMT-8 was also the most effective at reducing proteinuria in a Type I streptozotocin-induced model of diabetes (FIG. 9; from the

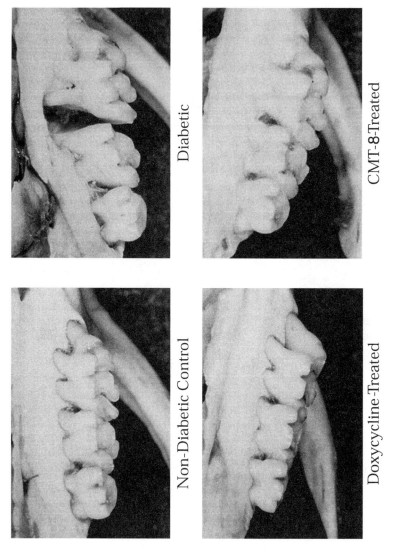

**FIGURE 7.** Representative de-fleshed maxillas from nondiabetic control, untreated, and tetracycline-treated diabetic rats. All of the jaws from these and other experimental groups were analyzed morphometrically for alveolar bone loss.

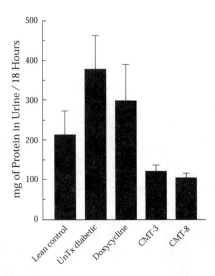

**FIGURE 8.** Urinary protein excretion by lean controls, untreated Type II diabetics, and diabetics treated with different tetracycline analogues.

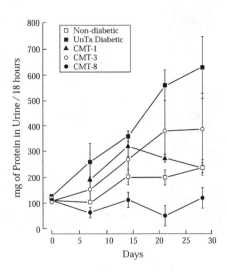

**FIGURE 9.** Urinary protein excretion by nondiabetic controls, untreated Type I diabetics, and diabetics treated with different tetracycline analogues.

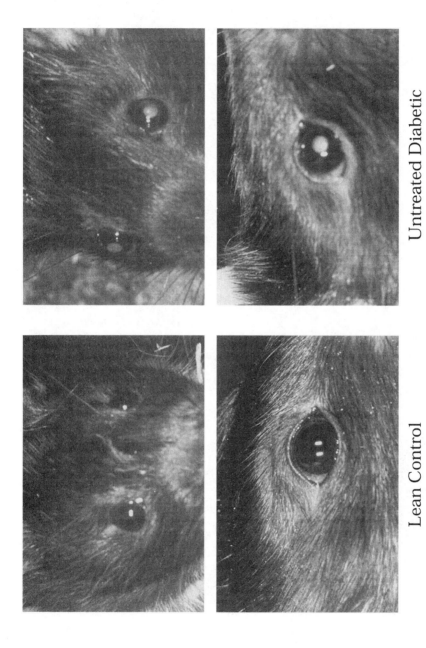

Untreated Diabetic

Lean Control

**FIGURE 10.** Cataract development in Type II ZDF/GMI diabetic rats.

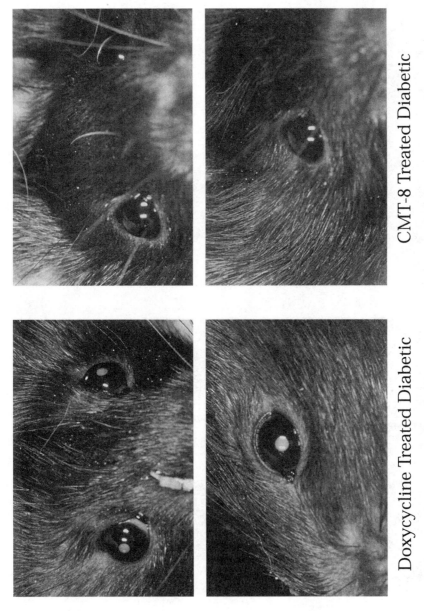

CMT-8 Treated Diabetic

Doxycycline Treated Diabetic

**FIGURE 11.** Prevention of cataract development in CMT-8- (but not doxcycline-) treated diabetic rats.

laboratory of Dr. Lenny Arbeit, Nephrology, SUNY at Stony Brook). The values for proteinuria seen in the Type I model at 30 days are similar to those seen in the Type II model at ~5 months, with comparable reductions observed after administration of the CMTs.

### *Cataracts*

Diabetics develop posterior subcapsular cataracts, characterized by a central opacity in the nucleus and cortex, and these were observed in the current studies of Type II ZDF animals, as shown in FIGURE 10. None of the lean controls developed clinically detectable cataracts, whereas 65% of the untreated Type II diabetic rats developed these ocular lesions (FIG. 10), an increase in incidence that was statistically significant ($p < 0.001$). Of extreme interest, treatment with CMT-8 appeared to prevent the development of these cataracts (FIG. 11) at 5 months while the antimicrobial parent compound doxycycline was not as effective. Treatment of the diabetic rats with CMT-8 produced a significant reduction, of about 37% ($p < 0.05$), in the incidence of cataract development; CMT-3 and doxy produced no significant beneficial effect. The mechanisms by which Type II (and Type I) diabetes causes the development of these cataracts are not yet understood.

### *Tooth Loss*

Tooth loss in the Type II ZDF model can be attributed to both caries and periodontal disease. Significant abscess formation, including life-threatening deep neck abscesses, were noted in the untreated, and, surprisingly, also in the doxycycline-treated diabetic animals. Of interest, these were the same groups that exhibited the

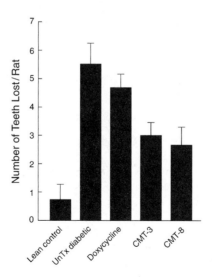

**FIGURE 12.** Tooth loss in untreated TypeII diabetic rats and diabetics treated with different tetracycline analogues at 7 months (5 months of daily administration of drug).

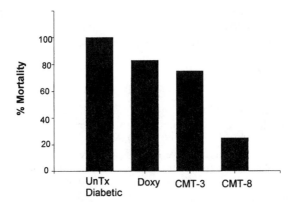

**FIGURE 13.** Percent mortality at 9 months of untreated Type II ZDF/GMI diabetic rats, and diabetics treated with different tetracycline analogues.

greatest tooth loss; up to 50% of the teeth were lost in the untreated diabetic animals by 7 months compared to <10% in the lean control group ($p < 0.001$) (FIG. 12). A 50% reduction ($p < 0.01$) in tooth loss was seen in the diabetic animals treated with CMT-3 and -8 at 7 months, but again, surprisingly, doxycycline therapy did <u>not</u> produce a significantly beneficial effect on this oral complication.

### *Mortality*

Initially, mortality was noted only in the doxy-treated ZDF rats beginning at 3–5 months of age; this was followed and eventually surpassed by increasing mortality rates in the untreated diabetics from 6–9 months. Only the lean controls and the CMT-8 treated animals survived beyond 12 months of age. CMT treatment not only resulted in a four-fold reduction in mortality (FIG. 13), but also increased longevity in the Type II ZDF diabetic animals, most likely by preventing the development of a number of the long-term complications of uncontrolled diabetes that eventually lead to the demise of these animals.

### DISCUSSION

The experiments described above were the first in a series of studies to investigate whether TC compounds, by non-antimicrobial mechanisms, could prevent complications of both Type I and Type II diabetes, associated with altered collagen metabolism, particularly excess MMP activity, and to identify in rat models of these diseases the most effective of the TC analogues.

Our initial interest was a pharmacologic approach to modulating the host response in a complication of Type I diabetes in the oral cavity, aggressive periodontal breakdown. Consistent with observations for skin extracts from the Type I diabetic animals, a four-fold (or greater) elevation in matrix metalloproteinase activity was noted in the gingival extracts of the untreated diabetic animals. We previously reported that both local (i.e., bacterial products such as endotoxins) and systemic factors, relevant to the pathogenesis of periodontal disease, were capable of increasing matrix-degrading enzyme activities in the gingival tissues of rats.[32] Systemic effects of Type I diabetes, producing elevations of MMP levels in both the skins (FIGS. 1–3) and gingival tissues (FIGS. 4–6), were observed in the streptozotocin-induced hyperglycemic rats. Sorsa *et al.*[33] reported that the cellular source of elevated MMP levels in the gingival crevicular fluid of Type I diabetic patients with periodontitis is the neutrophil, with 50–60% of the enzyme existing in the active form. These observations in humans are consistent with the data obtained from the gingival extracts; Western blot analyses demonstrated elevated leukocyte-type gelatinase (MMP-9) and collagenase (MMP-8) (FIG. 6) in these tissues of the Type I diabetic rat. Gelatin zymography revealed that the primary effect of diabetes on gelatinolytic activity was the enhancement of higher molecular weight forms possibly reflecting a mix of 92-kDa (MMP-9 monomer), 88-kDa (activated MMP-9), 220-kDa (MMP-9 homodimer), 125-kDa (heterodimer of MMP-9 complexed to lipocalin[34]) and 123-kDa (heterodimer of MMP-9 with TIMP-1.[35,32] It should be noted, however, that MMP-8 as well as MMP-9 can be produced by mesenchymal cells as well as neutrophils and the former cell type (e.g., under the influence of elevated cytokines and/or $PGE_2$ during diabetes) may be the source of these MMPs in the diabetic rat gingiva.[26]

The mechanisms for increased enzyme activity in the diabetic rats has not yet been elucidated. Elevated levels of iNOS, $PGE_2$ and certain cytokines have all been observed in diabetics, and any or all of these cell regulators could be responsible for elevated MMP expression during the hyperglycemic diabetic state. Furthermore, AGEs formed during long-term diabetes can interact with membrane-bound receptors on inflammatory and resident tissue cells to stimulate their production of a variety of cytokines. In particular, cytokines such as IL-1 and TNF can stimulate the synthesis of MMPs and other matrix-degrading enzymes by various cell types, a mechanism that is most likely responsible (at least in part) for the alveolar bone loss observed in the STZ-model of Type I diabetes. AGEs also have a propensity to form reactive oxygen intermediates, which are believed to activate at least some types of pro-MMPs.

Current therapies for the treatment of Type I and II diabetes include diet restrictions, insulin and/or oral medications (such as sulfonylureas, biguanides, alpha-glucosidase inhibitors and thiazolidinediones) to reduce the severity of hyperglycemia. However, despite advances in the treatment of both Type I and II diabetes, the long-term complications of this disease continue to exert a negative impact on the quality of life and longevity of these patients. To study these complications a Type II model of diabetes has been developed which can be studied for months, even up to a year. We studied the relative ability of two of the non-antimicrobial tetracyclines (CMTs-3 and -8, selected on the basis of the studies described) and a commercially available antimicrobial tetracycline (doxycycline or doxy) to prevent the development of some long-term complications of this disease, including tooth loss,[36] nephropathy,[37] and cataract development.

Diabetes is the single largest cause of end-stage kidney disease in the United States, and develops in 20–40% of patients with insulin-dependent diabetes and in many patients with NIDDM.[38,39] Effective medical therapy to halt the inexorable progression of renal disease due to diabetes, once the process is initiated, has been elusive. Diabetic nephropathy is a clinical syndrome of progressive renal dysfunction leading to hypertension, varying degrees of the nephrotic syndrome, and renal failure. Typically, proteinuria is the first manifestation of the syndrome and increases in severity with time.[40] Given that there are a limited number of animal models of spontaneous diabetes, and only a few of these develop clear-cut pathologic lesions in the kidney, further investigation using these animal model systems is desirable. Information on the nature and chronology of changes in renal functional status is important to assess the ability of pharmacologic therapies to retard the development of these lesions. In recent studies the ZDF/Gmi-*fa* animals, including their lean control litter mates, have been found to suffer from hydronephrosis. McCarthy *et al.*[37] demonstrated that albuminuria increases as more extensive sclerosis of the kidney glomeruli of these diabetic rats develops. Since treatment of the Type II diabetic rats with CMT-3 and CMT-8 reduced the severity of proteinuria by 74% (FIG. 8), compared to untreated diabetics (findings in the current study were consistent with the effects of tetracycline and CMTs on proteinuria in Type I diabetic rats [Ramsammy *et al.*,[41] Arbeit, personal communication, FIG. 9]), a rationale has been generated to justify future extensive histologic studies on the kidneys of these animals to determine whether the severity of glomerular sclerosis is also reduced. Such findings would indicate that these compounds may be useful in the prevention of diabetic nephropathy.

In human subjects, excretion of a significant amount of protein into the urine portends a shortened life expectancy,[40] unless dialysis or renal transplantation are carried out. The current study found that the life expectancy for the untreated ZDF/Gmi-*fa* Type II diabetic animals was less than 9 months. Of extreme interest, only treatment with CMT-8 was capable of extending the life span of the diabetics beyond 12 months. The reduction of proteinuria appeared to correlate with an increase in life expectancy, similar to the pattern seen in human subjects. Moreover, as one complication develops, the likelihood of additional complications developing increases.

An additional complication of diabetes is associated with swelling of the lens. This swelling is a result of the accumulation of fructose and sorbitol that increases the osmolality within the lens. As this process continues, lens protein becomes denatured and cataracts form.[40] Two types of cataracts have been described: (1) metabolic or juvenile cataracts, which are observed in children and young adults with uncontrolled diabetes, and (2) senile cataracts, which are more common than metabolic cataracts. The cataracts in most diabetics are similar to the senile cataracts observed in nondiabetic patients, but tend to occur at a younger age (e.g., the aging process is accelerated). The development of posterior subcapsular cataracts seen in humans, characterized by a central opacity in the nucleus and cortex, was observed, for the first time to our knowledge, in the current studies. CMT-8 administration to the Type II diabetic rats reduced the incidence of cataract development by 37% at 5 months. In fact, this therapeutic effect may have been underestimated since it was an unexpected finding and therefore was not documented at the earlier time points.

The prevention of alveolar bone loss in the Type I diabetic animals and the reduction in tooth loss in the Type II diabetic animals demonstrate the profound effects of the CMTs on the development of periodontal disease in these uncontrolled diabetic animals. Tooth loss in the Type II model can be attributed to both caries (previously unreported) and periodontal disease. Significant abscess formation, including life-threatening deep neck abscesses, was also noted for the first time in the untreated Type II animals and, *surprisingly*, also in the doxycycline-treated diabetic rats. Of interest, these were the same groups that exhibited the greatest tooth loss; up to 50% of the teeth were lost in the untreated diabetic animals by 7 months compared to <10% in the lean control group ($p < 0.001$). A 50% reduction ($p < 0.01$) in tooth loss was seen in the diabetic animals treated with CMTs-3 and -8 at 7 months, but again, surprisingly, doxycycline therapy did *not* produce a significantly beneficial effect on this oral complication (again demonstrating that the beneficial effects of tetracyclines in this model were *independent* of the drug's antibacterial efficacy).

A most surprising observation was that CMT treatment not only decreased mortality, but also increased longevity of the Type II ZDF diabetic animals, presumably by preventing the development of a number of long term life-threatening complications (e.g., nephropathy) of uncontrolled diabetes. These studies clearly indicate that CMTs, but not commercially available antibiotic tetracyclines, may have therapeutic applications for the long-term management of diabetes.

The results of the Diabetes Control and Complications Trial (DCCT) indicate that prolonged duration of elevated serum glucose concentrations in diabetics is involved in the pathogenesis of both retinopathy and nephropathy. While the results of the DCCT demonstrated that reductions in glucose concentrations significantly delayed the appearance of such complications, only 4% of the patients in the intensely treated cohort were able to consistently maintain normal glucose and glycated hemoglobin values.[42] Thus, in the large majority of patients with diabetes, the exposure to elevated glucose levels, over prolonged periods of time, cannot be avoided, and other means of preventing long-term complications must be developed. For this reason, considerable effort has been directed at identifying pharmacologic strategies that abrogate the deleterious effects of factors that act in concert with or independent of glycemic status in the pathogenesis of diabetic complications. The current studies, using animal models of both types of diabetes, suggest that the CMTs could be such a treatment, considering the ability of these newly developed tetracycline analogues to reduce the severity of kidney, eye, and periodontal complications despite blood glucose levels in the severely hyperglycemic range of 500–800 mg%.

## ACKNOWLEDGMENTS

We would like to acknowledge the efforts of Dr. Bob Greenwald and Ms. Susan Moak of Long Island Jewish Medical Center in the execution of the Type II diabetes experiments. The subsequent studies were supported by NIH Grant #s DE-00363, DE-03987 and CollaGenex Pharmaceuticals, Inc.

## REFERENCES

1. HERMAN, W. 1991. Prevention and Treatment of Complications of Diabetes: A Guide for Primary Care Practitioners. Centers for Disease Control. Atlanta, GA.
2. LOE, H. 1993. Periodontal disease: the sixth complication of diabetes mellitus. Diabetes Care **16** (Supple. 1): 329–334.
3. RUDERMAN, N. & C. HAUDENSCHILD. 1984. Diabetes as an atherogenic factor. Prog. Cardiovasc. Dis. **26:** 373–412.

4. FRANTZIS, T., C. REEVE & A. BROWN, JR. 1971. The ultrastructure of capillary basement membranes in the attached gingiva of diabetic and nondiabetic patients with periodontal disease. J. Periodontol. **42:** 406–411.
5. LISTGARTEN, M., F. RICKER, JR., L. LASTER, J. SHAPIRO & D. COHEN. 1974. Vascular basement lamina thickness in the normal and inflamed gingiva of diabetics and nondiabetics. J. Periodontol. **45:** 676–684.
6. MCNAMARA, T., J. KLINGSBERG, N. RAMAMURTHY & L. GOLUB. 1979. Crevicular fluid studies of a diabetic and her non-diabetic twin. J. Dent. Res. **58** (special issue A): 351.
7. KAPLAN, R., J. MULVIHILL, N. RAMAMURTHY & L. GOLUB. 1982. Gingival collagen metabolism in human diabetics. J. Dent. Res. **61:** 275.
8. MANOUCHEHR-POUR, M., P. SPAGNUOLO, H. RODMAN & N. BISSADA. 1981. Comparison of neutrophil chemotactic responses in diabetic patients with mild and severe periodontal disease. J. Periodontol. **52:** 410–414.
9. GOLUB, L., G. NICOLL, V. IACONO & N. RAMAMURTHY. 1982. In vivo crevicular leukocyte response to a chemotactic challenge: inhibition by experimental diabetes. Infect. Immun. **37:** 1013–1020.
10. CHASE, H., R. WILLIAMS & J. DUPONT. 1979. Increased prostaglandin synthesis in children with diabetes mellitus. J. Pediatrics. **94(2):** 185–189.
11. SALVI, G., B. YALDA, J. COLLINS, B. JONES, F. SMITH, R. ARNOLD & S. OFFENBACHER. 1997. Inflammatory mediator response as a potential risk marker for periodontal diseases in insulin-dependent diabetes mellitus patients. J. Periodontol. **68:** 127–135.
12. BROWNLEE, M., A. CERAMI & H. VLASSARA. 1988. Advanced glycosylation endproducts in tissue and the biochemical basis of diabetic complications. N. Engl. J. Med. **318:** 1315–1320.
13. BUCALA, R., Z. MAKITA, T. KOSCHINSKY, A. CERAMI & H. VLASSARA. 1993. Lipid advanced glycosylation: pathway for lipid oxidation in vivo. Proc. Natl. Acad. Sci. USA **90:** 6434–6438.
14. VLASSARA, H. 1991. Non-enzymatic glycosylation. Diabetes Annu. **6:** 371–389.
15. SCHMIDT, A.,S. YAN & D. STERN. 1995. The dark side of glucose. Nat. Med. **1(10):** 1002–1004.
16. SCHMIDT, A., O. HORI, J. CHEN, J. BRETT & D. STERN. 1995. AGE interaction with their endothelial receptors induce expression of VCAM-1: a potential mechanism for the accelerated vasculopathy of diabetes. J. Clin. Invest. **96:** 1375–1403.
17. VLASSARA, H. 1992. Receptor-mediated interactions of advanced glycosylation end products with cellular components within diabetic tissues. Diabetes **41:** 52–56.
18. SCHMIDT, A., M. HASU, D. POPOV, J. ZHANG, J. CHEN, S. YAN, J. BRETT, R. CAO, K. KUWABARA, C. GABRIELA, N. SIMIONESCU, M. SIMIONESCU & D. STERN. 1994. The receptor for advanced glycation endproducts (AGEs) has a central role in vessel wall interactions and gene activation in response to circulating AGE-proteins. Proc. Natl. Acad. Sci. USA **91:** 8807–8811.
19. GOLUB, L., M. SCHNEIR & N. RAMAMURTHY. 1978. Enhanced collagenase activity in diabetic rat gingiva: in vitro and in vivo evidence. J. Dent. Res. **57:** 520–525.
20. PETERSON, R. 1995. The Zucker Diabetic Fatty (ZDF) rat. In Advances in the Research of Diabetic Animals. Lessons from Animal Diabetes. E. Shafrir, Ed.: 1–7. Smith-Gordon. London.
21. RAMAMURTHY, N. & L. GOLUB. 1983. Diabetes increases collagenase activity in extracts of rat gingiva and skin. J. Periodont. Res. **18:** 23–30.
22. GOLUB, L.M., H.M. LEE, G. LEHRER, A. NEMIROFF, T.F. MCNAMARA, R. KAPLAN & N.S. RAMAMURTHY. 1983. Minocycline reduces gingival collagenolytic activity during diabetes: preliminary observations and a proposed new mechanism of action. J. Periodont. Res. **18:** 516–526.

23. YU, Z., M. LEUNG, N. RAMAMURTHY, T. MCNAMARA & L. GOLUB. 1992. HPLC determination of a chemically-modified non-antimicrobial tetracycline: biologic implications. Biochem. Med. **47:** 10–20.

24. MCCROSKERY, P.A., J.F. RICHARDS & E.D. HARRIS, JR. 1975. Purification and characterization of a collagenase extracted from rabbit tumours. Biochem. J. **152:** 131–142.

25. HEUSSEN, C. & E. DOWDLE. 1980. Electrophoretic analysis of plasminogen activators in polyacrylamide gels containing sodium dodecyl sulfate and copolymerized substrates. Anal. Biochem. **102:** 196–202.

26. SORSA, T., Y. DING, T. SALO, A. LAUHIO, O. TERONEN, T. INGMAN, H. OHTANI, N. ANDOH, S. TAKEHA & Y. KONTTINEN. 1994. Effects of tetracyclines on neutrophil, gingival and salivary collagenases: a functional and western blot assessment with special references to their cellular sources in periodontal diseases. Ann. N.Y. Acad. Sci. **732:** 112–131.

27. KLAUSEN, B., R. EVANS & C. SFINTESCU. 1989. Two complementary methods of assessing periodontal bone level in rats. Scand. J. Dent. Res. **97:** 494–499.

28. RYAN, M.E., N.S. RAMAMURTHY & L.M. GOLUB. 1998. Tetracyclines inhibit protein glycation in experimental diabetes. Adv. Dent. Res. **12:** 152–158.

29. AMIN, A., M. ATTUR, G. THAKKER, P.V. PATEL, P. PATEL, I. PATEL & S. ABRAMSON. 1996. A novel mechanism of action of tetracyclines: effects on nitric oxide synthases. Proc. Natl. Acad. Sci. USA **93:** 14014–14019.

30. TRACHTMAN, H., S. FUTTERWEIT, R. GREENWALD, S. MOAK, P. SINGHAL, N. FRANKI & A. AMIN. 1996. Chemically modified tetracyclines inhibit inducible nitric oxide synthase expression and nitric oxide production in cultured rat mesangial cells. Biochem. Biophys. Res. Commun. **229:** 243–248.

31. SODEK, J. & C. OVERALL. 1992. Matrix metalloproteinases in periodontal tissue remodelling. Matrix (Supple.): 352–362.

32. CHANG, K., M. RYAN, L. GOLUB, N. RAMAMURTHY & T. MCNAMARA. 1996. Local and systemic factors in periodontal disease increase matrix-degrading enzyme activities in rat gingiva: effects of minocycline therapy. Res. Commun. Mol. Path. Pharm. **91**(3): 303–318.

33. SORSA, T., T. INGMAN, K. SUOMALAINEN, S. HALINEN, H. SAARI, Y. KONTTINEN, V. UITTO & L. GOLUB. 1992. Cellular source and tetracycline inhibition of gingival crevicular fluid collagenase of patients with labile diabetes mellitus. J. Clin. Periodontol. **19:** 146–149.

34. KJELDSEN, L., D. BAINTON, H. SENGELOV & N. BORREGAARD. 1994. Identification of neutrophil gelatinase-associated lipocalin as a novel matrix protein of specific granules in human neutrophils. Blood **83:** 791–807.

35. TRIEBEL, S., J. BLASER, T. GOTE, G. PELZ, E. SCHUREN, M. SCHMITT & H. TSCHESCHE. 1995. Evidence for the tissue inhibitor of metalloproteinase-1 (TIMP-1) in human polymorphonuclear leucocytes. Eur. J. Biochem. **231:** 714–719.

36. SPROUL, P., F. RAHEMTULLA, C. PRINCE, M. JEFFCOAT & K. MCCARTHY. 1996. Diabetic periodontal disease: extracellular matrix alterations in an animal model. J. Dent. Res. **75,** (#1575).

37. MCCARTHY, K., D. ABRAHAMSON, K. BYNUM, P. ST. JOHN & J. COUCHMAN. 1994. Basement membrane-specific chondroitin sulfate proteoglycan is abnormally associated with the glomerular capillary basement membrane of diabetic rats. J. Histochem. Cytochem. **42:** 473–484.

38. NOTH, R. 1989. Diabetic nephropathy: hemodynamic basis and implications for disease management. Ann. Intern. Med. **110:** 795–813.

39. KROLEWSKI, A., J. WARRAM, A. CHRISTLIEB, E. BUSICK & C. KAHN. 1985. The changing natural history of nephropathy in Type I diabetes. Am. J. Med. **78:** 785–794.

40. LAVINE, R. 1990. Chronic complications of diabetes mellitus. *In* Internal Medicine for Dentistry. L. Rose & D. Kaye, Eds.: 1127–1130. C.V. Mosby. St Louis, MO.
41. RAMSAMMY, L., N. RAMAMURTHY, H. LEE, R. GREENWALD & L. GOLUB. 1989. Tetracyclines inhibit diabetic rat kidney type IV collagenolytic activity and proteinuria. Clin. Res. **37:** 459.
42. D.C.C.T. RESEARCH GROUP. 1993. The effect of intensive treatment of diabetes on the development and progression of long-term complications in insulin-dependent diabetes mellitus. N. Engl. J. Med. **329:** 977–986.

# Clinical Trials of a Matrix Metalloproteinase Inhibitor in Human Periodontal Disease

ROBERT A. ASHLEY[a] AND THE SDD CLINICAL RESEARCH TEAM

*CollaGenex Pharmaceuticals, Inc., Newtown, Pennsylvania 18940, USA*

**ABSTRACT:** After demonstration by Golub *et al.* of the ability of the tetracyclines to inhibit elevated collagenolytic activity in animal models of periodontal diseases, a clinical development program was initiated to demonstrate the potential of a subantimicrobial dose of doxycycline (SDD) to augment and maintain the improvements in clinical parameters of adult periodontitis (AP) afforded by conventional nonsurgical periodontal therapy. Clinical trials were carried out in which a number of different SDD dosing regimens and placebo were compared in patients administered a variety of adjunctive nonsurgical therapies. Measured parameters included levels of collagenase activity in gingival crevicular fluid (GCF) and gingival specimens, clinical attachment levels (cALv), probing pocket depths (PD), bleeding on probing (BOP), and subtraction radiographic measurements of alveolar bone height. When used as an adjunct to either scaling and root planing or supragingival scaling and dental prophylaxis, SDD was shown to reduce collagenase levels in both GCF and gingival biopsies, to augment and maintain cALv gains and PD reductions, to reduce BOP, and to prevent loss of alveolar bone height. These clinical responses arose in the absence of any significant effects on the subgingival microflora and without evidence of an increase in the incidence or severity of adverse reactions relative to the control groups. It is proposed that one of the mechanisms of action of SDD is as an inhibitor of pathologically elevated MMPs, including neutrophil and bone cell collagenases (MMP-8 and MMP-13), which are associated with the host response in chronic AP, and that SDD provides a novel systemic approach to the management of AP.

## INTRODUCTION

Periodontitis is the most common cause of adult tooth loss in the United States.[1] A recurring and site-specific condition, periodontitis involves inflammation of the gingiva together with loss of clinical attachment caused by the destruction of the periodontal support structures and alveolar bone.[2,3]

Although bacteria are necessary for initiating periodontitis, host responses are in large part responsible for the destruction of the periodontal support structures.[3] In patients with periodontitis, pathologic overactivity of host-derived matrix metalloproteinases (MMPs) occurs in response to the bacterial infection in the periodontal tissues. This leads to the excessive destruction of collagen, the primary structural component of the periodontal matrix.[4] In turn, MMP-mediated destruction of connective tissue collagen leads to gingival recession, pocket formation, and tooth mo-

[a]Address for correspondence: Robert A. Ashley, CollaGenex Pharmaceuticals, Inc., 301 South State Street, Newtown, Pennsylvania 18940. Phone, 215/579-7388; fax, 215/579-8577; e-mail, roba@collagenex.com

bility. In the absence of appropriate therapy, tooth loss may occur in advanced disease.

Commonly used therapies for treating periodontitis include mechanical procedures that control the localized bacterial infection by physically removing plaque and calculus. Although mechanical procedures are frequently effective in slowing the progression of periodontal disease, clinical outcomes may be suboptimal in some patients because these interventions may have only a limited effect on pathologic host responses. New strategies for managing destructive periodontal diseases aim to reduce the bacterial load while simultaneously suppressing the host responses that lead to tissue destruction. This "two-pronged" approach to treatment utilizes mechanical procedures as first-line strategies, with host-modulating pharmacotherapies as adjuncts to first-line treatments.

Recently, attention has focused on the tetracycline antibiotics and their unique ability to inhibit the activities of tissue-destructive MMPs. In a series of landmark studies, Golub and colleagues demonstrated that tetracyclines, such as doxycycline, inhibit collagenolytic activity in gingival tissues.[5,6] Significantly, this anticollagenolytic action of doxycycline occurs at doses below those required for antimicrobial effectiveness.[7–9] Doxycycline has been shown to inhibit MMP-8, the predominant MMP responsible for periodontal destruction, in extracts of inflamed human gingival tissue.[10] Treatment with doxycycline, 20 mg twice daily, reduced the excessive activity of MMPs and reduced the degradation of collagen in gingival crevicular fluid (GCF) taken from adult patients with periodontitis.[11] In a preliminary clinical study, doxycycline, 20 mg twice daily, inhibited collagenase activity and improved clinical attachment levels (cALv) and probing pocket depths (PD) when administered to periodontal patients periodically over a 6-month period compared with placebo.[12] Taken together, these studies suggest that subantimicrobial dose doxycycline (SDD) may have clinical utility as an adjunct to mechanical interventions in the treatment of periodontitis.

We describe here (1) a 12-week, dose-ranging, Phase II study evaluating the effect of SDD on GCF collagenase activity and selected clinical parameters and (2) long-term, Phase III clinical trials evaluating the efficacy and safety of SDD as a systemically administered adjunct to mechanical procedures in the treatment of adult periodontitis (AP). The effect of SDD on the dynamics of the periodontal microflora is also briefly summarized.

## METHODS

### Dose-Ranging Study

A Phase II clinical trial was conducted by Golub *et al.*[16] to select the optimal dosage of SDD that reduces GCF collagenase activity and improves attachment levels without inducing doxycycline resistance in microorganisms, and part of the study is summarized as follows. An institutional review board approved the study protocol; patients included in the study provided informed written consent. A total of 75 adult patients with active periodontitis (i.e., exhibiting pockets with repeatedly elevated GCF collagenase activity) were enrolled in a randomized, placebo-controlled, parallel-group, 12-week study. Patients were stratified according to levels of periodontal attachment and levels of GCF collagenase activity. At the baseline visit, patients received a dental scaling and prophylaxis and then were randomly assigned to receive various dosing regimens of SDD or placebo for 12 weeks.

Efficacy measures included the reduction in GCF collagenase activity (assessed at baseline and at weeks 2, 4, 8, and 12 of the treatment period) and the change in relative attachment levels (rALv, measured using a Florida Probe with disc attachment). rALv was assessed at baseline and at week 12 of the treatment period. Subgingival microbial samples were harvested at week 12, and sensitivity to doxycycline was determined by standard disk diffusion methodology using a doxycycline-impregnated (30-μg) disk.

Mean values for GCF collagenase activity were calculated using sample site as the unit of analysis. Mean values for rALv were calculated and used to compare changes occurring over the entire study for each treatment group. Changes from baseline and intergroup differences were tested for statistical significance using analysis of variance or covariance as appropriate.

## *Efficacy and Safety Studies*

To evaluate the safety and efficacy of adjunctive SDD in the clinical setting, four randomized, placebo-controlled, double-blind, parallel-group, Phase III clinical trials were conducted at multiple dental centers in the United States. An institutional review board at each dental center approved the protocols; patients included in the studies provided informed written consent. In all studies, patients received an oral pathology examination at the screening visit, at the baseline visit, and throughout the designated treatment period. Manual probing measurements of cALv and PD were conducted using a UNC-15 probe at screening, baseline, and throughout the treatment period. A single examiner with extensive experience in conducting clinical trials performed all manual probing measurements on a given patient. Duplicate measurements were compared statistically to ensure that individual examiners at each study site obtained acceptable percent reliability.

In three of the studies, the efficacy of SDD was evaluated in conjunction with supragingival scaling and dental prophylaxis (SSDP). A total of 437 patients with clinical evidence of periodontitis (i.e., at least two tooth sites with cALv and PD between 5 and 9 mm inclusive that bled on probing) were enrolled in the studies. Patients received SSDP (30-minute scaling to remove supragingival plaque and calculus followed by tooth polishing with prophylaxis paste and a rotary instrument) at the baseline visit and thereafter at 6-month intervals. Patients were randomly allocated to receive placebo ($n = 119$), SDD 10 mg once daily ($n = 80$), SDD 20 mg once daily ($n = 119$), or SDD 20 mg twice daily ($n = 119$) for 12 months. Patients were instructed to take study medication once in the morning and once in the evening, 1 hour before eating, at approximately 12-hour intervals. For patients receiving SDD once daily, morning medication contained SDD and evening medication contained placebo.

In a fourth study, the efficacy of SDD was evaluated in conjunction with scaling and root planing (SRP). A total of 190 adult patients with evidence of periodontitis (i.e., cALv and probing PD between 5 and 9 mm, inclusive, with bleeding on probing) in at least two tooth sites within each of two quadrants were enrolled in the study. Patients received SRP at the baseline visit. SRP was performed on the qualifying quadrants by the same therapist at each study center, with up to 1 hour allowed per quadrant, until the tooth and root surfaces were visually and/or tactilely free of all deposits. Patients were then randomly allocated to receive either placebo ($n = 94$) or SDD 20 mg twice daily ($n = 96$) for 9 months.

Patients were evaluated after 6 and 12 months of treatment (SSDP studies) or after 3, 6, and 9 months of treatment (SRP study). In all studies, efficacy measures included (1) the change in cALv from baseline, (2) the change in PD from baseline, and (3) the percentage of tooth sites with bleeding on probing (BOP). Efficacy endpoints were evaluated by manual probing at six sites around each tooth in the full mouth (SSDP studies) or within qualifying quadrants (SRP study). Tooth sites were stratified according to AAP criteria of disease severity at baseline (no disease, baseline PD 0 to 3 mm; mild-to-moderate disease, baseline PD 4 to 6 mm; severe disease, baseline PD $\geq 7$ mm). Baseline disease severity was determined as the average of duplicate measurements on a given tooth site by manual probing. If the difference between measurements was greater than 2 mm, a repeat measurement was made, and the two closest measurements were averaged to obtain the baseline value.

In the SSDP studies, tooth sites with rapidly progressing disease (e.g., attachment loss of 3 mm or more from baseline to month 6 as measured by manual probing) were discontinued from the study and subjected to mechanical therapy (typically SRP). Patients continued to receive adjunctive SDD or adjunctive placebo after mechanical therapy for the remainder of the treatment period.

Adverse events were recorded in patient diaries, reported at patient interviews, and reported in monthly phone calls. Laboratory tests were performed at the screening visit and throughout the treatment period.

For microbial assessments, subgingival plaque samples were collected at multiple time points using endodontic paper-points. Typical microscopic (e.g., darkfield microscopy) and culture-based techniques (e.g., enumeration of microbes on selective and nonselective media) were used to determine the proportion of distinct cellular morphotypes and to evaluate shifts in the normal periodontal flora, respectively. Agar dilution or agar gradient elution methods were used to assess the susceptibility of the periodontal microflora to doxycycline and other antibiotics.

All treatment group comparisons were performed on the intent-to-treat population (i.e., the population receiving study medication for at least 1 day and having at least one efficacy measurement). Treatment comparisons were performed using a last-observation-carried-forward algorithm (LOCF) to impute missing data at each time point. In the SSDP studies, patients were evaluated as a single population, with the results presented by treatment group. Treatment group comparisons of the per-site efficacy variables were carried out using generalized estimating equation regression techniques.[13,14] With respect to tooth sites with rapidly progressing disease, the data collected from sites after local therapy were evaluated separately. For the primary analysis, these data were replaced by imputed values based on LOCF techniques.

In the SRP study, treatment group comparisons were performed on per-patient variables. The least square means (or means expressed as percentages) and standard errors were calculated using an appropriate analysis-of-variance model. Microbial parameters were assessed using two-sample, unpaired $t$-tests.

For all studies, differences between treatment groups were considered statistically significant when the probability of a Type I error was less than 5% ($p < 0.05$).

## RESULTS

### *Dose-Ranging Study*

Over time, all treatment groups demonstrated reductions in GCF collagenase, with mean values approaching levels associated with healthy gingiva after 4 weeks of treatment.[15,16] At week 12, mean levels of GCF collagenase were significantly reduced from elevated baseline levels in patients receiving SDD 20 mg twice daily (Group 1; $p < 0.05$). For the group receiving SDD 20 mg twice daily (Group 1), the percent decrease in activity from baseline was 47.3% versus 29.1% for the group receiving placebo (FIG. 1).

The effect of SDD on rALv also is shown in FIGURE 1. Improvements in rALv were greatest in the treatment groups receiving doxycycline 20 mg once or twice daily, throughout the entire 12-week course of the study.

Treatment with SDD for 12 weeks did not lead to the emergence of doxycycline resistance of the subgingival microflora; post-treatment isolates remained susceptible to 3 μg/ml of doxycycline (a concentration approximating the levels of serum doxycycline following an antibiotic dosage [200 mg/day]).

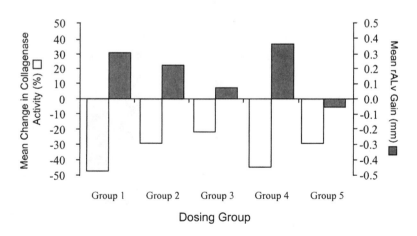

**FIGURE 1.** Effect of SDD on GCF collagenase activity and rALv. Patients received placebo or SDD as described in the Methods section. Group 1: SDD 20 mg twice daily × 12 weeks; Group 2: SDD 20 mg twice daily × 4 weeks then 20 mg once daily × 8 weeks; Group 3: SDD 20 mg twice daily × 4 weeks then placebo × 8 weeks; Group 4: SDD 20 mg once daily × 12 weeks; Group 5: placebo. SDD = subantimicrobial dose doxycycline; GCF = gingival crevicular fluid; rALv = relative attachment level.

## *Efficacy and Safety Studies*

### *Efficacy Results of SSDP Studies*

In all studies combined, no significant differences were demonstrated between the groups receiving either SDD 10 mg once daily or SDD 20 mg once daily and the group receiving placebo. However, significant treatment differences were demonstrated between the SDD 20 mg twice-daily group and the placebo group; these data are presented below.

Across all disease strata, improvements in cALv from baseline were demonstrated for both treatment groups, presumably owing to the course of SSDP administered at baseline and month 6. However, in tooth sites with mild-to-moderate disease (baseline PD 4 to 6 mm), attachment gains demonstrated with adjunctive SDD were significantly greater than attachment gains demonstrated with adjunctive placebo. After 12 months of treatment, the mean attachment gains were 0.67 mm with adjunctive SDD and 0.44 mm with adjunctive placebo ($p < 0.01$ versus placebo). In tooth sites with severe disease (baseline PD $\geq$ 7 mm), improvements in cALv also were significantly greater with adjunctive SDD than with adjunctive placebo; the mean attachment gains after 12 months of treatment were 1.27 mm with adjunctive SDD and 0.95 mm with adjunctive placebo ($p < 0.05$ versus placebo).

Improvements in PD from baseline were similar to improvements in cALv from baseline. In tooth sites with mild-to-moderate disease, mean reductions in PD were 0.71 mm following 12 months of treatment with adjunctive SDD 20 mg twice daily and 0.46 mm following 12 months of treatment with adjunctive placebo. The difference between the two groups was statistically significant ($p < 0.01$ versus placebo). In tooth sites with severe disease, treatment with adjunctive SDD 20 mg twice daily

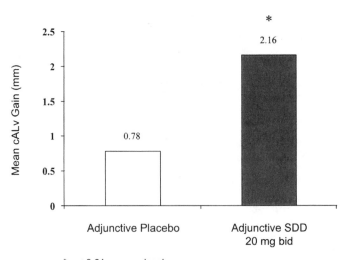

* $p < 0.01$ versus placebo

**FIGURE 2.** Effect of adjunctive SDD on cALv in tooth sites with rapidly progressing AP. Patients received placebo or SDD as described in the Methods section. SDD = subantimicrobial dose doxycycline; cALv = clinical attachment level.

reduced PD by 1.39 mm, whereas treatment with adjunctive placebo reduced PD by 0.96 mm ($p < 0.01$ versus placebo).

Attachment loss during the first 6 months of the study of 3 mm or more for which an investigator recommended SRP at the 6-month time point was demonstrated in 52 tooth sites in 9 patients receiving adjunctive placebo versus 14 tooth sites in 9 patients receiving adjunctive SDD. This corresponded to a 73% reduction in the incidence of rapid progression of periodontitis. In tooth sites with rapidly progressing disease that were subjected to SRP, the average attachment gain was significantly greater in the adjunctive SDD group than in the adjunctive placebo group (FIG. 2). At 12 months, the mean gain in cALv was 2.16 mm with adjunctive SDD and 0.78 mm with adjunctive placebo ($p < 0.01$ versus placebo).

With respect to BOP, significant improvements from baseline were demonstrated for the adjunctive SDD group after 12 months of treatment compared with the adjunctive placebo group in tooth sites with no disease (SDD [26.3%] versus placebo [31.4%]), in tooth sites with mild-to-moderate disease (SDD [52.3%] versus placebo [60.9%]), and in tooth sites with severe disease (SDD [69.1%] versus placebo, [79.5%]) (all $p < 0.05$).

**TABLE 1. Effect of adjunctive SDD on the mean per-patient average change in cALv from baseline (SRP study)[a,b]**

| | Mean change (± SEM) in cALv from baseline (mm) | | | | | |
|---|---|---|---|---|---|---|
| | Month 3 | | Month 6 | | Month 9 | |
| Baseline PD | Placebo | SDD | Placebo | SDD | Placebo | SDD |
| 0 to 3 mm | −0.14 | −0.22 | −0.21 | −0.25 | −0.20 | −0.25 |
| | (0.03) | (0.03) | (0.03) | (0.03) | (0.03) | (0.03) |
| | $n = 92$ | $n = 90$ | $n = 93$ | $n = 90$ | $n = 93$ | $n = 90$ |
| 4 to 6 mm | −0.71 | −0.86[c] | −0.83 | −0.98[c] | −0.86 | −1.03[c] |
| | (0.05) | (0.05) | (0.05) | (0.05) | (0.05) | (0.05) |
| | $n = 92$ | $n = 90$ | $n = 93$ | $n = 90$ | $n = 93$ | $n = 90$ |
| ≥ 7 mm | −0.91 | −1.38[d] | −1.14 | −1.59[c] | −1.17 | −1.55[c] |
| | (0.10) | (0.10) | (0.13) | (0.13) | (0.13) | (0.13) |
| | $n = 78$ | $n = 79$ | $n = 78$ | $n = 79$ | $n = 78$ | $n = 79$ |

[a]Negative change from baseline indicates attachment gain or improvement from baseline.
[b]Values represent the least-square means adjusted for investigator and baseline average.
[c]$p < 0.05$ versus placebo.
[d]$p < 0.01$ versus placebo.
ABBREVIATIONS: cALv = clinical attachment level; PD = probing depth; SDD = subantimicrobial doxycycline.

**TABLE 2. Effect of adjunctive SDD on the mean per-patient average change in PD from baseline (SRP study)[a,b]**

| | Mean change ($\pm$ SEM) in PD from baseline (mm) | | | | | |
| --- | --- | --- | --- | --- | --- | --- |
| | Month 3 | | Month 6 | | Month 9 | |
| Baseline PD | Placebo | SDD | Placebo | SDD | Placebo | SDD |
| 0 to 3 mm | $-0.03$ | $-0.12^c$ | $-0.07$ | $-0.16^c$ | $-0.05$ | $-0.16^c$ |
| | (0.02) | (0.02) | (0.02) | (0.02) | (0.02) | (0.02) |
| | $n = 92$ | $n = 90$ | $n = 93$ | $n = 90$ | $n = 93$ | $n = 90$ |
| 4 to 6 mm | $-0.60$ | $-0.82^c$ | $-0.68$ | $-0.91^c$ | $-0.69$ | $-0.95^c$ |
| | (0.05) | (0.05) | (0.05) | (0.05) | (0.05) | (0.05) |
| | $n = 92$ | $n = 90$ | $n = 93$ | $n = 90$ | $n = 93$ | $n = 90$ |
| $\geq 7$ mm | $-0.93$ | $-1.55^c$ | $-1.14$ | $-1.75^c$ | $-1.20$ | $-1.68^c$ |
| | (0.10) | (0.10) | (0.12) | (0.12) | (0.12) | (0.12) |
| | $n = 78$ | $n = 79$ | $n = 78$ | $n = 79$ | $n = 78$ | $n = 79$ |

[a]Negative change from baseline indicates PD decrease or improvement from baseline.
[b]Values represent the least-square means adjusted for investigator and baseline average.
[c]$p < 0.01$ versus placebo.
ABBREVIATIONS: PD = probing depth; SDD = subantimicrobial dose doxycycline.

## Efficacy Results of SRP Studies

In general, improvements in cALv from baseline were demonstrated for the adjunctive SDD treatment group and the adjunctive placebo group, presumably owing to the course of SRP administered at the baseline visit. However, the mean attachment gains were significantly greater with adjunctive SDD than with adjunctive placebo in tooth sites with mild-to-moderate disease (all $p < 0.05$) and in tooth sites with severe disease (all $p < 0.05$) after 3, 6, and 9 months of treatment. The mean per-patient average change in cALv from baseline for each disease stratum is shown in TABLE 1.

Reductions in PD from baseline also were demonstrated for both treatment groups, presumably owing to the course of SRP administered at baseline. However, the mean reductions in PD were significantly greater with adjunctive SDD than with adjunctive placebo at every time point in tooth sites with mild-to-moderate disease (all $p < 0.001$) and tooth sites with severe disease (all $p < 0.01$) after 3, 6, and 9 months of treatment (TABLE 2).

Adjunctive SDD reduced the incidence of BOP in tooth sites with no disease and in tooth sites with mild-to-moderate disease compared with adjunctive placebo (all $p < 0.05$). Significant treatment differences were demonstrated between adjunctive SDD and adjunctive placebo after 3, 6, and 9 months in tooth sites with no disease (all $p < 0.05$) and after 9 months in tooth sites with mild-to-moderate disease

($p < 0.05$). In sites with severe disease, a trend favoring adjunctive SDD over adjunctive placebo was noted.

*Safety Results of SSDP and SRP Studies*

A total of 428 patients (those patients receiving SDD 20 mg twice daily or placebo) were included in the safety analysis. Treatment with adjunctive SDD was well tolerated. The percentage of patients discontinuing treatment with adjunctive SDD due to adverse events of all causes was similar to that for placebo (6% versus 7%, respectively). The nature of adverse events leading to study discontinuation did not differ between the treatment groups.

The most frequent adverse events of all causes are shown in TABLE 3. Adverse events were generally transient and mild to moderate in severity. For both treatment groups, the most commonly reported adverse event was headache, followed by the common cold and flu symptoms. Patients randomized to placebo reported a slightly higher incidence of flu symptoms, toothache, and periodontal abscess than patients randomized to SDD. No clinically meaningful differences in the incidence of adverse events related to the gastrointestinal tract, urogenital tract, or the skin were noted between SDD and placebo. Between the two groups, no clinically meaningful differences in laboratory parameters were demonstrated.

TABLE 3. Incidence of the most common adverse events of all causes ($\geq 5\%$ for either treatment group)

| Adverse event | SDD 20 mg bid ($n = 213$) | Placebo ($n = 215$) |
|---|---|---|
| Headache | 26% | 26% |
| Common cold | 22% | 21% |
| Flu symptoms | 11% | 19% |
| Toothache | 7% | 13% |
| Periodontal abscess | 4% | 10% |
| Tooth disorder | 6% | 9% |
| Nausea | 8% | 6% |
| Sinusitis | 3% | 8% |
| Injury | 5% | 8% |
| Dyspepsia | 6% | 2% |
| Sore throat | 5% | 6% |
| Joint pain | 6% | 4% |
| Diarrhea | 6% | 4% |
| Sinus congestion | 5% | 5% |
| Coughing | 4% | 5% |

SDD = subantimicrobial dose doxycycline.

*Microbiology Results of SSDP and SRP Studies*

In general, the results of the studies demonstrate that treatment with adjunctive SDD 20 mg twice daily for up to 12 months did not result in a detrimental shift in the periodontal flora. The use of adjunctive SDD did not result in the colonization or overgrowth of periodontal flora by periodontal or opportunistic pathogens, including yeast and enteric microorganisms. Moreover, treatment with adjunctive SDD did not lead to the emergence of doxycycline resistance or multiantibiotic resistance of the subgingival microflora. No replacement or overgrowth of the periodontal flora by doxycycline-resistant bacteria or yeast was demonstrated.

## DISCUSSION

Periodontopathic bacteria and destructive host responses are involved in the initiation and progression of AP. Therefore, the successful long-term management of AP may require an approach to treatment that integrates therapies that address both etiologic components. In the studies described, SDD was used as a systemic adjunct to antimicrobial mechanical interventions in patients with AP. Improvements in periodontal parameters attributable to treatment with SDD likely arise, at least in part, from reductions in the activity of neutrophil collagenase (MMP-8), as shown previously,[12] and confirmed in the Phase II study conducted in patients with AP.

The results of the large-scale, Phase III studies demonstrate that adjunctive SDD improves the efficacy of mechanical interventions routinely used in treating AP in its initial stages. In the combined SSDP studies, treatment with adjunctive SDD resulted in significantly greater gains in attachment than did treatment with adjunctive placebo (all $p < 0.05$). Moreover, improvements in cALv were paralleled by similar improvements in PD; significant reductions in PD were demonstrated in the adjunctive SDD group compared with the adjunctive placebo group (all $p < 0.05$).

Likewise, in the SRP study, significantly greater improvements in cALv and PD were demonstrated with adjunctive SDD than with adjunctive placebo. Improvements from baseline were up to 52% greater for cALv and up to 67% greater for PD with adjunctive SDD than improvements with adjunctive placebo. Furthermore, improvements in clinical parameters were demonstrated with SDD after only 3 months of daily use, and attachment gains and reductions in pocket depth were maintained at 9 months of treatment with SDD.

In all Phase III studies, improvements in BOP were demonstrated with SDD 20 mg twice daily compared with placebo. Improvements in BOP are most likely attributable to improvements in the integrity of the collagen structure at the base of the periodontal pocket rather than to an anti-inflammatory effect.

The present studies also demonstrate that tooth sites with rapidly progressing disease benefit from treatment with adjunctive SDD. It is likely that patients who are highly susceptible to rapidly progressing periodontitis (e.g., those with dysfunctional host responses) and patients with tooth sites refractory to traditional therapies will respond favorably to adjunctive SDD. Evaluating the efficacy and safety of SDD as an adjunct to surgical interventions is a subject of future research.

In general, adjunctive SDD was well tolerated in these studies, with a low incidence of discontinuations due to adverse events. With respect to microbial parameters, treatment with SDD did not alter the dynamics of the subgingival microflora, nor did treatment with SDD lead to the emergence of doxycycline resistance or multiantibiotic resistance of the subgingival microflora.

To our knowledge, these studies represent the first demonstration of the clinical utility of chronic administration of an MMP inhibitor in a large, Phase III, patient population. These studies provided the basis for the approval of the U.S. Food and Drug Administration to market SDD in the United States under the trade name Periostat®. Periostat is the only MMP inhibitor currently approved for marketing in the United States and is the only product demonstrated to modulate host responses in chronic AP.

When used as an adjunct to mechanical procedures in patients with AP, SDD is a well-tolerated, systemic treatment that significantly improves several indices of periodontal health compared with placebo. SDD, an inhibitor of tissue-destructive MMPs in periodontal tissues, may have clinical utility in the treatment of destructive periodontal diseases.

## ACKNOWLEDGMENTS

The SDD Clinical Research Team include the following investigators: D. Adams, D.D.S., University of Oregon, OR; H.J. Baron, D.D.S., Ph.D., NJ; T. Blieden D.D.S., Eastman Dental Center, NY; J. Caton, D.D.S., Eastman Dental Center, NY; S. Ciancio, SUNY Buffalo, NY; R. Crout, D.M.D., West Virginia University, WV; L.M. Golub, D.M.D.., SUNY Stony Brook, NY; A. Hefti, D.D.S., University of Florida, FL; W. Killoy, D.D.S., University of Missouri, MO; B. Kohut, D.M.D., Warner Lambert Company, NJ; T.F. McNamara, Ph.D., SUNY Stony Brook, NY; R. Nagy, D.D.S., UCLA; R. O'Neal, D.D.S., University of Michigan, MI; G. Payonk, Ph.D., Johnson and Johnson Consumer Products, NJ; A. Polson, D.D.S., University of Pennsylvania, PA; C. Quinones, D.D.S., University of Texas at Houston, TX; T. Sipos, Ph.D., Digestive Care Inc., PA; E. Taggart, UCSF, CA; J. Thomas, Ph.D., West Virginia University, WV; M. Wolff, SUNY Stony Brook, NY; C. Walker, Ph.D., University of Florida, FL. The studies described were supported by CollaGenex Pharmaceuticals, Inc., Newtown, PA.

## REFERENCES

1. WILLIAMS, R.C. 1990. Medical progress: periodontal disease. N. Engl. J. Med. **332:** 373–382.
2. AMERICAN ACADEMY OF PERIODONTOLOGY. 1997. Treatment of gingivitis and periodontitis. J. Periodontol. **68:** 1246–1253.
3. AMERICAN ACADEMY OF PERIODONTOLOGY. 1996. Epidemiology of periodontal diseases. J. Periodontol. **67:** 935–945.
4. GOLUB, L.M., M.E. RYAN & R.C. WILLIAMS. 1998. Modulation of the host response in the treatment of periodontitis. Dent. Today **17:** 102–109.
5. GOLUB, L.M., H.M. LEE, G. LEHRER, *et al.* 1983. Minocycline reduces gingival collagenolytic activity during diabetes; preliminary observations and a proposed new mechanism of action. J. Periodont. Res. **18:** 516–524.

6. GOLUB, L.M., M. WOLFF, H.M. LEE, *et al.* 1985. Further evidence that tetracyclines inhibit collagenase activity in human crevicular fluid and from other mammalian sources. J. Periodont. Res. **20:** 12–23.

7. GOLUB, L.M., S. CIANCIO, N.S. RAMAMURTHY, *et al.* 1990. Low-dose doxycycline therapy: effect on gingival and crevicular fluid collagenase activity in humans. J. Periodont. Res. **25:** 321–330.

8. GOLUB, L.M., N.S. RAMAMURTHY, T.F. MCNAMARA, *et al.* 1991. Tetracyclines inhibit connective tissue breakdown: new therapeutic implications for an old family of drugs. Crit. Rev. Oral Biol. Med. **2:** 297–322.

9. GOLUB, L.M., M. WOLFF, S. ROBERTS, *et al.* 1994. Treating periodontal diseases by blocking tissue-destructive enzymes. JADA **125:** 163–169.

10. GOLUB, L.M., T. SORSA, H-M. LEE, *et al.* 1995. Doxycycline inhibits neutrophil (PMN)-type matrix metalloproteinases in human adult periodontitis gingiva. J. Clin. Periodontol. **22:** 100–109.

11. GOLUB, L.M., H.M. LEE, R.A. GREENWALD, *et al.* 1997. A matrix metalloproteinase inhibitor reduces bone-type collagen degradation fragments and specific collagenases in gingival crevicular fluid during adult periodontitis. Inflammation Res. **46:** 310–319.

12. CROUT, R.J., H.M. LEE , K. SCHROEDER K, *et al.* 1996. The "cyclic" regimen of low-dose doxycycline for adult periodontitis: a preliminary study. J. Periodontol. **67:** 506–514.

13. LIANG, K.Y. & S.L. ZEGER. 1986. Longitudinal data analysis using generalized linear models. Biometrika **73:** 13–22.

14. ZEGER, S.L. & K.Y. LIANG. 1986. Longitudinal data analysis for discrete and continuous outcomes. Biometrics **42:** 121–130.

15. CIANCIO, S., R. WAITE, T. SIPOS, *et al.* 1989. Gingival crevicular fluid collagenolytic activity in diagnosing periodontal disease [abstract]. J. Dent. Res. **68:** 334.

16. GOLUB, L.M., T.F. MCNAMARA, B. KOHUT, T. BLIEDEN *et al.* 1999. Adjunctive treatment with subantimicrobial doses of doxycycline: effects on gingival fluid collagenase activity and attachment loss in adult periodontitis. J. Periodontol. Submitted for publication.

# Effects of Chemically Modified Tetracycline, CMT-8, on Bone Loss and Osteoclast Structure and Function in Osteoporotic States

TAKAHISA SASAKI,[a,b] NOBUTAKA OHYORI,[a] KAZUHIRO DEBARI,[c]
NUMGAVARAM S. RAMAMURTHY,[d] AND LORNE M. GOLUB[d]

[a]Department of Anatomy and Cell Biology, and [c]Laboratory for Electron Microscopy,
Showa University Dental School, 1-5-8 Hatanodai, Shinagawa-ku,
Tokyo 142-8555, Japan

[d]Department of Oral Biology and Pathology, School of Dental Medicine,
Health Sciences Center, State University of New York at Stony Brook,
Stony Brook, New York 11794-8700, USA

ABSTRACT: We examined the effects of a nonantimicrobial tetracycline analogue, CMT-8, on bone loss and osteoclasts in ovariectomized (OVX) rats. Three-month-old female rats were OVX, and, one week later, distributed into three groups: sham-operated non-OVX controls, untreated OVX controls, and CMT-8–treated OVX rats. After 145 days of daily drug administration (p.o.), the femurs were dissected and examined histologically. Ovariectomy markedly decreased trabecular and cortical bone volume in the metaphyses compared to sham-operated controls. Treating the OVX rats with CMT-8 produced a significant inhibition of trabecular and cortical bone loss and induced new bone formation, in which connectivity of the trabecular struts was increased by bridging the adjacent longitudinal bone trabeculae. Ultrastructurally, CMT-8 reduced ruffled border formation in osteoclasts, while it caused no structural impairment in osteoblasts. To further evaluate the effects of CMT-8 on the resorbing activity of osteoclasts, osteoclasts were cultured on dentine slices pretreated with CMT-8 at concentrations of 2, 10, or 50 µg/ml, and resorption lacuna formation on the dentine surface was found to be reduced, dose-dependently, by the bound CMT-8. Our results suggest that CMT-8 therapy effectively inhibits post-ovariectomy bone loss not only by inducing new bone formation, but also by inhibiting osteoclastic bone resorption, and that CMT-8 binding to bone may provide a prolonged release delivery of this antiresorptive therapy.

## INTRODUCTION

Postmenopausal estrogen deficiency results in significant bone loss (osteoporosis) in humans and experimental animals.[1–6] The ovariectomized (OVX) mature- or aged-rat models have been widely used to explore various drug regimens to ameliorate this metabolic bone disease.[5,7–11] Recently these OVX rat models were reported

[b]Address for correspondence: T. Sasaki, Department of Anatomy and Cell Biology, Showa University Dental School, 1-5-8 Hatanodai, Shinagawa-ku, Tokyo 142-8555, Japan. Phone, 813-3784-8156; fax, 813-3781-0255.

to exhibit decreased severity of ovariectomy-induced osteoporosis when they were administered either antimicrobial tetracycline (TC)[12] or a nonantimicrobial TC analogue.[13] In the latter study, enhanced efficacy was observed when the chemically modified nonantimicrobial TC (CMT-1: 4-de-dimethylamino TC) was administered in combination with a nonsteroidal antiinflammatory drug, flurbiprofen.[13-15] We previously demonstrated that CMT-1 decreased the severity of diabetes-induced osteoporosis in rats.[16-19]

Regarding mechanisms, CMT-1 was found to be taken up by osteoblasts, to normalize the structure of atrophic osteoblasts, and to increase the production of $^3$H-proline-labeled bone collagen by these cells in diabetic rats.[17-20] In addition, the doxycycline administration over a prolonged time period appeared to reduce bone resorption, as indicated by a reduction in the urinary excretion of hydroxyproline and calcium.[16] However, the effects of TCs and CMTs on the bone-resorbing activity of osteoclasts requires further studies.[21]

The current study has attempted to determine the effects of long-term therapy (145 days) with newer CMT-8 (6α-deoxy hydroxy 4-de-dimethylamino TC) in the OVX rat model of osteoporosis. This CMT-8 was recently found to be a potent inhibitor of bone resorption and collagenase activities (including bone-type collagenase) in vitro and in vivo.[21-23] Here, we report that long-term therapy with CMT-8 markedly reduced the severity of estrogen-deficient osteoporosis in rats by enhancing new bone formation as well as by reducing bone resorption via impairment of the structure and function of osteoclasts.

## MATERIALS AND METHODS

### In vivo *Experiments*

Three-month-old female rats (Sprague-Dawley strain) served as the model of postmenopausal bone loss after they were ovariectomized (OVX). Anesthesia during the bilateral surgical procedure was accomplished by intraperitoneal injection of sodium nembutal. As experimental controls, rats were subjected to a sham surgical procedure. One week after ovariectomy, 15 rats (5 rats per experimental group) were assigned to the following three groups: sham-operated non-OVX controls, untreated OVX controls, and CMT-8-treated OVX rats; the latter received CMT-8 (5 mg/day/rat) suspended in 2% carboxymethylcellulose (CMC) by daily oral gavage. Untreated rats received CMC only. CMT-8 is a semi-synthetic TC (i.e., doxycycline) that has been chemically modified by removal of the dimethylamino group on carbon-4 of the "A" ring of the 4-ringed antibiotic molecule.[24,25] After 145 days of drug therapy, the rats were anesthetized with sodium nembutal and perfusion-fixed through the left ventricle with a mixture of 1% glutaraldehyde and 1% formaldehyde in 0.1 M sodium cacodylate buffer (pH 7.3).

The mid and distal portions of femurs were dissected and processed as follows: Longitudinal sections of femurs, including the metaphyses and epiphyseal growth plates, and cross sections of mid femurs were processed for backscattered electron microscopic (BSE) observations as described previously.[13] Quantitative analysis of BSE images was performed as described previously,[13] and the data were expressed as (1) the mean percent bone area of trabecular bone in the medullary cavity at the metaphysis and (2) the mean percent bone area of cortical bone in the mid-diaphysis portion. Five tissue sections were examined from each experimental group. Statistical differences between groups were evaluated by the two-tailed Student's *t* test. $p < 0.05$ was considered to be significant. Routine light and electron microscopic observations of femurs were also performed as described previously.[18]

## In vitro *Experiments*

As described previously,[26] mouse osteoclast-like multinucleated cells (OCLs) were prepared from ddY mice. Primary osteoblastic cells obtained from newborn mouse calvaria and bone marrow cells obtained from 7- to 9-week-old male mouse tibiae were cultured in $\alpha$–minimum essential medium containing 10% fetal bovine serum and 10nM $1\alpha,25(OH)_2D_3$ on culture dishes. OCLs were formed within 7 days of culture and collected by centrifugation at 250$g$ for 5 min. To assess the resorbing activity of OCLs, the OCL culture on dentine slices was examined by the method previously reported.[26,27] Prior to cell culture, dentine slices were either untreated or coated with CMT-8 by soaking in the solution containing CMT-8 at concentrations of 2, 10, or 50 $\mu$g/ml. OCL preparations were placed on these dentine slices and incubated for 20 hr. At the end of the culture period, the dentine slices were placed in 0.05M $NH_4OH$ for 30 min and cleaned by ultrasonication to remove adherent cells. These dentine slices were coated with carbon and their BSE and SEM images were examined as described previously.[28] The areas of resorption lacunae on BSE images were measured with an image analysis system, and the results were expressed as the mean ($\pm$ SD [standard deviation]) percentage of resorbed areas to the whole surface areas of dentine slices.

## RESULTS

### In vivo *Experiments*

Histomorphometric analysis of the BSE observations revealed statistically significant differences in bone areas between the different groups. In the sham-operated controls, the medullary cavities of the metaphyses were partially occupied by longitudinal and transverse-oriented bone trabeculae (mean bone area = 20.6%) (FIG. 1). Thin longitudinal bone trabeculae, generated from the epiphyseal growth plates (the primary trabecular bones) continued to form the secondary trabecular bones, which were connected to each other by the transverse bone trabeculae. These secondary trabecular bones then merged with the cortical bone. Wide plate-like trabecular bones, formed by convergence of the longitudinal and transverse bony trabeculae, were observed in the secondary trabecular bone area (FIG. 1).

In the untreated OVX rats, the trabecular bone area (mean bone area = 9.6%) was markedly reduced by about 46.6% ($p < 0.01$) compared to the sham-operated controls. This effect was primarily due to the loss of the transverse bone traveculae between longitudinal bone trabeculae, in which wide plate-like bone trabeculae were seldom observed (FIG. 2). Trabecular bone area in the CMT-8-treated OVX rats (mean bone area = 23.7%) was significantly ($p < 0.01$) increased by 147% compared to that observed in the untreated OVX rats, and was even slightly higher than that

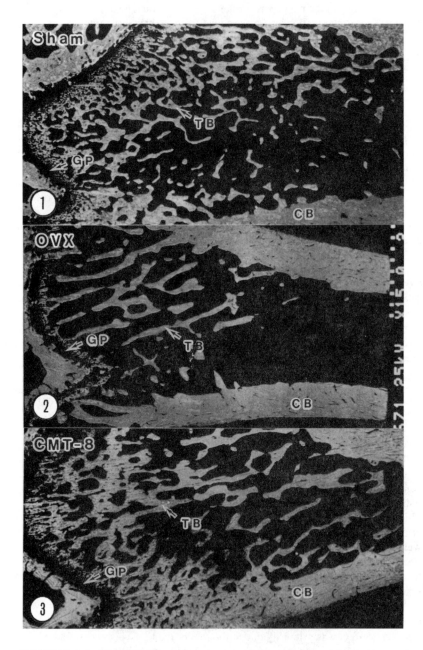

**FIGURES 1–3.** BSE micrographs showing longitudinal sections of the distal metaphyses of humeri in sham-operated control (FIG. 1), untreated OVX (FIG. 2), and CMT-8-treated OVX (FIG. 3) rats. GP: growth plate cartilage; TB: trabecular bone; CB: cortical bone. ×17.

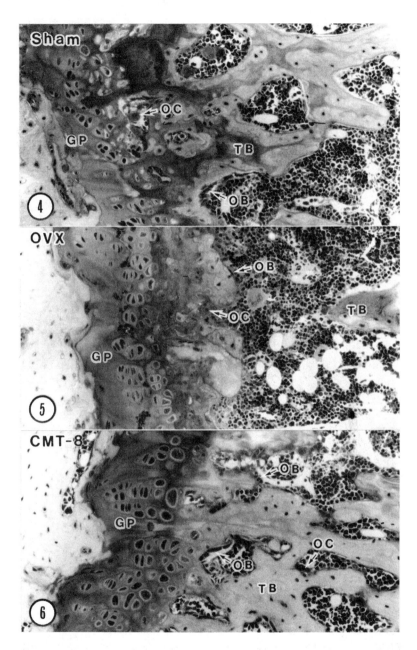

**FIGURES 4–6.** Light micrographs of the distal metaphyses of humeri in sham-operated control (FIG. 4), untreated OVX (FIG. 5), and CMT-8-treated OVX (FIG. 6) rats. GP: growth plate cartilage; TB: trabecular bone; OB: osteoblasts; OC: osteoclasts. ×175.

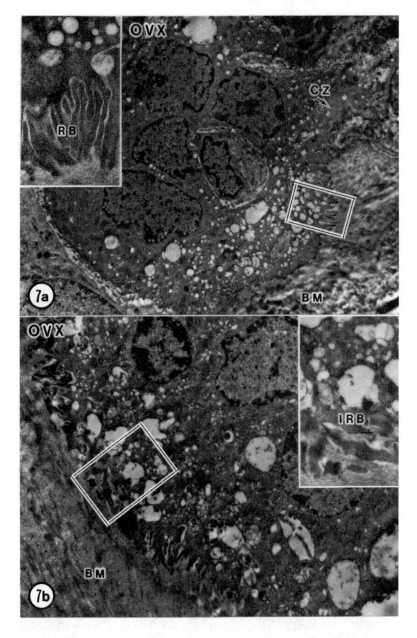

**FIGURES 7a and b.** TEM micrographs of the osteoclasts in CMT-8 treated OVX rats. FIGURE 7a shows an osteoclast with both a ruffled border (RB) and a clear zone (CZ) on the bone matrix (BM). FIGURE 7b shows an osteoclast with an irregularly formed ruffled border (IRB) and a clear zone facing the bone matrix (BM).

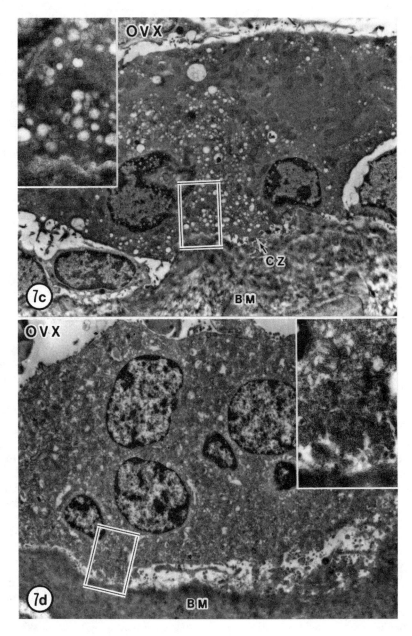

**FIGURES 7c and d.** FIGURE 7c shows an osteoclast with only a clear zone (CZ) on the bone matrix (BM). FIGURE 7d shows an osteoclast having neither a ruffled border nor a clear zone on the bone matrix (BM). Original magnification ×4,000 (**a, c, d**); ×5,000 (**b**). *Insets* show higher magnification views of the osteoclast-bone matrix border. ×16,000 (**a**); ×10,000 (**b**); ×12,000 (**c** and **d**).

seen in the sham-operated, non-OVX controls (FIG. 3). CMT-8 treatment of the OVX rats was found to induce many connections between the longitudinal and transverse bone trabeculae, which resulted in the appearance of wide plate-like trabecular bones. In addition, assessment of electron density of the CMT-8-treated OVX groups, which reflects the calcification rate of these newly-produced trabecular bones, was similar to that of pre-existing normal bones (FIG. 3).

In light microscopic observations on the sham-operated control rats, many bone trabeculae generated from the ossification zone of the metaphyseal growth plates were covered with thin osteoblast layers, and, to a much lesser extent, osteoclasts were observed in the ossification zone (FIG. 4). In the untreated OVX rats, the bone trabeculae were much fewer in number, and a decrease of transverse trabeculae between thin longitudinal trabeculae was evident. The osteoblast layers along the bone trabeculae were extremely thin, and many osteoclasts were observed in the ossification zone of reduced growth plates (FIG. 5). In addition, numerous adipocytes were distributed in the marrow cavities. In the CMT-8-treated OVX rats, longitudinal trabecular bones with numerous transverse connection were observed more frequently than those in sham-operated and untreated OVX rats (FIG. 6). Different from those in the untreated OVX rats, the bone trabeculae were covered with cuboidal osteoblasts, but osteoclasts were also observed in the ossification zone of reduced-growth plates and along the bone trabeculae. In marrow cavities, few adipocytes were observed (FIG. 6).

Ultrastructurally, osteoclasts located on bone surfaces in sham-operated and untreated OVX rats exhibited well-developed ruffled borders and clear zones facing the alveolar bone surfaces (data not shown). Of 10 randomly selected osteoclasts examined on bone surfaces in these rats, 9 osteoclasts had well-developed ruffled borders and one osteoclast exhibited an irregularly formed ruffled border. All osteoclasts exhibited clear zones and contained numerous pale vesicles and vacuoles in the cytoplasm (data not shown).

In 17 randomly selected osteoclasts examined in CMT-8-treated OVX rats, the following four structural variations were evident: (1) three osteoclasts had both ruffled borders and clear zones and contained numerous pale cytoplasmic vacuoles as in untreated controls (FIG. 7a); (2) one osteoclast possessed a clear zone and irregularly or poorly formed ruffled border, in which membrane infoldings of various sizes and configurations were seen (FIG. 7b); (3) six osteoclasts had only clear zones with various degrees of extension but no ruffled borders (FIG. 7c); and (4) seven osteoclasts showed the formation of neither ruffled borders nor clear zones (FIG. 7d). Neither necrotic nor apoptotic osteoclasts were observed.

Ultrastructural features of osteoblasts were also examined. The osteoblasts in untreated controls showed oval or cuboidal shape, contained the cytoplasmic organelles for protein synthesis such as the cisterns of rough-surfaced endoplasmic reticulum, and faced the osteoid layers of the trabecular bones. The osteoblasts in untreated OVX rats showed a flattened cell profile with few cytoplasmic organelles for protein synthesis. In CMT-8-treated OVX rats, the cuboidal-shaped osteoblasts with cytoplasmic organelles for protein synthesis were observed to face thick osteoid layers on the trabecular bones (data not shown), indicating increased osteoblastic activity and increased bone formation.

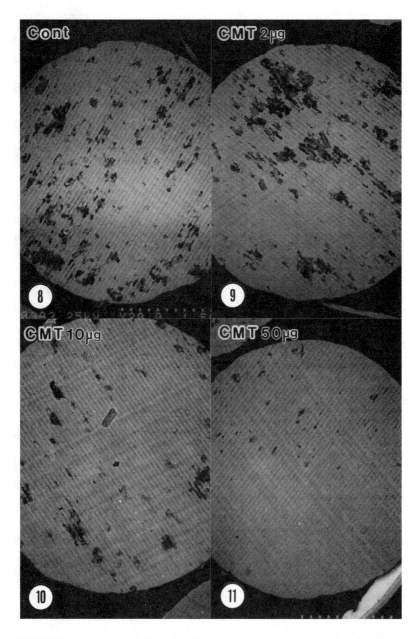

**FIGURES 8–11.** BSE micrographs showing black decalcified areas corresponding to resorption lacunae on dentine slices cultured with osteoclasts. The dentine slices are shown in a control culture (FIG. 8), coated with CMT-8 at 2 μg/ml (FIG. 9), coated with CMT-8 at 10 μg/ml (FIG. 10), and coated with CMT-8 at 50 μg/ml CMT-8 (FIG. 11). ×20.

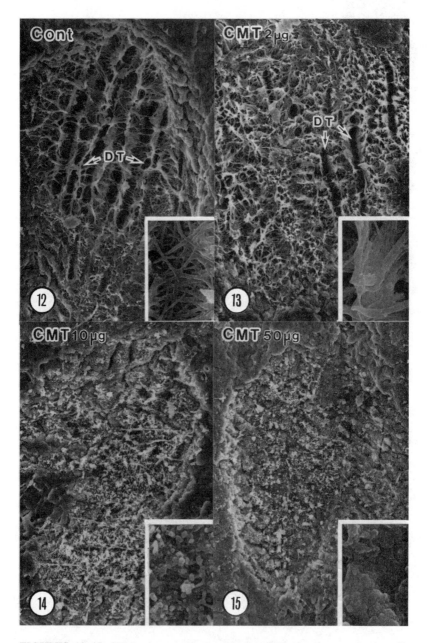

**FIGURES 12–15.** SEM micrographs showing resorption lacunae on dentine slices, which correspond to black decalcified areas in BSE images. The dentine slices are shown in a control culture (FIG. 12), coated with CMT-8 at 2 µg/ml (FIG. 13), coated with CMT-8 at 10 µg/ml (FIG. 14), and coated with CMT-8 at 50 µg/ml CMT-8 (FIG. 15). ×1,270; *insets* ×10,500.

## In vitro *Experiments*

To evaluate the effects of CMT-8 on the resorbing activity of osteoclasts, we cultured these cells on dentine slices which had been coated with CMT-8 at concentrations of 2, 10, or 50μg/ml; control cultures contained dentine slices that were not exposed to CMT-8. In BSE observations of these dentine slices, black decalcified areas corresponding to resorption lacunae were reduced by CMT-8 coating in a dose-dependent manner (FIGS. 8–11). Namely, the mean decalcified area in uncoated control dentine slices was approximately 15.3% of whole dentine surfaces (FIG. 8), which was similar to that (19.5%) in the dentine slices coated with CMT-8 at 2μg/ml (FIG. 9). The mean decalcified area was significantly and dose-dependently reduced to 8.4% at 10μg/ml CMT-8 (FIG. 10) and to 2.1% at 50μg/ml CMT-8 (FIG. 11). On the dentine surface coated with CMT-8 at 50μg/ml, decalcified area formation was inhibited almost completely (FIG. 11).

In SEM observations of these decalcified areas, owing to dissolution of apatite crystals in the peritubular and intertubular matrices, the enlarged openings of dentinal tubules and exposure of matrix collagen fibers were evident within round resorption lacunae on control dentine slices not exposed to CMT-8 (FIG. 12). Similar openings of dentinal tubules and exposure of matrix fibers were also observed over dentine slices coated with CMT-8 at 2μg/ml (FIG. 13), but rarely observed over dentine slices coated with CMT-8 at 10 or 50μg/ml (FIGS. 14 and 15). On the dentine surface coated with CMT-8 at 50μg/ml, even in the decalcified areas in BSE images, exposure of matrix fibers was hardly observed showing extremely solid surface topography (FIG. 15).

## DISCUSSION

Earlier studies using the diabetic rat model of osteoporosis demonstrated that TCs and CMTs could enhance osteoblastic bone formative activities.[17–19] TCs and CMTs have also been found to inhibit osteoclastic bone resorption in organ and cell cultures,[21,24,30] but they did not reduce osteoclast numbers.[31] Consistent with these *in vitro* studies, CMT and TC (minocycline) treatment of OVX rats did not significantly reduce osteoclast numbers, but did appear to inhibit bone resorption.[12,13] In the current study, CMT-8 was selected from among a variety of antimicrobial TCs (e.g., doxycycline) and nonantimicrobial CMTs 1-10, because: (1) cell and organ culture studies demonstrated that CMT-8 was the most potent TC compound to inhibit bone resorption *in vitro* ($IC_{50} \leq 1$ μg/ml),[21,22] and (2) an *in vivo* model of aggressive periodontal bone loss, induced by repeated injection of bacterial endotoxin directly into the gingival tissues, demonstrated that CMT-8 was more potent than other CMTs and doxycycline as an inhibitor of bone loss and bone-type collagenase (MMP-13) activity.[23]

As was evident in the current study, CMT-8 therapy markedly increased both the trabecular and cortical bone mass in the femurs of OVX rats. In this regard, CMT-8 apparently stimulated new bone formation in the metaphyseal trabecular bone area and produced plate-like wide bone trabeculae, and such bone-productive effect by CMT-8 is thought to be associated with the apparent uptake of CMT by osteoblasts and its ability to stimulate collagen production and alkaline phosphatase activity by these bone-lining cells.[17–20] It is, however, unlikely that CMT stimulates the proliferation and differentiation of osteoblastic cells. Rather, CMT is thought to directly act on the pre-existing osteoblasts for stimulation of cell activities.[17–19] In addition, because some CMTs were more effective in enhancing osteoblast activity during bone formation than were commercial TCs,[18,19] such proanabolic effects of the TC molecule are clearly independent of the drug's antimicrobial activity.

In a previous fluorescence microscopic study, new bone formation by CMT labeling was consistently observed on the trabecular bones and on the endosteal surfaces of the cortical bones in the CMT-treated OVX rats.[13] In the current study, CMT-8 therapy significantly increased trabecular bone mass in the OVX rats; in particular, CMT-8 appeared to markedly increase the connection of longitudinal and transverse bone trabeculae and wide plate-like bone trabeculae at the metaphyses of femurs.

In the current study, CMT-8 also appeared to inhibit bone resorption. Although, as reported by us previously,[13] CMT-8 did not appear to reduce osteoclast numbers, CMT-8 therapy impaired formation of the intact ruffled borders in osteoclasts. Because the ruffled border is the functional site of bone resorption,[27,32,33] those osteoclasts lacking the intact ruffled borders are thought to have reduced bone-resorbing activity. In addition, in the current osteoclast culture experiments, the CMT-8 coating reduced, significantly and dose-dependently, formation of decalcified areas and resorption lacunae on the dentine slices by cultured osteoclasts. Disappearance of matrix collagen exposure and enlargement of dentinal tubules by CMT-8 coating also suggest reduced demineralizing and degrading activities of osteoclasts. Rifkin et al.[34] already reported that TCs and CMT reduced acid production by isolated osteoclasts, but this effect was not due to a direct effect on the proton pump.

Inhibition of matrix collagen degradation by osteoclasts has been reported to be caused by cysteine proteinase inhibitors such as E-64.[28,33,35,36] It is well established that the inhibition of cysteine proteinases and/or MMPs results in the inhibition of osteoclastic bone resorption.[35,36] In this regard, Rifkin et al.[21] reported that TCs did not inhibit cysteine proteinases secreted by osteoclasts, but inhibited secretion of enzymes in a dose-dependent manner. It has been further reported that CMT-8 inhibited collagenolytic and gelatinolytic activities of MMPs in connective tissues.[22,23] Sorsa et al.[37] recently determined the functional sites of CMT-8 molecule ($Ca^{2+}/Zn^{2+}$ binding site at $C_{11}$ and $C_{12}$) which inhibit human and chicken MMPs. These in vitro studies appear to be consistent with our current study. Namely, reduction of demineralization and degradation of co-cultured dentine slices by CMT-8 treatment is thought to be due to reduced ruffled border formation in osteoclasts induced by CMT-8. However, it still remains to be clarified whether CMT-8 directly inhibits the enzymatic activities of cysteine proteinases and MMPs secreted by osteoclasts.

On the basis of the current study, the significant increase in trabecular bone mass in the CMT-8-treated OVX rats appears to reflect a synergy due to enhanced bone formation plus suppressed osteoclast function. It appears that long-term CMT-8 therapy normalizes the internal bone structures in OVX rats by (1) suppressing osteoclastic bone resorption at an early phase of estrogen deficiency and (2) by stimulating bone formation at a later phase post-ovariectomy; these early- and latephase effects may combine to prevent OVX-induced bone loss.

## ACKNOWLEDGMENTS

This study was supported in part by Grant R37 DE-03987 from the National Institute of Dental Research and by a grant-in-aid (No. 10671712) from the Ministry of Science, Education and Culture of Japan.

## REFERENCES

1. FAUGERE, M.C. *et al.* 1990. Bone changes occurring early after cessation of ovarian function in beagle dogs: a histomorphometric study employing sequential biopsies. J. Bone Min. Res. **5:** 263–272.
2. HELFRICH, M.D. *et al.* 1991. Morphologic features of bone in human osteopetrosis. Bone **12:** 411–419.
3. DANIELSEN, C.C. *et al.* 1993. Cortical bone mass, composition and mechanical properties in female rats in relation to age, long-term ovariectomy, and estrogen substitution. Calcif. Tissue Int. **52:** 26–33.
4. HIETALA, E.L. 1993. The effect of ovariectomy on periosteal bone formation and bone resorption in adult rats. Bone Min. **20:** 57–65.
5. MILLER, S.C. *et al.* 1991. Calcium absorption and osseous organ-, tissue-, and envelope-specific changes following ovariectomy in rats. Bone **12:** 439–446.
6. MOSEKILDE, L. *et al.* 1993. The effect of aging and ovariectomy on the vertebral bone mass and biomechanical properties of mature rats. Bone **14:** 1–6.
7. ISMAIL, F. *et al.* 1988. Serum bone Gla protein and vitamin D endocrine system in the oophorectomized rat. Endocrinology **122:** 624–630.
8. WRONSKI,T.J. *et al.* 1989. Time course of vertebral osteopenia in ovariectomized rats. Bone **10:** 295–301.
9. SHEN, V. *et al.* 1993. Loss of cancellous bone mass and connectivity in ovariectomized rats can be restored by combined treatment of PTH and estradiol. J. Clin. Invest. **91:** 2479–2487.
10. KALU, D.N. *et al.* 1993. Ovariectomy-induced bone loss and the hematopoietic system. Bone Min. **23:** 145–161.
11. TURNER, R.T. *et al.* 1993. Mechanism of action of estrogen on cancellous bone balance in tibiae of ovariectomized growing rats. J. Bone Min. Res. **8:** 359–366.
12. WILLIAMS, S. *et al.* 1996. Minocycline prevents the decrease in bone mineral density and trabecular bone in ovariectomized aged rats. Bone **19:** 637–644.
13. AOYAGI, M. *et al.* 1996. Tetracycline/flurbiprofen combination therapy modulates bone remodeling in ovariectomized rats: preliminary observations. Bone **19:** 629–635.
14. GREENWALD, R.A. *et al.* 1992. Tetracyclines suppress metalloproteinase activity in adjuvant arthritis, and in combination with flurbiprofen, ameliorate bone damage. J. Rheumatol. **19:** 927–938.
15. BAIN, S. *et al.* 1997. Tetracycline prevents cancellous bone loss and maintains near-normal rates of bone formation in streptozotocin diabetic rats. Bone **21:** 147–153.
16. GOLUB, L.M. *et al.* 1990. Tetracycline administration prevents diabetes-induced osteopenia in the rat: initial observations. Res. Commun. Chem. Pathol. Pharmacol. **68:** 24–40.
17. SASAKI, T. *et al.* 1990. Insulin-deficient diabetes impairs osteoblast and periodontal ligament fibroblast metabolism but does not affect ameloblasts and odontoblasts: response to tetracycline(s) administration. J. Biol. Buccale **18:** 215–226.
18. SASAKI, T. *et al.* 1991. Tetracycline administration restores osteoblast structure and function during experimental diabetes. Anat. Rec. **231:** 25–34.

19. SASAKI, T. *et al.* 1992. Tetracycline administration increases collagen synthesis in osteoblasts of streptozotocin-induced diabetic rats: quantitative autoradiographic study. Calcif. Tissue Int. **50:** 411–419.

20. SASAKI, T. *et al.* 1994. Bone cell and matrix bind chemically-modified non-antimicrobial tetracycline. Bone **15:** 373–375.

21. RIFKIN, B.R. *et al.* 1994. Modulation of bone resorption by tetracyclines. Ann. NY Acad. Sci. **732:** 165–180.

22. GREENWALD, R.A. *et al.* 1998. In vitro sensitivity of the three mammalian collagenases to tetracycline inhibition: relationship to bone and cartilage degradation. Bone **22:** 33–38.

23. RAMAMURTHY, N.S. *et al.* 1998. In vivo and in vitro inhibition of matrix metalloproteinases including MMP-13 by several chemically modified tetracyclines (CMTs). *In* Biological Mechanisms of Tooth Eruption, Reabsorption and Replacement by Implants. Z. Davidovitch & J. Mah, Eds.: 271–277. EBSCO Media. Birmingham, AL.

24. RYAN, M.E. *et al.* 1996. Potential of tetracyclines to modify cartilage breakdown in osteoarthritis. Curr. Opin. Rheumatol. **8:** 238–247.

25. RYAN, M.E. *et al.* 1996. Matrix metalloproteinases and their inhibition in periodontal treatment. Curr. Opin. Periodontol. **3:** 85–96.

26. NAKAMURA, I. *et al.* 1996. Osteoclast integrin $\alpha v\beta 3$ is present in the clear zone and contributes to cellular polarization. Cell Tissue Res. **286:** 507–515.

27. SASAKI, T. *et al.* 1994. Expression of vacuolar $H^+$-ATPase in osteoclasts and its role in resorption. Cell Tissue Res. **278:** 265–271.

28. DEBARI, K. *et al.* 1995. An ultrastructural evaluation of the effects of cysteine-proteinase inhibitors on osteoclastic resorptive functions. Calcif. Tissue Int. **56:** 566–570.

29. GOLUB, L.M. *et al.* 1984. Tetracyclines inhibit tissue collagenase activity. A new mechanism in the treatment of periodontal disease. J. Periodont. Res. **19:** 651–655.

30. GOMES, B.C. *et al.* 1984. Tetracyclines inhibit parathyroid hormone-induced bone resorption in organ culture. Experientia **40:** 1273–1275.

31. RIFKIN, B.R. *et al.* 1992. Effects of tetracyclines on rat osteoblast collagenase activity and bone resorption in vitro. *In* The Biological Mechanisms of Tooth Movement and Craniofacial Adaptation. Z. Davidovitch, Ed.: 85–90. EBSCO Media. Birmingham, AL.

32. VAANANEN, H.K. *et al.* 1990. Evidence for the presence of a proton pump of vacuolar $H^+$-ATPase type in the ruffled borders of osteoclasts. J. Cell Biol. **111:** 1305–1311.

33. SASAKI, T. 1996. Recent advances in the ultrastructural assessment of osteoclastic resorptive functions. Microsc. Res. Techn. **33:** 182–191.

34. RIFKIN, B.R. *et al.* 1992. Effects of tetracyclines on acid production by isolated osteoclasts [abstract]. J. Dent. Res. **71:** 235.

35. EVERTS, V. *et al.* 1992. Degradation of collagen in the bone-resorbing compartment underlying the osteoclast involves both cysteine-proteinases and matrix metalloproteinases. J. Cell Physiol. **150:** 221–231.

36. EVERTS, V. *et al.* 1998. Cysteine proteinases and matrix metalloproteinases play distinct roles in the subosteoclastic resorption zone. J. Bone Min. Res. **13:** 1420–1430.

37. SORSA, T. *et al.* 1998. Functional sites of chemically-modified tetracyclines: inhibition of the oxidative activation of human neutrophil and chicken osteoclast pro-matrix metalloproteinases. J. Rheumatol. **25:** 975–982.

# Specialized Surface Protrusions of Invasive Cells, Invadopodia and Lamellipodia, Have Differential MT1-MMP, MMP-2, and TIMP-2 Localization

WEN-TIEN CHEN[a] AND JAW-YUAN WANG

*Department of Medicine/Medical Oncology, State University of New York at Stony Brook, Stony Brook, New York 11794-8160, USA*

ABSTRACT: Surface protrusions, invadopodia, and analogous lamellipodia at the leading edge of an invasive cell, which make contact with the underlying extracellular matrix (ECM), are the main motor for cellular locomotion and invasion. Previous studies have demonstrated that invadopodia, but not lamellipodia, are sites of ECM degradation on the cell surface. Such degradative activity is in part due to the localization of latent matrix metalloproteinase–2 (MMP-2) and membrane type–1 MMP (MT1-MMP) to invadopodia, where MMP activation occurs. Although lamellipodia exhibit similar structure and mobility to invadopodia, lamellipodia, by virtue of their location at the cellular periphery, are readily accessible to the soluble tissue inhibitor of matrix metalloproteinase–2 (TIMP-2) and blood-borne inhibitors. We show here that TIMP-2 co-localizes with MT1-MMP and MMP-2 at lamellipodia but not with that of invadopodia. Thus, the MMP-TIMP localization at lamellipodia may be a key mechanism for the regulation of MMP activation on the cell surface, which in turn governs expression of the cell-invasive phenotype.

## MOBILITY OF SURFACE PROTRUSIONS: LAMELLIPODIA AND INVADOPODIA

Lamellipodia are 0.4–1.0-μm-thin, flat protrusions that form at the leading edge of the cultured cell migrating on a planar substratum (FIG. 1). They are considered as a motile organelle of the migratory cell. Mobility of lamellipodia was originally observed in the phenomenon of retraction-induced spreading of embryonic fibroblasts.[1] Retraction of the trailing edge of an embryonic fibroblast results in an abrupt increase in protrusive activity of its lamellipodia. A 10- to 30-fold increase in lamellipodial spreading occurs within 8 seconds after onset of retraction at the trailing edge and then decreases slightly so that by 1 minute the increase in spreading is five- to tenfold. During this period, there is a linear relationship between area increase at the leading edge and area decrease at the trailing edge. This increase was studied with time-lapse cinemicrography and scanning electron microscopy.[2] Increased spreading following retraction results primarily from an increase in the duration of

[a]Address for correspondence: Wen-Tien Chen, Ph.D., Department of Medicine/Medical Oncology, HSC T-16, Rm. 020, SUNY Stony Brook, Stony Brook, New York 11794-8160. Phone, 516/444-6948; 444-7594 (lab); fax, 516/444-2493; e-mail, wchen@mail.som.sunysb.edu

**A. side view**　　　　　　　　　**B. top view**

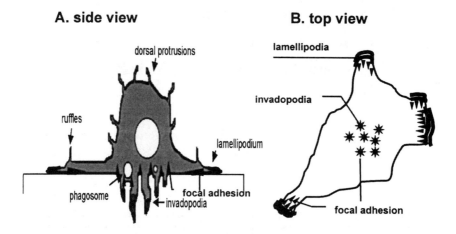

**FIGURE 1.** Highly schematic representation of the positional relationship between focal adhesions and lamellipodia or invadopodia, respectively, in a cell culture model by growing invasive cells on collagenous films. The surface protrusion lamellipodium is shown to extend from its base where there are dark streaks depicting focal adhesion at the leading edge of the invasive cell. The invadopodium forms directly under the cell body and invades the matrix film, where its base is surrounded by the fine streaks of focal adhesion. Upon malignant transformation, the cell becomes invasive: it increases its invadopodia and lamellipodia and decreases its focal adhesions. In cell culture models, lamellipodia are accessible to soluble proteins present in culture medium. It is proposed that soluble protease inhibitors may play an important role in downregulation of proteolytic activity of lamellipodia, hence the cell invasive phenotype. *Dark thick bands* indicate lamellipodia; *dark streaks,* focal adhesion at the base of lamellipodia; *dark dots,* invadopodia; *fine dark streaks,* focal adhesion surrounding invadopodia.

the extension phase of lamellipodial spreading. Much ruffling accompanies this increased spreading, particularly during its earliest phase. Upon retraction of the trailing edge, folds appear on the surface of the retracted tail and adjacent cell body; and, soon after, microvilli-like structures appear as well. Once the moving cell has fully respread, however, the upper surface is once again smooth and free of folds and microvilli. Like retraction of the tail, lamellipodial extension requires an active contraction, associated with a meshwork of microfilaments.[3, 4]

Invadopodia are fine cylindrical protrusions 0.4 μm in diameter that extend from the front of the cell body into the matrix layer (FIG. 1). They are also motile organelles of the invasive cell. Invadopodia were considered originally as membrane protrusions in Rous sarcoma virus (RSV)-transformed cells that drove the locomotion and invasion into the underlying extracellular matrix (ECM), as determined by interference reflection microscopy (IRM) and transmission electron microscopy.[4–6] They were initially termed "rosettes"[7] or "podosomes"[8] owing to their IRM morphology. The invadopodia of transformed fibroblasts form gray IRM rosette-shaped contact sites on planar fibronectin-gelatin substrata.[4–6] These rosette contacts change their relative IRM light intensity and size within 1-min intervals, and the cell surface at the rosettes is protruding and retracting perpendicular to the substratum. Some IRM spots could shift their positions relative to adjacent spots, thus suggesting that invadopodia are able to move laterally on the ventral surface of the cell. These activities often result, within 20 min, in dramatic changes in the morphology of the entire group of rosette contacts. In contrast, adhesive sites, such as focal and close contacts, retain their IRM images for 20 to 40 min, suggesting the stability of adhesion sites.

In contrast, lamellipodia form broad gray, IRM close contacts with the substratum. IRM combined with time-lapse cinemicrography had been used to examine the presence of lamellipodia-close contacts and the speeds of translocation for a variety of cell types.[3,4,9] Rapid translocation of amphibian leukocytes (average speed = 9.0 μm/min), amphibian epidermal cells (7 μm/min), teleost epidermal cells (7 μm/min), freshly seeded (2 hr) chick heart fibroblasts (moving 1–3 μm/min), RSV-transformed chick heart fibroblasts (moving 3 μm/min), the rapidly advancing (1–5 μm/min) margin of spreading human WI-38 fibroblasts, and isolated MDCK canine epithelial cells (0.5–1.0 μm/min), was found to correlate with the presence of lamellipodia-close contacts at the leading edge of the cell. Conversely, numerous dark streaks of focal contact were found associated with the slow rate of translocation displayed by older cultures (72 hr) of chick fibroblasts (less than 0.1 μm/min), well-spread WI-38 cells (less than or equal to 0.3 μm/min), and confluent MDCK cells (less than 0.01 μm/min).[4,9–13] It is concluded that close contacts formed under invadopodia and lamellipodia, but not focal contacts underlying adhesion sites, are associated with rapid cellular translocation, and that the build-up of focal contacts is associated with reduced cellular translocation and maintenance of the spread cell shape. Thus, invadopodia and lamellipodia are more labile than other ECM contact sites, such as focal adhesion and focal contact.

## STRUCTURE OF PROTRUSIONS:
## THEIR RELATIONSHIP WITH FOCAL ADHESIONS

Using a combination of approaches including IRM, frozen thin-sectioning, transmission electron microscopy, subcellular fractionation, and immunofluorescence, it was demonstrated that invadopodia and lamellipodia represent specialized surface protrusions that are filled with meshwork of microfilaments, while focal adhesions are stable membrane structures, which serve as termini of microfilament bundles.[4,10–18] However, protrusion and focal adhesion structures contain similar cytoskeleton-associated proteins including F-actin, vinculin, α-actinin, cortactin, tensin, and talin, as well as signal-transducing elements such as tyrosine kinases.[6,17,19]

Significantly, when specific anti-phosphotyrosine antibodies were used to detect tyrosine-phosphorylated proteins in frozen thin-sections of RSV-transformed cells actively invading fibronectin-coated gelatin beads, invasive transformed cells have seven-fold higher tyrosine phosphorylation than untransformed cells, which exhibit only focal adhesions.[17] Both tyrosine phosphorylation and motile activities of the invadopodia were inhibited by genistein, an inhibitor of tyrosine-specific kinases, but those of focal adhesions were not. To further analyze this process, the subcellular fractionation technique was developed to generate a surface membrane fraction contacting collagenous matrix that contains either highly enriched invadopodial or lamellipodial membranes from RSV-transformed chicken embryonic cells.[17] Four major tyrosine-phosphorylated proteins (150 kDa, 130 kDa, 81 kDa, and 77 kDa) were identified from isolated invadopodial fraction and are responsible for the intense tyrosine-phosphorylated protein labeling associated with invadopodia or lamellipodia extending into sites of matrix degradation. The 150-kDa protein may be specific for chicken cell invasiveness since it is approximately 3.6-fold enriched in the invadopodial fraction relative to the cell body fraction and is not observed in focal contacts. These results suggest that the enhanced phosphorylation of a specific tyrosine-containing protein by pp60$^{\text{v-src}}$ may directly contribute to the formation and function of invadopodia or lamellipodia.

Invasion of malignant human melanoma and breast carcinoma cells also involves formation of cell surface invadopodia and lamellipodia and reduction of focal adhesions.[18,20–22] Proteolytic fragments of ECM components have been implicated as a factor that promotes cell invasiveness. For example, one active peptide, IKVAV (Ile-Lys-Val-Ala-Val), of laminin, a major basement membrane glycoprotein, is located on the C-terminus of the long arm of the laminin α1 chain and promotes cellular attachment, migration, tumor growth, and metastasis.[23–26] Also, a 140-kDa matrix protein, ladsin, which appears to be identical to the laminin β2 chain, has been shown to promote scattering of carcinoma cells, stimulate cell migration, and bind to α3β1 integrin.[27,28] It is postulated that laminin can activate integrin-mediated cell adhesion and laminin fragments can transduce biochemical signals through integrin inside the cell to modulate the surface activities of degradation and invasion.[29,30]

Experimentation was designed to determine potential laminin fragments that may promote cell invasion[30]: a systematic screening using the cell invasion model was made with 113 overlapping synthetic peptide-beads covering the laminin α1 chain carboxyl-terminal globular domain (G domain amino acid residues 2111-3060).[18,31,32] It was demonstrated that the peptides designated AG-10 (NPWHSIY-ITRFG) and AG-32 (TWYKIAFQRNRK) ligated α6β1 integrin in a similar manner as anti-β1 and α6 integrin antibodies to promote melanoma invasiveness independently of the adhesion function of integrin receptors. Accompanying the induced invasiveness, there was an increased organization of seprase on invadopodia, but the expression levels of seprase, MMP-2, and β1 integrins were not altered.[30] Also, the anti-α3 antibodies and the laminin peptide HGD-6 activate the α3β1 integrin, which results in a downstream signaling cascade stimulating phagocytosis.[18] Recently, it was demonstrated that this induced invasion involves an increase in tyrosine phosphorylation of a 190-kDa signal component associated with GTPase-activating protein for Rho family members (p190RhoGAP; p190) and membrane-protrusive activities at invadopodia.[33] Activated Src may associate with the state of tyrosine phosphorylation of a major RasGAP-associated protein p190 RhoGAP (p190),[34–36] which in turn creates a binding site for the SH2 domains of other signaling molecules to promote actin cytoskeletal motility. p190 binds GTP and forms a stable association with GTPase-activating protein (GAP) that downregulates Rho and Rac activity.[35,37,38] It was concluded that activation of the α6β1 integrin signaling regulates the tyrosine phosphorylation state of p190, which in turn connects downstream signaling pathways through Rho family GTPases to actin cytoskeleton in invadopodia, thus promoting membrane-protrusive and -degradative activities necessary for cell invasion.[33]

On the basis of the results obtained from morphologic studies involving IRM, transmission electron microscopy, frozen thin-sectioning, and immunofluorescence, we have observed a close positional relationship between focal adhesions and lamellipodia or invadopodia, respectively, in invasive cells grown on collagenous films (FIG. 1). Lamellipodia appear to extend from streaks of focal adhesions at the leading edge of the cell migrating on the plane of the film, while invadopodia form directly under the cell body and invade into the film where their base is surrounded by fine streaks of focal adhesions. Although it remains to be tested, such close positioning may bring together cytoskeletal and signaling components necessary for formation of surface protrusions. Another possibility is the involvement of specific integrins for formation of focal adhesions and protrusions, respectively. The latter has been supported by a recent finding that $\alpha 3\beta 1$ integrin co-localizes with seprase in the same invadopodia and lamellipodia.[50] However, the $\alpha 5\beta 1$ integrin, a fibronectin receptor, localizes in focal adhesions and fine adhesion structures that surround the base of invadopodia (FIG. 1). The appearance of these fine adhesion structures containing $\alpha 5\beta 1$ was dependent upon the ability of LOX human melanoma cells to extend invadopodia, and thus was more prominent in cells cultured on thicker matrices. It is postulated that $\alpha 3\beta 1$ integrin is a protease-docking protein for directing seprase to invadopodia, and $\alpha 5\beta 1$ may participate in the adhesion process necessary for formation and extension of invadopodia.

## ECM-DEGRADING ACTIVITIES ON THE CELL SURFACE

An *in vitro* model has been established to identify invadopodia that involves growing cells on a cross-linked gelatin or collagen substratum: fluorescence-labeled or radio-labeled fibronectin covalently linked to the surface of a fixed gelatin film[4–6,39] or bead.[15–17,40] This method not only allows us to study cell surface structures during the initial stages of invasion (within 1–6 hours of cells making contact with the matrix), but also demonstrates the existence of cell surface proteases that may be involved in the local degradation of matrix components at cell contact sites by RSV-transformed cells and human cancer cell lines as well.[20,22,30,41] Invadopodia degrade a variety of immobilized substrates including fibronectin, laminin, type I collagen, type IV collagen, basement membrane-like material produced by tumor (Matrigel), and cell-free matrix produced by embryonic cells in a time- and cell-contact-dependent manner.[21,42,43] Several membrane-bound proteases are implicated in invadopodia, including high molecular weight, membrane-bound, serine- and metallo proteinases from RSV-transformed cells,[5,44] seprase,[20,21,42,43] membrane-bound MMP-2,[21,45] and MT1-MMP.[41] Thus, invadopodia have the biological activity of degrading ECM components.

Although lamellipodia share similar biological features with invadopodia as described above, the original cell culture model for invasion detected very little proteolytic activity at lamellipodia. However, when a moderately invasive RPMI7951 human melanoma cell line overexpresses MT1-MMP in lamellipodia, surface proteolytic activity has been identified both at lamellipodia and at invadopodia.[41] This is an important result, demonstrating that overbalancing by MT1-MMP of its inhibitor at lamellipodia can generate the observed ECM-degrading activity. In addition, it was demonstrated that the transmembranous/cytoplasmic domain of MT1-MMP ($TM_{MT\text{-}MMP,553}$AVGLAVFFFRRHGTPRRLLYCQRSLLDKV$_{582}$) was required for lamellipodial and invadopodial localization of the enzyme and therefore for directing cellular invasion, as neither occurred with MT-MMP mutants lacking the $TM_{MT\text{-}MMP}$ ($\Delta TM_{MT\text{-}MMP}$) or containing the $TM_{IL\text{-}2R}$ ($\Delta TM_{MT\text{-}MMP}/TM_{IL\text{-}2R}$). Furthermore, in the embryonic fibroblast line WI-38, which lacks invadopodia, the lamellipodia exhibited collagen-degrading activities and contained two serine integral membrane proteases, seprase and dipeptidyl peptidase IV (DPPIV).[51] Also, in invasive human breast carcinoma lines, MDA-MB-346 and Hs578T, MT1-MMP, seprase and DPPIV were found to concentrate at both invadopodia and lamellipodia.[52] We conclude that protease profiles in lamellipodia and in invadopodia are identical, but proteolytic activities at lamellipodia are suppressed in the cell culture model used.

## MMP INHIBITOR LOCALIZATION AT LAMELLIPODIA

Previous studies showed the localization of surface MMP including MT1-MMP and MMP-2 at invadopodia, but failed to identify any inhibitor at such site.[21,41] In contrast, lamellipodia are protrusions located at the cellular periphery and are readily accessible to soluble protease inhibitors. The possibility that tissue inhibitor of matrix metalloproteinase-2 (TIMP-2) is located on the lamellipodia surface was investigated using immunofluorescence on Hs578T cells growing on cross-linked gelatin films (FIG. 2). Hs578T cells invading the gelatin films were labeled with rabbit

---

**FIGURE 2.** Immunofluorescent localization of MT1-MMP, MMP-2, TIMP-2, and control membrane glycoprotein p90 at lamellipodia of invasive breast carcinoma cells growing on cross-linked gelatin films. *Arrows* in individual panels indicate sites of lamellipodia, and *open arrows* show other membrane folds at the leading edge of the cells. *Top panels:* phase-contrast image (*left*), MT1-MMP immunofluorescence (*middle*), and superimposed composite of the left and middle image (*right*) of Hs578T grown on the cross-linked gelatin film. Cells were stained for MT1-MMP using rabbit polyclonal antibody directed against MT1-MMP peptide, and areas of MT1-MMP distribution were visualized by epifluorescence. Both lamellipodia (*arrow in box*) and invadopodia (under the cell body) are stained with anti-MT1-MMP. *Bottom panels:* phase-contrast image and immunofluorescence of MMP-2 (*left*), TIMP-2 (*middle*), and p90 (*right*). Lamellipodia are brightly labeled with MMP-2, TIMP-2, and p90 (*arrows*), while other ruffling membranes are only stained with p90 but lack MMP-2 and TIMP-2 (*open arrows*). X600.

MATERIALS AND METHODS: Hs578T breast carcinoma cells were purchased from the American Type Culture Collection (Rockville, MD) and cultured on cross-linked gelatin films, fixed, and immunolabeled as previously described.[41] Rabbit polyclonal antibody raised against the peptide CDGNFDTVAMLRGEM (residues 310-333) of MT1-MMP was kindly provided by Dr. Hiroshi Sato (Kanazawa, Japan) and mouse mAb Ab45 from Bill Stetler-Stevenson (NIH, Bethesda). Mouse antibody T2-101 against TIMP-2 was purchased from Oncogene Research (Cambridge, MA). MAb C37 was directed against a p90 glycoprotein that was co-purified with human melanoma seprase during immunoaffinity purification, and it localizes exclusively on the plasma membrane.

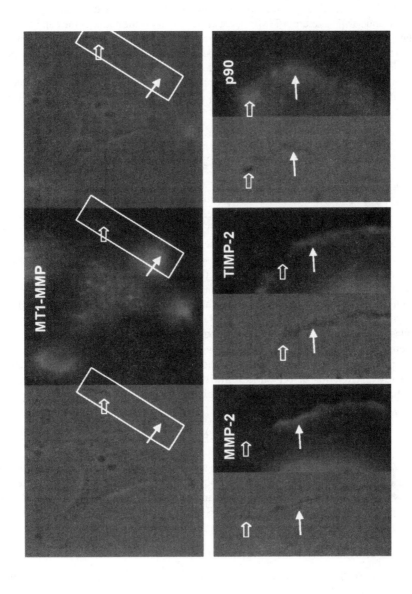

polyclonal antibodies directed against MT1-MMP, anti-MMP-2 mAb Ab45, anti-TIMP-2 mAb T2-101, or control anti-p90 mAb C37. FIGURE 2 shows that the lamellipodia are brightly labeled with MT1-MMP, MMP-2, TIMP-2, and p90 (*arrows*), while other ruffling membranes are only stained with p90 but lack MT1-MMP, MMP-2, and TIMP-2 (*open arrows*). In this study and those of others,[21,41] MT1-MMP and MMP-2, but not TIMP-2, are enriched on invadopodia of the cell invading the substratum. mAb C37 directed against a cell surface glycoprotein was not enriched on lamellipodia or invadopodia, but was enriched on dorsal membrane protrusions. We conclude that TIMP-2 co-localizes with MT1-MMP and MMP-2 at lamellipodia but not at invadopodia. Thus, the MMP-TIMP localization at lamellipodia may be a key mechanism for the regulation of MMP activation on the cell surface, which in turn governs expression of the cell-invasive phenotype.

## PROSPECTIVE: REGULATION OF THE CELL-INVASIVE PHENOTYPE BY SOLUBLE PROTEASE INHIBITORS

Cancer cells do not express all key proteases necessary for their invasive potential. In fact, carcinoma and melanoma cells in culture are able to recruit proteases and major protease inhibitors from serum.[45–47] Like the blood-clotting cascade, remodeling of the extracellular matrix at the invasion front is governed by highly regulated cascades of proteolytic enzymes and inhibitors present in tissue fluid and serum.[48,49] Serine proteases (e.g., uPA, tPA, plasmin) and metalloproteases (e.g., collagenases, stromelysins, gelatinases) as well as their respective endogenous inhibitors (e.g., PAI-1, PAI-2, and $\alpha$2-antiplasmin for the former, and TIMP-1, TIMP-2, and $\alpha$2-macroglobulin for the latter) are examples of such soluble molecules. Since surface-protrusion lamellipodia and invadopodia are both dynamic and proteolytic, they may recruit soluble collagenases, such as MMP-1 and MMP-2, via receptor integral membrane proteases and activate the matrix degradation. Subsequently, inactivation of cell surface MMP may involve soluble protease inhibitors including TIMP and serine protease inhibitors.

## ACKNOWLEDGMENTS

We are indebted to Leslie A. Goldstein and Stanley Zucker for helpful discussions and critical review of the manuscript and to Yunyun Yeh for volunteer assistance. This work was supported by USPHS Grant R01 CA-39077 and by a Susan Komen Breast Cancer Foundation Translational Research Award (to W.-T.C.).

## REFERENCES

1. CHEN, W.-T. 1979. Induction of spreading during fibroblast movement. J. Cell Biol. **81:** 684–691.
2. CHEN, W.-T. 1981. Surface changes during retraction-induced spreading of fibroblasts. J. Cell Sci. **49:** 1–13.
3. CHEN, W.-T. 1981. Mechanism of retraction of the trailing edge during fibroblast movement. J. Cell Biol. **90:** 187–200.

4. CHEN, W.-T. 1989. Proteolytic activity of specialized surface protrusions formed at rosette contact sites of transformed cells. J. Exp. Zool. **251:** 167–185.

5. CHEN, W.-T., K. OLDEN, B.A. BERNARD & F.F. CHU. 1984. Expression of transformation-associated protease(s) that degrade fibronectin at cell contact sites. J. Cell Biol. **98:** 1546–1555.

6. CHEN, W.-T., J.M. CHEN, S.J. PARSONS & J.T. PARSONS. 1985. Local degradation of fibronectin at sites of expression of the transforming gene product pp60src. Nature **316:** 156–158.

7. DAVID-PFEUTY, T. & S.J. SINGER. 1980. Altered distributions of the cytoskeletal proteins vinculin and alpha-actinin in cultured fibroblasts transformed by Rous sarcoma virus. Proc. Natl. Acad. Sci. USA **77:** 6687–6691.

8. TARONE, G., D. CIRILLO, F.G. GIANCOTTI, P.M. COMOGLIO & P.C. MARCHISIO. 1985. Rous sarcoma virus-transformed fibroblasts adhere primarily at discrete protrusions of the ventral membrane called podosomes. Exp. Cell Res. **159:** 141–157.

9. KOLEGA, J., M.S. SHURE, W.-T. CHEN & N.D. YOUNG. 1982. Rapid cellular translocation is related to close contacts formed between various cultured cells and their substrata. J. Cell Sci. **54:** 23–34.

10. CHEN, W.-T. & S.J. SINGER. 1982. Immunoelectron microscopic studies of the sites of cell- substratum and cell-cell contacts in cultured fibroblasts. J. Cell Biol. **95:** 205–222.

11. CHEN, W.-T., J.M. GREVE, D.I. GOTTLIEB & S.J. SINGER. 1985. Immunocytochemical localization of 140 kD cell adhesion molecules in cultured chicken fibroblasts, and in chicken smooth muscle and intestinal epithelial tissues. J. Histochem. Cytochem. **33:** 576–586.

12. CHEN, W.-T., E. HASEGAWA, T. HASEGAWA, C. WEINSTOCK & K.M. YAMADA. 1985. Development of cell surface linkage complexes in cultured fibroblasts. J. Cell Biol. **100:** 1103–1114.

13. Chen, W.-T., J. Wang, T. Hasegawa, S.S. Yamada & K.M. Yamada. 1986. Regulation of fibronectin receptor distribution by transformation, exogenous fibronectin, and synthetic peptides. J. Cell Biol. **103:** 1649–1661.

14. CHEN, W.-T. & S.J. SINGER. 1980. Fibronectin is not present in the focal adhesions formed between normal cultured fibroblasts and their substrata. Proc. Natl. Acad. Sci. USA **77:** 7318–7322.

15. MUELLER, S.C., T. KELLY, M.Z. DAI, H.N. DAI & W.-T. CHEN. 1989. Dynamic cytoskeleton-integrin associations induced by cell binding to immobilized fibronectin. J. Cell Biol. **109:** 3455–3464.

16. MUELLER, S.C. & W.-T. CHEN. 1991. Cellular invasion into matrix beads: localization of beta 1 integrins and fibronectin to the invadopodia. J. Cell Sci. **99:** 213–225.

17. MUELLER, S.C., Y. YEH & W.-T. CHEN. 1992. Tyrosine phosphorylation of membrane proteins mediates cellular invasion by transformed cells. J. Cell Biol. **119:** 1309–1325.

18. COOPMAN, P.J., D.M. THOMAS, K.R. GEHLSEN & S.C. MUELLER. 1996. Integrin $\alpha 3 \beta 1$ participates in the phagocytosis of extracellular matrix molecules by human breast cancer cells. Mol. Biol. Cell. **7:** 1789–1804.

19. BURRIDGE, K., K. FATH, T. KELLY, G. NUCKOLLS & C. TURNER. 1988. Focal adhesions: transmembrane junctions between the extracellular matrix and the cytoskeleton. Annu. Rev. Cell Biol. **4:** 487–525.

20. AOYAMA, A. & W.-T. CHEN. 1990. A 170-kDa membrane-bound protease is associated with the expression of invasiveness by human malignant melanoma cells. Proc. Natl. Acad. Sci. USA **87:** 8296–8300.

21. MONSKY, W.L., C.-Y. LIN, A. AOYAMA, T. KELLY, S.C. MUELLER, S.K. AKIYAMA & W.-T. CHEN. 1994. A potential marker protease of invasiveness, seprase, is localized on invadopodia of human malignant melanoma cells. Cancer Res. **54:** 5702–5710.

22. CHEN, W.-T., C.C. LEE, L. GOLDSTEIN, S. BERNIER, C.H. LIU, C.Y. LIN, Y. YEH, W.L. MONSKY, T. KELLY, M. DAI & S.C. MUELLER. 1994. Membrane proteases as potential diagnostic and therapeutic targets for breast malignancy. Breast Cancer Res. Treat. **31:** 217–226.

23. KANEMOTO, T., R. REICH, L. ROYCE, D. GREATOREX, S.H. ADLER, N. SHIRAISHI, G.R. MARTIN, Y. YAMADA & H.K. KLEINMAN. 1990. Identification of an amino acid sequence from the laminin A chain that stimulates metastasis and collagenase IV production. Proc. Natl. Acad. Sci. USA **87:** 2279–2283.

24. MACKAY, A.R., D.E. GOMEZ, A.M. NASON & U.P. THORGEIRSSON. 1994. Studies on the effects of laminin, E-8 fragment of laminin and synthetic laminin peptides PA22-2 and YIGSR on matrix metalloproteinases and tissue inhibitor of metalloproteinase expression [see comments]. Lab. Invest. **70:** 800–806.

25. STACK, M.S., R.D. GRAY & S.V. PIZZO. 1993. Modulation of murine B16F10 melanoma plasminogen activator production by a synthetic peptide derived from the laminin A chain. Cancer Res. **53:** 1998–2004.

26. YAMAMURA, K., M.C. KIBBEY & H.K. KLEINMAN. 1993. Melanoma cells selected for adhesion to laminin peptides have different malignant properties. Cancer Res. **53:** 423–428.

27. MIYAZAKI, K., Y. KIKKAWA, A. NAKAMURA, H. YASUMITSU & M. UMEDA. 1993. A large cell-adhesive scatter factor secreted by human gastric carcinoma cells. Proc. Natl. Acad. Sci. USA **90:** 11767–11771.

28. KIKKAWA, Y., M. UMEDA & K. MIYAZAKI. 1994. Marked stimulation of cell adhesion and motility by ladsin, a laminin-like scatter factor. J. Biochem. (Tokyo) **116:** 862–869.

29. SEFTOR, R.E.B., E.A. SEFTOR, W.G. STETLER-STEVENSON & M.J.C. HENDRIX. 1993. The 72 kDa type IV collagenase is modulated via differential expression of $\alpha_v\beta_3$ and $\alpha_5\beta_1$ integrins during human melanoma cell invasion. Cancer Res. **53:** 3411–3415.

30. NAKAHARA, H., M. NOMIZU, S.K. AKIYAMA, Y. YAMADA, Y. YEH & W.T. CHEN. 1996. A mechanism for regulation of melanoma invasion. Ligation of alpha6beta1 integrin by laminin G peptides. J. Biol. Chem. **271:** 27221–27224.

31. NOMIZU, M., W.H. KIM, K. YAMAMURA, A. UTANI, S.Y. SONG, A. OTAKA, P.P. ROLLER, H.K. KLEINMAN & Y. YAMADA. 1995. Identification of cell binding sites in the laminin alpha 1 chain carboxyl-terminal globular domain by systematic screening of synthetic peptides. J. Biol. Chem. **270:** 20583–20590.

32. MATTER, M.L. & G.W. LAURIE. 1994. A novel laminin E8 cell adhesion site required for lung alveolar formation in vitro. J. Cell Biol. **124:** 1083–1090.

33. NAKAHARA, H., S.C. MUELLER, M. NOMIZU, Y. YAMADA, Y. YEH & W.T. CHEN. 1998. Activation of beta1 integrin signaling stimulates tyrosine phosphorylation of p190RhoGAP and membrane-protrusive activities at invadopodia. J. Biol. Chem. **273:** 9–12.

34. CHANG, J.H., L.K. WILSON, J.S. MOYERS, K. ZHANG & S.J. PARSONS. 1993. Increased levels of p21ras-GTP and enhanced DNA synthesis accompany elevated tyrosyl phosphorylation of GAP-associated proteins, p190 and p62, in c-src overexpressors. Oncogene **8:** 959–967.

35. FOSTER, R., K.Q. HU, D.A. SHAYWITZ & J. SETTLEMAN. 1994. p190 RhoGAP, the major RasGAP-associated protein, binds GTP directly. Mol. Cell Biol. **14:** 7173–7181.

36. CHANG, J.H., S. GILL, J. SETTLEMAN & S.J. PARSONS. 1995. c-Src regulates the simultaneous rearrangement of actin cytoskeleton, p190RhoGAP, and p120RasGAP following epidermal growth factor stimulation. J. Cell Biol. **130:** 355–368.

37. BOUTON, A.H., S.B. KANNER, R.R. VINES, H.C. WANG, J.B. GIBBS & J.T. PARSONS. 1991. Transformation by pp60src or stimulation of cells with epidermal growth factor induces the stable association of tyrosine- phosphorylated cellular proteins with GTPase-activating protein. Mol. Cell Biol. **11:** 945–953.

38. SETTLEMAN, J., C.F. ALBRIGHT, L.C. FOSTER & R.A. WEINBERG.1992. Association between GTPase activators for Rho and Ras families. Nature **359:** 153–154.

39. CHEN, W.-T., Y. YEH & H. NAKAHARA. 1994. An *in vitro* cell invasion assay: determination of cell surface proteolytic activity that degrades extracellular matrix. J. Tiss. Cult. Meth. **16:** 177–181.

40. MUELLER, S.C. & J.-Y. ZHOU. 1994. Crosslinked gelatin beads: a culture method for the study of cell adhesion and invasion. J. Tiss. Cult. Meth. **16:** 183–188.

41. NAKAHARA, H., L. HOWARD, E.W. THOMPSON, H. SATO, M. SEIKI, Y. YEH & W.-T. CHEN. 1997. Transmembrane/cytoplasmic domain-mediated membrane type 1-matrix metalloprotease docking to invadopodia is required for cell invasion. Proc. Natl. Acad. Sci. USA **94:** 7959–7964.

42. KELLY, T., S.C. MUELLER, Y. YEH & W.-T. CHEN. 1994. Invadopodia promote proteolysis of a wide variety of extracellular matrix proteins. J. Cell Physiol. **158:** 299–308.

43. KELLY, T., S. KECHELAVA, T.L. ROZYPAL, K.W. WEST & S. KOROURIAN. 1998. Seprase, a membrane-bound protease, is overexpressed by invasive ductal carcinoma cells of human breast cancers. Mod. Pathol. **11:** 855–863.

44. CHEN, J.M. & W.-T. CHEN. 1987. Fibronectin-degrading proteases from the membranes of transformed cells. Cell **48:** 193–203.

45. MONSKY, W.L., T. KELLY, C.Y. LIN, Y. YEH, W.G. STETLER-STEVENSON, S.C. MUELLER & W.-T. CHEN. 1993. Binding and localization of M(r) 72,000 matrix metalloproteinase at cell surface invadopodia. Cancer Res. **53:** 3159–3164.

46. AZZAM, H.S., G. ARAND, M.E. LIPPMAN & E.W. THOMPSON. 1993. Association of MMP-2 activation potential with metastatic progression in human breast cancer cell lines independent of MMP- 2 production. J. Natl. Cancer Inst. **85:** 1758–1764.

47. SATO, H., T. TAKINO, Y. OKADA, J. CAO, A. SHINAGAWA, E. YAMAMOTO & M. SEIKI. 1994. A matrix metalloproteinase expressed on the surface of invasive tumour cells [see comments]. Nature **370:** 61–65.

48. BIRKEDAL-HANSEN, H., W.G.I. MOORE, M.K. BODDEN, L.J. WINDSOR, B. BIRKEDAL-HANSEN, A. DeCARLO & J.A. ENGLER. 1993. Matrix metalloproteinases: a review. Crit. Rev. Oral Biol. Med. **4:** 197–250.

49. WOESSNER, J.F., JR. 1995. Quantification of matrix metalloproteinases in tissue samples. Methods Enzymol. **248:** 510–528.

50. MUELLER, S.C., G. GHERSI, S. AKIYAMA, Q.-Z. SANG, H. NAKAHARA, L. HOWARD, M. PIÑEIRO-SÁNCHEZ, YUNYUN YEH & W.-T. CHEN. 1999. A novel protease-docking function of integrin. In preparation.

51. GHERSI, G., L.A. GOLDSTEIN, LARI HAKKINEN, HANNU LARJAVA, M. SALAMONE, J.-Y. WANG, Y. YEH & W.-T. CHEN. 1999. Wound-activated proteases: seprase and dipeptyl peptidase IV (DPPIV) are involved in cell migration and collagen degradation by human fibroblasts. In preparation.

52. WANG, J.-Y., L.A. GOLDSTEIN, T. YAMANE, Y. MORI, A.-K. NG, M. JONES, M. PIÑEIRO-SÁNCHEZ, Y. YEH & W.-T. CHEN. 1999. Seprase is a cell activation marker for invasiveness of both tumor and stromal cells in human breast ductal carcinoma. In preparation.

# Activation of ProMMP-9 by a Plasmin/MMP-3 Cascade in a Tumor Cell Model

## Regulation by Tissue Inhibitors of Metalloproteinases

ELIZABETH HAHN-DANTONA,[a] NOEMI RAMOS-DeSIMONE,[a] JOHN SIPLEY,[a] HIDEAKI NAGASE,[b] DEBORAH L. FRENCH,[c] AND JAMES P. QUIGLEY[a,d]

[a]Department of Pathology, State University of New York at Stony Brook, Stony Brook, New York 11794-8691, USA

[b]Department of Biochemistry and Molecular Biology, University of Kansas Medical Center, Kansas City, Kansas 66160-7421, USA

[c]Department of Medicine, The Mount Sinai Medical Center, New York, New York 10029-6574, USA

ABSTRACT: To examine MMP-9 activation in a cellular setting we employed cultures of human tumor cells that were induced to produce MMP-9 over a 200-fold concentration range (0.03 to 8.1 nM). The secreted levels of TIMPs in all the induced cultures remain relatively constant at 1–4 nM. Quantitation of the zymogen/active enzyme status of MMP-9 in the cultures indicates that even in the presence of potential activators, the molar ratio of endogenous MMP-9 to TIMP dictates whether proMMP-9 activation can progress. When the MMP-9/TIMP ratio exceeds 1.0, MMP-9 activation progresses, but only via an interacting protease cascade involving plasmin and stromelysin 1 (MMP-3). Plasmin, generated by the endogenous plasminogen activator (uPA), is not an efficient activator of proMMP-9. Plasmin, however, is very efficient at generating active MMP-3 from exogenously added proMMP-3. The activated MMP-3, when its concentration exceeds that of TIMP, becomes a potent activator of proMMP-9. Addition to the cultures of already-activated MMP-3 relinquishes the requirement for plasminogen and proMMP-3 additions and results in direct activation of the endogenous proMMP-9. The activated MMP-9 enhances the invasive phenotype of the cultured cells as their ability to transverse basement membrane is significantly increased following zymogen activation. That this enhanced tissue remodeling capability is due to the activation of MMP-9 is demonstrated through the use of a specific anti-MMP-9–blocking monoclonal antibody.

## INTRODUCTION

The extracellular matrix (ECM) is a complex combination of collagens, glycoproteins, glycosaminoglycans, and proteoglycans that forms barriers and provides structural support for tissues and organs in multicellular organisms.[1] Degradation of the ECM is a critical feature of cancer cell invasion, and there are several groups of

[d]Present address of corresponding author: James P. Quigley, Department of Vascular Biology, The Scripps Research Institute, 10550 North Torrey Pines Road, La Jolla, California 92037. Phone, 619/784-7108; e-mail, jquigley@scripps.edu

enzymes that are capable of modifying or degrading ECM components. One class of proteases that has the capability to cleave all the proteinaceous components of the ECM is the matrix metalloproteases (MMPs). Specific inhibitors of MMPs can block tumor cell invasion and metastasis in model systems.[2–4] These studies imply that MMPs play a role in tumor progression. However, the specific role for individual MMPs in this process is still unclear.

Most of the MMPs are secreted as inactive zymogens and can undergo proteolytic cleavage to become active enzymes. Therefore, MMP activation plays a crucial role in regulating MMP function. Activation of latent or proMMPs occurs through the disruption of the bond between a conserved unpaired cysteine in the amino terminal portion of the enzyme and the catalytic zinc ion in the active site.[5] This activation by proteolytic cleavage can occur intracellularly,[6] or at the cell surface.[7,8] However, the majority of the MMPs are capable of being activated by soluble proteolytic enzymes.[9–12] Once activated, the MMPs are inhibited very efficiently by the endogenous tissue inhibitors of metalloproteinases, TIMPs, with $K_i$ values in the range of $10^{-10}$ M.[13] TIMP expression is widespread and often the molar levels are equal or greater than that of MMPs.[14] ProMMP-2 and proMMP-9 are often found complexed to TIMP-2 and TIMP-1 respectively[15–17] adding another level of regulation to the catalytic ability of these enzymes.

Upregulation of MMP expression by tumor cells overcomes some of the regulatory controls that block MMP activation and activity. One MMP that is upregulated in many tumor cell systems is MMP-9.[14] Although structurally and catalytically similar to MMP-2, MMP-9 does not become activated under conditions where coexpressed MMP-2 is activated by cell-bound components.[18,19] The regulation of MMP-9 expression is also vastly different from that of MMP-2, as MMP-9 is regulated by cytokines[20,21] and growth factors,[22] and MMP-2 is often constituitively expressed and unresponsive to the same cytokines and growth factors. Upregulation of MMP-9 expression has been linked to many physiological and pathologic processes including bone resorption,[23] inflammation,[24,25] and tumor cell invasion.[26–29] More recent studies have shown that MMP-9 is directly involved in tumor metastasis[30,31] and in tumor cell intravasation.[32] The *in vivo* mechanism of MMP-9 activation has only recently begun to be defined. MMP-3 appears to be the most efficient activator of proMMP-9[10] and may be an *in vivo* activator. However, MMP-3 may not always be present in tissues and organs where MMP-9 is expressed. Even if MMP-3 is available, it is also produced as a zymogen and therefore requires activation. Furthermore, once MMP-3 is activated, the enzyme becomes susceptible to inhibition by tissue inhibitors of metalloproteinases (TIMPs). Therefore, the inhibitory potential of the TIMPs must be evaded to allow activation of MMP-9.

To demonstrate a possible tumor-associated mechanism for the activation of MMP-9 and the circumvention of TIMP-mediated control, we examined cultures of a breast carcinoma cell line, MDA-MB-231, which expresses MMP-9, TIMP-1 and TIMP-2. These cultures do not appear to express MMP-2, another potent gelatinase. Therefore, we could examine MMP-9 activation by monitoring the generation of specific gelatinase activity without any interference from active MMP-2. Using MDA-MB-231 cells, our laboratory has completed a study demonstrating that tumor cell–derived MMP-9 can be activated by MMP-3. This activation occurs through a proteolytic cascade beginning with uPA conversion of plasminogen to plasmin and subsequent activation of proMMP-3 by the newly generated plasmin. The activated MMP-9 contributes significantly to the basement membrane invasion exhibited by these tumor cells. The specificity of this mechanism is verified using neutralizing anti-MMP-9 monoclonal antibodies.

## MATERIALS AND METHODS

### Purified Proteins

Full-length proMMP-3 and the activated C-terminal deletion of human MMP-3 ($\Delta$C-MMP-3) was obtained as previously described.[33] Recombinant human TIMP-1 and proMMP-9 were generous gifts from Dr. Rafael Fridman (Department of Pathology and the Karmanos Cancer Institute, Wayne State University, Detroit, MI). Plasminogen was obtained from Boehringer Manneheim (Indianapolis, IL). Purified anticatalytic anti-human MMP-9 monoclonal antibodies were produced as described.[34]

### Cell Culture

MDA-MMP-9 cells were produced as described.[11] MDA-MB-231 and MDA-MMP-9 cells were cultured at 37°C in 7% $CO_2$ in DMEM (GIBCO BRL, Grand Island, NY) containing 10% heat-inactivated fetal bovine serum (FBS, Hyclone, UT) supplemented with 2 mM glutamine, 1 mM sodium pyruvate, 100 U/ml penicillin, and 100 µg/ml streptomycin (GIBCO BRL). MDA-MMP-9 cells were plated and grown to confluence in 100-mm dishes ($1.5–2 \times 10^7$ cells/plate) using the growth media described above. Each plate was washed twice with serum free DMEM with aprotinin (40 µg/ml) and once with serum-free DMEM alone. The cells were incubated in 5-ml serum-free DMEM with or without 100 ng/ml PMA, as indicated for 20 hours at 37°C. At the end of this time, the conditioned media were collected and centrifuged (3000 rpm $\times$ 10 min) to remove cells and cell debris.

### Basement Membrane Invasion Assay Using Matrigel

Matrigel-coated filter inserts (8-µm pore size) that fit into 24-well invasion chambers were obtained from Becton Dickinson (Bedford, MA). Cells were added to the upper compartment of the chamber as described.[11] After incubation the filter inserts were removed from the wells and the cells on the upper side of the filter were removed using cotton swabs. The filters were fixed, mounted, and stained according to the manufacturer's instructions (Becton Dickinson). The cells that had invaded through the Matrigel and were located on the underside of the filter were photographed and counted.

### Enzyme Activity in Solution

Gelatinase activity was measured in solution using heat-denatured [3]H-acetylated type I collagen purified from rat-tails.[35] Conditioned medium or purified reagents were incubated with the labeled gelatin (20 µg/ml, 2000 cpm/µg) in buffer (50 mM Tris, pH 7.5; 200 nM NaCl; 10 mM $CaCl_2$; 0.05% Brij-35; Sigma, St. Louis, MO) for 48 hours or as indicated followed by trichloroacetic acid precipitation, as previously described.[36]

## *Substrate Zymography*

Gelatin substrate zymography was performed using 8% SDS-polyacrylamide gels copolymerized with gelatin, as previously described.[19] Following electrophoresis, the gels were washed for 1 hour in 2.5% Triton X-100 (Sigma) and incubated overnight in buffer (50 mM Tris, pH 7.5; 200 nM NaCl; 10 mM $CaCl_2$; 0.05% Brij-35; 0.02% $NaN_3$). The gels were stained with Coomassie Brilliant Blue (Sigma) and clear zones marked the mobility of the zymogen and converted forms of MMP-9.

## RESULTS

To follow the activation of MMP-9 in a setting devoid of MMP-2, proMMP-9 production and conversion by the breast carcinoma cell line, MDA-MB-231, were examined in detail. These cells make 100-fold more TIMP than proMMP-9 (TABLE 1). In order to induce increased production of proMMP-9, MDA-MB-231 cells were stimulated with the phorbol ester, PMA. The conditioned media of untreated and PMA-treated MDA-MB-231 cells were examined by ELISA to determine the levels of MMP-9 and TIMP proteins in the cultures (TABLE 1). Even though PMA treatment caused an 8-fold increase in MMP-9 production, the TIMP protein level was still 20-fold greater than that of MMP-9. MMP-3 has been shown to activate proMMP-9 *in vitro*.[10,12] In order to determine whether proMMP-9 activation could be mediated by an MMP-3–dependent mechanism in a cell culture system, a proteolytic cascade that would yield active MMP-3 was initiated in the untreated and PMA-treated MDA-MB-231 cultures. Plasmin is a known activator of proMMP-3,[37] and MDA-MB-231 cells produce uPA at levels that are sufficient to catalyze the conversion of plasminogen to plasmin.[11] The proteolytic cascade was initiated by addi-

**TABLE 1: Levels of proMMP-9 and TIMPs in cultures of untreated and PMA-treated MDA-MB 231 cells**

| | MDA-MB-231 cells | |
|---|---|---|
| | Untreated | PMA-treated |
| ProMMP-9 | 0.03 nM | 0.25 nM |
| Total TIMP | 3.0 nM | 4.0 nM |
| ProMMP-9 conversion in the presence of proMMP-3 and plasminogen | None | None |

NOTE: The indicated cultures ($1 \times 10^5$ cells/$cm^2$) were incubated in serum-free medium containing plasminogen and proMMP-3 for 48 hours. The conditioned media were harvested and analyzed for proMMP-9, TIMP-1, and TIMP-2 by ELISA as described.[11] The values represent the mean of three separate determinations. Total TIMP represents the sum of the TIMP-1 and TIMP-2 values added together. Conversion of proMMP-9 in the conditioned media of PMA-treated and untreated MDA-MB-231 cells in the presence of proMMP-3 and plasminogen was monitored by gelatin zymography.

tion of plasminogen and proMMP-3 into untreated and PMA-treated MDA-MB-231 cultures. There was no detectable proMMP-9 conversion in the presence of plasminogen and proMMP-3 in the conditioned media from either untreated or PMA-treated MDA-MB-231 (TABLE 1).

MDA-MB-231 cells were transfected with a plasmid encoding the full-length coding region for MMP-9[11] in order to upregulate MMP-9 and achieve levels of proMMP-9 that exceed the TIMP level. A stable cell line was isolated (MDA-MMP-9), and it was determined that the transfected cells make 8 nM MMP-9, 4 nM TIMP and no detectable MMP-2. These cells were used to reinvestigate the activation of proMMP-9 through the plasmin/MMP-3 proteolytic cascade. MDA-MMP-9 cells were cultured in the presence of plasminogen and proMMP-3 either alone or in combination; samples of the culture supernatants were analyzed by zymography for conversion of proMMP-9, and also by solution phase gelatinase activity for the appearance of active enzyme. When MDA-MMP-9 cells were cultured in the presence of either plasminogen (2.5 µg/ml) or proMMP-3 (16 nM), an 86-kDa form of MMP-9 was detected using gelatin zymography (FIG. 1, zymograph lanes 2 and 3). The appearance of the 86-kDa MMP-9, however, was not associated with an increase in gelatinolytic activity above background levels present in the medium of cells cultured with no additions (FIG. 1, bar graph column 1 versus 2 and 3). When MDA-MMP-9 cells were cultured in the presence of both plasminogen (2.5 µg/ml) and proMMP-3 (16 nM), substantial gelatinase activity was generated (FIG. 1, column 4) and nearly all of the proMMP-9 was converted to the 82-kDa form of the enzyme (FIG. 1, zymograph lane 4). The generation of plasmin was essential for MMP-9 activation because the addition of the plasmin inhibitor, aprotinin, prevented the appearance of both 82-kDa MMP-9 and gelatinase activity (FIG. 1, condition 5). The activation of proMMP-9 was also blocked by the addition of TIMP-1 at concentrations in excess of MMP-3 and MMP-9 (condition 6). The most effective inhibitor of proMMP-9 processing and activation was an anti-MMP-9 monoclonal antibody that previously had been shown to block organomercurially mediated activation of MMP-9.[34] The purified antibody completely blocked the plasmin-plus proMMP-3-dependent activation of MMP-9 (condition 7), while control IgG allowed full conversion and activation (condition 8).

We then examined the functional relevance of proMMP-9 conversion by analyzing the transmigration of MDA-MMP-9 cells across a reconstituted basement membrane (Matrigel). MDA-MMP-9 cells were added to the upper compartment of Matrigel-coated invasion chambers. Cells that had transmigrated to the underside of the filter were fixed and stained (FIG. 2). The generation of MMP-9 gelatinase activity was monitored using samples of culture media collected from the upper compartment of the invasion chamber. In the presence of plasminogen alone (FIG. 2A), there is only a small increase in the number of MDA-MMP-9 cells that were able to invade through the reconstituted matrix compared to the untreated cells (data not shown). Gelatinase activity in the plasminogen-containing cultures was slightly above the background level (418 cpm verses 267 cpm, respectively). When both proMMP-3 and plasminogen are added to the system together (FIG. 2B), there is a substantial increase in the number of cells that transmigrate to the underside of the filter. This enhanced invasion was accompanied by conversion of proMMP-9 to the 82-kDa form of the enzyme (data not shown) and a substantial increase in gelatinolytic ac-

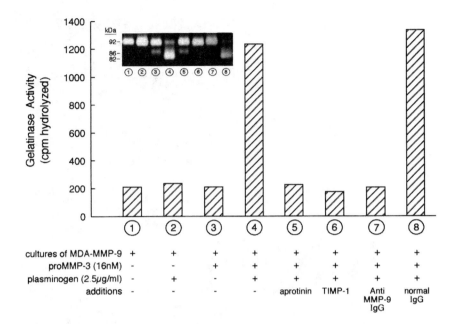

**FIGURE 1.** Activation of proMMP-9 in MDA-MMP-9 cultures supplemented with plasminogen and proMMP-3 and the effect of specific inhibitors on the activation cascade. Cultures of $1 \times 10^5$ MDA-MMP-9 cells were incubated for 48 hours in serum-free DMEM in the absence and presence of the indicated (+) components: plasminogen (2.5 μg/ml), proMMP-3 (16 nM), aprotinin (40 μg/ml), TIMP-1 (20 nM), anti-MMP-9 IgG (30 μg/ml), and normal mouse IgG (30 μg/ml). The conditioned media were collected and analyzed for proMMP-9 processing by gelatin substrate zymography (*inset*) and proMMP-9 activation by $^3$H-gelatinase activity in solution.

tivity (3056 cpm). Both the anti-MMP-9 monoclonal antibody (FIG. 2C) and TIMP-1 (FIG. 2D) were able to significantly inhibit this enhanced MDA-MMP-9 invasion. The gelatinase activities in these cultures (278 cpm and 342 cpm, respectively) were reduced to near-background level (267 cpm). These data indicate that activation of MMP-9 through a plasminogen/proMMP-3 cascade results in enhanced tumor cell invasion.

In the preceding experiments the presence of both plasminogen and proMMP-3 was necessary for proMMP-9 activation to occur (FIG. 1, condition **4**). To demonstrate that plasmin generated from plasminogen was necessary to activate the proMMP-3 and that MMP-3 alone could activate proMMP-9, a C-terminally truncated, stable, activated MMP-3 (ΔC-MMP-3; 33) was incubated with proMMP-9. MMP-9 conversion and activity were monitored by gelatin zymography (FIG. 3A) and $^3$H-gelatin degradation (FIG. 3C), respectively. As little as 0.5-nM active MMP-3 was able to generate both 82-kDa MMP-9 (FIG. 3A) and detectable amounts of gelatinolytic activity (FIG. 3C, 836 cpm). With increasing concentrations of ΔC-MMP-3, the amount of 92-kDa proMMP-9 decreased and the amount of the 82-kDa form

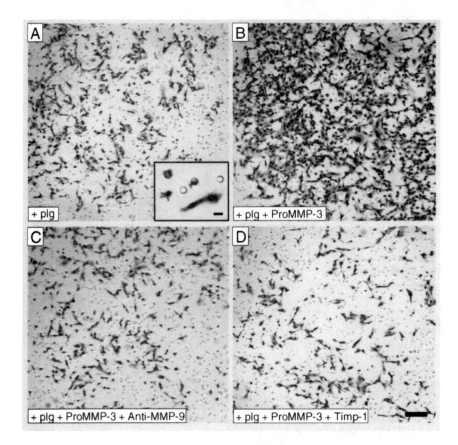

**FIGURE 2.** Invasion across a reconstituted basement membrane (Matrigel) is enhanced upon activation of proMMP-9. A suspension of $5 \times 10^4$ MDA-MMP-9 cells in serum-free DMEM was added to the upper compartments of invasion chambers fitted with Matrigel-coated, 8-μm-pore filter inserts. The cell suspension contained: plasminogen (2.5 μg/ml), **(A)**; plasminogen (2.5 μg/ml) + proMMP-3 (16 nM), **(B)**; plasminogen (2.5 μg/ml) + proMMP-3 (16 nM) + antiMMP-9 IgG (30 μg/ml), **(C)**; or plasminogen (2.5 μg/ml) + proMMP-3 (16 nM) + TIMP-1 (40 nM) **(D)**. After 48 hours the inserts were removed, fixed, and stained. The cells that had invaded and migrated to the underside of the filters were photographed. The *inset* shows individual cells migrating through the pores of the filter. The *bar* in panel D represents 200 μm.

of MMP-9 increased (FIG. 3A) with a corresponding increase in the level of gelatinolytic activity (FIG. 3C). However, when TIMP-1 was added to this system at a concentration of 4 nM, the approximate concentration found in the MDA-MB-231 cultures (TABLE 1), the activation of proMMP-9 by ΔC-MMP-3 was dampened considerably, and higher concentrations of ΔC-MMP-3 were required for effective activation. In the presence of TIMP-1 the 82-kDa MMP-9 and the accompanying gelatinolytic activity were detected only when active MMP-3 was present at concen-

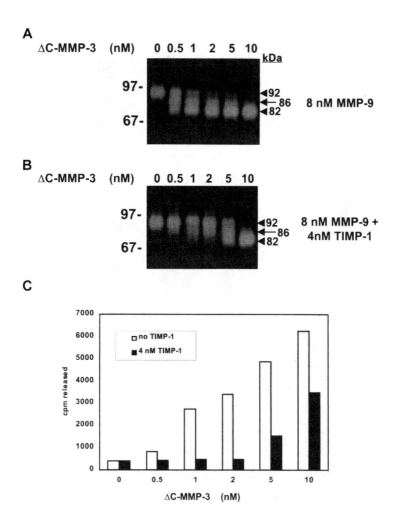

**FIGURE 3.** Conversion of purified proMMP-9 by active MMP-3 is dependent on exceeding the level of TIMP in the system. Purified proMMP-9 (8 nM) was incubated with radiolabeled gelatin and active MMP-3 (ΔC-MMP-3), as indicated, in the absence (**A**) or presence (**B**) of 4 nM TIMP-1. Samples were incubated at 37°C for 4 hours, at which time aliquots were removed for gelatin zymography (**A** and **B**, 1 μl) and for quantitation of ³H-gelatin degradation (**C**, 100 μl). The gelatinolytic activity of 10 nM ΔC-MMP-3 alone in this assay was indistinguishable from a buffer-alone control.

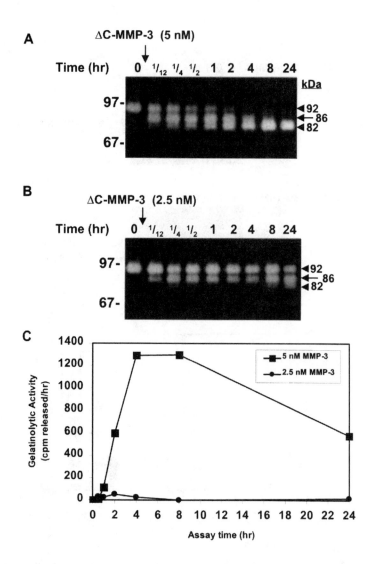

**FIGURE 4.** The time course of conversion of secreted ProMMP-9 by active MMP-3 is dependent on the balance of active MMPs to TIMP. Serum-free conditioned medium was collected after 24 hours from $1.5 \times 10^7$ MDA-MMP-9 cells, yielding secreted proMMP-9 at a concentration of 8-10 nM. Active MMP-3 [5 nM (**A**) or 2.5 nM (**B**) ΔC-MMP-3] was added to the medium along with radiolabeled gelatin, and the reactions were incubated at 37°C. At the indicated times, aliquots were removed for gelatin zymography (**A** and **B**, 1 μl) and for quantitation of [$^3$H]gelatin degradation (**C**, 100 μl).

trations greater than 5 nM (FIG. 3B and black bars in FIG. 3C). When the ΔC-MMP-3 concentration was below 5 nM, the gelatinolytic activity of the samples did not rise above the level of activity detected in the sample containing only proMMP-9 (392 cpm). These results indicate that a stoichiometric excess of TIMP-1 could inhibit MMP-3-mediated MMP-9 activation until the TIMP-1 was titered out by the active MMP-3.

To determine how TIMP in the tumor cell system effects the time course of activation of MMP-9 by active MMP-3, conditioned medium from the MDA-MMP-9 cells was incubated with 5 nM ΔC-MMP-3, a concentration above the endogenous level of TIMP (4 nM), or 2.5 nM ΔC MMP-3, a concentration below that of the endogenous TIMP. MMP-9 conversion and activation were monitored by gelatin zymography (FIG. 4A and B) and ³H-gelatin degradation (FIG. 4C), respectively. In the presence of 5 nM ΔC-MMP-3, some conversion of proMMP-9 was observed in less than 5 minutes, at which time the 86-kDa MMP-9 intermediate and small amounts of the 82-kDa form were detectable (FIG. 4A). With increasing times of incubation, there was an increase in the amount of the 82-kDa form of MMP-9 and a decrease in the amount of the 92-kDa zymogen, with maximum conversion occurring by 8 hours (FIG. 4A). After one hour of incubation, the gelatinolytic activity (FIG. 4C, solid squares) began to increase above the background levels and continued to increase steadily, peaking at 4–8 hours (1293 cpm/hr). Although 2.5 nM ΔC-MMP-3 was sufficient to generate the 86-kDa intermediate form of MMP-9, the activation progressed no further; no gelatinolytic activity was detected in the solution assay, even after 24 hours of incubation (FIG. 4C, solid circles).

When both plasminogen and proMMP-3 were added to the cellular invasion assay, we observed an MMP-9-dependent enhancement of basement membrane trans-

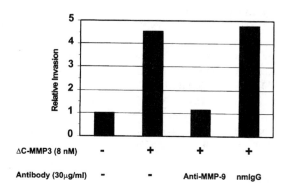

**FIGURE 5.** MDA-MMP-9 invasion of Matrigel-coated chambers in the presence of ΔC-MMP-3. A suspension of $1 \times 10^5$ MDA-MMP-9 cells was added to Matrigel-coated polycarbonate filters in the presence or absence of exogenous active MMP-3 (8 nM ΔC-MMP-3), anti-human MMP-9 monoclonal antibody (anti-MMP-9), or normal mouse IgG (nmIgG), as indicated. After incubation at 37°C, the cells that had invaded to the underside of the filters were stained and counted. The relative invasion of MDA-MMP-9 cells in the absence of active MMP-3 was arbitrarily set to 1.

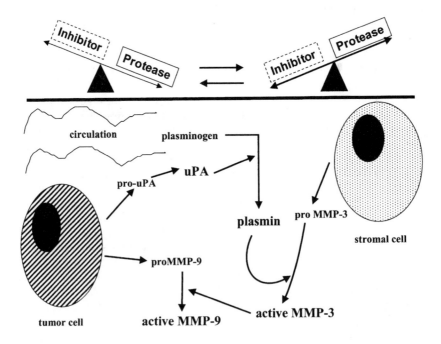

**FIGURE 6.** Model of proteolytic activation of proMMP-9 in a tumor-bearing tissue.

migration (FIG. 2). Since the addition of active MMP-3 alone could eliminate the requirement for plasminogen and proMMP-3 in the activation of proMMP-9, it was of interest to demonstrate that active MMP-3 alone could also mediate MMP-9-dependent cellular invasion. MDA-MMP-9 cells were placed on Matrigel-coated inserts in the absence or presence of 8 nM active MMP-3 (ΔC-MMP-3). In the presence of active MMP-3, MDA-MMP-9 invasion was increased 4.5-fold (FIG. 5), and the 82-kDa form of MMP-9 was detected in the conditioned medium of the upper compartment by gelatin zymography (data not shown). The increased invasion was inhibitable by a specific monoclonal antibody against human MMP-9, while a control antibody had no effect on invasion (FIG. 5). The results with the specific antibody confirm that active MMP-9 was the enzyme responsible for enhanced invasion in this tumor cell system.

## DISCUSSION

In this study, we have attempted to recapitulate some of the biochemical events that may occur during activation of proMMP-9 in an invasive tumor. We also have examined the functional involvement of active MMP-9 in a tumor invasion model using a highly specific, MMP-9-neutralizing antibody. Upregulation of MMP-9 expression by itself does not result in proMMP-9 activation. Addition of exogenous plasminogen and proMMP-3 to MDA-MMP-9 cultures resulted in proMMP-9 conversion and activation. Neither plasminogen (plasmin) nor proMMP-3 alone could activate proMMP-9: both components were necessary. The resulting increase in MMP-9 activity enhanced the ability of the cells to invade through a reconstituted basement membrane. A neutralizing monoclonal antibody specific for human MMP-9, purified TIMP-1, and the plasmin inhibitor aprotinin each were able to inhibit proMMP-9 conversion and activation. These specific inhibitors also blocked the MMP-9–dependent enhancement of basement membrane invasion. These experiments elucidate a possible mechanism for activation of MMP-9 *in vivo* (FIG. 6). In this model the tumor cell provides both uPA and proMMP-9, the plasminogen enters from the circulating plasma, and the host tissue provides proMMP-3. However, in order for the cascade to proceed, the natural inhibitor: protease balance must favor the proteases. When the balance favors the inhibitors and the inhibitors are in stoichiometric excess, the cascade will be arrested before active MMP-9 can be generated.

When a stable, activated MMP-3 ($\Delta$C-MMP-3) was added to the experimental systems, the requirement for exogenous plasminogen and proMMP-3 was eliminated, demonstrating that the plasminogen requirement was based on plasmin's ability to generate active MMP-3 from proMMP-3. In the biochemical studies performed using purified proMMP-9 (FIG. 3) or MDA-MMP-9–conditioned medium (FIG. 4), $\Delta$C-MMP-3 alone was able to mediate MMP-9 conversion and activation. The appearance of the 86-kDa form of MMP-9 depicted in FIGURES 3 and 4 did not coincide with gelatinolytic activity. However, the appearance of 82-kDa form of MMP-9 was coincident with the appearance of gelatinase activity. In the cell culture model for invasion (FIG. 5), exogenous $\Delta$C-MMP-3 was able to activate proMMP-9, and the resulting MMP-9 activity caused an increase in the invasive capacity of the cells.

We also have attempted to determine the role of the TIMPs in regulating the activation of proMMP-9 in a cell culture system. The experiments presented here demonstrate that TIMPs have at least three regulatory functions. First, the TIMPs can prevent the generation of active MMP-9 by inhibiting active MMP-3 or any other MMP activator of proMMP-9. Secondly, TIMP-1, by binding to proMMP-9, can dampen the activation. Thirdly, TIMPs directly inhibit the gelatinolytic activity of MMP-9. This study shows that the activation of MMP-9 by MMP-3 in culture can occur by a two-step process, verifying the results obtained in the test tube with purified reagents.[12] However, it appears that the first cleavage step, the generation of the 86-kDa form, is resistant to TIMP, even when the inhibitor is in molar excess (FIG. 4), implying that the kinetics for the MMP-3-mediated conversion of 92-kDa proMMP-9 to the 86-kDa form are highly favored over the inhibition of this conversion step by TIMP. However, the second step, which results in the generation of 82-kDa active MMP-9, is susceptible to TIMP inhibition, indicating that the second cleavage is much slower and, therefore, is the rate-limiting step in proMMP-9 activation.

While evidence exists, both in our studies and others,[10,12] implicating MMP-3 in proMMP-9 activation, it should be emphasized that MMP-3 may not be the sole activator of proMMP-9 *in vivo*. The homozygous MMP-3-deficient mice[38] do not seem to suffer from the same defects in bone development that are seen in the homozygous

MMP-9–deficient strain.[39] Furthermore, MMP-3 may not always be available in tissues where proMMP-9 activation occurs. Mazzeri et al.[40] examined activation of proMMP-9 and proMMP-2 in HT 1080 tumor cell cultures and concluded that plasmin could activate both gelatinases in the absence of proMMP-3. The plasmin that appeared to be responsible for this MMP activation was associated with the cell surface, and both gelatinases also were associated with cell surfaces.[40] Plasmin does not appear to activate proMMP-9 in our system. Employing a wide range of plasmin concentrations (0.2–20 µg/ml), we have been unable to observe any direct activation of proMMP-9.[11] However, the studies presented here and in that of Ramos-deSimone et al.[11] involve soluble MMP-9, which could be readily activated by soluble MMP-3. The apparent disparity between the data of Mazzeri et al.[40] and our data may reflect cell-associated activation versus solution phase activation. Another explanation for the conflicting experimental results could be the presence of specific cell surface receptors or proteases that are expressed in a cell-type–specific manner and are able to promote gelatinase activation, such as integrins or MT-MMPs. These potentially important molecules were not investigated in either study. Collagenase-3 (MMP-13) is another potential proMMP-9 activator because it exhibits favorable kinetics for proMMP-9 activation.[41] MMP-13 could be relevant in tissues where there is no active MMP-3, because the mechanism that activates proMMP-13 is distinct, as it involves MT1-MMP and MMP-2.[42]

The TIMPs are clearly important in regulating MMP-9 activation and the MMP-9–dependent enhancement of invasion in the MDA-MMP-9 system. The role of TIMPs in regulating the activation of MMP-9 is based mainly on stoichiometry. Thus, the ability of a tumor to circumvent TIMP control over MMP-9 activation *in vivo* may be due to the titering-out of the endogenous TIMP. In this model system, the TIMP was titered-out by the overexpression of MMP-9 and the addition of exogenous active MMP-3 (ΔC-MMP-3). *In vivo* saturation of TIMP could occur when total MMP expression is induced by cytokines or growth factors. This titering may only occur in focalized areas where ECM and basement membrane remodeling is necessary either for tissue differentiation, wound healing, or tumor invasion. This study demonstrates that titering of TIMPs must take place for initiation and progression of MMP-9 activation.

## REFERENCES

1. WERB, Z. 1997. ECM and cell surface proteolysis: regulating cellular ecology. Cell **91:** 439–442.
2. WATSON, S.A., T.M. MORRIS, G. ROBINSON, M.J. CRIMMIN, P.D. BROWN & J.D. HARDCASTLE. 1995. Inhibition of organ invasion by the matrix metalloproteinase inhibitor batimastat (BB-94) in two human colon carcinoma metastasis models. CANCER RES. **55:** 3629–3633.
3. RAY, J.M. & W.G. STETLER-STEVENSON. 1994. The role of matrix metalloproteases and their inhibitors in tumour invasion, metastasis and angiogenesis. Eur. Respir. J. **7:** 2062–2072.
4. WOJTOWICZ-PRAGA S., J. TORRI, M. JOHNSON, V. STEEN, J. MARSHALL, E. NESS, R. DICKSON, M. SALE, H.S. RASMUSSEN, T.A. CHIODO & M.J. HAWKINS. 1998. Phase I trial of Marimastat, a novel matrix metalloproteinase inhibitor, administered orally to patients with advanced lung cancer. J. Clin. Oncol. **16:** 2150–2156

5. SPRINGMAN, E.B., E.L. ANGLETON, H. BIRKEDAL-HANSEN & H.E. VANWART. 1990. Multiple modes of activation of latent human fibroblast collagenase: evidence for the role of a Cys73 active-site zinc complex in latency and a "cysteine switch" mechanism for activation. Proc. Natl. Acad. Sci. USA **87:** 364–368.

6. PEI, D. & S. WEISS. 1995. Furin-dependent intracellular activation of the human Stromelysin-3 zymogen. Nature **375:** 244–257.

7. SATO, H., T. TAKINO, Y. OKADA, J. CAO, A. SHINGAWA, E. YAMAMOTO & M. SEIKI. 1994. A matrix metalloproteinase expressed on the surface of invasive tumour cells. Nature **370:** 61–65.

8. STRONGIN, A.Y., I. COLLIER, G. BANNIKOV, B.L. MARMER, G.A. GRANT & G.I. GOLDBERG. 1995. Mechanism of cell surface activation of 72-kDa type IV collagenase. Isolation of the activated form of the membrane metalloprotease. J. Biol. Chem. **270:** 5331–5338.

9. HE, C., S.M. WILHELM, A.P. PENTLAND, B.L. MARMER, G.A. GRANT, A.Z. EISEN & G.I. GOLDBERG. 1989. Tissue cooperation in a proteolytic cascade activating human interstitial collagenase. Proc. Natl. Acad. Sci. USA **86:** 2632–2636.

10. OGATA, Y., Y. ITOH & H. NAGASE. 1995. Steps involved in activation of the promatrix metalloproteinase 9 (progelatinase B)-tissue inhibitor of metalloproteinases-1 complex by 4-aminophenylmercuric acetate and proteinases. J. Biol. Chem. **270:** 18506–18511.

11. RAMOS-DESIMONE, N., E. HAHN-DANTONA, J. SIPLEY, H. NAGASE, D.L. FRENCH & J.P. QUIGLEY. 1999. Activation of matrix metalloproteinase-9 (MMP-9) via a converging plasmin/stromelysin-1 cascade enhances tumor cell invasion. J. Biol. Chem. **274:** 13066–13076.

12. NAGASE, H., J.J. ENGHILD, K. SUZUKI & G. SALVENSEN. 1990. Stepwise activation mechanisms of the precursor of matrix metalloproteinase 3 (stromelysin) by proteinases and (4-aminophenyl)mercuric acetate. Biochemistry **29:** 5783–5789.

13. BIRKEDAL-HANSEN, H., W.G. MOORE, M.K. BODDEN, L.J. WINDSOR, B. BIRKEDAL-HANSEN, A. DECARLO & J.A. ENGLER. 1993. Matrix metalloproteinases: a review. Critical Rev. Oral Biol. Med. **4:** 197–250.

14. KOLKHORST V., J. STURZEBECHER & B. WIEDERANDERS. 1998. Inhibition of tumour cell invasion by protease inhibitors: correlation with the protease profile. J. Cancer Res. Clin. Oncol. **124:** 598–606.

15. GOLDBERG G.I., B.L. MARMER, G.A. GRANT, A.Z. EISEN, S. WILHELM & C.S. HE. 1989. Human 72-kilodalton type IV collagenase forms a complex with a tissue inhibitor of metalloproteases designated TIMP-2. Proc. Natl. Acad. Sci. USA **21:** 8207–8211.

16. HOWARD, E.W. & M. BANDA. 1991. Binding of tissue inhibitor of metalloproteinases 2 to two distinct binding sites on human 72-kDa gelatinase. J. Biol. Chem. **266:** 17972–17977.

17. GOLDBERG G.I., A. STRONGIN, I.E. COLLIER, L.T. GENRICH & B.L. MARMER. 1992. Interaction of 92-kDa type IV collagenase with the tissue inhibitor of metalloproteinases prevents dimerization, complex formation with interstitial collagenase, and activation of the proenzyme with stromelysin. J. Biol. Chem. **7:** 4583–4591.

18. BROWN, P.D., A.T. LEVY, I. MARGULIES, L.A. LIOTTA & W.G. STETLER-STEVENSON. 1990. Independent expression and cellular processing of Mr 72,000 type IV collagenase and interstitial collagenase in human tumorigenic cell lines. Cancer Res. **50:** 6184–6191.

19. WARD, R.V., S.J. ATKINSON, P.M. SLOCOMBE, A.J. DOCHERTY, J.J. REYNOLDS & G. MURPHY. 1991. Tissue inhibitor of metalloproteinases-2 inhibits the activation of 72 kDa progelatinase by fibroblast membranes. Biochem. Biophys. Acta **1079:** 242–246.

20. MOLL, U.M., G. YOUNGLEIB, K. ROSINSKI & J.P. QUIGLEY. 1990. Tumor promoter-stimulated Mr 92,000 gelatinase secreted by normal and malignant human cells: isolation and characterization of the enzyme from HT1080 tumor cells. Cancer Res. **50:** 6162–6170.

21. UNEMORI, E., M.S. HIBBS & E.P. ARMENTO. 1991. Constitutive expression of a 92-kD gelatinase (type V collagenase) by rheumatoid synovial fibroblasts and its induction in normal human fibroblasts by inflammatory cytokines. J. Clin. Invest. **88:** 1656–1662.

22. KONDAPAKA, S.B., R. FRIDMAN & K.B. REDDY. 1997. Epidermal growth factor and amphiregulin up-regulate matrix metalloproteinase-9 (MMP-9) in human breast cancer cells. Int. J. Cancer **70:** 722–726.

23. OKADA, Y., K. NAKA, K. KAWAMURA, I. NAKANISHI, M. FUJIMOTO, H. SATO & M. SEIKI. 1995. Localization of matrix metalloproteinase 9 (92-kilodalton gelatinase/type IV collagenase = gelatinase B) in osteoclasts: implications for bone resorption. Lab. Invest. **72:** 311–322.

24. HIBBS, M.S., K.A. HASTY, J.M. SEYER, A.H. KANG & C. MAINARDI. 1985. Biochemical and immunological characterization of the secreted forms of human neutrophil gelatinase. J. Biol. Chem. **260:** 2493–2500.

25. HEPPNER, K.J., L.M. MATRISIAN, R.A. JENSEN & W.H. ROGERS. 1996. Expression of most matrix metalloproteinase family members in breast cancer represents a tumor-induced host response. Am. J. Pathol. **149:** 273–282.

26. STETLER-STEVENSON, W.G., S. AZNAVOORIAN & L.A. LIOTTA. 1993. Tumor cell interactions with the extracellular matrix during invasion and metastasis. Annu. Rev. Cell Biol. **9:** 541–573.

27. MACDOUGALL, J.R. & L.M. MATRISIAN. 1995. Contributions of tumor and stromal matrix metalloproteinases to tumor progression, invasion and metastasis. Cancer Metast. Rev. **14:** 351–362.

28. MIGNATTI, P. & D.B. RIFKIN. 1993. Biology and biochemistry of proteinases in tumor invasion. Physiol. Rev. **73:** 161–195.

29. HIMELSTEIN, B.P., R. CONTE-SOLER, E.J. BERNHARD, D.W. DILKS & R.J. MUSCHEL. 1994. Metalloproteinases in tumor progression: the contribution of MMP-9. Invasion Metast. **14:** 246–258.

30. BERNHARD, E.J., S.B. GRUBER & R.J. MUSCHEL. 1994. Direct evidence linking expression of matrix metalloproteinase 9 (92-kDa gelatinase/collagenase) to the metastatic phenotype in transformed rat embryo cells. Proc. Natl. Acad. Sci. USA. **91:** 4293–4297.

31. HUA, J. & R.J. MUSCHEL. 1996. Inhibition of matrix metalloproteinase 9 expression by a ribozyme blocks metastasis in a rat sarcoma model system. Cancer Res. **56:** 5279–5284.

32. KIM, J., W. YU, K. KOVALSKI & L. OSSOWSKI. 1998. Requirement for specific proteases in cancer cell intravasation as revealed by a novel semiquantitative PCR-based assay. Cell **94:** 353–362.

33. SUZUKI, K., C. KAN, W. HUNG, M.R. GEHRING, K. BREW & H. NAGASE. 1998. Expression of human pro-matrix metalloproteinase 3 that lacks the N-terminal 34 residues in *Escherichia coli*: autoactivation and interaction with tissue inhibitor of metalloproteinase 1 (TIMP-1). J. Biol. Chem. **379:** 185–191.

34. RAMOS-DESIMONE, N, U.M. MOLL, J.P. QUIGLEY & D.L. FRENCH. 1993. Inhibition of matrix metalloproteinase 9 activation by a specific monoclonal antibody. Hybridoma **12:** 349–363.

35. MALLYA, S.K., K.A. MOOKTIAR & H.E. VANWART. 1986. Accurate, quantitative assays for the hydrolysis of soluble type I, II, and III 3H-acetylated collagens by bacterial and tissue collagenases. Anal. Biochem. **158:** 322–333.

36. AIMES, R.T. & J.P. QUIGLEY. 1997. Matrix metalloproteinase-2 is an interstitial colla-genase. Inhibitor-free enzyme catalyzes the cleavage of collagen fibrils and soluble native type I collagen generating the specific 3/4- and 1/4-length fragments. J. Biol. Chem. **270:** 5872–5876.

37. NAGASE, H., J.J. ENGHILD, K. SUZUKI & G. SALVENSEN. 1990. Stepwise activation mechanisms of the precursor of matrix metalloproteinase 3 (stromelysin) by protein-ases and (4-aminophenyl) mercuric acetate. Biochemistry. **29:** 5783–5789.

38. RUDOLPH-OWEN, L.A., D.L. HULBOY, C.L. WILSON, J. MUDGETT & L.M. MATRISIAN. 1997. Coordinate expression of matrix metalloproteinase family members in the uterus of normal, matrilysin-deficient, and stromelysin-1-deficient mice. Endocri-nology **138:** 4902–4911.

39. VU, T.H., J.M. SHIPLEY, G. BERGERS, J.E. BERGER, J.A. HELMS, D. HANAHAN, S.D. SHAPIRO, R.M. SENIOR & Z. WERB. 1998. MMP-9/gelatinase B is a key regula-tor of growth plate angiogenesis and apoptosis of hypertrophic chondrocytes. Cell **93:** 411–422.

40. MAZZIERI, R., L. MASIERO, L. ZANETTA, S. MONEA, M. ONISTO, S. GARBISA & P. MIGNATTI. 1997. Control of type IV collagenase activity by components of the urokinase-plasmin system: a regulatory mechanism with cell-bound reactants. EMBO J. **16:** 2319–2332.

41. KNAUPER, V., B. SMITH, C. LOPEZ-OTIN & G. MURPHY. 1997. Activation of progelati-nase B (proMMP-9) by active collagenase-3 (MMP-13). Eur. J. Biochem. **248:** 369–373.

42. KNAUPER, V., H. WILL, C. LOPEZ-OTIN, B. SMITH, S.J. ATKINSON, H. STANTON, R.M. HEMBRY & G. MURPHY. 1996. Cellular mechanisms for human procollage-nase-3 (MMP-13) activation. Evidence that MT1-MMP (MMP-14) and gelatinase A (MMP-2) are able to generate active enzyme. J. Biol. Chem. **271:** 17124–17131.

# Matrix Metalloproteinase Inhibition

## From The Jurassic To The Third Millennium

J. FREDERICK WOESSNER, JR.[a]

*Departments of Biochemistry and Molecular Biology, and Medicine,*
*University of Miami School of Medicine, Miami, Florida 33101, USA*

ABSTRACT: A brief historical introduction to the matrix metalloproteinase (MMP) field, which began in 1962, is followed by an overview of the inhibition of these proteases by natural inhibitors such as $\alpha_2$ macroglobulin and the TIMPs (tissue inhibitors of metalloproteinases) and by synthetic inhibitors, which are largely chelating agents. The latter include thiol, alkylcarbonyl, phosponamidate and hydroxamate compounds, as well as the tetracyclines. A review of the most recent progress concludes with prognostications as to where the field may be going next.

## INTRODUCTION

When this topic was assigned by Dr. Greenwald, my first thought was that the Jurassic was not nearly far enough into the past to capture the origins of the matrix metalloproteinases (MMPs). However, there has been a tradition of referring to early workers in this field as dinosaurs, so I assume he meant that I should refer back to the early portion of my career. From that point of view, the Jurassic period for the MMP field started in 1962. Jerome Gross and Charles Lapière[1] set out to establish how the metamorphosing tadpole of a frog could lose its tail, particularly the collagenous components, which were particularly resistant to proteolytic degradation. Their approach was to culture small explants of tail tissue on a substratum of reconstituted collagen fibrils. After several days, a clearing was noticed under the tissue and a collagenolytic enzyme could be recovered from the medium. The collagen molecules were cleaved very uniformly into two segments of 1/4 and 3/4 length.[2] Human collagenase (MMP-1) was purified eight years later[3] and shown to be a zymogen.[4]

Soon, additional MMPs began to come to light, and within 20 years of the original discovery some seven distinct types had been recognized (FIG. 1).

At this point, cloning techniques entered the scene and the rate of discovery accelerated rapidly so that by the end of 1997 an additional 12 MMPs were known. At the same time, MMPs were shown to be widely distributed in the animal kingdom down to C. *elegans* and in the plant kingdom in soybean and *Arabidopsis*. It was not long after the discovery of collagenase that efforts were made to inhibit the

[a]Address for correspondence: R-127, P.O. Box 016960, Miami, Florida 33101. Phone, 305/243-6510; fax, 305/243-3955; e-mail, fwoessne@mednet.med.miami.edu

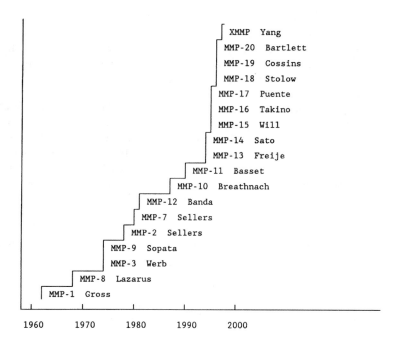

**FIGURE 1.** History of the discovery of the MMPs. References[1,5–21] are shown ascending in chronological order.

enzyme. Nagai *et al.*[22] inhibited tadpole collagenase with EDTA and cysteine. A number of simple chelators such as D-penicillamine and acetylcysteine were introduced by Berman,[23] who demonstrated that zinc was the catalytic metal ion. While these compounds were not suitable for treating animals, they were found to be useful in topical treatment of alkali burns of the cornea, in which excess collagenase action leads to destruction of the cornea.[24]

## ARE MMPs WORTH INHIBITING?

In recent years the MMPs have been implicated in a variety of important diseases including cancer (growth and metastasis), atherosclerosis, osteo- and rheumatoid arthritis, and emphysema.[25] However, the evidence has not been very strong, consisting largely of such correlations as higher rates of metastasis accompanied by higher levels of MMPs. Almost nothing is known about the natural substrates of the MMPs other than collagen;[26] that is, a great variety of proteins are cleaved *in vitro*, but there are few convincing demonstrations *in vivo* that a specific component of the matrix has been cleaved by a specific MMP. There do not appear to be any genetic diseases involving any of the MMPs. So one might justifiably ask whether there is a rational point to developing inhibitors to the MMPs with the goal of reversing disease processes. Recent work suggests more strongly that the answer is yes, based on the development of knockout mice for several of the MMPs[27] and on transfection experiments. Some of these are summarized in TABLE 1. In general, reducing MMP levels reduces tumorigenesis, invasiveness, emphysema, etc., whereas overexpression enhances these processes.

It may be argued that there would be significant medical applications for inhibition of MMPs, and indeed a great many pharmaceutical companies are engaged in such endeavors. In this review I will again follow the guidelines of Dr. Greenwald, who emphasizes the inhibitory nature of any mechanism that reduces the tissue levels of MMPs. For this reason the chapter is titled MMP *inhibition* rather than *inhibitors*. Taking this broader view, FIGURE 2 presents a schematic representation of various ways in which the tissue levels of MMPs might be reduced. The top half of the figure suggests ways in which to reduce the amount of active enzyme (this will be taken up later in the chapter) and the bottom half concerns what to do once the enzyme has lost its propeptide and become fully active (which will be taken up immediately).

**TABLE 1. Effect of over- and underexpression of MMPs**

| | Mouse | | Cancer cell line | |
|---|---|---|---|---|
| | Overexpress transfect[a] | Underexpress knockout[b] | Overexpress transfect[c] | Underexpress antisense[d] |
| MMP-1 | Emphysema, acanthosis | — | — | Invade gel ↓ |
| Collagen Noncleavable | | Fibrosis | — | — |
| MMP-2 | — | Tumor response ↓ | — | Inflammatory phenotype lost |
| MMP-3 | Precocious mammary development | MMP-7 ↑, MMP-10 ↑ Tumorigenesis ↓ | — | Tumorigenesis ↓ Arthritis ↓ |
| MMP-7 | Testicular changes | MMP-3 ↑, MMP-10 ↑ Tumorigenesis ↓ | Tumorigenesis ↑ Invasiveness ↑ | Tumorigenesis ↓ Invasiveness ↓ |
| MMP-9 | — | Abnormal growth plate development | — | Metastasis ↓ Invade gel ↓ |
| MMP-12 | — | Prevent emphysema | — | — |
| MMP-14 | — | — | Activate MMP-2 Invasiveness ↑ | — |

[a]References 28–31.
[b]References 32–38.
[c]References 39–41.
[d]References 42–47.

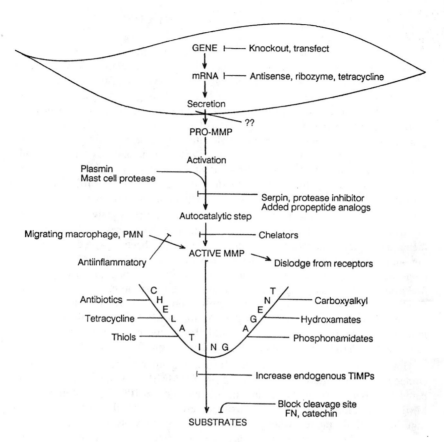

**FIGURE 2.** Approaches useful in reducing the level of MMPs in a tissue.

The vast majority of attention to blocking active MMPs has been given to chelating agents that bind to zinc at the active center, thereby inactivating the enzyme. However, there are also a few other possibilities. One of these is $\alpha_2$-macroglobulin, a large serum protein that offers the enzyme a bait region; when this is cut, the molecule changes shape and traps the MMP in a cage-like structure.[48] Another is to dislodge MMPs from their receptors. Some MMPs are associated with the cell surface, and others appear to be bound to the extracellular matrix—this prevents the MMP from diffusing away and keeps it under control of the cell. We have demonstrated that matrilysin and collagenase-3 are bound to heparan sulfate in the uterus and may be dislodged by heparin.[49] Gold salts are used to treat arthritis; these have been shown to work by the binding of a gold atom to a heavy metal site in the MMP distinct from that occupied by the catalytic zinc atom of the MMP.[50] Finally, there are compounds that are believed to inhibit MMPs by binding to cleavage sites on the substrate; an example of this is catechin.[51]

## INHIBITION OF ACTIVE MMPS BY CHELATORS

Mention was made above of early studies of chelators. However, these compounds had no specificity and would block any enzyme containing zinc. An early effort to specifically block MMP-1 was based on knowledge of the bond cleaved in collagen (Gly*Ile-Ala-Gly); a synthetic peptide including this sequence was readily cleaved.[52] The cleavage product was then modified to have an –SH group in place of the amino group; this compound had an $IC_{50}$ of 10 μM.[53] The Searle company introduced an inhibitor based on chelation by a carboxyalkyl moiety at the amino end of a dipeptide with $IC_{50} = 1.7$ μM.[54] This compound was shown to inhibit bone resorption in culture[55] and follicle rupture in perfused rat ovaries.[56] The same group developed an hydroxamate dipeptide with $IC_{50} = 20$ nM for MMP-1 and 4 nM for MMP-2.[57] Similar results were obtained with phosphonamidates.[58]

During the intervening 10 years there has been a burst of activity in the design of MMP inhibitors, with developments in all four of the classes of compounds just mentioned. A tremendous boost to inhibitor design has been provided by the working out of the X-ray structures of the major MMPs, starting with MMP-1.[59] Recent emphasis has been placed on maximizing specificity for a single MMP, optimizing oral availability for drug purposes, and minimizing possibilities of *in vivo* degradation by use of peptide mimetics. A reasonably good phosphonate compound was developed by Celltech[60] with $IC_{50} = 2.5$ nM for MMP-2. Carboxyalkyl tripeptide inhibitors for MMP-3 have $IC_{50} = 400$ pM.[61] Succinyl mercaptoketones and mercaptoalcohols have been developed with $IC_{50} = 140$ pM for MMP-9.[62]

However, most attention has been focused on improvements to hydroxamate compounds. Derivatives with $IC_{50} = 1–5$ nM for MMP-1, -3 and -7 have been prepared.[63] Dipeptide mimetic hydroxamates have $K_i = 100–500$ pM for various MMPs.[64] A sufonamide-based hydroxamate effectively blocks cartilage degradation in a rabbit model when given orally.[65] Various hydroxamates such as Marimastat™[66] and Trochate™[67] are now in clinical trial. At least six compounds are in clinical trial for cancer and arthritis.[68]

A completely different approach to inhibiting MMPs has been through a search for antibiotic compounds that would affect their activity. It was found that Actinomyces produced a powerful natural hydroxamate compounds, actinonin with $K_i = 1.4$ μM for MMP-1.[69] Actinomadura produces a series of hydroxamates known as *matlystatins*.[70]

Synthesis of modified matlystatins produces compounds with $IC_{50} = 1.2$ nM on MMP-9.[71] Nicotianamine from Streptomyces inhibits MMP-2 with $IC_{50} = 230$ nM.[72] Tetracyclines were shown in 1983 by Golub *et al.* to inhibit collagenase in gingival fluid and tissue;[73] this action is independent of the antibacterial activity of the drug.[74] Rather, the action is due to chelation through oxygen at C11 and C12.[75] However, the effects of tetracyclines *in vivo* appear more likely to be due to downregulation of collagenase messenger and protein expression, rather than by direct inhibition of MMP-1 or -8.[76,77] Finally, the propeptides of MMPs maintain the enzyme in latent form because cysteine in the peptide chelates zinc at the active center. This has led to the development of inhibitors based on the PRCGNPD sequence of the propeptides.[78] Such peptides were whittled down from 14 to 5 residues, with effective inhibition produced by Ac-RCGVP-NH₂.[79]

## TISSUE INHIBITORS OF METALLOPROTEINASES—TIMPs

TIMP was first noted in medium of human fibroblasts[80] and in serum where it appeared as a β1-serum protein.[81] The protein was purified from human fibroblasts[82] and rabbit bone[83] and sequenced in 1985.[84] This first form of TIMP is now designated TIMP-1. TIMP-2 was noted as an extra band when reverse zymography was first introduced.[85] It was purified from bovine vascular cells[86] and sequenced by Edman degradation.[87] TIMP-3 was first noticed as a novel protein in the extracellular matrix of chick embryo fibroblasts[88] but was not identified as the inhibitor ChiMP-3 until 1991.[89] It was cloned and sequenced in 1992.[90] Human TIMP-4 was recently identified by cloning.[91] A recent review of the TIMPs is presented by Gomez et al.[92]

All of the TIMPs are capable of inhibiting all of the MMPs, so far as has been tested. However, the interaction depends both on the TIMP and the MMP. Thus, MMPs lacking the hemopexin domain naturally (matrilysin) or by deletion bind less tightly than do full-length MMPs.[93] The N-terminal domain of the TIMPs appears to be the critical portion for inhibition,[94] although there is some enhancement of binding in the presence of the C-domain.[95] The exception to the general rule is that TIMP-1 appears to be a very poor inhibitor of MMP-14 attached to the cell surface.[96] We now know from the recent work of Gomis-Rüth et al.[97] on the crystal structure of the MMP-3/TIMP-1 complex that TIMP acts as a chelator: the amino $N$ and carbonyl oxygen of Cys-1 chelate the zinc of the active site, while Thr-2 and Val-4 fit the binding pockets at S1′ and S2′.

Attention has been given to altering the levels of TIMP in cells and tissues, for example, by transfection, with the thought of increasing the inhibition of MMPs. There is a certain logic in this approach, but one must not lose sight of the fact that the TIMPs also possess growth-stimulatory effects.[98] This effect is independent of the MMP-blocking effect.[99] Therefore, TIMP might be a two-edged sword, promoting growth where one was hoping to suppress it. A further consideration is that the TIMPs are able to inhibit the activation of many MMPs. This comes about if an MMP is first partially activated by the action of a second protease such as plasmin. This removes part of the propeptide, permitting the enzyme to undergo autocatalytic cleavage to the fully active form. It is this last step that is often blocked by TIMP.[100]

## BLOCKING THE PRODUCTION OF MMPs

So far we have considered methods of inhibiting or reducing the activity of active MMP. FIGURE 2 illustrates that there are many possibilities for inhibiting the production or activation of MMPs at earlier stages. The first step would be to introduce perturbations at the gene level—either knocking out the normal gene or transfecting it to increase the levels of MMPs (or more to the point, increasing TIMPs). This approach is briefly summarized in TABLE 1. At the next level, one might reduce the production of mRNA for a specific MMP (by use of antisense RNA or ribozymes) or a group of MMPs by agents that act on common promoter elements such as AP-1 and Jun. A few brief comments will be made on the use of ribozymes that can specifically affect the mRNA of a single MMP. Thus Flory *et al.*[101] delivered a stromelysin-specific ribozyme into the knee joints of rabbits. It was taken up into the synovium and reduced the expression of MMP-3 that followed interleukin-1 stimulation. An anti-MMP-2 ribozyme, introduced into glomerular mesangial cells, caused loss of inflammatory phenotype; addition of active, but not of pro-MMP-2, restored the inflammatory phenotype.[42] Hua and Muschel[46] introduced hammerhead ribozyme for MMP-9 into rat embryo sarcoma cells, which led to disappearance of message and protein. The cells remained tumorigenic but lost their ability to metastasize. Among more general agents that reduce mRNA levels of MMPs are the tetracyclines mentioned above.[76,77]

I know of no agents that specifically block the secretion of MMPs from the cell. However, now that we know several MMPs, such as stromelysin 3 and MT1-MMP, that are activated by furin,[102] it is possible that specific furin inhibitors might prevent activation within the cell leading to altered secretion or secretion of inactive enzyme.[103] The majority of the secreted MMPs are found in the tissue as proenzymes. These can be activated by a variety of agents such as organomercurials and trypsin, but it is likely that in vivo activation involves enzymes likely to be in the matrix such as plasmin, plasma kallikrein, cathepsin G, and mast cell chymases.[104] This suggests that one might be able to block the activation of MMPs by introducing specific inhibitors of these serine class proteases. There has been considerable attention given to the effects of knocking out the gene for plasminogen and plasminogen activator; mice lacking plasminogen did not produce detectable active form of MMP-9 in injured coronary arteries.[105]

It should not be overlooked that activation of proMMPs almost always involves a final autocatalytic step in which the enzyme cleaves off the remaining portion of its own propeptide. This was mentioned above in relation to TIMP's ability to block this final step. But it is also the case that many of the MMPs can activate other members of the group once they themselves become active. Some examples include MMP-2—proMMP-13, MMP-3—proMMP-1, MMP-7—proMMP-9, MMP-10—proMMP-1, etc.[104] Therefore, compounds useful for blocking MMPs should also prove useful for inhibiting autoactivation or activation of one MMP by another. While it requires the same concentrations of inhibitor to inhibit at the level of active enzyme or at the level of activation of proenzyme, it should be advantageous to inhibit before the proenzyme is fully converted to active enzyme or before the active enzyme has opportunity to activate other MMPs. Finally, one may maintain the tissue levels of MMPs at lower values if one blocks the entrance of cells such as macrophages and PMNs from the outside. This is usually done by the use of antiinflammatory drugs.

## INTO THE THIRD MILLENIUM

The field of MMPs and TIMPs has now grown very large, with more than 1,000 scientific papers published on the subject each year. The field may be said to have grown out of adolescence and reached a certain maturity. It is of interest to consider whether it will grow older—that is, whether there will be new directions and further development of the field. Having failed to predict the collapse of Russia or the development of the bear market, it is unlikely that I will have much more success in looking into the future of the inhibition of MMPs. Nonetheless, an outline of some possible developments is presented in TABLE 2.

First, we should consider whether we now have a clear idea of how many MMPs there are and what their functions are. FIGURE 1 suggests that MMPs were still being discovered at a significant rate. I am aware of several further additions to be made in the near future. However, the Human Genome Project should soon provide a good overview of the total population of MMPs, which can be recognized by their HEXGHXXGXXH motif plus the cysteine switch motif. With respect to the role of each MMP, a tremendous amount of work remains to be done: there are knockouts for only a few of the 11 known mouse MMPs, the mouse lacks the important MMP-1, and the true natural substrates for each MMP are also almost completely unknown.

With respect to synthetic inhibitors of MMPs, the pace of research and development is picking up and many new companies have entered the field. We can expect improved inhibitors with increased specificity based on X-ray structures. More structures remain to be completed. There are new developments in magnet research that promise improved NMR studies, which, coupled with enhanced algorithms for analyzing ever-larger protein structures, should soon lead to the ability to examine the structure of enzyme-inhibitor complexes in solution. Progress should continue in the development of peptide mimetics and similar derivatives that are stable *in vivo* and orally available. This may be aided by the use of combinatorial chemistry to screen vast numbers of potential inhibitors. While the focus at the moment is to find ways to selectively block each of the 15 or so MMPs, there is a long way to go to meet this goal. And when the goal is reached, it will probably

**TABLE 2. Future directions for the inhibition of MMPs**

1. Identify all human MMPs: Human Genome Project
   a. Determine their role by gene knockout experiments
   b. Identify critical substrates of the ECM for each MMP
2. Synthetic inhibitors of MMPs
   a. Improve specificity for individual MMPs based on X-ray structure
   b. Determine solution structure by advanced NMR techniques
   c. Develop peptide mimetics stable *in vivo*
   d. Use combinatorial chemistry to develop nonpeptide inhibitors
   e. Use combinations of inhibitors to block multiple MMPs
3. Modulation of tissue levels of MMPs
   a. Inactivate MMPs in specific targets by gene technology
   b. Target antisense RNA ribozymes to specific cells
   c. Increase production of tissue TIMPs
   d. Block pathways of MMP activation
   e. Develop drugs to regulate MMP mRNA synthesis

prove to be the case that inhibition of a single MMP is of little value therapeutically. This is because there are frequently multiple MMPs elevated in a disease process and when one MMP is knocked out or depressed, the cells tend to replace it with another MMP of similar activity. Therefore, it will be necessary to develop combinations of inhibitors to selectively inhibit multiple MMPs. This work will be facilitated by the development of chip technology that will permit rapid screening of all MMPs in a given tissue or even in a small cell cluster within the tissue.

With respect to modulating tissue levels of MMPs by other means, this area should benefit from the general advances in gene technology and gene therapy, a field just beginning to get under way. In most diseases involving MMPs it will be desirable to suppress gene activity for the MMP or to enhance TIMP gene expression. A start has already been made in the use of antisense RNA and specific ribozymes, but these have been applied mostly to isolated cells. Ways must be found to target these reagents to specific cells in the body. Because of the important role of the MMPs in normal physiological processes it is critical to selectively inhibit MMPs in the diseased tissue while affecting normal tissues as little as possible. This principle will limit the usefulness of synthetic inhibitors as well, except in very serious illnesses.

Little attention has been given to blocking the activation pathways of the MMPs. Because of the cascade nature of activation by natural proteases, the blocking of these enzymes that activate MMPs could be accomplished with much smaller doses of drugs than those required to block the MMPs after they become active. Also, inhibitors of other classes of proteases such as serine or cysteine enzymes could be brought into the picture, broadening the group of pharmaceutical agents available. There have been several studies of the ability of propeptides, and small regions of the propeptides, of MMPs to inhibit the enzyme activity. This area will no doubt be expanded with the development of peptide mimetics that are not rapidly metabolized. Such compounds could be thought of either as blocking active enzyme or as restoring a "natural" latency of the MMP.

Ways must be found to regulate the gene expression of MMPs through drugs acting on the promoter region. Retinoids and tetracyclines are among the compounds being explored for this purpose. In addition to treating gingival disease, tetracyclines have been shown to reduce levels of MMP protein in osteoarthritic cartilage,[106] melanoma cell,[107] and prostate cancer cells.[108] This is in addition to direct inhibition of MMPs owing to the chelating properties of tetracyclines.

Finally, one may mention the challenge to the field presented by the newly emerging ADAMs family of proteases. These enzymes contain *a d*isintegrin and *a m*etalloprotease domain, closely resembling MMPs in having the metzincin protein fold, but belonging to a distinct family of matrix metalloproteinases. Considerable attention has also been given lately to TACE, the TNFα-converting enzyme. This, and many related enzymes, are found on the cell surface, where they release various growth factors and cytokines. They are often referred to as "sheddases." These enzymes are inhibited by many of the same synthetic inhibitors, such as hydroxamates that block the MMPs,[109] although they are not blocked by TIMPs.[110] The challenge then will be to develop inhibitors not only specific for individual MMPs, but ones that will not affect the ADAMs. Of course, in some situations such as arthritis, it may be useful to block both activities, reducing inflammation as well as tissue destruction.[111]

In sum, while the MMP field has matured considerably, it is far from moribund and is just entering its prime of life. Many important disease processes may be controlled through regulation of MMPs and many important normal developmental processes will be better understood when the contribution of MMPs becomes clearer.

## ACKNOWLEDGMENT

This work was supported by Grant AR-16940 from the National Institutes of Health.

## REFERENCES

1. GROSS, J. & C.M. LAPIÈRE. 1962. Collagenolytic activity in amphibian tissues: a tissue culture assay. Proc. Natl. Acad. Sci. USA **48:** 1014–1022.
2. GROSS, J. & Y. NAGAI. 1965. Specific degradation of the collagen molecule by tadpole collagenolytic enzyme. Proc. Natl. Acad. Sci. USA **54:** 1197–1204.
3. BAUER, E.A., A.Z. EISEN & J.J. JEFFREY. 1970. Immunologic relationship of a purified human skin collagenase to other human and animal collagenases. Biochim. Biophys. Acta **206:** 152–160.
4. HARPER, E., K.J. BLOCH & J. GROSS. 1971. The zymogen of tadpole collagenase. Biochemistry **10:** 3035–3041.
5. LAZARUS, G.S., R.S. BROWN, J.R. DANIELS & H.M. FULLMER. 1968. Human granulocyte collagenase. Science **159:** 1483–1485.
6. WERB, Z. & J.J. REYNOLDS. 1974. Stimulation by endocytosis of the secretion of collagenase and neutral proteinase from rabbit synovial fibroblasts. J. Exp. Med. **140:** 1482–1497.
7. SOPATA, I. & A.M. DANCEWICZ. 1974. Presence of a gelatin-specific proteinase and its latent form in human leucocytes. Biochim. Biophys. Acta **370:** 510–523.
8. SELLERS, A., J.J. REYNOLDS & M.C. MEIKLE. 1978. Neutral metallo-proteinases of rabbit bone. Separation in latent forms of distinct enzymes that when activated degrade collagen, gelatin and proteoglycans. Biochem. J. **171:** 493–496.
9. SELLERS, A. & J.F. WOESSNER, JR. 1980. The extraction of a neutral metalloproteinase from the involuting rat uterus, and its action on cartilage proteoglycan. Biochem. J. **189:** 521–531.
10. BANDA, M.J., E.J. CLARK & Z. WERB. 1980. Limited proteolysis by macrophage elastase inactivates human alpha 1-proteinase inhibitor. J. Exp. Med. **152:** 1563–1570.
11. BREATHNACH, R., L.M. MATRISIAN, M.-C. GESNEL, A. STAUB & P. LEROY. 1987. Sequences coding for part of oncogene-induced transin are highly conserved in a related rat gene. Nucleic Acids Res. **15:** 1139–1151.
12. BASSET, P., J.P. BELLOCQ, C. WOLF, I. STOLL, P. HUTIN et al. 1990. A novel metalloproteinase gene specifically expressed in stromal cells of breast carcinomas. Nature **348:** 699–704.
13. FREIJE, J.M., I. DÍEZ-ITZA, M. BALBÍN, L.M. SÁNCHEZ, R. BLASCO et al. 1994. Molecular cloning and expression of collagenase-3, a novel human matrix metalloproteinase produced by breast carcinomas. J. Biol. Chem. **269:** 16766–16773.
14. SATO, H., T. TAKINO, Y. OKADA, J. CAO, A. SHINAGAWA et al. 1994. A matrix metalloproteinase expressed on the surface of invasive tumour cells. Nature **370:** 61–65.
15. WILL, H. & B. HINZMANN. 1995. cDNA sequence and mRNA tissue distribution of a novel human matrix metalloproteinase with a potential transmembrane segment. Eur. J. Biochem. **231:** 602–608.

16. TAKINO, T., H. SATO, A. SHINAGAWA & M. SEIKI. 1995. Identification of the second membrane-type matrix metalloproteinase (MT-MMP-2) gene from a human placenta cDNA library. MT-MMPs form a unique membrane-type subclass in the MMP family. J. Biol. Chem. **270:** 23013–23020.

17. PUENTE, X.S., A.M. PENDÁS, E. LLANO, G. VELASCO & C. LÓPEZ-OTÍN. 1996. Molecular cloning of a novel membrane-type matrix metalloproteinase from a human breast carcinoma. Cancer Res. **56:** 944–949.

18. STOLOW, M.A., D.D. BAUZON, J. LI, T. SEDGWICK, V.C. LIANG et al. 1996. Identification and characterization of a novel collagenase in *Xenopus laevis*: possible roles during frog development. Mol. Biol. Cell **7:** 1471–1483.

19. KARLSON, J., A.K. BORG-KARLSON, R. UNELIUS, M.C. SHOSHAN, N. WILKING et al. 1996. Inhibition of tumor cell growth by monoterpenes *in vitro*: evidence of a Ras-independent mechanism of action. Anticancer Drugs **7:** 422–429.

20. BARTLETT, J.D., J.P. SIMMER, J. XUE, H.C. MARGOLIS & E.C. MORENO. 1996. Molecular cloning and mRNA tissue distribution of a novel matrix metalloproteinase isolated from porcine enamel organ. Gene **183:** 123–128.

21. YANG, M.Z., M.T. MURRAY & M. KURKINEN. 1997. A novel matrix metalloproteinase gene (XMMP) encoding vitronectin-like motifs is transiently expressed in *Xenopus laevis* early embryo development. J. Biol. Chem. **272:** 13527–13533.

22. NAGAI, Y., C.M. LAPIÈRE & J. GROSS. 1966. Tadpole collagenase. Preparation and purification. Biochemistry **5:** 3123–3130.

23. BERMAN, M.B. & R. MANABE. 1973. Corneal collagenases: evidence for zinc metalloenzymes. Ann. Ophthalmol. **5:** 1193–1209.

24. BROWN, S.I., S. AKIYA & C.A. WELLER. 1969. Prevention of the ulcers of the alkali-burned cornea. Preliminary studies with collagenase inhibitors. Arch. Ophthalmol. **82:** 95–97.

25. PARKS, W.C. & R.P. MECHAM. 1998. Matrix Metalloproteinases. Academic Press. San Diego, CA.

26. WOESSNER, J.F., JR. 1998. The matrix metalloproteinase family. *In* Matrix Metalloproteinases. W.C. Parks & R.P. Mecham, Eds. :1–14. Academic Press. San Diego, CA.

27. MECHAM, R.P., T.J. BROEKELMANN, C.J. FLISZAR, S.D. SHAPIRO, H.G. WELGUS et al. 1997. Elastin degradation by matrix metalloproteinases—cleavage site specificity and mechanisms of elastolysis. J. Biol. Chem. **272:** 18071–18076.

28. D'ARMIENTO, J., S.S. DALAL, Y. OKADA, R.A. BERG & K. CHADA. 1992. Collagenase expression in the lungs of transgenic mice causes pulmonary emphysema. Cell **71:** 955–961.

29. D'ARMIENTO, J., T. DICOLANDREA, S.S. DALAL, Y. OKADA, M.T. HUANG et al. 1995. Collagenase expression in transgenic mouse skin causes hyperkeratosis and acanthosis and increases susceptibility to tumorigenesis. Mol. Cell. Biol. **15:** 5732–5739.

30. SYMPSON, C.J., R.S. TALHOUK, C.M. ALEXANDER, J.R. CHIN, S.M. CLIFT et al. 1994. Targeted expression of stromelysin-1 in mammary gland provides evidence for a role of proteinases in branching morphogenesis and the requirement for an intact basement membrane for tissue- specific gene expression. J. Cell Biol. **125:** 681–693.

31. RUDOLPH-OWEN, L.A., P. CANNON & L.M. MATRISIAN. 1998. Overexpression of the matrix metalloproteinase matrilysin results in premature mammary gland differentiation and male infertility. Mol. Biol.Cell **9:** 421–435.

32. LIU, X., H. WU, M. BYRNE, J. JEFFREY, S. KRANE et al. 1995. A targeted mutation at the known collagenase cleavage site in mouse type I collagen impairs tissue remodeling. J. Cell Biol. **130:** 227–237.

33. ITOH, T., M. TANIOKA, H. YOSHIDA, T. YOSHIOKA, H. NISHIMOTO et al. 1998. Reduced angiogenesis and tumor progression in gelatinase A-deficient mice. Cancer Res. **58:** 1048–1051.

34. RUDOLPH-OWEN, L.A., D.L. HULBOY, C.L. WILSON, J. MUDGETT & L.M. MATRISIAN. 1997. Coordinate expression of matrix metalloproteinase family members in the uterus of normal, matrilysin-deficient, and stromelysin-1-deficient mice. Endocrinology 138: 4902–4911.

35. MASSON, R., O. LEFEBVRE, A. NOEL, M. EL FAHIME, M.P. CHENARD et al. 1998. In vivo evidence that the stromelysin-3 metalloproteinase contributes in a paracrine manner to epithelial cell malignancy. J. Cell Biol. 140: 1535–1541.

36. WILSON, C.L., K.J. HEPPNER, P.A. LABOSKY, B.L. HOGAN & L.M. MATRISIAN. 1997. Intestinal tumorigenesis is suppressed in mice lacking the metalloproteinase matrilysin. Proc. Natl. Acad. Sci. USA 94: 1402–1407.

37. VU, T.H., J.M. SHIPLEY, G. BERGERS, J.E. BERGER, J.A. HELMS et al. 1998. MMP-9/ gelatinase B is a key regulator of growth plate angiogenesis and apoptosis of hypertrophic chondrocytes. Cell 93: 411–422.

38. HAUTAMAKI, R.D., D.K. KOBAYASHI, R.M. SENIOR & S.D. SHAPIRO. 1997. Requirement for macrophage elastase for cigarette smoke-induced emphysema in mice. Science 277: 2002–2004.

39. WITTY, J.P., S. MCDONNELL, K.J. NEWELL, P. CANNON, M. NAVRE et al. 1994. Modulation of matrilysin levels in colon carcinoma cell lines affects tumorigenicity in vivo. Cancer Res. 54: 4805–4812.

40. YAMAMOTO, H., F. ITOH, Y. HINODA & K. IMAI. 1995. Suppression of matrilysin inhibits colon cancer cell invasion in vitro. Int. J. Cancer 61: 218–222.

41. DERYUGINA, E.I., G.X. LUO, R.A. REISFELD, M.A. BOURDON & A. STRONGIN. 1997. Tumor cell invasion through matrigel is regulated by activated matrix metalloproteinase-2. Anticancer Res. 17: 3201–3210.

42. TURCK, J., A.S. POLLOCK, L.K. LEE, H.P. MARTI & D.H. LOVETT. 1996. Matrix metalloproteinase 2 (gelatinase A) regulates glomerular mesangial cell proliferation and differentiation. J. Biol. Chem. 271: 15074–15083.

43. LOCHTER, A., A. SREBROW, C.J. SYMPSON, N. TERRACIO, Z. WERB et al. 1997. Misregulation of stromelysin-1 expression in mouse mammary tumor cells accompanies acquisition of stromelysin-1-dependent invasive properties. J. Biol. Chem. 272: 5007–5015.

44. DRAPER, K.G., P. PAVCO, J. MCSWIGGEN, J. GUSTOFSON, & D.T. STINCHCOMB, inventors; Ribozyme Pharmaceuticals, assignee. 1997. Ribozymes that cleave human stromelysin mRNA and can be used as inflammation inhibitors or for arthritis treatment. US Patent 5,612,215, pp.1–241. Application date: February 17, 1995.

45. HAYAKAWA, T., N. FUJIMOTO, R.V. WARD & K. IWATA. 1994. Interaction between progelatinase A and TIMP-2. Ann. N. Y. Acad. Sci. 732: 389–391.

46. HUA, J. & R.J. MUSCHEL. 1996. Inhibition of matrix metalloproteinase 9 expression by a ribozyme blocks metastasis in a rat sarcoma model system. Cancer Res. 56: 5279–5284.

47. MELONG, R.K. & P.S. MILLER. 1996. Inhibition of human collagenase activity by antisense oligonucleoside methylphosphonates. Antisense Nucleic Acid Drug Dev. 6: 273–280.

48. NAGASE, H., Y. ITOH & S. BINNER. 1994. Interaction of alpha 2-macroglobulin with matrix metalloproteinases and its use for identification of their active forms. Ann. N. Y. Acad. Sci. 732: 294–302.

49. YU, W.-H. & J.F. WOESSNER, JR. 1997. Binding of matrilysin to glycosaminoglycan [abstract]. FASEB J. 9: A1227.

50. MALLYA, S.K. & H.E. VAN WART. 1989. Mechanism of inhibition of human neutrophil collagenase by gold(I) chrysotherapeutic compounds. Interaction at a heavy metal binding site. J. Biol. Chem. 264: 1594–1601.

51. WAUTERS, P., Y. EECKHOUT & G. VAES. 1986. Oxidation products are responsible for the resistance to the action of collagenase conferred on collagen by (+)-catechin. Biochem. Pharmacol. **35:** 2971–2973.

52. MASUI, Y., T. TAKEMOTO, S. SAKAKIBARA, H. HORI & Y. NAGAI. 1977. Synthetic substrates for vertebrate collagenase. Biochem. Med. **17:** 215–221.

53. GRAY, R.D., H.H. SANEII & A.F. SPATOLA. 1981. Metal binding peptide inhibitors of vertebrate collagenase. Biochem. Biophys. Res. Commun. **101:** 1251–1258.

54. MCCULLAGH, K., H. WADSWORTH, & M. HANN, inventors; G.D. Searle & Co., assignee. 1984. Carboxyalkyl peptide derivatives. Eur. Patent No. 126,974, pp. 1-111. Date of application: April 26, 1983.

55. DELAISSÉ, J.M., Y. EECKHOUT, C. SEAR, A. GALLOWAY, K. MCCULLAGH et al. 1985. A new synthetic inhibitor of mammalian tissue collagenase inhibits bone resorption in culture. Biochem.Biophys.Res.Commun. **133:** 483–490.

56. BRÄNNSTRÖM, M., J.F. WOESSNER, JR., R.D. KOOS, C.H. SEAR & W.J. LEMAIRE. 1988. Inhibitors of mammalian tissue collagenase and metalloproteinases suppress ovulation in the perfused rat ovary. Endocrinology **122:** 1715–1721.

57. BUTLER, T.A., C. ZHU, R.A. MUELLER, G.C. FULLER, W.J. LEMAIRE & J.F. WOESSNER, JR. 1991. Inhibition of ovulation in the perfused rat ovary by the synthetic collagenase inhibitor SC 44463. Biol. Reprod. **44:** 1183–1188.

58. GALARDY, R.E., D. GROBELNY, Z.P. KORYLEWICZ & L. PONCZ. 1992. Inhibition of human skin collagenase by phosphorus-containing peptides. Matrix Suppl. **1:** 259–262.

59. LOVEJOY, B., A. CLEASBY, A.M. HASSELL, K. LONGLEY, M.A. LUTHER et al. 1994. Structure of the catalytic domain of fibroblast collagenase complexed with an inhibitor. Science **263:** 375–377.

60. MORPHY, J.R., N.R.A. BEELEY, B.A. BOYCE, J. LEONARD, B. MASON et al. 1994. Potent and selective inhibitors of gelatinase A. 2. Carboxylic and phosphonic acid derivatives. Bioorg. Med. Chem. Lett. **4:** 2747–2752.

61. ESSER, C.K., N.J. KEVIN, N.A. YATES & K.T. CHAPMAN. 1997. Solid-phase synthesis of a N-carboxyalkyl tripeptide combinatorial library. Bioorg. Med. Chem. Lett. **7:** 2639–2644.

62. LEVIN, J.I., J.F. DIJOSEPH, L.M. KILLAR, M.A. SHARR, J.S. SKOTNICKI et al. 1998. The asymmetric synthesis and in vitro characterization of succinyl mercaptoalcohol and mercaptoketone inhibitors of matrix metalloproteinases. Bioorg. Med. Chem. Lett. **8:** 1163–1168.

63. CHEN, J.J., Y.P. ZHANG, S. HAMMOND, N. DEWDNEY, T. HO et al. 1996. Design, synthesis, activity & structure of a novel class of matrix metalloproteinase inhibitors containing a heterocyclic P- 2'-P-3' amide bond isostere. Bioorg. Med. Chem. Lett. **6:** 1601–1606.

64. LEVY, D.E., F. LAPIERRE, W.S. LIANG, W.Q. YE, C.W. LANGE et al. 1998. Matrix metalloproteinase inhibitors - a structure-activity study. J. Med. Chem. **41:** 199–223.

65. MACPHERSON, L.J., E.K. BAYBURT, M.P. CAPPARELLI, B.J. CARROLL, R. GOLDSTEIN et al. 1997. Discovery of CGS 27023A, a non-peptidic, potent, and orally active stromelysin inhibitor that blocks cartilage degradation in rabbits. J. Med. Chem. **40:** 2525–2532.

66. WOJTOWICZ-PRAGA, S., J. LOW, J. MARSHALL, E. NESS, R. DICKSON et al. 1996. Phase I trial of a novel matrix metalloproteinase inhibitor batimastat (BB-94) in patients with advanced cancer. Invest. New Drugs **14:** 193–202.

67. LEWIS, E.J., J. BISHOP, K.M.K. BOTTOMLEY, D. BRADSHAW, M. BREWSTER et al. 1997. Ro 32-3555, an orally active collagenase inhibitor, prevents cartilage breakdown in vitro and in vivo. Br. J. Pharmacol. **121:** 540–546.

68. BROWN, P.D. 1998. Synthetic inhibitors of matrix metalloproteinases. *In* Matrix Metalloproteinases. W.C. Parks and R.P. Mecham, Eds. :243-261. Academic Press. San Diego, CA.

69. FAUCHER, D.C., Y. LELIÈVRE & T. CARTWRIGHT. 1987. An inhibitor of mammalian collagenase active at micromolar concentrations from an actinomycete culture broth. J. Antibiot. (Tokyo) **40**: 1757–1761.

70. OGITA, T., A. SATO, R. ENOKITA, K. SUZUKI, M. ISHII *et al.* 1992. Matlystatins, new inhibitors of typeIV collagenases from Actinomadura atramentaria. I. Taxonomy, fermentation, isolation, and physico-chemical properties of matlystatin-group compounds. J. Antibiot. (Tokyo) **45**: 1723–1732.

71. TAMAKI, K., K. TANZAWA, S. KURIHARA, T. OIKAWA, S. MONMA *et al.* 1995. Synthesis and structure-activity relationships of gelatinase inhibitors derived from matlystatins. Chem. Pharm. Bull. (Tokyo) **43**: 1883–1893.

72. SUZUKI, K., K. SHIMADA, S. NOZOE, K. TANZAWA & T. OGITA. 1996. Isolation of nicotianamine as a gelatinase inhibitor. J. Antibiot. (Tokyo) **49**: 1284–1285.

73. GOLUB, L.M., H.M. LEE, G. LEHRER, A. NEMIROFF, T.F. MCNAMARA *et al.* 1983. Minocycline reduces gingival collagenolytic activity during diabetes. Preliminary observations and a proposed new mechanism of action. J. Periodont. Res. **18**: 516–526.

74. GOLUB, L.M., T.F. MCNAMARA, G. D'ANGELO, R.A. GREENWALD & N.S. RAMAMURTHY. 1987. A non-antibacterial chemically-modified tetracycline inhibits mammalian collagenase activity. J. Dent. Res. **66**: 1310–1314.

75. SORSA, T., N.S. RAMAMURTHY, A.T. VERNILLO, X. ZHANG, Y.T. KONTTINEN *et al.* 1998. Functional sites of chemically modified tetracyclines—inhibition of the oxidative activation of human neutrophil and chicken osteoclast pro-matrix metalloproteinases. J. Rheumatol. **25**: 975–982.

76. UITTO, V.-J., J.D. FIRTH, L. NIP & L.M. GOLUB. 1994. Doxycycline and chemically modified tetracyclines inhibit gelatinase A (MMP-2) gene expression in human skin keratinocytes. Ann. N. Y. Acad. Sci. **732**: 140–151.

77. JONAT, C., F.Z. CHUNG & V.M. BARAGI. 1996. Transcriptional downregulation of stromelysin by tetracycline. J. Cell. Biochem. **60**: 341–347.

78. STETLER-STEVENSON, W.G., J.A. TALANO, M.E. GALLAGHER, H.C. KRUTZSCH & L.A. LIOTTA. 1991. Inhibition of human type IV collagenase by a highly conserved peptide sequence derived from its prosegment. Am. J. Med. Sci. **302**: 163–170.

79. FOTOUHI, N., A. LUGO, M. VISNICK, L. LUSCH, R. WALSKY *et al.* 1994. Potent peptide inhibitors of stromelysin based on the prodomain region of matrix metalloproteinases. J. Biol. Chem. **269**: 30227–30231.

80. BAUER, E.A., G.P. STRICKLIN, J.J. JEFFREY & A.Z. EISEN. 1975. Collagenase production by human skin fibroblasts. Biochem. Biophys. Res. Commun. **64**: 232–240.

81. WOOLLEY, D.E., D.R. ROBERTS & J.M. EVANSON. 1975. Inhibition of human collagenase activity by a small molecular weight serum protein. Biochem. Biophys. Res. Commun. **66**: 747–754.

82. WELGUS, H.G., G.P. STRICKLIN, A.Z. EISEN, E.A. BAUER, R.V. COONEY *et al.* 1979. A specific inhibitor of vertebrate collagenase produced by human skin fibroblasts. J. Biol. Chem. **254**: 1938–1943.

83. CAWSTON, T.E., W.A. GALLOWAY, E. MERCER, G. MURPHY & J.J. REYNOLDS. 1981. Purification of rabbit bone inhibitor of collagenase. Biochem. J. **195**: 159–165.

84. DOCHERTY, A.J.P., A. LYONS, B.J. SMITH, E.M. WRIGHT, P.E. STEPHENS *et al.* 1985. Sequence of human tissue inhibitor of metalloproteinases and its identity to erythroid-potentiating activity. Nature **318**: 66–69.

85. HERRON, G.S., M.J. BANDA, E.J. CLARK, J. GAVRILOVIC & Z. WERB. 1986. Secretion of metalloproteinases by stimulated capillary endothelial cells. II. Expression of collagenase and stromelysin activities is regulated by endogenous inhibitors. J. Biol. Chem. **261:** 2814–2818.

86. DeCLERCK, Y.A. 1988. Purification and characterization of a collagenase inhibitor produced by bovine vascular smooth muscle cells. Arch. Biochem. Biophys. **265:** 28–37.

87. STETLER-STEVENSON, W.G., H.C. KRUTZSCH & L.A. LIOTTA. 1989. Tissue inhibitor of metalloproteinase (TIMP-2). A new member of the metalloproteinase inhibitor family. J. Biol. Chem. **264:** 17374–17378.

88. BLENIS, J. & S.P. HAWKES. 1983. Transformation-sensitive protein associated with the cell substratum of chicken embryo fibroblasts. Proc. Natl. Acad. Sci. USA **80:** 770–774.

89. STASKUS, P.W., F.R. MASIARZ, L.J. PALLANCK & S.P. HAWKES. 1991. The 21-kDa protein is a transformation-sensitive metalloproteinase inhibitor of chicken fibroblasts. J. Biol. Chem. **266:** 449–454.

90. PAVLOFF, N., P.W. STASKUS, N.S. KISHNANI & S.P. HAWKES. 1992. A new inhibitor of metalloproteinases from chicken: ChIMP-3. A third member of the TIMP family. J. Biol. Chem. **267:** 17321–17326.

91. GREENE, J., M. WANG, Y.E. LIU, L.A. RAYMOND, C. ROSEN et al. 1996. Molecular cloning and characterization of human tissue inhibitor of metalloproteinase 4. J. Biol. Chem. **271:** 30375–30380.

92. GOMEZ, D.E., D.F. ALONSO, H. YOSHIJI & U.P. THORGEIRSSON. 1997. Tissue inhibitors of metalloproteinases - structure, regulation and biological functions. Eur. J. Cell. Biol. **74:** 111–122.

93. WINDSOR, L.J., D.L. STEELE, S.B. LeBLANC & K.B. TAYLOR. 1997. Catalytic domain comparisons of human fibroblast-type collagenase, stromelysin-1, and matrilysin. Biochim. Biophys. Acta **1334:** 261–272.

94. MURPHY, G., A. HOUBRECHTS, M.I. COCKETT, R.A. WILLIAMSON, M. O'SHEA et al. 1991. The N-terminal domain of tissue inhibitor of metalloproteinases retains metalloproteinase inhibitory activity. Biochemistry **30:** 8097–8102.

95. CRABBE, T., S.M. KELLY & N.C. PRICE. 1996. An analysis of the conformational changes that accompany the activation and inhibition of gelatinase A. FEBS Lett. **380:** 53–57.

96. D'ORTHO, M.P., H. STANTON, M. BUTLER, S.J. ATKINSON, G. MURPHY et al. 1998. MT1-MMP on the cell surface causes focal degradation of gelatin films. FEBS Lett. **421:** 159–164.

97. GOMIS-RÜTH, F.X., K. MASKOS, M. BETZ, A. BERGNER, R. HUBER et al. 1997. Mechanism of inhibition of the human matrix metalloproteinase stromelysin-1 by TIMP-1. Nature **389:** 77–81.

98. HAYAKAWA, T., K. YAMASHITA, E. OHUCHI & A. SHINAGAWA. 1994. Cell growth-promoting activity of tissue inhibitor of metalloproteinases-2 (TIMP-2). J. Cell Sci. **107:** 2373–2379.

99. CHESLER, L., D.W. GOLDE, N. BERSCH & M.D. JOHNSON. 1995. Metalloproteinase inhibition and erythroid potentiation are independent activities of tissue inhibitor of metalloproteinases-1. Blood **86:** 4506–4515.

100. NAGASE, H., K. SUZUKI, Y. ITOH, C.C. KAN, M.R. GEHRING et al. 1996. Involvement of tissue inhibitors of metalloproteinases (TIMPS) during matrix metalloproteinase activation. Adv. Exp. Med. Biol. **389:** 23–31.

101. FLORY, C.M., P.A. PAVCO, T.C. JARVIS, M.E. LESCH, F.E. WINCOTT et al. 1996. Nuclease-resistant ribozymes decrease stromelysin mRNA levels in rabbit synovium following exogenous delivery to the knee joint. Proc. Natl. Acad. Sci. USA **93:** 754–758.

102. PEI, D. & S.J. WEISS. 1995. Furin-dependent intracellular activation of the human stromelysin-3 zymogen. Nature **375:** 244–247.

103. MAQUOI, E., A. NOEL, F. FRANKENNE, H. ANGLIKER, G. MURPHY *et al.* 1998. Inhibition of matrix metalloproteinase 2 maturation and HT1080 invasiveness by a synthetic furin inhibitor. FEBS Lett. **424:** 262–266.

104. NAGASE, H. 1997. Activation mechanisms of matrix metalloproteinases. Biol. Chem. **378:** 151–160.

105. LIJNEN, H.R., B. VANHOEF, F. LUPU, L. MOONS, P. CARMELIET *et al.* 1998. Function of the plasminogen/plasmin and matrix metalloproteinase systems after vascular injury in mice with targeted inactivation of fibrinolytic system genes. Arterioscl. Thromb. Vasc. Biol. **18:** 1035–1045.

106. SMITH, G.N., L.P. YU, K.D. BRANDT & W.N. CAPELLO. 1998. Oral administration of doxycycline reduces collagenase and gelatinase activities in extracts of human osteoarthritic cartilage. J. Rheumatol. **25:** 532–535.

107. SEFTOR, R.E.B., E.A. SEFTOR, J.E. DE LARCO, D.E. KLEINER, J. LEFERSON *et al.* 1998. Chemically modified tetracyclines inhibit human melanoma cell invasion and metastasis. Clin. Exp. Metast. **16:** 217–225.

108. FIFE, R.S., G.W. SLEDGE, B.J. ROTH & C. PROCTOR. 1998. Effects of doxycycline on human prostate cancer cells in vitro. Cancer Lett. **127:** 37–41.

109. BLACK, R.A., C.T. RAUCH, C.J. KOZLOSKY, J.J. PESCHON, J.L. SLACK *et al.* 1997. A metalloproteinase disintegrin that releases tumour-necrosis factor-alpha from cells. Nature **385:** 729–733.

110. BLACK, R.A., F.H. DURIE, C. OTTEN-EVANS, R. MILLER, J.L. SLACK *et al.* 1996. Relaxed specificity of matrix metalloproteinases (MMPS) and TIMP insensitivity of tumor necrosis factor-alpha (TNF-alpha) production suggest the major TNF-alpha converting enzyme is not an MMP. Biochem. Biophys. Res. Commun. **225:** 400–405.

111. DIMARTINO, M., C. WOLFF, W. HIGH, G. STROUP, S. HOFFMAN et al. 1997. Anti-arthritic activity of hydroxamic acid-based pseudopeptide inhibitors of matrix metalloproteinases and TNF-alpha processing. Inflamm. Res. **46:** 211–215.

# C-Telopeptide Pyridinoline Cross-Links

## Sensitive Indicators of Periodontal Tissue Destruction

WILLIAM V. GIANNOBILE[a]

*Department of Periodontics/Prevention/Geriatrics, University of Michigan, 1011 North University Avenue, Ann Arbor, Michigan 48109-1078, USA*

ABSTRACT: C-telopeptides and related pyridinoline cross-links of bone Type I collagen are sensitive markers of bone resorption in osteolytic diseases such as osteoporosis and osteoarthritis. We have studied the release of C-telopeptide pyridinoline crosslinks of Type I collagen as measures of bone destruction in periodontal disease. Studies in preclinical animal models and humans have demonstrated the relationship between radiographic bone loss and crevicular fluid C-telopeptide levels. We have recently found that C-telopeptide levels correlate strongly with microbial pathogens associated with periodontitis and around endosseous dental implants. Host-modulation of bone-related collagen breakdown has been shown by studies in humans demonstrating that MMP inhibition blocks tissue destruction and release of C-telopeptides in patients with active periodontal disease.

Periodontal disease is initiated by microbial pathogens that elicit a host immune response with subsequent tissue destruction of the periodontal structures.[1] Patients afflicted with periodontitis experience breakdown of alveolar bone, periodontal ligament, and tooth root cementum. Uncontrolled periodontal tissue destruction eventually leads to tooth loss. A challenge in periodontology is the accurate and early assessment of tissue breakdown in patients with periodontal disease. Current methods of disease detection lack diagnostic sensitivity and specificity. Periodontal diagnostic procedures address several important functions, which include: screening; diagnosis of specific periodontal diseases; identification of sites or subjects at increased risk of experiencing progression of periodontal destruction; treatment planning; and monitoring of therapy.[2]

This paper focuses on the ability of the class of molecules known as the pyridinoline cross-links to identify tooth sites or subjects at an increased risk of experiencing periodontal tissue destruction. Research will be reviewed in preclinical and clinical investigations that illustrate the relationship among pyridinoline cross-links, periodontal pathogens, and alveolar bone loss. The paper concludes with recent developments in the inhibition of periodontal tissue destruction by matrix metalloproteinase (MMP) inhibitors such as doxycycline.

[a] Address for telecommunication: Phone, 734/764-1562; fax, 734/763-5503; e-mail, wgiannob@umich.edu

## PYRIDINOLINE CROSS-LINKS:
## BIOCHEMISTRY AND CLINICAL CORRELATIONS

Pyridinoline cross-links have emerged as very promising biomarkers of bone resorption in an array of osteolytic diseases. Pyridinoline (hydroxylysl pyridinoline or Pyr) and deoxypyridinoline (lysyl pyridinoline or Dpy), N-telopeptides, and C-telopeptides have been the best studied members of this class of collagen-degradative molecules.[3] Newly formed collagen fibrils deposited in the extracellular matrix of bone are stabilized by mature cross-links formed by lysyl oxidase on lysine and hydroxylysine residues in the N- and C-terminal regions of collagen chains. This process results in the formation of divalent collagen cross-links that by further condensation yield trivalent Pyr and Dpy.

Osteoclastic bone resorption initiates the release of cross-linked immunoreactive telopeptides Pyr and Dpy. Urinary levels of Pyr and Dpy correlate with histological parameters of bone turnover from bone biopsies[4] as well as by radiopharmaceutical uptake at sites of bone resorption.[5] In bone turnover diseases such as osteoporosis,[6] rheumatoid arthritis,[7] and Paget's disease,[8] increased levels of Pyr and Dpy are found in the circulation. In patients with metastatic bone disease, bisphosphanate therapy potently decreased circulating levels of Pyr and Dpy.[9] Furthermore, postmenopausal osteoporotic subjects experienced significant decreases in Pyr and Dpy after bisphosphanate[10] or estrogen[11] therapy, coincident with increases in spinal bone mineral density.

## PYRIDINOLINE CROSS-LINKS AND BONE RESORPTION
## IN PERIODONTITIS

Given the important contribution of Pyr cross-links in systemic osteolytic diseases, over the past half of a decade researchers have begun to explore the ability of these molecules to detect bone resorption in the periodontium. The majority of diagnostic agents currently target the use of gingival crevicular fluid (GCF), which is an exudate that can be harvested from the gingival sulcus or periodontal pocket. As GCF traverses the inflamed tissue, it appears to carry molecules involved in the destructive process (FIG. 1).[12] Therefore, GCF offers great potential as a source for factors that may be associated with active tissue destruction. Several investigators have assessed the roles of various connective tissue breakdown products or enzymes involved in bone metabolism from GCF (reviewed in Refs. 13 and 14). However, most of these molecules are not bone-specific markers of tissue breakdown, and thus are simply mediators of combined soft and hard tissue inflammatory events (FIG. 2). Host factors in the GCF associated with the anatomic events of periodontitis and peri-implantitis may be useful as markers for identifying and predicting future disease progression. A biochemical marker specific for bone degradation may be useful to differentiate the presence of gingival inflammation from active periodontal and peri-implant bone destruction.[14]

Recently, several studies have been conducted examining C-telopeptides in GCF. Talonpoika and Hämäläinen demonstrated strong correlations between C-telopeptide

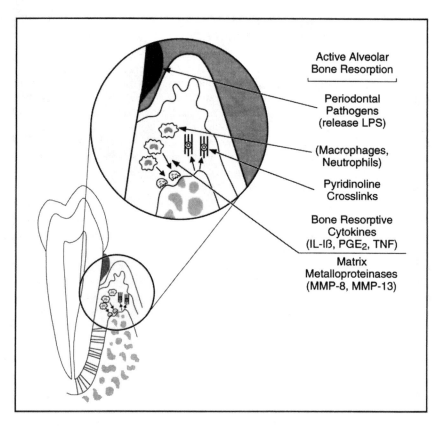

**FIGURE 1.** Release of pyridinoline cross-links during active alveolar bone resorption in periodontitis. Periodontitis is initiated by bacterial plaque and endotoxin, which cause a host immune response with recruitment of inflammatory cells such as neutrophils to the periodontal lesion. Neutrophils release MMP-8, which causes alveolar bone, periodontal ligament, and cementum destruction. Pyridinoline cross-links such as Dpy and C-telopeptide are released into the GCF soon after the initiation of osteoclastic bone resorption in periodontitis.

levels and clinical parameters of periodontal tissue destruction such as radiographic bone level and pocket depth.[15] Furthermore, they showed that in a small subset of patients provided periodontal therapy, dramatic reduction of GCF C-telopeptide levels resulted as soon as 2 days after treatment. In a ligature-induced experimental periodontitis study in dogs, the GCF levels of pyridinoline cross-links (C-telopeptide and deoxypyridinoline) significantly increased during the development of attachment loss and osteoclastic bone resorption[16,17] (Fig. 3). As early as 3 days after the initiation of experimental disease, tartrate-resistant acid phosphastase (TRAP+) mononuclear and TRAP+ multinucleated cells could be noted at the alveolar crest and gingival connective tissue.[17] The TRAP+ cells related strongly with elevated levels of Dpy in the serum, urine and GCF, suggesting that these markers are sensitive enough to be detected not only locally in the GCF but also systemically in the circulation. Moreover,

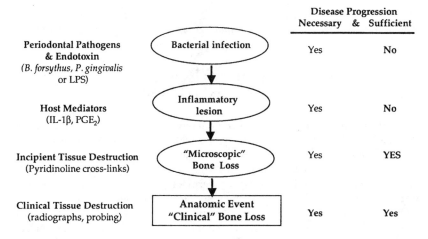

FIGURE 2. Checkpoints of periodontal disease progression. This diagram demonstrates examples of pathogens, cytokines, tissue breakdown components, and clinical indices used to detect stages of active periodontitis. The pathway of pathogenic infection by bacteria and endotoxin leading to a host inflammatory response results in tissue destruction as measured by bone-degradative products (e.g., pyridinoline cross-links). The anatomic event of alveolar bone loss demonstrated on a radiograph or periodontal probing gives the patient's diagnosis of periodontitis. Interventions identifying tissue destruction prior to significant anatomic events are goals of periodontal diagnostic tests.

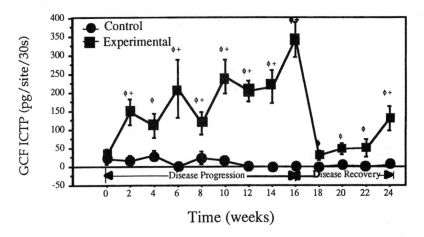

Time (weeks)

FIGURE 3. Crevicular fluid C-telopeptide levels in response to experimental periodontitis in beagles. Note significant increases in GCF C-telopeptide throughout the entire disease progression phase following the initiation of periodontal disease by silk ligatures. Once ligatures were removed, concomitant with scaling and root planing, GCF telopeptide levels decreased significantly compared to baseline for the first 6 weeks of the disease recovery phase followed by another significant increase in C-telopeptide at week 24. *Error bars* represent ± SE from a total of 36 samples/time point. [+]$p < 0.05$ compared to control; [φ]$p < 0.05$ compared to baseline. (From Giannobile *et al.*[16] Reproduced by permission.)

C-telopeptide levels in GCF were highly sensitive and specific for predicting future alveolar bone loss as measured by computer-assisted digitizing radiography.[16]

## RELATIONSHIP BETWEEN PERIODONTAL PATHOGENS AND C-TELOPEPTIDES

Destructive periodontal diseases can be thought of as a series of infections that affect individual or multiple periodontal sites within an individual.[1] A large body of literature supports an etiologic role of selected subgingival species such as *Bacteroides forsythus, Porphyromonas gingivalis,* and *Treponema denticola* in periodontal diseases.[18] These species may promote periodontal tissue destruction, including bone resorption, by the expression of a multitude of virulence factors.

Our group has recently demonstrated the relationship between GCF C-telopeptide levels and the associated subgingival microbiota in patients with periodontal disease.[19] Thirty-six periodontal subjects were evaluated for C-telopeptides in relation to a panel of subgingival species in subjects exhibiting various clinical presentations such as health, gingivitis, and periodontitis. Subgingival plaque and GCF samples were taken from tooth sites in each of 36 subjects. The presence and amounts of 40 subgingival taxa were determined in plaque samples using whole genomic DNA probes and checkerboard DNA-DNA hybridization, and GCF C-telopeptide levels were quantified by RIA. Clinical assessments made at the same sites included pocket depth and attachment level. Relationships between C-telopeptide levels and clinical parameters as well as subgingival species were determined by regression analysis. The results demonstrated significant differences among disease categories for GCF C-telopeptide levels for healthy persons ($1.1 \pm 0.6$ pg/site [mean $\pm$ SEM]) and patients

**TABLE 1. Relationship between prevealence of specific periodontal pathogens and mean subject C-telopeptide levels**

| Species | r(s)* |
|---|---|
| C. rectus | 0.55 |
| B. forsythus | 0.54 |
| F. periodonticum | 0.53 |
| C. showae | 0.52 |
| P. intermedia | 0.49 |
| F. nucleatum ss nucleatum | 0.48 |
| P. nigrescens | 0.47 |
| E. nodatum | 0.46 |
| P. gingivalis | 0.42 |
| F. nucleatum ss polymorphum | 0.42 |
| T. denticola | 0.40 |

*$p < 0.01$.

with gingivitis ($14.8 \pm 6.6$ pg/site) and periodontitis ($30.3 \pm 5.7$ pg/site). C-telopeptide levels related modestly to several clinical parameters. Regression analysis indicated that C-telopeptide levels correlated strongly with mean subject levels of several periodontal pathogens including *B. forsythus, P. gingivalis, P. intermedia, P. nigrescens,* and *T. denticola* ($p < 0.01$) (TABLE 1). The data indicate that there is a positive relationship between C-telopeptides and periodontal pathogens.

## RELATIONSHIP BETWEEN PERIODONTAL PATHOGENS WITH C-TELOPEPTIDES AT ORAL IMPLANT FIXTURES

The treatment of edentulous and partially edentulous individuals with endosseous oral titanium implants is a predictable treatment option for tooth replacement.[20] However, a small portion of the population experiences peri-implantitis, an inflammatory condition characterized by both connective tissue and bone destruction that can result in dental implant failure. Current methods used to assess peri-implant and periodontal health include probing and radiography, which present a historical perspective, but give no information regarding current disease activity. Oringer and co-workers recently reported the application of the DNA-DNA hybridization technique on the enumeration of subgingival organisms associated with endosseous dental implant fixtures in 22 human subjects.[21] GCF and plaque samples were collected at implant and tooth sites. Radioimmunoassay techniques were utilized to determine GCF C-telopeptide levels, and plaque samples were analyzed utilizing checkerboard DNA-DNA hybridization. C-telopeptide levels and subgingival plaque composition were not significantly different between implants and teeth. Implant sites colonized by *Prevotella intermedia, Capnocytophaga gingivalis, Fusobacterium nucleatum ss vincentii,* and *Streptococcus gordonii* exhibited odds ratios of 12.4, 9.3, 8.1, and 6.7, respectively, of detecting C-telopeptide. These results suggest a relationship between elevated C-telopeptide levels at implant sites and some subgingival organisms associated with disease progression. Future studies should examine levels of C-telopeptide and other putative pyridinoline cross-links at implant fixtures exhibiting clinical signs of failure. Finally, longitudinal studies are necessary to examine the predictive ability of GCF C-telopeptide to identify the development of peri-implant bone loss.

## BLOCKING OF PERIODONTAL TISSUE DESTRUCTION BY MATRIX METALLOPROTEINASES

Studies performed by Golub and co-workers have helped to elucidate the ability of MMP inhibitors to block tissue destruction in patients with periodontal disease (reviewed in Refs. 22 and 23). We recently reported results from eighteen patients with moderate to severe periodontitis entered into an open-labeled study to examine the ability of systemically administered doxycycline (20 mg b.i.d.) on the reduction of tissue destruction.[24] At baseline, and at 1- and 2-month appointments, GCF samples were analyzed for C-telopeptide and osteocalcin by RIA, and collagenolytic enzyme activity and MMP species were studied by Western blotting. The results from this investigation revealed that C-telopeptide levels decreased nearly 70% after the first month of administration, while collagenase activity decreased more than 30% from baseline (FIG. 4). The reductions in GCF C-telopeptide and collagenase levels were sustained for an additional month. Interestingly, osteocalcin, a marker of bone turnover did not relate to C-telopeptides and did not predict response to therapy. Levels of matrix metalloproteinases (MMPs) -8 and -13 were decreased concomitant with C-telopeptide reductions during the 2-month observation period. These data suggest that MMP inhibitors, such as low-dose doxycycline, may block attachment and alveolar bone loss resulting from host-response modifiers such as MMPs.

## CONCLUSIONS

Pyridinoline cross-links are important markers of bone resorption in a variety of bone metabolic diseases including periodontitis. These molecules may become important indicators of active tissue destruction in the accurate diagnosis of alveolar

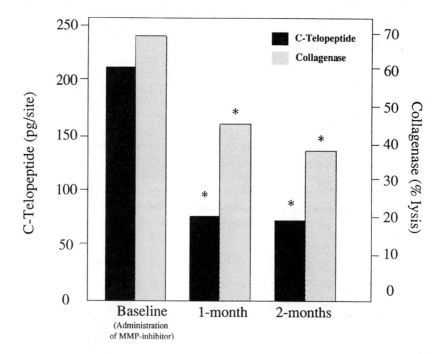

**FIGURE 4.** Doxycycline therapy inhibits collagenolytic breakdown of periodontal tissues. GCF C-telopeptide (by RIA) and collagenase (by % [$^3$H-methyl] collagen α components degraded to $α^A$ fragments during incubation with 1.0 mM APMA at 22°C) in GCF of subjects with severe periodontitis before and during a 2-month regimen of 20 mg b.i.d. doxycycline. Each value represents the mean levels from 28 sites in 7 subjects. ($*p < 0.05$ as compared to baseline.) (From Weinberg and Bral.[23] Reproduced by permission.)

bone loss. Furthermore, the use of blockers of collagen breakdown such as matrix metalloproteinase inhibitors may prove useful in the control of periodontal tissue destruction as monitored by pyridinoline cross-links in periodontal patients.

## ACKNOWLEDGMENTS

The collaborations and discussions with Drs. Sig Socransky, Anne Haffajee, Rich Oringer, Michael Palys, and Larry Golub are appreciated. These studies have been supported by grants from the NIH (Grants DE 04881 and DE 11814) and the Institute of Molecular Biology, Inc.

## REFERENCES

1. SOCRANSKY, S.S. & A.D. HAFFAJEE. 1997. The nature of periodontal diseases. Ann. Periodontol. **2:** 3–10.
2. ARMITAGE, G.C. 1996. Periodontal diseases: diagnosis. Ann. Periodontol. **1:** 37–215.
3. CALVO, M.S., D.R. EYRE & C.M. GUNDBERG. 1996. Molecular basis and clinical application of biologic markers of bone turnover. Endocrine Rev. **17:** 333–368.
4. DELMAS, P.D., A. SCHLEMMER, E. GINEYTS, B. RIIS & C. CHRISTIANSEN. 1991. Urinary excretion of pyridinoline crosslinks correlates with bone turnover measured on iliac crest biopsy in patients with vertebral osteoporosis. J. Bone Mineral Res. **6:** 639–644.
5. EASTELL, R., L. HAMPTON, A. COLWELL, J.R. GREEN, A.M.A. ASSIRI, R. HESP & R.G.G. RUSSELL. 1990. Urinary collagen crosslinks are highly correlated with radioisotopic measurements of bone resorption. *In* Osteoporosis. C. Christiansen & K. Overgaard, Eds.: 469–470. Osteopress ApS. Copenhagen.
6. EASTELL, R., S.P. ROBINS, T. COLWELL, A.M.A. ASSIRI, B.L. RIGGS & R.G.G. RUSSELL, 1993. Evaluation of bone turnover in type I osteoporosis using biochemical markers specific for both bone formation and bone resorption. Osteoporosis Int. **3:** 255–260.
7. BLACK, D., M. MARABANI, R.D. STURROCK & S.P. ROBINS. 1989. Urinary excretion of the hydroxypyridinium cross links of collagen in patients with rheumatoid arthritis. Ann. Rheum. Dis. **48:** 641–644.
8. UEBELHART, D., E. GINEYTS, M.-C. CHAPUY & P.D. DELMAS. 1990. Urinary excretion of pyridinium crosslinks: a new marker of bone resorption in metabolic bone disease. Bone Mineral **8:** 87–96.
9. KYLMALA, T., T.L. TAMMELA, L. RISTELI, J. RISTELI, M. KONTTURI & I. ELOMAA. 1995. Type I collagen degradation product (ICTP) gives information about the nature of bone metastases and has prognostic value in prostate cancer. Br. J. Cancer **71:** 1061–1064.
10. GARNERO, P., W.J. SHIH, E. GINEYTS, D.B. KARPF & P.D. DELMAS. 1994. Comparison of new biochemical markers of bone turnover in late postmenopausal osteoporotic women in response to alendronate treatment. J. Endocrinol. Metab. **79:** 1693-1700.
11. YASUMIZU, T., K. HOSHI, S. IIJIMA & A. ASAKA. 1998. Serum concentration of the pyridinoline cross-linked carboxyterminal telopeptide of type I collagen (ICTP) is a useful indicator of decline and recovery of bone mineral density in lumbar spine: analysis in Japanese postmenopausal women with or without hormone replacement. Endocrine Journal **45:** 45–51.
12. CIMASONI, G. 1983. Crevicular Fluid Updated. Karger. Basel.
13. LAMSTER, I.B. 1997. Evaluation of components of gingival crevicular fluid as diagnostic tests. Ann. Periodontol. **2:** 123–37.

14. GIANNOBILE, W.V. 1997. Crevicular fluid biomarkers of oral bone loss. Curr. Opinion Periodontol. **4:** 19–30.

15. TALONPOIKA, J.T. & M.M. HÄMÄLÄINEN. 1994. Type I carboxyterminal telopeptide in human gingival crevicular fluid in different clinical conditions and after periodontal treatment. J. Clin. Periodontol. **21:** 21–34.

16. GIANNOBILE, W.V., S.E. LYNCH, R.G. DENMARK, D.W. PAQUETTE, J.P. FIORELLINI & R.C. WILLIAMS. 1995. Crevicular fluid osteocalcin and pyridinoline cross-linked carboxyterminal telopeptide of type I collagen (ICTP) as markers of rapid bone turnover in periodontitis. J. Clin. Periodontol. **22:** 904–910.

17. SHIBUTANI, T., Y. MURAHASHI, E. TSUKADA, Y. IWAYAMA & J.N.M. HEERSCHE. 1997. Experimentally induced periodontitis in beagle dogs causes rapid increases in osteoclastic resorption of alveolar bone. J. Periodontol. **68:** 385–391.

18. SOCRANSKY, S.S., A.D. HAFFAJEE, M.A. CUGINI, C. SMITH & R.L. KENT JR. 1998. Microbial complexes in subgingival plaque. J. Clin. Periodontol. **25:** 134–144.

19. PALYS, M.D., A.D. HAFFAJEE, S.S. SOCRANSKY & W.V. GIANNOBILE. 1998. Relationship between C-telopeptide pyridinoline cross-links (ICTP) and putative periodontal pathogens in periodontitis. J. Clin. Periodontol. **25:** 865–871.

20. FIORELLINI, J.P., G. MARTUSCELLI & H.P. WEBER. 1998. Longitudinal studies on implant systems. Periodontology 2000 **17:** 125–131.

21. ORINGER, R.J., M.D. PALYS, A. IRANMANESH, J.P. FIORELLINI, A.D. HAFFAJEE, S.S. SOCRANSKY & W.V. GIANNOBILE. 1998. C-telopeptide pyridinoline cross-links (ICTP) and periodontal pathogens associated with endosseous oral implants. Clin. Oral Implants Res. **9:** 365–373.

22. RYAN, M.E., S. RAMAMURTHY & L.M. GOLUB. 1996. Matrix metalloproteinases and their inhibition in periodontal treatment. Curr. Opinion Periodontol. **3:** 85–96.

23. WEINBERG, M.A. & M. BRAL. 1998. Tetracycline and its analogues: A therapeutic paradigm in periodontal diseases. Crit. Rev. Oral Biol. Med. **9:** 322–332.

24. GOLUB, L. M., H.M. LEE, R.A. GREENWALD, M.E. RYAN, T. SORSA, T. SALO & W.V. GIANNOBILE. 1997. A matrix metalloproteinase inhibitor reduces bone-type collagen degradation fragments and specific collagenases in gingival crevicular fluid during adult periodontitis. Inflamm. Res. **46:** 310–319.

# Thirty-six Years in the Clinic without an MMP Inhibitor

## What Hath Collagenase Wrought?

ROBERT A. GREENWALD

*Long Island Jewish Medical Center, New Hyde Park, New York 11040, USA*

*Biology is much more complex than chemistry—especially combinatorial chemistry!*
—Observation attributed to Inhibitus of Matrixin, ancient pharmacologist

ABSTRACT: Vertebrate collagenase was discovered in 1962, and within a few short years, several inhibitors had been identified. At one time or another, virtually every major drug company has had an MMP inhibitor program, but in 1999, there is only one such product on the market. With a potential market for lifelong therapy in rheumatoid arthritis, osteoarthritis, periodontal disease, osteoporosis, and cancer, this is certainly puzzling. The problem is that the chemistry appears to have outstripped the biology. *In vitro*, there are many inhibitors with nanomolar or picomolar efficacy, but *in vivo* efficacy in animal models does not always follow. There is also a conceptual problem regarding broad-spectrum vs. highly specific inhibitors. Designing human trials to demonstrate MMP inhibition and clinical efficacy is a daunting problem, especially if one seeks to distinguish anti-MMP activity from anti-inflammatory effect. Adult periodontal disease may be the best available human disease model for development of an MMPI.

Collagenase 1 (MMP-1), the prototypic MMP, was discovered, as most workers know, by Gross and Lapierre[1] observing the resorption of the metamorphosing tadpole. Within a few short years, excess collagenase (CGase) had been identified in several human disease states, including dystrophic epidermolysis bullosa, periodontal disease, and rheumatoid arthritis. By 1969, several potential inhibitors had been characterized. In the ensuing 30 years, literally hundreds of MMP-inhibitory compounds have been developed, and the potential market for such drugs is immense. Why then did it take 36 years after the discovery of MMP-1 for the first product (low-dose doxycycline, approved in October 1998, just prior to this symposium) to reach the market for the express purpose of inhibiting pathologically excessive collagenase?

At one time or another, virtually every major pharmaceutical company has had an MMP-inhibitor (MMPI) program. At this meeting, there were presentations from Agouron, Bayer, Roche/Syntex, Novartis, CollaGenex, British Biotech, DuPont Merck, Parke Davis, Glaxo, Proctor and Gamble, Amersham, Lilly, and Wyeth-Ayerst. Many of these companies are old hands at this game, and good inhibitors have been on the shelf for years. Why are they not in the clinic?

It is said that at any one time, there are 100,000 people worldwide involved in drug development, and yet, on average, only about 12 to 16 new drugs are introduced in a year, many of which are "me-too" agents. R. Hirschmann[2] has noted, "Activity in relevant *in vitro* and *in vivo* assays does not mean that a compound is a drug. Rather, a drug is a substance approved by a regulatory agency in a medically sophisticated country. Many a promising compound has failed in safety studies, or in the clinic, because of poor bioavailability, rapid metabolism, species difference, etc. Discovering an active compound is relatively easy, but discovering an important new drug remains unbelievably difficult."

## PLUSES AND MINUSES

Surely lack of potential market share is not the reason for the paucity of commercial MMPIs. The list of diseases characterized by excessive collagenase includes rheumatoid arthritis (RA), osteoarthritis (OA), osteoporosis, periodontal disease (PD), metastatic tumor and neoplastic growth, various ocular disorders, aneurysm and possibly atherosclerosis. Any one of these alone could support the development of a suitable drug and, in most cases, the therapy would be lifelong. No greater potential market exists.

Feasibility is not the issue either. There are hundreds, if not thousands, of suitable inhibitors, and the introduction of combinatorial chemistry has increased the numbers by several orders of magnitude. Admittedly, many of the early molecules had poor bioavailability, but that problem has been addressed and the newer agents are capable of better tissue penetration.

The advent of the knockout mouse has, of course, introduced a conceptual problem. No greater inhibitor exists than the organism with gene deletion for the enzyme in question. It is said that Merck abandoned their stromelysin inhibitor program when they discovered that models of arthritis were unchanged in the matching knockout mice. One more reminder of the complexity and redundancy of biologic systems!

## CHOOSING THE RIGHT TARGET

There is great debate in the anti-MMP camp about broad spectrum vs. highly targeted inhibitors. Most of the older inhibitors (e.g., galardin, batimistat, tetracyclines) hit more than one MMP. Some agents inhibit one or more varieties of the same basic enzyme (e.g., the three collagenases) without any effect on related enzymes such as stromelysin or gelatinase (e.g., Trocade). Others are more multi-talented. Using advanced X-ray crystallography, chemists have been able to refine the "fit" of potential inhibitor molecules so as to achieve nanomolar or even picomolar inhibition parameters for specific enzymes, notably MMP-13. Since some MMPs are constitutive and others are inducible (especially in areas of inflammation), it can be argued that inhibitors should be targeted to the latter, comparable to the situation with Cyclooxygenase (COX)-1 and -2. It is not at all clear, however, that such an approach will be more useful clinically than a broad-spectrum agent.

Several strategies are available for inhibiting an MMP. Inhibition can be accomplished by intervening in expression, activation, or enzyme activity, and a method has been suggested whereby MMP activity is reduced by upregulation of TIMP. It is not at all clear that any *one* approach is intrinsically better than another. In tissue and/ or cell culture, dexamethasone is a wonderful inhibitor of MMP expression, but one need only remember that most patients with rheumatoid arthritis are treated with steroids for many years if not decades and still demonstrate progressive loss of collagen-dependent tissues, not to mention worsening osteoporosis.

## A SHORT HISTORY OF INHIBITOR DEVELOPMENT

Alpha$_2$-macroglobulin was the first natural MMP inhibitor identified. This protein, along with alpha$_1$-antiprotease, is recognized as the first line of defense, albeit clearly overwhelmed in most disease processes. The therapeutic use of proteins of this class, and of TIMPs and related molecules, suffers from the same restrictions affecting all protein treatments—the need for parenteral administration, short half-life, great expense in purification, and the need to create safe controls. (It has long been my belief, for example, that if you are testing a TIMP, the control should be the same TIMP inactivated by a molecular change rather than saline or albumin. But to do that experiment, both proteins must be demonstrated to be equivalent in safety, greatly increasing the cost of drug development.) Various tissue extracts (cartilage, vitreous, aorta, gingiva, tumors, fibroblasts) have been described that inhibit either CGase or angiogenesis or both, but their practical application remains uncertain; the active components would have to be purified and expressed as recombinant proteins, with the problems described above.

The first low molecular weight inhibitors were either chelating agents (e.g., EDTA or ortho-phenanthroline), which are unusable medically, or sulfhydryl reagents, such as DTT and mercapto-compounds, including thiol peptides. Wyeth, Squibb, ICI, and American Home Products were among the companies that obtained patents on such compounds in the mid-80s. The test systems used to screen for MMPI activity have included skin CGase, IL-1-induced cartilage degradation, the rat air pouch, cotton pellet granuloma, ovulation, animal models of inflammatory arthritis, and Matrigel invasion. Gold salts, widely used for treatment of rheumatiod arthritis, were said at one time to inhibit CGase, and eriochrome black T was noted to be selective for CGase but not gelatinases. Galardin was effective in corneal systems, and peptide inhibitors of gelatinase were described. The tetracyclines have a long history of utility as MMP inhibitors, and recently, some bisphosphonates have been noted to have anti-MMP properties. Thus there is no shortage of potential MMPIs.

## *IN VIVO* VERITAS!

The MMP inhibitor field—like many other problems in medicinal chemistry—is plagued by the disparity between *in vitro* results and what might/does happen in a whole organism. First, only old-fashioned doctors like myself still measure CGase by grinding up the tissue and digesting the substrate. In the modern era, one uses Northern blotting, DNA chips, and the like, to detect changes in expression. Many labs use automated screening assays for proteinase activity with peptidic substrates conjugated to fluorometrically detectable side groups. Clearly, such screens do not deal with the problems of bioavailability, etc., and the cation concentrations usually employed are not those of the physiologic milieu where the target enzyme is expected to be active.

One need only consider the fact that many compounds with $IC_{50}$ values in the nanomolar or even picomolar range have been reported which are ineffective in animal models, In contrast, the tetracyclines, whose activity is comparatively unimpressive *in vitro* (inhibitory parameters that are low micromolar or high nanomolar), are nevertheless highly effective in animal models and in human disease. It is a far step from a 96-well microtiter plate to which you add an octapeptide substrate and purified, recombinant enzyme to an organ or tissue where a blizzard of pathologic events is occurring simultaneously.

Analogy can be made to the development of the selective COX-2 (cyclooxygenase) inhibitors. Most drugs classified as nonsteroidal anti-inflammatory drugs inhibit both the constitutive enzyme, COX-1, which maintains normal physiology in the stomach and kidneys, as well as COX-2, an inducible enzyme that engenders prostaglandin production at the site of disease, thereby causing pain and inflammation. By manipulation of enzyme source, substrate concentration, and reaction conditions, it is often possible to show that any given drug inhibits either or both enzymes. Ultimately, the drug industry adopted *in vivo* human data to resolve this question. Blood samples are taken after oral administration, and serum thromboxane $A_2$ levels (a product of COX-1) are compared to $PGE_2$ levels after LPS stimulation of peripheral blood monocytes (a function of COX-2).

A good collagenase example is the streptozotocin diabetic rat. In a diabetic rat, skin collagenase levels rise dramatically, correlated with loss of body mass and loss of skin collagen content. Since rats do not make MMP-1, this collagenase is almost certainly MMP-13, which is currently the favored target for human disease (and which appears to be inducible as in the manner of COX-2). By simply gavaging the test agent by mouth for 3–4 weeks and harvesting the skin for collagen/collagenase assay, *in vivo* documentation of MMP inhibition can readily be observed; most usable tetracyclines are effective in this model. This is an old-fashioned but inexpensive screening model which is currently in use at some pharmaceutical houses.

Thus we come to a formidable problem: how do you prove that a drug that is designed to prevent connective tissue degradation actually does so in human disease? In animal models of rapidly progressive disease, where large samples of tissue can be harvested *ad lib*, proving efficacy is relatively easy. In human rheumatoid arthritis (RA) or osteoarthritis (OA), for example, where tissues cannot be readily sampled, this is a very difficult proposition. The gold standard (i.e., radiologic progression), requires serial studies over time spans of two or more years combined with labor-intensive interpretations by trained observers. We know from three decades of experience that proving analgesic and/or anti-inflammatory effectiveness, which can be done over a short time frame, will have virtually no predictive value regarding progression of disease. An investigator faced with the goal of proving that a new drug can truly prevent joint damage faces the daunting prospect of a frighteningly expensive multiyear trial involving large numbers of patients. I suggest that one of the major reasons for the lack of current availability of an MMP inhibitor is that although a potential drug may pass the *in vitro* and animal screens, the problems of proving efficacy in humans are so great that such agents are often shelved prematurely.

Two factors might mitigate this problem: measurement of collagen cross-links, and studies of periodontal disease.

## HUMAN STUDY DESIGN

It is a basic tenet of the CGase field that triple-helical collagen can only be degraded by a CGase. If excessive collagen breakdown occurs, elevated levels of breakdown products should become apparent in serum and/or urine, and specificity is enhanced if the breakdown products contain moieties known to be collagen-specific. *It follows that if a collagenase inhibitor is administered therapeutically in a situation in which excessive collagenase activity transpires, there should then be a decrease in the amount of collagen-unique end product(s) of digestion that can be detected.* Since there are now assays available for at least five different collagen cross-link degradation products, these appear to be the appropriate surrogates for demonstration of efficacy *in vivo* for human disease.[3] I believe that this is where the primary outcome for MMP inhibitors lies.

There is in fact only one study in the entire literature in which a collagenase inhibitor was administered orally to a group of patients suffering from enhanced connective tissue degradation due to increased CGase levels, and it was shown that *both* the pathologically elevated enzyme levels were normalized *and* that excess collagen breakdown was also reduced. This was a study of low-dose doxycycline given to a small cohort of subjects with periodontal disease.[4] Gingival CGase and GCF ICTP (a collagen-degradation marker) were both normalized by the MMP inhibitory treatment. Interestingly, it appears that MMP-13 was the major target in these tissues.

If a pharmaceutical company approached me today and asked me—as has happened—how can we do a human study to show that our MMP inhibitor is effective in RA (or OA) without committing to several thousand patients to be followed radiologically for three years, I would reply as follows. First, do the rat models of diabetic skin collagenase and either diabetic or endotoxin periodontal disease to show *in vivo* efficacy in both inhibiting collagenase and preventing connective tissue breakdown (especially bone).

Then do a 3-month proof-of-concept study in human periodontal disease. There are ample patients, tissue sampling is straightforward, and the same enzymes and markers are relevant. In fact, the parallels between PD and RA with regard to pathogenesis are striking. (A full discussion of this subject will appear elsewhere.[5]) Not only have the same mediators (e.g., cytokines, enzymes, and local immunoglobulin production) been implicated in both diseases, but both disorders respond to nonsteroidal anti-inflammatory drugs as well as to MMP inhibitors. A successful study in PD would set the stage for a similar trial in RA.

The third study would be a 3-month RA trial in which only systemic collagen cross-link levels are measured. Any drug which is capable of reducing elevated collagen cross-link excretion by 50% or more after 3 months of therapy would be a candidate for a true "DMARD" effect, that is preventing progressive radiologic damage. If the test agent also inhibited proteoglycan release from IL-1 treated cartilage, studies of serum and/or urine proteoglycan markers should also be included.

There are dozens of antirheumatic drugs on the market, and for virtually all of them, the FDA-approved claims pertain to pain, swelling, stiffness, etc. In my opinion, the anti-MMP drugs should be completely divorced from anti-inflammatory action. We must recall that years of conventional therapy with NSAIDs, DMARDs, and/or steroids generally does not stop the progression of RA. We have no real need for another drug to reduce the parameters of inflammation; the holy grail is to stop progression. I would also argue that background therapy is irrelevant. In most RA trials, there are restrictions on which drugs can be taken just prior to and during the study. *If you accept the premise that no existing drugs stop RA progression, and that the parameters to be measured with the test agent are those of collagen breakdown and radiologic change, then it is not necessary to restrict the initial treatment.* In fact, it is also quite permissible for the patient to change conventional drugs during the trial. Of course, the FDA must be convinced of this admittedly radical view.

In my opinion, anti-collagenolytic therapy for RA, if such becomes available, will have to be combined with aggressive anti-inflammatory therapy for maximal benefit. One study has shown that anti-inflammatory therapy allows greater ingress of MMP inhibitors into tissue.[6] An MMP inhibitor should be an "add-on," superimposed on background treatment, and should be evaluated by methods likely to demonstrate the unique aspects of such a therapeutic approach rather than the conventional measures of efficacy in RA. I would not expect an MMP inhibitor to have a substantive short-term effect on walk time or joint tenderness; measures of collagen cross-link release would be the primary outcome measures in any such study. Such an agent should then be considered for a placebo-controlled, long-term trial (2 to 3 years) in which radiologic parameters, functional assessment, and progression to surgery would be the primary outcome measures.

## CONCLUSION

Periodontal disease is a target for MMPIs by itself; it also can serve as a surrogate for RA. Both animal and human PD studies are easier and less costly than arthritis trials. Before committing large sums to a prolonged RA trial, proof-of-concept studies in PD should be carefully evaluated.

## REFERENCES

1. GROSS, J. & C. LAPIERE. 1962. Collagenolytic activity in amphibian tissues: a tissue culture assay. Proc. Natl. Acad. Sci. USA **48:** 1014–1022.
2. HIRSCHMANN, R. 1991. Medicinal chemistry in the gold age of biology: lessons from steroid and peptide research and review. Angew Chem. Int. Ed. Engl. **30:** 1278–1301
3. GREENWALD, R.A. 1996. Monitoring collagen degradation in arthritic patients: the search for suitable surrogates. Arthritis and Rheumatism **39:** 1455–1465
4. GOLUB L.M, H.M. LEE, R.A. GREENWALD *et al.* 1997. A matrix metalloproteinase inhibitor reduces bone-type collagen degradation fragments and bone-type collagenase in gingival crevicular fluid during adult periodontitis. Inflammation Res. **46:** 310–319

5. GREENWALD, R.A. & K. KIRKWOOD. 1999. Adult periodontitis as a model for rheumatoid arthritis. J. Rheumatol. In press.
6. LEUNG M., R.A. GREENWALD, N.S. RAMAMURTHY *et al.* 1995. Effect of NSAIDs on the uptake of 4-dedimethylaminotetracycline and the inhibition of metalloproteinase activities in the inflamed joint of the adjuvant arthritic rat. J. Rheumatol. **22:** 1726–1731.

# Physical and Biological Regulation of Proteoglycan Turnover around Chondrocytes in Cartilage Explants

## Implications for Tissue Degradation and Repair

THOMAS M. QUINN,[a,b] ADRIAN A. MAUNG,[a] ALAN J. GRODZINSKY,[a,c] ERNST B. HUNZIKER,[b] AND JOHN D. SANDY[d]

[a]Continuum Electromechanics Group, Center for Biomedical Engineering, Department of Electrical Engineering and Computer Science, Massachusetts Institute of Technology, Cambridge, Massachusetts 02139, USA

[b]M.E. Mueller Institute for Biomechanics, University of Bern, Switzerland

[d]Shriners Hospitals for Children, Tampa, Florida, USA

ABSTRACT: The development of clinical strategies for cartilage repair and inhibition of matrix degradation may be facilitated by a better understanding of (1) the chondrocyte phenotype in the context of a damaged extracellular matrix, and (2) the roles of biochemical and biomechanical pathways by which matrix metabolism is mediated. Using methods of quantitative autoradiography, we examined the cell-length scale patterns of proteoglycan deposition and turnover in the cell-associated matrices of chondrocytes in adult bovine and calf cartilage explants. Results highlight a rapid turnover in the pericellular matrix, which may indicate spatial organization of PG metabolic pools, and specific biomechanical roles for different matrix regions. Subsequent to injurious compression of calf explants, which resulted in grossly visible tissue cracks and caused a decrease in the number of viable chondrocytes within explants, cell-mediated matrix catabolic processes appeared to increase, resulting in apparently increased rates of proteoglycan turnover around active cells. Furthermore, the influences of cell-stimulatory factors such as IL-1β appeared to be delayed in their effects subsequent to injurious compression, suggesting interactions between biomechanical and biochemical pathways of PG degradation. These results may provide a useful reference point in the development of *in vitro* models for cartilage injury and disease, and hint at possible new approaches in the development of cartilage repair strategies.

## INTRODUCTION

### Turnover of Cartilage Extracellular Matrix

The macromolecules that constitute the extracellular matrix of cartilage are continually being degraded, disassembled, synthesized anew, and reassembled in a process of turnover. Understanding of the mechanisms by which this cell-mediated

[c]Address for telecommunication: Phone, 617/253-4969; fax, 617/258-5239; e-mail, alg@mit.edu

process is affected by mechanical forces and biochemical factors has profound implications for the understanding of tissue remodeling, response to injury, and progression of degradative diseases.[1,2]

*In vitro* studies have been undertaken in an effort to characterize the baseline turnover processes of cartilage proteoglycans (PGs) in culture.[3,4] In serum-free explant culture, the half-life of aggregating PGs in the cartilage ECM was found to vary from ~6 days in adult bovine tissue[3,5] to ~14–21 days in calf tissue.[3,6] In the presence of cycloheximide or at 4°C, these half-lives were significantly increased,[3,5] suggesting that cell activity was required for PG catabolism. Specific enzyme action has been confirmed by the identification of characteristic PG degradation fragments[4] and core protein cleavage sites,[4,7] most notably between the G1 and G2 domains and in sparsely glycosylated areas of the CS-attachment domain.[4,7] Very few of the GAG-rich C-terminal cleavage products have been observed remaining in the ECM,[3,4] although the N-terminal products remain bound in the tissue[7] through stabilizing interactions between G1, link protein, and hyaluronan.[8] Furthermore, the PG released to cartilage/chondrocyte culture media *in vitro*[9,10] and to synovial fluid *in vivo*[11,12] is specifically degraded; PG release is therefore thought to be a rapid cell-mediated process directed by regulated protease action.[7]

The identification of distinct pools of aggregating PGs which turn over at different rates has suggested higher levels of structural organization within the PG compartment.[13] *De novo* ECM synthesized by chondrocytes in alginate gel exhibited more rapid PG turnover in the pericellular matrix than in the further-removed matrix.[14]

In degenerative diseases of cartilage, the physiological balance between ECM synthesis and degradation is altered in favor of the latter. In the specific context of osteoarthritis, this appears to be due to a cell-mediated upregulation of the normal degradative processes,[12] in combination with the synthesis of poorly assembled matrix.[15] Factors that may contribute to the onset of such diseases are many, and remain largely uncertain.[1] Strong clinical correlations exist, however, between abnormal patterns of mechanical loading and the later onset of osteoarthritis.[2]

### Effects of Cell-Stimulatory Factors on Turnover

*In vitro* culture systems provide a means for the selective investigation of physiologically important biochemical factors which may be important to cartilage extracellular matrix turnover.[3] Several recent studies have been conducted on the effects of growth factors, which appear to be important stimuli for cartilage ECM synthesis and assembly; these have included insulin-like growth factor,[5,6,16] fibroblast growth factor-2 (FGF-2),[17] and transforming growth factor-β (TGFβ).[13,18] In the context of cartilage ECM degradation, cytokines and other biochemical factors which have received particular attention recently include interleukin-1 (IL-1),[13,14,16,20] tumor necrosis factor (TNF),[13] and retinoic acid (RA).[18,21–23] The intracellular and extracellular pathways of ECM turnover which are induced by these individual biochemical factors are beginning to be elucidated,[4,7,20,24] together with combined effects[13,16,18] and the development of working hypotheses regarding cytokine/biochemical factor-mediated ECM turnover regulation.

Interleukin 1-β (IL-1β) is an 18-kDa cytokine produced by chondrocytes and other cells in synovial joints.[25] It has been implicated as an important mediator of car-

tilage degeneration in rheumatoid arthritis,[26] and has been shown to be involved in
the development of OA-like conditions subsequent to mechanical trauma.[27] IL-1β
interacts with chondrocytes via cell surface receptors[28] and induces the degradation
of extracellular PGs and collagen,[20] leading to the alteration of tissue physical properties and loss of function.[21] Matrix metalloproteinase (MMP) mRNA and protein
expression levels are elevated by IL-1β,[29,30] and tissue inhibitor of MMP (TIMP)
activity is modulated[25]; however, the PG degradation products which are induced by
IL-1β are not typical of MMP-mediated cleavage,[7] but cleavage by a glutamate-
specific proteinase which has been termed aggrecanase. Indeed an aggrecanase
(ADMP-1) has now been cloned[31] which appears to represent the cartilage activity
and it belongs to a novel class of metalloproteinases termed ADAM-ts (for *a d*isintegrin *a*nd *m*etalloproteinase-*t*hrombo*s*pondin type). Effects of IL-1β are ameliorat-
ed by TIMPs,[20,32] or by the addition of some cytokines.[13]

Retinoic acid (RA) is a derivative of vitamin A which induces cell-mediated[23]
matrix degeneration in cartilage through both increased rates of loss[23] and decreased
rates of synthesis[22] of PGs and collagen. It is normally present in the cartilage ECM
at concentrations of ~ 0.1–10 nM, and may be part of an endogenous biochemically
mediated control system for PG turnover.[18] RA binds directly to sites in the cell nucleus, causing altered DNA transcription[33] and cytoskeletal modifications.[34] RA
causes an increase in the synthesis of some MMPs,[20] but a decrease in others, notably those upregulated by IL-1.[35] The PG degradation products which result from RA
stimulation are also atypical of MMPs, but typical of those generated by
aggrecanase[7] (ADMP-1); indeed both IL-1 and RA-mediated matrix degradation are
inhibited by broad-spectrum MMP inhibitors (hydroxamates)[20] which have activity
against both the MMPs and the ADAM-ts group since all are $Zn^{2+}$-dependent metalloendopeptidases. RA-induced matrix degradation has also been shown to be inhibited by serum[22,23] or IGF-1.[18]

Ample evidence[5] suggests that insulin-like growth factor-1 (IGF-1) is the major
factor in serum which stimulates PG synthesis in cartilage explant culture.[36] It is
normally present in articular cartilage at ~ 1–50 ng/g tissue,[37,38] and interacts with
chondrocytes via cell surface receptors.[39,40] IGF-1 induces increased PG and collagen synthesis in adult and immature cartilage,[5,6,18] and results in decreased PG
loss in adult tissue.[6] Overall, the effects of IGF-1 appear to be to increase the half-
lives of major matrix macromolecules,[6] and the maintenance of tissue mechanical
function.[17]

### *Effects of Mechanical Loading on Turnover*

In addition to biochemical factors, mechanical loading plays an important role in
the mediation of cartilage ECM macromolecule synthesis,[41–43] assembly,[44] degradation,[45,46] and turnover.[41,45] Normal physiological stresses on cartilage range from
~0–20 MPa,[47–49] applied at characteristic frequencies of ~0–1 Hz. *In vitro* studies
have shown that the synthesis rates of PG, collagen, and other matrix macromolecules are reduced in a dose-dependent manner by static compression to a fixed
thickness,[42,43,50] and the acquisition of the highly aggregatable form of newly synthesized PGs is delayed.[44] Low frequency (~0.001 Hz) alternation between levels of
static compression, or "cyclic" compression, has been observed to significantly decrease the half-lives of PGs and collagen in the ECM.[45] However, PG and collagen

synthesis were upregulated by low-amplitude, sinusoidal, dynamic thickness perturbations at frequencies between ~0.01 Hz and 1 Hz.[42] The mechanisms by which
such normal levels of applied mechanical compression influence ECM macromolecule synthesis, assembly, degradation, and release are not well understood, and may
involve a confluence of physicochemical[51–53] and biomechanical[54,65] factors, including interactions with cytokine/biochemical factor-mediated regulatory pathways.[27,41,43]

Applied mechanical stresses that exceed the normal amplitude and/or frequency
ranges may give rise to significant tissue injury *in vitro*,[46,56] providing reasonable
models for *in vivo* cartilage mechanical trauma.[27,57] The short- and long-term responses of chondrocytes to such mechanical injury of cartilage are thought to be associated with the development of degradative cartilage diseases such as
osteoarthritis.[58] Impact loading of cartilage characteristically induced visible damage to tissue explants,[56,57] accompanied by tissue swelling *in vitro*,[46] suggesting
damage to the collagen network and changes in tissue material properties. Such mechanical injuries may significantly alter cell viability,[46,56] and result in increased
rates of turnover of ECM macromolecules,[45] and expression of degradative factors.[27] These results provide significant weight to the hypothesis that cartilage injury
may initiate a series of events that have long-term consequences for cell-mediated
tissue degeneration.[58] Therefore, the effects of cartilage injury on biochemically
and/or biomechanically mediated regulatory pathways of cartilage ECM maintenance are of significant interest.[1,2]

## Objectives

The goals of the current study involved the study of proteoglycan deposition and
turnover within the cell-associated matrices of chondrocytes in adult bovine and calf
articular cartilage. With a view toward issues that may be important to the development of repair strategies for damaged or diseased cartilage, proteoglycan turnover
was assessed under control conditions, and also subsequent to injurious mechanical
compression. In order to identify possible interactions between biomechanical and
biochemical pathways of cell-mediated PG metabolism, the additional influences of
IL-1β, RA, and IGF-1 on PG turnover in control and injured cartilage were also investigated.

## METHODS

### Tissue Culture

#### Preliminary Culture and Radiolabel

Groups of six 3-mm-diameter × 0.6-mm-thick-explants were obtained as previously described[42] from the femoropatellar grooves of adult steers (~2 years old).
Tissue acquisition methods resulted in explants made of middle-zone cartilage (i.e.,
600 μm of tissue thickness below the superficial ~100 μm). For one week of preliminary culture, explants were maintained in tissue culture media (0.25 mL/disk,
changed daily) consisting of DMEM (containing 10 mM HEPES, 1 g/L glucose,
2 mM L-glutamine, and 110 mg/L sodium pyruvate; Gibco BRL) supplemented with

0.1 mM nonessential amino acids (Sigma), 0.4 mM L-proline, 100 U/mL penicillin, 100 g/mL streptomycin, 20 g/mL ascorbate and 20% fetal bovine serum (Hyclone). Explants were radiolabeled for the last 6 hours of the 7-day preliminary culture with 40 μCi/mL Na$_2$-[$^{35}$S]sulfate (for radiolabeling of sulfated proteoglycans).[42] Radiolabeling was terminated by 4 × 0.25 mL/disk × 15-min washes in DMEM.

Groups of six 3-mm-diameter × 1-mm-thick cartilage explant disks were also similarly obtained from the femoropatellar grooves of freshly slaughtered veal calves (~3 weeks old). Conditions of preliminary culture and $^{35}$S-sulfate label were identical to those for steer explants, except that 10% FBS was used. Under similar conditions, calf and adult bovine cartilage has previously been observed to obtain steady-state rates of matrix synthesis and release of ECM macromolecules to culture media after ~5 days.[36]

## Chase Conditions

After radiolabeling, explants were chased in culture for up to 3 weeks. Culture media (0.25 mL/[disk · day]) was changed after day 1 of the chase and then once every two days thereafter. Steer explants were chased in supplemented DMEM with 20% FBS; selected groups of explants were removed from culture immediately after the wash (day 0) and on days 1, 2, 4, 10, and 20 of chase. Calf explants were chased in supplemented DMEM without FBS; selected groups of explants were removed from culture on days 3, 7, and 13 of chase. The change from culture in 10% serum prior to radiolabel to no serum during chase was expected to promote degradative conditions in calf explants.[3,6]

The effects of injurious mechanical compression of calf explants were also explored through the application of a compression protocol similar to one previously employed in cartilage ECM turnover studies.[45] Selected explant disks were cycled between their free-swelling (FSW) thicknesses (~1.2–1.4 mm) and 0.5 mm using custom-made manually operated polysulfone compression/culture chambers similar to those previously described.[42,45] Disks were alternately left at FSW conditions or compressed to 0.5 mm thickness for 1 hour each; a single compression-release cycle therefore took two hours. Compression was applied in the axial direction with impermeable platens and without radial confinement. The transition between FSW and 0.5 mm thickness was applied in ~15 sec at a constant compression velocity of ~0.07 mm/sec. A total of six (6) compression-release cycles were applied over a 12-hour period on day 1 of chase. No compression was applied at any other time in culture.

For calf explants, conditions during the chase period were further modified for investigation of the effects of biochemical cell-stimulatory factors. Individual groups of calf explants were chased in the presence of one of 100 ng/mL interleukin-1β (IL-1β), 10 nM retinoic acid (RA), or 100 ng/mL insulin-like growth factor-1 (IGF-1).

### Histological Analysis

## Quantitative Autoradiography

After removal from culture, selected explants were chemically fixed in 0.05 M sodium cacodylate with 2.5% cetylpyridinium chloride and 2% glutaraldehyde, and prepared for light microscopy with toluidine blue staining and histologic autoradiography by established methods.[59] As previously described,[60] semi-automated image analyses were performed on 1-μm-thick, explant-bisecting vertical cross-sections (e.g., FIG. 1). One section was analyzed per explant disk. Digitized high-power light microscope (Olympus Vanox) images, 100 μm × 75 μm in total area, with a resolution of 6 pixels/μm, were captured using a CCD color video camera (Sony), frame grabber (RasterOps XLTV), and microcomputer (Macintosh). Systematic random samples[61] of ~20 locations within each explant cross-section were examined for cells with a normal histologic appearance and the presence of a well-defined

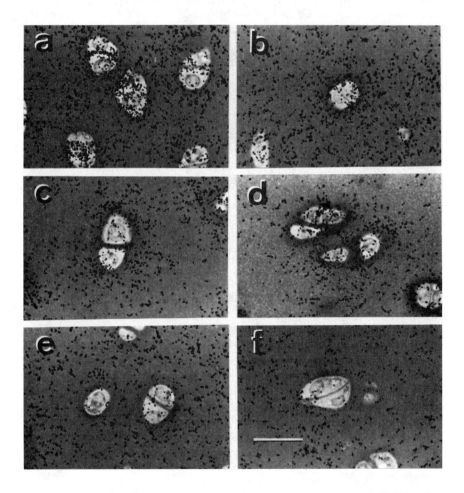

**FIGURE 1.** Autoradiographic histologic appearance of chondrocytes and ECM within adult bovine articular cartilage explants (toluidine blue stain). Explants were chemically fixed immediately after a wash following $^{35}$S-sulfate radiolabeling **(a)**, and after chase periods of 1 day **(b)**, 2 days **(c)**, 4 days **(d)**, 10 days **(e)**, and 20 days **(f)**. *Bar:* 20 μm.

nucleus–cytoplasm interface. These features were taken to be indicative of cells that had been sectioned approximately through the "middle," so that distances from the cell membrane as seen on sections were representative of the actual distances in three dimensions. Color images for analysis were then centered on whichever suitable cell section profile had a nucleus with geometric center closest to the middle of the sampled location. Using an image-processing program (IPLab Spectrum, Signal Analytics Corp.), a human user traced the cell-matrix boundary of the centrally located cell and autoradiography grains were identified by blue or green intensity thresholding. The physical space represented within each image was then parameterized in terms of radial position relative to the traced cell-matrix boundary. Previously developed methods of calculating autoradiography grain density[59] were then employed within regions of space defined by concentric annuli, 1 μm in breadth and conforming to individual cell shapes with an angular resolution of $\pi/12$ rad, at increasing distances from the cell membrane.[60]

For the measurement of tissue-average grain densities, without regard for location relative to any cell membranes, grain density was measured over entire 100-μm × 75-μm images, independently sampled. Traced cell-matrix interfaces acquired during grain density measurements were also used for the estimation of cell volume with the nucleator sizing principle, using the average pixel location of the trace as reference point and sine-weighted directional probes for isotropic sampling within vertical sections.[62] Cell volume fraction ($V_V$) was estimated by point-counting[61] of independently sampled images. Matrix volume per cell was then estimated from the combined cell volume and cell volume fraction data.

## Biochemical Analysis

### Radiolabel Release to Culture Media

Spent tissue culture media was immediately frozen after use and stored at −20°C throughout the culture period. At the end of the chase, selected explant disks were frozen prior to overnight extraction in 4 M guanidine HCl as previously described,[41] and subsequent papain digestion.[42] Assay results from guanidine-HCl extracts and papain digests were summed for the determination of total quantities for each explant. Radioactivity of media and tissue samples was measured by liquid scintillation counting (LKB Wallac 1211 Rackbeta with Ecolume scintillation fluid). For the determination of total amounts of radioactivity incorporated during the label, each explant radioactivity was summed together with the radioactivities of all of its chase media aliquots.

### Determination of Macromolecular Content

Sulfated glycosaminoglycan (GAG) contents of media and tissue samples were assessed by the DMMB dye method.[63] Gross size characterization of radiolabeled macromolecules was performed on selected samples with Sephadex G25 chromatography (MWCO ~5 kDa).

### Characterization of Proteoglycan Degradation Fragments

Sepharose CL-2B chromatography (in the presence of excess hyaluronan and link protein) was used for the examination of size and aggregability of aggrecan fragments lost to medium. Western analysis of selected media fractions was also performed to assess the size distributions of G1-containing aggrecan fragments.[7]

### Statistical Analysis

Data are reported as mean ± SEM. Trends within data sets (e.g., as functions of chase time or distance from cell membrane) were identified by one-way analysis of variance followed by Tukey's paired comparison procedure. Differences between injuriously compressed samples and controls were identified using two-tailed Student's *t*-tests for distributions with unequal variances; differences were considered to be significant for $p < 0.05$. Comparisons between tissue-average grain densities and those measured within localized extracellular matrix zones were performed using Dunnett's method for comparisons to a standard.[64] Half-lives ($\tau_{1/2}$) of macromolecules within individual extracellular matrix zones were estimated by modeling autoradiography grain density ($\rho$) versus time ($t$) data with the exponential function $\rho(t) = \rho_0 \exp(-t/\tau)$ where $\rho_0 = \rho(t = 0)$ and $\tau = \tau_{1/2}/\ln 2$ was the exponential time constant. Linear least-squares fits of $\ln \rho(t)$ versus $t$ provided estimates of $\rho_0$ and $\tau_{1/2}$ as functions of distance from the cell membrane, within 1-μm intervals. Best-fit values of $\tau_{1/2}$ were averaged within defined spatial regions for the identification of macromolecular half-lives with distinct matrix compartments.

### RESULTS

Our histologic preparation and image analysis methods result in constant proportionalities between autoradiography grain densities and the concentrations of macromolecules synthesized during radiolabel[59] and deposited into the ECM. Increases of matrix water content associated with tissue swelling after explantation were expected to have stopped after ~1 day, which was well prior to radiolabeling after one week of preliminary culture. Therefore, in explants not subjected to any mechanical compression, decreases in autoradiography grain densities with time in chase were interpreted as due to decreases in numbers of radiolabeled macromolecules present in defined extracellular matrix regions (FIG. 1).

Characteristic patterns of extracellular proteoglycan deposition in steer cartilage were evident in explants fixed on day 0. At this early time of chase, [35]S-proteoglycans were most concentrated adjacent to cells, within ~2 μm from the cell membrane (FIG. 2; $p < 0.01$ in 26 of 26 comparisons). By day 4 of chase, [35]S-proteoglycan grain densities were decreased from the day 0 and 1 values for all distances from the cell membrane; these differences were most significant in the cell-associated matrix up to 8 μm from the cell membrane ($p < 0.05$ in 15 of 16 comparisons). This was consistent with the interpretation that a process of proteoglycan degradation was dominating the changes in grain density, as opposed to transport of PGs within the extracellular matrix several days after synthesis. In general, rates of decrease of [35]S-proteoglycan grain density were largest next to the cell membrane and became smaller with increasing distance from the cell (FIG. 2). The half-life of proteoglycans within the matrix region 5 μm from the cell membrane was 10.5 ± 1.2 days, which was significantly shorter ($p < 0.01$) than for proteoglycans localized between 10 and 15 μm from the cell membrane (51.7 ± 8.0 days; FIG. 2).

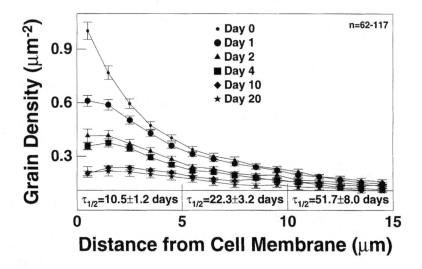

**FIGURE 2.** Evolution of autoradiography grain density in the extracellular matrices around individual chondrocytes within adult bovine articular cartilage explants. Explants were pulse radiolabeled with $^{35}$S-sulfate and subsequently chased for 0 days (•), 1 day (●), 2 days (▲), 4 days (■), 10 days (◆), or 20 days (★). Area densities of histologic autoradiography grains versus distance from the cell membrane were determined for several (~20) cells sampled from each of 6 identically treated explant disks (mean ± SEM, $n \approx 100$). Half-lives ($\tau_{1/2}$) of radiolabeled proteoglycans were determined within three 5-μm spatial intervals (mean ± SEM, $n = 5$).

Similar trends for spatial dependence of proteoglycan deposition and turnover under control conditions were also observed in calf explants (FIG. 3a). $^{35}$S-proteoglycan deposition was most concentrated immediately adjacent to the cells, within ~2 μm from the cell membrane ($p < 0.01$ in 24 of 26 comparisons). Autoradiography grain densities subsequently remained highest in this pericellular region until approximately day 3 (FIG. 3a). The rate of decrease of $^{35}$S-PG grain density was also relatively high in the pericellular region, however, so that by days 7–13 the positions of the peak grain densities were located ~3–5 μm from the cell membrane (FIG. 3a). In general, the apparent rates of PG deposition and turnover as evidenced by the evolution of $^{35}$S-PG grain densities were largest next to the cell membrane and became smaller with increasing distance from the cell (FIG. 3a).[65]

Injurious compression of calf cartilage explants had several effects on cell and matrix morphology and cell-associated PG matrix synthesis. Tissue cracks appeared to indicate mechanical failure of the collagen matrix and may have been associated with changes in tissue material properties.[65] Indeed, subsequent studies[66] have confirmed that such levels of injurious compression cause a significant decrease in tissue stiffness in unconfined compression, indicative of failure of the collagen network. Confined compression stiffness also decreases, though less pronouncedly. Injury was also associated with the appearance of a large number of apparently inactive

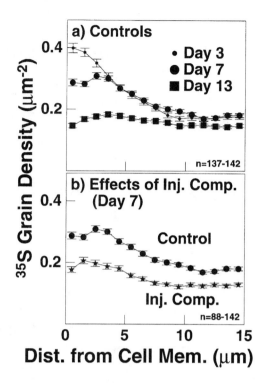

**FIGURE 3. a)** Evolution of autoradiography grain density in the extracellular matrices around individual chondrocytes within calf articular cartilage explants. Explants were pulse-radiolabeled with $^{35}$S-sulfate and subsequently chased for 3 days (•), 7 days (●), or 13 days (■). Area densities of histologic autoradiography grains versus distance from the cell membrane were determined for several (~20) cells sampled from each of 6 identically treated explant disks (mean ± SEM, $n \approx 100$). **b)** $^{35}$S-autoradiography grain densities around viable chondrocytes, which represented labeled proteoglycans remaining in the matrix, showed significant decreases ($p < 0.01$) in injuriously compressed explants (★) compared to controls (●) on day 7 of chase. Grain densities versus distance from the cell membrane were determined for several (~15) cells sampled from each of 6 identically treated explant disks (mean ± SEM, $n \approx 90$).

cells interspersed among viable chondrocytes, such that the volume fraction of viable cells changed from 8.4 ± 0.3 to 3.4 ± 0.7% ($p < 0.01$) and the volume of matrix per viable cell increased from ~11,000 to ~42,000 μm$^3$.[67] Furthermore, viable chondrocytes were larger than normal (1570 ± 120 μm$^3$; $p < 0.01$ compared to cells from control explants, which had a volume of 970 ± 50 μm$^3$), and were associated with decreased $^{35}$S-PG grain densities throughout the cell-associated matrix as compared with uncompressed controls on day 7 of chase (FIG. 3b). By day 13, significant effects of injurious compression on $^{35}$S-autoradiography grain density around metabolically active cells were confined to extracellular regions ≥4 μm from the cell membrane.[65]

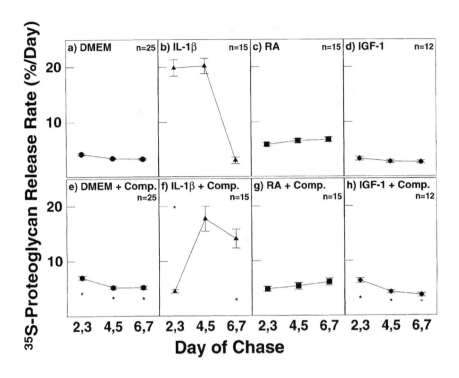

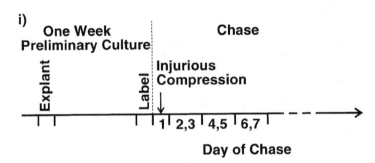

**FIGURE 4. a–h)** Rates of release of $^{35}$S-proteoglycans to culture media from calf cartilage explants over the chase period (mean ± SEM). Chase culture media consisted of DMEM (●; **a, e**), or DMEM supplemented with 100 ng/mL IL-1β (▲; **b, f**), 10 nM RA (■; **c, g**), or 100 ng/mL IGF-1 (◆; **d, h**). Data were acquired from several repeated experiments; in all cases injuriously compressed explant disks (**e, f, g, h**) were matched to otherwise identically treated uncompressed explants (**a, b, c, d**) taken from the same region of the calf femoropatellar groove. Differences between injuriously compressed and matched uncompressed explants are represented by stars (☆, $p < 0.05$; ★, $p < 0.01$; two-tailed Student's $t$-test) plotted at the positions of the uncompressed means (**e, f, g, h**). **i)** Experimental timeline for calf cartilage explant preliminary culture, radiolabeling, injurious compression, and chase.

Release of radiolabel to the media of control explants was initially relatively high on day 1 of chase, during which ~15% of the initial $^{35}$S-label was lost.[67] This release rate subsequently decreased to ~3–4% per day for the remainder of the chase (FIG. 4a). Sephadex G25 chromatography revealed that ~30% of the $^{35}$S-label released by control explants on day 1 of chase was associated with proteoglycans; this proportion was subsequently > 80% for the remainder of the chase.[67]

Release of $^{35}$S-PGs was significantly increased by the presence of 100 ng/mL IL-1β (FIG. 4b). On days 2–5 of chase, ~20% of the initial $^{35}$S-label was lost per day, resulting in almost total loss of $^{35}$S-radioactivity from explants by day 7 of chase (FIG. 4b). Sephadex G25 chromatography indicated that more than 60% of this released $^{35}$S-label was associated with PGs or PG fragments on day 1; this proportion was subsequently >80% for the remainder of the chase. In the presence of 10 nM RA, the kinetics of the release of $^{35}$S-PG was similar to that of control explants, with an apparent slightly increased release rate of ~5% per day on days 2–7 of chase (FIG. 4c). Explants cultured in 100 ng/mL IGF-1 exhibited $^{35}$S-PG release kinetics, which were similar to those of controls (FIG. 4d).

The effects of injurious mechanical compression on the kinetics of $^{35}$S-PG release in control explants were similar to those previously observed by Sah *et al.* in the context of high-amplitude cyclic compression.[45] A marked increase in release rate was observed the day the compression protocol was applied (day 1 of chase),[67] followed by a sustained increased rate of release until at least day 7 of chase (FIG. 4e). This was accompanied by an increase in the fraction of released radiolabel which was associated with proteoglycans (by G25 chromatography).[67] The extent of the increase in $^{35}$S-PG release due to injurious compression on day 1 of chase was not as marked in the presence of either 100 ng/mL IL-1β or 10 nM RA.[67] Furthermore, the very large release rates of $^{35}$S-PGs and PG fragments observed in uncompressed IL-1β-treated explants appeared to be delayed by ~2 days in compressed explants (FIGS. 4b and f). No effects of injurious compression were observed with respect to $^{35}$S-PG release in the presence of 10 nM RA from days 2–7 of chase (FIG. 4g).

Kinetics of release of total PG to culture media, as evidenced by GAG content, was similar to the kinetics of recently synthesized $^{35}$S-PG release. In control, RA, and IGF-1 samples, PG release was maintained at similar steady-state levels throughout the chase period.[67] IL-1β–treated explants, however, exhibited a dramatic increase in PG release rates over the other uncompressed conditions, particularly during days 2–5 of chase (FIGS. 5a and c). Mechanical compression in control explants caused a markedly increased PG release rate on the day of compression (day 1 of chase) followed by a sustained increased rate of PG release on days 2–5.[67] CL-2B chromatography and Western analysis of media indicated that the increase in PG release associated with injurious compression was accompanied by elevated release rates of high molecular weight, G1-bearing species which formed stable aggregates with HA (FIG. 6). Effects of compression on PG release from IL-1β-treated explants were not evident on day 1.[67] As with the $^{35}$S-label release, however, injurious compression appeared to delay the onset of the dramatic loss of PGs from explants treated with IL-1β (FIGS. 5b and d). In the presence of RA, the marked increase in PG release on day 1 was only sustained subsequently over days 2 and 3; similar results were observed in the presence of IGF-1.[67]

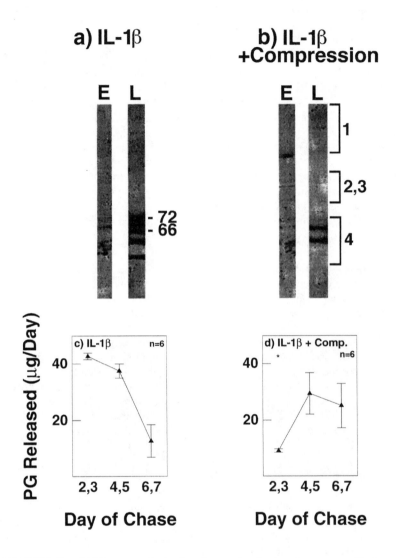

**FIGURE 5. a, b)** Western analysis of media samples from explants cultured in the presence of 100 ng/mL IL-1β. Pooled (*n* = 5) media from early (E; days 1–3) and late (L; days 4–7) in the chase, for both uncompressed (**a**) and injuriously compressed (**b**) explants, were subjected to analysis with anti-ATEGQV, which binds to the G1 (HA-binding) domain of aggrecan core protein. Release of small MW G1-bearing species is commonly interpreted as a final stage in chondrocyte-mediated PG catabolism.[7] **c, d)** Release of total proteoglycan to culture media from calf explants chased in the presence of IL-1β. Differences between injuriously compressed (**d**) and matched uncompressed (**c**) explants are represented by stars (★, *p* < 0.01; two-tailed Student's *t*-test) plotted at the positions of the uncompressed means (**d**).

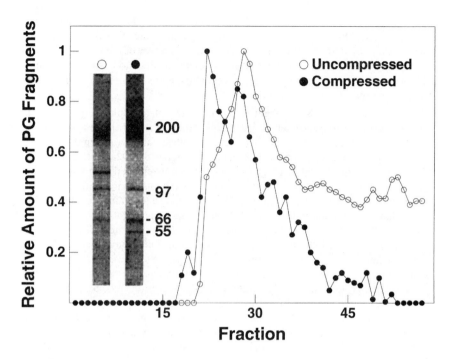

**FIGURE 6.** Sepharose CL-2B chromatography and Western analysis of media samples from calf explants cultured in DMEM only. Pooled ($n = 5$) media from days 4–7 of chase of uncompressed (○) and injuriously compressed (●) explants was CL-2B fractionated in the presence of excess HA and link protein. Western blots were obtained using anti-ATEGQV, which binds to the G1 (HA-binding) domain of aggrecan core protein.[7]

## DISCUSSION

In both adult bovine and calf control explants, newly synthesized PGs were distributed throughout the ECM around individual cells, but with particularly high concentrations in the pericellular matrix (FIG. 1). Similarly, PG turnover was also most rapid in the pericellular matrix, in both adult bovine (FIG. 2) and calf explants (FIG. 3a). The rapid turnover of proteoglycans in the pericellular matrix that we observed was consistent with findings in previous studies,[68] including chondrocyte cultures in alginate gel[14] and agarose (work in progress). Together, these observations suggest that cell-length scale spatial organization of proteoglycan deposition and turnover is a highly conserved feature of the cartilage ECM which may have a fundamental physiological significance. Whereas the farther-removed matrix likely fulfills the organ-level biomechanical function of articular cartilage as a load-bearing, low-friction tissue in synovial joints, cell-matrix interactions in the pericellular matrix are ultimately the means by which physical signals associated with tissue deformations are communicated to chondrocytes.[60] Therefore, manipulation of the composition and physical properties of the immediate pericellular environment may be one means by which chondrocytes can amplify or attenuate microphysical stimuli which accompany tissue loading. In addition, rapid turnover of the pericellular matrix may be one means by which chondrocytes can contribute to tissue remodelling and repair in the face of continued joint-level demands for biomechanical functionality from the further-removed matrix.

Matrix proteoglycans are thought to be composed of multiple metabolic pools which turn over at different rates.[69,70] Moreover, these pools of tissue aggrecan are probably not homogeneous in that there are molecules that are full-length and also those that are C-terminally truncated. Recently[70–72] it has been shown that a considerable proportion of normal cartilage matrix aggrecan is represented by the ADMP-1-generated core which terminates at Glu1687 (bovine numbering with 19 residue leader peptide).[73] Release of this species may indeed explain the observation (FIG. 6) that compression induced the release of a large aggregating species with a core size of about 200 kDa, which is clearly C-terminally truncated. Such compression-induced release might therefore be the result of an enhanced degradation of the hyaluronan network resulting in the release of aggregating species, rather than an accelerated proteolysis, for which there was no clear evidence (FIG. 6). On the other hand increased proteolysis clearly results from IL-1 addition (FIG. 5) and this was accompanied by the enhanced release of the G1 domain of aggrecan generated by interglobular domain cleavage. This process, however, appeared to be delayed by mechanical injury. It is therefore clear that an understanding of the interplay between cytokine and mechanically induced degradation of cartilage matrix will require further delineation of the aggrecan degradative pathway which can involve both C-terminal and N-terminal proteolysis in addition to hyaluronan cleavage.[74] A schematic of the likely sequence of these events and the steps which appear to be sensitive to mechanical injury is shown in FIGURE 7.

In addition, pathways of PG catabolism may be spatially localized around chondrocytes[68,75] owing to concentration distributions of secreted or membrane-associated degradative enzymes.[77] Since spatial organization of PG turnover in the cell-associated matrix appears to be a highly conserved aspect of chondrocyte phenotypic expression, these distinct metabolic pools of matrix proteoglycans may represent an important link between matrix biomechanical function and biochemical processing.

Tissue injury was associated with a significant ($>50\%$) reduction in cell viability and a concomitant increase in matrix volume per viable cell.[65] Recent studies suggest that a subpopulation of such injuriously compressed cells has been induced to undergo apoptosis.[66] Viable chondrocytes were also significantly larger than cells from control explants and appeared to mediate an acceleration of matrix PG degradation and turnover. A significant decrease in the density of $^{35}$S-PG autoradiography grains around viable chondrocytes (FIG. 3b) was evident until at least day 13 of chase.[65] This was consistent with observations of elevated rates of $^{35}$S-PG release and tissue swelling secondary to injurious compression. Random sampling of tissue-length scale $^{35}$S-autoradiography grain density in injuriously compressed explants indicated that the tissue mean grain density was $0.163 \pm 0.006 \, \mu m^{-2}$ on day 7 and $0.142 \pm 0.006 \, \mu m^{-2}$ on day 13 of chase ($n = 60$). These tissue means were significantly greater ($p < 0.01$) than extracellular grain densities between ~8 μm and 15 μm

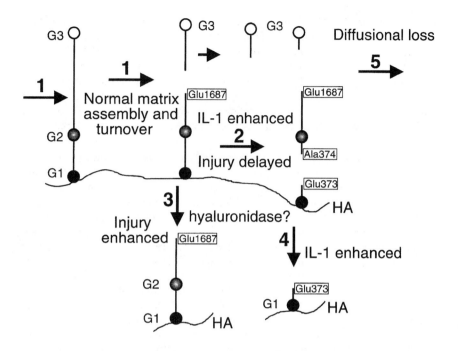

**FIGURE 7.** Schematic of the cartilage aggrecan turnover pathway and possible interplay of injury with cytokine-dependent events. The hypothetical scheme describes a single pool of aggrecan, which is produced (steps 1) by normal synthesis followed by controlled and limited C-terminal truncation at Glu1687 by aggrecanase (ADMP-1). This process generates free G3-bearing fragments, which are lost by diffusion (step 5). In the presence of IL-1 (steps 2 and 4) there is accelerated degradation at Glu1687 and initiation of cleavage at Glu373 by aggrecanase (ADMP-1). Step 4 also appears to involve hyaluronan catabolism. Step 2 generates the product Ala374-Glu1687, which is lost by diffusion (step 5). Mechanically induced injury appears to delay step 2. On the other hand injury appears to accelerate the loss of large aggregating species (step 3) by a process that may also involve hyaluronan catabolism.

from the cell membranes of metabolically active cells in these explants, which were $0.130 \pm 0.006\ \mu m^{-2}$ ($n = 88$) on day 7 (7 of 7 differences significant at $p < 0.002$) and $0.122 \pm 0.006\ \mu m^{-2}$ ($n = 116$) on day 13 (5 of 7 differences significant at $p < 0.05$; FIG. 3b). For a typical matrix volume per cell of $11,000\ \mu m^3$,[67] this extracellular matrix zone was expected to include the cell-associated matrices of several nearby cells, and therefore to be representative of the local tissue mean around the sampled cells. Together, these observations indicate that PG turnover in injured explants was most rapid in the vicinity of the viable cells which remained. Since measured rates of PG release from injuriously compressed explants were already significantly elevated over control levels (FIG. 4e), it would appear that metabolically active cells in injured explants were mediating an abnormally rapid rate of proteoglycan degradation.

Such activity may have been due to direct mechanical disruption of cytoskeletal architecture or cell-matrix interactions, or part of a regulated cell response involving morphological changes and matrix remodeling within a damaged ECM. These effects of acute *in vitro* injury on cell and matrix morphology and PG turnover bear compelling similarities to changes which are known to occur in osteoarthritic cartilage.[58] While direct associations to long-term *in vivo* disease may clearly not be drawn, the phenotypic expression of cells that survive mechanical injury may be of significant interest for the development of appropriate strategies of articular cartilage repair.

The effects of IL-1β on release of GAG and $^{35}$S-PGs from calf explants appeared to have been delayed by ~2 days by mechanical injury (FIGS. 4b and f). Furthermore, Western analysis indicated that the onset of massive release of small MW G1-bearing fragments (a final stage in IL-1β–mediated PG degradation[7]) was similarly delayed by injury (FIG. 5). These observations indicated an interaction between biomechanical and biochemical pathways by which PG matrix catabolism is mediated. One possible explanation might be that injurious compression represented a significant "shock" to cells, even if they remained viable, such that their normal responses to some biochemical stimuli were compromised. However, no interference with the effects of RA on PG matrix degradation was evident, although these would have been more difficult to quantify since the effects of 10 nM RA were not as dramatic as those of 100 ng/mL IL-1β (FIGS. 4b and f). In any case, these results suggest that (1) cartilage injury can significantly alter the response of chondrocytes to biochemical stimuli; and (2) mechanical injury, and mechanical compression in general, may be reasonably thought of as a "physical cytokine" for chondrocytes, the effects of which can be equally important to a successful cartilage repair strategy as many biochemical factors.

## ACKNOWLEDGMENTS

This work was supported by NIH Grants AR33236 and AR45779, AO/ASIF Grant 96-Q76, and a fellowship from the Arthritis Society of Canada (T.M.Q.). We thank Sophia T. Kung and Greg Allen for helpful discussions, and Elke Berger, Veronique Gaschen, Eva Kapfinger, Prasanna Perumbuli, and Michael Ponticiello for technical contributions.

## REFERENCES

1. DIEPPE, P. 1995. Osteoarthritis and molecular markers. A rheumatologist's perspective. Acta Orthop. Scand. **66** (Suppl 266): 1–5.
2. RADIN, E.L. 1995. Osteoarthrosis—the orthopedic surgeon's perspective. Acta Orthop. Scand. **66** (Suppl 266): 6–9.
3. CAMPBELL, M.A., C.J. HANDLEY, V.C. HASCALL, R.A. CAMPBELL & D.A. LOWTHER. 1984. Turnover of proteoglycans in cultures of bovine articular cartilage. Arch. Biochem. Biophys. **234**(1): 275–289.
4. ILIC, M.Z., C.J. HANDLEY, H.C. ROBINSON & M.T. MOK. 1992. Mechanism of catabolism of aggrecan by articular cartilage. Arch. Biochem. Biophys. **294**(1): 115–122.

5. McQuillan, D.J., C.J. Handley, M.A. Campbell, S. Bolis, V.E. Milway & A.C. Herington. 1986. Stimulation of proteoglycan biosynthesis by serum and insulin-like growth factor-1 in cultured bovine articular cartilage. Biochem. J. **240:** 423–430.

6. Sah, R.L.Y., A.C. Chen, A.J. Grodzinsky & S.B. Trippel. 1994. Differential effects of bFGF and IGF-1 on matrix metabolism in calf and adult bovine cartilage explants. Arch. Biochem. Biophys. **308**(1): 137–147.

7. Sandy, J.D., A.H.K. Plaas & T.J. Koob. 1995. Pathways of aggrecan processing in joint tissues: implications for disease mechanism and monitoring. Acta. Orthop. Scand. **66** (Suppl 266): 26–32.

8. Morgelin, M., D. Heinegard, J. Engel & M. Paulsson. 1994. The cartilage proteoglycan aggregate: assembly through combined protein-carbohydrate and protein-protein interactions. J. Biophys. Chem. **50**(1–2): 113–128.

9. Sandy, J.D., P.J. Neame, R.E. Boynton & C.R. Flannery. 1991. Catabolism of aggrecan in cartilage explants. Identification of a major cleavage site within the interglobular domain. J. Biol. Chem. **14:** 8683–8685.

10. Lark, M.W., J.T. Gordy, J.R. Weidner, J. Ayala, J.H. Kimura, H.R. Williams, R.A. Mumford, C.R. Flannery, S.S. Carlsson, M. Iwata & J.D. Sandy. 1995. Cell-mediated catabolism of aggrecan: evidence that cleavage at the aggrecanase site (Glu 373 - Ala 374) is a primary event in proteolysis of the interglobular domain. J. Biol. Chem. **270:** 2550–2556.

11. Sandy, J.D., C.R. Flannery, P.J. Neame & S.F. Lohmander. 1992. The structure of aggrecan fragments in osteoarthritic synovial fluid. Evidence for the involvement in osteoarthritis of a novel proteinase which cleaves the Glu 373 - Ala 374 bond in the interglobular domain. J. Clin. Invest. **89**(2): 1512–1516.

12. Lohmander, L.S., P.J. Neame & J.D. Sandy. 1993. The structure of aggrecan fragments in human synovial fluid. Evidence that aggrecanase mediates cartilage degradation in inflammatory joint disease, joint injury and osteoarthritis. Arth. Rheum. **36:** 1214–1222.

13. Hardingham, T.E., M.T. Bayliss, V. Rayan & D.P. Noble. 1992. Effects of growth factors and cytokines on proteoglycan turnover in articular cartilage. Br. J. Rheum. **31**(1): 1–6.

14. Mok, S.S., K. Masuda, H.J. Hauselmann, M.B. Aydelotte & E.J.M.A. Thonar. 1994. Aggrecan synthesized by mature bovine chondrocytes suspended in alginate. J. Biol. Chem. **269**(52): 33021–33027.

15. Mankin, H.J. & K.D. Brandt. 1992. Biochemistry and metabolism of articular cartilage in osteoarthritis. *In* Osteoarthritis: Diagnosis and Medical/Surgical Management, 2nd ed. R.W. Moskowitz *et al.* Eds.: 109–154. W.B. Saunders. Philadelphia, PA.

16. Fosang, A.J., J.A. Tyler & T.A Hardingham. 1991. Effect of interleukin-1 and insulin like growth factor-1 on the release of proteoglycan components and hyaluronan from pig articular cartilage in explant culture. Matrix **11:** 17–24.

17. Sah, R.L.Y., S.B. Trippel & A.J. Grodzinsky. 1996. Differential effects of serum, IGF-1, and FGF-2 on the maintenance of cartilage physical properties during long-term culture. J. Orth. Res. **14:** 44–52.

18. Morales, T.I. 1994. Transforming growth factor-( and insulin-like growth factor-1 restore proteoglycan metabolism of bovine articular cartilage after depletion by retinoic acid. Arch. Biochem. Biophys. **315**(1): 190–198.

19. Arner, E.C. & M.A. Pratta. 1989. Independent effects of interleukin-1 on proteoglycan breakdown, proteoglycan synthesis, and prostaglandin E2 release from cartilage in organ culture. Arth. Rheum. **32**(3): 288–297.

20. BUTTLE, D.J., C.J. HANDLEY, M.Z. ILIC, J. SAKLATVALA, M. MURATA & A.J. BARRETT. 1993. Inhibition of cartilage proteoglycan release by a specific inactivator of cathepsin B and an inhibitor of matrix metalloproteinases: evidence for two converging pathways of chondrocyte-mediated proteoglycan degradation. Arth. Rheum. **36**(12): 1709–1717.

21. BONASSAR, L.J., K.A. JEFFRIES, C.G. PAGUIO & A.J. GRODZINSKY. 1995. Cartilage degradation and associated changes in biomechanical and electromechanical properties. Acta. Orthop. Scand. **66** (Suppl 266): 38–44.

22. CAMPBELL, M.A. & C.J. HANDLEY. 1987. The effect of retinoic acid on proteoglycan biosynthesis in bovine articular cartilage cultures. Arch. Biochem. Biophys. **253**(2): 462–474.

23. CAMPBELL, M.A. & C.J. HANDLEY. The effect of retinoic acid on proteoglycan turnover in bovine articular cartilage cultures. Arch. Biochem. Biophys. **258**(1): 143–155.

24. BUTTLE, D.J., J. SAKLATVALA, M. TAMAI & A.J. BARRETT. 1992. Inhibition of interleukin 1-stimulated cartilage proteoglycan degradation by a lipophilic inactivator of cysteine endopeptidases. Biochem. J. **281:** 175–177.

25. SHINGU, M., Y. NAGAI, T. ISAYAMA, T. NAONO & M. NOBUNAGA. 1993. The effects of cytokines on metalloproteinase inhibitors (TIMP) and collagenase production by human chondrocytes and TIMP production by synovial cells and endothelial cells. Clin. Exp. Immunol. **94:** 145–149.

26. DARLING, J.M., L.H. GLIMCHER, S. SHORTKROFF, B. ALBANO & E.M. GRAVALLESE. 1994. Expression of metalloproteinases in pigmented villonodular synovitis. Hum. Pathol. **25:** 825–830.

27. PICKVANCE, E.A., T.R. OEGEMA & R.C. THOMPSON. 1993. Immunolocalization of selected cytokines and proteases in canine articular cartilage after transarticular loading. J. Orthop. Res. **11:** 313–323.

28. CHANDRASEKHAR, S. & A.K. HARVEY. 1989. Induction of interleukin-1 receptors on chondrocytes by fibroblast growth factor: a possible mechanism for modulation of interleukin-1 activity. J. Cell Physiol. **138:** 236–246.

29. KANDEL, R.A., K.P.H. PRITZKER, G.B. MILLS & T.F. CRUZ. 1990. Fetal bovine serum inhibits chondrocyte collagenase production: interleukin 1 reverses this effect. Biochim. Biophys. Acta **1053:** 130–134.

30. HUTCHINSON, N.I., M.W. LARK, K.L. MACNAUL, C. HARPER, L.A. HOERNER, J. MCDONNELL, S. DONATELLI, V.L. MOORE & E.K. BAYNE. 1992. In vivo expression of stromelysin in synovium and cartilage of rabbits injected intraarticularly with interleukin-1 beta. Arth. Rheum. **35:** 1227–1233.

31. ARNER, E.C., T. BURN, M.A. PRATTA, R. LIU, J.M. TRZASKOS, R.C. NEWTON, C.P. DECICCO et al. 1999. Isolation and identification of "aggrecanase": a novel aggrecan-degrading metalloproteinase (ADMP). Trans. Orth. Res. Soc. **45:** 38.

32. SEED, M.P., S. ISMAIEL, C.Y. CHEUNG, T.A. THOMSON, C.R. GARDNER, R.M. ATKINS & C.J. ELSON. 1993. Inhibition of interleukin 1 beta induced rat and human cartilage degradation in vitro by the metalloproteinase inhibitor u27391. Ann. Rheum. Dis. **52:** 37–43.

33. NOJI, S., T. YAMAAI, E. KOYAMA, T. NOHNO & S. TANIGUCHI. 1989. Spatial and temporal expression pattern of retinoic acid receptor genes during mouse bone development. FEBS Lett. **257:** 93–96.

34. BROWN, P.D. & P.D. BENYA. 1988. Alterations in chondrocyte cytoskeletal architecture during phenotypic modulation by retinoic acid and dihydrochalasin B-induced reexpression. J. Cell Biol. **106:** 171–179.

35. LAFYATIS, R., S.J. KIM, P. ANGEL, A.B. ROBERTS, M.B. SPORN, M. KARIN & R.L. WILDER. 1990. Interleukin-1 stimulates and all-trans-retinoic acid inhibits collagenase gene expression through its 5´ activator protein-1-binding site. Mol. Endocrinol. **4**(7): 973–980.

36. HASCALL, V.C., C.J. HANDLEY, D.A. MCQUILLAN, G.K. HASCALL, H.C. ROBINSON & D.A. LOWTHER. 1983. The effect of serum on biosynthesis of proteoglycans by bovine articular cartilage in culture. Arch. Biochem. Biophys. **224**(1): 206–223.

37. SCHALKWIJK, J., L.A.B. JOOSTEN, W.B. VAN DEN BERG, J.J. VAN WYK & L.B.A. VAN DE PUTTE. 1989. Insulin-like growth factor stimulation of chondrocyte proteoglycan synthesis by human synovial fluid. Arth. Rheum. **32**(1): 66–71.

38. SCHNEIDERMAN, R., N. ROSENBERG, Y. HISS, P. LEE & A. MAROUDAS. 1995. Concentration and size distribution of IGF-1 in human normal and osteoarthritic synovial fluid and cartilage. Acta Orthop. Scand. **66** (Suppl 266): 75–76.

39. TRIPPEL, S.B., J.J. VAN WYK & H.J. MANKIN. 1986. Localization of somatomedin-C binding to bovine growth-plate chondrocytes in situ. J. Bone Joint Surg. [Am.] **68**: 897–903.

40. TRIPPEL, S.B., S.D. CHERNAUSEK, J.J. VAN WYK, A.C. MOSES & H.J. MANKIN. 1988. Demonstration of type I and type II somatomedin receptors on bovine growth plate chondrocytes. J. Orthop. Res. **6**: 817–826.

41. BEHRENS, F., E.L. KRAFT & T.R. OEGEMA. 1989. Biochemical changes in articular cartilage after joint immobilization by casting or external fixation. J. Orthop. Res. **7**: 335–343.

42. SAH, R.L.Y., Y.J. KIM, J.Y.H. DOONG, A.J. GRODZINSKY, A.H.K. PLAAS & J.D. SANDY. 1989. Biosynthetic response of cartilage explants to dynamic compression. J. Orthop. Res. **7**: 619–636.

43. KIM, Y.J., A.J. GRODZINSKY & A.H.K. PLAAS. 1996. Compression of cartilage results in differential effects on biosynthetic pathways for aggrecan, link protein, and hyaluronan. Arch. Biochem. Biophys. **328**(1): 331–340.

44. SAH, R.L.Y., A.J. GRODZINSKY, A.H.K. PLAAS & J.D. SANDY. 1990. Effects of tissue compression on the hyaluronate-binding properties of newly synthesized proteoglycans in cartilage explants. Biochem. J. **267**: 803–808.

45. SAH, R.L.Y., J.Y.H. DOONG, A.J. GRODZINSKY, A.H.K. PLAAS & J.D. SANDY. 1991. Effects of compression on the loss of newly synthesized proteoglycans and proteins from cartilage explants. Arch. Biochem. Biophys. **286**(1): 20–29.

46. JEFFREY, J.E., D.W. GREGORY & R.M. ASPDEN. 1995. Matrix damage and chondrocyte viability following a single impact load on articular cartilage. Arch. Biochem. Biophys. **322**(1): 87–96.

47. AFOKE, N.Y.P., P.D. BYERS & W.C. HUTTON. 1987. Contact pressures in the human hip joint. J. Bone Joint Surg. **69B**: 536–541.

48. HODGE, W.A., R.S. FIJAN, K.L. CARLSON, R.G. BURGESS, W.H. HARRIS & R.W. MANN. 1986. Contact pressures in the human hip joint measured in vivo. Proc. Natl. Acad. Sci. USA **83**: 2879–2883.

49. HUBERTI, H.H. & W.C. HAYES. 1984. Patellofemoral contact pressures: the influence of q-angle and tendofemoral contact. J. Bone Joint Surg. **66A**: 715–724.

50. GRAY, M.L., A.M. PIZZANELLI, R.C. LEE, A.J. GRODZINSKY & D.A. SWANN. 1989. Kinetics of the chondrocyte biosynthetic response to compressive load and release. Biochim. Biophys. Acta **991**: 415–425.

51. GRAY, M.L., A.M. PIZZANELLI, A.J. GRODZINSKY & R.C. LEE. 1988. Mechanical and physicochemical determinants of the chondrocyte biosynthetic response. J. Orthop. Res. **6**: 777–792.

52. GARCIA, A.M., A.C. BLACK & M.L. GRAY. 1994. Effects of physicochemical factors on the growth of mandibular condyles in vitro. Calcif. Tissue Int. **54**: 499–504.

53. KIM, Y.J., R.L.Y. SAH, A.J. GRODZINSKY, A.H.K. PLAAS & J.D. SANDY. 1994. Mechanical regulation of cartilage biosynthetic behavior: Physical stimuli. Arch. Biochem. Biophys. **311**(1): 1–12.

54. KIM, Y.J., L.J. BONASSAR & A.J. GRODZINSKY. 1995. The role of cartilage streaming potential, fluid flow, and pressure in the stimulation of chondrocyte biosynthesis during dynamic compression. J. Biomech. **28**(9): 1055–1066.

55. BUSCHMANN, M.D., E.B. HUNZIKER, Y.J. KIM & A.J. GRODZINSKY. 1996. Altered aggrecan synthesis correlates with cell and nucleus structure in statically compressed cartilage. J. Cell Sci. **109**: 499–508.

56. REPO, R.U. & J.B. FINLAY. 1977. Survival of articular cartilage after controlled impact. J. Bone Joint Surg. **59-A(8)**: 1068–1076.

57. THOMPSON, R.C., M.J. VENER, H.J. GRIFFITHS, J.L. LEWIS, T.R. OEGEMA & L. WALLACE. 1993. Scanning electron-microscopic and magnetic resonance-imaging studies of injuries to the patellofemoral joint after acute transarticular loading. J. Bone Joint Surg. **75-A(5)**: 704–713.

58. HOWELL, D.S., B.V. TREADWELL & S.B. TRIPPEL. 1992. Etiopathogenesis of osteoarthritis. *In* Osteoarthritis: Diagnosis and Medical/Surgical Management, 2nd ed. R.W. Moskowitz *et al.*, Eds.: 233–252. W.B. Saunders. Philadelphia, PA.

59. BUSCHMANN, M.D., A.M. MAURER, E. BERGER & E.B. HUNZIKER. 1996. A method of quantitative autoradiography for the spatial localization of proteoglycan synthesis rates in cartilage. J. Histochem. Cytochem. **44**(5): 423–431.

60. QUINN, T.M., A.J. GRODZINSKY, M.D. BUSCHMANN, Y.J. KIM & E.B. HUNZIKER. 1998. Mechanical compression alters proteoglycan deposition and matrix deformation around individual cells in cartilage explants. J. Cell Sci. **111**(5): 573–583.

61. GUNDERSON, H.J.G. & E.B. JENSEN. 1987. The efficiency of systematic sampling in stereology and its prediction. J. Microsc. **147**(3): 229–263.

62. GUNDERSON, H.J.G. 1988. The nucleator. J. Microsc. **151**(1): 3–21.

63. FARNDALE, R.W., D.J. BUTTLE & A.J. BARRETT. 1986. Improved quantitation and discrimination of sulphated glycosaminoglycans by use of dimethylmethylene blue. Biochim. Biophys. Acta. **883**: 173–177.

64. FISHER, L.D. & G. VAN BELLE. 1993. Biostatistics: A Methodology for the Health Sciences. John Wiley and Sons, New York, NY.

65. QUINN, T.M., A.J. GRODZINSKY, E.B. HUNZIKER & J.D. SANDY. 1998. Effects of injurious compression on matrix turnover around individual cells in calf articular cartilage explants. J. Orth. Res. **16**: 490–499.

66. LOENING, A.M., M.E. LEVENSTON, I.E. JAMES, M.E. NUTTALL, H.K. HUNG, M. GOWEN, A.J. GRODZINSKY & M.W. LARK. 1999. Injurious compression of bovine articular cartilage induces chondrocyte apoptosis before detectable mechanical damage. Trans. Orth. Res. Soc. **45**: 42.

67. QUINN, T.M. 1996. Articular cartilage: matrix assembly, mediation of chondrocyte metabolism, and response to compression. Ph.D. Thesis, MIT, Cambridge, MA.

68. WINTER, G.M., C.A. POOLE, M.Z. ILIC, J.M. ROSS, H.C. ROBINSON & C.J. HANDLEY. 1998. Identification of distinct metabolic pools of aggrecan and their relationship to type VI collagen in the chondrons of mature bovine articular cartilage explants. Conn. Tiss. Res. **37**(3–4): 277–293.

69. LOHMANDER, S. 1977. Turnover of proteoglycans in guinea pig costal cartilage. Arch. Biochem. Biophys. **180**(1): 93–101.

70. ILIC, M.Z., H.C. ROBINSON & C.J. HANDLEY. 1998. Characterization of aggrecan retained and lost from the extracellular matrix of articular cartilage. Involvement of carboxyl-terminal processing in the catabolism of aggrecan. J. Biol. Chem. **273**(28): 17451–17458.

71. SANDY, J.D., P.J. ROUGHLEY, R.A. MUMFORD & M.W. LARK. 1996. Evidence for aggrecanase-mediated cleavage at the E1714-G1715 bond of the CS-attachment region of human aggrecan in vivo. Trans. Orth. Res. Soc. **42:** 145.

72. SANDY, J.D., V.T. THOMPSON & C. VERSCHAREN. 1999. Chondrocyte-mediated catabolism of aggrecan: kinetic analysis of aggrecanase action identifies the G1-bearing product of cleavage at the Glu-Glu-Glu sequence in the CS2 attachment region as a stable intermediate in aggrecan turnover. Submitted for publication.

73. HERING, T.M., J. KOLLAR & T.D. HUYNH. 1997. Complete coding sequence of bovine aggrecan: comparative structural analysis. Arch. Biochem. Biophys. **345**(2): 259–270.

74. BONASSAR, L.J., J.D. SANDY, M.W. LARK, A.H.K. PLAAS, E.H. FRANK & A.J. GRODZINSKY. 1997. Inhibition of cartilage degradation and changes in physical properties induced by IL-1α and retinoic acid using matrix metalloproteinase inhibitors. Arch. Biochem. Biophys. **344**(2): 404–412.

75. LARK, M.W., E.K. BAYNE, J. FLANAGAN, C.F. HARPER, L.A. HOERRNER, N.I. HUTCHINSON, I.I. SINGER, S.A. DONATELLI, J.R. WEIDNER, H.R. WILLIAMS, R.A. MUMFORD & L.S. LOHMANDER. 1997. Aggrecan degradation in human cartilage: evidence for both matrix metalloproteinase and aggrecanase activity in normal, osteoarthritic, and rheumatoid joints. J. Clin. Invest. **100**(1): 93–106.

76. HUGHES, C.E., F.H. BUTTNER, B. EIDENMULLER, B. CATERSON & E. BARTNIK. 1997. Utilization of a recombinant substrate rAgg1 to study the biochemical properties of aggrecanase in cell culture systems. J. Biol. Chem. **272**(32): 20269–20274.

# Adamalysins

## A Family of Metzincins Including TNF-α Converting Enzyme (TACE)

LORAN KILLAR,[a,d] JUDITH WHITE,[b] ROY BLACK,[c] AND JACQUES PESCHON[c]

[a]Oncology/Immunoinflammatory Diseases, Wyeth-Ayerst Research,
Princeton, New Jersey 08543, USA

[b]Department of Cell Biology, University of Virginia,
Charlottesville, Virginia 22908, USA

[c]Immunex Corporation, Seattle, Washington 98101, USA

ABSTRACT: The adamalysins are a family of proteins in the metzincin super-family of metalloproteases, which also includes the matrix metalloproteinases. There are two subfamilies of adamalysins: the *s*nake *v*enom *m*etalloproteases (SVMPs) and the ADAMs (proteins containing *a d*isintegrin *a*nd *m*etallopro-tease domain). At least 23 ADAMs have been identified to date. The ADAMs are expressed by a wide variety of cell types, and are involved in functions as diverse as sperm-egg binding, myotube formation, neurogenesis, and pro-teolytic processing of cell surface proteins. An overview of the ADAM family and their functions will be presented. TACE is a unique member of the ADAM family that cleaves membrane-bound TNF-α to generate soluble TNF-α. Mice lacking proteolytically active TACE have been generated and characterized. The TACE knock-out results in perinatal lethality. Cells from the TACE-deficient mice release 80–90% less soluble TNF-α than do wild-type cells. Irradiated mice that are reconstituted with TACE knock-out hematopoeitic stem cells have markedly reduced levels of serum TNF-α following LPS challenge, compared to irradiated mice reconstituted with wild-type cells, suggesting that TACE is the major TNF-α converting enzyme *in vivo*. TACE-deficient cells are compromised in the generation of other soluble proteins that are produced as the result of cleavage of a membrane precursor form, suggesting that TACE is involved in multiple shedding events.

## STRUCTURE AND POTENTIAL FUNCTIONS OF ADAMS

ADAMs are a family of cell surface proteins containing *a d*isintegrin *a*nd *m*etal-loprotease domain, which combine features of adhesion molecules and proteases.[1] Twenty-three members have been disclosed to date (TABLE 1). The first ADAMs to be described were shown to be involved in reproductive functions, particularly sper-matogenesis and sperm-egg fusion.[2–4] Other members of this family of proteins may have key roles in development.[2] Recently, certain ADAMs have been shown to have key proteolytic functions,[5] and those will be the focus of this overview.

[d] Address for telecommunication: Phone, 732/274-4046; fax, 732/274-4129;
e-mail, killarl@war.wyeth.com

**TABLE 1. The ADAMS**[a]

| ADAM No. | Other name | ADAM No. | Other name |
|---|---|---|---|
| ADAM 1[a] | fertilin α | ADAM 13[c] | — |
| ADAM 2 | fertilin β | ADAM 14[c] | adm-1 |
| ADAM 3 | cyritestin, tMDC I | ADAM 15 | metargidin, MDC 15 |
| ADAM 4[b] | tMDC V | ADAM 16[c] | MDC 16 |
| ADAM 5[b] | tMDC II | ADAM 17 | TACE |
| ADAM 6 | tMDC IV | ADAM 18[b] | tMDC III |
| ADAM 7 | EAP I | ADAM 19[b] | meltrin β |
| ADAM 8 | MS2 | ADAM 20 | — |
| ADAM 9 | MDC 9, meltrin γ | ADAM 21 | — |
| ADAM 10 | MADM, kuzbanian | ADAM 22 | MDC 2 |
| ADAM 11 | MDC | ADAM 23 | MDC 3 |
| ADAM 12 | meltrin α | | |

NOTE: Shading indicates potential protease activity.
[a]For an updated list of ADAMs, contact the website for Dr. White's Laboratory at http://www.med.virginia.edu/~jag6m/whitelab.html
[b]Pseudogene in human.
[c]No human homologue to date.
[d]No mammalian homologue to date, potential protease activity.

Structurally, the ADAMs are most closely related to the NIII SVMPs.[6,7] Together SVMPs and ADAMs belong to a superfamily of Zn-dependent metalloproteases known as the metzincins, which also includes the matrix metalloproteinases (MMPs).[8] The domain structure of the SVMPs and ADAMs is shown in FIG 1. The most notable difference between the two subfamilies is that all of the SVMPs are secreted, whereas all of the ADAMs described to date have a membrane-bound form. Alternatively spliced soluble forms have been described for at least two ADAMs,[9,10] and soluble forms of additional ADAMs may be identified in the future. Both families contain a pro domain, which renders the protease domain inactive until the pro domain is proteolytically removed. The pro domain employs a cysteine switch mechanism similar to that used by the MMPs.[11–13] There is a putative furin cleavage site between the pro and protease domains in most ADAMs, and cleavage by furin or furin-like enzymes is believed to be the physiological mechanism of activation of ADAM zymogens.[12,13] The protease domains contain a conserved sequence (HEXGHXXGXXHD) which includes the three histidine residues that coordinate the catalytic $Zn^{2+}$.[5,8] A significant difference between the SVMPs and ADAMs is that all SVMPs are active proteases, whereas about half of the ADAMs have alterations in the sequence of the $Zn^{2+}$ coordinating cassette that is predicted to result in the loss of protease activity.[5] The disintegrin domain was so named in the SVMPs because it was shown to disrupt integrin functions, particularly those involved in platelet aggregation.[6,7] In a number of the ADAMs, the distintegrin domain has been

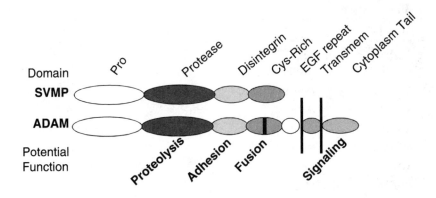

**FIGURE 1.** Organization and potential functions of adamalysin domains comparing a NIII SVMP to ADAMs. The *dark bar* in the Cys-rich region of the ADAM represents a candidate fusion peptide present in certain ADAMs. (Adapted from Black and White.[5])

shown to mediate cell–cell interactions via integrin molecules.[14–17] The Cys-rich domain in some ADAMs may be involved in cell fusion events.[18,19] There is an EGF repeat domain of unknown function next to the cell membrane, a transmembrane region, and a cytoplasmic domain. In some of the ADAMs, particularly those with protease activity, the cytoplasmic domain contains potential phosphorylation sites and SH3 binding domains, suggesting that these ADAMs may have signaling activity.[5] On the basis of this domain structure, ADAMs have the potential to mediate a variety of functions including proteolysis, adhesion, fusion, and signaling.

## ADAM PROTEASES

TABLE 2 lists those ADAMs with intact $Zn^{2+}$ binding motifs in their protease domains. ADAMs 10, 12, and 17 have been demonstrated to actually have protease activity.[13,20–23] For those ADAMs that are demonstrable proteases, their activity can be inhibited by zinc chelators (e.g., EDTA or 1,10-phenanthroline) and certain small molecule matrix metalloproteinase inhibitors.[13,20–23] In addition, TIMP-3 has been shown to inhibit substrate cleavage by ADAM-17.[2]

Recently, physiological roles for the protease activity of ADAMs 10 and 17 have been elucidated, and will be described in more detail below. Potential functions for most of the other protease ADAMs are based hypothetically on their expression patterns. ADAM-1 (fertilin α) may be involved in spermatogenesis in a number of mammals, and is found as a heterodimer with ADAM-2 (fertilin β) on mature sperm

**TABLE 2. ADAMS with HEXGHXXGXXHD boxes**

| ADAM No. | Other name | Potential role |
|---|---|---|
| ADAM 1 | fertilin α | spermatogenesis |
| ADAM 8 | MS2 | immune function |
| ADAM 9 | MDC 9, meltrin γ | myogenesis, osteogenesis |
| ADAM 10 | MADM, kuzbanian | neurogenesis |
| ADAM 12 | meltrin α | myogenesis, osteogenesis |
| ADAM 13 | — | neural crest cell migration |
| ADAM 15 | metargidin, MDC 15 | blood vessel function |
| ADAM 16 | MDC 16 | spermatogenesis |
| ADAM 17 | TACE | cleaves pro-TNF-α |
| ADAM 19 | meltrin β | myogenesis, osteogenesis |
| ADAM 20 | — | spermatogenesis |

(bovine and guinea pig).[3,19] The function of the protease domain in ADAM-1 is not clearly defined, but the fact that a form of ADAM-1 expressing the protease domain is found on immature sperm suggests it may have a role in the development of sperm.[3] On mature sperm, both ADAM-1 and 2 have been processed such that both the pro domain and the protease domains have been cleaved off.[3,19] The disintegrin domain of ADAM-2 is involved in sperm-egg binding (via $\alpha_6\beta_1$ integrin) and there is evidence to suggest that the Cys-rich domain of ADAM-1 may mediate sperm–egg fusion.[18,19] As ADAM-1 is a pseudogene in humans,[2,25] it is hypothesized that one of the newly identified human testis-specific ADAMs (e.g., ADAM-20 or 21) may act as the equivalent of fertilin α in humans.[26] ADAM-8 is found on monocytes and macrophages and is upregulated when the cells are activated, suggesting that it may have a role in immune or inflammatory responses, although its exact function is unknown.[27,28] ADAMs-9, 12, and 19 (meltrins γ, α, β) appear to have a role in myogenesis and/or osteogenesis.[10,13,29–31] ADAM-12 is interesting in that a soluble form resulting from alternative splicing of the gene has been indentified and shown to induce myogenesis.[10] ADAM-12 has been shown to cleave the bait region of α-2 macroglobulin, but the physiological role of its protease activity remains to be elucidated.[13] ADAM-13 is proposed to be involved in neural crest cell migration and/or somitogenesis in *Xenopus*.[32] A unique feature of human ADAM-15 is that it has an RGD sequence in its disintegrin domain which specifically mediates adhesion to the integrin $\alpha_V\beta_3$.[16] The importance of $\alpha_V\beta_3$ in the interaction of vascular endothelial cells with each other and with other cell types suggests that the disintegrin domain of ADAM-15 may have a role in blood vessel function, but its activity as a protease (if any) is unknown.[33–39]

A potential role for the protease ADAMs in pathogenic processes is suggested by the observation that a number of these molecules are upregulated in certain diseases, including those for which matrix metalloproteinases are targets for therapeutic intervention. For example, expression of ADAMs 10, 12, and 15 is upregulated in certain tumor cells; and ADAMs 10, 12, 15, and 17 have been shown to be expressed by chondrocytes from articular cartilage.[40–43]

## ADAM-10

One ADAM for which a physiological protease function has been strongly implicated is ADAM-10. ADAM-10 is a homologue of a protease first described in fruit flies called Kuzbanian or *kuz*, which is involved in a developmental process known as lateral inhibition, a process that results in signalling cells to stop differentiating.[44] Lateral inhibition is an important regulatory process in the development of the nervous system, muscle, and hematopoietic cells.[44,45] Initial reports suggested that *kuz* is responsible for the processing of a protein called Notch. A proposed mechanism for lateral inhibition is that Notch signals the cell to stop differentiating when the proteolytically processed form of Notch on the cell is activated by binding to a Notch ligand on an adjacent cell.[44,45] A recent report, however, suggests that *kuz* actually processes the Notch ligand Delta, rather than Notch, and the *kuz*-processed, soluble form of Delta binds to and activates Notch.[46] Regardless of the mechanism (which requires further elucidation), *kuz* appears to have a role in lateral inhibition. Expression of *kuz* constructs lacking a functional protease domain exerts a dominant-negative effect on lateral inhibition in *Drosophila* and *Xenopus* neurogenesis.[45] A similar construct of murine ADAM-10 (with a nonfunctional protease domain) also acts as a dominant negative and blocks lateral inhibition in those species, indicating evolutionary conservation of this enzyme's structure and function.[45] This evidence suggests that ADAM-10 has an important role in development.

ADAM-10 was shown to be capable of cleaving membrane-bound TNF-α,[47,48] but subsequent studies indicate that ADAM-10 is probably not a physiological TNF-α convertase, given that ADAM-10 is expressed normally on TACE knock-out cells that are compromised in their ability to shed TNF-α (unpublished observations). ADAM-10 has been shown to cleave myelin basic protein and type IV collagen *in vitro*.[20,21] Although it is not yet known whether these are physiological substrates *in vivo*, this data warrants further investigation for a potential role of ADAM-10 in the pathogenesis of demyelinating diseases such as multiple sclerosis and/or conditions involving compromise of the basement membrane such as extracellular remodeling in renal pathology or tumor metastasis. As mentioned above, ADAM-10 is highly expressed in some types of tumor cells.[40] ADAM-10 expression is upregulated on arthritic chondrocytes, raising the question of whether ADAM-10 degrades cartilage matrix components and thus has a role in the pathogenesis of arthritis.[41,42] There is recent evidence that ADAM-10 can activate proMMP-2 *in vitro*, which has interesting implications for potential metzincin cascades in physiology and disease.[49] There is still much to learn about the physiological and potential pathological activities of ADAM-10.

## ADAM-17/TNF-α CONVERTING ENZYME (TACE)

Tumor necrosis factor-α (TNF-α) is expressed as a membrane-bound homotrimer, which is proteolytically cleaved at a specific site (between alanine 76 and valine 77) to generate soluble TNF-α.[22,23] TNF-α is a pleotrophic cytokine with many proinflammatory activities, and has been implicated in numerous disease processes.[50,51] The role of TNF-α as a key pathological mediator in rheumatoid arthritis and Crohn's disease has been validated by the dramatic clinical efficacy of soluble TNF-α receptor chimeras and/or anti-TNF-α antibodies.[52–57] These biological agents are an important first step in targeting TNF-α in disease. Protein therapeutics have certain disadvantages, however, including their expense, requirements for parenteral administration, and the potential for developing neutralizing immune responses against them. The identification of orally active, small molecule inhibitors of TNF-α for treating chronic conditions is therefore highly desirable. Numerous approaches are being pursued for identifying small-molecules that inhibit the production or activity of TNF-α.[58] Relevant to a discussion of ADAM proteases is the targeting of ADAM-17 or TNF-α converting enzyme (TACE), the enzyme that is responsible for the generation of soluble TNF-α.

A number of years ago, the observation was made that the generation of soluble TNF-α could be inhibited both *in vitro* and *in vivo* by certain small-molecule inhibitors of the matrix metalloproteinases.[59–61] The search ensued for the metalloenzyme that mediated TNF-α processing, and in 1997, investigators at Immunex and GlaxoWellcome published reports describing the isolation and cloning of ADAM-17 as TACE.[22,23] TACE displays the typical domain structure of the ADAMs, but, along with ADAM-10, is evolutionarily distant from the other ADAMs.[5,22,23] The crystal structure of the protease domain of TACE has been determined.[62] Like most other ADAMs, TACE has a furin cleavage site between the pro and catalytic domains. TACE is ubiquitously expressed in its active form. In most cell types, TACE is not upregulated upon activation of the cells.[22] There appear to be higher levels of TACE expressed on chondrocytes from arthritic cartilage compared to chondrocytes from normal cartilage, but the reason for increased expression on chondrocytes from diseased tissue is not known.[43] TACE has been shown to be the primary enzyme that processes membrane-bound TNF-α. Cultured cells that are homozygous for a mutated TACE that lacks a functional catalytic domain, cannot efficiently generate soluble TNF-α.[22] Irradiated mice reconstituted with hematopoietic stem cells expressing mutant proteolytically inactive TACE express much lower levels of serum TNF-α in response to LPS challenge than do irradiated mice reconstituted with wild-type cells (unpublished data).

A surprising finding is that generation of homozygous TACE knockout mice using the same mutant TACE construct (deletion of the $Zn^{2+}$ binding region of the catalytic domain) leads to perinatal lethality, with the majority of the embryos dying between e17.5 and one day after birth.[63] Heterozygous animals with the TACE mutation are normal.[63] Knocking out TNF-α itself, or either or both TNF receptors, results in homozygous animals that are phenotypically normal, implying that TACE is doing more than processing TNF-α.[64–66] To demonstrate that lethality in the TACE knockout is not due to aberrant responses to membrane-bound TNF-α, the TACE knockouts were crossed with mice lacking both TNF-α receptors. The TACE muta-

**TABLE 3. Effect of TACE$^{\Delta Zn/\Delta Zn}$ mutation in cells and mice**

| |
|---|
| *Cellular shedding defects* |
| TNF-α. L-Selectin, TNFR-p75, TGF-α |
| |
| *Mouse phenotype* |
| Perinatal lethality |
| Open eyelids after e16.5 |
| Hair and coat abnormalities |
| Skin abnormalities |
| Epithelial abnormalities in multiple organs |

tion proves lethal in these animals as well.[63] Isolated cells from TACE mutant embryos, or the rare TACE mutant mouse that survives for a few weeks, are compromised in shedding a number of cell surface proteins[63] (TABLE 3). The phenotype displayed by the TACE knock-out animals resembles that of TGF-α knockout animals, and TGF-α shedding is deficient in TACE knockout cells[63] (TABLE 3). The TGF-α knockout is not lethal, however.[67] Knocking out the receptor for TGF-α, the EGF receptor, results in embryonic lethalilty.[68,69] Because other EGFR ligands are shed proteins, the hypothesis has been raised that shedding of these other EGFR ligands which are required for growth and development is compromised in the TACE knockouts.[63]

These studies raise numerous questions about the role of TACE in development and in the mature animal; how TACE activity is regulated, the function of the other domains of TACE, whether TACE is involved in the processing of other cell surface proteins, and the validity of TACE as a therapeutic target for specifically modulating TNF-α activity.

## SUMMARY

The ADAMs are a comparatively recently identified family of molecules that may function as mediators of adhesion, fusion, proteolysis, and signaling. There are currently 23 members of the family, and it is likely that more will be identified in the future. Relatively little is known about the role these molecules play in physiology and pathology, and how they function mechanistically. Future investigation of these interesting molecules may reveal pivotal roles for ADAMs in normal physiology and disease. ADAMs may prove to be useful targets for therapeutic manipulation in a variety of pathological conditions.

## REFERENCES

1. WOLFSBERG, T.G., P.D. STRAIGHT, R.L. GERENA *et al.* 1995. ADAM, a widely distributed and developmentally regulated gene family encoding membrane proteins with *A Disintegrin And Metalloprotease* domain. Dev. Biol. **169:** 378–383.

2. WOLFSBERG, T.G. & J.M. WHITE. 1996. ADAMs in fertilization and development. Dev. Biol. **180:** 389–401.

3. MYLES, D.G. & P. PRIMAKOFF. 1997. Why did the sperm cross the cumulus? To get to the oocyte. Functions of the sperm surface proteins PH-20 and fertilin in arriving at, and fusing with, the egg. Biol. Reprod. **56:** 320–327.

4. CHO, C., D.O. BUNCH, J.-E. FAURE *et al.* 1998. Fertilization defects in sperm from mice lacking fertilin β. Science **281:** 1857–1859.

5. BLACK, R.A. & J.M. WHITE. 1998. ADAMS—focus on the protease domain. Curr. Opin. Cell Biol. **10:** 654–659.

6. PAINE, M.J., H.P. DESMOND, R.D. THEAKSTON *et al.* 1992. Purification, cloning, and molecular characterization of a high molecular weight hemorrhagic metalloprotease, jararhagin, from *Bothrops jararaca* venom. Insights into the disintegrin gene family. J. Biol. Chem. **267:** 22869–22876.

7. HUANG, T.F. 1998. What have snakes taught us about integrins? Cell. Molec. Life Sci. **54:** 527–540.

8. STOCKER, W., F. GRAMS, U. BAUMANN *et al.* 1995. The metzincins—topological and sequential relations between the astacins, adamalysins, serralysins, and matrixins (collagenases) define a superfamily of zinc-peptidases. Protein Sci. **4:** 823-840.

9. KATAGIRI T., Y. HARADA, M. EMI *et al.* 1995. Human metalloprotease/disintegrin-like (MDC) gene: exon-intron organization and alternative splicing. Cytogenet. Cell Genet. **68:** 39–44.

10. GILPIN, B.J., F. LOECHEL, M.-G. MATTEI *et al.* 1998. A novel, secreted form of human ADAM 12 (Meltrin α) provokes myogenesis *in vivo.* J. Biol. Chem. **273:** 157–166.

11. NAGASE, H. 1997. Activation mechanisms of matrix metalloproteinases. Biol. Chem. **378:** 151–160.

12. MILHIET, P.E., S. CHEVALLIER, D. CORBEIL *et al.* 1995. Proteolytic processing of the alpha-subunit of rat endopeptidase-24.18 by furin. Biochem. J. **309:** 683–688.

13. LOECHEL, F., B.J. GILPIN, E. ENGVALL *et al.* 1998. Human ADAM 12 (meltrin α) is an active metalloprotease. J. Biol. Chem. **273:** 16993–16997.

14. ALMEIDA, E.A., A.P. HUOVILA, A.E. SUTHERLAND *et al.* 1995. Mouse egg integrin $\alpha_6\beta_1$ functions as a sperm receptor. Cell **81:** 1095–1104.

15. YUAN, R., P. PRIMAKOFF & D.G. MYLES. 1997. A role for the disintegrin domain of cyritestin, a sperm surface protein belonging to the ADAM family, in mouse sperm-egg plasma membrane adhesion and fusion. J. Cell Biol. **137:** 105–112.

16. ZHANG, X.P., T. KAMATA, K. YOKOYAMA *et al.* 1998. Specific interaction of the recombinant disintegrin-like domain of MDC-15 (metargidin, ADAM-15) with integrin $\alpha_v\beta_3$. J. Biol. Chem. **273:** 7345–7350.

17. CHEN, M.S., E.A.C. ALMEIDA, A.-P. J. HUOVILA *et al.* 1999. Evidence that distinct states of the integrin $\alpha_v\beta_1$ interact with laminin and an ADAM. J. Cell Biol. **144:** in press.

18. HUOVILA, A.-P.J., E.A.C. ALMEIDA & J.M. WHITE. 1996. ADAMs and cell fusion. Curr. Opin. Cell Biol. **8:** 692–699.

19. WATERS, S.I. & J.M. WHITE. 1997. Biochemical and molecular characterization of bovine fertilin α and β (ADAM 1 and ADAM 2): a candidate sperm-egg binding/fusion complex. Biol. Reprod. **56:** 1245–1254.

20. HOWARD, L., X. LU, S. MITCHELL *et al.* 1996. Molecular cloning of MADM: A catalytically active mammalian disintegrin-metalloprotease expressed in various cell types. Biochem. J. **317:** 45–50.

21. MILLICHIP, M.I., D.J. DALLAS, E. WU *et al.* 1997. The metallo-disintegrin ADAM10 (MADM) from bovine kidney has type IV collagenase activity in vitro. Biochem. Biophys. Res. Commun. **245:** 594–598

22. BLACK, R.A., C.T. RAUCH, C.J. KOZLOSKY *et al.* 1997. A metalloproteinase disintegrin that releases tumour-necrosis-α from cells. Nature **385:** 729–733.

23. Moss, M.L., S.-L. C. Jin, M.E. Milla *et al.* 1997. Cloning of a disintegrin metallo-proteinase that processes precursor tumour-necrosis factor-α. Nature **385:** 733–736.

24. Amour, A., P.M. Slocombe, A. Webster *et al.* 1998. TNF-α converting enzyme (TACE) is inhibited by TIMP-3. FEBS Lett. **435:** 39–44.

25. Jury, J.A., J. Frayne, & L. Hall. 1997. The human fertilin alpha gene is nonfunctional: implications for its proposed role in fertilization. Biochem. J. **321:** 577–581.

26. Vanhuijsduijnen, R.H. 1998. ADAM 20 and 21 - Two novel human testis-specific membrane metalloproteases with similarity to fertilin-α. Gene. **206:** 273–282.

27. Yoshida, S., M. Setoguchi, Y. Higuchi *et al.* 1990. Molecular cloning of cDNA encoding MS2 antigen, a novel cell surface antigen strongly expressed in murine monocytic lineage. Internatl. Immunol. **2:** 585–591.

28. Schluesener, H.J. 1998. The disintegrin domain of ADAM 8 enhances protection against rat experimental autoimmune encephalomyelitis, neuritis and uveitis by a polyvalent autoantigen vaccine. J. Neuroimmunol. **87:** 197–202.

29. Inoue, D., M. Reid, L. Lum *et al.* 1998. Cloning and initial characterization of mouse meltrin beta and analysis of the expression of four metalloprotease-disintegrins in bone cells. J. Biol. Chem. **273:** 4180–4187.

30. Kurisaki, T., A. Masuda, N. Osumi *et al.* 1998. Spatially- and temporally-restricted expression of meltrin α (ADAM12) and β (ADAM19) in mouse embryo. Mechan. Develop. **73:** 211–215.

31. Yagami-hiromasa, T., T. Sato, T. Kurisaki *et al.* 1995. A metalloprotease- disintegrin participating in myoblast fusion. Nature. **377:** 652–656

32. Alfandari, D., T.G. Wolfsberg, J.M. White *et al.* 1997. ADAM 13: a novel ADAM expressed in somitic mesoderm and neural crest cells during *Xenopus laevis* development. Dev. Biol. **182:** 314–330.

33. Weerasinghe, D., K.P. McHugh, F.P. Ross *et al.* 1998. A role for the $\alpha_v\beta_3$ integrin in the transmigration of monocytes. J. Cell Biol. **142:** 595–607.

34. Scatena, M., M. Almeida, M.L. Chaisson *et al.* 1998. NF-κB mediates $\alpha_v\beta_3$ integrin-induced endothelial cell survival. J. Cell Biol. **141:** 1083–1093.

35. Ruegg, C., A. Yilmaz, G. Bieler *et al.* 1998. Evidence for the involvement of endothelial cell integrin $\alpha_v\beta_3$ in the disruption of the tumor vasculature induced by TNF and IFN-γ. Nature Med. **4:** 408–414.

36. Eliceiri, B.P., R. Klemke, S. Stromblad *et al.* 1998. Integrin $\alpha_v\beta_3$ requirement for sustained mitogen-activated protein kinase activity during angiogenesis. J. Cell. Biol. **140:** 1255–1263.

37. Woodard, A.S., G. Garcia-Cardena, M. Leong *et al.* 1998. The synergistic activity of $\alpha_v\beta_3$ integrin and PDGF receptor increases cell migration. J. Cell Sci. **111:** 469–478.

38. Bombeli, T., B.R. Schwartz & J.M. Harlan. 1998. Adhesion of activated platelets to endothelial cells: evidence for a GPIIbIIIa-dependent bridging mechanism and novel roles for endothelial intercellular adhesion molecule 1 (ICAM-1), $\alpha_v\beta_3$ integrin, and GPIbα. J. Exp. Med. **187:** 329–339.

39. Herren, B., E.W. Raines & R. Ross. 1997. Expression of a disintegrin-like protein in cultured human vascular cells and in vivo. FASEB J. **11:** 173–180.

40. Wu, E., P.I. Croucher & N. McKie. 1997. Expression of members of the novel membrane linked metalloproteinase family ADAM in cells derived from a range of haematological malignancies. Biochem. Biophys. Res. Commun. **235:** 437–442.

41. McKie, N., T. Edwards, D.J. Dallas *et al.* 1997. Expression of members of a novel membrane linked metalloproteinase family (ADAM) in human articular chondrocytes. Biochem. Biophys. Res. Commun. **230:** 335–339.

42. Chubinskaya, S., G. Cs-Szabo & K.E. Kuettner *et al.* 1997. ADAM-10 message is expressed in human articular cartilage. J. Histochem. Cytochem. **46:** 723–729.

43. PATEL, I.R., M.G. ATTUR, R.N. PATEL *et al.* 1998. TNF-α convertase enzyme from human arthritis-affected cartilage: isolation of cDNA by differential display expression of the active enzyme, and regulation of TNF-α. J. Immunol. **160:** 4570–4579.
44. NYE, J.S. 1997. Developmental signaling: Notch signals Kuz it's cleaved. Curr. Biol. **7:** 716–720.
45. PAN, D. & G.M. RUBIN. 1997. Kuzbanian controls proteolytic processing of Notch and mediates lateral inhibition during Drosophila and vertebrate neurogenesis. Cell. **90:** 271–280.
46. QI, H., M.D. RAND, X. WU *et al.* 1999. Processing of the Notch ligand Delta by the metalloprotease Kuzbanian. Science **283:** 91–94.
47. LUNN, C.A., X. FAN, B. DALIE *et al.* 1997. Purification of ADAM 10 from bovine spleen as a TNFα convertase. FEBS Lett. **400:** 333–335.
48. ROSENDAHL, M.S., S.C. KO, D.L. LONG *et al.* 1997. Identification of characterization of a pro-tumor necrosis factor-α-processing enzyme from the ADAM family of zinc metalloproteases. J. Biol. Chem. **272:** 24588–24593.
49. MCKIE, N. 1999. Molecular and biochemical characterization of ADAM-10. In press.
50. SHERRY, B. & A. CERAMI. 1988. Cachectin/tumor necrosis factor exerts endocrine, paracrine, and autocrine control of inflammatory responses. J. Cell Biol. **107:** 1269–1277.
51. BAZZONI, F. & B. BEUTLER. 1996. The tumor necrosis factor ligand and receptor families. N. Engl. J. Med. **334:** 1717–1725.
52. KALDEN, J.R. & B. MANGER. 1998. Biologic agents in the treatment of inflammatory rheumatic diseases. Curr. Opin. Rheumatol. **10:** 174–178.
53. MORELAND, L.W. 1998. Soluble tumor necrosis factor receptor fusion protein (ENBREL) as a therapy for rheumatoid arthritis. Rheum. Dis. Clin. N. Amer. **24:** 579–591.
54. RUTGEERTS, P. 1998. Medical therapy of inflammatory bowel disease. Digestion **59:** 453–469.
55. Centocor's Infliximab Launched in US, Scrip 10/14/98, **#2378:** 20
56. First Approval for Immunex's Enbrel, Scrip 11/06/98, **#2385:** 14
57. Remicade Effective in Rheumatoid Arthritis, PharmaMarketletter 11/16/98, **25:** 18.
58. EIGLER, A., B. SINHA, G. HARTMANN *et al.* 1997. Taming TNF: strategies to restrain this proinflammatory cytokine. Immunol. Today **18:** 487–492.
59. MOHLER, K.M., P.R. SLEATH, J.N. FITZNER *et al.* 1994. Protection against a lethal dose of endotoxin by an inhibitor of tumour necrosis factor processing. Nature **270:** 218–220.
60. MCGEEHAN, G.M., J.D. BECHERER, R.C. BAST, JR. *et al.* 1994. Regulation of tumour necrosis factor-α processing by a metalloproteinase inhibitor. Nature **370:** 558–561.
61. GEARING, A.J., P. BECKETT, M. CHRISTODOULOU *et al.* 1994. Processing of tumour necrosis factor-α precursor by metalloproteinases. Nature **370:** 555–557.
62. MASKOS, K., C. FERNANDEZ-CATALAN, R. HUBER *et al.* 1998. Crystal structure of the catalytic domain of human tumor necrosis factor-α-converting enzyme. Proc. Natl. Acad. Sci. USA **95:** 3408–3412.
63. PESCHON, J.J., J.L. SLACK, P. REDDY *et al.* 1998. An essential role for ectodomain shedding in mammalian development. Science **282:** 1281–1284.
64. PASPARAKIS, M., L. ALEXOPOULOU, V. EPISKOPOU *et al.* 1996. Immune and inflammatory responses in TNF α-deficient mice: a critical requirement for TNF α in the formation of primary B cell follicles, follicular dendritic cell networks and germinal centers, and in the maturation of the humoral immune response. J. Exp. Med. **184:** 1397–1411.
65. MARINO, M.W., A. DUNN, D. GRAIL *et al.* 1997. Characterization of tumor necrosis factor-deficient mice. Proc. Natl. Acad. Sci. USA **94:** 8093–8098.

66. PESCHON, J.J., D.S. TORRANCE, K.L. STOCKING *et al.* 1998. TNF receptor-deficient mice reveal divergent roles for p55 and p75 in several models of inflammation. J. Immunol. **160:** 943–952.
67. LUETTEKE, N.C., T.H. QIU, R.L. PEIFFER *et al.* 1993. TGF α deficiency results in hair follicle and eye abnormalities in targeted and waved-1 mice. Cell **73:** 263–278.
68. MIETTINEN, P.J., J.E. BERGER, J. MENESES *et al.* 1995. Epithelial immaturity and multiorgan failure in mice lacking epidermal growth factor receptor. Nature **376:** 337–341.
69. SIBILIA, M. & E.F. WAGNER. 1995. Strain-dependent epithelial defects in mice lacking the EGF receptor. Science **269:** 234–238.

# MMP Inhibition and Downregulation by Bisphosphonates

OLLI TERONEN,[a,e,h] PIA HEIKKILÄ,[d] YRJÖ T. KONTTINEN,[c,e]
MINNA LAITINEN,[g] TUULA SALO,[f] ROELAND HANEMAAIJER,[i]
ANNELI TERONEN,[d] PÄIVI MAISI,[e] AND TIMO SORSA[b]

*Department of Oral and Maxillofacial Surgery,[a] Periodontology,[b] and Anatomy,[c]
Institute of Dentistry,[d] and Helsinki University Central Hospital,[e]
University of Helsinki, Helsinki, Finland*

[f]*Department of Oral Pathology, Institute of Dentistry, University of Oulu, Oulu, Finland*

[g]*Institute of Medical Technology, Surgery, University of Tampere, Tampere, Finland*

[e]*Faculty of Veterinary Medicine, University of Helsinki, Helsinki, Finland*

[i]*Gaubius Laboratory, TNO-PG, PO Box 2215, 2301 CE Leiden, the Netherlands*

ABSTRACT: Bisphosphonates are a group of drugs capable of inhibiting bone resorption, and are thus used for the treatment of bone diseases, such as Paget's disease, osteoporosis, and for bone metastases of malignant tumors. Their primary cellular target is considered to be the osteoclast. The molecular mechanisms responsible for the downregulation of bone resorption by bisphosphonates have remain unclear. We have discovered that various matrix metalloproteinases (MMPs) are inhibited *in vitro* by several bisphosphonates. This novel finding may, in part, explain the efficacy of bisphosphonates in their current indications in humans. In enzyme activity tests using purified and recombinant enzymes, we have observed the inhibition of MMP-1, -2, -3, -7, -8, -9, -12, -13, and -14 by clodronate, alendronate, pamidronate, zolendronate, nedrinate, and clodrinate. The $IC_{50}$s range from 50 to 150 μM. We have also shown that clodronate can downregulate the expression of MT1-MMP protein and mRNA in several cell lines. Additionally, several bisphosphonates decrease the degree of invasion of malignant melanoma (C8161) and fibrosarcoma (HT1080) cells through artificial basement membrane (Matrigel) in cell cultures at $IC_{50}$s of 50–150 μM and below. Having low toxicity and proven to be well tolerated after several years in human use, bisphosphonates have the potential to become one of the main MMP-inhibitors for MMP-related human soft and hard tissue–destructive diseases in the near future.

It has been clear for some time that inhibition of MMPs by pharmacologic agents in certain pathologic conditions would be beneficial to downregulate the tissue-destructive course in certain inflammatory and malignant diseases. Therefore the development of new MMP inhibitors to drugs it is hoped will be useful in the future is of great interest. Although the therapeutic utility of a collagenase/MMP inhibitor had already been recognized briefly after the initial discovery of vertebrate-type col-

[h]Address for correspondence: Dr. Olli Teronen, Department of Oral and Maxillofacial Surgery, Institute of Dentistry, P.O. Box 41, FIN-00014 University of Helsinki, Finland. Phone, (358)-9-191 27 257; fax, (358)-9-191 27 265; e-mail, olli.teronen@helsinki.fi

lagenase or MMP-1,[1] it is somewhat surprising that no safe pharmacologic product without adverse side-effects is currently available for therapeutic use as an MMP inhibitor. The discovery that tetracyclines can control tissue-eroding MMPs was made by Golub *et al.* more than 15 years ago.[2] In clinical use, low nonmicrobial doses of tetracyclines, especially doxycycline, have proved to be effective in the treatment of periodontal disease and arthritides.[3–9] Tetracycline derivatives lacking antimicrobial properties also called chemically modified tetracyclines (CMTs), have been produced and tested for the treatment of periodontitis and arthritis in animal models. It remains to be seen whether CMTs and tetracyclines (doxycycline) develop undesirable side effects, such as gastrointestinal disturbances and potential antibiotic resistance, during long-term treatment.

Some active side-targeted or peptidomimetic MMP inhibitors have been designed and tested.[10] Although very effective *in vitro,* at even nanomolar concentrations, these new "designer" inhibitors require higher doses than expected *in vivo,* and still have a long way to go to become clinically ready and safe pharmaceutical products owing to their early stage of development and the multiple steps a new drug has to undergo before approval for clinical use.[10]

Several studies during recent years have described beneficial effects of bisphosphonates in malignant human diseases such as breast cancer and multiple myeloma.[11–14] Other effects of bisphosphonates include inhibition of breast cancer cell adhesion *in vitro,*[15] inhibition of tooth-root resorption in rats,[16] reduction of the human metastastic cancer burden in nude mice,[17] and a more effective decrease in osteolytic bone metastasis with a combination of TIMP-2 and bisphosphonate ibandronate than with either alone.[18] Finally, treatment with bisphosphonates halted periodontal tissue destruction and the progression of bone destruction during the course of periodontitis in monkeys, beagle dogs, and humans.[19–21] These phenomena strongly mimic the outcome following use of any matrix metalloproteinase inhibitor.

The bisphosphonates are a class of drugs developed in the past three decades for use against various diseases involving disturbances in bone and calcium metabolism. Bisphosphonates are synthetic compounds that exhibit strong affinity for the hydroxyapatite crystal of bone. The P–C–P bond of bisphosphonates is stable in heat and most chemical reagents, and completely resistant to enzymatic hydrolysis, but can be hydrolyzed in solution by ultraviolet light. The major effect of pharmacologically active bisphosphonates has so far been their recognized ability to inhibit bone resorption. Bisphosphonates are generally well tolerated and have low toxicity.[22,23] At first, the antiosteolytic action of bisphosphonates was supposed to be due to the diminished dissolution of the hydroxyapatite crystals to which bisphosphonates are attached. Today, the mechanisms of this inhibition have not been completely clarified, but several lines of evidence indicate that the inhibition of bone resorption by bisphosphonates is a result of one or more of the following mechanisms: (a) direct inhibition of osteoclast function; (b) physicochemical incorporation into the skeletal matrix, thereby interfering with the actual process of bone resorption; or (c) direct inhibition of osteoblast-mediated cytokine production. Rodan and Fleisch[24] conclude that although the detailed mechanism of action of bisphosphonates has not been elucidated, it is clear that at the tissue level all bisphosphonates inhibit bone resorption, bone turnover, and therefore, bone loss. At the cellular level bisphosphonates are considered to act directly and/or indirectly to inactivate osteoclastic bone resorption.[24] Vacuolar ATPase, squalene synthetase, and protein tyrosine phosphatase are inhibited with certain bisphosphonates, but conclusions as to the impact of these findings on bone resorption or other possible biological effects *in vivo* remains to be established.[24–27]

In this work we describe the inhibition of catalytic activities of several genetically distinct, but structurally related MMP family members by various bisphosphonates such as clodronate, alendronate, pamidronate, and zolendronate. This inhibition at least partly explains the effects of bisphosphonates when used for treatment of diseases involving significant bone and soft tissue destruction. This therefore suggests that certain MMPs may indeed play an important role in bone resorption. Furthermore, our findings may lead to new medical indications and use of these safe, familiar drugs in numerous tissue-destructive inflammatory and malignant diseases.

## MATERIALS

### Chemicals

Pamidronate and zolendronate were donated by manufacturers for this purpose as was alendronate (Merck, Sharp & Dohme, West Point, PA, USA). Clodronate was purchased from Leiras, Turku, Finland.

### Matrix Metalloproteinases (MMPs)

For enzymatic assays of MMP activities in human biological fluids and extracts, saliva, gingival crevicular fluid (GCF), peri-implant sulcus fluid, and jaw cyst as well as gingival tissue extracts were used. The various types of MMPs were studied by Western blotting with specific poly- and monoclonal antibodies.[28–31] Purified MMPs from human peripheral blood neutrophils as well as human gingival fibroblasts and keratinocytes (MMP-1, -2, -8, and -9) were used as described in earlier papers.[28–33] Recombinant MMPs (MMP-1, -3, -7, -12, -13, -14, and -20), kindly provided by Drs. G. Murphy, V. Knäuper, and C. López-Otín were also used in the experiments.

## METHODS

### Cell Cultures

Human osteosarcoma MG-63, U2-OS, human melanoma C8161, and human fibrosarcoma HT1080 cell lines were purchased from American Type Culture Collection (ATCC, Rockville, MD). Cells were cultured in a humid atmosphere of 5% $CO_2$ and 95% air at 37°C in Dulbecco's modified eagle's medium with 1000 mg glucose/L and sodium bicarbonate (DMEM, Sigma Chemical Co, St Louis, MO) supplemented with 10% heat-inactivated fetal bovine serum (FBS) (Gibco BRL, Paisley, Scotland), 2.2 mM to which were added L-glutamine, 1% non-essential amino acids (cat. no. 11140-035, Gibco BRL), penicillin (500 IU/ml)(Gibco BRL), and streptomycin (50 mg/ml)(Gibco BRL). After reaching confluence, the cells were harvested by trypsinization with 0.25% trypsin-EDTA (Gibco-BRL) and reseeded for further use.

## Invasion Assay

Tumor cell invasion was studied using 6.4-mm-diameter Boyden chambers precoated with Matrigel (Becton Dickinson, Belford, MA); the layer of Matrigel matrix serves as reconstituted basement membrane *in vitro*. This uniform layer occludes the pores of the membrane, blocking noninvasive cells from migrating through the membrane. By contrast, invasive cells are able to detach themselves from and migrate through the Matrigel matrix-treated membrane.

## Quantification of Matrigel Invading Cells

Cells were preincubated at 37° C in a 5% $CO_2$ atmosphere for 2 hr in the presence of different concentrations of bisphosphonate (0–500 μM) in the media. Each well was plated with 100,000 cells in a 500-μl volume, in fresh serum-containing media. After cells were cultured for 16–20 hr, they were fixed in methanol, washed, and stained in toluidene blue. Cells were removed from the upper surface of the membrane with a cotton swab and the cells migrated in the membrane were quantitated by calculating under microscope.

## Cell Culture and Gelatin Zymography

For MMP assays cells were trypsinized and plated at the desired density ($1 \times 10^5$ cells per well) in the 24-well tissue culture plate (Greiner GmbH, Frickenhausen, Germany). Confluent cell layers were washed with fresh medium and cultured for 24 hours with fresh medium containing $1 \times 10^{-7}$ M phorbol 12–myristate 13–acetate (PMA) (Sigma). The cell layers were then washed twice with serum-free medium and re-fed with serum-free medium containing 0, 0.1, 1.0, or 2.5 mM clodronate (Leiras, Turku, Finland). Cultures were incubated for 24 hours, and the conditioned media harvested and analyzed for MMPs by zymography and immunoassays.

## Gelatin Zymography

Zymograms were used for the detection of distinct molecular forms of MMP-2 in cell culture media in varying concentrations of bisphosphonates reflecting the proteolytic activation of proMMP-2 from higher molecular form to lower ones. Samples were incubated in Laemmli's sample buffer for 30 minutes at 22°C, after which 20 μl of the sample was loaded onto each lane of a polyacrylamide (10% sodium dodecyl sulfate) gel containing gelatin. A separate lane was run with molecular weight standards (high and low molecular weight prestained standards from BioRad Lab., Richmond, CA, USA). Electrophoresis was carried out in precooled pool buffer (50 mM Tris; 0.38 M glysine; 0.1% SDS; pH 8.3) at a continuous voltage of 110 V. Gels were washed three times for 10 minutes in 50 mM Tris, buffer, pH 7.5, and supplemented with 2.5% Tween and 0.02% $NaN_3$; then three times for 10 minutes in the same buffer supplemented with 1 μM $ZnCl_2$ and 5 mM $CaCl_2$; and overnight at 37°C in 50 mM Tris buffer with 5 mM $CaCl_2$, 1 μM $ZnCl_2$, and 0.02% $NaN_3$, pH 7.5. The reaction was stopped with Coomassie Brilliant Blue R 250 staining followed by destaining in 5% acetic acid and 10% methanol in water. The gelatinases from the sample solubilize their substrate in the gel and therefore were visualized as white bands against the blue-stained surrounding gel, where the substrate is intact and heavily stained.

### RNA Extraction and Nothern Blot Analysis

Total RNA was prepared from $5 \times 10^7$ cells treated with $10^{-7}$ PMA using Quick-Prep® total RNA extraction kit (Pharmacia Biotech, Uppsala, Sweden). Northern analysis of MT-1-MMP expression was carried out as described.[53]

### Measurement of Collagenase Activity

Samples (biological fluids, tissue extracts, purified human MMPs, recombinant human MMPs, or cell culture media) with or without 1 mM APMA-treatment (an optimal organomercurial proMMP activator) were incubated with pure type I collagen in the presence of varying concentrations of bisphosphonates. Subsequently, the samples were separated by gel electrophoresis, and the type I collagen degradation was quantitated by densitometry.[28]

### Measurement of β-caseinolytic Activity

Samples were incubated with 52 μM β-casein for 1 hr at 22°C.[34] The enzyme reaction was then halted, and samples were electrophoresed and analyzed using a densitometer (BioRad Model GS-700 Imiging Densitometer). The disappearance of 21-kDa β-casein was regarded as MMP activity and quantitated.[34]

### Colorimetric Assay for Matrix Metalloproteinases Using Modified Pro-urokinase as Substrate

The recombinant catalytic domain of human MT1-MMP and other pure human MMPs[35,36] was incubated with 0–2000 μM clodronate and other bisphosphonates (1 hour preincubation) and assayed for MT1-MMP and other MMPs' activity according to Verheijen *et al.*,[36] with the results expressed as activity units, in which one unit is $1000 \times \Delta A_{405}/hr$.[2]

### Measurement of Gelatinase Activity by Soluble Radioactive Type I Gelatin Substrate

Gelatinolytic activities of the samples and purified MMPs were assayed with [125]I-labeled gelatin as a substrate.[37] Specimens were incubated with [125]I-labeled gelatin for 1 hour at 37°C, after which the undegraded gelatin was precipitated with 20% TCA. The supernatants and precipitates were counted separately by a gamma scintillation counter.[38]

**TABLE 1. The effects of various bisphosphonates on MMP activities, activation, production, and malignant cell invasion**

| Downregulation or inhibition of: | Clondronate | Pamidronate | Alendronate | Zolendronate | Nedrinate | Clodrinate | Tilundronate |
|---|---|---|---|---|---|---|---|
| MMP-1 activity | +++ | +++ | | | | | +++ |
| MMP-2 activity | +++ | + | | | | | |
| MMP-3 activity | +++ | +++ | +++ | | | | +++ |
| MMP-8 activity | +++ | +++ | +++ | | | | |
| MMP-9 activity | +++ | +++ | ++ | | | | |
| MMP-12 activity | | | +++ | +++ | | | |
| MMP-13 activity | | +++ | +++ | | | | |
| MMP-14 activity | +++ | | | | | | |
| MMP-20 activity | | | | +++ | | | |
| U2-OS MMP-2 activation | +++ | | | | | | |
| MG63 MMP-2 activation | ++ | | | | | | |
| MG63 MT1-MMP expression | +++ | | | | | | |
| MG63 MT1-MMP production | +++ | | | | | | |
| HT1080 cell invasion | +++ | | +++ | | +++ | +++ | |
| C8161 cell invasion | +++ | | +++ | | +++ | +++ | |

## RESULTS

TABLE 1 summarizes the effects of distinct bisphosphonates on MMP activities, MMP expression, and MMP production. We have also observed dose-dependent decrease in invasion of malignant cell lines through experimental basement membrane (Matrigel) in the presence of bisphosphonates.

### *Inhibition of MMP Activities, Expression, and Production by Bisphosphonates*

Clodronate inhibited neutrophil MMP-8–mediated degradation both of β-casein and of type I collagen substrates. Inhibition of type I collagen as well as β-casein degradation were both dose-dependent (TABLE 1).[29] Furthermore, not only was the purified neutrophil MMP-8 inhibited, but the corresponding enzyme in complex biological/inflammatory fluids such as gingival crevicular fluid (GCF) and peri-implant sulcular fluid (PISF) was inhibited as well.[29] Clodronate at 200 and 400 μM concentrations efficiently inhibited recombinant MMP-1.[30] Although significant and specific degradation of type I collagen by buffer-treated MMP-1 from jaw cyst and human fibroblast sources occurred, no formation of the specific αA (3/4) cleavage products of type I collagen occurred after pre-treatment of MMP-1 with clodronate at therapeutically attainable levels.[30] Enzymatic activities of MMP-9 and MMP-2 are also inhibited by clodronate (TABLE 1).

Pamidronate and alendronate, each in the concentration range of 250 to 1000 μM, dose-dependently inhibited the degradation of 21-kDa β-casein by human recombinant MMP-13 (TABLE 1).[39] We further found that alendronate and pamidronate can also inhibit human recombinant stromelysin (MMP-3),[39] which has been closely related to various malignancies. Additionally, pamidronate and alendronate inhibit human MMP-1, -2, -8, -9, -12, and MMP-20 at corresponding concentrations (TABLE 1).[39] Zolendronate inhibited MMP-3, -12, -13, and -20 dose-dependently.

Clodronate, in the concentration range of 200 to 2000 μM, inhibited the activity of the catalytic domain of human recombinant MT1-MMP (TABLE 1).[39] The $IC_{50}$ was extrapolated to be 80–120 μM.

Zymography showed that, PMA-induced MG-63 and U2-OS cell culture media cell lines contained MMP-2 in both proform (72 kDa) and proteolytically activated form (68 kDa). When the concentration of clodronate was increased, the active (68 kDa) in relation to the proform (72 kDa) decreased significantly (TABLE 1; FIG. 1A and B).

Immunologic staining of the MG-63 monolayer in glass slides showed clear decrease in the amount of pericellular MT1-MMP immunoreactivity. Further, Northern blotting showed reduced MT1-MMP mRNA in clodronate-treated PMA-induced HT1080 fibrosarcoma cells compared with non–clodronate-treated ones (data not shown).

When cultured on Matrigel, human melanoma C8161 and human fibrosarcoma HT1080 cell lines efficiently invade the Matrigel membrane. Invasion assays showed that, in the presence of human normal serum, bisphosphonates (clodronate, pamidronate, alendronate, nedrinate, and clodrinate) constantly and dose-dependent reduced the invasion of these malignant cell lines; the $IC_{50}$s for bisphosphonate-induced reductions of *in vitro* invasion of human malignant cell lines studied were 30–80 μM (FIG. 2).

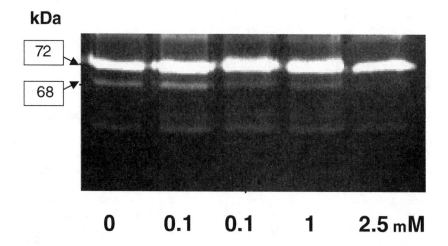

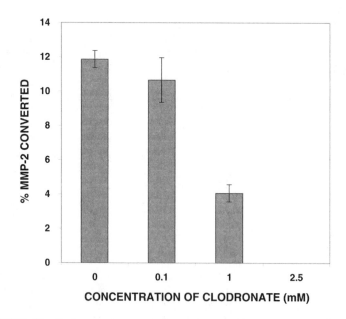

**FIGURE 1A.** Zymogram of PMA-induced U2-OS osteosarcoma cell culture media. 72-kDa gelatinolytic band represents proMMP-2 and 68-kDa its porteolytically converted from; both are confirmed to be MMP-2 by Western blotting. Clodronate (concentrations indicated below lanes) dose-dependently decreases the processing of proMMP to the 68-kDa form.

**FIGURE 1B.** Clodronate decreases MMP-2 conversion in U2-OS osteosarcoma cell culture media. Seven parallel experiments of the percentage of the 68-kDa form of total MMP-2 activities in zymography were calculated by scanning. Block graph shows the mean and standard deviation of these calculations.

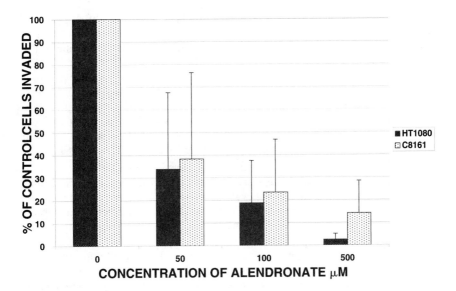

**FIGURE 2.** Block graph presents mean and standard deviation of five parallel experiments in invasion assay. Melanoma (C8161) and fibrosarcoma (HT1080) cells were cultured on Matrigel membrane. Invasion of the cells in the membrane were calculated in the presence of different concentrations of alendronate. The amount of cells invaded were calculated under the microscope and compared to the identical control without alendronate as 100% invasion.

## DISCUSSION

Nakaya and co-workers have independently confirmed our original findings,[29,30] reporting on the inhibition of MMP-1 and MMP-3 by tilundronate.[40,41] Overall, the mechanism of the MMP inhibition may involve the ability of bisphosphonates to act as cation-chelators. For the MMP inhibition by tetracyclines and CMTs, the same mechanism has been suggested.[2,3,32,42–44] Noteworthy is the fact that this inhibition occurs at a substantially low concentration (about 100 μM), obtained in targeted areas by *in vivo* medication, and bisphosphonates are known to seek in the bone and especially under osteoclasts.[23,48] Nevertheless, the specific inhibition of distinct matrix metalloproteinases seems to differ with various bisphosphonates, a finding very similar to that observed for TCs and CMTs.[3,45]

From the the viewpoint of pharmacologic intervention, of interest is the predominant role and apparent imbalances between MMPs and tisssue inhibitors of matrix metalloproteinases (TIMPs) in malignant and inflammatory tissue-destructive diseases. Finding safe MMP inhibitors for long-term clinical use has become a major goal in the design of optimal treatment procedures and medications to halt ongoing pathologic tissue-loss in such diseases. In addition to the classic involvement of MMPs in cancer and metastatic tissue degradation, the role of MMPs have been shown in recent studies to play a part in molecular steps involving the growth of the primary tumor, in angiogenesis, in the initiation of growth at an ectopic site, and in the sustained growth of metastatic foci which are becoming clinically detectable tumors.[46] Membrane-type MMPs have been under growing interest in the recent years and their role in malignant cell invasion has been well established owing to their ability to target proteolytic action at specific direction through activation, as have other MMPs, at cell surface upon their secretion. The suprisingly good results in the treatment of multiple myeloma and breast cancer with bisphosphonates may thus in part be explained by the inhibition of MMPs, and especially inhibition of proMMP-2 activation and MT1-MMP activity/expression at a cellular level, as shown in our experiments. Furthermore, the invasive potential of the several different types of human malignant cell lines studied could be downregulated by bisphosphonates.

The pharmacologic MMP inhibition can be beneficial in arthritides, in various kinds of ulceration, several skin diseases, in oral aphtous lesions, in periodontal diseases, dental peri-implantitis, in jaw cysts, and in prevention of the loosening of total hip replacement (and other) prostheses/implants. The impact of such a clinically safe inhibitor in the long-term treatment of these soft- and hard-tissue-destructive diseases would be immense. The increasing interest in MMP inhibitors has led to the development of various new active-site-targeted peptide inhibitors (anti-MMP peptidomimietics) in addition to the previously known tetracyclines and their non-antimicrobial CMT-derivatives.[47] Despite the promising prospects of these effective peptidomimetic inhibitors, they are burdened with long testing periods prior to possible approval for their use clinically.[10] It is noteworthy that these anticollagenolytic peptidomimetics are very efficient *in vitro* MMP inhibitors at pico- and nanomolar levels, but their beneficial anti-tissue-destructive effects *in vivo* have been obtained at micromolar serum levels, thus, making them comparable to bisphosphonates and tetracyclines.[2,3,10,48]

Bisphosphonates have been shown to be beneficial in the treatment of bone destruction in malignancies with increased metastatic potential as well as in the treatment of inflammatory tissue-destructive diseases in humans.[11–13,49,50] The molecular mechanism(s) of the ability of bisphosphonates to prevent and inhibit tissue destruction and cancer dissemination have, however, remained unclear.[24] Most recently, it has been shown that alendronate reverses the ability of TIMP-2 to inhibit MMP-2 from degradation by plasmin *in vitro*.[51] The inhibition of MMPs by bisphosphonates should be considered potentially one of the main mechanisms explaining, at least partially, their antitumor effects, such as reduction of the actual tumor burden and the spread not only to bony tissues but also to soft tissues.[49] In fact, recent studies by Ramamurthy *et al.*[52] show that in the endotoxin (LPS)-induced rat periodontitis model, when compared to single use of doxycycline or bisphosphonate, combination treatment with these broad-spectrum MMP-inhibitors (doxycycline/CMTs with bisphosphonate) lead in synergistically enhanced reduction of periodontal soft- and bone-tissue destruction, associated with inhibition and downregulation of gingival tissue MMPs and serine proteinases.[52]

## SUMMARY

Bisphosphonates have the ability to inhibit MMPs at several levels starting from expression of MMPs to cellular events known to be related to MMPs' actions. New findings show the novel potential of bisphosphonates to inhibit major tissue-destructive enzymes and may help to clarify the beneficial molecular mechanisms of these agents in inhibiting and preventing hard- and soft-tissue destruction. These observations can be useful in future development of drugs as clinically relevant MMP inhibitors/downregulators alone and/or in different combinations in the treatment of diseases involving pathologic excessive extracellular matrix degradation.

Bisphosphonates are thus among the prime candidates for pharmacologic use as MMP inhibitors locally or systemically for conditions such as periodontal disease, dental peri-implantitis, rheumatoid arthritis and other arthitides, loosening of total hip replacement prostheses, several skin diseases, oral aphtous ulcers, and even for cancer. Bisphosphonates which are already well tolerated in humans, may thus soon be used in clinical medicine, sooner than will the use of inhibitors currently at the developmental stage. Having low toxicity and proven to be well tolerated after several years in human use, bisphosphonates have the potential to become one of the main MMP inhibitors for MMP related human diseases in the near future.

## ACKNOWLEDGMENTS

This study was supported by grants from the Finnish Dental Society, the Academy of Finland, and by the EVO Clinical Research Grant (TKILO19).

## REFERENCES

1. GROSS, J. & C.M. LAPIÉRE. 1962. Collagenolytic activity in amphibian tissue culture assay. Proc. Natl. Acad. Sci. USA **54:** 1197–1204.
2. GOLUB, L.M. *et al.* 1983. Minocycline reduces gingival collagenolytic activity during diabetes. Preliminary observations and a proposed new mechanism of action. J. Periodont. Res. **18:** 516–526.
3. GOLUB, L.M. *et al.* 1992. Host modulation with tetracyclines and their chemically modified analogues. Curr. Opin. Dent. **2:** 80–90.
4. GOLUB, L.M. *et al.* 1997. A matrix metalloproteinase inhibitor reduces bone-type collagen degradation fragments and specific collagenases in gingival crevicular fluid during adult periodontitis. Inflamm. Res. **46:** 310–319.
5. INGMAN, T. *et al.* 1993. Tetracycline inhibition and the cellular source of collagenase in gingival crevicular fluid in different periodontal diseases. J. Periodontol. **64:** 82–88.
6. LAUHIO, A. *et al.* 1994. Reduction of matrix metalloproteinase 8-neutrophil collagenase levels during long-term doxycycline treatment of reactive arthritis. Antimicrob. Agents Chemother. **38:** 400–402.
7. LAUHIO, A. *et al.* 1994. In vivo inhibition of human neutrophil collagenase (MMP-8) activity during long-term combination therapy of doxycycline and NSAID in acute reactive arthritis. Clin. Exp. Immunol. **98:** 21–28.
8. LAUHIO, A. *et al.* 1995. Tetracyclines in the treatment of rheumatoid arthritis. Lancet. **340:** 645–646.
9. NORDSTRÖM, D. *et al.* 1998. Anti-collagenolytic mechanism of action of doxycycline treatment in rheumatoid arthritis. Rheumatol. Int. **5:** 175–180.
10. BROWN, P.D. & R. GIAVAZZI. 1995. Matrix metalloproteinase inhibition: a review of anti-tumour activity. Ann. Oncol. **6:** 967–974.
11. DELMAS, P.D. 1996. Bisphosphonates in the treatment of bone diseases. N. Engl. J. Med. **335:** 1836–1837.

12. HORTOBAGYI, G.N. *et al.* 1996. Efficacy of pamidronate in reducing skeletal complications in patients with breast cancer and lytic bone metastases. Protocol 19 Aredia Breast Cancer Study Group. N. Engl. J. Med. **335:** 1785–1791.
13. BANKHEAD, C. 1997. Bisphosphonates spearhead new approach to treating bone metastases. J. Natl. Cancer Inst. **89:** 115–116.
14. DIEL, I.J. *et al.* 1998. Reduction in new metastases in breast cancer with adjuvant clodronate treatment. N. Engl. J. Med. **339:** 357–363.
15. VAN DER PLUIJM, G. *et al.* 1996. Bisphosphonates inhibit the adhesion of breast cancer cells to bone matrices in vitro. J. Clin. Invest. **98:** 698–705.
16. IGARASHI, K. *et al.* 1996. Inhibitory effect of the topical administration of a bisphosphonate (risedronate) on root resorption incident to orthodontic tooth movement in rats. J. Dent. Res. **75:** 1644–1649.
17. SASAKI, A. *et al.* 1995. Bisphosphonate risendronate reduces metastatic human breast cancer burden in bone in nude mice. Cancer Res. **55:** 3551–3557.
18. YONEDA, T. *et al.* 1997. Inhibition of osteolytic bone metastasis of breast cancer by combined treatment with the bisphosphonate ibandronate and tissue inhibitor of the matrix metalloproteinase-2. J. Clin. Invest. **99:** 2509–2517.
19. JEFFCOAT, M.K. & M.S. REDDY. 1996. Alveolar bone loss and osteoporosis: Evidence for a common mode of therapy using the bisphosphonate alendronate. *In*: Biological Mechanisms of Tooth Movement and Craniofacial Adaptation. Z. Davidowitch & L.A. Norton, Eds.: 365–373. Harvard Society for Advancement of Orthodontics, Boston, MA.
20. REDDY, M.S. *et al.* 1995. Alendronate treatment of naturally occuring periodontitis in beagle dogs. J. Periodontol. **66:** 211–217.
21. WEINREB, M. *et al.* 1994. Histomorphometrical analysis of the effects of the bisphosphonate alendronate on bone loss caused by experimental periodontitis in monkeys. J. Periodont. Res. **29:** 35–40.
22. FLEISCH, H. *et al.* 1991. Bisphosphonates. Pharmacology and use in the treatment of tumour-induced hypercalcaemic and metastatic bone diseases. Drugs. **42:** 919–944.
23. FLEISCH, H. *et al.* 1997. Bisphosphonates in bone disease. From laboratory to the patient. The Parthenon Publishing Group. New York.
24. RODAN, G.A. & H.A. FLEISCH. 1996. Bisphosphonates: mechanisms of action. J. Clin. Invest. **97:** 2692–2696.
25. FELIX, R. *et al.* 1976. The effect of several diphosphonates on acid phosphohydrolases and other lysosomal enzymes. Biochim. Biophys. Acta **429:** 429–438.
26. SKOREY, K. *et al.* 1997. How does alendronate inhibit protein-tyrosine phosphatases? J. Biol. Chem. **272:** 22472–22480.
27. SMITH, A. *et al.* 1996. Protein-tyrosine phosphatase activity regulates osteoclast formation and function: inhibition by alendronate. Proc. Natl. Acad. Sci. USA **93:** 3068–3073.
28. TERONEN, O. *et al.* 1995. Characterization of interstitial collagenases in jaw cyst walls. Eur. J. Oral Sci. **103:** 141–147.
29. TERONEN, O. *et al.* 1997. Inhibition of matrix metalloproteinase-1 by dichloromethylene bisphosphonate (clodronate). Calcif. Tissue Int. **61:** 59–61.
30. TERONEN, O. *et al.* 1997. Human neutrophil collagenase MMP-8 in peri-implant sulcus fluid and its inhibition by clodronate. J. Dent. Res. **76:** 1529–1537.
31. SORSA, T. *et al.* 1994. Effects of tetracyclines on neutrophil, gingival, and salivary collagenases. A functional and western-blot assessment with special reference to their cellular sources in periodontal diseases. Ann. N.Y. Acad. Sci. **732:** 112–131.
32. SORSA, T. *et al.* 1997. Activation of type IV procollagenenase by human tumor-associated trypsin-2. J. Biol. Chem. **272:** 21067–21074.

33. TERONEN, O. *et al.* 1995. Identification and characterization of gelatinases/type IV collagenases in jaw cysts. J. Oral Pathol. Med. **24:** 78–84.

34. DECARLO, A. 1994. Matrix metalloproteinase activation and induction in keratinocytes by a purified thiol-proteinase from Porphyromonas gingivalis. PhD thesis, University of Alabama at Birmingham, Alabama, USA.

35. LICHTE, A. *et al.* 1996. The recombinant catalytic domain of membrane-type matrix metalloproteinase-1 (MT1-MMP) induces activation of progelatinase A and progelatinase A complexed with TIMP-2. FEBS Lett. **397:** 277–282.

36. VERHEIJEN, J.H. *et al.* 1997. Modified proenzymes as artificial substrates for proteolytic enzymes: colorimetric assay of bacterial collagenase and matrix metalloproteinase activity using modified pro-urokinase. Biochem J. **323:** 603–609.

37. RISTELI, L. & J. RISTELI. 1987. Analysis of extracellular matrix proteins in biological fluids. Meth. Enzymol. **145:** 391–411.

38. MÄKELÄ, M. *et al.* 1994. Matrix metalloproteinases (MMP-2 and MMP-9) of oral cavity: cellular origin and relationship to periodontal status. J. Dent. Res. **73:** 1397–1406.

39. TERONEN, O. 1998. Jaw cyst matrix metalloproteinases (MMPs) and inhibition of MMPs by bisphosphonates. PhD thesis, University of Helsinki, Helsinki, Finland.

40. NAKAYA, H. *et al.* 1997. Effects of bisphosphonate on MMP-1 activity and mRNA expression. J. Dent. Res. **76:** A2725.

41. NAKAYA, H. *et al.* 1997. Effects of bisphosphonates on MMP-3 activity and mRNA expression [abstract]. Annual meeting of the American Academy of Periodontology.

42. GOLUB, L.M. *et al.* 1985. Further evidence that tetracyclines inhibit collagenase activity in human crevicular fluid and from other mammalian sources. J. Periodont. Res. **20:** 12–23.

43. RYAN, M.E. *et al.* 1996. Matrix metalloproteinases and their inhibition in periodontal treatment. Curr. Opin. Periodont. **3:** 85–96.

44. JONAT, C. *et al.* 1996. Transcriptional downregulation of stromelysin by tetracycline. J. Cell. Biochem. **60:** 341–347.

45. SUOMALAINEN, K. *et al.* 1992. Tetracycline inhibition identifies the cellular origin of interstitial collagenases in human periodontal diseases in vivo. Oral Microbiol. Immunol. **7:** 121–123.

46. CHAMBERS, A.F. & L.M. MATRISIAN. 1997. Changing views of the role of matrix metalloproteinases in metastasis. J. Natl. Cancer Inst. **89:** 1260–1270.

47. SEFTOR, R.E. *et al.* 1998. Chemically modified tetracyclines inhibit human melanoma cell invasion and metastasis. Clin. Exp. Metastasis. **16:** 217–225.

48. SATO, M. *et al.* 1991. Bisphosphonate action. Alendronate localization in rat bone and effects on osteoclast ultrastructure. J. Clin. Invest. **88:** 2095–2105.

49. SIRIS, E.S. 1997. Breast cancer and osteolytic metastases: Can bisphosphonates help? Nat. Med. **2:** 151–152.

50. DELMAS, P.D. 1996. Bisphosphonates in the treatment of bone diseases. N. Engl. J. Med. **335:** 1836–1837.

51. FARINA, A.R. *et al.* 1998. Tissue inhibitor of metalloproteinase-2 protection of matrix metalloproteinase-2 from degradation by plasmin is reversed by divalent cation chelator EDTA and the bisphosphonate alendronate. Cancer Res. **58:** 2957–2960.

52. RAMAMURTHY, N.S. *et al.* 1999. A suboptimal CMT-8 clodronate combination therapy inhibits alveolar bone resorption in LPS induced periodontitis in rats. Ann. N.Y. Acad. Sci. This volume.

53. HANEMAAIJER, R. *et al.* 1997. Matrix metalloproteinase-8 is expressed by rheumatoid synovial fibroblasts and endothelial cells. J. Biol. Chem. **272:** 31504–31509.

# Retinoid-Mediated Suppression of Tumor Invasion and Matrix Metalloproteinase Synthesis

MATTHIAS P. SCHOENERMARK,[a,b] TERESA I. MITCHELL,[a] JONI L. RUTTER,[a]
PETER R. RECZEK,[c] AND CONSTANCE E. BRINCKERHOFF[a]

[a]Dartmouth Medical School, Hanover, New Hampshire 03755, USA

[c]Bristol-Myers Squibb, Buffalo, New York 14213, USA

ABSTRACT: Cancer mortality usually results from the tumor invading the
local environment and metastasizing to vital organs, e.g. liver, lung, and brain.
Degradation of the extracellular matrix is, therefore, the *sine qua non* of tumor
cell invasion. this degradation is mediated mainly by MMPs, and thus, inhibition of MMP synthesis is a target for anticancer agents. Tumor cells must
traverse both the basement membrane (type IV collagen) and the interstitial
stroma (type I collagen). Therefore, we used scanning electron microscopy to
examine the invasive behavior of several aggressive tumor cell lines, A2058
melanoma cells, and SCC and FaDu squamous cell carcinomas through these
matrices; and we monitored the ability of all-*trans* retinoic acid and several
RAR-specific ligands to block invasion. We demonstrate that several retinoids,
which are specific RAR α, β, or γ agonists/antagonists, selectively inhibited
MMP synthesis in the three tumor cell lines. However, there was not a common
pattern of MMP inhibition by a particular retinoid. For instance, a RARα antagonist suppressed MMP-1 and MMP-2 synthesis in the melanoma cell line,
but not in the FaDu or SCC-25 cells. On the other hand, synthesis of MMP-1
and MMP-9 by the FaDu cells was affected hardly at all, while a RARγ antagonist reduced the levels of MMP-2. Only all-*trans* retinoic acid reduced MMP-1
synthesis in these cells. We postulate that the differences may be related to a
differential pattern of RAR expression in each of these cells, and that the RARs
expressed by each cell line may not be targets of these RAR specific compounds. All-*trans* retinoic acid is a pan ligand, binding to all three RARs and,
therefore, may modulate gene expression more generally. We conclude that the
power of these new ligands lies in their specificity, which can be directed towards modulating expression of certain RARs and, thus, of certain MMPs. By
blocking MMP synthesis, retinoids may be effective in cancer therapy by decreasing tumor invasiveness.

Degradation of the extracellular matrix is an essential step in the process of tumor
invasion and metastasis. This degradation is mediated principally by matrix metalloproteinases, a family of enzymes that, collectively, destroys matrix macromolecules.[1-4] The type IV collagen in basement membrane is degraded primarily by
MMP-2 and MMP-9, while the stromal collagens (types I and III) are destroyed by
the interstitial collagenases, of which there are three: MMP-1, MMP-8, and

[b]Present address: Department of Otolaryngology, Hannover Medical School, Carl-Neuberg-
Strasse 1, D-30625 Hannover/Germany.

MMP-13. Of these, MMP-1 is the most ubiquitously expressed, and is produced by stromal fibroblasts as well as by some tumor cells, thereby providing several mechanisms for facilitating collagen breakdown and the invasive ability of tumor cells.[1–4]

Because matrix degradation is so central to the invasive/metastatic behavior of tumors, and because it is usually this invasive/metastatic ability that is life-threatening, therapeutic strategies designed to inhibit invasion and keep a tumor localized are attractive. Given the link between high levels of MMP gene expression and tumor invasion, it is not surprising that considerable attention has been devoted to inhibiting enzyme *activity,* and hence the ability to degrade the extracellular matrix around tumor cells. Indeed, a number of synthetic inhibitors of MMP activity have been developed.[5] These compounds are potent, but reversible, inhibitors that bind to the active site of MMPs. Although clinical trials are in progress, some compounds are very insoluble and have poor bioavailability when administered orally,[5] making their success problematic. In addition, they may be associated with undesirable side effects.[5]

Another approach is the inhibition of MMP *synthesis* by the vitamin A analogues, retinoids. Retinoids mediate their effects on gene expression by two classes of nuclear hormone receptors, the retinoic acid receptors (RARs) and retinoid X receptors (RXRs), both of which belong to the steroid hormone receptor super family and have $\alpha$, $\beta$ and $\gamma$ subtypes (reviewed in Refs. 4 and 6). RARs and RXRs often heterodimerize and act through a variety of RAR element (RARE) motifs that resemble the sequence AG[G/T]TCA.[4,6] However, with the exception of MMP-11 (stromelysin-3),[7] MMP promoters do not contain a classical RARE. Instead, repression of MMP transcription by retinoids is mediated by several mechanisms that include (a) upregulation of RAR mRNAs, (b) downregulation of Fos and Jun mRNAs, (c) sequestration of Fos and Jun proteins, and (d) the interaction of RAR/RXR heterodimers with the AP-1 site, indirectly, via AP-1 binding proteins.[6] Oral administration of all-*trans*-retinoic acid and synthetic retinoids has shown some promise against a variety of kinds of cancer.[8,9] The therapeutic effects of retinoids occur through several mechanisms, including inhibition of cell proliferation, induction of apoptosis, enhanced cell differentiation,[8,9] and, significantly, suppression of tumor invasion.[8–11] While the decrease in invasiveness may be partly due to retinoid-mediated repression of tumor cell motility,[10,11] it is also attributable to suppression of MMP production.[3,4,6,10]

Malignant melanoma represents one of the most aggressive and invasive cancers, and is notoriously recalcitrant in responding to traditional chemotherapies.[12,13] Squamous cell carcinoma of the head and neck is also aggressive, often requiring mutilating surgery and has a high recurrence rate.[4,15] Thus, like melanoma, these tumors also have a poor prognosis. In both kinds of cancer, the most important prognostic factors are lymph node metastasis and invasive depth of the tumor, both of which are determined by the activity of MMPs that are produced by the tumor cells or by the surrounding stromal tissues.[1,2,12,13] These invasive tumors are, therefore, among the most promising as targets for retinoid therapy, where inhibition of MMP synthesis may ablate the invasive process. While retinoids have been somewhat effective in the treatment of head and neck carcinoma,[14,15] their success in melanoma has been limited, similar to results with many other chemotherapeutic agents.[12,13] Nonetheless, recent reports suggest that RAR/RXR specific ligands are effective against melanoma[16,19] and against specific MMPs,[7] including MMP-1-specific compounds AM-80, CD666,[7] thus sparking renewed interest in retinoids for treating melanoma.[7,16–19,20]

Because of this renewed interest in retinoids as chemopreventive and chemotherapeutic agents for these aggressive forms of cancer, we have analyzed a panel of ligand-specific retinoids for their ability to repress the profile of MMPs expressed by A2058 melanoma cells, and two head and neck cancer cell lines, FaDu and SCC-25 cells. We also used a sensitive and quantitative *in vitro* invasion assay[21] to measure the ability of the tumor cell lines to invade a reconstituted matrix of type IV collagen (Matrigel®) or type I collagen, and the ability of retinoids to block this invasion.

## MATERIALS AND METHODS

### *Retinoids*

All *trans*-retinoic acid was purchased from Sigma (St. Louis, MO) and was prepared as previously described. RAR-specific retinoids were provided by Bristol Myers Squibb, and were as follows: RAR $\alpha$ agonist #194753; RAR $\beta$ agonist #185411; RAR $\alpha$ antagonists #195614 and 189532; RAR $\gamma$ antagonist #191681; RAR $\alpha$, $\beta$, $\gamma$ antagonist #189453. Stocks of each compound were prepared at $10^{-3}$ M in DMSO and stored at $-70°C$. All compounds were protected from light. Cells were treated with retinoids for 24 hours or 48 hours under serum-free conditions, at $10^{-6}$ M, unless otherwise noted.

### *Cell Culture*

A2058 cells were grown in Dulbecco's Modified Eagles Medium (DMEM) with 10% fetal calf serum. FaDu cells were grown in Eagle's Minimum Essential Medium with 10% FCS. For SCC-25 cells, the medium (1:1 of Ham's F-12 and DMEM) was supplemented with 0.04 $\mu g/ml$ hydrocortisone. All stocks were grown in the presence of penicillin/streptomycin (37°C in 5% $CO_2$). After 3 to 4 days, when the cells were confluent, they were passaged 1:7 with 0.25% trypsin.[22,23] Cultures of human foreskin fibroblasts were prepared as described previously.[22]

### *Northern Blot Analysis*

Northern analysis was used to measure levels of mRNA for MMPs. At confluence, all cells were washed three times with Hanks' Balanced Salt Solution (HBSS) to remove traces of serum, and then cultured in serum-free DMEM, with or without retinoid ($10^{-6}$ M) for 24 hours. Total RNA was then harvested, and 10 $\mu g$ were assayed as described.[22,23] Blots were probed sequentially with cDNAs for human MMP-1, MMP-3, MMP-2, MMP-9 (a generous gift of Dr. Harald Tschesche), MMP-13, and MT1-MMP.[23] Autoradiographs were exposed for varying lengths of time, from several hours, 24 hours, or as long as 4 days. The longest exposure times were required to detect those MMPs that were expressed at a low level.

## In vitro *Invasion Assay with Scanning Electron Microscopy and Confocal Laser Scan Microscopy*

The assay utilizes a modified Boyden chamber[21] with the two compartments that are separated by a nitrocellulose filter.[21,23] Autoclaved filters were coated with Matrigel®[21] or collagen type I (1 mg/ml; Sigma), which was diluted in sterile DMEM in the presence of 1% antibiotics and successively applied to the membrane ($3 \times 150$ µl, $1 \times 550$ µl, allowed to gel at 37°C for 30 min, air dried for 1hr at room temperature). Single cell suspensions of $10^5$ tumor cells alone, or co-cultures of tumor cells and normal skin fibroblasts ($5 \times 10^4$ of each cell type) were washed and resuspended in serum-free media. Cell viability was monitored by trypan blue exclusion. The lower chamber was filled with serum-free media, and $10^5$ cells in 1 ml were added to the upper chamber. Chambers were incubated at 37°C and 5% $CO_2$ for 24 hours or 48 hours, and cellular invasion was assessed by scanning electron microscopy (SEM).

### Quantitation of Invasion with Confocal Laser Scan Microscopy (CLSM)

A quantitative assay that uses confocal laser microscopy[21,23] measured the invasive ability of the tumor cells through Matrigel® and type I collagen. Confocal laser scan microscopy (CLSM) was performed on the invasion chamber membranes.[21,23] Filters were washed in PBS, fixed in 2-propanol, and then dried and made transparent by soaking in 100% xylene. The transparent filters were dried again, mounted on a microscope slide, and subjected to CLSM in the reflection mode to allow for three dimensional analysis. The $z$-dimension was scanned in 2-µm steps, and the fluorescent signal of each layer was represented as a TIFF-file, and saved in Microsoft Excel©.[21,23] Invasion of each cell was plotted as the percentage of cells migrating through each 2-µm layer of the membrane.

## RESULTS

### Profile of MMP Expression and Inhibition by Retinoids

A2058 melanoma cells and FaDu and SCC-25 squamous carcinoma cells were examined for their profile of MMP expression. We measured the interstitial collagenases (MMP-1, MMP-13), the gelatinases/type IV collagenases (MMP-2, MMP-9), stromelysin (MMP-3), and membrane-bound MT1-MMP. This profile of MMPs was chosen because it represents the most commonly expressed MMPs, and because, taken together, these enzymes utilize most of the substrates present in the extracellular matrix and the basement membrane. The results of these studies are presented in FIGURES 1-4, and are summarized in TABLE 1.

Figure 1 (lane 1) shows the pattern of constitutive expression of MMPs by A2058 melanoma cells. These cells express MMP-1, MMP-2, MT1-MMP, and low levels of MMP-9. Neither stromelysin (MMP-3) nor collagenase-3 (MMP-13) was detectable by Northern analysis. (FIG. 2, lane 1). In contrast, FIGURE 2 illustrates that FaDu cells express both MMP-3 and MMP-13 (lane 2). These cells also express MMP-1

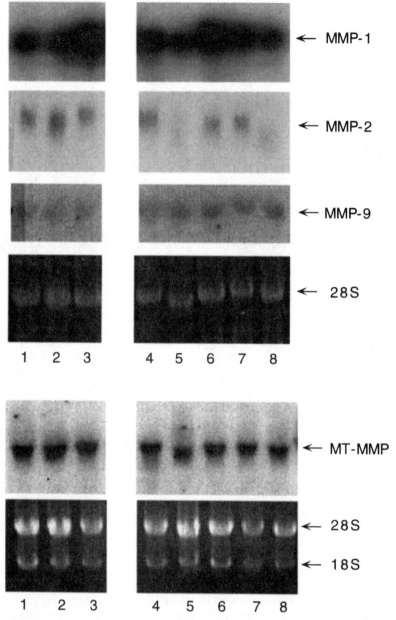

**FIGURE 1.** Profile of MMP expression by A2058 melanoma cells and inhibition by retinoids. Confluent cultures of A2058 melanoma cells were washed with Hank's Balanced
Salt Solution and placed in serum-free DMEM without or with retinoids ($10^{-6}$ M) for 24
hours. Total RNA was harvested and analyzed for expression of MMPs by Northern blot
analysis. *Lane 1:* untreated; *lane 2:* α agonist #194753; *lane 3:* β agonist #185411; *lane 4:*
α antagonist #195614; *lane 5:* α antagonist #189532; *lane 6:* γ antagonist #191681; *lane 7:*
α, β γ antagonist #189453; *lane 8:* all-*trans*-retinoic acid.

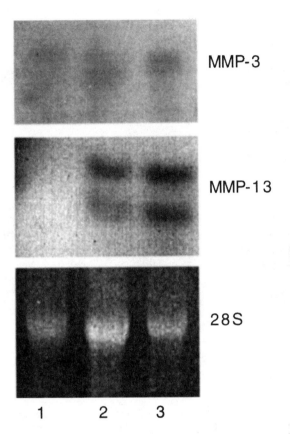

**FIGURE 2.** Profile of MMP expression by A2058 melanoma cells and by FaDu and SCC-25 squamous carcinoma cells. Confluent cultures of cells were washed with Hank's Balanced Salt Solution and placed in serum-free DMEM without or with retinoids ($10^{-6}$ M) for 24 hours. Total RNA was harvested and analyzed for expression of MMP-3 (stromelysin-1) or MMP-13 (collagenase-3) by Northern blot analysis. *Lane 1:* A3058 melanoma cells; *lane 2:* FaDu cells; *lane 3:* SCC-25 cells.

and MMP-9 constitutively (FIG. 3, lane 1), although the levels of MMP-1 are not as great as those seen in the A2058 cells. FaDu cells do not express MMP-2 or MT1-MMP (TABLE 1 and data not shown). Finally, the SCC-25 cells express MMP-3 and MMP-13 (FIG. 2, lane 3), along with substantial amounts of MMP-1, MMP-2, MT1-MMP, and low levels of MMP-9 (FIG. 4, lane 1). Thus, each of these three tumor cell lines expresses a number of MMPs, which gives them the potential to degrade both stromal collagens and basement membrane.

We next tested the ability of a series of specific RAR/RXR agonist/antagonists to affect MMP expression. Since all-*trans*-retinoic acid ($10^{-6}$ M) has inhibited MMP production by both normal and tumor cells,[4,6] this compound was included as a pos-

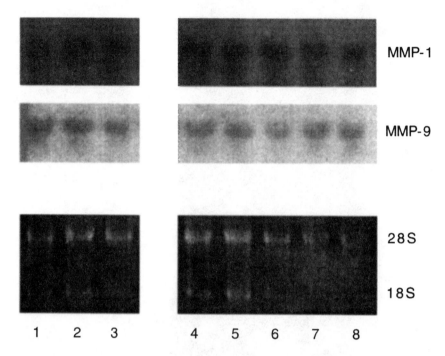

**FIGURE 3.** Profile of MMP expression by FaDu cells and inhibition by retinoids. Confluent cultures of FaDu cells were washed with Hank's Balanced Salt Solution and placed in serum-free DMEM for 24 hours. Total RNA was harvested and analyzed for expression of MMPs by Northern blot analysis. *Lane 1:* untreated; *lane 2:* α agonist #194753; *lane 3:* β agonist #185411; *lane 4:* α antagonist #195614; *lane 5:* α antagonist #189532; *lane 6:* γ antagonist #191681; *lane 7:* α, β γ antagonist #189453; *lane 8:* all-*trans*-retinoic acid.

**TABLE 1. Summary of MMP gene expression**

| MMP | Cells | | | |
|---|---|---|---|---|
| | A2058 | FaDu | SCC-25 | HFF |
| MMP-1 | + | + | + | + |
| MMP-2 | + | − | + | + |
| MMP-3 | − | ± | ± | + |
| MMP-9 | + | + | + | + |
| MMP-13 | − | + | + | − |
| MT1-MMP | + | − | + | + |

NOTE: Total mRNA was isolated from A2058, FaDu, or SCC-25 cells, and 10 µg/lane was analyzed by Northern blot with cDNAs specific for MMP mRNAs. + = mRNA is strongly expressed; ± = weaker expression; − = no expression detected.

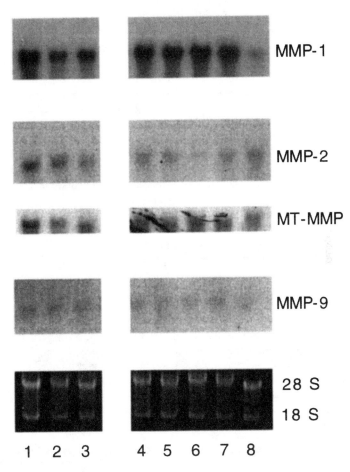

**FIGURE 4.** Profile of MMP expression by SCC-25 cells and inhibition by retinoids. Confluent cultures of SCC-25 cells were washed with Hank's Balanced Salt Solution and placed in serum-free DMEM without or with retinoids ($10^{-6}$ M) for 24 hours. Total RNA was harvested and analyzed for expression of MMPs by Northern blot analysis. *Lane 1:* untreated; *lane 2:* α agonist #194753; *lane 3:* β agonist #185411; *lane 4:* α antagonist #195614; *lane 5:* α antagonist #189532; *lane 6:* γ antagonist #191681; *lane 7:* α, β γ antagonist #189453; *lane 8:* all-*trans*-retinoic acid.

itive control. A series of six BMS retinoids ($10^{-6}$ M) were tested for their ability to inhibit the panel of MMPs expressed by A2058 melanoma cells, and by SCC-25 and FaDu squamous cell carcinoma cells.

As shown in FIGURE 1, most of the retinoids tested had no effect on the level of expression of MMPs by A2058 melanoma cells. However, RAR α antagonist, #189532, inhibited expression of MMP-1 and MMP- 2, and the level of inhibition seen with this compound paralleled that seen with all-*trans*-retinoic acid (lane 8).

Both the β agonist (lane 3) and the γ antagonist (lane 6) appear to increase levels of MMP-1, but not of the other MMPs.

In contrast, MMP production by FaDu cells was not inhibited by either the BMS compounds or all-*trans*-retinoic acid (FIG. 3). However, the retinoids displayed a different pattern of MMP inhibition when tested on the SCC-25 cells (FIG. 4). The α and β agonists (lanes 2 and 3, respectively) reduced MMP-1 (collagenase-1) expression, but were not as effective as all-*trans*-retinoic acid (lane 8). All of the compounds, including all-*trans*-retinoic acid, inhibited expression of MMP-2, with the γ antagonist (#191681, lane 6) perhaps being the most potent. MT1-MMP expression was slightly reduced by most compounds, although the α agonist (#194753, lane 2) appears to be the most effective, showing a level of inhibition similar to that seen with all-*trans*-retinoic acid (lane 8). Expression of MMP-9 was weak, and was not substantially affected by any of the compounds. Finally, MMP-13 (collagenase-3) expression was not inhibited by all-*trans*-retinoic acid (data not shown). Thus, no one compound or class of compounds is universally effective in the SCC-25 cells. Nonetheless, the agonists appear to inhibit the collagenases, while the antagonists seem to be effective against the gelatinase/type IVase.

### *Invasive Ability of Tumor Cells through Matrigel and Type I Collagen and Inhibition by Retinoids*

We have previously demonstrated that A2058 cells invade Matrigel®, but do not invade type I collagen unless they are cultured in the presence of conditioned medium derived from normal fibroblasts.[23] Furthermore, invasion is blocked if the tumor cells are treated with all-*trans*-retinoic acid ($10^{-6}$ M). Importantly, there are no apoptotic effects on the cells from retinoic acid.[23] Because BMS #189532 was as effective as all-*trans*-retinoic acid in reducing MMP-1 mRNA (FIG. 1), we tested its

**TABLE 2. Quanitation of invasion by FaDu and SCC25 squamous cell carcinomas**

| Cell Type | Matrix | Treatment | Noninvasive fraction | Invasive fraction |
|-----------|--------|-----------|----------------------|-------------------|
| FaDu | Matrigel | — | 82% | $18 \pm 7\%$ |
| FaDu | Matrigel | retinoic acid | 99% | $0.8 \pm 0.6\%$ |
| FaDu | collagen I | — | 90% | $10.4 \pm 2.8\%$ |
| FaDu | collagen I | retinoic acid | 100% | 0% |
| SCC25 | Matrigel | — | | $52.3 \pm 8.1\%$ |
| SCC25 | Matrigel | retinoic acid | 72% | $28 \pm 2.3\%$ |
| SCC25 | collagen I | — | 100% | 0% |
| SCC25 | collagen I | retinoic acid | 100% | 0% |

NOTE: Invasion through Matrigel or type I collagen by FaDu or SCC-25 cells, in the presence or absence of all-*trans*–retinoic acid ($10^{-6}$ M), was quantified by confocal laser scanning microscopy. Data are expressed as the percent of cells invading the matrices to a depth of 10 μm.

ability to block invasion of A2058 melanoma cells. Untreated A2058 cells co-cultured with fibroblasts invade the collagen matrix (FIG. 5A). After 24 hours of treatment with BMS #189532 at $10^{-6}$ M, invasion is blocked but the A2058 cells are beginning to show signs of toxicity, as seen by the presence of "blebs" (FIGS. 5B and C). However, after 48 hours co-culture with a lower concentration of this retinoid ($10^{-8}$ M), invasion by the tumor cells is still blocked (FIG. 5D). The fibroblasts appear healthy, but the A2058 cells are dying. Thus, a low concentration of this retinoid, the $\alpha$ antagonist #189532, appears (a) to block invasion of type I collagen by the melanoma cells and (b) to be selectively toxic for these cells.

Since this compound also reduced levels of MMP-2 mRNA in A2058 cells, we tested its ability to block invasion through Matrigel, and found that it could not do so (data not shown). Thus, the $\alpha$ antagonist #189532 is a selective inhibitor of MMP invasiveness in A2058 melanoma cells, preventing invasion through type I collagen, but not through Matrigel. The fact that this compound has differential effects on invasion through these two matrices, even though it can suppress both the synthesis of MMP-1 and MMP-2, suggests that different mechanisms are mediating invasion through these matrices (see the Discussion section).

Next we examined the ability of the squamous cell carcinoma line, FaDu, to invade Matrigel and type I collagen (FIG. 6). After 24 hours, the cells have begun to invade Matrigel (FIG. 6A), and by 48 hours, there is more invasion (FIG. 6B). However, treatment with all-*trans*-retinoic acid for 24 hours blocks the invasive ability of FaDu cells on Matrigel (FIG. 6C). These cells also produce MMP-1 and MMP-13 (FIGS. 2 and 3), either of which may mediate their ability to invade collagen (FIG. 6D), and this invasion is prevented by treatment with retinoic acid ($10^{-6}$ M) (TABLE 2). Quantitation of invasion by FaDu cells through Matrigel and type I collagen revealed that approximately 18% or 10% invaded these matrices, respectively, to a depth of 10 µM (TABLE 2). In the presence of retinoic acid ($10^{-6}$ M for 24 hours), about 1% of the cells invaded to this depth with either matrix. The ability of all-*trans*-retinoic acid to block invasion is somewhat puzzling, since this compound does not reduce expression of MMP-1 or MMP-9 (FIG. 3). Perhaps there is some indirect effect, such as inhibiting the synthesis of a proteinase that activates latent MMPs[23] or perhaps retinoic acid is inhibiting MMP-13 production (TABLE 1) in these cells.

Another head and neck carcinoma cell line, SCC-25, was examined for its ability to invade these matrices. FIGURE 7A shows that after 24 hours, the SCC-25 cells have invaded Matrigel, and FIGURE 7B illustrates that treatment with all-*trans*-retinoic acid ($10^{-6}$ M) can partially block this invasion. Indeed, 52% of untreated cells invaded Matrigel, compared to 28% for cells treated with all-*trans*–retinoic acid ($10^{-6}$ M) (TABLE 2). Thus, there are some cells that continue to invade, even in the presence of the retinoid, suggesting the presence of a subpopulation of cells that is resistant to the effects of all-*trans*–retinoic acid (see the Discussion section). In contrast, these SCC-25 cells did not invade type I collagen (FIG. 7C) despite production of several interstitial collagenases (FIG. 4 and TABLE 1). Perhaps, like the A2058 melanoma cells, they require the presence of stromal factor(s).[23]

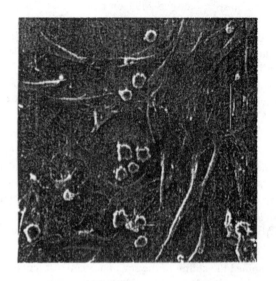

B.

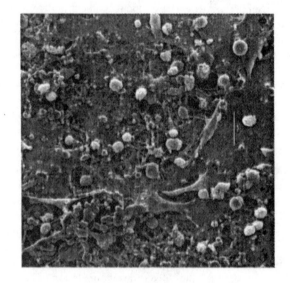

A.

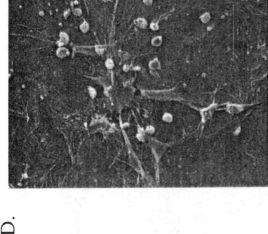

**FIGURE 5.** Retinoid-mediated inhibition of invasion of type I collagen by A2058 melanoma cells. Single cell suspensions in 1 ml medium, consisting of tumor cells and human foreskin fibroblasts in equal numbers of $5 \times 10^4$ each, were co-cultured on a matrix of type I collagen. The lower chamber was filled with serum-free media, and $10^5$ cells in 1 ml were seeded on the collagen matrix in the upper chamber in the absence or presence of the RAR $\alpha$ antagonist #189532 for 24 or 48 hours. Cells were visualized by scanning electron microscopy. **(A)** Co-culture of A2058 melanoma cells and human skin fibroblasts on type I collagen. Note that the A2058 cells degrade the collagen matrix, as seen by the "halo" of digested collagen around the tumor cells. The fibroblasts remain on the surface. **(B)** Co-culture of A2058 melanoma cells and human skin fibroblasts on type I collagen in the presence of the $\alpha$ antagonist #189532 at $10^{-6}$ M for 24 hours. Invasion is blocked, and the A2058 cells show "blebs," which may indicate cytotoxicity. **(C)** Co-culture of A2058 melanoma cells and human skin fibroblasts on type I collagen in the presence of the $\alpha$ antagonist #189532 at $10^{-6}$ M for 48 hours. Again, invasion is blocked, and both the A2058 cells and the fibroblasts are affected by the retinoid. Most cells are dead or show toxic changes. **(D)** Co-culture of A2058 melanoma cells and human skin fibroblasts on type I collagen in the presence of the $\alpha$ antagonist #189532 at $10^{-8}$ M for 48 hours. Invasion is blocked. The fibroblasts remain normal, while the A2058 melanoma cells are affected by the retinoid.

D.

C.

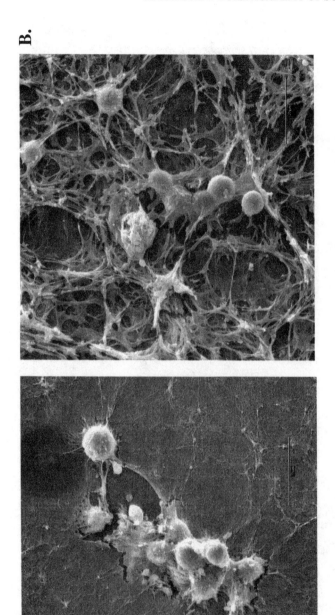

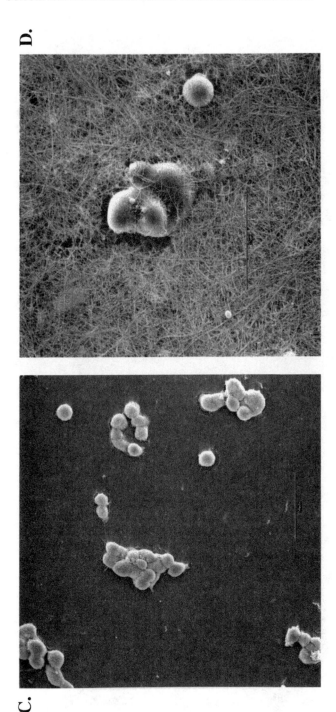

**FIGURE 6.** Invasion of Matrigel and type I collagen by FaDu cells in inhibition by retinoic acid. Single cell suspensions, consisting of $10^5$ FaDu tumor cells in 1 ml, were seeded on Matrigel or type I collagen in the upper chamber in the absence or presence of all-*trans*–retinoic acid ($10^{-6}$ M) for 24 or 48 hours. The lower chamber was filled with serum-free media. Cells were visualized by scanning electron microscopy. (**A**) After 24 hours on Matrigel, there is an interaction of the FaDu cells with the matrix, and the cells are beginning to invade the Matrigel. (**B**) By 48 hours, invasion is more pronounced. (**C**) Invasion through Matrigel is blocked by treatment with all-*trans*–retinoic acid. (**D**) After 24 hours on type I collagen, FaDu cells begin to invade the matrix. Note the "halo" of digested matrix around the cells.

B.

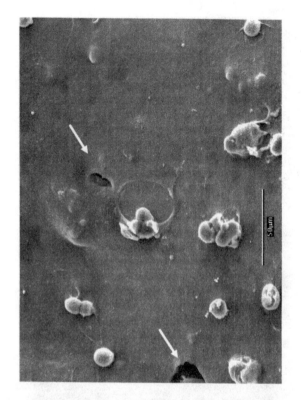

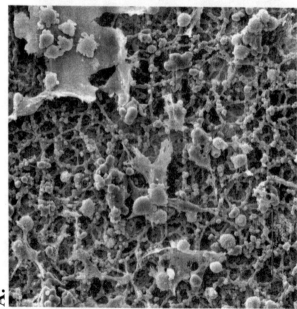

A.

**FIGURE 7.** Retinoid-mediated inhibition of invasion of Matrigel by SCC-25 cells. Single cell suspensions, consisting of $10^5$ SCC-25 tumor cells, were washed three times in HBSS, resuspended in serum-free media, and counted. The lower chamber was filled with serum-free media, and $10^5$ cells in 1 ml were seeded on Matrigel or type I collagen in the upper chamber in the absence or presence of all-*trans*–retinoic acid ($10^{-6}$ M) for 24 hr. Cells were visualized by scanning electron microscopy. (**A**) After 24 hours on Matrigel, the SCC-25 cells have degraded the matrix, and the underlying filter is exposed. (**B**) After 24 hours on Matrigel, in the presence of retinoic acid invasion is partially blocked. Some cells have invaded the matrix (*white arrows*). (**C**) After 24 hours on type I collagen, SCC-25 cells have not degraded and invaded the collagen.

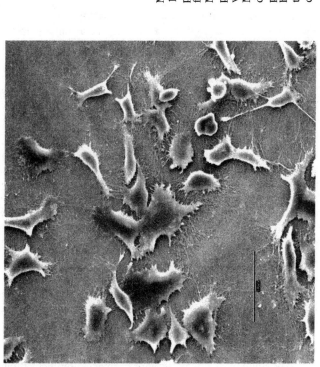

C.

## DISCUSSION

In this paper, we demonstrate the ability of several retinoids, which are specific RAR $\alpha$, $\beta$, or $\gamma$ agonists/antagonists, to selectively inhibit MMP synthesis in three different tumor cell lines. Perhaps the most striking observation is the fact that there is not a common pattern of MMP inhibition by a particular retinoid. For example, the $\alpha$ antagonist #189532 suppressed MMP-1 and MMP-2 synthesis in the melanoma cell line, but not in the FaDu or SCC-25 cells. In contrast, synthesis of MMP-1 and MMP-9 by the FaDu cells was affected hardly at all, while the $\gamma$ antagonist #191681 reduced the levels of MMP-2. Only all-*trans*–retinoic acid reduced MMP-1 synthesis by these cells. Perhaps these discrepancies are due to a differential pattern of RAR expression in each of these cells,[6,8,9,14,15] and further, the RARs expressed by each cell line may not be targets of these RAR-specific compounds.[16,17] On the other hand, all-*trans*–retinoic acid is a pan ligand, binding to all three RARs and modulating gene expression more generally.[6,8] Thus, the power of these new ligands resides in their specificity, which can be targeted to mediating the expression of certain RARs and, therefore, of certain MMPs. However, before these compounds become effective therapeutic agents, it may be necessary to first characterize the pattern of RAR/RXR expression by each cell type so that the appropriate ligand can be used.

A2058 cells are highly invasive through both Matrigel and type I collagen.[23] Presumably, invasion through Matrigel is mediated by MMP-2, which can be activated by MT1-MMP. However, invasion through collagen occurs only in the presence of a soluble factor(s) secreted by normal fibroblasts.[23] Conceivably, this factor is another MMP (i.e., stromelysin) which can activate latent collagenase and which is produced by the fibroblasts, but not by the melanoma cells (FIG. 2).[3,4,23] The BMS $\alpha$ antagonist #189532 is particularly effective in suppressing MMPs expressed by the A2058 cells and in blocking invasion through collagen. Perhaps, it also suppresses MMP production by the fibroblasts[4,6] in this co-culture system, thereby amplifying the inhibitory effect on invasion.[23]

In contrast, FaDu cells invade both Matrigel and type I collagen, and they migrate through these matrices without the need for co-culture with fibroblasts, suggesting that the cells have a mechanism for activating latent MMPs. It is interesting that invasion is blocked by retinoic acid, but that synthesis of MMP-1 or MMP-9 is not affected. Possibly, MMP-13 (collagenase-3) contributes to collagen degradation by these cells (TABLE 1), but the effect of BMS retinoids and of all-*trans*–retinoic acid on the expression of this MMP has not yet been measured.

Somewhat surprisingly, we noted that the SCC-25 cells do not invade type I collagen within 24 hours, despite production of MMP-1, MMP-13 and MT1-MMP, all of which can degrade interstitial collagen. Furthermore, recent data obtained by co-culture of SCC-25 cells and fibroblasts in serum-free conditions demonstrate (a) proliferation of the tumor cells, (b) architectural rearrangement of the tumor cells into clusters surrounded by the stromal cells, and (c) possible invasion of the collagen matrix by the tumor cells (data not shown). Perhaps, then, similar to the A2058 melanoma cells,[23] co-culture of SCC-25 cells with stromal cells may facilitate tumor invasiveness.

Finally, SCC-25 cells produce a variety of MMPs, and these cells readily invade Matrigel. It is interesting that all-*trans*–retinoic acid only partially blocks this invasive ability, suggesting the existence of a retinoid-resistant subpopulation of cells.[24] Many tumor cells exhibit aberrant/defective expression of RARs and/or RXRs, which is accompanied by a loss of retinoid sensitivity.[8,9,12] For example, MDA231 breast cancer cells are insensitive to the growth-inhibitory effects of RA unless either RAR α or β is transfected into the cells.[24] Similarly, in several head and neck cancer cell lines, RAR β is either not expressed, or expression is dysregulated, and the cell lines are retinoid-resistant.[8–10] Here, too, however, expression of wild-type receptor restores retinoid-mediated growth suppression.[8–10,24] The ability to transfect functional RAR/RXR receptors into cells and to restore retinoid responsiveness suggests a novel strategy for future gene therapies,[25] where endogenous RAR/RXR expression has gone awry.

However, our current data indicate that in RAR/RXR-responsive cells, retinoids are effective inhibitors of MMP synthesis and of tumor cell invasion, suggesting that these compounds might be therapeutically effective in cancer. Indeed, pharmacologic doses of retinoids can exert therapeutic effects in a number of disorders, such as acne and psoriasis and leukemia, as well as some kinds of head and neck cancers,[8–10,26] and it is possible that at least some of these effects are due to inhibition of MMP synthesis.

Over the last decade, our understanding of how retinoids exert their effects on gene expression has exploded.[26–31] Retinoic acid receptors (RARs), which bind all-*trans*– and 9-*cis*–retinoic acid and retinoid X receptors (RXRs), which bind 9-*cis*–retinoic acid, and their specific subtypes α, β and γ, were identified. A role for vitamin A and its derivatives as hormones with profound effects on gene expression was confirmed. In most cases, retinoids affect their target genes through RAR/RXR interaction with specific retinoic acid response elements (RAREs). However, retinoids repress MMP gene expression primarily by interacting with the Fos and Jun transcription proteins that bind to the AP-1 site located in the proximal promoter of most MMPs, including the collagenases.[4,6]

The recent development of RAR/RXR-specific ligands that affect the expression of certain genes represents an important step in the possibility of actualizing directed gene expression.[16,20,25] These "second generation" compounds have the ability to regulate certain genes, thereby potentially avoiding some of the more general side effects associated with the use of parent retinoids. Possibly, these RAR/RXR-specific ligands will, like their original counterparts, act synergistically with glucocorticoids to reduce MMP production,[32] thereby providing targeted effects on the expression of certain genes. Further, because of this synergism, lower doses of each drug can be used, further subverting certain effects associated with conventional doses.

The next steps will be the identification of additional retinoids that are effective in suppressing MMP gene expression. In addition, the *in vitro* invasion assay described here provides a valuable model system for assessing the invasive potential of tumor cells through extracellular matrices, and for evaluating the ability of specific retinoids to subvert this invasive behavior, and hence to stop metastatic disease. Finally, by extending these studies to include other melanoma cell lines and other head and neck carcinomas, we can determine whether paradigms of MMP expression and inhibition by ligand-specific retinoids become evident for certain types of cancer.

## ACKNOWLEDGMENTS

This work was supported by NIH Grant AR-26599, the RGK Foundation, and Bristol-Myers Squibb (C.E.B), the Mildred-Scheel-Foundation for Cancer Research (M.P.S), and NIH Grant ST32-CA-09658 (J.L.R.)

## REFERENCES

1. MacDougall, J.R. & L.M. Matrisian. 1995. Contributions of tumor and stromal matrix metalloproteinases to tumor progression, invasion and metastasis. Cancer Metastasis Rev. **14:** 351–362.
2. Chambers, A.F. & L.A. Matrisian. 1997. Changing views on the role of matrix metalloproteinases in metastasis. J. Natl. Cancer Inst. **89:** 1260–1270.
3. Borden, P. & R.A. Heller. 1997. Transcriptional control of matrix metalloproteinases and the tissue inhibitors of matrix metalloproteinases. Crit. Rev. Eukaryotic Gene Expression **7:** 159–178.
4. Vincenti, M.P., L.A. White, D.J. Schroen, U. Benbow & C.E. Brinckerhoff. 1996. Regulating expression of the gene for matrix metalloproteinase-1 (collagenase): mechanisms that control enzyme activity, transcription and mRNA stability. Crit. Rev. Eucaryotic Gene Expression **6:** 391–411.
5. Wojowicz-Praga, S., J. Low, J. Marshall, E. Ness, R. Dickson, J. Barter, M. Sale, P. McCann, J. Moore, A. Cole & M.J. Hawkins. 1996. Phase I trial of a novel matrix metelloproteinase inhibitor Batimastat (BB-94) in patients with advanced cancer. Investigational New Drugs **14:** 193–202.
6. Schroen, D.J., & C.E. Brinckerhoff. 1997. Nuclear hormone receptors inhibit matrix metalloproteinase (MMP) gene expression through diverse mechanisms. Gene Expression **6:** 197–207.
7. Guerin, E., M.-G. Ludwig, P. Basset & P. Anglard. 1997. Stromelysin-3 induction and interstitial collagenase repression by retinoic acid. J. Biol. Chem. **272:** 11088–11095.
8. Smith, M.A., D.R. Parkinson, B.D. Cheson & M.A. Friedman. 1992. Retinoids in cancer therapy. J. Clin. Oncol. **10:** 839–864.
9. Khuri, F.R., S.M. Lippman, M.R. Spitz, R. Lotan & W.K. Hong. 1997. Molecular epidemiology and retinoid chemoprevention of head and neck cancer. J. Natl. Cancer Inst. **89:** 199–211.
10. Hendrix, M.J.C., W.R. Wood, E.A. Seftor, D. Lotan, M. Nakajima, R.L. Misiorowski, R.E.B. Seftor, W.G. Stetler-Stevenson, S.J. Bavacqua, L.A. Liotta, M.E. Sobel, A. Raz & R. Lotan. 1990. Retinoic acid inhibition of human melanoma cell invasion through a reconstituted basement membrane and its relation to decreases in the expression of proteolytic enzymes and motility factor receptor. Cancer Res. **50:** 4121–4130.
11. Helige, C., J. Smolle, G. Zellnig, E. Hartman, R. Fink-Puches, H. Kerl & H.A. Tritthart. 1993. Inhibition of K1735-M2 melanoma cell invasion in vitro by retinoic acid. Clin. Exp. Metastasis **11:** 409–418.
12. Morton, D.L., E.R. Essner, J.M. Kirkwood & R.G. Parker. 1997. Malignant melanoma. *In* Cancer Medicine. J.F. Holland, R.C. Bast, D.L. Morton, E. Frei, D.W. Kufe & R.R. Weichselbaum, Eds.: 2467–2499. Williams & Wilkins. Baltimore.
13. Rigel, D.S., R.J. Friedman, A.W. Kopf & M.K. Silverman. 1991. Factors influencing survival in melanoma. Dermatol. Clin. **9:** 631–642.

14. LOTAN, R. 1996. Retinoids and their receptors in modulation of differentiation, development, and prevention of head and neck cancers. Anticancer Res. **16:** 15–19.

15. ISSING, W.J. & T.P. WUSTROW. 1996. Expression of retinoic acid receptors in squamous cell carcinoma and their possible implications for chemoprevention. Anticancer Res. **16:** 2373–2377.

16. SPANJAARD, R.A., M. IKEDA, P.J. LEE, B. CHARPENTIER, W.W. CHIN & T.J. EBERLEIN. 1997. Specific activation of retinoic acid receptors (RARs) and retinoid X receptors reveals a unique role for RAR γ in induction of differentiation and apoptosis in S91 melanoma cells. J. Biol. Chem. **272:** 18990–18999.

17. SCHADENDORF, D., M.A. KERN, M. ARTUC, H.L. PAHL, T. ROSENBACH, I. FICHTNER, W. NURNBERG, S. STUTING, E. VON STEBUT, M. WORM, A. MAKKI, K. JURGOVSKY, G. KOLDE & B.M. HENZ. 1996. Treatment of melanoma cells with the synthetic retinoid CD437 induces apoptosis via activation of AP-1 in vitro, and causes cell growth inhibition in xenografts in vivo. J. Cell Biol. **135:** 1889–1898.

18. SCHADENDORF, D., M. WORM, K. JURGOVSKY, E. DIPPEL, U. REICHERT & B.M. CZARNETZKI. 1995. Effects of various synthetic retinoid on proliferation and immunophenotype of human melanoma cell in vitro. Recent Results Cancer Res. **139:** 183–193.

19. SOBALLE, P.W. & M. HERLYN. 1994. Cellular pathways leading to melanoma differentiation: therapeutic implications. Melanoma Res. **4:** 213–223.

20. ZHANG, X.K., Y. LIU & M.O. LEE. 1996. Retinoic receptors in human lung cancer and breast cancer. Mutation Res. **350:** 267–277.

21. SCHOENERMARK, M.P., O. BOCK, A. BUCHNER, R. STEINMEIR, U. BENBOW & T. LENARZ. 1997. Quantification of tumor cell invasion using confocal laser scan microscopy (CLSM). Nature Med. **3:** 1167–1171.

22. RUTTER, J.L., U. BENBOW, C.I. COON & C.E. BRINCKERHOFF. 1997. Cell-type specific regulation of human interstitial collagenase-1 gene expression by Interleukin-1β (IL-1β) in human fibroblasts and BC-8701 breast cancer cells. J. Cell. Biochem. **66:** 322–336.

23. BENBOW, U., M.P. SCHOENERMARK, T.I. MITCHELL, J.L. RUTTER, H. NAGASE & C.E. BRINCKERHOFF. 1998. Novel host/tumor cell interactions mediate invasive and non-invasive behavior of melanoma cells through type I collagen. Submitted.

24. SHEIKH, M.S., Z.M. SHAO, X.S. LI, M. DAWSON, A.M. JETTEN, S. WU, B.A. CONLEY, M. GARCIA, H. ROCHEFORT & J.A. FONTANA. 1994. Retinoic-resistant estrogen receptor-negative human breast carcinoma cells transfected with retinoic acid receptor-alpha acquire sensitivity to growth inhibition by retinoids. J. Biol. Chem. **269:** 21440–21447.

25. BLAU, H. & P. KHAVARI. 1997. Gene therapy: Progress, problems, prospects. Nature Med. **3:** 612.

26. SPORN, M.B., A.B. ROBERTS & D.S. GOODMAN, Eds. 1994. The Retinoids. Raven Press. New York.

27. GUIGUERE, V., E.S. ONG, P. SEGUI & R.M. EVANS. 1987. Identification of a receptor for the morphogen retinoic acid. Nature **33:** 624–629.

28. ZALENT, A., A. KRUST, M. PETKOVICH, P. KASTNER & P. CHAMBON. 1989. Cloning of murine α and β and a novel receptor γ predominantly expressed in skin. Nature **339:** 714–717.

29. MANGELDORF, D.J., U. BORGMEYER, R.A. HEYMAN, J.Y. ZHOU, E.S. ONG, A.E. ORO, A. KAKIZUKA & R.M. EVANS. 1992. Characterization of three RXR genes that mediate the action of 9-cis retinoic acid. Genes & Dev. **6:** 329–344.

30. LEID, M., P. KASTNER, R. LYONS, H. NAKSHATRI, M. SAUNCERS, T. ZACHAREWSKI, J.Y. CHEN, A. STAUB, J.M. GARNIER, S. MADER & P. CHAMBON. 1992. Purification, cloning and RXR identity of the HeLa cell factor with which RAR or TR heterodimerizes to bind to target sequences efficiently. Cell **68:** 377–395.

31. ALLENBY, G., M-T. BOCQUEL, M. SAUNDERS, S. KAZMER, J. SPECK, M. ROSENBERGER, A. LOVEY, P. KASTNER, J. GRIPPO, P. CHAMBON & A. LEVIN. 1993. Retinoic acid receptors and retinoid X receptors: interactions with endogenous retinoic acids. Proc. Natl. Acad.Sci., **90:** 30–34. 1981.

32. BRINCKERHOFF, C.E. & E.D. HARRIS, JR. 1981. Modulation by retinoic acid and corticosteroids of collagenase production by rabbit synovial fibroblasts treated with phorbol myristate acetate or polyethylene glycol. Biochim. Biophys. Acta **677:** 424–432.

# Determination of Gelatinase-A (MMP-2) Activity Using a Novel Immunocapture Assay

STEPHEN J. CAPPER,[a,b] JAN VERHEIJEN,[c] LYNNE SMITH,[a] MIKE SULLY,[a] HETTY VISSER,[c] AND ROELAND HANEMAAIJER[c]

[a]Nycomed Amersham, Cardiff Laboratories, Forest Farm Estate, Whitchurch, Cardiff CF4 7YT, Wales, UK

[c]Gaubius Laboratory, TNO-PG, PO Box 2215, 2301 CE Leiden, the Netherlands

It is important to be able to measure the amount of matrix metalloproteinase (MMP) produced, as well as how much of this becomes activated, to evaluate the contribution of a given MMP to disease development. A novel technology is described for determining the activity of MMPs using a colorimetric assay. We have already applied this technique to develop a specific MMP-9 assay.[1,2] Here we have validated the MMP-2 assay with cell culture and serum/plasma samples. In addition, the assay has been applied to measure MMP-2 activity levels in tumor homogenates, urine, saliva, and synovial fluids.

## METHODS

The assay is based on a modified pro-urokinase, where the activation sequence, normally recognized by plasmin (Pro-Arg-Phe-Lys ⇕ Ile-Ile-Gly-Gly), was replaced by a sequence that is specifically recognized by MMPs (Arg-Pro-Leu-Gly ⇕ Ile-Ile-Gly-Gly). A chromogenic peptide substrate for urokinase is then used to measure the active urokinase, generated through MMP activation of the modified urokinase. The assay uses an MMP-2 specific antibody–coated 96-well microtiter plate to confer MMP-2 specificity. MMP-2 is captured overnight from biological fluids or cell culture media in the microtiter plate wells. Immobilized latent MMP-2 is then activated using APMA to measure total MMP-2. Differential activation with APMA enables the analysis of the activity of both already-active and latent forms of MMP-2. The assay protocol consists of capture of 100 µl sample or standard (proMMP-2) overnight at 4°C. Following washing, 50 µl APMA (0.5 mM) or buffer (for measuring total or active MMP-2, respectively) is added with 50 µl detection reagent (pro-urokinase and substrate) and incubated at 37°C. The standard range is 0.75–12 ng/ml for a 1.5-hr incubation, and this is increased to 0.19–3 ng/ml for 4-hr incubation. The sensitivities for these protocols are 0.5 ng/ml and 0.19 ng/ml, respectively. The assay is specific for MMP-2; cross-reaction with pro- and active MMP-2 is 100%, proMMP-2/TIMP-2 complex 43%, and MMP-9, 1, 3 < 1%.

[b]Corresponding author. Address for telecommunication: Phone, 441/222-526480; fax, 441/222-526230; e-mail, stephen.capper@eu.apbiotech.com

## RESULTS

Serum/plasma levels of total MMP-2 were: serum $222 \pm 54$ ng/ml; heparin plasma $139 \pm 25$ ng/ml; EDTA plasma $85 \pm 21$ ng/ml; serum $211 \pm 39$ ng/ml (all $n = 5$). The results for the other samples are shown (FIGS. 1 and 2).

The serum and plasma results show measurable levels. EDTA samples are much lower, probably due to the inhibition of activity by chelation. MMP-2 can be successfully measured in cell culture as mostly pro-form. In saliva from normals and urine from patients, significant levels of active MMP-2 can be detected in some cases. The breast tumor homogenates show predominantly proMMP-2, although active

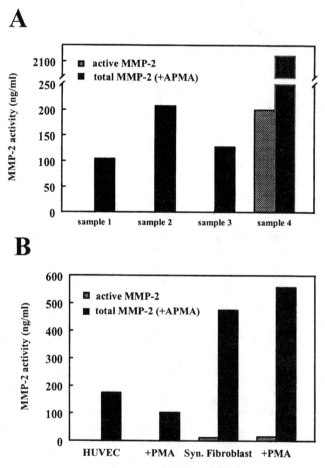

**FIGURE 1.** (A) Total and active MMP-2 levels in synovial fluid samples from arthritis patients measured by the MMP-2 activity assay. Each sample is a different patient. (B) Total and active MMP-2 levels in cell culture media from HUVECs and human synovial fibroblasts with and without PMA stimulation.

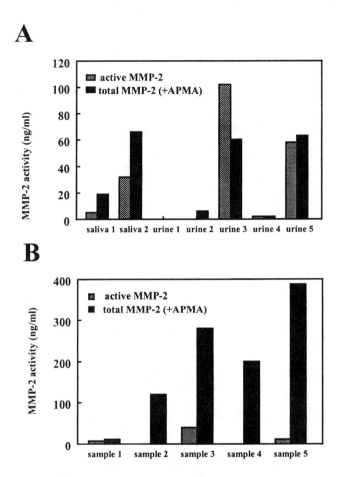

**FIGURE 2.** (A) Total and active MMP-2 levels in saliva from normal volunteers and urine from bladder cancer patients. Each sample is a different individual. (B) Total and active MMP-2 levels in breast tumor homogenates. Each sample is a different patient.

MMP-2 can clearly be seen in some samples. Synovial fluid samples from arthritis patients similarly show proMMP-2; however, one patient showed very significant amounts of active enzyme.

## DISCUSSION

A simple, reliable assay for MMP-2 activity has been developed and used to measure latent/total MMP-2 and active MMP-2 simultaneously. Initial results have shown the utility of the assay in distinguishing samples that have elevated levels of active MMP-2.

## REFERENCES

1. VERHEIJEN, J.H. *et al.* 1997. Biochem. J. **323:** 603–609.
2. HANEMAAIJER, R. *et al.* 1998. Matrix Biol. **17:** 335–347.

# Analytical Aspects regarding the Measurement of Metalloproteinases

MICHAEL LEIN,[a] KLAUS JUNG, LARS NOWAK, DIETMAR SCHNORR, AND
STEFAN A. LOENING

*Department of Urology, University Hospital Charité,*
*Humboldt University, Berlin, Germany*

## OBJECTIVE

MMPs and TIMPs are interesting new diagnostic tools in oncology, liver diseases, and rheumatoid arthrithis.[1] There is evidence that changes of MMPs in tissue are also manifest in biological body fluids like blood or urine.[2] These changes might offer the possibility to determine MMP concentrations by simple, noninvasive tests. Commercial assays for the determination of MMPs and TIMPs have been introduced.[3] The objective of this study was to demonstrate the importance of a standard preanalytic and analytic procedure for the measurement of components.

## DESIGN AND METHODS

In order to investigate the analytical reliability of new commercial ELISA tests, BIOTRAK™ test kits (Amersham Int.) were used for the determination of MMP-1, 2, 3, 9; TIMP-1,2; and the MMP-1/TIMP-1 complex in blood. Blood samples from healthy male volunteers ($n = 10$) were simultaneously collected into plastic tubes for preparation of serum samples and into potassium EDTA-coated tubes as well as into lithium heparin-coated plastic tubes for preparation of plasma samples (Sarstedt, Nümbrecht, Germany; Monovette systems 03.1589, 03.1528, 05.1167). The tubes were stored at room temperature and centrifuged within 60 min after venipuncture at $1600\,g$ for 15 min. The supernatant was then carefully removed and stored at −80°C until the actual analysis was performed.

## RESULTS

The detection limits and the precision data were in the usual ranges of ELISA tests so that concentrations found in blood could be determined reliably. The time between venipuncture and preparation of the sample (30 min, 1, 2, 3 hours) has no effect on the measured sample concentration. MMP-1, TIMP-1, and the MMP-1/TIMP-1 complex were lower in heparin plasma than in serum or EDTA plasma. MMP-2, MMP-9, and TIMP-2 measured in serum, heparin plasma, and EDTA plas-

[a]Address for correspondence: Michael Lein, M.D., Department of Urology, University Hospital Charité, Humboldt University, Schumannstr. 20/21, 10098 Berlin, Germany. Phone, ++49 30/2802 4714; fax, ++49 30/2802 5615.

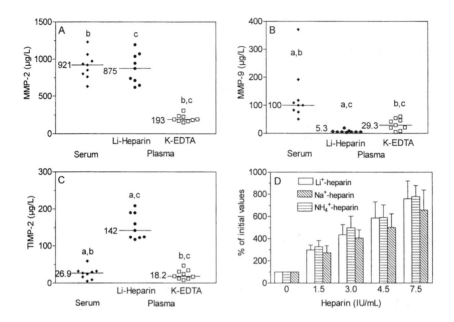

**FIGURE 1.** MMP-2 (**A**), MMP-9 (**B**), and TIMP-2 (**C**) concentrations in dependence on sample processing. Individual values and the median are given. a, Significant difference between serum and heparin plasma; b, significant difference between serum and EDTA plasma; c, significant difference between heparin plasma and EDTA plasma. (**D**) The influence of heparin on the TIMP-2 measurement. TIMP-2 concentrations in serum samples were measured after the addition of different concentrations of lithium, potassium, and ammonium heparin. Data are given as percentages of the arithmetic means ± SD.

ma were markedly different (FIG. 1 A, B, C). MMP-2 values did not differ in serum and heparin plasma but were lower in EDTA plasma. MMP-9 concentration was ~ 3- to 20-fold higher in serum than in heparin and EDTA plasma. TIMP-2 values were 5- to 8-fold higher in heparin plasma than in serum and EDTA plasma. To realize the possible inhibitory or stimulatory effect of EDTA and heparin on the determination of these components, EDTA or heparin was added to serum samples ($n = 4$), getting concentration equivalents. MMP-2 and MMP-9 concentrations were 96–106% of the initial values without additives (no interference by EDTA and heparin). Heparin but not EDTA produced increased TIMP-2 values dependent on the concentration but not on the kind of heparin (FIG. 1 D).

## CONCLUSION

The clinician should be aware that the commutability of MMP and TIMP values measured by ELISA tests are impossible if different kinds of specimens are used. To avoid clinical misinterpretations, at least consistent specimens should be determined. There is an urgent need to solve the possibly common problems of the proper specimen for measurement of MMPs and TIMPs.

## REFERENCES

1. ISHIGURO, N. *et al.* 1996. Determination of stromelysinn-1, 72 and 92 kDa type IV collagenase, tissue inhibitor of metalloproteinase-1 (TIMP-1), and TIMP-2 in synovial fluid and serum from patients with rheumatoid arthritis J. Rheumatol. **23:** 1599–1604.
2. ZUCKER, S. *et al.* 1995. Plasma assay of gelatinase B: Tissue inhibitor of metalloproteinase complexes in cancer. Cancer **76:** 700–708.
3. JUNG, K. *et al.* 1996. Role of specimen collection in preanalytical variation of metalloproteinases and their inhibitors [Technical Letter] Clin. Chem. **42:** 2043–2044.

# Hyper-resistance to Infection in TIMP-1–Deficient Mice Is Neutrophil Dependent but Not Immune Cell Autonomous

KEITH OSIEWICZ, MICHAEL McGARRY, AND PAUL D. SOLOWAY[a]

*Department of Molecular and Cellular Biology, Roswell Park Cancer Institute, Elm and Carlton Streets, Buffalo, New York 14263, USA*

The TIMP/MMP axis has been postulated to influence both acquired and innate immune responses *in vivo*.[1] Among the experimental results consistent with this possibility are the observations that macrophages from mice deficient for MMP-12 have reduced kinetics of migration both *in vitro* and *in vivo*[2] and that the complement factor C1 inhibitor can be degraded *in vitro* by MMP-8 and -9.[3] However, reports documenting a direct influence of the TIMP/MMP axis on responses to infection have been lacking. We have observed that loss of TIMP-1 had a dramatic effect on the course of corneal infection by *Pseudomonas aeruginosa*. Infections became established and bacteria underwent a burst of growth identically in both groups of mice. However, coincident with massive neutrophil accumulation in corneas between 24 and 48 hours postinfection, the bacterial burden in mutant mice dropped to a level 1,600-fold lower than in wild-type mice. This was shown to be reversible by treatment of mutant mice with BB-94. Furthermore, the mechanism underlying the hyper-resistance to infection in mutant mice was shown to be complement dependent since depletion of the complement system from mutant mice suppressed their phenotype (Osiewicz *et al.*, submitted for publication).

The complement branch of the innate immune system consists of approximately 30 plasma proteins that can be activated by a proteolytic cascade to generate two bactericidal activities. Bacteria may be killed directly by the formation of a pore in bacterial membranes by the membrane attack complex of complement. Alternatively, activated complement factor C3b can bind to bacterial surfaces targeting bacteria for phagocytosis by complement receptor-containing neutrophils.

In order to study the contribution of neutrophils to the hyper-resistance phenotype seen in TIMP-1–deficient mice, infections were done in neutropenic mice, and the bacterial burdens were measured 48 hours after infection. Neutropenia was induced by cyclophosphamide administered intraperitoneally on day 0 and day 5 at a dose of 150 µg/kg. Infections were done on day 4 using an inoculum of $3 \times 10^6$ CFU/eye, and bacterial burdens were measured on day 6 by homogenization of whole eyes followed by plate count assays. During the course of the infection, cyclophosphamide treatment reduced the number of circulating neutrophils by approximately 90% (data not shown). Macrophage numbers were reduced by 50%; however, these cells were very rare in infected eyes during the first 48 hours of infection and are unlikely to be significant to the mutant phenotype. The results showed that in neutropenic mice, the

[a]Corresponding author. Phone, 716/845-5843; fax, 716/845-8389;
e-mail, psoloway@mcbio.med.buffalo.edu

**TABLE 1. Corneal infections in neutropenic mice**

| Cyclophosphamide treatment | WT CFU/eye | Mut CFU/eye | Ratio WT/Mut | $p$ |
|:---:|:---:|:---:|:---:|:---:|
| − | $5.6 \times 10^6$ | $3.6 \times 10^3$ | 1600 | <0.05 |
| + | $3.7 \times 10^7$ | $4.5 \times 10^7$ | 0.82 | >0.5 |

NOTE: Cyclophosphamide-induced neutrophil depletion and infections were performed as described in the text. Bacterial burden is reported as colony-forming units of bacteria (CFU) per eye by plate-count assays of whole-eye homogenates. All tests included a minimum of four wild-type (WT) or TIMP-1–deficient (Mut) strain 129SvJae mice aged 3–6 weeks.

bacterial burden was identical in wild-type and TIMP-1–deficient animals at 48 hours after infection, indicating that hyper-resistance to infection in the mutants was neutrophil dependent (TABLE 1).

To determine if there was a cell-autonomous alteration of neutrophils caused by the *timp-1* mutation that enhanced their bactericidal activity, infections were repeated in mice that had been engrafted with either autologous or heterologous bone marrow. Female mice were irradiated with 900 rads from a $Co^{60}$ source, then intravenously injected with $10^7$ bone marrow cells isolated from the femurs and tibia of male donors. Two weeks after adoptive transfer, corneal infections were established and bacterial burdens were assayed as above. At the time of sacrifice, DNA was isolated from blood and the success of adoptive transfer confirmed by a quantitative PCR assay for X- and Y-chromosome–specific sequences (not shown).

In the control experiments, irradiated mice that were reconstituted with bone marrow from donors of the same genotype responded to infections in the same way as mice that were not subjected to bone marrow transplants—mutant recipients of mutant bone marrow were hyper-resistant to infections relative to wild-type recipients of wild-type bone marrow by a factor of 2,300 ($p < 0.0001$, TABLE 2). In the heterologous transfer experiments in which wild-type recipients received mutant bone marrow and mutant recipients received wild-type bone marrow, the wild-type recipients maintained the wild-type phenotype, while the mutant recipients acquired the wild-type phenotype (TABLE 2). The results indicated that while hyper-resistance to infection in TIMP-1–deficient mice is neutrophil dependent, it is not due to cell-

**TABLE 2. Corneal infections in bone marrow transplant recipients**

| Recipient | Donor | CFU/Eye | $p$ |
|:---:|:---:|:---:|:---:|
| +/+ | +/0 | $1.0 \times 10^5$ | |
| +/+ | −/0 | $3.9 \times 10^4$ | >0.5 |
| −/− | −/0 | $4.3 \times 10^1$ | <0.01 |
| −/− | +/0 | $1.5 \times 10^3$ | |

NOTE: Wild-type (+/+) or mutant (−/−) female mice were given bone marrow transplants using marrow from wild-type (+/0) or mutant (−/0) male donors. Engraftment, confirmation of adoptive transfer, infections, and measurement of bacterial burden were done as described in the text. All tests included a minimum of four strain 129SvJae mice aged 3–6 weeks.

autonomous changes in the mutant neutrophil population. Furthermore, the presence of TIMP-1 in mice, regardless of the tissue producing it, was sufficient to suppress the *timp-1* mutant phenotype.

## REFERENCES

1. GOETZL, E.J. *et al.* 1996. Matrix metalloproteinases in immunity. [Review]. J. Immunol. **156:** 1–4.
2. SHIPLEY, J.M. *et al.* 1996. Metalloelastase is required for macrophage-mediated proteolysis and matrix invasion in mice. Proc. Nat. Acad. Sci. USA **93:** 3942–3946.
3. KNAUPER, V. *et al.* 1991. Inactivation of human plasma C1-inhibitor by human PMN leucocyte matrix metalloproteinases. FEBS Lett. **290:** 99–102.

# Shed Membrane Vesicles and Selective Localization of Gelatinases and MMP-9/TIMP-1 Complexes

VINCENZA DOLO,[a] ANGELA GINESTRA,[a] DONATA CASSARÁ,[a] GIULIO GHERSI,[a] HIDEAKI NAGASE,[b] AND MARIA LETIZIA VITTORELLI[a,c]

[a]Dipartimento di Biologia Cellulare e dello Sviluppo,
Università di Palermo, 90128 Palermo, Italy

[b]Department of Biochemistry and Molecular Biology,
University of Kansas Medical Center, Kansas City, Kansas 66160, USA

Shedding of membrane vesicles is a vital phenomenon frequently observed in tumor cells and suggested to be involved in several aspects of tumor progression.[1] As we have recently shown, in 8701-BC breast carcinoma cells, vesicle shedding is modulated by extracellular signal; and several molecules, probably involved in tumor progression, are specifically densified on shed vesicles.[2] One of the clustered molecules is gelatinase B (MMP-9), which is present in both active and proenzyme forms. Western blot analyses of vesicles shed by 8701-BC breast carcinoma cells demonstrate that most MMP-9 molecules bound to vesicle membranes are in association with TIMP-1, as components of high $M_r$ complexes.[2] ProMMP-9/TIMP-1 complexes have been detected in vesicles shed by all other analyzed cell lines. Other proteolytic enzymes, on the contrary, were detected in vesicles shed by some cell lines and not in others. The urokinase type of plasminogen activator (uPA), while not detected in vesicles shed by 8701-BC and MCF-7 cells, is present in MDA MB-231. In vesicles shed by HT-1080 fibrosarcoma, we find not only MMP-9 and uPA, but also gelatinase A (MMP-2).[3] We observe a positive correlation between the quantity of shed vesicles, the amount of lytic enzymes they carry, and the in vitro invasive capability of the different cell lines. These findings indicate that vesicle production might represent an important mechanism by which tumor cells induce the localized degradation of the extracellular matrix and acquire metastatic capabilities.

We have no evidence of which membrane vesicle components are involved in binding of proMMP-9/TIMP-1 complexes. Since Olson et al. recently reported that, in breast epithelial cells, proMMP-9 binds to α2 (IV) chain of collagen IV,[4] we tested 8701-BC shed vesicles for the presence of this molecule. As shown by FIGURE 1, when Western blot analyses were performed in nonboiled, nonreducing conditions, antibodies against the α2 (IV) chain of human collagen IV, recognized two components. One of them, having an electrophoretic mobility of 190 kDa, corresponds to the α2 (IV) chain of collagen IV, the molecule identified as the proMMP-9 receptor

[c]Address for correspondence: M. Letizia Vittorelli, Dipartimento di Biologia Cellulare e dello Sviluppo, Viale delle Scienze, 90128 Palermo, Italy. Phone, 0039/91-425081; fax, 0039/91-420897; e-mail, mlvitt@mbox.unipa.it

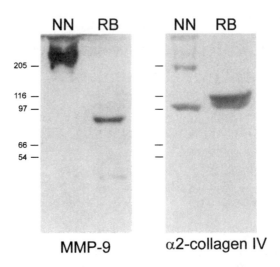

**FIGURE 1.** Immunological identification of MMP-9 and α2 (IV) chain of collagen IV in vesicles shed by 8701-BC breast carcinoma cells. Western blot analyses of vesicles shed by 8701-BC breast carcinoma cells with anti-MMP-9 polyclonal antibodies (*left panel*); and monoclonal AB 1910 antibodies against human α2 (IV) chain of collagen IV (*right panel*). 40 μg of proteins/lane, 10% SDS-PAGE. NN, nonreduced nonboiled samples; RB, reduced and boiled samples (1% β-mercaptoethanol and 3 min. at 100°C).

on the MCF-10A cell surface.[4] The second component has an electrophoretic mobility of about 100 kDa and probably corresponds to a fragment of the same molecule. These results demonstrate the presence of relatively large amounts of α2 (IV) chain of collagen IV in vesicles shed by 8701-BC cells. The 190-kDA component is absent in samples boiled and treated with reducing agents. In these conditions antibodies recognized a doublet of components, running with an electrophoretic mobility of about 100 kDA. Those components are likely to be fragments created by partial degradation of α2 (IV) chain of collagen IV molecules. These fragments were not observed by Olson *et al.*[4] on the cell surface of MCF-10A cells. Degrading activity of MMP-9 molecules on the α2 (IV) chain of collagen IV was demonstrated, however,[4] and active MMP-9 molecules are present in vesicles shed by 8701-BC.[2]

In our experiments, interactions between α2 (IV) chain of collagen IV and proMMP-9 molecules were not detected. It is, however, possible that large complexes, existing *in vivo*, are destroyed during blot procedures. We will therefore analyze vesicles for interactions of proMMP-9/TIMP-1 complexes with the α2 (IV) chain of collagen IV, using different techniques such as cross-linking and/or immunoprecipitation.

## ACKNOWLEDGMENT

We thank the Italian Association for Cancer Research (AIRC) for its financial support of this work.

## REFERENCES

1. TAYLOR, D.D. & P.H. BLACK. 1986. Shedding of plasma membrane fragments. Neo-plastic and developmental importance. *In* Developmental Biology Vol. 3. M. Stein-berg, Ed: 33–57, Plenum Press. New York.
2. DOLO, V. *et al.* 1998. Selective localization of MMP-9, β-1 integrins and HLA-I mole-cules on membrane vesicles shed by 8701-BC breast carcinoma cells. Cancer Res. **58:** 4227–4232.
3. GINESTRA, A. *et al.* 1997: Urokinase plasminogen activator and gelatinases are associ-ated with membrane vesicles shed by human HT-1080 fibrosarcoma cells. J. Biol. Chem. **27:** 17216–17222.
4. OLSON, M.W. *et al.* 1998. High affinity binding of latent MMP-9 to the $\alpha2$ (IV) chain of collagen IV. J. Biol. Chem. **273:** 10672–10681.

# Tissue Inhibitor of Metalloproteinase-1 Is Not an Acute-Phase Protein

PETER M. TIMMS,[a,b] STEWART CAMPBELL,[c,d] PAUL MAXWELL,[e]
DEREK L. SHIELDS,[e] AND BOOTH J.DANESH[c]

[a]Immunoassay Laboratory, Department of Clinical Biochemistry,
51-53 Bart's Close, St. Bartholomew's Hospital, London EC1A 7BE, UK

[c]Department of Gastroenterology, Stobhill NHS Trust, Glasgow G21 3UW, UK

[d]Department of Human Nutrition, Glasgow Royal Infirmary, Glasgow, UK

[e]Department of Clinical Biochemistry, Stobhill NHS Trust, Glasgow, UK

## INTRODUCTION

The amount of collagen in tissues is tightly controlled by the rate of collagen synthesis and the rate of collagen degradation. Collagen degradation is governed by the hydrolytic actions of matrix metalloproteinases (MMP). The activity of MMP is inhibited by a group of antiproteinases called *tissue inhibitor of metalloproteinase* (TIMP).[1] TIMP-1, the best characterized of these antiproteinases, has a molecular weight of 28,000 and is glycosylated at two sites.

TIMP-1 is an important factor controlling the development of hepatic fibrosis.[2,3] The activity of TIMP-1 is upregulated by IL6,[4] which is also involved in the acute-phase response. Since an acute-phase protein response[5] is also found in patients with alcoholic liver disease, we investigated a group of alcoholic patients who were undergoing acute alcohol withdrawal to see if TIMP-1 correlates with a variety of established acute-phase proteins.

## PATIENTS AND METHODS

Alcoholic patients (24: 22 males and 2 females) were recruited consecutively to an alcohol and drug rehabilitation unit. All had previously consumed more than 150 g of alcohol daily for greater than one year. The patients had a variable degree of liver disease as judged by clinical and biochemical criteria. Screening tests for other causes of liver disease (autoimmune hepatitis; hepatitis A, B, or C; and hemochromatosis) were negative.

There is no data on the half-life of TIMP-1, so we compared TIMP-1 levels with the acute-phase proteins prealbumin, albumin, transferrin, orosomucoid, ferritin, and CRP, all of which have different half-lives.

Plasma samples were obtained within three days of admission. Plasma TIMP-1 was assayed using an ELISA sandwich assay from Amersham. The acute-phase proteins were assayed using immunoturbidometric assays.

[b]Address for telecommunication: Phone, 0171/601 8254; fax, 0171/796 4676;
e-mail, p.m.timms@mds.qmw.ac.uk

Data was normalized either by log transformation to the base$_{10}$ (CRP and orosomucoid) or by taking the square root of TIMP-1. All statistical calculations were performed using Minitab Statistical Software Minitab (Pennsylvania, USA).

## RESULTS

Most alcoholic subjects had TIMP-1 levels elevated above the normal range. TABLE 1 shows the levels of TIMP-1 and acute-phase proteins found in our study. TABLE 2 shows no significant correlation between TIMP-1 and the other acute-phase proteins measured, although the acute phase proteins did correlate with each other significantly. There were significant correlations between $\log_{10}$ orosomucoid and albumin ($r = -0.45$), $\log_{10}$ orosomucoid and $\log_{10}$ CRP ($r = 0.54$), $\log_{10}$ CRP and prealbumin ($r = -0.61$).

**TABLE 1. Range of TIMP-1 and acute phase proteins in our alcoholic subjects**

|  | Range in alcoholic subjects | | Normal range |
| --- | --- | --- | --- |
|  | Median | Range | |
| TIMP-1 ng/ml | 1100 | 208–2349 | 61–325 |
| Ferritin µg/L | 276 | 13–1181 | 6–260 |
| CRP mg/L | 4 | 0–102 | < 20 |
| Orosomucoid g/L | 1.27 | 0.83–2.24 | 0.55–1.4 |
| Albumin g/L | 43 | 34–49 | 36–52 |
| Prealbumin g/L | 0.32 | 0.12–0.44 | 0.2–0.4 |
| Transferrin g/L | 2.50 | 1.77–3.36 | 2–4 |

**TABLE 2. Pearson correlation coefficients between TIMP-1 and the acute phase proteins**

|  | Ferritin µg/L | Albumin g/L | Log CRP mg/L | Log Oros[a] g/L | Transferrin g/L | Prealbumin g/L |
| --- | --- | --- | --- | --- | --- | --- |
| $\sqrt{\text{TIMP}}$ ng/ml | $r = 0.013$ $p = 0.95$ | $r = -0.12$ $p = 0.57$ | $r = 0.35$ $p = 0.11$ | $r = 0.29$ $p = 0.16$ | $r = -0.03$ $p = 0.89$ | $r = -0.12$ $p = 0.59$ |

[a]Log Oros $\log_{10}$ orosomucoid.

## DISCUSSION

TIMP-1 concentration is increased in fibrosis associated with liver disease, left ventricular hypertrophy secondary to hypertension,[6] and after wound healing. The acute-phase response is associated with interleukin-1, -6, and tumor necrosis factor release, which have been documented to stimulate TIMP-1 release. We assayed CRP, prealbumin, transferrin, ferritin, and orosomucoid because they cover a wide range of half-lives (from hours for CRP to 20 days for albumin). The lack of correlation between TIMP-1 and CRP observed in this study has previously been reported by Plumpton,[7] who did not investigate a group of alcoholic patients. We found no correlation between any of the measured acute-phase proteins and TIMP-1. There were correlations between CRP and prealbumin, CRP and orosomucoid, and orosomucoid and albumin. This provides further evidence that this study was powerful enough to identify a correlation between known acute-phase proteins, suggesting that the lack of correlations seen with TIMP-1 and the acute-phase proteins was not just due to lack of power of the study. The elevated TIMP-1 and the correlations between the established acute-phase proteins suggest that there is a different control mechanism governing TIMP-1 and acute-phase protein modulation in these patients. Theoretically the release of TIMP-1 as part of a generalized acute-phase response would be physiologically undesirable since an acute-phase response could be associated with increased collagen deposition.

## ACKNOWLEDGMENTS

We thank Amersham for the kits for the TIMP-1 assay.

## REFERENCES

1. GOMEZ, D.E., D.F. ALONSO, H. YOSHIJI & U.P. THORGEIRSSON. 1997. Tissue inhibitors of metalloproteinases: structure, regulation and biological function. Eur. J. Cell Biol. **74:** 111–122.

2. WALSH, K., P.M. TIMMS & S. CAMPBELL. 1998. Plasma levels of MMP-2 and TIMP-1 and 2 as non-invasive markers of liver disease in chronic hepatitis C: comparison using ROC analysis. Dig. Dis. Sci. In press.

3. LI, J., A. ROSMAN, M. LOE & C.S. LIEBER. 1994. Tissue inhibitor of metalloproteinase is increased in the serum of precirrhotic and cirrhotic alcoholic patients and can serve as a marker of fibrosis. Hepatology **19:** 1418–1423.

4. BIRKEDAL-HANSEN, W.K. MOORE & M.K. BODDEN. 1993. Matrix metalloproteinases: a review. Crit. Rev. Oral Biol. Med. **4**(2): 197–250.

5. LIPPI, G., S. FEDI & M. GRASSI. 1992. Acute phase proteins in alcoholics with or without liver injury. Ital. J. Gastroenterol. **24**(7): 383–385.

6. TIMMS, P.M. & V. SRIKANTHAN. 1998. Hypertension causes an increase in tissue inhibitor of metalloproteinase-1. Am. J. Hypertens. **11:** (4pt2) 1A.

7. PLUMPTON, T.A., I.M. CLARK, C. PLUMPTON et al. 1995. Development of an enzyme-linked immunosorbent assay to measure total TIMP-1 and measurement of TIMP-1 and CRP in serum. Clin. Chim. Acta **240:** 137–154.

# Regulation of TIMP-1 Expression by Hypoxia in Kidney Fibroblasts

JILL T. NORMAN,[a,b] IAN M. CLARK,[c] AND PATRICIA L. GARCIA[a]

[a]Department of Medicine, Royal Free and University College Medical School, London, WC1E 6JJ, UK

[c]School of Biological Sciences, University of East Anglia, Norwich, Norfolk, NR4 7TJ, UK.

TIMP-1 is overexpressed in fibrotic kidney diseases,[1] in which extracellular matrix accumulation and obliteration of the microvasculature are likely to lead to tissue hypoxia,[2] suggesting that hypoxia may be a stimulus for TIMP-1 expression. Interstitial fibroblasts are the major effectors of matrix accumulation; our previous *in vitro* studies showed that in human kidney fibroblasts (KF), hypoxia (1% $O_2$) induced a specific, time-dependent increase in TIMP-1 mRNA and protein that is dependent on new RNA and protein synthesis[3] (FIG.1) but is independent of hypoxia-induced, autocrine mediators. Transduction of the signal of a fall in ambient oxygen to changes in TIMP-1 expression appears to involve a heme-protein $O_2$-sensor and activation of protein kinase C– and tyrosine kinase–mediated signaling pathways.[3]

Hypoxia can alter mRNA levels by altering gene transcription, via hypoxia-response elements, and/or by altering mRNA stability.[4] Transient transfections with human TIMP-1 promoter-CAT reporter constructs[5] were used to examine basal and hypoxia-induced regulation of TIMP-1 promoter activity in KF. In normoxic cells, the most active region of the promoter was −738/+95, with distal negative regulatory elements between −1800/−738. Deletion of either 5´ or 3´ sequences from the −738/+95 promoter reduced activity, although the −59/+95 construct, which is inactive in skin fibroblasts,[5] retained some basal activity in KF, suggesting cell-type differences in regulation. Hypoxia stimulated activity of the −738/+95 promoter, indicating that hypoxia increases TIMP-1 gene transcription. 5´- and 3´-deletions[5] of the promoter localized a hypoxia-inducible element (HRE) to between −59/+8. Hypoxic regulation of a number of genes has been shown to be mediated by hypoxia-inducible factor-1 (HIF-1), a heterodimeric $\alpha\beta$ transcription factor, that binds to the sequence 5´-CGTG-3´.[6] The putative HRE in the hTIMP promoter contains a HIF-binding motif (−26/−23), suggesting a role for this factor in hypoxia-induced TIMP-1 gene transcription. HIF-1$\alpha$ mRNA is constitutively expressed in KF; hypoxia induced an early increase in HIF-1$\alpha$ mRNA and nuclear HIF-1$\alpha$ protein, which preceded the increase in TIMP-1mRNA. Mutation of the −26/−23 CGTG sequence suppressed hypoxia-inducibility of the promoter, indicating HIF binding is required for activity (FIG.2).

[b]Address for correspondence: Jill Norman, Ph.D., Department of Medicine, The Rayne Institute, 5 University Street, London WC1E 6JJ, Great Britain. Phone, 44 171/209 6179; fax, 44 171/209 6211; e-mail, rmhajtn@ucl.ac.uk

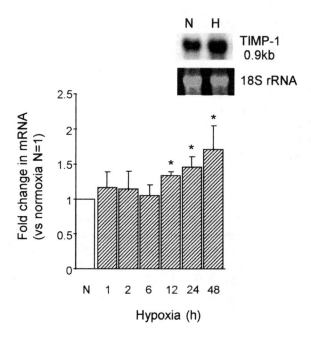

**FIGURE 1.** Hypoxia ($1\% O_2$) induces a time-dependent increase in TIMP-1 mRNA in kidney fibroblasts. Northern blots of TIMP-1 mRNA expression were quantitated by densitometry and normalized for loading by comparison to ethidium bromide–stained rRNA. Normoxic cells (N; 21% oxygen) were assigned an arbitrary value of 1 and the fold change with hypoxia (H) calculated. *$p < 0.05$ vs. N, $n = 4$ experiments. *Inset* shows a representative Northern blot for 48-hour samples.

These data show that hypoxia stimulates TIMP-1 mRNA via HIF-binding to a HRE, a novel regulatory mechanism for this family of inhibitors that may be relevant to changes in TIMP-1 expression in diseases associated with reduced tissue oxygenation. The various TIMPs expressed by KF (TIMP-1, -2, -3) are differentially regulated by hypoxia; mRNAs for TIMP-1 and -3 are increased, while TIMP-2 mRNA expression is unaffected.

## ACKNOWLEDGMENT

These studies were supported by a British Heart Foundation Grant # PG/96045 to J.N. and an Arthritis Research Campaign fellowship to I.M.C.

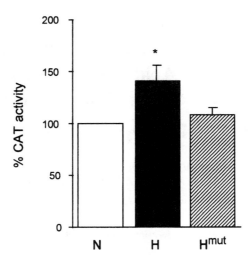

**FIGURE 2.** Hypoxia stimulates TIMP-1 promoter activity. Hypoxia increases CAT activity in fibroblasts transfected with the −102/+95 hTIMP-CAT reporter construct. Mutation of the sequence 5′-CGTG-3′ at −26/−23 to 5′-AAAC-3′ (H$^{mut}$) suppresses hypoxia-induced activity (H). Reporter gene activity is corrected for efficiency of transfection and for total cell protein, and data are presented as percent increase in CAT activity above normoxia (N; 100%). $n = 3$ experiments.

## REFERENCES

1. EDDY, A. 1996. Molecular insights into renal fibrosis. J. Am. Soc. Nephrol. **7:** 2495–2508.
2. FINE, L.G., C. ORPHANIDES & J. NORMAN. 1998. Progressive renal disease: The chronic hypoxia hypothesis. Kidney. Int. **53** Suppl. **65:** S74–S78.
3. NORMAN, J. 1997. Mechanisms of hypoxia induced changes in collagen-I (coll-I) and tissue inhibitor of metalloproteinase-1 (TIMP-1) gene expression in human renal fibroblasts [abstract]. J. Am. Soc. Nephrol. **8:** 523.
4. BUNN, H.F. & R. POYNTON. 1996. Oxygen sensing and molecular adaptation to hypoxia. Physiol. Rev. **76:** 839–885.
5. CLARK, I.M., A.D. ROWAN, D.R. EDWARDS *et al.* 1997. Transcriptional activity of the human tissue inhibitor of metalloproteinases 1 (TIMP-1) gene in fibroblasts involves elements in the promoter, exon 1 and intron 1. Biochem J. **324:** 611–617.
6. WOOD, S.M. & P. RATCLIFFE. 1997. Mammalian oxygen sensing and hypoxia-inducible factor 1. Int. J. Biochem. Cell. Biol. **29:** 1419–1432.

# Induction of Human Tissue Inhibitor of Metalloproteinase–1 Gene Expression by All-*trans* Retinoic Acid in Combination with Basic Fibroblast Growth Factor Involves both p42/44 and p38 MAP Kinases

HEATHER F. BIGG,[a] RONNIE McLEOD,[b,c] JASMINE WATERS,[b]
TIM E. CAWSTON,[d] JOHN F. NOLAN,[e] AND IAN M. CLARK[b]

[a]*Department of Oral, Medical and Surgical Sciences, University of British Columbia,
Vancouver, British Columbia V16 1Z3, Canada*

[b]*School of Biological Sciences, University of East Anglia, Norwich,
Norfolk, NR4 7TJ, UK*

[d]*Department of Rheumatology, University of Newcastle, Newcastle, NE2 4HH, UK*

[e]*Norfolk & Norwich Health Care NHS Trust, Norwich, NR1 3SR, UK*

Matrix metalloproteinases (MMPs) are a family of neutral zinc-dependent enzymes that have a diverse spectrum of biological activity within the human body. Processes that are known to require the activity of MMPs include angiogenesis and dermal wound healing, embryogenesis, uterine involution, and pericellular activation cascades.[1] The activity of the MMPs is controlled at the levels of synthesis and secretion, activation, and inhibition. Diseased states whereby the control of MMP expression is perturbed include rheumatoid arthritis and osteoarthritis, corneal ulceration, tumor invasion, metastasis, and liver fibrosis.[2] Inhibition of the MMPs is finely controlled by a family of specific MMP inhibitors, the tissue inhibitors of metalloproteinases (TIMPs). These TIMPs have been shown to inhibit all of the active MMPs.[3]

It has previously been reported by ourselves[4] and others that all-trans retinoic acid (ATRA) stimulates TIMP-1 production from fibroblasts. This study demonstrates that all-trans retinoic acid interacts synergistically with basic fibroblast growth factor (bFGF) to superinduce the production of TIMP-1, and begins to explore the mechanism by which this occurs. Additionally we have begun to investigate whether similar mechanisms exist within osteoblasts.

The synergistic stimulation of TIMP-1 protein by ATRA and bFGF increased over 72 hours. A short (1–12-hour) incubation with bFGF alone followed by ATRA alone gave a synergistic induction of TIMP-1 protein similar to that seen with both agents together; whereas a longer incubation (24–72 hours) with bFGF alone followed by ATRA alone abrogated the effect. Steady state levels of TIMP-1 mRNA were induced 14–40-fold above control by ATRA and bFGF. In contrast, TIMP-2

[c]Corresponding author. Phone, 01603/593796; fax, 01603/592250; e-mail, r.mcleod@uea.ac.uk

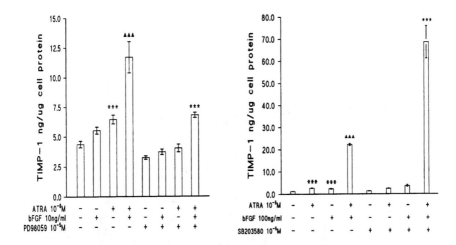

**FIGURE 1.** Role of p38 and p42/44 MAP kinases on the expression of TIMP-1 in human skin fibroblasts. Human skin fibroblasts ($5 \times 10^4$/well) were plated out in 24-well plates, then stepped down into MEM/1% AT-FCS for 48 hr. The growth factors and inhibitors were then added to the cells in MEM/1% AT-FCS and incubated at 37°C for 72 hr. The media was harvested, made 0.02% with respect to Na Azide, and batch tested for TIMP-1 within 2 weeks using a TIMP-1 ELISA developed in our laboratory.

mRNA was not induced by ATRA and bFGF. Control TIMP-1 mRNA was stable in the presence of the transcriptional inhibitor DRB for 48 hours; no differences were seen with bFGF, ATRA, or both reagents.

The induction of TIMP-1 mRNA by ATRA and bFGF was greatly diminished by cycloheximide and therefore required new protein synthesis. ATRA suppressed bFGF-stimulated collagenase (MMP-1) mRNA levels. The tyrosine kinase inhibitor genistein caused a dose-dependent inhibition of TIMP-1 protein induction by ATRA and bFGF. A MEK-1 inhibitor (PD98059) inhibited both basal and induced levels of TIMP-1, while specific p38 MAP kinase inhibitors (SB203580, SB202190) further enhanced the synergistic stimulation of TIMP-1 by ATRA and bFGF. (FIG. 1).

Preliminary data has been collected with human osteoblasts derived from sequentially digested cancellous bone (from femoral shaft) that may demonstrate differences in species/tissue/age-specific osteoblast TIMP-1 expression. Osteoblasts (verified by alkaline phosphatase staining)[6] treated with ATRA over a 72-hr time point (FIG. 2) show an inverted dose-response curve compared to that originally reported by Overall[5] for fetal rat calvarial osteoblasts, with ATRA inducing TIMP-1 expression rather than suppressing TIMP-1 expression as reported. Studies to investigate differences in ATRA-induced/repressed TIMP-1 expression in human versus rat, adult versus fetal, and long bone versus flat bone, respectively, are ongoing at present.

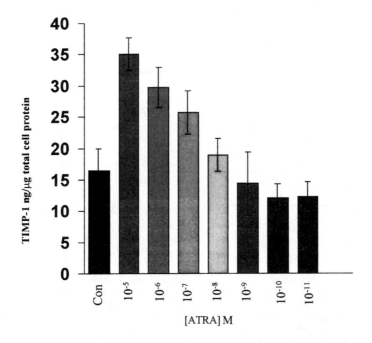

**FIGURE 2.** Stimulation of TIMP-1 expression by ATRA in human long bone osteo-blasts. Human osteoblasts prepared from the sequential digestion of cancellous bone removed during prosthetic hip replacement were cultured in DMEM-F12/10%FCS. Cells were stepped down into media containing 1% AT-FCS for 48 hr prior to the addition of ATRA at the respective concentration. Cells were incubated for 72 hr in the presence of ATRA at 37°C. The media was harvested, made 0.02% with respect to Na azide, and batch tested for TIMP-1 within 2 weeks using a TIMP-1 ELISA method developed in the laboratory.

## ACKNOWLEDGMENTS

This work was supported by the Arthritis Research Campaign (UK).

## REFERENCES

1. MURPHY, G., A.G.P. DOCHERTY, R.M. HEMBRY & J.J. REYNOLDS. 1991. Metallopro-teinases and tissue damage. Brit. J. Rheumatol. **30:** 25–31.
2. GREENWALD, R.A. & L.M. GOLUB, EDS. 1994. Inhibition of Matrix Metalloprotein-ases: Therapeutic Potential. Ann. N.Y. Acad. Sci. **732.** New York.
3. DENHARDT, D.T., B. FENG, D.R. EDWARDS, E.T. COCUZZI & U.M. MALYANKAR. 1993. Tissue inhibitors of metalloproteinases (TIMP, aka EPA)—structure, control of expression and biological functions. Pharmacol. Ther. **59:** 329–341.
4. BIGG, H.F. & T.E. CAWSTON. 1995. All-trans retinoic acid interacts synergistically with basic fibroblast growth factor and epidermal growth factor to stimulate the pro-duction of tissue inhibitor of metalloproteinase from fibroblasts. Arch. Biochem. Biophys. **319:** 74–83.

5. OVERALL, C.M. 1995. Repression of tissue inhibitor of matrix metalloproteinase expression by all-trans retinoic acid in rat bone cell populations: comparison with transforming growth factor-$\beta$1 J. Cell. Physiol. **164:** 17–25.

6. WEISS, M.J., P.S. HENTHORN, M.A. LAFFERTY, C. SLAUGHTER, M. RADUCHA & H. HARRIS. 1986 Isolation and characterisation of a cDNA encoding a human-liver bone kidney-type alkaline phosphatase. Proc. Natl. Acad. Sci. USA **83:** 7182–7186.

# Transcriptional Regulation of the Human Tissue Inhibitor of Metalloproteinases–1: Mapping Transcriptional Control in Intron-1

GREG DEAN[a] AND IAN M. CLARK

*Department of Biological Sciences, University of East Anglia,*
*University Plain, Norwich, Norfolk, NR4 7TJ, England*

The matrix metalloproteinases (MMPs) are a family of enzymes involved in the turnover and degradation of extracellular matrix. The active form of all MMPs are inhibited by a family of specific inhibitors, the tissue inhibitors of metalloproteinases (TIMPs). Inhibition of MMP activity by TIMPs represents an important control mechanism. Aberrant matrix turnover is involved in a number of pathologies including rheumatoid arthritis and osteoarthritis, tumor invasion and metastasis, and liver fibrosis. Therefore, an understanding of TIMP-1 gene regulation would be useful in developing therapies for these diseases.

A 2.9-kb clone of human TIMP-1 was sequenced and determined to contain 1.7 kb upstream of exon-1, exon-1, intron-1, exon-2, and the start of intron-2.[1] Further studies showed the importance of an Ap-1 and PEA3 site located just upstream of exon-1 in basal transcription.[1] Initial studies have suggested that intron-1 of the human TIMP-1 gene contains elements involved in transcriptional regulation. The presence of intron-1 in promoter reporter fusion studies greatly reduces reporter expression relative to constructs lacking the intron-1, indicating that regulatory elements are present in intron-1 of human TIMP-1. Corresponding work with mouse TIMP-1 by Flenniken and Williams[2] showed that the entire intron-1 was required in reporter constructs for correct developmental expression. 3′ deletions of human TIMP-1 intron-1 were created and subcloned into a chloramphenicol O-acetyltransferase (CAT) reporter plasmid.[3] Transient transfections were carried out with these constructs into primary human skin fibroblast cells using the FuGENE 6 transfection reagent (Boehringer Mannheim). Subsequent analysis of CAT expression was determined using a CAT ELISA kit (Boehringer Mannheim) and showed that there are at least three positive and three negative elements present in intron-1 of human TIMP-1 (FIG. 1).

Using the data from this crude map, DNase1 footprinting of intron-1 using the SureTrack Footprinting kit (Pharmacia Biotech) is being undertaken. One fragment towards the 3′ of intron-1 covering the deletion fragment +684/+748 (putative negative element) showed areas of protection from DNase I digestion, suggesting that protein was bound to this region. On analysis of this sequence with a transcription factor database, consensus sequences were found for c-Myb, Sp1, and c-Ets/PEA3, in and overlapping the protected region (see FIG. 2).

[a]Address for telecommunication: Phone, 01603/593796; fax, 01603/592250; e-mail, g.dean@uea.ac.uk

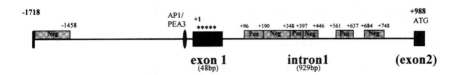

FIGURE 1. Positive and negative elements present in intron-1 of human TIMP-1.

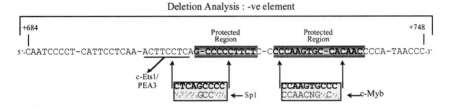

FIGURE 2. Consensus sequences found for c-Myb, Sp1, and c-Ets/PEA3 in and over-lapping the protected regions.

In order to confirm the authenticity of these sites, electrophoretic mobility shift assays (EMSA) and "super shifts" will be needed, with the aim of introducing mutations back into reporter constructs to analyze the functional significance of these sites *in vivo*.

## ACKNOWLEDGMENT

This work was supported by the Arthritis Research Campaign.

## REFERENCES

1. CLARK, I.M., A.D. ROWAN, D.R. EDWARDS, T. BECH-HANSEN, D.A.MANN, M.J. BAHR & T.E. CAWSTON. 1997. Transcriptional activity of the human tissue inhibitor of metalloproteinases 1 (TIMP-1) gene in fibroblasts involves elements in the promoter, exon-1 and intron-1. Biochem. J. 324: 611–617.
2. FLENNIKEN, A.M. & B.R.G. WILLIAMS. 1990. Developmental expression of the endogenous TIMP gene and a TIMP-*lacZ* fusion gene in transgenic mice. Genes Dev. 4: 1094–1106.
3. BRUNO L. & G. SCHUTZ. 1987. CAT constructions with multiple unique restriction sites for the functional analysis of eukaryotic promoters and regulatory elements. Nucleic Acids Res. 15: 5490.

# Balance between MMP-9 and TIMP-1 Expressed by Human Bronchial Epithelial Cells: Relevance to Asthma

P.M. YAO,[a] H. LEMJABBAR,[a] M.P. D'ORTHO,[a] B. MAITRE,[a] P. GOSSETT,[b] B. WALLAERT,[b] AND C. LAFUMA[a,c]

[a]INSERM U492, Faculté de Médecine, Créteil, France

[b]INSERM U416, Institut Pasteur, Lille, France

Human bronchial epithelial cells (HBECs) may play an important role in normal growth and development as well as in normal extracellular matrix turnover, thereby contributing in the maintenance of the structural and functional integrity of the lung.[1] HBECs may also be involved in responses to bronchial tree insults during inflammatory remodeling or wound healing. Injury to the bronchial epithelial surface, including mechanical trauma and toxin or inflammatory mediator exposure,[2] can result in sloughing of epithelial cells, leading to partial exposure of the basement membrane. The actual concept is that the wound repair process following epithelial denudation is achieved by a predictable sequence of resting epithelial cell spreading, migration, and proliferation. This process is probably affected by matrix metalloproteinases (MMPs) well known to degrade most matrix macromolecular components[3] under the control of their specific inhibitors, TIMPs.[4] More particularly the gelatinase subgroup comprising MMP-9 (gelatinase B or 92-kDa gelatinase) and MMP-2 (gelatinase A or 72-kDa gelatinase) specifically degrades basement membrane type IV collagenase and anchoring fibril type VII collagen.

The aim of the work presented here was to investigate: (1) the ability of cultured HBECs from human explants to express gelatinases and TIMPs in basal conditions and after stimulation by LPS endotoxin or inflammatory cytokines IL-1β and TNF-α; (2) the modulation of gelatinase and TIMP production by cell-matrix interactions; and (3) the contribution of gelatinases and TIMPs in human bronchial inflammation during asthma and status asthmaticus.

All of the immunological, enzymatic, and quantitative RT-PCR data show that:

1. The primary cultures of HBECs constitutively express major MMP-9, minor MMP-2, as well as mainly TIMP-1, suggesting that these enzymes and their inhibitor may be involved in the turnover and controlled degradation of the subepithelial basement membrane, and in epithelial cell-cell interactions.[5,6]

2. The marked upregulation and activation of MMP-9, the unchanged expression or restrictive modulation of TIMP-1, and the absence of detectable

[c]Address for correspondence: Dr. Chantal Lafuma, Inserm U492, Faculté de Medécine, 8, rue du Général Sarrail, 94010 Créteil, France. Phone, 33 1 4981 3523; fax, 33 1 4898 1777, e-mail, lafuma@im3.inserm.fr

TIMP-2 and TIMP-3 production by HBECs in response to inflammatory conditions (IL-1β, TNF-α, and LPS) may induce an imbalance between MMP-9 and TIMP- I in favor of the metalloproteinase, thus promoting extensive degradation of specific macromolecular molecular components of the extracellular matrix underlying HBECs, as well as detachment of the cells.[6]

3.  Primary cultures of HBECs exhibit differential gelatinase expression and unchanged TIMP-1 expression in response to the collagen type used as a matrix substrate. The same differences are evidenced under basal conditions and during exposure to LPS or proinflammatory cytokines. The restrictive modulation of MMP-9 expression as well as the loss of both MMP-9 and MMP-2 activation in the presence of type IV collagen strongly suggest that this type of collagen is associated with homeostatic HBEC phenotype, and limits the ability of HBECs to degrade the matrix. By contrast, the persistent activation of both gelatinases and the MMP-9 production in greater amount by HBECs plated on types I + III collagen support the concept of an HBEC resorptive phenotype with this substrate.[7]

4.  The presence of acutely elevated levels of MMP-9 in the bronchial airways of patients with status asthmaticus (SA) is partly counterbalanced by high TIMP-1 levels.[8] However, the presence of free metallogelatinolytic activity in status asthmaticus strongly suggests imbalance between MMP-9 and TIMP-1 in favor of MMP-9. The concomitant overproduction of stromelysin-l–activated forms supports the possibility that MMP-9 may be activated by stromelysin-1 during SA. In contrast, no or little MMP-9, nor stromelysin-1, nor TIMP-1 were detected during mild asthma, pointing out that the mechanism of airway inflammation in SA may be quite distinct from those in mild asthma. Several arguments allowed us to propose that cell origin of such elevated MMP-9 levels in SA may be shared between numerous activated chemoattracted neutrophils and activated bronchial epithelial cells *in situ*, in response to lung injury. This acute enhanced secretion of MMP-9 only partially counterbalanced by TIMP-1 overproduction in bronchial airways during SA appear to be partly responsible for edema, increased bronchial lung permeability, and possibly some destruction of airways.

## REFERENCES

1. SHOJI, S., K.A. RICHARD, R.F. ERTL, R.A. ROBBINS, J. LINDER & S.I. RENNARD. 1989. Am. J. Respir. Cell Mol. Biol. **1:** 13–20.
2. SNIDER, G.L., E.C. LUCEY, T.G. CHRISTENSEN, P.J. STONE, J.D. CALORE, A. CANTANESE & C. FRANZBLAU. 1984. Am. Rev. Respir. Dis. **129:** 155–160.
3. BIRKEDAL-HANSEN, H., W.G.I. MOORE, M.K. BODDEN, L.J. WINDSOR, B. BIRKEDAL-HANSEN, A. DECARLO & J.A. ENGLER. 1993. Crit. Rev. Oral Biol. Med. **4**(2): 197–250.
4. GOMEZ, D.E., D.F. ALONSO, H. YOSHIJI & U.P. THORGEIRSSON. 1997. Eur. J. Cell Biol. **74:** 111–122.

5. YAO, P.M., J.M. BUHLER, M.P. D'ORTHO, F. LEBARGY, C. DELCLAUX, A. HARF & C. LAFUMA. 1996. J. Biol. Chem. **271:** 15580–15589.
6. YAO, P.M., B. MAITRE, C. DELACOURT, J.M. BUHLER, A. HARF & C. LAFUMA. 1997. Am. J. Physiol. **273** (Lung Cell. Mol. Physiol. **17**): L866–L874.
7. YAO, P.M., C. DELCLAUX, M.P. D'ORTHO, B. MAITRE, A. HARF & C. LAFUMA. 1998. Am. J. Respir. Cell Mol. Biol. **18:** 813–822.
8. LEMJABBAR, H., P. GOSSET, C. LAMBLIN, I. TILLIE, D. HARTMANN, B. WALLAERT, A. TONNEL & C. LAFUMA. 1999. Am. J. Respir. Crit. Care Med. **159:** 1298–1307.

# Use of Encapsulated Cells Secreting Murine TIMP-2 Ameliorates Collagen-Induced Arthritis in Mice

NICOLE RENGGLI-ZULLIGER,[a] JEAN DUDLER,[a] NOBORU FUJIMOTO,[b]
KAZUSHI IWATA,[b] AND ALEXANDER SO[a,c]

[a]Department of Rheumatology, CHUV, 1011 Lausanne, Switzerland

[b]Biopharmaceutical Department, Fuji Chemical Industries, Ltd.,
Takaoka, Toyama 933-8511, Japan

## INTRODUCTION

Arthritis is characterized by an irreversible process of connective tissue breakdown in the joints, leading to permanent loss of function and disability. If most therapies are efficient at relieving pain and inflammation, we still are unable to control the underlying process of destruction. It is believed that an altered equilibrium between matrix metalloproteinases (MMPs) and tissue inhibitors of metalloproteinases (TIMPs) in the joints leads to the pathologic destruction of cartilage and bone.[1]

Based on this assumption, several studies in various pathologies have investigated the therapeutic potential of restoring the balance between enzymes and inhibitors by increasing the amount of circulating inhibitors.[2,3]

In this work, prevention of cartilage breakdown was evaluated *in vivo* in a murine collagen-induced arthritis model (CIA). Inhibition of MMPs was realized by the subcutaneous implantation of two 1-cm-long semipermeable hollow fibers (capsules) containing cells overexpressing murine TIMP-2 (mTIMP-2).

## MATERIAL AND METHODS

### Induction of Collagen-Induced Arthritis (CIA)

Male DBA/1J (H-2q) mice between 8–10 weeks of age were obtained from BRL/RCC Biotechnology & Animal Breeding (Füllinsdorf, Switzerland). 100 μg of native chicken type II collagen (M.M. Griffith; Utah, USA) emulsified in Complete Freund's Adjuvant (Difco; Basel, Switzerland) was injected intradermally at base of tail. At day 24 after the first injection, a booster injection of 100 μg of native chicken type II collagen was done in Incomplete Freund's Adjuvant. Clinical scoring was done blindly as described elsewhere.[4]

[c]Corresponding author: Professor Alexander So, Department of Rheumatology, Centre Hospitalier Universitaire Vaudois, CH-1011 Lausanne, Switzerland. Phone, +41 21 314 14 49; fax, +41 21 314 15 33; e-mail, aso@chuv.hospvd.ch

## Capsules

Murine myoblasts cells (C2C12) transfected with mTIMP-2 (kindly obtained from D. Edwards; Alberta, Canada) subcloned into the pPI-dn-DHFR vector[5] were injected (1.8 µl/capsule; 270,000 cells/1-cm-long capsule) into 500-µm inner diameter poly(ether-sulfone)(PES) fibers from AKZO-Fiber Nobel AG (Wupperthal, Germany). This semipermeable membrane allows the bidirectional exchange of both nutrients and the overexpressed protein secreted by encapsulated cells. Before implantation cell differentiation into myotubes was induced and two capsules per mice were implanted subcutaneously.

## One-Step Sandwich Enzyme Immunoassay for mTIMP-2

Measurement of murine TIMP-2 in mouse sera was performed using the human TIMP-2 ELISA kit (Fuji Chemical Industries Ltd., Japan) according to manufacturer's instructions,[6] but using a recombinant mTIMP-2 purified from *P. pastoris* as standard.

## RESULTS

### Delayed Onset of Collagen-Induced Arthritis in Mice Bearing Capsules Secreting mTIMP-2

Therapeutic potential of mTIMP-2 was tested in the CIA model. Two capsules overexpressing mTIMP-2 were implanted one day before arthritis induction in the treatment group, whereas the control group received no capsules. Previous experiments already demonstrated that the course of CIA was not significantly affected by implantation of capsules containing untransfected C2C12 cells compared to no implantation.

The survival of the encapsulated cells after explantation (day 44) was demonstrated histologically. Survival of the cells was homogenous throughout the capsule with no necrosis and a high rate of cells differentiated into myotubes.

Encapsulated cells were able to increase the circulating levels of mTIMP-2, from $27 \pm 9$ ng/ml in the control mice ($n = 6$) to $41 \pm 10$ ng/ml in treated mice ($n = 6$) at day 44, equivalent to a 1.5-fold increase (mean $\pm$ SEM) (FIG. 1). Mice bearing capsules ($n = 6$) showed a delayed onset compared to control mice ($n = 8$), which was significant at day 30 using Wilcoxon rank sum test ($p \leq 0.05$). In the same conditions, no effect on the incidence or the severity of disease was observed. Once declared, arthritis seemed to progress at the same speed as in the control mice (FIG. 2). This was corroborated by the histology of the paws, where severe arthritis was present in both groups, with synovitis, cartilage proteoglycan depletion, cartilage, and bone erosions.

## DISCUSSION

Increased serum level of mTIMP-2 obtained with our encapsulated cells correlated with the observed delayed onset of arthritis, but had no influence on the severity

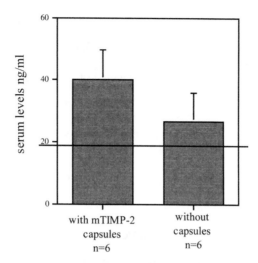

**FIGURE 1.** Comparison of serum levels at day 44 after CIA induction of mice with capsules secreting TIMP-2 or compared to control mice without capsules. The sera level of mice with TIMP-2 capsules is 1.5 times increased compared to mice without capsules (mean ± SEM). *Horizontal line* shows serum levels before immunization, 19.6 ± 3.6 ng/ml.

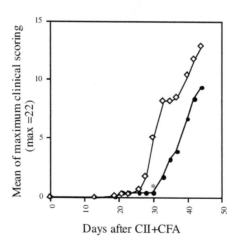

**FIGURE 2.** Mean of maximum clinical arthritic scores of mice with capsules secreting TIMP-2 (*circles*, $n = 6$) compared to control mice without capsules (*diamonds*, $n = 8$). A delayed onset is observed for the mice with capsules that is statistically significant at day 30 $p \leq 0.05$ using Wilcoxon's rank sum test.*

or the incidence of the disease. Although the protein activity was not demonstrated, a 1.5-fold increase may well not be sufficient to inhibit on the long term the highly expressed MMPs. The observed delay of arthritis onset is probably the result of an amelioration of the clinical picture through a decrease of inflammation. Further experiments, using capsules with a higher secretion rate, will enable the evaluation of cartilage breakdown inhibition using this *ex vivo* gene therapy.

## ACKNOWLEDGMENTS

We thank the laboratory of Gene Therapy Center and Surgical Research, CHUV, Lausanne, for their technical assistance, especially Vivianne Padrun and Liliane Schnell. The work was supported by the Jean and Linette Warnery Foundation.

## REFERENCES

1. MARTEL-PELLETIER, J., *et al.* 1994. Excess of metalloproteases over tissue inhibitor of metalloprotease may contribute to cartilage degradation in osteoarthritis and rheumatoid arthritis. Lab. Invest. **70:** 807–815.
2. CARMICHAEL, D.F. *et al.* 1989. Systemic administration of TIMP in the treatment of collagen-induced arthritis in mice. Agents Actions **27:** 378–379.
3. CONWAY, J.G. *et al.* 1995. Inhibition of cartilage and bone destruction in adjuvant arthritis in the rat by a matrix metalloproteinase inhibitor. J. Exp. Med. **182:** 449–457.
4. GREENWALD, R.A. *et al.* Handbook of animal models for rheumatic diseases. CRC Press. Boca Raton, FL.
5. RINSCH, C. *et al.* 1997. A gene therapy approach to regulated delivery of erythropoietin as a function of oxygen tension. Hum. Gene. Ther. **8:** 1881–1889.
6. FUJIMOTO, N. *et al.* 1993. A one-step sandwich enzyme immunoassay for tissue inhibitor of metalloproteinases-2 using monoclonal antibodies. Clin. Chim. Acta **220:** 31–45.

# TIMP-1 and TIMP-2 Perform Different Functions *in Vivo*

ZHIPING WANG AND PAUL D. SOLOWAY[a]

*Department of Molecular and Cellular Biology, Roswell Park Cancer Institute,*
*Elm and Carlton Streets, Buffalo, New York 14263, USA*

TIMPs-1 and -2 have been shown to share common activities *in vitro* including erythroid-potentiating,[1,2] MMP-inhibiting,[3,4] and growth-promoting activities.[5,6] Differences between the TIMPs have been observed as well; for example, TIMP-2 but not TIMP-1 has been shown to collaborate with MMP-14 in the activation of proMMP-2 on cell surfaces *in vitro*.[7,8] In order to identify the *in vivo* functions of TIMPs-1 and -2 and to characterize any differences between them, mice deficient for these TIMPs were developed (Ref. 9; Osiewicz *et al.*, this volume; Osiewicz *et al.*, submitted for publication; Wang *et al.*, in preparation).

TIMP-1–deficient mice were shown to be hyper-resistant to corneal infections with *Pseudomonas aeruginosa* by a complement-dependent mechanism (Osiewicz *et al.*, this volume and Osiewicz *et al.*, submitted). Infections became established and bacteria underwent a burst of growth identically in wild-type and TIMP-1–deficient mice. However, coincident with massive neutrophil accumulation in corneas between 24 and 48 hours postinfection, the bacterial burden in mutant mice dropped to a level 1,600-fold lower than in wild-type mice. This phenotype was suppressed by BB-94 treatment and depletion of the complement system. To determine if loss of TIMP-2 also resulted in hyper-resistance to infection, corneal infections were established in wild-type and TIMP-2–deficient mice using $10^7$ colony forming units (CFU) per eye of *P. aeruginosa*. Twenty-four hours after infection, eyes were removed and homogenized, and the bacterial burdens were measured by plate count. The results showed that in contrast to TIMP-1–deficient mice, mice lacking TIMP-2 were not hyper-resistant to infections; instead, the bacterial burdens were indistinguishable between wild-type and TIMP-2–deficient mice (TABLE 1). This revealed that although TIMP-1 can potently regulate complement-dependent responses to bacterial infections *in vivo*, TIMP-2 does not have this effect.

TIMP-2 has been shown to play a role in the activation of proMMP-2 *in vitro*[7,8] and is required for efficient activation *in vivo* (Wang *et al.*, submitted). To determine if TIMP-1 performs this function *in vivo*, tissues from wild-type and mutant mice were assayed by gelatin zymography for the various forms of MMP-2. Lysates of mouse lung were prepared by homogenizing tissue pieces with a Wheaton homogenizer in RIPA buffer (150 mM NaCl, 1.0% NP40, 0.5% deoxycholate, 0.1% sodium dodecyl sulfate, 50 mM Tris-Cl, pH 8.0) at 4°C using 1 ml buffer per 3 mg tissue. Lysates were centrifuged at $14,000 \times g$ for 15 minutes, and the insoluble, ECM-enriched fractions were washed in RIPA buffer, aliquoted, and stored at $-80°C$ or

[a]Corresponding author. Phone, 716/845-5843; fax, 716/845-8389;
e-mail, psoloway@mcbio.med.buffalo.edu

**TABLE 1. Corneal infections in *Timp* mutant mice**

| Experiment | *Timp* Genotype | CFU/Eye | Mut/WT | $p$ |
|---|---|---|---|---|
| 1 | WT | $5.6 \times 10^6$ | 1600 | $< 0.05$ |
| | t1/0 | $3.6 \times 10^3$ | | |
| 2 | WT | $3.8 \times 10^6$ | 0.6 | 0.5 |
| | t2/t2 | $2.2 \times 10^6$ | | |

NOTE: Corneal infections were established in wild-type (WT) mice or mice deficient for TIMP-1 (t1/0) or TIMP-2 (t2/t2) in two separate experiments. Twenty-four hours after infection with $10^7$ CFU/eye of *P. aeruginosa,* eyes were removed and homogenized, and the bacterial burden (CFU/eye) was measured by plate count. The ratio of the bacterial burdens in eyes of mutant (Mut) mice relative to WT are shown, as are the $p$ values for the burdens. All tests included a minimum of four mice aged 3–6 weeks.

immediately analyzed by gelatin zymography. Lysates containing an equivalent of 3 mg wet tissue were loaded onto each lane (FIG. 1).

The results showed that latent proMMP-2 underwent normal activation to its active lower molecular weight form in wild-type and TIMP-1–deficient mice, and, in contrast to TIMP-2, is not relevant to proMMP-2 activation.

Collectively, the results of these two experiments highlight key differences between the *in vivo* functions of TIMP-1 and TIMP-2. TIMP-1 but not TIMP-2 influences complement-dependent responses to bacterial infection, and TIMP-2 but not TIMP-1 influences activation of proMMP-2.

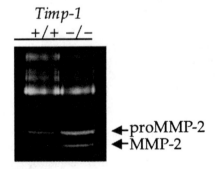

**FIGURE 1.** Gelatin zymography of lung extracts from TIMP-deficient mice. Lung extracts from wild-type (+/+) or *Timp*-mutant mice (–/–) were prepared as described in the text and analyzed by gelatin zymography. The positions of latent proMMP-2 and active MMP-2 are shown.

## REFERENCES

1. DOCHERTY, A.J., A. LYONS, B.J. SMITH, E.M. WRIGHT, P.E. STEPHENS, T.J. HARRIS, G. MURPHY & J.J. REYNOLDS. 1985. Nature 318(6041): 66–69.
2. STETLER-STEVENSON, W.G., N. BERSCH & D.W. GOLDE. 1992. FEBS Lett. 296(2): 231–234.
3. MERCER, E., T.E. CAWSTON, M. DE SILVA & B.L. HAZLEMAN. 1985. Biochem. J. 231(3): 505–510.
4. STETLER-STEVENSON, W.G., H.C. KRUTZSCH & L.A. LIOTTA. 1989. J. Biol. Chem. 264(29): 17374–17378.
5. HAYAKAWA, T., K. YAMASHITA, K. TANZAWA, E. UCHIJIMA & K. IWATA. 1992. FEBS Lett. 298(1): 29–32.
6. HAYAKAWA, T., K. YAMASHITA, E. OHUCHI & A. SHINAGAWA. 1994. J. Cell Sci. 107(Part 9): 2373–2379.
7. STRONGIN, A.Y., I. COLLIER, G. BANNIKOV, B.L. MARMER, G.A. GRANT & G.I. GOLDBERG. 1995. J. Biol. Chem. 270(10): 5331–5338.
8. WILL, H., S.J. ATKINSON, G.S. BUTLER, B. SMITH & G. MURPHY. 1996. J. Biol. Chem. 271(29): 17119–17123.
9. SOLOWAY, P.D., C.M. ALEXANDER, Z. WERB & R. JAENISCH. 1996. Oncogene 13: 2307–2314

# Tissue Inhibitor of Metalloproteinase-2 Induces Apoptosis in Human T Lymphocytes

MEGAN S. LIM,[a,b] LILIANA GUEDEZ,[c] WILLIAM G. STETLER-STEVENSON,[c] AND MARY ALICE STETLER-STEVENSON[c]

[a]Department of Laboratory Medicine, Sunnybrook and Women's College Health Science Center, University of Toronto, 2075 Bayview Avenue, Toronto, Ontario, M4N 3M5, Canada

[c]Laboratory of Pathology, National Cancer Institute, National Institutes of Health, Bethesda, Maryland 20892, USA

TIMPs are multifunctional proteins with metalloproteinase-inhibitory properties as well as growth-modulatory activities.[1] These proteins are also expressed in normal and malignant hematologic cells, and thus may play a role in their physiology. Our detailed analysis of expression of MMPs and TIMPs in cells of the lymphoid system demonstrated specific patterns of expression in cells of B- and T-cell lineage.[2] TIMP-2 is expressed at variable levels by T lymphoblastic lymphoma cells, while resting peripheral blood T cells express minimal levels. Furthermore, stimulation of peripheral blood T cells induces expression of TIMP-2 late in activation, when apoptosis is prominent. We therefore studied the effect of TIMP-2 on T-cell apoptosis using morphologic analysis as well as flow cytometric quantitation of Annexin V-FITC–stained cells. Our studies demonstrate that rTIMP-2 increased apoptosis in activated peripheral blood T cells, whereas unstimulated T cells were not susceptible to TIMP-2–mediated apoptosis. This effect was specific to TIMP-2 and was not observed with TIMP-1. Recombinant TIMP-2 increased the percentage of cells undergoing apoptosis in a dose-dependent manner. Recombinant TIMP-2 also induced apoptosis of Tsup and Jurkat T lymphoblastic lymphoma cell lines. The metalloproteinase-inhibitory function of TIMP-2 appears important in this process in that synthetic metalloproteinase inhibitors BB94, GM6001, and KB8301 also increased activation-induced apoptosis. A TIMP-2 peptide lacking the N-terminal domain, which is critical for MMP inhibition, did not induce apoptosis. A neutralizing antibody to TIMP-2 inhibited this process and resulted in slowing down the rate of apoptosis. This effect was not observed with anti–TIMP-1 antibody or an isotype control antibody. Molecules involved in ligand-mediated pathways of T-cell apoptosis are the Fas/Fas ligand and TNF receptor/TNF system. Many cell surface proteins are processed by synthetic MMP inhibitors.[3,4] Proteolysis of membrane-associated proteins is important for the conversion of latent receptor ligands to active forms, for regulation of signaling pathways via cleavage of transmembrane receptors, and for altering adhesion molecule interactions with the extracellular matrix. The potential role of TIMP-2 in inhibition of cleavage of Fas ligand was studied. Analysis of cell

[b]Address for correspondence: Dr. Megan S. Lim, The Department of Laboratory Medicine, Sunnybrook and Women's College Health Science Center, University of Toronto, 2075 Bayview Ave., Toronto, Ontario, M4N 3M5, Canada. Phone, 416-480-4600; e-mail, megan.lim@sunnybrook.on.ca

surface expression of Fas ligand, and determination of soluble forms of Fas ligand in cells grown in the presence or absence of TIMP-2 and BB94 support the hypothesis that these molecules inhibit the cleavage of Fas ligand and thus increase the rate of T-cell apoptosis. The use of blocking antibodies to Fas also demonstrated that soluble forms of Fas ligand are nonfunctional in signaling cell death as compared to membrane-bound forms. These data indicate that MMPs and TIMPs can modulate the growth and immune response of hematologic cells.

## REFERENCES

1. CORCORAN, M.L. & W.G. STETLER-STEVENSON. 1995. Tissue inhibitor of metellopro-teinases-2 (TIMP-2) stimulates fibroblast proliferation via a cyclic adenosin 3′,5′-mono-phosphate (cAMP)–dependent mechanism. J. Biol. Chem. **270:** 13453–13459.
2. STETLER-STEVENSON, W.G., M.A. MANSOOR, M.S. LIM *et al.* 1997. Expression of matrix metalloproteinases and tissue inhibitors of metalloproteinases (TIMPs) in reactive and neoplastic lymphoid cells. Blood **89:** 1708–1715.
3. BENNETT, T.A., E.B. LYNAM, L.A. SKLAR & S. ROGELJ. 1996. Hydroxamate-based metalloprotease inhibitor blocks shedding of L-selectin adhesion molecule from leu-kocytes. Functional consequences for neutrophil aggregation. J. Immunol. **156:** 3093–3097.
4. CROWE, P.D. *et al.* 1995. A metalloprotease inhibitor blocks shedding of the 80-kD TNF receptor and TNF processing in T lymphocytes. J. Exp. Med. **181:** 1205–1210.

# Analysis of the Interaction of TIMP-2 and MMPs: Engineering the Changes

MIKE HUTTON,[a,b] GEORGINA S. BUTLER,[a] BETH A. WATTAM,[c]
FRANCES WILLENBROCK,[d] RICHARD A. WILLIAMSON,[e]
AND GILLIAN MURPHY[a]

[a]School of Biological Sciences, University of East Anglia, Norwich, NR4 7TJ, UK

[c]Thrombosis Research Institute, Emmanuel Kaye Building,
Manresa Road, London, SW3 6LR, UK

[d]Department of Biochemistry, Queen Mary and Westfield College,
University of London, London, E1 4NS, UK

[e]Department of Biosciences, University of Kent, Canterbury, Kent, CT2 7NJ, UK

The tissue inhibitors of metalloproteinases (TIMPs) regulate the activation and pro-teolytic activity of the matrix metalloproteinases (MMPs). An imbalance in the rel-ative concentrations of MMP and TIMP has been observed in many degradative diseases.[1] Four members of the family (TIMPs 1–4) have been identified and cloned from a number of species. The TIMPs share 40% sequence similarity with consider-ably higher structural similarity. The TIMPs consist of a six-loop structure formed by 12 conserved cysteine residues. The first three loops form the N-terminal domain, which is highly conserved and has been shown to bind directly into the active site of MMPs. The C-terminal regions, composed of the final three loops, are more diver-gent and may be responsible for the selectivity of inhibition and binding efficiency of TIMPs to MMPs.

We have initiated a study examining interactions of the N-terminal domain of TIMP-2 (N–TIMP-2) with MMPs, using isolated N-terminal domains expressed in *E. coli*. The rationale for this study was initially based on the NMR solution structure of N–TIMP-2,[2] which showed that the core of the N–TIMP-2 protein is a closed five-stranded β-barrel homologous to the oligosaccharide/oligonucleotide–binding pro-tein family. Further information has been obtained from the crystal structure of the catalytic domain of MMP3–TIMP-1 complex,[3] NMR chemical shift perturbation studies on the interaction of MMP3 and N–TIMP-2,[4] and most recently the crystal structure of MT1-MMP and TIMP-2.[5] This body of work has identified three major regions of the N–TIMP-2 protein as being important in the interaction with MMPs: $Cys^1$- $Ser^4$, and $Ala^{70}$-$Gly^{73}$, which together form a "ridge" on the surface of the TIMP and interact directly with the active site cleft of MMP-3; and $Glu^{28}$-$Lys^{41}$, which is at the apex of the AB loop forming a β-hairpin. We have made site-directed mutants in two of these regions and investigated the effect on the association rate with several MMPs.

[b]Address for correspondence: Dr. Mike Hutton, School of Biological Sciences, University of East Anglia, Norwich, NR4 7TJ, UK. Phone, 00441603 593828; fax, 00441603 593874; e-mail, m.hutton@uea.ac.uk

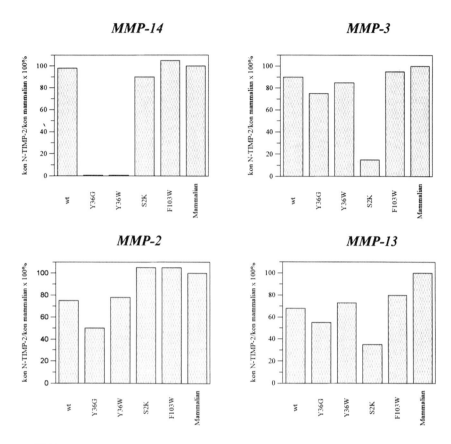

**FIGURE 1.** Association rate measurements for N-TIMP-2 mutants. Association rate constants ($k_{on}$) were measured using the appropriate concentrations of enzymes and inhibitors in a Perkin Elmer LS50B spectrofluorimeter using 1 µM quenched fluorescent substrates.[9,10] Values expressed are relative to mammalian N–TIMP-2, which has been set to 100%.

The N-terminal domain of wild-type (A21T) and mutant human TIMP-2 were prepared by expression as inclusion bodies using the pET vector (pET23d) in *E. coli* BL21 (DE3). Point mutations were introduced using the overlap extension method,[6] and the products were sequenced. The expressed proteins were prepared by solubilization and refolding.[7] N–TIMP-2 and MMP concentrations were determined by active site titration[8]; in order to accurately determine the concentration of N–TIMP-2 mutants, they were titrated against several MMPs. The association rates ($k_{on}$) were determined as described in FIGURE 1. The association data (FIG. 1) show that each mutant has a $k_{on}$ similar to wild-type with at least one of the MMPs used in the analysis. This suggests that the mutants have been correctly refolded from inclusion bodies, and this has been confirmed by nonreduced HPLC tryptic digest studies. The changing of residue Tyr[36] to Gly severely alters the rate of association with MMP-14, while having much less effect with other MMPs. Mutating Ser[2] to a charged lysine has a marked effect on MMP-3 and MMP-13 association, while not affecting the other MMPs used.

Thus these data demonstrate that greater specificity can be introduced into the TIMP-2 molecule by site-directed mutagenesis, with the mutation of a single amino acid affecting the binding of N–TIMP-2 to certain MMPs, while leaving others unaffected. This preliminary mutagenic work indicates that combining mutagenesis with further structural studies will allow us to produce more specific TIMPs where the selective inhibition of particular MMPs is desirable; and that this could be put to therapeutic use in the treatment of several disease processes.

## ACKNOWLEDGMENTS

This work is supported by the Arthritis Research Campaign and the Wellcome Trust, UK.

## REFERENCES

1. NAGASE, H., S.K. DAS, S.K. DEY, J.L. FOWLKES, W. HUANG & K. BREW. 1997. *In* Inhibitors of Metalloproteinases in Development and Disease. S.P. Hawkes, D.R. Edwards & R. Khokha, Eds. Harwood Academic Publishing. Lausanne. In press.
2. WILLIAMSON, R.A., G. MARTORELL, M.D. CARR, G. MURPHY, A.J.P. DOCHERTY, R.B. FREEDMAN & J. FEENEY. 1994. Solution structure of the active domain of tissue inhibitor of metalloproteinases-2. A new member of the OB fold protein family. Biochemistry 33: 11745–11759.
3. GOMIS-RÜTH, F.-X., K. MASKOS, M. BETZ, A. BERGNER, R. HUBER, K. SUZUKI, N. YOSHIDA, H. NAGASE, K. BREW, G.P. BOURENKOV, H. BARTUNIK & W. BODE. 1997. Mechanism of inhibition of the human matrix metalloproteinase stromelysin-1 by TIMP-1. Nature 389: 77–79.
4. MUSKETT, F.W., T.A. FRENKIEL, J. FEENEY, R.B. FREEDMAN, M.D. CARR & R.A. WILLIAMSON. 1998. High resolution structure of the N-terminal domain of Tissue Inhibitor of Metalloproteinase-2 and characterisation of its interaction site with Matrix Metalloproteinase-3. J. Biol. Chem. 273: 21736–21743.
5. FERNANDEZ-CATALAN, C., W. BODE, R. HUBER, D. TURK, J.J. CALVETE, A. LICHTE, H. TSCHESCHE & K. MASKOS. 1998. Crystal structure of the complex formed by the membrane type 1–matrix metalloproteinase with the tissue inhibitor of metalloproteinase-2, the soluble progelatinase A receptor. EMBO J. 17: 5238–5248.
6. HO, S.N., H.D. HUNT, R.M. HORTON, J.K. PULLEN & L.R. PEASE. 1989. Site-directed mutagenesis by overlap extension using the polymerase chain reaction. Gene 77: 51–59.
7. WILLIAMSON, R.A., D. NATALIA, C.K. GEE, G. MURPHY, M.D. CARR & R.B. FREEDMAN. 1996. Chemically and conformationally authentic active domain of human tissue inhibitor of metalloproteinase-2 refolded from bacterial inclusion bodies. Eur. J. Biochem. 241: 476–483.
8. MURPHY, G. & F. WILLENBROCK. 1995. Tissue inhibitors of matrix metalloendopeptidases. Methods Enzymol. 248: 496–510.
9. WILLENBROCK, F., T. CRABBE, P.M. SLOCOMBE, C.W. SUTTON, A.J.P. DOCHERTY, M.I. COCKETT, M. O'SHEA, K. BROCKLEHURST, I.R. PHILLIPS & G. MURPHY. 1993. The activity of the tissue inhibitors of metalloproteinases is regulated by C-terminal domain interactions: a kinetic analysis of the inhibition of gelatinase A. Biochemistry 32: 4330–4337.

10. NGUYEN, Q., F. WILLENBROCK, M.I. COCKETT, M. O'SHEA, A.J.P. DOCHERTY & G. MURPHY. 1994. Different domain interactions are involved in the binding of tissue inhibitors of metalloproteinases to stromelysin-1 and gelatinase A. Biochemistry **33:** 2089–2095.

# Murine TIMP-2 Gene–Targeted Mutation

J. CATERINA,[a,b] N. CATERINA,[a] S. YAMADA,[a] K. HOLMBÄCK,[a]
G. LONGENECKER,[a] J. SHI,[a] S. NETZEL-ARNETT,[a] J. ENGLER,[c]
A. YERMOVSKI,[c] J. WINDSOR,[c] AND H. BIRKEDAL-HANSEN[a]

[a]National Institute of Dental and Craniofacial Research, National Institutes of Health, Bethesda, Maryland 20892, USA

[c]University of Alabama at Birmingham, Birmingham, Alabama 35294, USA

Incorrectly regulated extracellular matrix (ECM) remodeling has been implicated in a number of physiologically aberrant conditions. Tissue inhibitors of metalloproteinases (TIMPs) are thought to play a critical role in the regulation of ECM remodeling resulting from their ability to inhibit matrix metalloproteinases (MMPs), whose various substrate specificities allow for degradation of the majority of ECM components. Although there is a high degree of homology among the four TIMP family members presently identified and some functional overlap in their inhibitory ability, they exhibit divergent expression patterns and some degree of preferential complex formation with the various MMPs.[1] In order to gain a better understanding of the role these inhibitors play individually and as a family, we have cloned the murine TIMP-2 gene locus, used homologous recombinatory gene-targeted mutagenesis technology to introduce a large chromosomal deletion into the TIMP-2 gene locus in the murine embryonic stem (ES) cell line HM-1,[2] and created mice lacking either one or both of their intact TIMP-2 gene loci (FIG. 1).

The mutant locus resulting from the homologous recombination event retains the ability to transcribe a form of the TIMP-2 message without the regions corresponding to exons 2 and 3 of the wild-type message. Furthermore, splicing of exon 1 to exon 4 retains the correct reading frame. The protein product of this message gives rise to a peptide of the expected size with only residual activity as assayed by reverse zymography (data not shown).

The homozygous mutant mice appear to be physiologically normal. No reduction in growth rate is seen compared to wild-type littermates. They are fertile and produce healthy offspring. Average life span is comparable to wild-type littermates as well.

Cultures of primary fibroblasts were prepared from both homozygous mutant and wild-type mice. Fibroblast culture medium was subjected to organomercurial activation. Aliquots of activation reactions were stopped with 1,10-phenathroline at various time points and assayed by gelatin zymography (FIG. 2). Gelatinase A (MMP2) was converted to its active form much more quickly from cultures of homozygous mutant fibroblasts. This activation could be slowed to the rate seen in cultures from wild-type fibroblasts by the addition of purified recombinant TIMP-2.

Fibroblast cultures stimulated by conconavolin A were also assayed by gelatin zymography. In contrast to results seen with *in vitro* MMP2 activation, culture me-

[b]Address for correspondence: NIDCR/NIH, 9000 Rockville Pike, Building 30, Room 4A400, Bethesda, Maryland 20892-4380. Phone, 301/496-3923; fax, 301/594-1253; e-mail, jc239h@nih.gov

**FIGURE 1.** Primers were designed to the 5′ and 3′ untranslated regions of the mouse TIMP-2 cDNA, and PCR of those probes was used to screen a P1 mouse genomic DNA library (Genome Systems) to obtain the entire TIMP-2 locus. Three such clones were obtained, but due to the extensive size of intron 1 (>50 kb), they have not been fully characterized. The exon (*upper case*)/intron (*lower case*) border sequences are also shown with the previous and next fully coded amino acid of each border identified and numbered. The TIMP-2 gene-targeting vector was constructed using a 2-kb BamHI 5′ homology region and a 6-kb EcoRI/HindIII 3′ homology region; the phosphoglycerate kinase (pgk) promoter was used to drive expression of both the positive selection drug resistance gene neomycin (NEO) and the homologous recombination enriching negative selection gene herpes simplex virus thymidine kinase (TK). The targeting construct was linearized and electroporated into HM-1 ES cells. Electroporated cells were subjected to Geneticin and gancyclovir selection. Surviving colonies were isolated and genotyped by both PCR and Southern blot analyses. (E: *Eco* RI, X: *Xba* I, B: *Bam* HI)

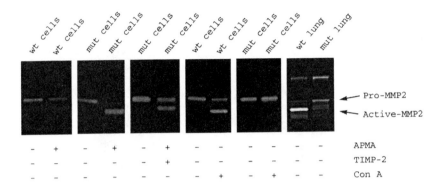

**FIGURE 2.** Gelatin zymography was performed to determine the effects of loss of TIMP-2 on the activation of gelatinase A *in vitro* by organomercurial and *in vivo* in serum-free medium from cultured primary fibroblasts and in lung tissue homogenates.

dium from mutant fibroblasts exhibited greatly reduced conversion of MMP2 to the active form compared to that from wild-type fibroblasts. This same phenomenon was seen when extracts from mouse lung tissue were examined, providing physiological evidence for the theoretical requirement for TIMP-2 in MMP2 activation.

## REFERENCES

1. BIRKEDAL-HANSEN, H. *et al.* 1993. Matrix metalloproteinases: a review. Crit. Rev. Oral Biol. Med. **4**(2): 197–250.
2. MAGIN, T.M. *et al.* 1992. A new mouse embryonic stem cell line with good germ line contribution and gene targeting frequency. Nucleic Acids Res. **20**(14): 3795–3796.

# The Expression of TIMP-3 in Hepatoma Cell Lines

YOSHIAKI KATANO,[a] YOSHIHIDE FUKUDA, ISAO NAKANO,
HIDENORI TOYODA, MIEKO EBATA, KEN-ICHI NAGANO,
KIYOSHI MORITA, SHOUICHI YOKOZAKI, MAMIKO TAKEUCHI,
KAZUHIKO HAYASHI, AND TETSUO HAYAKAWA

*Second Department of Internal Medicine, Nagoya University School of Medicine,
Nagoya 466-8550, Japan*

Tissue inhibitors of metalloproteinases (TIMPs) constitute a family of proteins, of which four members have so far been identified.[1–4] TIMP-1 or -2 has been studied for liver fibrosis[5] or hepatoma invasion/metastasis.[6] TIMP-3, which has a similar structure to TIMP-1 and -2, has an ability to inhibit matrix metalloproteinases. It has been reported that TIMP-3 expression was increased in the tissues of breast cancer.[3] As regards liver tissues, although it has been demonstrated that TIMP-3 expression is not detected in normal liver tissue,[3] TIMP-3 expression in hepatoma has not been cleared. In this study, we examined TIMP-3 expression in hepatoma cell lines.

## MATERIALS AND METHODS

HepG2 cell line was obtained from American Type Culture Collection; Huh7, PLC/PRF/5, HLE, and HLF cell lines were obtained from Health Science Research Resources Bank. Each cell line was incubated in semiconfluent state in serum-free RPMI1640 for 48 hours. After further incubation for 24 hours with or without 100 unit/ml of tumor necrosis factor-$\alpha$ (TNF-$\alpha$) or 50 ng/ml of 12-o-tetradecanoylphorbol 13-acetate (TPA), total RNA from each cell line was extracted with RNAzol B. The cDNA was synthesized from 1 µg of total RNA using AMV reverse transcriptase with random hexamers. The gene of TIMP-1, -2, or -3 was amplified by PCR method. Moreover, TIMP-3 gene was amplified by second PCR. The specific sets of PCR primers for TIMP-1, -2, and -3 were designed as follows; for TIMP-1 5´-CTG GCT TCT GGC ATC CTG TTG TTG-3´ (sense, nt 81–104) and 5´-CAT GGC GGG GGT GTA GAC GAA-3´ (antisense, nt 329–09), for TIMP-2 5´-AAA GCG GTC AGT GAG AAG GAA GTG-3´ (sense, nt 412–435) and 5´-CCG GGG AGG AGA TGT AGC A-3´ (antisense, nt 778–760), for TIMP-3 outer primer pairs 5´-AGC TGG AGC CTG GGG GAC TG-3´ (sense, nt 315–334) and 5´-CTT GCG CTG GGA GAG GGT GAG-3´ (antisense, nt 677–657), and for TIMP-3 inner primer pairs 5´-CTC GCC CAG CCA CCC CCA GGA C-3´ (sense, nt 356–377) and 5´-AGC CCC GTG TAC

[a]Address for correspondence: 2nd Department of Internal Medicine, Nagoya University School of Medicine, 65 Tsurumai-cho, Showa-ku, Nagoya 466-8550, Japan. Phone, 81-52-741-2111(ext.2169); fax, 81-52-744-2178; e-mail, ykatano@tsuru.med.nagoya-u.ac.jp

ATC TTG CCA TCA-3′ (antisense, nt 628–605). Amplified products were separated by agarose gel electrophoresis or confirmed the sequences by direct sequencing method.

## RESULTS AND DISCUSSION

We studied the expression of TIMP-1, -2, or -3 gene in five hepatoma cell lines: HepG2, Huh7, PLC/PRF/5, HLE, and HLF. In all hepatoma cell lines, the expressions of TIMP-1 and -2 were detected under serum-free condition. In HepG2, Huh7, HLE, and HLF cell lines, TIMP-3 expression was also recognized under untreated condition. However, the detection of TIMP-3 gene needed second PCR (FIG. 1). Although the TIMP-3 gene was detected only by first PCR in HLE cell line, the intensity was weaker compared with TIMP-1 or -2 gene; and TIMP-3 expression was increased by stimulating with TNF-α or TPA in HLE cell line (FIG. 2). On the other hand, TIMP-3 expression in PLC/PRF/5 cell line was not detected either under untreated condition or by stimulating with TNF-α, but was recognized in the presence of TPA (FIG. 1). It

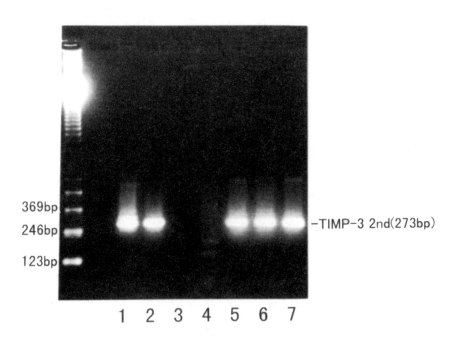

**FIGURE 1.** Expression of TIMP-3 gene in hepatoma cell lines. Second PCR products from hepatoma cell lines were applied. *Lane 1*, untreated HepG2; *lane 2*, untreated Huh7; *lane 3*, untreated PLC/PRF/5; *lane 4*, PLC/PRF/5 in presence of TNF-α; *lane 5*, PLC/PRF/5 in presence of TPA; *lane 6*, untreated HLE; *lane 7*, untreated HLF. *Size markers* are indicated to the *left* of the gel. The position of TIMP-3 band is shown to the *right* of the gel.

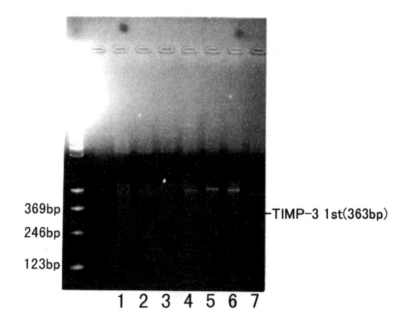

**FIGURE 2.** Analysis of TIMP-3 expression in hepatoma cell lines. First PCR products from hepatoma cell lines were applied. *Lane 1,* untreated HepG2; *lane 2,* untreated Huh7; *lane 3,* untreated PLC/PRF/5; *lane 4,* untreated HLE; *lane 5,* HLE in presence of TNF-α; *lane 6,* HLE in presence of TPA; *lane 7,* untreated HLF. *Size markers* are indicated to the *left* of the gel. The position of TIMP-3 band is shown to the *right* of the gel.

is possible that TIMP-3 plays a role regarding hepatoma cells, but further study, including its relationship with matrix metalloproteinases, is necessary.

## REFERENCES

1. CARMICHAEL, D.F., A. SOMMER *et al.* 1983. Primary structure and cDNA cloning of human fibroblast collagenase inhibitor. Proc. Natl. Acad. Sci. USA **83:** 2407–2411.
2. STETLER-STEVENSON, W.G., P.D. BROWN *et al.* 1990. Tissue inhibitor of metalloproteinase-2 (TIMP-2) mRNA expression in tumor cell lines and human tumor tissues. J. Biol. Chem. **265:** 13933–13938.
3. URIA, J.A., A.A. FERRANDO *et al.* 1994. Structure and expression in breast tumors of human TIMP-3, a new member of the metalloproteinase inhibitor family. Cancer Res. **54:** 2091–2094.
4. GREEN, J., M. WANG *et al.* 1996. Molecular cloning and characterization of human tissue inhibitor of metalloproteinase 4. J. Biol. Chem. **271:** 30375–30380.
5. BENYON, R.C., J.P. IREDALE *et al.* 1996. Expression of tissue inhibitor of metalloproteinases 1 and 2 is increased in fibrotic human liver. Gastroenterology **110:** 821–831.
6. NAKATSUKASA, H., K. ASHIDA *et al.* 1996. Cellular distribution of transcripts for tissue inhibitor of metalloproteinases 1 and 2 in human hepatocellular carcinoma. Hepatology **24:** 82–88.

# *In Situ* Activity of Gelatinases during Lewis Lung Carcinoma Progression

GORDON C. TUCKER,[a] GILLES FERRY, STÉPHANE PROVENT,
AND GHANEM ATASSI

*Institut de Recherches Servier, Department of Experimental Oncology,
11 rue des Moulineaux, 92150 Suresnes, France*

In recent years, a number of techniques, including immunohistochemical analyses of tissues from experimental or clinical tumor samples, perturbation experiments with endogenous or pharmacological inhibitors, and knockout mouse models all point to the contribution of matrix metalloproteinases (MMPs) to tumor progression. However, MMPs require activation for extracellular activity, and endogenous inhibitors can bind to the active forms; immunohistochemistry with the majority of available antibodies does not allow distinction between these pro-, activated, and inhibited MMPs *in situ*. Thus, it is not clear when, where, or to what extent MMPs are capable of activity *in situ*. This information is crucial in order to understand the exact mechanism of action of pharmacological MMP inhibitors (MMPIs) in terms of putative targets. Therefore, we assayed the activity status of gelatinases *in situ* in a mouse syngenic tumor model sensitive to MMPIs. Following intramuscular injection of Lewis lung carcinoma cells (LLC), a highly vascularized tumor is produced with concomitant local invasion and lung metastases formation, thus reproducing crucial steps thought to involve MMPs. Treatment with some MMPIs can lead to a significant, although modest, decrease of the primary LLC tumor growth, and to various effects on the spontaneous pulmonary metastases.[1–3]

As revealed by zymography, in contrast to normal muscle extracts displaying only low amounts of MMP-2 (72-kDa gelatinase A) and absence of MMP-9 (92-kDa gelatinase B), extracts from the primary LLC tumor contained all of pro-, intermediate, and activated MMP-2 and MMP-9 forms (FIG. 1). In addition, only the tumor extracts were able to cleave native type IV collagen, and this activity was specifically inhibited by the MMPI Ro-31-9790,[4] thus demonstrating the presence of at least one subgroup of uninhibited activated MMPs, the gelatinases. We next investigated the localization of MMP activities directly on histological frozen sections, by following the emergence of fluorescence *in situ* after degradation of an initially quenched fluorescent pseudopeptide substrate of MMPs. By using image analysis and the MMPI Ro-31-9790, we demonstrated the presence of an intense MMP-dependent signal in the LLC primary tumor, whereas tissues distant from the tumor or muscle from control animals exhibited low levels of fluorescence (FIG. 2, upper panels). Occasionally, brightly reactive areas were observed at the front of tumor cell invasion into the muscle layers and in the stroma around the tumor (FIG. 2, lower panels).

[a]Address for correspondence: Dr. Gordon C. Tucker, Institut de Recherches Servier, Department of Experimental Oncology, 11 rue des Moulineaux, 92150 Suresnes, France. Phone, 33 1 41 18 25 00; fax, 33 1 41 18 24 40.

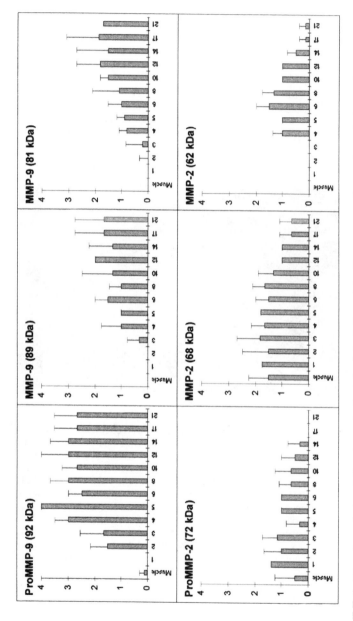

**FIGURE 1.** Time-dependent variations in production of gelatinolytic proteins from extracts of control muscle (tumor-free animals) and of LLC tumors. *Numbers on the abscissa* indicate time (in days) after intramuscular injection of tumor cells. *Numbers on the ordinate axis* correspond to arbitrary units reflecting the amount of gelatinase produced as inferred from visual estimation of the clear lysis band on gelatin-containing gels after SDS-PAGE zymography. Equal amounts of total tissue mass were loaded on each lane of the zymograms. Results are the means ± SEM of four individual experiments carried out with two animals per condition. Three distinct species of each gelatinase (MMP-2 and MMP-9) were detected in the extracts: a proform and the intermediate and active forms *(from left to right)*.

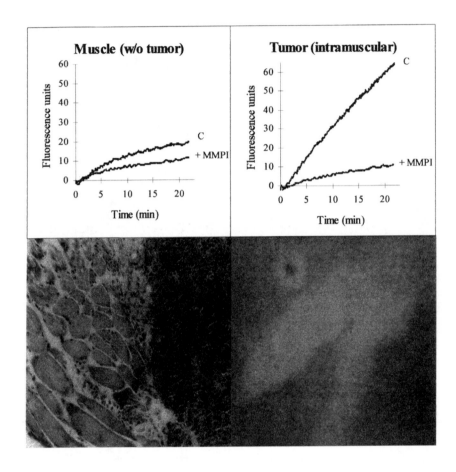

**FIGURE 2.** Fluorescence of the reporter substrate for active MMPs. The substrate Dnp-Pro-β-cyclohexyl-Ala-Gly-Cys(Me)-His-Ala-Lys(N-Me-Abz)-NH$_2$ (Bachem, Switzerland), which is cleaved between amino acids Gly and Cys,[6] was used to reveal activated MMPs *in situ*. Ten-micrometer-thick frozen sections of tumor-free muscle (*upper left*) or primary LLC tumor samples (*upper right*) were placed between two coverslips in 100 μl of 50 mM Tris, 200 mM NaCl, 5 mM CaCl$_2$, 0.1% Brij35, pH 7.5 buffer containing 100 μM of the substrate in the presence (+MMPI) or not (C) of 100 μM of Ro-31-9790. Time-lapse fluorescence imaging was performed at room temperature and high magnification (×20 objective) on a Nikon microscope equipped with fluorescence filters (excitation 340 nM, emission 440 nm) coupled to a CCD camera linked to an image analysis system (Magiscan station from Joyce Loebl equipped with the Magical software). Images were then captured at lower magnification at the level of areas corresponding to the tumor-muscle interface (*lower left panel*, hematoxylin and eosin staining of a paraffin section of a tumor sample eight days after intramuscular implantation, muscle on the *left*). An intense fluorescence is occasionally observed on frozen sections (*lower right panel*) at this interface containing invading tumor cells, fibroblastic cells, and neutrophils (the muscle layers are on the *upper left corner* and the tumor mass, separated by stromal tissue, is located *below*).

In conclusion, these findings demonstrate the presence of active and uninhibited MMPs, in particular in restricted stromal areas, during the invasive phase of LLC tumor progression. The use of more specific substrates for revealing the activity and of more selective MMPIs will clarify whether or not the corresponding MMPs, active *in situ*, are gelatinases. In addition, analysis by fluorescence of the electrophoretic patterns of tissue samples in the presence of an overlay containing the substrate should allow a more correct identification (in terms of molecular weights) of the active MMPs revealed *in situ* in the species of interest. The methodology presented permits a rapid evaluation of MMP activity *in situ* in less than an hour as compared to at least a day for *in situ* gelatin zymography with photographic emulsion or for detection of enzymatic activity using fluorescent casein or gelatin.[5] It would also be interesting to inject small fluorogenic substrates *in vivo* in or near the tumors and assay local *in situ* activity with confocal and intravital fluorescence microscopy.

## REFERENCES

1. ANDERSON, I.A. *et al.* 1996. Combination therapy including a gelatinase inhibitor and cytotoxic agent reduces local invasion and metastasis of murine Lewis lung carcinoma. Cancer Res. **56:** 715–718.
2. CONWAY, J.G. *et al.* 1996. Effect of matrix metalloproteinase inhibitors on tumor growth and spontaneous metastasis. Clin. Exp. Metastasis **14:** 115–124.
3. BULL, C. *et al.* 1998. Activity of the biphenyl matrix metalloproteinase inhibitor BAY 12-9566 in murine *in vivo* models [Abstr. #2062]. Proc. Am. Assoc. Cancer Res. **39:** 302.
4. BROADHURST, D.J. *et al.* 1992. European Patent Application, EP-497, 192-A.
5. GALIS, Z.S. *et al.* 1995. Microscopic localization of active proteases by in situ zymography : detection of matrix metalloproteinase activity in vascular tissue. FASEB J. **9:** 974–980.
6. BICKETT, D.M. *et al.* 1993. A high throughput fluorogenic substrate for interstitial collagenase (MMP-1) and gelatinase (MMP-9). Anal. Biochem. **212:** 58–64.

# Targeting Matrix Metalloproteinases in Human Prostate Cancer

QINGXIANG AMY SANG,[a,b] MARTIN A. SCHWARTZ,[a] HUI LI,[a]
LELAND W.K. CHUNG,[c] AND HAIYEN E. ZHAU[c]

[a]*Department of Chemistry, Florida State University, Tallahassee, Florida 32306, USA*

[c]*Department of Urology, University of Virginia, Charlottesville, Virginia 22908, USA*

## INTRODUCTION

The natural history of human prostate cancer progression has its origin as multi-focal lesions from the peripheral zone, extends through the prostatic capsule into the seminal vesicles, and disseminates to the lymph nodes, the bone, and the visceral organs.[1] Identification of new markers for diagnosis and novel targets for therapeutic intervention are critical steps in the process of eradicating prostate cancer. To metastasize, prostatic cancer cells must break down the basement membrane underneath the epithelial prostate cancer cells and the endothelial cells of blood vessels using matrix-degrading proteinases. Matrix metalloproteinases (MMPs) are a family of hydrolases that dissolve the macromolecules of connective tissues such as collagens, laminins, and fibronectin.[2] MMPs are important molecules used by invading cells to facilitate invasive growth and spread.

## MATRIX METALLOPROTEINASES ARE PRODUCED BY PROSTATE CANCER CELL LINES

Androgen-independent prostate carcinoma cell lines C4-2 and ARCaP were derived from men with metastatic prostate cancers.[1] The androgen-repressed human prostate carcinoma cell line (ARCaP) resembles closely the behaviors of the advanced stage of human prostate cancer. We have examined MMPs produced by those cell lines.[1] Gelatin and casein zymograms and a continuous fluorimetric MMP assay with a synthetic substrate are used to detect the MMP activity. Large amounts of latent and active forms of the gelatinase A (MMP-2) and gelatinase B (MMP-9), and some other MMPs are produced by ARCaP cells.[1] Moderate amounts of MMP-2 and MMP-9 were also produced by the C4-2 cells. MMPs are present predominantly in the cell culture media. Membrane type 1 matrix metalloproteinase was also detected in the plasma membrane enriched fraction (data not shown).

[b]Address for correspondence: Professor Q.X. Amy Sang, Department of Chemistry, 203 Dittmer Laboratory of Chemistry Building, Florida State University, Tallahassee, Florida 32306-4390. Phone, 850/644-8683; fax, 850/644-8281; e-mail, sang@chem.fsu.edu

## INHIBITION OF MMP ACTIVITIES BY
## RATIONALLY DESIGNED MMP INHIBITORS

Novel and potent MMP inhibitors have been designed, synthesized, and characterized by Drs. M.A. Schwartz and H.E. Van Wart's research groups[3,4] (FIG. 1 and TABLE 1). The inhibition of prostate cancer MMP activities by four of those inhibitors are tested. The total net MMP activities in C4-2 and ARCalP cell media were determined, and the $IC_{50}$ values ($IC_{50}$, the inhibitor concentration that inhibits 50% of the enzyme activity) of four MMP inhibitors against the MMP activities in the media were measured using a fluorescence substrate (TABLE 1). As shown in FIGURE 1 and TABLE 1, MAG-182 (a sulfhydryl inhibitor) and YLL-224 (a sulfodiimine inhibitor) have the highest inhibitory activity against MMPs that are present as crude mixtures in the culture media obtained from the ARCaP and C4-2 cells, respectively. ARCaP culture medium contained 20-fold higher net MMP activities than that of C4-2 culture medium. The net MMP activity in culture medium correlates with the metastatic behavior of the prostate cancer cell line *in vivo*. The effects of MMP inhibitors on the growth of C4-2 and ARCaP cell lines were tested. $IC_{50}$ values of MAG-181 on C4-2 cells and YLL-224 on ARCaP cells were found to be 220 nM and 74 nM, respectively. Other inhibitors tested, MAG-182, YLL-49, and the control compound N-acetylcysteine were ineffective up to 10 μM. No visible cytotoxicity was detected up to 50 μM of the inhibitors. Preliminary study shows that MAG-182 (30 mg/kg) has growth-retarding activity against ARCaP tumor in athymic nude mice (data not shown).

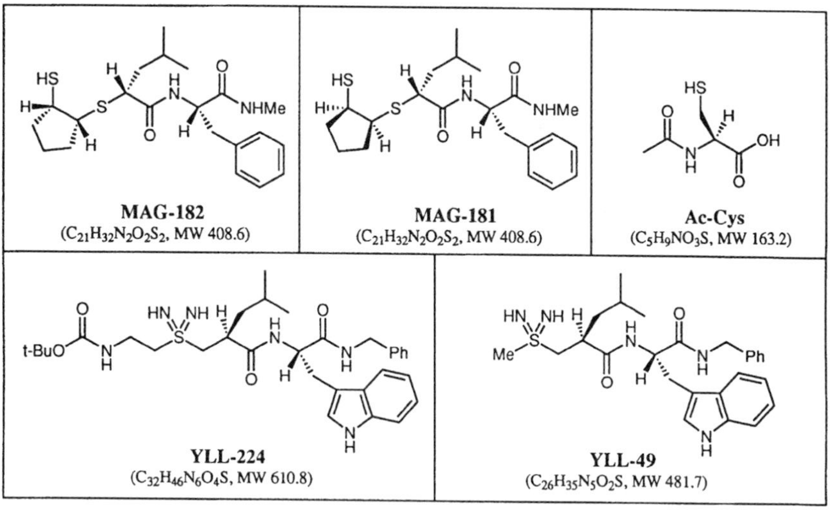

**FIGURE 1.** Structures of synthetic matrix metalloproteinase inhibitors. The structures, formula, and molecular weights of two sulfhydryl inhibitors (MAG-182 and MAG-181) and two sulfodiimine inhibitors (YLL-224 and YLL-49), as well as a control compound N-acetylcysteine (Ac-Cys) are shown.

**TABLE 1. Summary of IC$_{50}$ values (in nM) of four synthetic matrix metalloproteinase inhibitors (MAG-182, MAG-181, YLL-224, YLL-49) and a control compound (Ac-Cys) on five purified matrix metalloproteinases (HNC, HFC, HNG, HFG, HFS, MLN) and two conditioned cell culture media from two prostate cancer cell lines (C4-2 and ARCaP) in the absence/presence of 5α-dihydrotestosterone$^a$**

| Inhibitor | Purified enzymes | | | | | | Cell culture media | | | |
|---|---|---|---|---|---|---|---|---|---|---|
|  | HNC | HFC | HNG | HFG | HFS | MLN | C− | C+ | A− | A+ |
| MAG-182 | 0.89 | 49 | 0.72 | 1.1 | 470 | 56 | 890 | 3200 | 50 | 170$^c$ |
| MAG-181 |  | 680 | 44 | 85 | 2500 | 680 | 1200 | 1900 | 320 | 1300$^c$ |
| Ac-Cys |  |  | >9 mM | >9 mM |  |  | 4 mM$^c$ | 8 mM | 2 mM$^c$ | 1 mM$^c$ |
| YLL-224 | 5.9 | 180 | 44 | 63 | 4500 | 210 | 330$^c$ | 240 | 300 | 200$^c$ |
| YLL-49 | 21 | 320 | 200 | 130 | 14000 | 1300 | 380$^c$ | 370 | 440 | 500$^c$ |

$^a$See References 3 and 4 for details on enzyme sources and assay methods.
$^b$The cell culture media were concentrated 5–8 fold by Amicon Centriprep-10.
$^c$Assays did not go to 0% activity, even at high inhibitor concentrations.
ABBREVIATIONS: HNC, human neutrophil collagenase, matrix metalloproteinase-8, MMP-8; HFC, human fibroblast collagenase, MMP-1; HNG, human neutrophil gelatinase B, MMP-9; HFS, human fibroblast stromelysin, MMP-3; MLN, matrilysin, MMP-7; C-/C+, conditioned serum-free media from C4-2 cell culture in the absence/presence of 100 pM 5α-dihydrotestosterone; A-/A+, conditioned serum-free media from ARCaP cell culture in the absence/presence of 100 pM 5α-dihydrotestosterone.

## ACKNOWLEDGMENTS

This work was supported in part by grants from the Elsa U. Pardee Foundation and from NIH, 1R29CA78646 to QX.A.S.; and by NIH grants CA56307/DK47596 to L.W.K.C. and CA57361 to HY.E.Z. We thank Drs. Mohammad A. Ghaffari and Yi-Lin Luo for synthesizing the metalloproteinase inhibitors; Ms. Meiqin Chen and Mr. Delbert D. Bauzon for their excellent technical assistance; and Drs. Harold E. Van Wart, Henning Birkedal-Hansen, and Hideaki Nagase for their reagents and support.

## REFERENCES

1. ZHAU, H.E., S.-M. CHANG, B.-Q. CHEN, Y. WANG, H. ZHANG, C. KAO, Q.A. SANG, S.J. PATHAKS & L.W.K. CHUNG. 1996. Androgen-repressed phenotype in human prostate cancer. Proc. Natl. Acad. Sci. USA **93:** 15152–15157.
2. SANG, Q.A. & D.A. DOUGLAS. 1996. Computational sequence analysis of matrix metalloproteinases. J. Protein Chem. **15:** 137–160.
3. SCHWARTZ, M.A. & H.E. VAN WART. 1995. Mercaptosulfide metalloproteinase inhibitors. U.S. Patent 5455262.
4. SCHWARTZ., M.A. & H.E. VAN WART. 1995. Sulfoximine and sulfodiimine matrix metalloproteinase inhibitors. U.S. Patent 5470834.

# Expression of Matrix Metalloproteinases in Gastric Carcinoma and Possibility of Clinical Application of Matrix Metalloproteinase Inhibitor *in Vivo*

YOSHIHIDE OTANI,[a,b] TETSURO KUBOTA,[a] YOSHIHIKO SAKURAI,[a] NAOKI IGARASHI,[a] TAKEYOSHI YOKOYAMA,[a] MASARU KIMATA,[a] NORIHITO WADA,[a] KAORI KAMEYAMA,[c] KOICHIRO KUMAI,[a] YASUNORI OKADA,[c] AND MASAKI KITAJIMA[a]

*Departments of [a]Surgery and [c]Pathology, School of Medicine, Keio University, Tokyo, Japan*

## INTRODUCTION

Peritoneal dissemination of carcinoma cells is the cause of poor prognosis of patients with T3-4 gastric carcinoma. The mechanism of spread of carcinoma cells through the gastric wall and formation of peritoneal dissemination nodules is in urgent need of analysis. The extracellular space of the stomach wall is composed of a number of macromolecules. Of these, collagen types I and III compose the major structural element of the extracellular matrix. Although MMP-1 is capable of degrading these collagens, little is known about the role of MMP-1 in invasion and metastasis of carcinoma. In order to investigate the expression of the MMP-1 gene in carcinoma tissue, an *in situ* hybridization study was carried out. Therapeutic effect of the MMP inhibitor R-94138 was also evaluated using a peritoneal dissemination model of nude mice.

## MATERIALS AND METHODS

*Samples.* Tissue samples were obtained from surgically resected clinical specimens.

*In Situ Hybridization.* The hybridization procedure employed in this study was essentially the same as that described previously.[1] The hybridization was performed using [35]S-labeled MMP-1 cDNA probe at 45°C for 16 hr. After hybridization, the slides were dipped into emulsion, exposed, developed, and then stained with hematoxylin.

*Nude Mice Peritoneal Dissemination Model.* A gastric carcinoma cell line, TMK-1 ($5 \times 10^5$/body) was injected intraperitoneally into Bulb/c nude mice (male, 5 months). Subsequently, R-94138 (Sankyo Company, Japan) (30 mg/kg) was ad-

---

[b]Address for correspondence: Yoshihide Otani, M.D., Department of Surgery, School of Medicine, Keio University, 35 Shinanomachi, Shinjuku-ku, Tokyo, 160-8582 Japan. Phone, +81-3-3353-1211 ext. 62334; fax, +81-3-3355-4707; e-mail, otaniy@med.keio.ac.jp

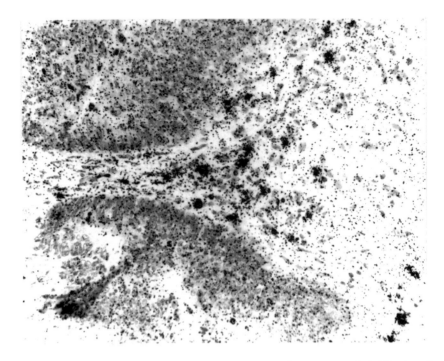

**FIGURE 1.** MMP-1 mRNA expression in gastric carcinoma. *In situ* hybridization of MMP-1 mRNA reveals that MMP-1 mRNA is expressed in the stromal cells, including fibroblasts, macrophages, and various inflammatory cells. However, the signal is not found in carcinoma cells. Original magnification × 240, reduced by 75%.

ministered into the mice. They were sacrificed 5 weeks after the injection, and therapeutic efficacy was evaluated by counting the number of peritoneal nodules.

*MMP Inhibitor.* Synthetic MMP inhibitor R-94138 is structurally similar to matlystatin, which is isolated from *Actinomadura atramentaria*.[2]

## RESULTS

*MMP-1 Expression in Gastric Carcinoma.* Overexpression of MMP-1 was observed in stromal cells, including fibroblasts, macrophages, and various inflammatory cells, which were closely associated with carcinoma nests, especially at the margin of invasion of carcinoma tissue. However, signal was not found within carcinoma cells (FIG. 1).

*Preventive Effect of R-94138 for the Formation of Peritoneal Dissemination Nodules.* R-94138 has shown the inhibitory effect for the formation of peritoneal dissemination nodules in nude mice ($28.0 \pm 11.9$ nodules/body vs. $12.3 \pm 1.7$ nodules/body, $p < 0.05$) (TABLE 1).

**TABLE 1. Inhibitory effect of R-94138 for the formation of peritoneal nodules**

| Treatment | n | Nodules/body |
|---|---|---|
| Control | 4 | $28.0 \pm 11.9^a$ |
| R-94138 (30 mg/body) | 4 | $12.3 \pm 1.7$ |

$^a$Mean $\pm$ SD, $p < 0.05$.

## DISCUSSION

Interaction between carcinoma cells and stromal cells seems one of the key path-ways of MMP-1 production. It is speculated that carcinoma cells produce several stimuli that induce the gene expression of MMPs in the stromal cells, and that the stromal cell–derived MMPs are activated at the surface of carcinoma cells, which causes degradation of extracellular matrix and tumor invasion. The preventive effect of R-94138 for peritoneal dissemination may give hope for the cure of patients with T3-4 gastric carcinoma.

## REFERENCES

1. OTANI, Y. *et al.* 1994. Gene expression of interstitial collagenase (matrix metallopro-teinase 1) in gastrointestinal tract cancers. J. Gastroenterol **29:** 391–397.
2. TAMAKI K. *et al.* 1995. Synthesis and structure-activity relationships of gelatinase inhibitors derived from matlystatins. Chem. Pharmacol. Bull. **43:** 1883–1893.

# Metalloproteinases and Tissue Inhibitors of Matrix-Metalloproteinases in Plasma of Patients with Prostate Cancer and in Prostate Cancer Tissue

MICHAEL LEIN,[a] LARS NOWAK, KLAUS JUNG, CHRISTIAN LAUBE, NORBERT ULBRICHT, DIETMAR SCHNORR, AND STEFAN A. LOENING

*Department of Urology, University Hospital Charité,*
*Humboldt University, Berlin, Germany*

## OBJECTIVE

Matrix metalloproteinases (MMP) form a group of enzymes with the common ability to degrade various components of the extracellular matrix (collagen, elastin, gelatin). It could be shown that increased levels of MMPs are associated with the invasive and metastatic potential in several human malignant tumors—e.g., in breast, colon, lung cancer.[1] Low TIMP expression correlates with enhanced invasive properties of human tumors. MMPs are controlled by various mechanisms (enzyme synthesis, activation, TIMP). The balance between MMPs and TIMPs as both positive and negative modulators of the invasive and metastatic processes appears to be decisive.[2] MMPs and TIMPs in blood have been recommended as diagnostic markers. The objectives were to evaluate their potential usefulness in patients with prostate cancer (PCa) and to prove the biological significance of MMPs and TIMPs in PCa both in blood and tissue.

## DESIGN AND METHODS

MMP-1, MMP-3, TIMP-1 and the complex MMP-1/TIMP1 were measured in plasma by ELISA tests (BIOTRAK™, Amersham Int., UK) in healthy controls ($n = 35$), in patients with benign prostate hyperplasia (BPH; $n = 29$), and in patients with PCa without metastasis ($n = 29$) and with metastasis ($n = 18$). Plasma–MMP-2 determination was performed (BIOTRAK, Amersham Int., UK) in 38 healthy men, 30 BPH-patients, and in PCa-patients without metastasis ($n = 54$) and with metastasis ($n = 31$). Prostate tissue samples ($n = 9$) were obtained from the cancerous and noncancerous parts of the same prostates (homogenization and extraction with 0.25% Triton X-100 and heat at 60°C). Total MMP activity and TIMP-1 in prostate tissue were determined with a continuous fluorimetric assay (Spectrofluorimeter LS 50B, Perkin-Elmer, Germany) and the ELISA-test of BIOTRAK, respectively.

[a]Address for correspondence: Michael Lein, M.D., Department of Urology, University Hospital Charité, Humboldt University, Schumannstr. 20/21, 10098 Berlin, Germany. Phone, ++49 30/2802 4714; fax, ++49 30/2802 5615.

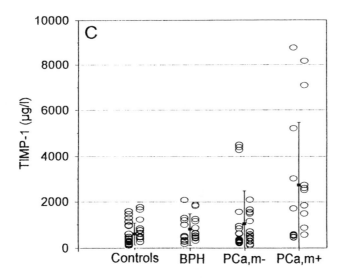

**FIGURE 1.** Plasma concentration of TIMP-1 in controls, in patients with benign prostate hyperplasia (BPH) and in patients with prostate cancer (PCa) without metastasis (PCa, m−) and with metastasis (PCa, m+).

**TABLE 1. MMP activity and TIMP-1 value in human prostate**

|  | Normal tissue | Tumor tissue |
|---|---|---|
| MMP |  |  |
| MU/g wet tissue | 88.8 (30.7–173) | 50.8 (41.4–86.1)* |
| MU/g protein | 2497 (984–5524) | 1580 (1127–2447)* |
| TIMP-1 |  |  |
| μg/g wet tissue | 12.4 (5.05–57.8) | 4.49 (1.23–17.1)* |
| μg/g protein | 237.8 (104–1248) | 96.7 (25.7–411.8)* |
| MMP/TIMP |  |  |
| Wet tissue basis | 4.0 (1.4–30.6) | 11.1 (2.5–47.9)* |
| Protein basis | 5.3 (1.9–40.9) | 15.5 (3.4–60.9)* |

*Significant differences.

## RESULTS

The mean values of MMP-1 and complex MMP-1/TIMP-1 were not different among the four groups studied. The mean MMP-3 and TIMP-1 concentrations were significantly higher in PCa-patients with metastases compared to controls, BPH, and PCa patients without metastases. Ten of these 18 patients had TIMP-1 concentrations higher than the upper 95% reference limit (FIG. 1). The correlation between MMP parameters and staging, grading, and PSA were calculated. TIMP-1 concentrations correlate with tumor staging (Spearman rank correlation: 0.52). PCa patients without and with metastasis were not characterized by higher MMP-2 concentrations in blood.

The median values of MMPs and TIMP-1 on wet weight basis and on protein basis were significantly lower in cancerous tissue samples than in their normal counterparts. The ratio MMP/TIMP-1 on wet weight basis and on protein basis was significantly higher in cancerous tissue samples than in the normal counterparts (TABLE 1).

## CONCLUSION

MMP-3 and TIMP-1 were significantly increased in plasma of patients with advanced cancer. Results point towards plasma TIMP-1 concentration as a potential marker for progression of PCa. PCa patients were not characterized by higher MMP-2 concentration in blood. MMP and TIMP tissue extraction procedure was demonstrated as a suitable quantitative method. The decreased ratio can be interpreted as an indicator of the imbalance between MMP and TIMP. Imbalance could be characteristic of carcinoma tissue.

### REFERENCES

1. MURRAY, G.I. et al. 1996. Matrix metalloproteinase-1 is associated with poor prognosis in colorectal cancer. Nature 2: 461–462.
2. ANDERSON, I.C. et al. 1995. Stromelysin-3 is overexpressed by stromal elements in primary non-small cell lung cancers and regulated by retinoic acid in pulmonary fibroblasts. Cancer Res. 55: 4120–4126.

# MMP Inhibition Reduces Intimal Hyperplasia in a Human Vein Graft Stenosis Model

I.M. LOFTUS,[a] K. PORTER, M. PETERSON, J. BOYLE, N.J.M. LONDON,
P.R.F. BELL, AND M.M. THOMPSON

*Department of Surgery, Leicester University, Leicester, UK*

## INTRODUCTION

More than 30% of vein bypass grafts develop stenoses in the first year after operation.[1] The underlying pathological process is intimal hyperplasia, characterized by the early migration and proliferation of smooth muscle cells.[2] This is dependent upon the degradation of the extracellular matrix by matrix metalloproteinases (MMPs), secreted predominantly by the smooth muscle cells. In particular, gelatinase A and B (MMP-2 and MMP-9) efficiently degrade type IV collagen, the major structural component of the basement membrane.

Previous animal studies have shown that MMP inhibition prevents the migration and proliferation of smooth muscle cells. Although work using human tissue has been limited, we have shown, using a well-validated organ culture model of human intimal hyperplasia,[3,4] that increased production of MMP-9 occurs during neointima formation.[5]

The hypothesis in this study was that raised levels of MMPs may play a significant role in intimal hyperplasia and that inhibition of these enzymes may prevent intimal hyperplasia and offer a potential therapeutic strategy for the prevention of vein graft stenoses. The aim was to investigate whether marimastat (a nonselective MMP inhibitor) or doxycycline (an antibiotic of the tetracycline family known to inhibit MMPs) reduce neointima formation in cultured long saphenous vein.

## METHODS

Segments of long saphenous vein were obtained from consecutive patients undergoing arterial bypass grafting for critical ischemia and prepared for culture as previously described.[3] Vein was divided equally, and individual segments were cultured in control conditions or in medium supplemented with marimastat or doxycycline at therapeutic concentrations ($10^{-6}$ mol/l and 10 μg/ml, respectively) based on the mean plasma trough levels achieved *in vivo*. Culture medium and drug was replaced every 2 days, and at the end of the 14-day period the segments were divided into two equal halves for:

*Measurement of Neointimal Thickness.* Vein was fixed in 4% paraformaldehyde for 18 hours, embedded in paraffin, and cut in 4-μm sections. Paraffin sections were

[a]Address for correspondence: Mr. Ian Loftus, The University Department of Surgery, Clinical Sciences Building, Leicester Royal Infirmary, P.O. Box 65, Leicester LE2 7LX, UK. Phone, 00 44 116 252 2140; fax, 00 44 116 252 3179; e-mail, ianloftus@aol.com

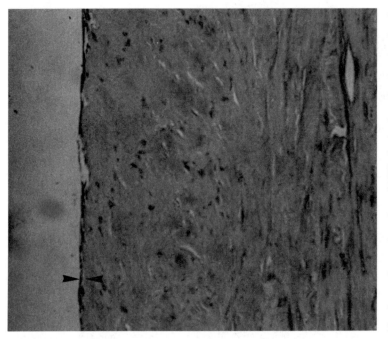

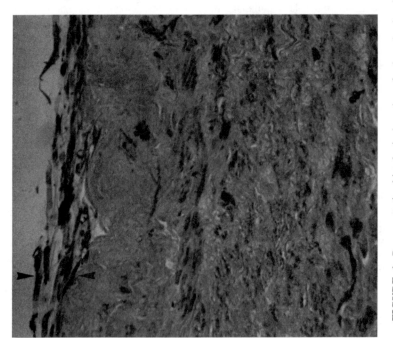

**FIGURE 1.** Representative histological section of cultured vein with a combined smooth muscle actin and Miller's elastin stain, showing development of neointima (*arrows*). Vein segments treated with marimastat (*right*) and doxycycline at a therapeutic concentration developed significantly less neointima than control veins (*left*).

stained using a monoclonal anti–smooth muscle actin superimposed with Miller's elastin stain to identify the layers of the vein wall. Measurements of neointimal thickness were made using a computerized image analysis system (Improvision, Coventry, UK), taking a minimum of 30 measurements across the entire length of two consecutive sections of each vein segment.

*Quantification of Gelatinase Levels.* Tissue was homogenized, dialyzed and protein standardized prior to gelatin zymography. The levels of MMP-2 and MMP-9 were determined by densitometric analysis of the bands detected on gelatin zymography, and, to allow for variation between gels, presented as a ratio to a positive control (HT-1080 sarcoma cell line).

Results are expressed as median and 95% confidence intervals. Differences between groups were analyzed using the Wilcoxon paired rank test with significance assumed at the 95% level.

## RESULTS

FIGURE 1 shows representative histological sections of vein cultured with and without marimastat. During the 14-day culture period, all control veins developed a significant neointima with a median thickness of 22 μm (range 17–29 μm). Both marimastat and doxycycline significantly attenuated neointima formation (median 4.5/5.5 μm, range 0–8/3–11 μm, $p = 0.006$ and $0.02$, respectively).

Gelatin zymography revealed that the levels of MMP-9 and MMP-2 were significantly reduced in vein cultured with marimastat ($p < 0.01$ and $0.05$, respectively). Doxycycline significantly reduced the level of MMP-9 ($p < 0.05$), but the observed reduction in MMP-2 failed to reach significance (FIG. 2).

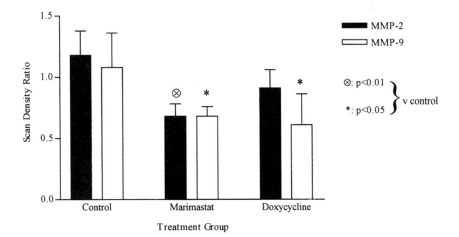

**FIGURE 2.** Protease activity of vein homogenates cultured in control conditions or with medium supplemented with marimastat or doxycycline. Results are the median (95% CI) of the scan density as a ratio to an HT-1080 control.

## CONCLUSION

This study has demonstrated significant inhibition of neointima formation by both marimastat and doxycycline in parallel with reduced gelatinase activity. The pharmacological profile and bioavailability make both attractive for clinical trials investigating the effect of MMP inhibition on the development of vein graft stenoses.

## ACKNOWLEDGMENTS

The authors thank British Biotechnology Pharmaceuticals Ltd. (Oxford, UK) for the marimastat. Mr. I.M. Loftus is a Royal College of Surgeons of England Research Fellow.

## REFERENCES

1. VARTY, K., K. ALLEN, P.R.F. BELL & N.J.M. LONDON. 1993. Infrainguinal vein graft stenosis. Br. J. Surg. **80:** 825–833.
2. CLOWES, A.W. & S.M. SCHWARTZ. 1985. Significance of quiescent smooth muscle migration in the injured rat carotid artery. Circ. Res. **56:** 139–145.
3. PORTER, K.E., K. VARTY, L. JONES, P.R.F. BELL & N.J.M. LONDON. 1996. Human saphenous vein organ culture: a useful model of intimal hyperplasia? Eur. J. Vasc. Endovasc. Surg. **11:** 48–58.
4. ALLEN, K.E., K. VARTY, L. JONES, R.D. SAYERS, P.R.F. BELL & N.J.M. LONDON. 1994. Human venous endothelium can promote intimal hyperplasia in a paracrine manner. J. Vasc. Surg. **19:** 577–584.
5. PORTER, K.E., M.M. THOMPSON, I.M. LOFTUS, E. MCDERMOTT, L. JONES, M. CROWTHER et al. 1998. Production and inhibition of the gelatinolytic matrix metalloproteinases in a human model of vein graft stenosis. Eur. J. Vasc. Endovasc. Surg. In press.

# Increased MMP-9 Activity in Acute Carotid Plaques: Therapeutic Avenues to Prevent Stroke

I.M. LOFTUS,[a] S. GOODALL, M. CROWTHER, L. JONES, P.R.F. BELL, A.R. NAYLOR, AND M.M. THOMPSON

*Department of Surgery, Leicester University, Leicester, UK*

## INTRODUCTION

It is now generally accepted that acute changes in atherosclerotic plaques are a prelude to the onset of clinical syndromes such as transient ischemic attacks, stroke,[1] angina, or myocardial infarction.[2] Each phase of the atherosclerotic process involves modification of the extracellular matrix, with the integrity of the plaque depending upon the relative degradation and synthesis of the individual matrix components. Unstable plaques, at risk of acute disruption and subsequent thrombosis, are characterized by an inflammatory infiltration and increased matrix degrading activity in their most vulnerable regions.[3] The major physiological regulators of the ECM are the matrix metalloproteinases (MMPs), and release of these enzymes, particularly by inflammatory cells, has been suggested as a mechanism for plaque disruption.[4]

The hypothesis in this study was that a localized imbalance of MMPs within carotid plaques may be responsible for acute plaque disruption and the onset of ischemic events. Should MMP activity be specifically related to acute plaque changes, then it may be possible to modify plaques by the administration of an MMP inhibitor. The aim was to establish the character, nature, expression, and production of MMPs in carotid plaques and to correlate MMP activity with clinical status, spontaneous embolization on transcranial Doppler (indicative of plaque instability), plus histological evidence of plaque disruption.

## MATERIALS AND METHODS

A consecutive series of 75 patients undergoing carotid endarterectomy (CEA) were entered into this study. All patients underwent a thorough preoperative assessment including a detailed history to establish the nature and duration of ischemic events and ultrasound assessment of the carotid plaque. Patients were allocated to one of four symptom groups (group 1 = asymptomatic; group 2 = no symptoms for > 6 months; group 3 = symptoms between 1 and 6 months preoperatively; group 4 = symptoms within 1 month of endarterectomy). All patients underwent preoperative

[a]Address for correspondence: Mr. Ian Loftus, The University Department of Surgery, Clinical Sciences Building, Leicester Royal Infirmary, P.O. Box 65, Leicester LE2 7LX, UK. Phone, 00 44 116 252 2140; fax, 00 44 116 252 3179; e-mail, ianloftus@aol.com

MCA monitoring with transcranial Doppler for 30 minutes to identify those with ongoing microembolization.

Carotid plaques, plus control tissue from a nondiseased part of the proximal common carotid artery and serum, were obtained from all patients at the time of surgery. Plaques were divided into three longitudinal portions and:

1. fixed in 4% paraformaldehyde prior to histological examination,
2. homogenized for quantification of MMP levels, and
3. homogenized for RNA extraction and quantification of MMP expression.

*MMP Activity.* The activity of the major MMP subtypes and their naturally occurring inhibitors (TIMPs) were quantified using a combination of gelatin zymography (followed by densitometric analysis and comparison to HT-1080 sarcoma cell line as positive control), plus ELISA (Amersham, UK). To determine the site of MMP activity histological sections were stained for the major MMP subtypes and inhibitors.

*MMP Production and Expression.* Following extraction of messenger RNA from plaque homogenate using coated magnetic beads, the reverse transcription polymerase reaction was performed to produce cDNA which was amplified and quantified. To establish the site of MMP production, *in situ* hybridization of paraformaldehyde fixed tissue was performed using digoxigenin-labeled oligonucleotide probes.

*Plaque Histology.* All plaques were fixed and paraffin embedded prior to sectioning and staining. An independent, experienced histopathologist graded each plaque as to the presence of plaque rupture, cap thinning, intraplaque hemorrhage, thrombus, necrosis, and other complex features.

## RESULTS

The allocation of all patients to one of four symptom groups was as follows: group 1 ($n = 20$), group 2 ($n = 16$), group 3 ($n = 18$), group 4 ($n = 21$). There were no significant differences between the symptom groups in age, sex, smoking habit, or the other independent vascular risk factors. The incidence of plaque rupture was significantly higher in group 4 (52% vs. <20% for the other three groups, $p < 0.05$), as was the rate of spontaneous embolization (57% vs. <20% for the other groups, $p < 0.01$), justifying the allocation to symptom groups.

Gelatin zymography detected significantly higher levels of both active and latent MMP-9 in the most highly symptomatic plaques compared to the other three groups (FIG. 1; $p = 0.001$ for both forms, Kruskall Wallis). There was no difference in the MMP-2 activity. These results were confirmed by ELISA, which demonstrated a fourfold increase in the concentration of MMP-9 in group 4 compared to the other three groups (median 128 ng/ml vs. < 35 ng/ml for the other groups). This was highly significant on statistical analysis ($p = 0.003$, Kruskall Wallis).

Furthermore, the level of MMP-9 was significantly higher in those plaques with evidence of ongoing spontaneous embolization detected on TCD monitoring and plaques with histological evidence of plaque rupture ($p = 0.019$ and 0.03, respectively, Mann Whitney U-Test). There was no significant difference between the groups in the plaque levels of the other major subtypes (MMPs 1, 2, or 3) or the tissue inhibitors of MMPs (TIMPs 1 and 2), nor in the serum levels of all subtypes including MMP-9.

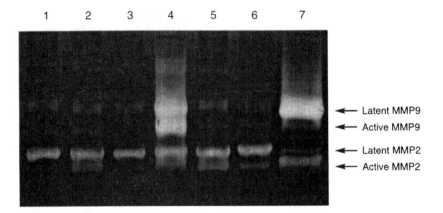

**FIGURE 1.** Representative gelatin zymogram of equal protein loads of plaque homogenates (samples 1–4, one plaque from each symptom group; 5–6, control vascular tissue; 7, HT-1080 positive control). The level of MMP9, both active and latent, can be seen to be higher in the recently symptomatic plaque.

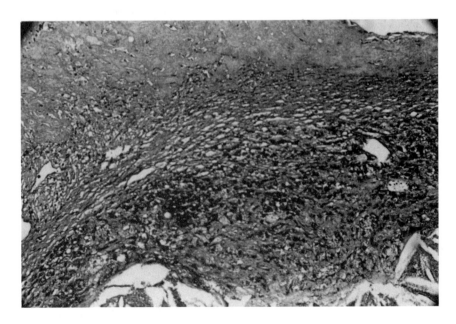

**FIGURE 2.** *In situ* hybridization for MMP-9 demonstrating intense staining around the shoulder and lipid core of a highly symptomatic carotid plaque associated with an intense inflammatory infiltrate.

RT-PCR and *in situ* hybridization revealed an increase in the expression of MMP-9 in the most highly symptomatic plaques (FIG. 2). The most intense staining was seen surrounding the lipid-rich core, particularly at the shoulder of the plaque, the area most vulnerable to acute disruption and matching the area of intense macrophage infiltration.

## CONCLUSION

In the most unstable carotid plaques, based on patient symptomatology, ongoing spontaneous embolization and histological evidence of plaque rupture, there is significantly more MMP-9 in both latent and active forms. This high concentration of MMP-9 is mirrored by an increase in its expression in the vulnerable region of the plaque, corresponding to the area of macrophage infiltration. There is no difference in the other major subtypes, nor in the serum levels of all MMP subtypes. Furthermore, there is a localized imbalance between this high level of MMP-9 and its naturally occurring inhibitors. This imbalance may be responsible for the acute disruption that leads to the onset of ischemic events, and thus represents an attractive target for pharmacotherapy aimed at stabilizing atherosclerotic plaques.

## REFERENCES

1. CARR, S., A. FARB, W.H. PEARCE, R. VIRMANI & J.S.T. YAO. 1996. Atherosclerotic plaque rupture in symptomatic carotid artery stenosis. J. Vasc. Surg. **23:** 756–765.
2. MORENO, P.R., V.H. BERNADI, J. LOPEZ-CUELLAR, A.M. MURCIA, I.F. PALACIOS, H.K. GOLD *et al.* 1996. Macrophages, smooth muscle cells and tissue factor in unstable angina. Circulation **94:** 3090–3097.
3. GALIS, Z.S., G.K. SUKHOVA, M.W. LARK & P. LIBBY. 1994. Increased expression of matrix metalloproteinases and matrix degrading activity in vulnerable regions of human atherosclerotic plaques. J. Clin. Invest. **94:** 2493–2503.
4. BROWN, D.L., M.S. HIBBS, M. KEARNEY, C. LOUSHIN & J.M. ISNER. 1994. Identification of 92-kD gelatinase in human coronary atherosclerotic lesions. Circulation **91:** 2125–2131.

# Enhanced Expression of Matrix Metalloproteinase-3, -12, and -13 mRNAs in the Aortas of Apolipoprotein E–deficient Mice with Advanced Atherosclerosis

ARCO Y. JENG,[a] MARY CHOU, WILBUR K. SAWYER, SHARI L. CAPLAN, JEAN VON LINDEN-REED, MICHAEL JEUNE, AND MARGARET FORNEY PRESCOTT

*Novartis Institute for Biomedical Research, Summit, New Jersey 07901, USA*

## INTRODUCTION

Increased expression of matrix metalloproteinases (MMPs) has been reported at sites of advanced atherosclerosis and aneurysm in humans.[1,2] An animal model that has been reported to develop advanced atherosclerosis and medial elastin degradation with progression to ectasia and aneurysm is the apolipoprotein E–deficient (apoE$^{-/-}$) mouse.[3–5] Using immunohistochemistry, a variety of MMPs have been demonstrated at sites of atherosclerosis and elastin degradation in apoE$^{-/-}$ mice on a high-fat diet.[5] However, little information is available on MMP expression in apoE$^{-/-}$ mice on a normal diet or in mice that are not as susceptible to atherosclerosis on either normal or high-fat diet. The present study compares mRNA expression of MMP-3, -9, -12, and -13 in mice that are highly susceptible (apoE$^{-/-}$), moderately susceptible (C57BL/6J and B6129F1/J), or resistant (C3H/HeJ and Balb/cJ) to diet-induced atherosclerosis.[6]

## MATERIALS AND METHODS

All mouse strains were obtained from Jackson Laboratories (Bar Harbor, ME) at 4–6 weeks of age. After one week of acclimation, groups of 6 animals from each strain were fed with either normal diet or a high-fat diet (1% cholesterol, 0.5% cholic acid, and 18% butter fat) ad lib for 16 weeks. Mice were then sacrificed, and aortas were perfused with saline *in situ,* carefully removed, cleaned of adventitial fat, and promptly snap frozen in liquid $N_2$.

Three aortas from each group were pooled (2 pools per group) and total RNA was extracted using TRI Reagent (Molecular Research Center, Cincinnati, OH). Samples containing 10 µg of total RNA each were resolved by electrophoresis in a 1.2% agarose/formaldehyde gel, transferred to a nitrocellulose membrane, and hybridized with a [$^{32}$P]-labeled cDNA probe. Autoradiography was carried out using Hyperfilm ECL (Amersham, Arlington Heights, IL), and signals were quantified by densitom-

[a]Address for correspondence: Dr. Arco Y. Jeng, LSB-2227, Novartis Pharmaceuticals Corporation, 556 Morris Avenue, Summit, NJ 07901. Phone, 908/277-5924; fax, 908/277-4756; e-mail, arco.jeng@pharma.novartis.com

etry. The hybridized cDNA probe on membranes was removed by boiling in 0.5% SDS, and the same membranes were used for hybridization with different probes.

The cDNA probes for MMP-9, -12, and -13 were obtained by PCR amplification of a mouse macrophage cDNA library (Stratagene, La Jolla, CA). The 5´ and 3´ primers used were 5´-GAATTCTGTTCAGCAAGGGGCGTGTC-3´ and 5´-GAAT-TCAAACAGTCCAACAAGAAAGGAC-3´, respectively, for MMP-9; 5´-CTC-GAGGAAGCTTCCTGGGAGTCCAG-3´ and 5´-CTCGAGCCCTGAGCATAGA-GTGAATATG-3´, respectively, for MMP-12; and 5´-CTCGAGCATGCTTCCT-GATGATGACGTT-3´ and 5´-CTCGAGCCCCACCCCATACATCTGAAA-3´, respectively, for MMP-13. The MMP-3 cDNA probe was amplified by PCR from mouse liver Quick-Clone cDNA (Clontech, Palo Alto, CA) using 5´ and 3´ primers 5´-GAATTCTGGGCTATACGAGGGCACGA-3´ and 5´-GAATTCGCACTTCCTT-TCACAAAGACTC-3´, respectively. The resulting PCR products were subcloned into pCR2.1 (Invitrogen, Carlsbad, CA), and the inserts were confirmed by DNA sequencing. The cDNA probes corresponding to MMP-3 (432 bp) and MMP-9 (459 bp) were excised by digestion with *Eco*RI and those corresponding to MMP-12 (565 bp) and MMP-13 (943 bp) by *Xho*I. The mouse GADPH cDNA probe was purchased from Ambion (Austin, TX).

## RESULTS AND DISCUSSION

Formation of advanced atherosclerotic lesions overlying areas of medial elastin destruction was observed solely in apoE$^{-/-}$ mice on a high-fat diet. The apoE$^{-/-}$ mice on a normal diet exhibited only early fatty-streak type lesions covering less than 5% of the aorta, and no elastin degradation was noted within the media. No other mouse

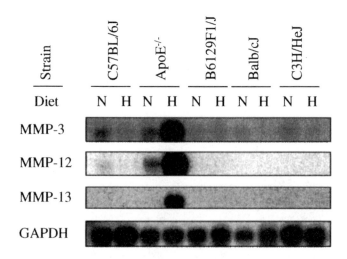

**FIGURE 1.** Autoradiograph showing mRNA expression of MMP-3, -12, and -13 in the aortas of different strains of mice on a normal (N) or high-fat (H) diet.

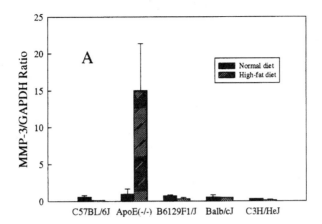

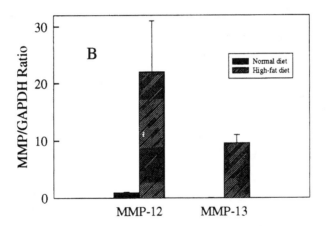

**FIGURE 2.** Normalized mRNA expression of MMPs. Signals in Northern blots shown in FIGURE 1 were quantified by densitometry and normalized with respect to that of GADPH. (**A**) MMP-3 expression in the aortas of various strains of mice on a normal or high-fat diet for 16 weeks. (**B**) MMP-12 and MMP-13 mRNA expression in the aortas of apoE$^{-/-}$ mice on a normal or high-fat diet for 16 weeks. Three aortas from the same group of animals were pooled for mRNA extraction. Each *bar* represents mean value with a range of error from two independent determinations.

strains on a high-fat diet developed atherosclerotic lesions or medial elastin degradation (data not shown). The mRNA expression of MMP-3 (1.8 kb), MMP-12 (1.8 kb), and MMP-13 (2.9 kb) was either not seen or was very low in all strains on a normal diet and in all strains on a high-fat diet excluding apoE$^{-/-}$ mice (FIG. 1). Analysis by densitometry showed that the mRNA levels of MMP-3, -12, and -13 were increased by 21 ($\pm$ 9)-, 24 ($\pm$ 9)-, and 190 ($\pm$ 29)-fold (mean $\pm$ range of error, $n = 2$ pools, with 3 animals in each pool), respectively, in apoE$^{-/-}$ mice on a high-fat diet compared with those on normal diet (FIG. 2). No significant mRNA expression of MMP-9 was detected in any strain on either diet (results not shown).

The results are consistent with immunohistochemical demonstration of MMP-3, -12, and -13 localized to areas of atherosclerosis and elastolysis in apoE$^{-/-}$ mice fed with a high-fat diet as reported by Carmeliet et al.[5] However, these authors also observed staining for MMP-9, whereas no significant changes in the mRNA levels of MMP-9 were found in the present report. While it remains to be investigated whether or not mRNA expression of MMPs is age dependent, the current study supports the potential role of MMPs in atherosclerosis and elastin degradation in apoE$^{-/-}$ mice on a high-fat diet.

## REFERENCES

1. GALIS, Z.S., G.K. SUKHOVA, M.W. LARK & P. LIBBY. 1994. Increased expression of matrix metalloproteinases and matrix degrading activity in vulnerable regions of human atherosclerotic plaques. J. Clin. Invest. 94: 2493–2503.
2. NEWMAN, K.M., A.M. MALON, R.D. SHIN, J.V. SCHOLES, W.G. RAMEY & M.D. TILSON. 1994. Matrix metalloproteinases in abdominal aortic aneurysm: characterization, purification, and their possible sources. Connect. Tissue Res. 30: 265–276.
3. PLUMP, A.S., J.D. SMITH, T. HAYEK, K. AALTO-SETÄTÄ, A. WALSH, J.G. VERSTUYFT, E.M. RUBIN & J.L. BRESLOW. 1992. Severe hypercholesterolemia and atherosclerosis in apolipoprotein E-deficient mice created by homologous recombination in ES cells. Cell 71: 343–353.
4. NAKASHIMA, Y., A.S. PLUMP, E.W. RAINES, J.L. BRESLOW & R. ROSS. 1994. ApoE-deficient mice develop lesions of all phases of atherosclerosis throughout the arterial tree. Arterioscler. Thromb. 14: 133–140.
5. CARMELIET, P., L. MOONS, R. LIJNEN, M. BAES, V. LEMAÎTRE, P. TIPPING, A. DREW, Y. EECKHOUT, S. SHAPIRO, F. LUPU & D. COLLEN. 1997. Urokinase-generated plasmin activates matrix metalloproteinases during aneurysm formation. Nature Genet. 17: 439–444.
6. NISHINA P.M., J. WANG, W. TOYOFUKU, F.A. KUYPERS, B.Y. ISHIDA & B. PAIGEN. 1993. Atherosclerosis and plasma and liver lipids in nine inbred strains of mice. Lipids 28: 599–605.

# Defects in Matrix Metalloproteinase Inhibitory Stoichiometry and Selective MMP Induction in Patients with Nonischemic or Ischemic Dilated Cardiomyopathy

MYTSI L. COKER, JAMES L. ZELLNER, ARTHUR J. CRUMBLEY, AND FRANCIS G. SPINALE[a]

*Division of Cardiothoracic Surgery, Medical University of South Carolina, Charleston, South Carolina 29425, USA*

## INTRODUCTION

The left ventricular (LV) myocardial collagen matrix has been proposed to participate in the maintenance of LV geometry. With several cardiac disease states such as dilated cardiomyopathy (DCM), alterations in extracellular matrix composition and structure have been reported to occur that may facilitate LV remodeling.[1-3] An important family of enzymes responsible for collagen remodeling are the matrix metalloproteinases (MMPs).[4,5] Several species of MMPs have been detected in the LV myocardium as well as the tissue inhibitors of MMPs, or TIMPs.[6] However, the molecular basis for changes in MMP/TIMP expression, MMP/TIMP stoichiometry, and the relation to MMP activity with nonischemic DCM and ischemic DCM remains unknown. Accordingly, the goal of the present study was to examine MMP/TIMP species expression and MMP activity in patients with end-stage nonischemic and ischemic DCM.

## PATIENTS/METHODS

The LV myocardial samples were collected from DCM patients at the time of transplant (nonischemic, $n = 21$; ischemic, $n = 16$) and from normal trauma patients ($n = 13$). LV myocardial samples were subjected to substrate zymography in order to measure MMP activity (pixels).[6] LV myocardial content (ng/g) of interstitial collagenase (MMP-1), tissue MMP inhibitor-1 (TIMP-1), specific MMP-1/TIMP-1 complex formation (COMP), 72-kDa gelatinase (MMP-2), stromelysin (MMP-3), and 92-kDa gelatinase (MMP-9) were quantitated by internally validated ELISA.

[a]Address for correspondence: Francis G. Spinale, M.D., Ph.D., Medical University of South Carolina, 770 MUSC Complex, Suite 625, Strom Thurmond Research Building, Charleston, South Carolina 29425. Phone, 843/953-3498; fax, 843/953-3499.

## RESULTS

Abundant zymographic activity was observed in LV myocardial samples, and this activity corresponded to molecular weights consistent with MMP species (FIG. 1).[4,5] LV MMP activity was measured using densitometric methods and is summarized in FIGURE 1. LV MMP activity was increased approximately threefold with both forms of DCM. LV myocardial MMP-1 content was decreased with both forms of DCM, while levels of TIMP-1, the primary inhibitor of MMP-1, remained unchanged (TABLE 1). The levels of the MMP-1/TIMP-1 complex decreased with both forms of DCM. LV myocardial MMP-9 levels were increased approximately twofold in both forms of DCM. LV myocardial MMP-2 and MMP-3 content were increased with nonischemic DCM, but not with ischemic DCM.

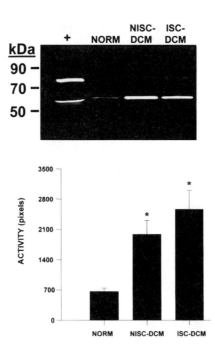

**FIGURE 1.** *(Top)* Zymographic activity in LV myocardial extracts was examined in normal (NORM), nonischemic DCM (NISC-DCM), and ischemic DCM (ISC-DCM) groups using gelatin as a proteolytic substrate. Conditioned media from an HT1080 cell line incubated in the presence (+) of PMA were included in all zymograms. In the LV myocardial extracts, proteolytic activity was observed at the 50–90-kDa region. With both NISC-DCM and ISC-DCM, LV myocardial zymographic activity appeared increased. *(Bottom)* Total LV myocardial zymographic activity for this gelatin substrate was quantitated using densitometric methods. LV myocardial zymographic activity was increased over threefold with NISC-DCM and ISC-DCM when compared to NORM (*$p < 0.05$).

**TABLE 1. Human LV myocardial MMP and TIMP levels in dilated cardiomyopathy**[a]

| Species | Normal (n = 13) | Nonischemic DCM (n = 21) | Ischemic DCM (n = 16) |
|---------|-----------------|--------------------------|-----------------------|
| MMP-1   | $780 \pm 148$   | $380 \pm 45^{b}$         | $294 \pm 35^{b}$      |
| TIMP-1  | $535 \pm 93$    | $549 \pm 36$             | $447 \pm 39$          |
| COMP    | $1920 \pm 172$  | $1369 \pm 154^{b}$       | $871 \pm 154^{b,c}$   |
| MMP-2   | $394 \pm 91$    | $675 \pm 61^{b}$         | $365 \pm 67^{c}$      |
| MMP-3   | $664 \pm 77$    | $1099 \pm 139^{b}$       | $682 \pm 95^{c}$      |
| MMP-9   | $71 \pm 9$      | $159 \pm 21^{b}$         | $130 \pm 22^{b}$      |

[a]Values expressed in ng/g LV.
[b]$p < 0.05$ vs. normal.
[c]$p < 0.05$ vs. nonischemic DCM.

## DISCUSSION

MMPs have been implicated in a number of tissue remodeling processes.[4,5] The present study demonstrated by *in vitro* MMP zymography increased MMP activity in patients with DCM due to nonischemic or ischemic etiologies. However, the present study suggests that the molecular basis for this increased MMP activity may be different in these two etiologies of DCM. Specifically, with nonischemic DCM, the content of several MMPs species is increased in the LV myocardium, which likely contributed to increased MMP activity. However, with ischemic DCM, a differential pattern of MMP expression was observed where only MMP-9 was increased from normal levels. Therefore, this differential LV myocardial MMP expression suggests that regulatory processes and/or external stimuli may be different between these two forms of DCM.

The findings of the present study also suggest that a loss of endogenous MMP inhibitory control may occur within the LV myocardium with DCM. Evidence to support this possibility includes the reduced expression of LV myocardial MMP-1 along with no change in TIMP-1 levels with DCM. Furthermore, the actual LV myocardial MMP-1/TIMP-1 complexes formed with both forms of DCM were reduced. Thus, alterations in MMP/TIMP stoichiometry within the LV myocardium may also contribute to the increased MMP activity with DCM. In conclusion, the results from this study suggest that the contributory factors for the LV remodeling process with DCM include increased MMP activity and defects in MMP inhibitory control.

This study builds upon past reports[2,3,6] by identifying the relationship between MMP expression and MMP activity to specific forms of DCM. Thus, potential therapeutic targets for modulating MMP expression and activity may be different for these two forms of cardiomyopathic disease.

## ACKNOWLEDGMENTS

Supported by National Institutes of Health Grants HL-45024 and HL-56603 (FGS), American Heart Association Grant-in Aid (F.G.S.). F.G.S. is an Established Investigator of the American Heart Association.

## REFERENCES

1. WEBER, K., R. PICK, J. JANICKI, G. GADODIA & J. LAKIER. 1988. Inadequate collagen tethers in dilated cardiomyopathy. Am. Heart J. **116**(6): 1641–1646.
2. GUNJA-SMITH, Z., A. MORALES, R. ROMANELLI & J. WOESSNER. 1996. Remodeling of human myocardial collagen in idiopathic dilated cardiomyopathy. Am. J. Pathol. **148**(5): 1639–1648.
3. SPINALE, F., M. TOMITA, J. ZELLNER, J. COOK, F. CRAWFORD & M. ZILE. 1991. Collagen remodeling and changes in LV function during development and recovery from supraventricular tachycardia. Am. J. Physiol. **261**: H308–H318.
4. BIRKEDAL-HANSEN, H., W. MOORE, M. BODDEN, L. WINDSOR, B. BIRKEDAL-HANSEN, A. DECARLO & J. ENGLER. 1992. Matrix metalloproteinases: a review. Crit. Rev. Oral Biol. Med. **4**(2): 197–250.
5. WERB, Z. & C. ALEXANDER. 1993. Proteinases and matrix degradation. *In* Textbook of Rheumatology. W.N. Kelly et al., Eds.: 248–268. W.B. Saunders. New York.
6. THOMAS, C., M.L. COKER, J.L. ZELLNER, J.R. HANDY, A.J. CRUMBLEY & F.G. SPINALE. 1998. Increased matrix metalloproteinase activity and selective upregulation in LV myocardium from patients with end-stage dilated cardiomyopathy. Circulation **97**: 1708–1715.

# Modulation of Matrix Metalloproteinases in Trophoblast Cell Lines

MARIA MORGAN AND SUSAN McDONNELL[a]

*School of Biological Sciences, Dublin City University, Dublin 9, Ireland*

Human fetal development depends on the embryo rapidly gaining access to the maternal circulation. This hurdle is overcome by trophoblast cells, which form the fetal portion of the human placenta, by transiently exhibiting an invasive phenotype. Thus during early pregnancy fetal cytotrophoblast cells invade the uterus and its arterial network. This invasive activity peaks during the 12th week of pregnancy and declines rapidly thereafter. In contrast to tumor cell invasion, trophoblast invasion is precisely regulated, being confined spatially to the uterus and temporally to early pregnancy. The invasive properties manifested by trophoblasts are made possible by the secretion of proteolytic enzymes that can degrade components of the extracellular matrix (ECM). A number of investigators have shown that the matrix metalloproteinases (MMPs) are important mediators of trophoblast invasion.[1,2] In this paper we examine the expression and regulation of MMPs in a series of trophoblast continuous cell lines referred to as $ED_{27}$, $ED_{31}$, $ED_{77}$, and a choriocarcinoma cell line BeWo. These cell lines have been previously characterized using several biochemical markers and have been shown to express MMPs and their inhibitors.[3] The trophoblast cell lines produced both MMP-2 and MMP-9, while the BeWo produced only MMP-2. An *in vitro* invasion assay demonstrated that the trophoblast cell lines were capable of invading through a Matrigel-coated filter, while the BeWo were unable to invade.

An understanding of the complex regulatory mechanisms controlling expression of MMP-2 and MMP-9 is clearly important for understanding the process of invasion that takes place during pregnancy. The aim of this study was to investigate the modulation of MMP expression in these trophoblast cell lines by cytokines, ECM components, hormones, and hypoxic conditions. We examined the effect of three cytokines, IL-1β, EGF, and TGF-β on trophoblast MMP production by RT-PCR and zymography. Of these cytokines tested, only IL-1β had a significant effect, upregulating expression of MMP-9 at the mRNA and protein level and increasing the *in vitro* invasive ability of the cells. The effect of IL-1β on MMP-9 mRNA levels was also found to be dose and time dependent, with a concentration of 5 ng/ml IL-1β and a treatment time of 8 hr sufficient for maximal induction. We also examined the effect of cytokines on MMP-2 expression in the trophoblast and BeWo cell lines and found that MMP-2 expression levels could not be modulated by any of the cytokines tested in this study.

Cellular invasion of the basement membrane is a multistep process in which cells recognize integral components of this structure and attach to them. Attachment to the

[a]Corresponding author. Address for telecommunication: Phone, 353-1-7045244; fax, 353-1-7045412; e-mail, susan.mcdonnell@dcu.ie

basement membrane triggers expression of genes necessary for the invasive cells to detach from the basement membrane, degrade it, and migrate through it. Many integrins are known to be capable of signal transduction, mediating the signal between the ECM and the cell interior via its receptor.[4] We decided to investigate this by culturing the trophoblast cell lines on various matrices and looking at MMP expression. The matrices examined included plastic, laminin, fibronectin, type IV collagen, and type I collagen. Although the repertoire of MMPs expressed by the trophoblast cells was not influenced by adhesion of the cells to the various substrates, depending on the nature of the substrate, regulatory effects were seen. MMP-9 expression was upregulated by laminin, but there was no effect on MMP-2.

Hormonal regulation of MMPs is an area of research that has been the focus of much interest recently, particularly in the reproductive system. However, we found that progesterone, even in the presence of β-estradiol, had no effect on MMP-9 mRNA expression levels in the trophoblast cell lines. We also looked at the effect of hypoxia on MMP expression by varying the oxygen content of the culture atmosphere. The rationale for this experiment was the dramatic changes in oxygen content of the placental environment that occurs during early gestation. However, our results showed that hypoxia did not alter the expression of MMPs in these cells.

Using trophoblast continuous cell lines as a model system of invasion during pregnancy, we have investigated the modulation of MMP expression by cytokines, ECM components, and hormones. Of these factors tested only IL-1β and laminin had a significant effect on MMP-9 expression. Considering that MMP-2 and MMP-9 are developmentally regulated during placentation and also the differential response of these two proteinases to cytokine treatment, it is tempting to speculate that cytokines and extracellular matrix components, in particular IL-1β and laminin, are involved in controlling MMP-9 activity throughout placentation.

## REFERENCES

1. BISCHOF, P. et al. 1991. Expression of extracellular matrix-degrading metalloproteinases by cultured human cytotrophoblast cells: effects of cell adhesion and immunopurification. Am. J. Obstet. Gynecol. 165: 1791–1801.
2. LIBRACH, C.L. et al. 1991. 92-kD type IV collagenase mediates invasion of human cytotrophoblasts. J. Cell Biol. 113: 437–449.
3. MORGAN, M. et al. 1998. Expression of metalloproteinases and their inhibitors in human trophoblast continuous cell lines. Exp. Cell Res. 242: 18–26.
4. SHIMIZU, Y. et al. 1990. Co-stimulation of proliferative responses of resting CD4 + T cells by the interaction of VLA-4 and VLA-5 with fibronectin or VLA-6 with laminin. J. Immunol. 145: 59–67.

# Effect of Relaxin on Tissue Inhibitor of Metalloproteinase-1 and -2 in the Porcine Uterus and Cervix

J.A. LENHART,[a] K.M. OHLETH,[a] P.L. RYAN,[a] S.S. PALMER,[b] AND C.A. BAGNELL[a,c]

[a]Department of Animal Sciences, Rutgers University, 84 Lipman Drive, New Brunswick, New Jersey 08901-8525, USA

[b]R.W. Johnson Pharmaceutical Research Institute, Raritan, New Jersey 08869, USA

Relaxin (RLX) changes the mechanical properties of reproductive tissues, stimulating growth and remodeling in the porcine uterus and cervix. While the mechanism for RLX-induced remodeling is not fully defined, tissue restructuring during the cycle, pregnancy, and postpartum involution is regulated through the activity of matrix metalloproteinases (MMPs). MMP regulation is complex, occurring at the level of protein synthesis, enzyme activation, and inhibition of active protease by specific tissue inhibitors of metalloproteinases (TIMPs). We have shown that RLX increases MMP-2 protein expression and activity in uterine secretions, while inhibiting tissue-associated uterine and cervical MMP-2 activity.[1] Whether these differences in MMP-2 activity are due to RLX-induced changes in uterine or cervical TIMPs is the focus of this study. To determine the effect of RLX on TIMP protein in the porcine uterus and cervix, prepubertal pigs were treated with RLX to induce uterine and cervical growth. Immunoblot analysis, using bovine TIMP-1 and human TIMP-2 monoclonal antibodies (0.25 µg/ml; Calbiochem), revealed that immunoreactive TIMP proteins were expressed in the uterus and cervix of both control and RLX-treated animals. RLX enhanced immunoreactive TIMP-1 and TIMP-2 in the uterus and uterine cervix, but had no effect on TIMP expression in the vaginal cervix. In addition, the ability of protein extracts from the uterus and cervix to inhibit the activity of both MMP-2 and MMP-9 (Biogenesis) was investigated. Gelatinase degradation of a fluorescently labeled, gelatinase-specific peptide substrate (RWJ-PRI), in the presence or absence of uterine flushes (10 µl) or protein extracts (30 µg) from control and RLX-treated animals was assessed. MMP-2 activity was attenuated in the presence of extracts of uterine and uterine cervical tissue. Inhibitory activity was greater ($p < 0.05$) in tissues from RLX-treated animals when compared to controls. Additionally, there was no evidence for inhibition of MMP-9 activity in the presence of extracts from either control or RLX-treated animals. In light of the dramatic effects of relaxin on porcine reproductive tissue remodeling, the reason for an increase in TIMP proteins and associated activity in response to relaxin is unclear. However, relaxin-induced reproductive tissue remodeling involves increases in the activity of other connective tissue enzymes responsible for matrix reorganization, such as col-

[c]Corresponding author. Address for telecommunication: Phone, 732/932-9095; fax, 732/932-6996; e-mail, bagnell@aesop.rutgers.edu

lagenase,[2] proteoglycanase,[2] and plasminogen activator.[3] Furthermore, while TIMP-1 and TIMP-2 have well-established roles in connective tissue remodeling, TIMPs have other biological functions. The increase in porcine TIMP expression/activity in response to relaxin may be involved in these other TIMP-mediated actions such as growth[4,5] or angiogenesis.[6]

## REFERENCES

1. LENHART, J.A. *et al.* 1998. Effect of relaxin on matrix metalloproteinase-2 in the porcine uterus and cervix. Biol. Reprod. **58**(1): 326.
2. TOO, C.K.L. *et al.* 1986. The effect of oestrogen and relaxin on uterine and cervical enzymes: collagenase, proteoglycanase and B-glycuronidase. Acta Endocrinol. **111:** 394.
3. WANG-LEE, J.L. *et al.* 1998. Regulation of urokinase- and tissue-type plasminogen activator by relaxin in the uterus and cervix of the prepubertal gilt. J. Reprod. Fertil. **114:** 119.
4. HAYAKAWA, T. *et al.* 1994. Cell growth promoting activity of tissue inhibitor of metalloproteinases-2 (TIMP-2). J. Cell Sci. **107:** 2372.
5. HAYAKAWA, T. *et al.* 1992. Growth promoting activity of tissue inhibitor of metal loproteinases-1 (TIMP-1). FEBS Lett. **298:** 29.
6. MURPHY, A.N. *et al.* 1993. Tissue inhibitor of metalloproteinases-2 inhibits bFGF-induced human microvascular endothelial cell proliferation. J. Cell Physiol. **157:** 351.

# Matrix Metalloproteinases Exhibit Different Expression Patterns in Inflammatory Demyelinating Diseases of the Central and Peripheral Nervous System

B.C. KIESEIER,[a,b] J.M. CLEMENTS,[c] A.J.H. GEARING,[c] AND H.-P. HARTUNG[a]

[a]Department of Neurology, Karl-Franzens-Universität, Graz, Austria

[c]British Biotech Pharmaceuticals Limited, Oxford, UK

Inflammatory demyelinating diseases of the nervous system, such as multiple sclerosis (MS) or the Guillain-Barré syndrome (GBS), represent common neurologic disorders. The understanding of the pathogenesis implicated in these conditions is incomplete. It is commonly accepted that these diseases are immune mediated. An emerging body of evidence suggests that matrix metalloproteinases (MMPs) may be of paramount importance in mediating inflammatory demyelinating diseases of the central (CNS) and peripheral nervous system (PNS). Primarily, this evidence is based on studies in animal models, such as experimental autoimmune encephalomyelitis (EAE), an animal model for inflammatory demyelination of the CNS, and experimental autoimmune neuritis (EAN), a model for inflammatory demyelination of the PNS. We investigated the temporospatial expression pattern of various MMPs in both experimental models in an attempt to elucidate the role of these proteases in the pathogenesis of these disabling diseases.

## MMPs IN EXPERIMENTAL AUTOIMMUNE ENCEPHALOMYELITIS

EAE is an inflammatory disorder of the CNS and is commonly used as an animal model to investigate pathophysiologic mechanisms in MS and their therapeutic modulation. In rodents EAE can be actively induced by immunization with CNS proteins, or adoptively transferred by injection of activated encephalitogenic T cells specific for these antigens. Broad-spectrum inhibition of MMP activity suppressed the development of and reversed clinical EAE in a dose-dependent way. These results were paralleled by restoration of the damaged blood-brain barrier in the inflammatory phase of the disease, and a significant reduction in gelatinase activity in the cerebrospinal fluid.[1,2] We studied the temporospatial expression pattern of various MMPs during the clinical course of EAE and found that MMPs in the CNS do not respond to inflammatory stimuli in an all-or-none fashion but rather are differentially

[b]Address for correspondence: Bernd C. Kieseier, M.D. Department of Neurology, Karl-Franzens-Universität, Auenbruggerplatz 22, 8036 Graz, Austria. Phone, +43-316-385-2981; fax, +43-316-325520; e-mail, bc.kieseier@kfunigraz.ac.at

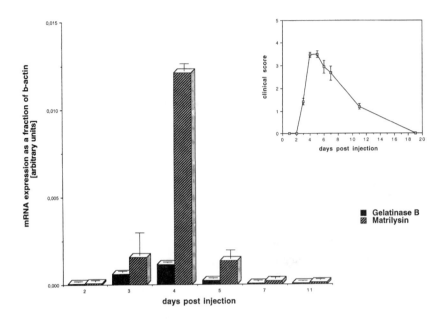

**FIGURE 1.** MMP mRNA expression during T cell–mediated experimental autoimmune encephalomyelitis: mRNA levels are expressed as a fraction of β-actin at different time points during the clinical course of the disease, which is demonstrated in the graph in the *upper right corner*.

regulated. Especially mRNAs for matrilysin and gelatinase B were significantly upregulated coincident with peak disease severity. Increased mRNA expression of gelatinase B was associated with enhanced proteolytic activity, as demonstrated by gelatin zymography; and immunohistochemistry localized this MMP to infiltrating mononuclear cells (FIG. 1).[3]

## MMPs IN EXPERIMENTAL AUTOIMMUNE NEURITIS

EAN is an acute inflammatory demyelinating disease of the PNS, commonly accepted as a suitable experimental model of the GBS. In susceptible animals it can be induced by active immunization with whole peripheral nerve homogenate, myelin, the myelin proteins P2 and P0, or peptides thereof, or by adoptive transfer of neuritogenic T cells. The broad-spectrum MMP inhibitor BB-1101 is capable of attenuating disease severity when given from symptom onset.[4] We found that during the entire clinical course of the disease mRNAs for collagenase-3, gelatinase A, stromelysin-1 and -3, and MT-MMP-1 and –3 were constitutively expressed. In contrast, mRNAs for matrilysin and gelatinase B were found to be upregulated during the initial phase of the disease, with peak levels coincident with maximum clinical disease

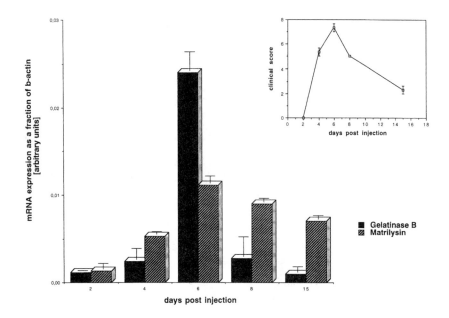

**FIGURE 2.** MMP mRNA expression during T cell–mediated experimental autoimmune neuritis, expressed as a fraction of β-actin at different time points during the course of the disease, which is demonstrated in the graph in the *upper right corner.*

severity. The expression of matrilysin remained upregulated throughout the recovery phase of AT-EAN, implicating a potential role of this metalloprotease in restoring the integrity of the PNS (FIG. 2). Matrilysin and gelatinase B expression could also be demonstrated in nerve biopsies from patients with GBS.[5]

## CONCLUSION

Important pathologic mechanisms in the genesis of demyelination, such as leukocyte recruitment, blood-brain, or nerve barrier breakdown, myelin destruction, and release of the disease promoting cytokine tumor necrosis factor–α, are considered to be MMP-dependent processes. Gelatinase B and matrilysin are both selectively upregulated during inflammatory diseases of the CNS and PNS. Differences in quantitative amounts as well as in the temporospatial expression point to different mechanisms of the response to inflammation in the CNS and the PNS. Regulation of MMP activity appears more likely to be a system-specific answer to an inflammatory stimulus, and thus MMPs may contribute to the pathogenesis of inflammatory demyelinating disorders of the CNS and PNS in different ways.

## REFERENCES

1. Gijbels, K. *et al.* 1994. Reversal of experimental autoimmune encephalomyelitis with a hydroxamate inhibitor of matrix metalloproteinases. J. Clin. Invest. **94:** 2177–2182.
2. Hewson, A.K. *et al.* 1995. Suppression of experimental allergic encephalomyelitis in the Lewis rat by the matrix metalloproteinase inhibitor Ro31-9790. Inflamm. Res. **44:** 345–349.
3. Kieseier, B.C. *et al.* 1998. Matrix metalloproteinase-9 and -7 are regulated in experimental autoimmune encephalomyelitis. Brain **121:** 159–166.
4. Hughes, P.M. *et al.* 1998. Matrix metalloproteinase expression during experimental autoimmune neuritis. Brain **121:** 481–494.
5. Kieseier, B.C. *et al.* 1998. Matrix metalloproteinases MMP-9 and MMP-7 are expressed in experimental autoimmune neuritis and the Guillain-Barré Syndrome. Ann. Neurol. **43:** 427–434.

# Matrix Metalloproteinases in Sterile Corneal Melts

G. GEERLING,[a,b,c] A.M. JOUSSEN,[d] J.T. DANIELS,[b] B. MULHOLLAND,[a,b]
P.T. KHAW,[a,b] AND J.K.G. DART[a]

[a]Moorfields Eye Hospital, 162 City Road, London EC 1V 2PD, UK

[b]Institute of Ophthalmology, Bath Street, London EC IV 9EL, UK

[d]Department of Ophthalmology, University of Heidelberg,
INF 400, D-69120 Heidelberg, Germany

## INTRODUCTION

Corneal ulceration is a rare but serious complication of local and systemic diseases of infectious (e.g., pseudomonas aeruginosa[1]) or noninfectious (e.g., rheumatoid arthritis[2]) origin. The clinical picture ranges from a central melt in an uninflamed eye to peripheral ulceration associated with scleritis.[3,4] The disease may progress to perforation of the globe, and can recur after corneal grafting. The initiating mechanism often remains unclear. In some cases trauma or tear deficiency can lead to epithelial breakdown following which an infectious agent or inflammatory cells (polymorphonuclear leukocytes [PMN] and macrophages) may then infiltrate and destroy corneal stroma by protease action.[4] However, in other cases the ulceration may be sterile and clinically free of inflammatory infiltration but continues to be progressive. In previous studies MMP-1 (collagenase) was found in corneal disease associated with rheumatoid arthritis, and MMP-2 and -9 (gelatinases A and B) were reported to be increased in the tears of patients after corneal grafting.[5] This study aimed to determine the pattern of MMP-2 and MMP-9 in tears of patients with sterile corneal melts.

## MATERIALS AND METHODS

*Tear Sampling.* Following local ethics committee approval, tears were sampled from age-matched healthy volunteers (group I: 26 eyes/16 subjects, mean age $72 \pm 11$ years), patients with connective tissue disease but no ocular disease (group II: 28 eyes/16 patients), and patients with healed or active sterile corneal melts (group III: 31 eyes/15 patients, mean age $77 \pm 8$ years). Samples were collected with capillaries and sponges to determine the influence of the collecting device on the result. After topical anesthesia with oxybuprocaine a sterile surgical sponge (Sugi, $1 \times 3 \times 4$ mm) was placed for 5 minutes into the inferior conjunctival fornix. The sponge was then centrifuged for 5 minutes at 13,000 rpm and the retrieved tear sample volume determined with a micropipette and stored at $-70°C$ until further use.

[c]Address for correspondence: Gerd Geerling, Moorfields Eye Hospital, 162 City Road, London EC IV 2PD, United Kingdom. Phone, 0044-171-5662823; e-mail: geerlingg@aol.com

*Gelatin Zymography.* After defrosting, a twofold serial dilution starting from 1 : 8 to 1 : 2048 was performed with phosphate-buffered saline. MWs were detected using gelatin zymography analysis following a recently described protocol (PAGE gel/ SDS-gel electrophoresis Novex R&D Systems, Oxon, UK).[6] Gelatinolytic activity appeared as clear bands on Coomassie blue stained background.

*Calculation of the Molecular Weight of Bands of Activity.* The molecular weight of each band in the sample was determined using prestained protein standards of known molecular weights to plot a standard graph of log(molecular weight). The highest dilution at which a band was detectable was noted.

*Inhibition of Gelatinolytic Activity.* To confirm that the bands of gelatinolytic activity resulted from MMP activity, zymography was repeated in the presence of the broad-spectrum inhibitor 1; 10-phenanthroline prepared at concentrations of 2 nM, 200 nM, and 20 mM in dimethyl sulfoxide.

*Activation of Matrix Metalloproteinases.* To provide further evidence for the identity of the MMPs in the tear samples, zymography was repeated after incubation with 2 mM aminophylmercuric acetate (APMA. Sigma), which activates the secreted inactive proenzyme. The results were compared with those obtained from pure human MMP-2 and MMP-9 (Biogenesis) treated in the same way.

*Statistical Analysis.* The Wilcoxon-Rank-Sum-Test (Mann-Whitney) for nonparametric data was used to compare the results of the three groups, and the sign rank test was used to compare the paired data on MMP-2 and -9 within each group.

## RESULTS

The device to collect the tear sample (sponge or capillary) did not influence the level of MMPs detected (TABLE 1, $p = 0.55$). Tears of healthy controls showed inactive MMP-9 at dilutions down to 1 : 128 and latent MMP-2 to 1 : 32. In tears of patients with active or previous corneal ulcers latent MMP-9 could be detected to a dilution of 1 : 2048 and MMP-2 to 1 : 1024. Incubation with APMA resulted in a band shift of the latent MMP-2 (72 kDa) and MMP-9 (92 kDa) to the MW of the active form of enzyme (MMP-2 = 65 kDa MMP-9 = 82 kDa). Incubation with 1,10-phenanthroline resulted in a dose-dependent inhibition of the gelatinolytic activity.

Patients with systemic connective tissue disease but no ocular disease had higher levels of MMP-2 and -9 than normals ($p < 0.00001$) but were not different from patients with a healed corneal ulcer. Patients with rheumatoid arthritis but no ocular disease had significantly higher levels of MMPs than patients with other connective tissue diseases (MMP-9, $p = 0.026$; MMP-2, $p = 0.012$).

Tears of patients with a history of active or inactive corneal melts showed significantly ($p < 0.00001$) more MMP-2 and -9 than tears of normals. No difference was found for the quantity of MMP-9 whether or not the ulcerative process was active. However, significantly more MMP-2 was detected in eyes with an active ulcer than in eyes with a healed ulcer ($p = 0.0068$). The level of MMP-9 did not differ significantly from MMP-2 except in tears of normal eyes, which contained more MMP-9 than MMP-2 ($p = 0.0002$).

**TABLE 1. Median and range (minimum and maximum) of dilution up to which MMP-2 or -9 was detectable in the zymogram**

| | $n$ (Eyes) | MMP-2 (71 kDa) | MMP-9 (92 kDa) |
|---|---|---|---|
| Group I: Normal control | 26 | 0 (0–32)[a,b] | 16 (0–128)[a,b,c] |
| Collected with a sponge | 20 | 0 (0–32)[d] | 16 (0–128)[c,d] |
| Collected with a capillary | 6 | 8 (0–16)[d] | 16 (16–32)[c,d] |
| Group II: Systemic rheumatic disease | 28 | 128 (16–512)[a,e] | 64 (32–512)[a,e] |
| Rheumatoid arthritis | 10 | 128 (64–521)[f] | 128 (64–512)[f] |
| Other connective diseases[i] | 18 | 64 (16–256)[f] | 64 (32–256)[f] |
| Group III: Corneal melts | 31 | 256 (32–1024)[b] | 256 (64–2024)[b] |
| Active disease | 12 | 512 (64–1024)[g,h] | 512 (64–1024)[g,h] |
| Inactive disease (previous melt) | 19 | 512 (32–512)[e,h] | 256 (64–2048)[e,g] |
| Rheumatoid arthritis | 20 | 256 (32–1024) | 256 (64–2048) |
| Other diseases[j] | 10 | 192 (128–512) | 256 (64–256) |

[a]Significant difference between groups I and II ($p < 0.00001$).
[b]Significant difference between groups II and III ($p < 0.00001$).
[c]Level of MMP-2 and -9 differed significantly in healthy normals only ($p = 0.0002$).
[d]No significant difference between sampling devices ($p = 0.55$).
[e]No significant difference between samples of groups II and III (MMP-2: $p = 0.06$; MMP-9: $p = 0.18$).
[f]Significant difference between patients of group II with rheumatoid arthritis and other connective tissue diseases ($p < 0.0001$).
[g]No significant difference for MMP-9 between active and inactive corneal ulcers ($p > 0.05$).
[h]Significant difference for MMP-2 between active and inactive corneal ulcers ($p > 0.0002$).
[i]Systemic lupus erythematodes: $n = 5$; Wegener's granulomatosis: $n = 4$; dermatomyositis: $n = 6$; Kabayashi syndrome: $n = 2$.
[j]Mooren's ulcer: $n = 2$; systemic lupus erythematodes: $n = 4$; dry eye: $n = 4$.

## DISCUSSION

MMPs are produced by inflammatory cells and bacteria but also by corneal keratocytes themselves. Fini *et al.* demonstrated their central role in the highly regulated turnover of corneal extracellular matrix.[7] There are good reasons for believing that overexpression of some of them may be an essential pathophysiological part in corneal ulceration. First, increased collagenolytic activity has been found in epithelium taken from corneal ulcers of various types.[8] Second, MMP-2 has been reported to be increased in corneal tissue and tears of patients with other corneal diseases such as keratoconus and following corneal surgery (corneal grafting).[5,9] Finally, as in rheumatoid disease and other autoimmune diseases, MMPs play a role in the pathogenesis of the systemic manifestations.[10]

Our data indicate, that MMP-9 (produced by epithelial cells) is a regular component of normal tears. Patients with rheumatoid arthritis show higher activities, which may be due to subclinical ocular surface disease and/or inflammation. In active corneal ulcers—when the epithelial and stromal wound healing is upregulated—MMP-9 and MMP-2 (produced by keratocytes) are present in tears in much higher concentrations than in inactive disease. Further work will determine whether this is a pathological overexpression or a normal wound healing reaction.

## ACKNOWLEDGMENT

This work was supported by Grants DFG Ge 895/4-1 and LORS 486.

## REFERENCES

1. TWINING, S.S., S.D. DAVIS & R.A. HYNDIUK. 1986. Relationship between proteases and descemetocele formation in experimental Pseudomonas keratitis. Curr. Eye Res. **5:** 503–510.
2. PFISTER, R.R. & G.E. MURPHY. 1980. Corneal stromal ulceration and perforation associated with Sjoegren's syndrome. Arch. Ophthalmol. **98:** 89–94.
3. EIFERMAN, R.A., D.J. CAROTHERS & J.A. YANKEELOV. 1979. Peripheral rheumatoid ulceration and evidence for conjunctival collagenase production. Am. J. Ophthalmol. **87:** 703–709.
4. BUMETT, J.M., L.E. SMITH, J.W. PRAUSE & K.R. KENYON. 1981. Acute inflammatory cells and collagenase in tears of human melting corneas. Invest. Ophthalmol. Vis. Sci. **20:** S173.
5. BARRO, C.D., J.P. ROMANET, A. FDILI, M. GUILLOT & F, MOREL. 1998. Gelatinase concentration in tears of corneal-grafted patients. Curr. Eye Res. **17:** 174–182.
6. KON, C.K., N.L. OCCLESTON, D. CHARTERIS, J.T. DANIELS, G.W. AYLWARD & P.T. KHAW. 1998. A prospective study of matrix metalloproteinases in proliferative vitreoretinopathy. Invest. Ophthalmol. Vis. Sci. **39:** 1524–1529.
7. FINI, M.E. & M. GIRARD. 1990. Expression of collagenolytic/gelatinolytic metalloproteinases by normal cornea. Invest. Ophthalmol. Vis. Sci. **31:** 1779–1788.
8. SLANSKY, H.H., U.L. GNADINGER, M. ITOI & C.H. DOHLMAN. 1969. Collagenase in corneal ulcerations. Arch. Ophthalmol. **82:** 108–111.
9. SMITH, V.A., B.B. HOH, M. LITTLETON & D.L. EASTY. 1995. Over-expression of a gelatinase A activity in keratoconus. Eye **9:** 429–433.
10. MURPHY, G. & R.M. HEMBRY. 1992. Proteinases in rheumatoid arthritis. J. Rheumatol. **32:** S62–64.

# Chemokine-Induced Extravasation of MonoMac 6 Cells: Chemotaxis and MMP Activity

CHRISTIANE M. KLIER AND PETER J. NELSON[a]

AG Klinische Biochemie, Medizinische Poliklinik,
Ludwig-Maximilians University of Munich, Germany

## INTRODUCTION

Chemokines are *chemo*tactic cyto*kines* that control the trafficking of immune effector cells into sites of inflammation.[1] The migration of leukocytes from the peripheral circulation through the basal lamina and ECM into tissue spaces is facilitated by the secretion of matrix metalloproteases (MMPs). MMPs modulate the turnover of the extracellular matrix.[2] The process of extravasation and transmigration is thought to be closely linked to the regulation of active MMP expression.[3] Our work examines chemokine regulation of MMP activity during the extravasation process.

## RESULTS AND DISCUSSION

### Expression of Chemokine Receptors and MMPs

A study of monocyte extravasation was conducted using the monocyte-like cell line MonoMac 6.[4] MonoMac 6 cells respond to monocyte chemoattractant protein (MCP) 1 through the chemokine receptor (CCR) 2 and show dose-dependent chemotaxis through fibronectin-coated filters (FIG. 1a). The expression of MMPs and *t*issue *i*nhibitor of *m*atrix *m*etalloproteinases (TIMPs) in MonoMac 6 cells was characterized using *f*luorescence *a*ctivated *c*ell *s*orting (FACS) analysis. In this assay, cells were fixed and either surface stained or stained internally following permeabilization of the cells (FIX and PERM, Caltag) using primary antibodies directed against various MMP and TIMP proteins (Calbiochem). Bound antibodies were detected with FITC-labeled goat anti-mouse IgG (Jackson Immunoresearch Laboratories). MonoMac 6 cells were found to constitutively express MMP-2, MMP-3, MMP-9, and TIMP-1 intracellularly and on the cell surface (TABLE 1, data not shown). As a functional assay of MMP activity linked to chemokine-induced migration, filter inserts (Costar, 5-micron pores, 6.5 mm) were coated with Matrigel,[3] and migration of MonoMac 6 cells was induced using 20 ng/ml MCP-1. The time course shown in FIGURE 1b revealed that significant migration of MonoMac 6 cells was observed after 6 (5%) and 8 (15%) hours, respectively. This process is considerably slower than

[a]Corresponding author: AG Klinische Biochemie, Medizinische Poliklinik, Ludwig Maximilians-Universität, Schillerstr. 42, 80336 München, Germany. Phone, +49 89 5996 844; fax, +49 89 5996 860; e-mail, nelson@med.poli.med-uni.muenchen.de

a)                                              b)

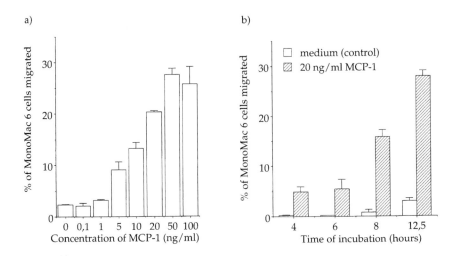

FIGURE 1. (a) MonoMac 6 cells migrate through fibronectin-coated membranes in response to MCP-1 in a dose-dependent manner. Filter inserts were coated with fibronectin,
and 450,000 cells were placed in the upper well. After 45 minutes the number of cells that
had transmigrated was determined using a cell counter (Casy I, Schärfe systems). (b) Kinetics of MonoMac 6 migration through a model basement membrane in response to MCP-1.
Filter inserts were coated with Matrigel (provided by Hynda Kleinman, NIH). The number
of MonoMac 6 cells that transmigrated at the indicated times in response to 20 ng/ml MCP-
1 or medium controls was counted. Data were obtained in duplicates and are expressed as
percent of total number of cells.

that seen through fibronectin-coated filter inserts (17%, 45 min, data not shown) suggesting an upregulation of MMP expression and activity during MCP-1–induced migration through Matrigel.

### MMP Expression during Migration through Matrigel

Changes in MMP and TIMP protein expression in MonoMac 6 cells following
MCP-1–induced migration through Matrigel were analyzed. Nonmigrated cells, as
well as cells that migrated through the Matrigel-coated filters, were collected and

TABLE 1. Internal expression of MMPs/TIMPs in MonoMac 6 cells during
MCP-1–induced migration through Matrigel-coated filter inserts

|                      | MMP-1 | MMP-2 | MMP-3 | MMP-9 | TIMP-1 | MT1-MMP |
|----------------------|-------|-------|-------|-------|--------|---------|
| Nonmigrated cells    | +     | +     | +     | +     | +      | −       |
| Transmigrated cells  | +     | ++    | ++    | ++    | +      | +       |

NOTE: MonoMac 6 cells were subjected to migration as described in FIGURE 1(b). After 10
hours of incubation, cells from the top of the filter insert (nonmigrated) as well as transmigrated
cells (representing 18% of total number of cells) were harvested and stained for FACS analysis.

subjected to FACS analysis. FACS analysis indicates an upregulation of internal stores of MMP-2, MMP-3, and MMP-9 after MCP-1–induced migration (TABLE 1). MMP-2 and MMP-9 proteins were released into the culture medium during transmigration as assessed by gelatine zymography[3] (data not shown). Further, small amounts of MMP-3 protein were detected in the culture supernatant as determined in immunoblot (data not shown). This finding supports the hypothesis that chemokines are involved in the regulation of matrix metalloproteinases during transmigration.

## REFERENCES

1. NELSON, P.J. & A.M. KRENSKY. 1998. Chemokines, lymphocyte biology and viruses: what goes around, comes around. Curr. Opin. Immunol. **10:** 265–270.
2. NAGASE, H. 1997. Activation mechanisms of matrix metalloproteinases. Biol. Chem. **378:** 151–160.
3. XIA, M. *et al.* 1996. Stimulus-specificity of matrix metalloproteinase–dependence of human T cell migration through a model basement membrane. J. Immunol. **156:** 160–167.
4. ZIEGLER-HEITBROCK *et al.* 1988. Establishment of a human cell line (Mono Mac 6) with characteristics of mature monocytes. Int. J. Cancer **41:** 456–461.

# Elevated Plasma Gelatinase A (MMP-2) Activity Is Associated with Quiescent Crohn's Disease

A.E. KOSSAKOWSKA,[a,b] S.A.C. MEDLICOTT,[a] D.R. EDWARDS,[c] L. GUYN,[c] A.L. STABBLER,[a] L.R. SUTHERLAND,[c] AND S.J. URBANSKI[a]

*Departments of [a]Pathology and [c]Medicine and Pharmacology, Foothills Hospital, University of Calgary, Alberta T2N 2T9, Canada*

Crohn's disease (CD) is an idiopathic inflammatory condition characterized by an unpredictable clinical course and frequent relapses. Stenosis, fissures, and fistulae, seen in 20–40% of patients, likely result from abnormal extracellular matrix (ECM) metabolism. Such complications may necessitate an alteration of medical therapy or surgical intervention. Many assays for markers of CD relapse have been investigated, but at present no test is implemented clinically.

MMPs are a family of zinc-dependent enzymes that degrade structural proteins of the ECM. These enzymes are involved in a variety of processes including atrophy (e.g., mammary gland involution), inflammatory disease (e.g., rheumatoid arthritis), tissue remodeling (e.g., liver cirrhosis, wound healing), and neoplasia (tumor invasion). Specific MMPs have been studied at the site of active bowel inflammation in CD, with immunofluorescence demonstrating a focal overexpression of MMP-9. Plasma MMP and tissue inhibitors of metalloproteinases (TIMPs) activity, however, have not been investigated in CD.

The goal of this study was to measure the activity of free (unbound) plasma MMP-2, -9 and their inhibitors, TIMPs-1, -2, -3 in patients with CD. More specifically, we sought to determine if plasma activity of unbound MMP-2, -9 correlates with a clinical index of disease activity (CDAI).

## METHODS

Plasma from 64 patients, collected on two separate visits, was analyzed by forward and reverse gelatin zymography, as well as by Western blot. The normal range of plasma MMP-2, -9 and inhibitory TIMP-1, -2, -3 activity was established from 39 healthy volunteers. Remission was defined as a CDAI score of $\leq 150$.

[b]Corresponding author: Department of Pathology and Laboratory Medicine, Foothills Medical Centre, 1403 29 Street NW, Calgary, Alberta T2N 2T9, Canada. Phone, 403/670-4759; fax, 403/670-4748; e-mail, anna.kossakowska@CRHA-Health.ab.ca

## RESULTS AND DISCUSSION

This study represents the first comprehensive investigation of MMP and TIMP functional activities in plasma of patients with inflammatory bowel disease. We have made four key observations:

First, median plasma MMP-2 activities were decreased in patients with active disease, compared to those with quiescent disease and controls. This distribution achieved statistical significance at visit one ($p = 0.009$ and $p = 0.003$, respectively), and a similar though nonsignificant trend was apparent at visit two (FIG. 1). Reduced MMP-2 levels in active CD was also evident in those 15 patients who changed disease status from visit one to visit two.

Second, median MMP-9 activities were elevated in all CD patients versus controls, reaching statistical significance at visit two ($p = 0.002$). This increased MMP-9 activity was not related to CDAI status of disease.

Third, the activity of TIMP-2 and -3, was reduced in all CD patients compared to controls. Regardless of CDAI ratings of disease status, TIMP-2 and -3 showed a marked reduction in activity at visits one and two (visit one, $p = 0.0006$, $p = 0.0001$ and visit two, $p = 0.005$, $p = 0.006$ for TIMP-2 and -3, respectively) (FIG. 2). TIMP-1 showed no significant variation between CD patients and healthy volunteers or between CD patients with opposing CDAI status.

Fourth, plasma gelatinases and TIMP levels failed to show significant variation relative to disease anatomic location or to tobacco, corticosteroid, or 5-ASA use. Thus, MMPs and TIMPs may be effectors of pathophysiological mechanisms of CD,

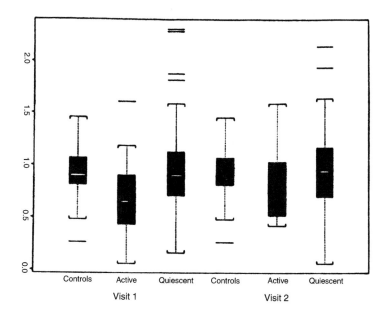

**FIGURE 1.** Distributions of MMP-2 activity in quiescent versus active Crohn's disease.

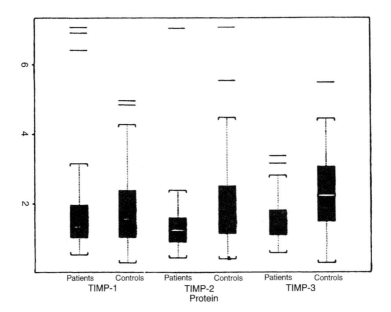

**FIGURE 2.** Comparisons of TIMP activities with Crohn's disease versus controls.

unaffected by therapeutic intervention with steroids or 5-ASA. In regard to the effectiveness of plasma MMPs as an assay for CD activity, MMP-2 appears the most promising but is limited by a prominent overlap with control levels.

The existence of MMP/TIMP complexes may contribute, in part, to dissimilar levels of gelatinases A and B in our study. Elevated MMP-9 in all CD patients detected at visit two is accompanied by normal TIMP-1 levels, compared to controls. In contrast, reduced levels of MMP-2 and TIMP-2 in patients with active disease may be indicative of a greater proportion of gelatinase A complexed to TIMP-2 and a net loss of MMP-2 gelatinolytic activity. The complete explanation for reduced levels of TIMP-2 and -3 in CD patients (reduced production and/or complex formation) is yet to be elucidated. Possibly, MMP-2 is inactivated during ongoing inflammation, then activated in the healing stage, which is characterized by granulation tissue formation.

## ACKNOWLEDGMENT

This work was supported by a grant from Foothills Hospital Research and Development Committee, Protocol #1661/8299.

# Regulation of Matrix Metalloproteinases in Human Intestinal Mucosa

SYLVIA L.F. PENDER,[a] CATRIONA McKENZIE, AZHAR SHAIDA, AND THOMAS T. MacDONALD

*Department of Paediatric Gastroenterology, St. Bartholomew's and the Royal London School of Medicine and Dentistry, London, EC1A 7BE, UK*

In inflammatory bowel disease (IBD) the concentrations of proinflammatory cytokines such as IL-1β and TNF-α are markedly increased.[1] There is a wealth of data to show that cytokines can increase matrix metalloproteinase (MMP) production in a variety of cell types.[2] It has been shown that MMPs play an important role in IBD. In particular, interstitial collagenase and stromelysin-1 mRNA are high in granulation tissue associated with Crohn's disease, ulcerative colitis, and gastric ulcers.[3–5]

We have previously demonstrated that MMPs play a major part in T cell–mediated tissue injury in human gut.[6–8] Elevated local concentrations of cytokines in the mucosa rapidly upregulate MMP-1 and MMP-3 production by resident stromal cells, and there is subsequent degradation of lamina propria extracellular matrix and basement membrane, and epithelial shedding.[9]

In this study, we have investigated the regulation of MMPs and TIMP-1 by cytokines and contact-dependent pathways in isolated gut stromal cells. Human fetal small intestine was obtained within 2 hours of surgical termination. Mesenchymal cells were isolated from the mucosa of human fetal small intestine and were maintained in MEM + 10% FCS. Cells from passage 4 were used in all experiments, as at this stage the number of epithelial and HLADR-positive cells is minimal. Mesenchymal cells were stimulated with TNF-α, IL-1β, IFN-γ and/or IL-10, anti-VLA-4 or VCAM-1 fusion protein for 2 days in the absence of serum.

By immunostaining we showed that mesenchymal cell lines isolated from human fetal small intestine were 99% smooth muscle cell actin positive, 45% desmin positive, only 2% DR positive, and less than 1% cytokeratin positive; none were CD3 or IL-2R positive. MMP-1 and MMP-3 production were rapidly upregulated when these cells were stimulated with IL-1β or TNF-α but not IFN-γ in serum-free medium. TIMP-1 production was unchanged.[6] The immunosuppressive cytokine IL-10, which is reported to increase MMP-3 production in skin fibroblasts, has no effect on gut stromal cells. The only biological activity we could detect for IL-10 was a slight synergy with TNF-α to increase gelatinase production. About 80% of gut stromal cells express the alpha 4 beta 1 integrin. By ligating VLA-4 with antibodies or a VCAM-1 fusion protein, MMP-3 production by stromal cells was slightly increased; however the production of activated gelatinase A was dramatically increased.

[a]Address for correspondence: Sylvia L.F. Pender, Department of Paediatric Gastroenterology, St. Bartholomew's and the Royal London School of Medicine and Dentistry, Suite 31, 3rd Floor, Dominion House, 59 Bartholomew's Close, London EC1A 7BE, UK. Phone, +44 (0) 171-601 8172 or 171-601 8489; fax, +44 (0) 171-600 5901; e-mail: s.pender@mds.qmw.ac.uk

Our results indicate that gut stromal cells are a potent source of MMPs and that they can be activated by contact-dependent and cytokine-driven mechanisms to increase MMP production. Ongoing studies are aimed at identifying the function of MMPs in the gut.

## ACKNOWLEDGMENTS

This work was supported by the Wellcome Trust, the Joint Research Board of St. Bartholomew's Hospital, and the Crohn's in Childhood Research Association. This study received ethical approval from the Hackney and District Health Authority, London.

## REFERENCES

1. RUGTVEIT, J., E.M. NILSEN, A. BAKKA, H. CARLSEN, P. BRANDTZAEG & H. SCOTT. 1997. Cytokine profiles differ in newly recruited and resident subsets of mucosal macrophages from inflammatory bowel disease. Gastroenterology 112: 1493–1505.
2. MURPHY, G. & J.J. REYNOLDS. 1993. Extracellular matrix degradation. In Connective Tissue and Its Heritable Disorders: Molecular, Genetic, and Medical Aspects. P.M. Royce & B. Steinmann, Eds.: 287–316. Wiley-Liss, New York.
3. BAILEY, C.J., R.M. HEMBRY, A. ALEXANDER, M.H. IRVING, M.E. GRANT & C.A. SHUTTLEWORTH. 1994. Distribution of the matrix metalloproteinases stromelysin, gelatinases A and B, and collagenase in Crohn's disease and normal intestine. J. Clin. Pathol. 47: 113–116.
4. SAARIALHO-KERE, U.K., M. VAALAMO, P. PUOLAKKAINEN, K. AIROLA, W.C. PARKS & M.L. KARJALAINEN-LINDSBERG. 1996. Enhanced expression of matrilysin, collagenase, and stromelysin- 1 in gastrointestinal ulcers. Am. J. Pathol. 148: 519–526.
5. VAALAMO, M., M.L. KARJALAINEN-LINDSBERG, P. PUOLAKKAINEN, J. KERE & U. SAARIALHO-KERE. 1998. Distinct expression profiles of stromelysin-2 (MMP-10), collagenase-3 (MMP-13), macrophage metalloelastase (MMP-12), and tissue inhibitor of metalloproteinases-3 (TIMP-3) in intestinal ulcerations. Am. J. Pathol. 152: 1005–1014.
6. PENDER, S.L., S.P. TICKLE, A.J. DOCHERTY, D. HOWIE, N.C. WATHEN & T.T. MACDONALD. 1997. A major role for matrix metalloproteinases in T cell injury in the gut. J. Immunol. 158: 1582–1590.
7. PENDER, S.L., J.M. FELL, S.M. CHAMOW, A. ASHKENAZI & T.T. MACDONALD. 1998. A p55 TNF receptor immunoadhesin prevents T cell–mediated intestinal injury by inhibiting matrix metalloproteinase production. J. Immunol. 160: 4098–4103.
8. PENDER, S.L., E.J. BREESE, U. GUNTHER, D. HOWIE, N.C. WATHEN, D. SCHUPPAN & T.T. MACDONALD. 1998. Suppression of T cell–mediated injury in human gut by interleukin-10: role of matrix metalloproteinases. Gastroenterology 115: 573–583.
9. PENDER, S.L., P. LIONETTI, S.H. MURCH, N. WATHAN & T.T. MACDONALD. 1996. Proteolytic degradation of intestinal mucosal extracellular matrix after lamina propria T cell activation. Gut 39: 284–290.

# MMP-13 and MMP-1 Expression in Tissues of Normal Articular Joints

SUE A. YOCUM,[a] LORI L. LOPRESTI-MORROW, LISA M. REEVES, AND
PETER G. MITCHELL

*Pfizer—Central Research Division, Inflammation Research Group, Eastern Point Road,
Groton, Connecticut 06340, USA*

In the pathogenesis of rheumatoid arthritis and osteoarthritis, a major feature is the degradation and loss of type II collagen with associated fibrillation of the cartilage. The vertebrate collagenases, collagenase-I (MMP-1), collagenase-2 (MMP-8), and collagenase-3 (MMP-13) represent the three known members of the matrix metalloproteinase (MMP) family capable of cleaving fibrillar collagen and are distinguished by their tissue expression and ability to degrade the different types of collagen. MMP-1, the first of these enzymes to be identified, is expressed in an array of connective tissues as well as in fibroblasts, macrophages, and some tumors; it preferentially cleaves type III collagen. MMP-8 is expressed primarily in polymorphonuclear phagocytes and is most efficient against type I collagen. Originally cloned from a breast carcinoma, human MMP-13 has been found in chondrocytes and osteoarthritic cartilage and has greatest activity against type II collagen.[1,2] MMP-1 and MMP-13 message expression is variable in osteoarthritic tissue, but cytokine stimulation of osteoarthritic cartilage results in both MMP-1 and MMP-13 induction.[1]

Although the relative contributions of each of these enzymes in normal tissue processes or in the arthritic disease process is unclear, inhibitor studies suggest MMP-13 plays a key role in collagen degradation. While mice and rats appear to have only MMP-13, the rabbit has an MMP-1 homologue as well and may provide a more relevant animal model for arthritic diseases.[3] In the present study we have examined the expression of MMP-1 and MMP-13 in rabbit joint connective tissues and normal human joint connective tissues in order to have a better understanding of the expression of these enzymes in the transition from normal to the diseased state.

## MATERIALS AND METHODS

Normal human cartilage was obtained within 24 hours postmortem. New Zealand White rabbit articular cartilage was obtained from freshly euthanized specimens. Articular cartilage was removed from the underlying bone under aseptic conditions and placed in DMEM containing antimycotic/antibiotic and gentamicin. Connective tissues representing tendon, synovium, ligament, and meniscus were aseptically removed and cultured as above. Dermal cuttings from New Zealand White rabbits were rinsed in 70% ethanol followed by several rinses in DMEM with antimycotic/

[a]Address for telecommunication: Phone, 860/715-0458; fax, 860/441-5719;
e-mail, s_a_yocum@groton.pfizer.com

antibiotic and gentamicin. Cuttings were incubated in 20% FBS in DMEM in 100-mm$^2$ dishes for outgrowth of skin fibroblasts Data represent cells within the first five passages. Human foreskin fibroblasts were prepared in manner similar to that of rabbit skin fibroblasts and were used within the first ten passages. Total RNA was prepared as described.[1] Northern blot preparation and analysis with human MMP-1 and human MMP-13 probes were essentially as previously described.[1]

## RESULTS

An examination of the mRNA expression of MMP-1 and MMP-13 in connective tissues from the articular joints of New Zealand White rabbits indicates distinct differences between MMP-1 and MMP-13 in IL-1 induction (FIG. 1). Low levels of MMP-1 message were detectable in some of the connective tissues, and treatment with IL-1 for 24 hours greatly increased message in all of the tissues examined (articular cartilage, synovium, meniscal cartilage, ligament, tendon, and skin fibroblasts). MMP-13 message expression, however, was limited to only the articular cartilage. Fibroblasts prepared from rabbit synovial tissue do express low amounts of MMP-13 following 24 hours of IL-1 treatment (data not shown).

An examination of normal human connective tissues also results in a distinct profile of expression. In human articular cartilage, synovium, and meniscal cartilage,

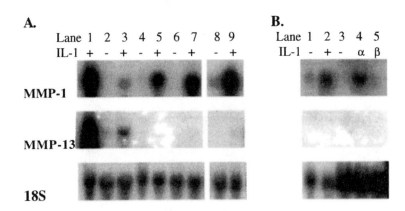

**FIGURE 1.** MMP-1 and MMP-13 expression in articular joint connective tissues of New Zealand white rabbits. (**A**) Total RNA (15 μg per lane) was prepared from tissues obtained from freshly euthanized rabbits. Each type of tissue was divided into two aliquots and cultured for 24 hours either in the presence of human recombinant IL-1α or in the absence of IL-1. *Lane 1*, IL-1 treated human osteoarthritic cartilage as a positive control; *lanes 2 and 3*, articular cartilage; *lanes 4 and 5*, synovium; *lanes 6 and 7*, meniscal cartilage; *lanes 8 and 9*, ligament. (**B**) 10 μg of total RNA from tendon was prepared as in A (above). Skin fibroblasts were produced from outgrowth of dermal cuttings in DMEM with 20% fetal bovine serum. 15 μg of total RNA was prepared from cultures stimulated for 24 hours ± 100 ng/ml human IL-1α or β. *Lanes 1 and 2*, tendon; *lanes 3–5*, skin fibroblasts.

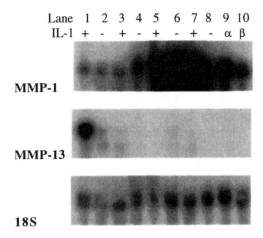

Lane 1 2 3 4 5 6 7 8 9 10
IL-1 + - + - + - + - α β

MMP-1

MMP-13

18S

FIGURE 2. MMP-1 and MMP-13 expression in normal human articular joint connective tissues. Total RNA (15 µg per lane) was prepared from tissues obtained postmortem. Each type of tissue was divided into two aliquots and treated for 24 hours ± 5 ng/ml human recombinant IL-1α (unless otherwise indicated) in serum-free DMEM. *Lane 1,* IL-1 treated porcine chondrocytes as a positive control; *lanes 2 and 3,* articular cartilage; *lanes 4 and 5,* synovium; *lanes 6 and 7,* meniscal cartilage; *lanes 8–10,* human foreskin fibroblasts.

MMP-1 message is evident prior to IL-1 treatment (FIG. 2). IL-1 stimulation for 24 hours yielded a significant induction of MMP-1 message, while articular cartilage did not show any substantial increase in message. MMP-13 message was faintly visible in articular cartilage and meniscal cartilage with IL-1 stimulation yielding no apparent increase in signal. Thus, IL-1 stimulation of normal articular cartilage explants did not induce either MMP-1 or MMP-13 message after 24 hours of treatment. MMP-13 mRNA was not detectable in synovium or human foreskin fibroblasts. As with rabbit, fibroblasts prepared from human synovium expressed low levels of MMP-13 mRNA at 8–24 hours following IL-1 stimulation (data not shown).

## CONCLUSIONS

The profiles of MMP-1 and MMP-13 mRNA expression in tissues from normal articular joints is distinct. Examination of rabbit articular joint connective tissues and rabbit skin fibroblasts demonstrates IL-1 induction of MMP-1 mRNA, whereas MMP-13 expression is limited to the articular cartilage. In normal human articular joint connective tissues, MMP-13 message is restricted to articular cartilage and meniscal cartilage and is expressed at relatively low levels. IL-1 stimulation of normal human articular cartilage for 24 hours did not yield detectable induction of MMP-1 and MMP-13 mRNA. This data further establishes the limited tissue expression of MMP-13, and suggests that cytokine induction of MMP-13 in normal versus osteoarthritic cartilage may differ.

## REFERENCES

1. MITCHELL, P.G, H.A. MAGNA, L.M. REEVES, L.L. LOPRESTI-MORROW, S.A. YOCUM, P.J. ROSNER, K.F. GEOGHEGAN & J.E. HAMBOR. 1996. Cloning, expression, and type II collagenolytic activity of matrix metalloproteinase-13 from human osteoarthritic cartilage. J. Clin. Invest. **97:** 761–768.
2. KNAUPER, V., C. LOPEZ-OTIN, B.J. SMITH, C.G. KNIGHT & G. MURPHY. 1996. Biochemical characterization of human collagenase-3. J. Biol. Chem. **271:** 1544–1550.
3. VINCENTI, M.P, C.I. COON, J.A. MENGSHOL, S.A. YOCUM, P.G. MITCHELL & C.E. BRINCKERHOFF. 1998. Cloning of the gene for interstitial collagenase-3 (matrix metalloproteinase-13) from rabbit synovial fibroblasts: differential expression with collagenase-1 (matrix metalloproteinase-1). Biochem. J. **331:** 341–346.

# Expression and Localization of TIMP-1, TIMP-2, MMP-13, MMP-2, and MMP-9 in Early and Advanced Experimental Lung Silicosis

ANNIE PARDO,[a] JULIA PÉREZ-RAMOS, LOURDES SEGURA-VALDEZ, REMEDIOS RAMÍREZ, AND MOISÉS SELMAN

*Facultad de Ciencias, Universidad Nacional Autónoma de México,*
*Universidad Autónoma Metropolitana, Unidad Xochimilco,*
*Instituto Nacional de Enfermedades Respiratorias, México*

## INTRODUCTION

Chronic exposure to crystalline silica particles results in macrophage-lymphocytic granulomatous lung inflammation, which is followed by abnormal and progressive accumulation of extracellular matrix.[1,2] However, the pathogenic mechanisms and the sequence of the pathological events leading to the fibrotic response have not been well defined. In this context, there is evidence suggesting an upregulation of a variety of fibrogenic cytokines such as tumor necrosis factor alpha (TNF-α) and transforming growth factor beta (TGF-β), with increased synthesis and secretion of lung extracellular matrix components.[3,4] However, studies on matrix degradation are scanty.

Extracellular matrix degradation involves the matrix metalloproteinases (MMPs), a conserved family with a zinc binding site in the catalytic domain, and an amino terminal domain responsible for the zymogen inactive state. MMP family include the collagenases, which degrade fibrillar collagens; the stromelysins, which cleave proteoglycans and some glycoproteins; the gelatinases A and B, which degrade basement membrane type IV collagen; and the membrane-type metalloproteinases, which are able to activate progelatinase A.[5,6] MMPs activity is regulated at different levels including the transcriptional level, the proenzyme activation, and the inhibition of active enzymes by a family of tissue inhibitors of metalloproteinases (TIMPs).[7] Here we determined the temporal pattern of expression and localization of collagenase-3 (MMP-13), gelatinases A and B (MMP-2 and MMP-9), and TIMP-1 and TIMP-2 during the evolution of rat experimental silicosis.

[a]Address for correspondence: Annie Pardo, Ph.D., Facultad de Ciencias, U.N.A.M., Apartado Postal 21-630, Coyoacán, México DF, CP 04000, México. Fax, 525/622-4910; e-mail, aps@hp.fciencias.unam.mx

## MATERIAL AND METHODS

Lung silicosis was induced in adult Wistar rats by a single intratracheal administration of 50 mg of quartz dust in sterile saline. Eight rats were sacrificed at 15, 45, and 60 days after silica instillation, and eight normal animals instilled with saline were used as controls. Animals were anesthetized, and the lungs were instilled with 4% paraformaldehyde and used for histology, *in situ* hybridization, and immunohistochemistry, as described elsewhere.[8,9] Additionally, bronchoalveolar lavage (BAL) was performed in six controls and six silica-exposed rats at 15, 45, and 60 days, and aliquots of 8 µl of fluid were used to analyze gelatinase activity in gelatin substrate SDS gel as previously described.[8]

## RESULTS

A significant increase in total inflammatory cells was observed in BAL from silicotic rats. The inflammatory response was characterized by an increment of lymphocytes and neutrophils at 15 and 45 days, and also by macrophages at 60 days. Zymography of BAL fluid from silica-exposed rats revealed increased gelatinolytic activities of progelatinase A and its activated form when compared with controls. Additionally, silicotic rats also showed bands with ~ MW of 95 and 86 kDa representing progelatinase B and its activated form.

By *in situ* hybridization and immunohistochemistry, younger silicotic granulomas exhibited intense staining for MMP-2, MMP-9, MMP-13, TIMP-1, and TIMP-2. Labeling was usually restricted to the granulomas and surrounding areas. By contrast, older granulomas, characterized by the presence of concentric layers of hyaline fibers in the center, displayed similar staining for TIMPs, but MMP signaling was markedly reduced. Saline-treated animals showed scattered positive cells. A semiquantitative evaluation is shown in TABLE 1.

Collagenase-3 transcript and protein was detectable in alveolar epithelial cells, macrophages, and fibroblasts. MMP-2 mRNA was observed mainly in mesenchymal cells and macrophages, while MMP-9 mRNA was expressed by macrophages, type 2 pneumocytes, and neutrophils. TIMP-1 and TIMP-2 were expressed by macrophages and mesenchymal cells.

TABLE 1. Semiquantitative evaluation of MMPs and TIMPs expression during the evolution of silicosis

| Time after silica exposure | MMP-13 | MMP-2 | MMP-9 | TIMP-1 | TIMP-2 |
|---|---|---|---|---|---|
| 15 Days | +++ | +++ | ++ | +++ | +++ |
| 45 Days | ++ | ++ | ++ | +++ | +++ |
| 60 Days | + | + | + | ++ | +++ |

## DISCUSSION

The findings of this study suggest that in early inflammatory silicotic granulomas there is a marked upregulation of collagenase-3 and gelatinases A and B, which decrease when the lesions evolve to fibrosis. By contrast, TIMP-1 and TIMP-2 also display a considerable increase from the early phases but show a more discrete reduction in the fibrotic granulomas. These results support the notion of an imbalance in the MMP/TIMP ratio during the evolution of experimental silicosis that could enhance the fibrotic response. Excessive initial gelatinolytic and collagenolytic activities may participate in basement membrane disruption, matrix remodeling, and growth factor release. Decreased collagenolytic activity in advanced phases may contribute to collagen accumulation and the development of progressive fibrosis.

## ACKNOWLEDGMENTS

This work was supported by PUIS & PAPIT: IN202697 (UNAM); CONACYT F643-M9406.

## REFERENCES

1. HEPPLESTON, A.G. 1984. Pulmonary toxicology of silica, coal and asbestos. Environ. Health Perspect. **55:** 111–127.

2. RAMOS, C.M. MONTAÑO, G. GONZÁLEZ, F. VADILLO & M. SELMAN. 1988. Collagen metabolism in experimental lung silicosis. A trimodal behavior of collagenolysis. Lung **166:** 347–353.

3. PIGUET, P.F., M.A. COLLART, G.E. GRAU, A.P. SAPPINO & P. VASSALLI. 1990. Requirement of tumour necrosis factor for development of silica-induced pulmonary fibrosis. Nature **344:** 245-247.

4. MARIANI, T.J., J.D. ROBY, R.P. MECHAM, W.C. PARKS, E. CROUCH & R.A. PIERCE. 1996. Localization of type I procollagen gene expression in silica-induced granulomatous lung disease and implication of transforming growth factor-β as a mediator of fibrosis. Am. J. Pathol. **148:** 151-164.

5. BIRKEDAL-HANSEN, H.W., G.I. MOORE, M.K. BODDEN, L.J. WINDSOR, B. BIRKEDAL-HANSEN, A. DeCARLO & J.A. ENGLER. 1993. Matrix metalloproteinases: a review. Crit. Oral Biol. Med. **4:** 197-250.

6. WOESSNER, JR. J.F. 1994. The family of matrix metalloproteinases. Ann. N.Y. Acad. Sci. **732:** 11–21.

7. GÓMEZ, D., D. ALONSO, H. YOSHIJI & U.P. THORGEIRSSON. 1997. Tissue Inhibitors of metalloproteinases: structure, regulation and biological functions. Eur. J. Cell Biol. **74:** 111–122.

8. PARDO, A., M. SELMAN, K. RIDGE, R. BARRIOS & I.J. SZNAJDER. 1996. Increased expression of gelatinases and collagenase in rat lungs exposed to 100% oxygen. Am. J. Respir. Crit. Care Med. **154:** 1067–1075.

9. BARRIOS, R., A. PARDO, C. RAMOS, M. MONTAÑO, R. RAMÍREZ & M. SELMAN. 1997. Upregulation of acidic fibroblast growth factor (FGF-1) during the development of experimental diffuse lung fibrosis. Am. J. Physiol. **273:** L451–L458.

# Type II Collagen Peptide Release from Rabbit Articular Cartilage

B.R. FELICE,[a] C.O. CHICHESTER,[a,b] AND H.-J. BARRACH[c]

[a]Department of Biomedical Sciences, University of Rhode Island,
Kingston, Rhode Island 02881, USA

[c]Department of Orthopaedics, Brown University-Rhode Island Hospital,
Providence, Rhode Island 02903, USA

Osteoarthritis (OA) is characterized by the destruction of articular cartilage resulting in the loss of two major components, proteoglycan and type II collagen. The collagen fibers in articular cartilage provide a supporting network that, when damaged, is thought to irreversibly destroy the matrix. The degradation of type II collagen and other extracellular matrix components is controlled by the activity of the matrix metalloproteinases (MMPs).[1] MMPs 1, 8, and 13 are the only known mammalian enzymes capable of cleaving triple helical fibrillar collagen.[2] Immunohistological studies of human osteoarthritic articular cartilage and explants from bovine cartilage showed extensive cleavage of collagen type II (CII) in the tissue.[3] The morphological findings were supported by the quantification of partially degraded denatured CII extracted from human articular cartilage using epitope-specific antibodies.[4,5]

This study evaluated the metalloproteinase-dependent release of CII peptides from rabbit articular cartilage using both *in vivo* and *in vitro* models. CII degradation peptides were quantitated using monoclonal antibodies directed against epitopes found in the cyanogen bromide cleavage peptide 9.7 (CB 9.7) region in type II collagen. The antibodies were used in a sandwich format to quantify peptide release *in vivo* or in an inhibition ELISA to examine the release of these epitopes in an explant system. Antibody 14:7:D8 was generated against a synthetic peptide 15 amino acids in length (GPQGPRGDKGEAGEP) coupled to KLH. Antibody 18:6:D6 was made against CB 9.7 coupled to ovalbumin. The antibodies react with epitopes in CB 9.7 and similar repeated epitopes on CB 11.

We used a surgically induced model of OA in rabbits (Hulth-Telhag) to determine if quantifiable amounts of CII fragments are released from articular cartilage into synovial fluid. The right knee joint was surgically altered by severing the medial collateral ligament and the anterior and posterior cruciate ligaments, and by removing the medial meniscus. A sham operation was performed on the left knee in which only the joint capsule was opened. Synovial fluid from the operated knee and lavage fluid from the sham-operated knee were collected at intervals up to 8 weeks after surgery. The CII peptide concentrations were measured using a magnetic bead sandwich assay employing the two monoclonal antibodies reactive against CII peptides. Carboxy-methyl cellulose–purified type II collagen cyanogen bromide cleavage

[b]Corresponding author: Clinton Chichester, Ph.D., Department of Biomedical Sciences, University of Rhode Island, Kingston, RI 02881. Phone, 401/874-5034; fax, 401/874-5048; e-mail, chichester@uri.edu

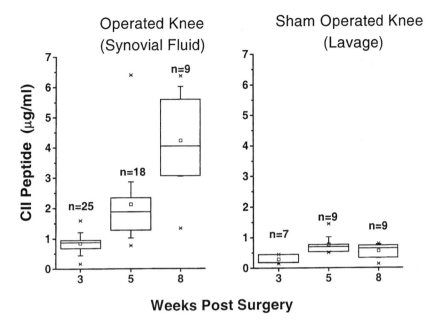

**FIGURE 1.** Concentration of CII peptides in the synovial fluid and lavage fluid of rabbits with either surgically induced OA (Hulth-Telhag model) or sham operation. Animals were sacrificed at 3, 5, and 8 weeks after surgery. The CII peptide concentration, expressed as μg/ml, was measured using a magnetic bead sandwich ELISA employing monoclonal antibodies 14:7:D8 and 18:6:D6. Statistical analysis was performed using the Kruskal-Wallis nonparametric ANOVA test.

peptides from bovine articular cartilage were employed as standard. The synovial fluid of the operated knee showed a gradual increase in CII peptide concentrations over time (FIG. 1). The boxed values are 25th and 75th percentile, with the mean and medium values inside the box. The values for the week 8 group are significantly different from those of the week 3 group ($p < 0.001$). The lavage fluid of the control knee showed insignificant levels of CII peptides

The MMPs are induced by cytokines, particularly IL-1α, but the enzymes exist in the inactive proenzyme form until extracellular activation.[6] It has been proposed that plasmin may be a physiological activator of the proenzymes.[4] Saito and coworkers[7] designed an *in vitro* model of OA employing both IL-1α and plasminogen in rabbit articular cartilage explant cultures. Cultures incubated with IL-1α and plasminogen exhibited MMP induction and activation, which lead to a significant amount of hydroxyproline release into the supernatant. This system, with modifications, was used to evaluate release of the same epitopes quantitated in the *in vivo* model.

Slices of articular cartilage were obtained from the knee joints of 4-kg rabbits and cultured as previously described,[7] with the exception of the substitution of 50 μg/ml bovine serum albumin for 1% fetal bovine serum. On the second day of culture the medium was replaced with culture medium containing IL-1α (1 ng/ml) or plasminogen (100 μg/ml) or IL-1α and plasminogen. At 3- or 4-day intervals the culture medium was collected for analysis and fresh medium containing the appropriate test compounds was added. IL-1α or plasminogen alone had little effect on release of the 14:7:D8 or 18:6:D6 epitope. The combination of IL-1α and plasminogen led to significant CII peptide release with both epitopes maximally released at day 18. Media

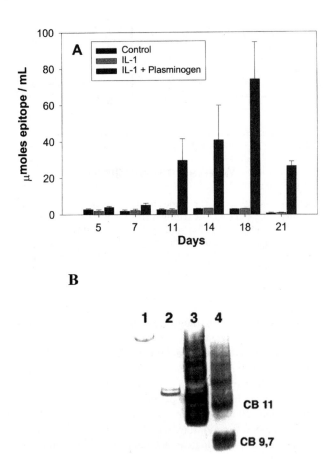

**FIGURE 2.** Release of CII peptides in rabbit articular cartilage explants measured with antibody 18:6:D6. Rabbit articular cartilage slices (50 mg/ml) were cultured according to the method of Saito et al.[7] On the second day media were replaced with culture medium containing IL-1α (1 ng/ml) or IL-1α (1 ng/ml) and plasminogen (100 μg/ml). At the times indicated, the media were taken for analysis and replaced with fresh media. (**A**) Concentration of 18:6:D6 epitopes in culture media quantitated by inhibition ELISA. (**B**) Western blot of culture supernatants using alkaline phosphatase-labeled 18:6:D6. (1) Control, (2) IL-1, (3) IL-1+plasminogen, (4) CII CNBR peptides.

concentrations of the 18:6:D6 epitope are shown in FIGURE 2A. The release of 18:6:D6 epitope was consistently higher than that of the 14:7:D8 epitope. On day 18 the 18:6:D6 epitope concentration was 104-fold greater than that of the 14:7:D8 epitope. Western blot analysis of the day 21 culture supernatants using antibody 18:6:D6 showed that there was relatively little immunoreactive material in the control culture (FIG. 2B). The IL-1α culture medium revealed a band at a slightly higher molecular weight than CB 11 (25.2 kDa). The combination of IL-1α and plasminogen dramatically increased amounts of this peptide. In the same culture this band was also identified by antibody 14:7:D8 (data not shown). Because of its molecular weight this band may represent the collagenase-generated TCA fragment that has undergone significant degradation or a relatively intact TCB fragment.

OA is characterized by a complex induction and activation of proteolytic activities. In the MMP cascade there are multiple sites that may be targeted for therapeutic manipulation. The CII peptide(s) identified in this study may have utility in evaluating the importance of each of these particular targets for inhibiting articular cartilage breakdown. The models described in this study represent viable screening methodologies for evaluating metalloproteinase inhibitors. The *in vitro* rabbit explant model coupled with the measurement of CII degradation peptides is amenable to high-volume screening, while the Hulth-Telhag model demonstrates rapid elevation in metalloproteinase activity resulting in considerable CII peptide release.

## REFERENCES

1. MURPHY, G. *et al.* 1992. The matrix metalloproteinases and their inhibitors. Am. J. Respir. Cell Biol. **7:** 120–135.
2. PROCKOP, D.J. *et al.* 1995. Collagens: molecular biology, diseases, and potentials for therapy. Ann. Rev. Biochem. **64:** 403–434.
3. DODGE, G.R. & A.R. POOL. 1989. Immunohistochemical detection and immunochemical analysis of type II collagen degradation in human normal, rheumatoid, and osteoarthritic cartilages and in explants of bovine articular cartilage cultured with interleukin 1. J. Clin. Invest. **83:** 647–661.
4. HOLLANDER, A.P. *et al.* 1994. Increased damage to collagen type II collagen in osteoarthritic articular cartilage detected by a new immunoassay. J. Clin. Invest. **93:** 1722–1732.
5. BILLINGHURST, R.C. *et al.* 1997. Enhanced cleavage of type II collagen by collagenases in osteoarthritic articular cartilage. J. Clin. Invest. **99:** 1534–1545.
6. DOCHERTY, A.J.P. *et al.* 1992. The matrix metalloproteinases and their natural inhibitors: prospects for treating degenerative tissue diseases. Trends Biotechnol. **10:** 200–207.
7. SAITO, S. *et al.* 1997. Collagen degradation induced by the combination of Il-1α and plasminogen in rabbit articular explant culture. J. Biochem. **122:** 49–54.

# Inhibition of Articular Cartilage Degradation in Culture by a Novel Nonpeptidic Matrix Metalloproteinase Inhibitor

R. CLARK BILLINGHURST,[a,b] KEVIN O'BRIEN,[a] A. ROBIN POOLE,[c]
AND C. WAYNE McILWRAITH[a]

[a]Equine Orthopaedic Research Laboratory, Department of Clinical Sciences,
Colorado State University, Fort Collins, Colorado 80523, USA

[c]Joint Diseases Laboratory, Shriners Hospitals for Children, Department of Surgery,
McGill University, Montreal, Quebec, Canada H3G 1A6

## INTRODUCTION

The integrity of articular cartilage is dependent, in large part, upon its two main structural components, aggrecan and type II collagen. The loss of aggrecan from the extracellular matrix and, more importantly, the subsequent breakdown of the collagen framework are believed to signal the irreversible stage of cartilage degradation that is a characteristic of many human arthritides.[1] The normal turnover of these extracellular matrix molecules in healthy cartilage and their accelerated breakdown in disease are the direct result of the activity of proteolytic enzymes produced within the cartilage by chondrocytes and/or from exogenous cellular sources within bone, synovium, and synovial fluid. The matrix metalloproteinases (MMPs) are a family of zinc-dependent neutral endoproteinases that are capable of degrading most of the protein components of the extracellular matrix of articular cartilage.[2] Catabolic cytokines, such as interleukin-1 (IL-1) and tumor necrosis factor–alpha (TNF-$\alpha$), are potent stimulators of MMP gene expression.

The inhibition of MMPs has been targeted as a potentially significant therapeutic approach in the management of joint disease.[3] This can take place at the gene level by inhibiting the known mediators of MMP expression, such as IL-1 and TNF-$\alpha$, or at the protein level by either preventing the activation of these enzymes that are secreted as zymogens or by directly inhibiting the mature activated enzymes. Tissue inhibitors of metalloproteinases (TIMPs) are normally produced to balance levels of active MMPs present in tissues and body fluids. The development of inhibitors of the activated MMPs is also the area that has seen the most activity in pharmaceutical research during recent years, in large part due to the involvement of MMPs in cancer as well as arthritis. The first-generation inhibitors were synthetic peptides, designed to mimic the cleavage site of the natural MMP substrate. They were coupled to chelating agents that would bind to the enzyme's active site catalytic zinc atom, thereby preventing the cleavage of the enzyme's natural substrate. Although proved to be quite effective *in vitro*, these inhibitors had low bioavailability, necessitating

[b]Address for correspondence: R. Clark Billinghurst, D.V.M., Ph.D., Equine Orthopaedic Research Laboratory, Department of Clinical Sciences, Colorado State University, Fort Collins, CO 80523. Phone, 970/491-4593; fax, 970/491-4138; e-mail, rbilli@lamar.colostate.edu

very large doses or alternatives to oral dosing, and resulting in increased side effects during clinical trials. The new generation of MMP inhibitors has addressed these problems,[4] and this study examines the effects of one such inhibitor in an *in vitro* model of cytokine-induced articular cartilage degradation.

## RESULTS AND DISCUSSION

The inhibitory constants of the nonpeptidic synthetic MMP inhibitor BAY 12-9566 for the collagenases MMP-8 and MMP-13 were determined to be 2.2 $\mu$M and 66 $\mu$M, respectively, using purified equine type II collagen, recombinant human enzymes, and the recently described COL2-3/4C$_{short}$ immunoassay[5] for detecting collagenase-cleaved type II collagen fragments. These IC$_{50}$s, along with the previously determined IC$_{50}$ of >5 $\mu$M for the collagenase MMP-1, indicate that BAY 12-9566 is much less effective in inhibiting the mammalian collagenases than the gelatinases MMP-2 (17 nM) and MMP-9 (512 nM), stromelysin-1 (MMP-3: 165 nM), or type I matrix metalloproteinase (MT1-MMP: 414 nM).

Articular cartilage was collected from the metacarpophalangeal joints of six horses (age = 2 years) and sectioned into 25–50-mg pieces. These explants were incubated for 20 days in serum-free culture medium with recombinant human IL-1$\alpha$

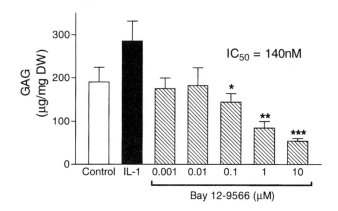

**FIGURE 1.** Effect of the MMP inhibitor BAY 12-9566 on glycosaminoglycan (GAG) release from articular cartilage explants. Equine articular cartilage was incubated for 20 days in serum-free control media or 10 ng/ml of IL-1$\alpha$ in the absence or presence of BAY 12-9566. Conditioned media were collected every other day and analyzed for proteoglycan breakdown by measuring sulfated GAG concentrations using the dimethylmethylene blue dye assay.[13] Shown are the cumulative amounts ($\mu$g) of GAG released per milligram of cartilage dry weight over the 20 days of culture, expressed as the means of six replicates for each test condition (*error bars* = SEM). The percent inhibition of IL-1–stimulated GAG release was plotted against inhibitor concentration (not shown) to determine the IC$_{50}$ value shown in the figure. Statistically significant differences between IL-1–induced GAG release totals for each concentration of inhibitor and GAG release totals for IL-1–stimulated explants cultured without inhibitor are shown as * for $p < 0.05$, ** for $p < 0.01$, and *** for $p < 0.001$.

(10 ng/ml), in the absence or presence of 10-fold increasing concentrations (1 nM to 10 µM) of BAY 12-9566. Coincubation with this inhibitor significantly reduced in a dose-dependent manner the catabolic effects of IL-1α on proteoglycan (FIG. 1) and type II collagen (FIG. 2) degradation and release in the cartilage cultures. In addition, endogenous unstimulated release was inhibited. The most significant overall positive effects were noted for 100 nM of BAY 12-9566 in terms of decreased collagenase-cleaved type II collagen fragment generation and release ($p < 0.01$), decreased proteoglycan release ($p < 0.05$), and increased proteoglycan synthesis and DNA content ($p < 0.05$; not shown). Although the higher concentrations of 1 µM and 10 µM resulted in statistically greater reductions in the IL-1–induced generation/release of cleaved type II collagen ($p < 0.01$) and proteoglycan turnover ($p < 0.001$), both concentrations negatively affected proteoglycan synthesis and DNA content (data not shown). This, along with the determination of $IC_{50}$s for the inhibition of cleaved type II collagen (7 nM) and proteoglycan (140 nM) release, support the use of BAY 12-9566 at submicromolar levels in this system.

This study suggests a significant role for stromelysin-1, the gelatinases, and/or MT1-MMP in the generation and/or release of cleaved type II collagen from IL-1–stimulated articular cartilage. This is based on the observations that the $IC_{50}$ for

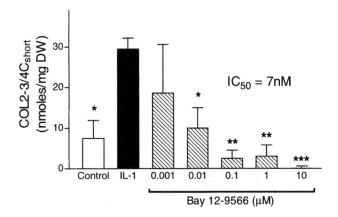

**FIGURE 2.** Effect of the MMP inhibitor BAY 12-9566 on collagenase-cleaved type II collagen release from articular cartilage explants. Equine articular cartilage was incubated for 20 days in serum-free control media or 10 ng/ml of IL-1α in the absence or presence of BAY 12-9566. Conditioned media were collected every other day and analyzed for type II collagen breakdown by measuring the amount of collagenase-cleaved type II collagen using the recently described COL2-3/4C$_{short}$ immunoassay.[5] Shown are the cumulative amounts (nmoles) of type II collagen fragments bearing the collagenase-generated COL2-3/4C$_{short}$ neoepitope released per milligram of cartilage dry weight over the 20 days of culture, expressed as the means of six replicates for each test condition (*error bars* = SEM). The percent inhibition of IL-1–stimulated cleaved type II collagen release was plotted against inhibitor concentration (not shown) to determine the $IC_{50}$ value shown in the figure. Statistically significant differences between IL-1–induced cleaved type II collagen totals for each concentration of inhibitor and cleaved collagen totals for IL-1–stimulated explants cultured without inhibitor are shown as * for $p < 0.05$, ** for $p < 0.01$, and *** for $p < 0.001$.

cleaved collagen release (7 nM) was approximately 300- to 10,000-fold lower than the inhibitory constants of BAY 12-9566 for MMP-1, -8, and -13, and only 2- to 70-fold lower than $IC_{50}$s for MMP-2, MMP-3, MMP-9, and MT1-MMP. Stromelysin has been shown to significantly enhance collagenase activity,[6,7] so inhibition of MMP-3 may lead indirectly to lower levels of cleaved triple helical type II collagen through a subsequent reduction in collagenase activity. MT1-MMP can digest native fibrillar type II collagen[8] and can induce progelatinase and procollagenase activation cascades.[9] The $IC_{50}$ for proteoglycan degradation (140 nM) approximates the $IC_{50}$ for MMP-3, supporting involvement of this enzyme in aggrecan catabolism,[10] although inhibition of the recently characterized proteinase "aggrecanase" by this MMP inhibitor cannot be ruled out.[11,12] One must also consider the potential for more concentrated tissue levels of the inhibitor due to the high protein-binding characteristics of BAY 12-9566.

## ACKNOWLEDGMENTS

This work was supported by funding provided by Bayer AG, Leverkusen, Germany.

## REFERENCES

1. JUBB, R.W. & H.B. FELL. 1980. The breakdown of collagen by chondrocytes. J. Path. **130:** 159–167.
2. BIRKEDAL-HANSEN, H. *et al.* 1993. Matrix metalloproteinases: a review. Crit. Rev. Oral Biol. Med. **4:** 197–250.
3. CAWSTON, T.E. 1996. Metalloproteinase inhibitors and the prevention of connective tissue breakdown. Pharmacol. Ther. **70:** 163–182.
4. WHITE, A.D. *et al.* 1997. Emerging therapeutic advances for the development of second generation matrix metalloproteinase inhibitors. Curr. Pharmaceut. Design **3:** 45–58.
5. BILLINGHURST, R.C. *et al.* 1997. Enhanced cleavage of type II collagen by collagenases in osteoarthritic articular cartilage. J. Clin. Invest. **99:** 1534–1545.
6. MURPHY, G. *et al.* 1987. Stromelysin is an activator of procollagenase. Biochem. J. **248:** 265–268.
7. BRINCKERHOFF, C.E. *et al.* 1990. Rabbit procollagenase synthesized and secreted by a high yield mammalian expression vector requires stromelysin (matrix metalloproteinase-3) for maximal activation. J. Biol. Chem. **265:** 22262–22269.
8. OHUCHI, E. *et al.* 1997. Membrane type 1 matrix metalloproteinase digests interstitial collagens and other extracellular matrix macromolecules. J. Biol. Chem. **272:** 2446–2451.
9. COWELL, S. *et al.* 1998. Induction of matrix metalloproteinase activation cascades based on membrane-type 1 matrix metalloproteinase: associated activation of gelatinase A, gelatinase B and collagenase 3. Biochem. J. **331:** 453–458.
10. FLANNERY, C.R. *et al.* 1992. Identification of a stromelysin cleavage site within the interglobular domain of human aggrecan. J. Biol. Chem. **267:** 1008–1014.
11. ARNER, E.C. *et al.* 1998. Cytokine-induced cartilage proteoglycan degradation is mediated by aggrecanase. Osteoarthritis Cart. **6:** 214–228.
12. ARNER, E.C. *et al.* 1999. Generation and characterization of aggrecanase. J. Biol. Chem. **274:** 6594–6601.
13. FARNDALE, R.W. *et al.* 1986. Improved quantitation and discrimination of sulphated glycosaminoglycans by use of dimethylmethylene blue. Biochim. Biophys. Acta **883:** 173–177.

# Collagen-PVP Decreases Collagen Turnover in Synovial Tissue Cultures from Rheumatoid Arthritis Patients

J. FURUZAWA-CARBALLEDA,[a,b] J. ALCOCER-VARELA,[c]
AND L. DÍAZ DE LEÓN[a]

[a]Department of Cellular Biology, Instituto de Investigaciones Biomédicas,
U.N.A.M. Ciudad Universitaria, P.O. Box 70228, Mexico City 04510, Mexico

[c]Department of Immunology and Rheumatology,
Instituto Nacional de la Nutrición Salvador Zubirán,
Vasco de Quiroga 15, CP 14000, Mexico City, Mexico

Rheumatoid arthritis (RA) is a chronic systemic disorder. Affected joints exhibit inflammation, abnormal immune response, and synovial hyperplasia. The perpetuation of the inflammatory process is mediated by the increased expression of cell adhesion molecules (CAM), such as members of the selectin (ELAM-1) and integrin families ($\alpha_{1-6}$, $\alpha_d$, $\beta_{1-4}$, etc.), and of the immunoglobulin gene superfamily (ICAM-1 and VCAM-1), prior to activation by proinflammatory cytokines.[1] CAM regulate cartilage and bone destruction favoring cellular influx that (concomitantly) overexpress matrix metalloproteinases (MMPs). MMPs are secreted by cells in an inactive form, and they are activated in the extracellular space by the action of other proteinases. The regulation of these enzymes is tightly controlled via cytokines and interactions with tissue inhibitors of the metalloproteinases (TIMPs).[2] Our aim was to evaluate total collagen content, the relative percentage of types I and III collagen, calcium-dependent and -independent collagenolytic activity, as well as TIMP-1 and adhesion molecule expression in synovial tissue (ST) cultures treated with a γ-irradiated mixture of pepsinized porcine type I collagen and polyvinylpyrrolidone (collagen-PVP). This biocompound has been shown to improve skin wound repair and bone fractures in rats.[3] Also, intralesional injection of collagen-PVP in hypertrophic scars diminished pruritus, pain, erythema, volume, inflammatory infiltrates; and improved tissue architecture. Moreover, the biodrug has been demonstrated to modulate extracellular matrix turnover, mainly types I and III collagen, and to downregulate the expression level of IL-1β, TNF-α, PDGF, and VCAM-1.[4] Besides, intradermal administration of collagen-PVP to scleroderma lesions has been shown to improve skin texture, appearance, and tissue architecture and to downmodulate IL-1β and ELAM-1 expressions.[5] Collagen-PVP has the advantage of being a biologic drug with minimal risks, because no side effects have been determined in healthy volunteers and hypertrophic scar patients treated for long periods with the biodrug. This was evaluated by clinical and laboratory tests. Based on the above-mentioned observations, this study included five RA patients with mean

[b]Corresponding author. Address for telecommunication: Phone, 525/6 22 38 19; fax, 525/6 22 38 97; e-mail, furuzawa@servidor.unam.mx

age of $41.0 \pm 17.0$ years and a mean disease duration of $9.6 \pm 5.1$ years. All of them fulfilled the American College of Rheumatology criteria for the diagnosis of RA. Disease-modifying antirheumatic drugs and nonsteroidal antiinflammatory drugs were prescribed for all patients before total knee joint replacement surgery. ST fragments of approximately $3–5 \text{ mm}^3$ were obtained and incubated during 7 days with 500 µl of RPMI-1640 with 10% FCS and with or without 1% collagen-PVP (Fibroquel). Supernatants and ST were collected after 3 and 7 days of culture. Total collagen content from ST was quantified by colorimetric reaction of the Woessner micromethod,[6] and the results were normalized by µg of DNA. The relative percentage of types I and III collagen was evaluated from cultured ST. Duplicate samples were analyzed by interrupted gel electrophoresis and densitometric analysis. Calcium-dependent and -independent collagenolytic activity was measured in the supernatants of tissue cultures by a biochemical assay, determining the degradation of rat N-[propionate-2,3-$^3$H]–labeled type I collagen in $CaCl_2$ or EDTA buffer, and the results were normalized by µg of DNA. Supernatant concentration of TIMP-1 was assessed by one-step sandwich enzyme-linked immunosorbent assay (ELISA) system, according to the manufacturer's instructions (Amersham, UK). Immunohistochemical procedures to detect ICAM-1 and VCAM-1 were carried out according to Krötzsch-Gómez *et al.*[4] Adhesion molecule expression was assessed by estimating positively staining cells in blood vessels and mesenchyma, and it was reported as the percentage of immunoreactive cells. All experiments were performed at least in duplicate. Statistical analysis was performed by paired Student *t*-test. Data were expressed as the mean $\pm$ SEM (standard error of the mean); $p < 0.05$ was considered significant. Histological data indicated that rheumatoid synovium from control cultures presented a variable content of inflammatory cells and fibrosis. ST *in vitro* treatment with 1% collagen-PVP diminished inflammatory infiltrates and increased type III collagen (data not shown). Densitometric analysis of interrupted gel electrophoresis showed a two-fold increase of type III collagen in treated synovium, in a time-dependent fashion versus control cultures ($p = 0.001$, treated compared to untreated cultures; FIG. 1a). However, treated ST did not display changes in total collagen content compared to control cultures, except at 3 days of culture ($p = 0.02$; FIG. 1b). On the other hand, calcium-dependent collagenolytic activity in supernatants from ST-treated cultures exhibited slightly lower levels compared to untreated supernatants, but they were not statistically significant (FIG. 1c). In contrast, calcium-independent collagenolytic activity from treated cultures diminished in a time-dependent manner at statistically significant levels compared to untreated cultures ($p = 0.02$ and $p = 0.04$ for 3 and 7 days of culture, respectively, treated versus untreated; FIG. 1d). TIMP-1 levels were threefold lower in ST supernatants from the collagen-PVP treated group than in those from control samples. Values were statistically significant at 7 days of culture ($p = 0.03$, treated versus untreated cultures; FIG. le). ICAM-1 and VCAM-1 were detected in ST in both blood vessels and spread cells. Significant differences between control and treated groups were found, where ST collagen-PVP treated cultures downregulated ICAM-1 and VCAM-I expression. Values were statistically significant in blood vessel cells (ICAM-1, $p = 0.04$ at 7 days of culture and VCAM-1, $p = 0.05$ and $p = 0.01$ at 3 and 7 days of culture respectively, treated versus untreated cultures; FIGS. 2a and b). Also, spread cells exhibited the same pattern as blood vessel cells (ICAM-1, $p = 0.03$ and VCAM-1,

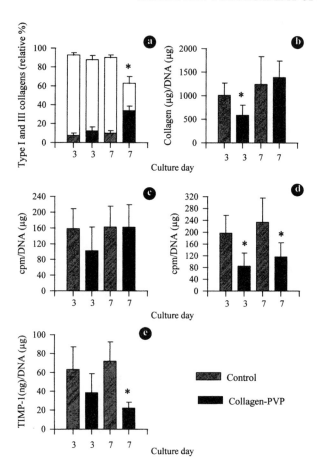

**FIGURE 1.** Collagen-PVP effect on collagen turnover in ST cultures. **(a)** Types I and III collagen relative percentage. Homogenates from ST were treated with pepsin for 72 hr and then dialyzed in 5 mM acetic acid. Twenty μl of ST homogenates were applied in a 0.1% SDS, 7.5% polyacrylamide gel. Gel electrophoresis was interrupted in the middle, then 5 μl of 5% β-mercaptoethanol solution was applied. Gel was incubated for 15 min, and then the system was switched on. Gel was stained for proteins with silver nitrate, and densitometric analysis was performed. Differences between control and biodrug-treated groups were observed at 7 days of culture (*$p = 0.001$). *Open bars* represent type I collagen percentage, *hatched bars* represent type III collagen in control cultures, and *solid bars* represent type III collagen in treated cultures. **(b)** Total collagen content in ST cultures with or without 1% collagen-PVP treatment. Collagen concentration was measured by colorimetric reaction in acid hydrolyzates to determine hydroxiproline content. Comparison between control and treated cultures showed differences at 3 days of culture with *$p = 0.02$. **(c)** Calcium-dependent collagenolytic activity. To measure the degradation of rat N-[propionate-2,3-$^3$H]–labeled type I collagen, the former was incubated with 140 μl of supernatants from 3 and 7 days of culture and 5 mM CaC1$_2$ buffer, pH 7.4, for 24 hr at 35°C, in a total volume of 500 μl. Reaction was stopped by the addition of 50 μl of 80 mM $o$-phenantroline and incubated for 30 minutes. Then proteins were precipitated with 500 μl of 1,4-dioxan, and the tubes were centrifuged.

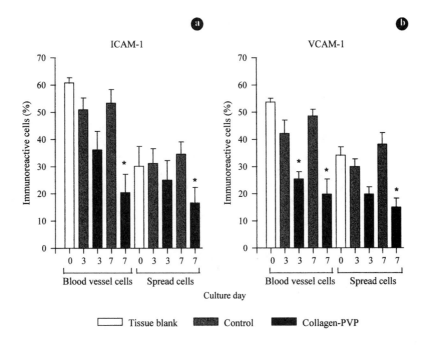

**FIGURE 2.** Collagen-PVP effect on ICAM-1 and VCAM-1 expression in ST culture. The percentage of adhesion molecule–expressing immunoreactive cells was determined in blood vessels and spread cells of ST cultures with or without 1% collagen-PVP treatment during 7 days, **(a)** ICAM-1 values were different in blood vessels and spread cells between treated and control cultures, with the following statistical levels: blood vessel cells, *$p = 0.04$ at 7 days of culture; spread cells, *$p = 0.03$ at 7 days of culture. **(b)** VCAM-1 values in blood vessels: *$p = 0.05$ and *$p = 0.01$ at 3 and 7 days of culture, respectively. In spread cells, *$p = 0.03$ at 7 days of culture. Results represent the mean ± SEM of at least two tissue sections from each patient.

---

**FIGURE 1/continued.** Radioactivity in 150 μl of each supernatant was measured in a liquid scintillation counter. **(d)** Calcium-independent collagenolytic activity. Measurement was performed as described above, but 10 mM EDTA buffer was added instead of 5 mM $CaCl_2$ buffer. Collagen-PVP treated groups were compared with untreated controls, these data were statistically significant with *$p = 0.02$ for 3 days of culture and *$p = 0.04$ for 7 days of culture. **(e)** TIMP-1 production in ST cultures with or without collagen-PVP treatment. The levels (ng/DNA μg) were measured in supernatants of cultures diluted 1:5 to 1:20. TIMP-1 expression in biodrug-treated cultures was significantly lower relative to that in control cultures (*$p = 0.03$ at 7 days of culture, treated versus untreated cultures). Results represent the mean ± SEM of at least two tissue fragments from each patient.

$p = 0.03$ at 7 days of culture, treated versus untreated control cultures; FIGS. 2a and b). Based on these results, we suggest that collagen-PVP modulates collagen turnover, since it decreases collagenolytic activity, as well as TIMP-1 production and increases the amount of type III collagen similar to levels observed in healthy ST controls (data not shown). Possibly the chronic inflammatory process is altered by collagen-PVP action, due to ICAM-1 and VCAM-1 downmodulation. This may be related to the downregulation of IL-1 and TNF-$\alpha$, since both proinflammatory cytokines are capable of inducing proliferation and migration of synovial cells via CAM,[1] as well as of inducing the expression and activation of collagenolytic enzymes.[2] This correlates with previous data in hypertrophic scars and scleroderma lesions treated with the biodrug, where these proinflammatory cytokines and adhesion molecules diminished with treatment.[4] In conclusion, collagen-PVP induces a downmodulation but not an inhibition of inflammatory parameters and probably allows a gradual and better recovery of homeostasis in ST of RA patients.

## ACKNOWLEDGMENT

We acknowledge Dr. Edgar Krötzsch for his critical review; Alejandro Quintana Díaz, B.Sc., for technical assistance; and Ana Luisa Weckmann, M.Sc., for correcting the English version of the manuscript. This work was partially supported by Grants LDL-94 provided by ÁSPID S.A. de C.V., PADEP 030308 and 030352 (UNAM), and PUIS (UNAM).

## REFERENCES

1. PALEOLOG, E.M., M. HUNT et al. 1996. Deactivation of vascular endothelium by monoclonal anti-tumor necrosis factor $\alpha$ antibody in rheumatoid arthritis. Arthritis Rheum. **39:** 1082–1091.
2. BRENNAN, F.M., K.A. BROWNE et al. 1997. Reduction of serum MMP-1 and MMP-3 in rheumatoid arthritis patients following anti-tumour necrosis factor-$\alpha$ (cA2) therapy. Br. J. Rheumatol. **36:** 643–650.
3. ALMAZÁN DIAZ, A., J.C. DE LA CRUZ GARCIA et al. 1996. Investigación experimental de la regeneración ósea en fémures de rata después de la aplicación de colágena I polimerizada: estudio radiológico, histológico e histoquímico. Rev. Mex. Ortop. Traum. **10:** 142–152.
4. KRÖTZSCH-GÓMEZ, F.E., J. FURUZAWA-CARBALLEDA et al. 1998. Cytokine expression is downregulated by collagen-polyvinylpyrrolidone in hypertrophic scars. J. Invest. Dermatol. **111:** 828–834.
5. BARILE, L., J. FURUZAWA-CARBALLEDA et al. 1998. Comparative study of collagen-polyvinylpyrrolidone vs. triamcinolone acetate in systemic sclerosis. Clin. Exp. Rheumatol. **16:** 370.
6. WOESSNER, J.F. 1961. The determination of hydroxyproline in tissue and protein samples containing small proportions of this iminoacid. Arch. Biochem. Biophys. **93:** 440–447.

# Matrix Metalloproteinase Inhibitors Block Osteoclastic Resorption of Calvarial Bone but not the Resorption of Long Bone

V. EVERTS,[a,b,c] W. KORPER,[a] A.J.P. DOCHERTY,[d] AND W. BEERTSEN[b]

[a]Academic Medical Centre, Department of Cell Biology, University of Amsterdam,
P.O. Box 22700, 1100 DE Amsterdam, the Netherlands

[b]Department of Periodontology, Academic Centre for Dentistry Amsterdam (ACTA),
University of Amsterdam, Amsterdam, the Netherlands

[d]Celltech, Slough, UK

## INTRODUCTION

It is generally accepted that cysteine proteinases (CPs) play an essential role in the degradation of bone matrix by osteoclasts. With respect to another class of proteinases, the matrix metalloproteinases (MMPs), contradictory results have been presented.[1,2] Since in these studies osteoclasts from different bone types were used, we hypothesized that functional differences exist between osteoclasts that occupy different sites of the body. In this study we compared osteoclastic bone degradation of calvarial and long bones and analyzed (i) the effect of selective CP- and MMP-inhibitors, and (ii) the resorption of bone slices by isolated calvarial and long bone osteoclasts and the effect of proteinase inhibitors thereupon.

## MATERIAL AND METHODS

Calvariae and long bones were dissected from 5-day-old mice and cultured for 24 hr in M-199 with 5% fetal calf serum in the presence or absence of the CP-inhibitor E-64 (40 μM), or one of the MMP-inhibitors CT166, CT1399, CT1746, or CT1847 (each at a final concentration of 10 μM). At the concentrations used the inhibitors block all enzymes of their respective class. Following the culture period the explants were processed for electron microscopic analysis.[2]

Calvariae and long bones were dissected from 5-day-old rabbits, and osteoclasts were isolated and seeded on bone slices obtained from bovine cortical bone. The cells were cultured for 48 hr in the presence or absence of the proteinase inhibitors as indicated above. After culturing the bone slices were cleaned of cells and processed for scanning electron microscopic analysis.

[c]Address for correspondence: V. Everts, Academic Medical Centre, University of Amsterdam, Department of Cell Biology and Histology. P.O. Box 22700, 1100 DE Amsterdam, the Netherlands. Phone, 31 20 5664720; fax, 31 20 6974156; e-mail, v.everts@amc.uva.nl

## RESULTS

### *Effect of Proteinase Inhibitors on Osteoclastic Bone Degradation*

*Calvaria*

Inhibition of the activity of CPs or MMPs resulted in the occurrence of large demineralized areas of nondigested bone matrix adjacent to actively resorbing osteoclasts.[2] These data indicate that mineral dissolution continued, whereas resorption of the matrix was strongly inhibited. Resorption of this matrix appears to depend on both CP- and MMP-activity.

*Long Bone*

Inhibition of the activity of CPs resulted, as with calvarial bone, in the occurrence of demineralized matrix in the subosteoclastic resorption zone.[3] Such an effect was not found in the presence of either MMP inhibitor (FIG. 1), thus suggesting that re-

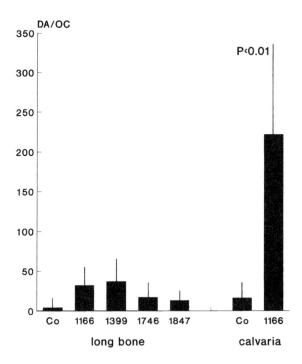

**FIGURE 1.** The effect of the MMP-inhibitors CT1166, CT1399, CT1746, and CT1847 on the volume density of demineralized bone adjacent to actively resorbing osteoclasts in cultures of long bones and calvariae. Bone explants were cultured for 24 hr in the absence or presence of the inhibitors and subsequently morphometrically analyzed. Data are expressed as demineralized area per osteoclast (DA/OC).

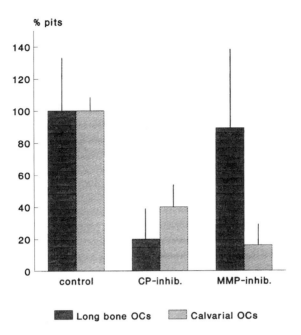

**FIGURE 2.** Number of resorption lacunae of bone slices incubated with isolated osteo-clasts obtained from long bones or calvariae. Osteoclasts were cultured in the absence or presence of the CP inhibitor E-64 or the MMP-inhibitor CT1166.

sorption of bone matrix of long bones depends on the activity of CPs but not on MMP activity.

### Resorption by Isolated Osteoclasts

Culturing of isolated osteoclasts seeded on bone slices resulted in the appearance of resorption pits. Bone resorption by osteoclasts isolated from both sources (calvaria and long bone) was inhibited by the CP inhibitor (FIG. 2). Inhibition of MMP activity had no effect on resorption by isolated long bone osteoclasts, whereas an inhibitory effect was seen with calvarial osteoclasts (FIG. 2).

## CONCLUSIONS

1. Osteoclastic degradation of calvarial bone depends on the activity of CPs and MMPs.
2. Osteoclastic degradation of long bone depends on the activity of CPs. MMPs do not play an important role in this process.
3. Osteoclasts constitute functionally different subpopulations.

## REFERENCES

1. DELAISSÉ, J.M. *et al.* 1987. The effects of inhibitors of cysteine proteinases and collagenase on the resorptive activity of isolated osteoclasts. Bone **8:** 305–313.
2. EVERTS, V. *et al.* 1998. Cysteine proteinases and matrix metalloproteinases play distinct roles in the subosteoclastic resorption zone. J. Bone Miner. Res. **13:** 1420–
   1430.
3. EVERTS, V. *et al.* 1988. Effects of the proteinase inhibitors leupeptin and E-64 on
   osteoclastic bone resorption. Calcif. Tissue Int. **43:** 172–178.

# Matrix Metalloproteinase Inhibitor CGS 27023A Protects COMP and Proteoglycan in the Bovine Articular Cartilage but not the Release of Their Fragments from Cartilage after Prolonged Stimulation *in Vitro* with IL-1α

VISHWAS GANU,[a,b] RICHARD MELTON,[a] WEIGWANG WANG,[a]
AND DON ROBERTS[c]

[a]*Arthritis and Bone Metabolism, Novartis Institute for Biomedical Research,
556 Morris Avenue, Summit, New Jersey 07901, USA*

[c]*Tulane University, Covington, Louisiana 70433, USA*

## INTRODUCTION

In arthritis, the metalloproteinases (MP) such as stromelysin-1, collagenase-1 and -3, MT1-MMP, 92-kDa gelatinase, and as yet unidentified MP aggrecanase are thought to play a role in the degradation of proteoglycans (PG), cartilage oligomeric matrix protein (COMP), and type II collagen, the components of articular cartilage. It is not yet clear which of these MP plays a dominant role in the degradation of cartilage. The matrix metalloproteinase inhibitors CGS 27023A[1] and BB-94[2] are potent inhibitors of several MPs, but only BB-94 inhibits aggrecanase activity.[3] Here, we investigated whether inhibition of these two activities is needed in protecting the cartilage components PG and COMP. We used them in an IL-1α–induced bovine cartilage degradation assay[4] and determined their effect on the release of these cartilage components from cartilage and also their retention in the cartilage after prolonged IL-1 stimulation of cartilage.

## MATERIALS AND METHODS

### Reagents

The supplies for tissue culture were obtained either from Sigma (St. Louis, MO) or GIBCO BRL (Gaithersburg, MD). Rabbit anti-peptide antibody to the C-terminal domain of COMP was prepared, and bovine articular cartilage (BAC) was isolated, not in full thickness, from tarsal joints of young calves and cultured as previously

[b]Address for correspondence: Vishwas Ganu, Ph.D., Novartis Institute for Biomedical Research, 556, Morris Ave., Bldg130.3207, Summit, NJ 07901. Phone, 908/277-7956; fax, 908/ 277-2577; e-mail, vishwas.ganu@pharma.novartis.com

described.[5] Matrix metalloproteinase inhibitors CGS 27023A (N-hydroxy-2(R)-[[4-methoxysulfonyl](3-picolyl)amino]-3-methylbutaneamide hydrochloride and BB-94 ((2S, 3R)-5-methyl-thienylthio) methyl]hexanohydroxamic acid were synthesized in house by Drs. D. Parker and J. Skiles (Chemistry research, Novartis Institute for Biomedical Research, Summit, NJ).

## Effect of Drug Treatment on Proteoglycan and COMP in Cartilage Stimulated with IL-1α

Two BAC slices, not in full thickness, weighing ~80 mg/slice/well, in 2 ml of serum-free medium (DMEM containing 50 µg/ml bovine serum albumin) were cultured in a 24-well flat-bottom tissue culture plate. The following experimental groups, in replicates of 3, were arranged: unstimulated control, cartilage stimulated with 10 ng/ml recombinant human IL-1α (rhIL-1α, Cistron, Pinebrook, NJ), and cartilage stimulated with 10 ng/ml rh IL-1α in the presence of 3 µM and 10 µM of each of CGS 27023A and BB-94. The fresh IL-1 and inhibitors were added on day 3 and day 7, and the media were harvested on days 3, 7, and 11. On day 11, all BAC pieces were suspended in 0.2% FCS in DMEM. On Day 14, the BAC chips were blotted on Kim wipes, weighed, and fixed in STF tissue fixative for histological analysis. The proteoglycan in the media was assayed as previously described.[4] The harvested media was concentrated, and COMP fragments in the media were analyzed by Western analysis using rabbit anti-C-terminal COMP peptide antibody.[5] The immunoreactive bands were quantitated using Sigma gel software.[5]

## Histological Assessment of Cartilage Explants

To detect proteoglycans in the cartilage, samples were processed using a routine paraffin procedure, sectioned at 3–6 µm, and all were stained with safranin O-fast green using a hematoxylin counterstain. COMP was detected by immunohistology using an AEC staining kit (Dako) and the rabbit-antipeptide antibody to the C-terminal domain of COMP as the primary antibody (1:100 dilution).

## RESULTS AND DISCUSSION

FIGURE 1 shows that the IL-1α stimulation of BAC resulted in the release of proteoglycan into the tissue culture media, and that this release was not inhibited at any time by CGS 27023A. BB-94 inhibited the proteoglycan release on day 3 but not on day 7 or day 11. On the basis of these results, we expected about the same level of safranin O staining for proteoglycan in the inhibitor-treated cartilage and in the cartilage treated with IL-1 alone. But contrary to these predictions, when compared with IL-1–treated cartilage, we found moderate safranin O staining for proteoglycan in CGS 27023A- and BB-94-treated cartilage specimens with pronounced staining in the pericellular region of chondrocytes. This was totally unexpected since IL-1–treated cartilage had severe loss of safranin O staining for proteoglycan throughout the matrix, and the results of the analysis of the culture fluids had suggested that the two inhibitors were not protecting the proteins from degradation. The unstimulated

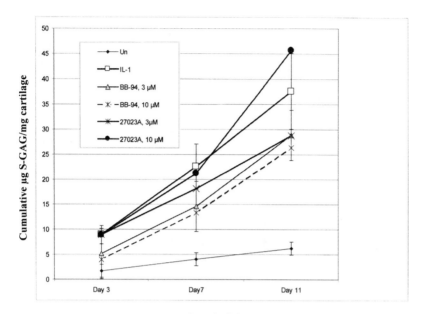

**Days in Culture**

**FIGURE 1.** Effect of CGS 27023A and BB-94 on IL-1–induced proteoglycan released from bovine articular cartilage in culture. Cartilage pieces in 24-well plates were cultured in serum-free DMEM containing 50 µg/ml bovine serum albumin for 11 days without any stimuli (Un) or with 10 ng/ml rh IL-1α alone (IL-1) or with 10 ng/ml rh IL-1α together with 3 or 10 µM CGS 27023A or BB-94. The amount of PG lost into the incubation media was determined after the media changes, done on days 3, 7, and 11, by dimethylmethylene blue dye binding assay. Each graph is cumulative, and each time point is a mean from replicates of 6 (IL-1 and Un groups) or 3 (IL-1 and inhibitors). Only the 10 µM BB-94 treatment group inhibited the proteoglycan released into the media on day 3 to an extent that was statistically significant ($p < 0.05$); all other treatment groups did not significantly inhibit the proteoglycan released into the media measured on days 3, 7, and 11.

cartilage, when compared with other experimental groups, had the most intense safranin O staining throughout the matrix. Therefore, our results suggest that non–MP-dependent mechanisms are also involved in the release of proteoglycan into the media, but inhibition of MP can still play a role in retaining newly synthesized proteoglycan in the cartilage.

FIGURE 2 shows the densitometric analysis of 67–80-kDa and 180-kDa COMP fragments that were detected in the tissue culture media. Both metalloproteinase inhibitors arrested their formation in the media on day 3, BB-94 being slightly more effective with regards to the formation of 67–80-kDa fragment. But these inhibitors appear to lose their effectiveness in inhibiting fragment release into the media on day 7 and day 11. Immunohistochemical analysis of the cartilage treated with IL-1, in

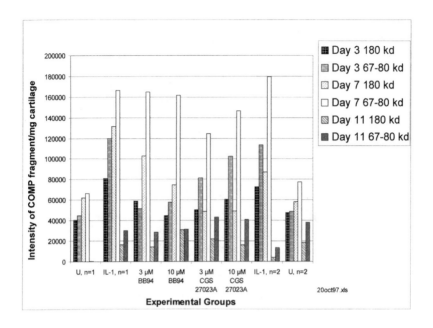

**FIGURE 2.** Densitometric analysis of 67–80-kDa and 180-kDa COMP fragments released into the media after IL-1α stimulation of bovine articular cartilage in culture. Cartilage pieces in 24-well plates were cultured in serum-free DMEM containing 50 µg/ml bovine serum albumin for 11 days without any stimuli (U, unstimulated, $n = 6$) or with 10 ng/ml rh IL-1α alone (IL-1, $n = 6$) or with 10 ng/ml rh IL-1α together with 3 or 10 µM CGS 27023A or BB-94 ($n = 3$). The incubation media from each group was pooled, concentrated using Centricon-10, and 15 µl loaded onto a 4–12% nonreducing SDS-PAGE. After electrophoresis, the samples were transferred onto a nitrocellulose membrane, and immunoreactive bands detected using an antibody to the C-terminal domain of COMP. The 67–80-kDa and 180-kDa immunoreactive bands were quantitated using Sigma gel software. The intensity corresponding to the band was normalized for the volume of the media and the total wet weight of the cartilage and then plotted against the experimental group. Unstimulated cartilage, and IL-1–stimulated cartilage groups were run in duplicate ($n = 3$ in 2 groups). The values of COMP fragment intensities shown above are the averages of the values obtained for two individual groups.

the presence of either of these two drugs, revealed higher staining for COMP in the pericellular region of chondrocytes. This was almost absent in IL-1–treated control cartilage specimens. It is likely, therefore, that the non–MP-dependent mechanisms responsible for the release of proteoglycan may also control the release of COMP fragments into the media. Nevertheless, histology of drug-treated samples also suggests that inhibition of MP is likely to play a favorable role in retaining COMP in the cartilage as well.

## REFERENCES

1. MACPHERSON, L.J. *et al.* 1997. Discovery of CGS 27023A, a non-peptidic, potent, and orally active stromelysin inhibitor that blocks cartilage degradation in rabbits. J. Med. Chem. **40:** 2525–2532.
2. BOTOS, I. *et al.* 1996. Batimastat, a potent matrix metalloproteinase inhibitor, exhibits an unexpected mode of binding. Proc. Natl. Acad. Sci. USA **93:** 2749–2754.
3. BUTTLE, D.J. *et al.* 1993. Inhibition of cartilage proteoglycan release by a specific inactivator of cathepsin B and an inhibitor of matrix metalloproteinases. Arthritis Rheum. **12:** 1709–1717.
4. GOLDBERG, R.L. *et al.* 1995. Time dependent release of matrix components from bovine cartilage after IL-1α treatment and the relative inhibition by matrix metallo-proteinase inhibitors (abstr.) Trans. Orthop. Res. Soc. **20:** 125.
5. GANU, V.S. *et al.* 1998. Inhibition of IL-1α induced COMP degradation in bovine articular cartilage by matrix metalloproteinase inhibitors: potential role for matrix metalloproteinases in the generation of COMP fragments in arthritic synovial fluid. Arthritis Rheum. **41:** 2143–2151.

# Modulation of the Synthesis and Activation of Matrix Metalloproteinases in IL-1–Treated Chondrocytes by Antirheumatic Drugs

T. SADOWSKI[a] AND J. STEINMEYER[b]

*Institute for Pharmacology and Toxicology, University of Bonn,*
*Reuterstr. 2b, D-53113 Bonn, Germany*

The destruction of articular cartilage during osteoarthritis and rheumatoid arthritis is characterized by the degradation and loss of collagen and proteoglycans. Matrix metalloproteinases (MMP) such as collagenase (MMP-1) and stromelysin (MMP-3) belong to the key enzymes of this proteolytic destruction. They are secreted as inactive proforms and must be activated before they can degrade their substrates. The serine protease plasmin, generated from the proenzyme plasminogen by the plasminogen activators u-PA (urokinase type) and t-PA (tissue type), is involved in the activation of these MMPs. Within the extracellular matrix, control over this enzymatic cascade is exerted by specific inhibitors, called tissue inhibitor of metalloproteinases (TIMP) and plasminogen activator inhibitor (PAI). Identification of agents that might inhibit or slow down the degradation of extracellular matrix *in vivo* has long been a therapeutic goal. Our study was therefore designed to determine the potential of some commercially available antirheumatic drugs to reduce the IL-1–mediated increase in proteolytic activities by either inhibiting the biosynthesis of the enzymes MMP-1, MMP-3, t-PA, and u-PA and/or by stimulating the biosynthesis of the specific inhibitors TIMP-1 and PAI-1.

## METHODS

Chondrocytes from the metacarpophalangeal joints of 18–24 month-old steers were isolated by enzymatic digestion and encapsulated in alginate beads. Cultures were maintained for 8 days at 37°C, 5% $CO_2$ and 95% humidity. During the final 48 hours cells were treated with 0.5 ng/ml human recombinant IL-1α in the presence or absence of drugs. MMP activity within the media was determined by using immobilized collagen[1] or [$^3$H]proteoglycan monomers[2] as substrates. Latent MMPs were first activated by the addition of trypsin or APMA. The plasminogen activator activity was measured by a coupled photometric assay.[3] The activity of the inhibitor TIMP was quantitated by its ability to inhibit gelatinase.[4] The mRNA expression of the enzymes and their inhibitors was determined by RT-PCR. Amplification was performed in the presence of digoxigenin(DIG)-11-dUTP. DIG-labeled PCR products were then analyzed by PCR-ELISA. TIMP-1 and MMP-3 protein synthesis was de-

[a]Address for telecommunication: Phone/fax, +49-228-735445.
[b]Present address: Orthopaedic Clinic of the University of Giessen, Paul-Meimberg-Str. 3, D-35385 Giessen, Germany.

termined by radiolabeling the chondrocytes with [$^{35}$S]methionin/cystein during the last 18 hours of the culture period followed by immunoprecipitation using specific antibodies. Precipitates were subsequently subjected to SDS gel electrophoresis. Gels were dried and analyzed by measuring the radioactivity of bands with a linear analyzer (Berthold). PAI-1 protein was determined by ELISA. The detection limit was 0.5 ng/ml PAI-1.

# RESULTS

Treatment of chondrocytes with IL-1 resulted in an increased production of proteases. No effect could be seen on the activity of TIMP or on the synthesis of PAI-1 protein. All drugs were initially tested at a concentration of $10^{-5}$ M. Since the resulting effects could mainly be traced back to alterations in mRNA expression, we further investigated the effects of additional concentrations of drugs on the transcription of these proteins. IL-1–stimulated *MMP-1 mRNA expression* was reduced to nearly baseline by the glucocorticoids dexamethasone and triamcinolone acetonide. This inhibition was maintained even at a concentration of $10^{-7}$ M. Glycosaminoglycan polysulfate (GAGPS) dose dependently inhibited MMP-1 mRNA expression with an $IC_{50}$ of 2.24 µM. Indomethacin, meloxicam, and naproxen also significantly reduced this mRNA expression. However, at a concentration lower than $10^{-5}$ M no effect could be determined. None of the other drugs tested such as acetylsalicylic acid, ademetionine, diacerein, diclofenac-Na, oxaceprol, and tiaprofenic acid proved to be effective. Evaluation of the *mRNA expression of MMP-3* yielded similar results. The glucocorticoids significantly reduced the expression over the whole concentration range tested. GAGPS and meloxicam also reversed the IL-1 stimulatory effect on MMP-3 synthesis with an $IC_{50}$ value of 13.96 µM and 4.51 µM, respectively. Indomethacin and tiaprofenic acid were ineffective at concentrations lower than $10^{-5}$ M, whereas none of the other drugs tested displayed any inhibitory effects. Dexamethasone and triamcinolone acetonide significantly reduced *TIMP-1 mRNA levels,* whereas ademetionine increased the expression of TIMP-1 even at a concentration of $5 \times 10^{-6}$ M. Nearly all of the tested NSAIDS and glucocorticoids as well as GAGPS significantly reduced the IL-1–mediated increased activity of *plasminogen activators,* which appears to be due to the inhibition of mRNA expression during the treatment period. Analysis by RT-PCR revealed differential effects of the drugs on the isoforms t-PA and u-PA. Dexamethasone and triamcinolone acetonide strongly inhibited t-PA mRNA expression even at a concentration of $10^{-7}$ M, whereas the latter glucocorticoid showed only a weak reduction of u-PA expression. Similar results were also obtained for GAGPS. All of the tested NSAIDs dose dependently inhibited the mRNA expression of t-PA. In addition, indomethacin and tiaprofenic acid also reduced the expression of u-PA. The biosynthesis of the inhibitor PAI-1 was enhanced by GAGPS and by all NSAIDs and glucocorticoids tested. Analysis by RT-PCR revealed that these drugs did not increase the *mRNA expression of PAI-1* as measured after 48 hours of treatment. It remains to be determined whether these drugs act posttranscriptionally or enhance the mRNA expression only at the beginning of the treatment period.

## DISCUSSION

Our findings indicate that the tested antirheumatic drugs display differential effects on the biosynthesis and activation of MMPs and the physiological inhibitors TIMP-1 and PAI-1. We observed a significant inhibitory or stimulatory effect on the biosynthesis of enzymes and their inhibitors brought about by several drugs. Only limited information about the *in vivo* concentrations of these drugs within articular cartilage are available.Taking into account the already known serum and/or synovial fluid levels of these drugs and assuming that they are not accumulating within cartilage, we conclude that our *in vitro*–determined effects may be of therapeutic relevance for only few drugs.

## REFERENCES

1. YOSHIOKA, H., I. OYAMADA & G. USUKU. 1987. An assay of collagenase activity using enzyme-linked immunosorbent assay for mammalian collagenase. Anal. Biochem. **166:** 172–177.
2. NAGASE, H. & J.F. WOESSNER, JR. 1980. An improved assay for proteases and polysaccharides employing a cartilage proteoglycan substrate entrapped in polyacrylamide particles. Anal. Biochem. **107:** 385–392.
3. VERHEIJEN, J.H. 1988. Tissue-type plasminogen activator and fast-acting plasminogen activator inhibitor in plasma. Methods Enzymol. **163:** 302–309.
4. KORITSAS, V.M. & H.J. ATKINSON. 1995. An assay for detecting nanogram levels of proteolytic enzymes. Anal. Biochem. **227:** 22–26.

# Effects of 1α,25DihydroxyvitaminD$_3$ on Matrix Metalloproteinase Expression by Rheumatoid Synovial Cells and Articular Chondrocytes in Vitro

LYNNE C. TETLOW AND DAVID E. WOOLLEY[a]

*University Department of Medicine, Manchester Royal Infirmary,*
*Oxford Road, Manchester, M13 9WL, UK*

The biologically active form of vitamin D, 1α,25dihydroxyvitamin D$_3$ (1,25D$_3$), through its interaction with intracellular vitamin D receptors (VDR), is reported to effect a variety of anabolic and catabolic events. 1,25D$_3$ effects changes in calcium and phosphorus metabolism as well as cell proliferation, cell differentiation, and immune function.[1,2] 1,25D$_3$ also modulates chondrocytic function such as proteoglycan and collagen synthesis, and effects changes in matrix metalloproteinase (MMP) expression by rabbit chondrocytes and human mononuclear phagocytes.[1,3,4] As MMPs, synoviocytes, and chondrocytes are implicated in cartilage degradation in rheumatoid arthritis (RA),[5,6] we have examined the distribution of VDR at sites of cartilage erosion and the *in vitro* effects of 1,25D$_3$ on MMP production by synovial fibroblasts and articular chondrocytes derived from rheumatoid tissues.

## RESULTS

Carnoy's fixed tissues from rheumatoid joints were examined for the presence of VDR by immunolocalization. VDR staining of cells was observed in all specimens of rheumatoid synovium ($n = 18$), but to variable extents (range ~5 to 50% of cells). Cell types positive for VDR included macrophages, fibroblasts, and endothelial cells (FIG. 1a), and some lymphocytic cells. Specimens of cartilage-pannus junction ($n = 10$) were also shown to contain VDR-positive cells, including chondrocytes close to the erosive lesion (FIG. 1b). By contrast, negligible staining was observed for chondrocytes in normal, healthy cartilage ($n = 10$), indicating that VDR expression appears to be upregulated in "rheumatoid" cartilage and the overlying pannus tissue.

Rheumatoid synovial fibroblasts (RSF) or human articular chondrocytes (HAC) were grown to confluence in 12-well culture dishes in Dulbecco's Modified Eagles Medium (DMEM) + 10% fetal calf serum. Fresh DMEM containing 2% FCS ± 1,25D$_3$, ± IL-1, or ± IL-1 + 1,25D$_3$ was added for 48 hours, after which the medium was collected and assayed for MMPs-1, -3, and -9 by ELISA methodology. 1,25D$_3$ had no effect on cell numbers during the course of the experiments.

[a]Address for telecommunication: Phone, 0044 0161-276-4240; fax, 0044 0161-274-4833; e-mail, david.woolley@mri.cmht.nwest.nhs.uk

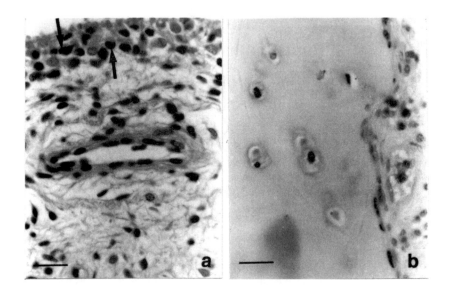

**FIGURE 1.** Immunolocalization of vitamin D receptors (VDR) in the rheumatoid lesion. (a) VDR expression in rheumatoid synovial tissue showing a proportion of synovial lining cells (*arrows*), endothelial cells, and sublining cells positively stained (*black*). *Bar* = 30 μm. (b) VDR expression by chondrocytes and pannus cells at a cartilage-pannus junction. *Bar* = 30 μm. VDR demonstration using a rat monoclonal antibody and detection by the alkaline phosphatase-avidin/biotin system with New Fuchsin substrate.

$1,25D_3$ had no effect on the basal production of MMPs-1, -3, and -9 by RSF; but when these cells were stimulated with interleukin-1β (IL-1), the resultant increase in production of MMPs-1, -3, and -9 was significantly suppressed when cocultured with $1,25D_3$ (TABLE 1). By contrast, the basal production of MMP-1 and MMP-3 by HAC was increased slightly by $1,25D_3$. Although the IL-1 stimulation of MMP-1 production by HAC was unaffected by $1,25D_3$, the IL-1 stimulation of MMP-3 production was significantly enhanced by the addition of $1,25D_3$.

## DISCUSSION

Cytokines play an important role in the regulation of the MMP phenotype for both RSF and HAC cultures, especially the proinflammatory mediators IL-1β and TNF-α. Vitamin D metabolites regulate the level of IL-1 in growth plate cartilage,[3] and both cytokines and MMPs have been demonstrated within the rheumatoid lesion.[6] Macrophages from synovial fluids can synthesize $1,25D_3$, and elevated levels of vitamin D metabolites have been reported in human arthritic synovial fluids. However, the factors that regulate VDR expression by specific cell types remain uncertain.

**TABLE 1.** Effects of 1,25D$_3$ on MMP-1, -3, and -9 production by rheumatoid synovial fibroblasts (RSF) and articular chondrocytes (HAC) *in vitro,* with and without stimulation by interleukin-1β

|  | MMP-1 | MMP-3 | MMP-9 |
|---|---|---|---|
| RSF |  |  |  |
| Control | 195 ± 5 | 18 ± 2 | 37 ± 6 |
| + 1,25D$_3$ | 148 ± 9 | 20 ± 3 | 42 ± 3 |
| + IL-1 | 3129 ± 227 | 893 ± 87 | 143 ± 12 |
| + IL-1 + 1,25D$_3$ | 2034 ± 159 | 280 ± 34 | 72 ± 5 |
| HAC |  |  |  |
| Control | 154 ± 8 | 34 ± 5 | — |
| + 1,25D$_3$ | 171 ± 6 | 46 ± 4 | — |
| + IL-1 | 421 ± 7 | 456 ± 25 | — |
| + IL-1 + 1,25D$_3$ | 453 ± 8 | 764 ± 50 | — |

NOTE: Data given are representative samples from several experiments using RSF ($n = 4$) and HAC ($n = 3$) cultures, each with qualitative similarities for the effects of 1,25D$_3$ and IL-1β. Values represent ngMMP/ml culture medium/$10^6$ cells/48 hr, ± SEM, as determined by ELISA methodology.

## CONCLUSIONS

The microfocal nature of VDR expression demonstrated in the rheumatoid lesion, the availability of vitamin D metabolites in arthritic joints, and the disparate effects of 1,25D$_3$ on the modulation of IL-1–stimulated MMP expression by RSF and HAC support the concept that vitamin D metabolites may contribute to MMP regulation in the pathophysiological processes associated with RA.

## ACKNOWLEDGMENT

This work was supported by Grant No. W0541 from the Arthritis Research Campaign, UK.

## REFERENCES

1. WALTERS, M.R. 1992. Newly identified actions of the vitamin D endocrine system. Endocrine Rev. **13:** 719–764.
2. BINDERUP, L. 1992. Immunological properties of vitamin D analogues and metabolites. Biochem. Pharmacol. **43:** 1885–1892.
3. DEAN, D.D., SCHWARTZ et al. 1997. Interleukin 1α and β in growth plate cartilage are regulated by vitamin D metabolites. J. Bone Miner. Res. **12:** 1560–1569.

4. GOLDRING, M.B. 1993. Degradation of articular cartilage in culture: regulatory factors. *In* Joint Cartilage Degradation—Basic and Clinical Aspects. J.F. Woessner & D.S. Howell, Eds.: 281–345. Marcel Dekker, New York.
5. TETLOW, L.C., M. LEES *et al.* 1993. Differential expression of gelatinase B (MMP-9) and stromelysin 1 (MMP-3) by rheumatoid synovial cells in vitro and in vivo. Rheumatol. Int. **13:** 53–59.
6. TETLOW, L.C. & D.E. WOOLLEY. 1995. Mast cells, cytokines and metalloproteinases at the rheumatoid lesion Ann. Rheum. Dis. **54:** 896–903.

# Synovial Cytokine and Growth Factor Regulation of MMPs/TIMPs: Implications for Erosions and Angiogenesis in Early Rheumatoid and Psoriatic Arthritis Patients

U. FEARON, R. REECE, J. SMITH, P. EMERY, AND D.J. VEALE[a]

*The Rheumatology Rehabilitation Research Unit, University of Leeds, Leeds, LS2 9NZ, UK*

## INTRODUCTION

The two most common types of arthritis—rheumatoid arthritis (RA) and psoriatic arthritis (PsA)—manifest different clinical and radiological features. Initial synovial membrane (SM) studies in PsA demonstrated vascular abnormalities on electron microscopy;[1] immunohistology showed increased vascularity, reduced hyperplasia, and E-selectin expression.[2,3] We have described a distinct macroscopic vascular morphology in PsA.[4] NFκB is an important transcription factor in the induction of cytokine genes[5,6] including TNF-α, IL-1, IL-6, LIF, and OSM, which in turn modulate production of MMP/TIMPs at the site of joint erosion.[7,8] Angiogenesis is under control of growth factors (TGF-β and VEGF) that may be produced in response to cytokines[9] and may play an important role in the increased vascular pattern observed macroscopically and histologically in PsA. This study examines the expression of cytokines (OSM, LIF), transcription/inhibitor factors (NFκB, IκBα), growth factors (TGFβ-1, VEGF) and their production in SM, synovial fluid (SF), and SM lining layer cells (LLC) from patients with early RA (ERA), early PsA (EPsA), and osteoarthritis (OA).

## PATIENTS AND METHODS

Twenty-seven patients (ERA = 9; EPsA = 9 [< 6 months]; OA = 9) underwent arthroscopic study, approved by local ethics committee and after informed consent was obtained. Paired SM biopsies, SF, and unique LLC were obtained under direct visualization. SM biopsies were sectioned (4–7 μm) and stained for NFκB, IκBα, and VEGF using a routine three-stage immunoperoxidase technique. SM was analyzed using a semiquantitative scoring method (0–4). Synovial LLCs were obtained by gentle agitation of the SM lining layer, collected carefully with a syringe and prepared as cytospins for immunofluorescent staining for NFκB and IκBα. Paired SF were analyzed for LIF, OSM, and TGFβ-1 by ELISA (R&D Systems, Abingdon).

[a]Address for correspondence: Dr. Douglas J. Veale, Department of Rheumatology, Leeds General Infirmary, Great Georges Street, Leeds, LS1 3EX, UK. Phone, +44 113 392 3956; fax, +44 113 392 3804; e-mail, rrrdjv@leeds.ac.uk

**TABLE 1. Comparison of VEGF-1, IκBα, and NFκB expression in synovial tissue from patients with ERA and EPsA**

|  | ERA ($n = 10$) | EPsA ($n = 10$) |
| --- | --- | --- |
| VEGF-1 | $0.85 \pm 0.34$ | $2.4 \pm 0.38$ |
| IκBα |  |  |
| (SS) | $1.25 \pm 0.13$ | $1.28 \pm 0.13$ |
| (LL) | $1.25 \pm 0.008$ | $2.08 \pm 0.19^*$ |
| NFκB |  |  |
| (SS) | $1.25 \pm 0.26$ | $1.02 \pm 0.21$ |
| (LL) | $1.75 \pm 0.21$ | $0.67 \pm 0.18^{**}$ |

NOTE:　SS = substroma; LL = lining layer. Values are expressed as mean ± SEM. Level of significance: $p < 0.05$. * $p = 0.008$. ** $p = 0.0094$.

## RESULTS

Increased expression of NFκB : IκBα was observed in ERA SM lining layer compared to EPsA (TABLE 1); no difference was observed in stromal cells. NFκB nuclear translocation was confirmed in isolated synovial ERA LLCs on immunofluorescence staining. Paired SF cytokines levels—LIF and OSM—were significantly higher in ERA vs. EPsA (LIF: 33[8] vs. 11[9] pg/ml, $p = 0.05$; OSM: 47[8] vs. 19[12] pg/ml, $p = 0.06$), respectively. Increased VEGF expression was found in EPsA SM blood vessels compared to ERA SM, associated with increased levels of SF TGFβ-1 in EPsA compared to ERA SF (mean[sem]: 1128[157] and 574[108] ng/ml, $p < 0.02$), respectively. Preliminary results suggest TNF-α mRNA levels in synovial LLC are increased in ERA cells (data not shown). In addition, a distinct and highly vascular pattern was observed in EPsA by videoarthroscopy.

## DISCUSSION

Increased NFκB : IκBα ratio and NFκB nuclear translocation in ERA SM lining layer indicates differential signaling in SM lining layer cells. NFκB plays a major role in upregulation of several cytokines, chemokines, and adhesion molecules that result in MMP production. In this study OSM and LIF levels reflected NFκB expression with increased levels in ERA. OSM and LIF belong to the IL-6 family of cytokines[10] and have previously been shown to modulate production of MMPs and their inhibitors TIMPs.[7,8] These results suggest that the increased activation of NFκB in ERA patients probably leads to increased cytokines, lining layer hyperplasia, and increased MMPs associated with synovial cartilage invasion and joint erosion. Inhibition of NFκB activation by corticosteroids[11,12] is believed to increase apoptosis, which may explain their dramatic clinical response in RA and relative inefficacy in PsA. In contrast, in EPsA there were lower levels of NFκB, OSM, and

LIF, but angiogenic factors VEGF and TGFβ-1 were increased. These results support the observation of a distinct vascular morphology observed on videoarthroscopy[4] and the previously reported increased vascularity of PsA SM histology. This may provide a rationale for the distinct immunohistological pattern found in SM of PsA.[1–3] This is the first report of specific molecular and immunohistocellular factors that may explain the clinical and radiological patterns in PsA and RA. Finally, that divergent patterns occur at early stages of disease provide strong evidence for different pathogeneses.

## REFERENCES

1. ESPINOZA, L.R., F.B. VASEY, C.G. ESPINOZA, T.S. BOCANEGRA & B.F. GERMAIN. 1982. Vascular changes in psoriatic synovium. Arthritis Rheum. **25:** 677–684.
2. VEALE, D., G. YANNI, S. ROGERS, L. BARNES, B. BRESNIHAN & O. FITZGERALD. 1993. Reduced synovial macrophage numbers, ELAM-1 expression, and lining layer hyperplasia in psoriatic arthritis compared to rheumatoid arthritis. Arthritis Rheum. **36**(7): 893–900.
3. JONES, S.M., J. DIXEY, N.D. HALL & N.J. MCHUGH. 1997. Expression of the cutaneous lymphocyte antigen and its counter-receptor E-selectin in the skin and joints of pateints with psoriatic arthritis. Br. J. Rheumatol. **36**(7): 748–757.
4. REECE, R., J.D. CANETE, P. EMERY & D.J. VEALE. 1998. Inter-observer reliability of arthroscopic synovitis grading in inflammatory arthritis (abstr.) Br. J. Rheumatol. **37:** 31.
5. HANDEL, M.L., L.B. MCMORROW & E.M. GRAVALLESE. 1995. Nuclear factor-κB in rheumatoid synovium: localisation of p50 and p65. Arthritis Rheum. **38:** 1762–1770.
6. BLAKE, D.R., P.G. WINYARD & R. MAROK. 1994.The contribution of hypoxic-reperfusion injury to inflammatory synovitis: the influence of reactive oxygen intermediates on the transcriptional control of inflammation. Ann. N.Y. Acad. Sci. **723:** 308–317.
7. LOTZ. M. & P.-A. GUERNE. 1991. Interleukin-6 enhances the synthesis of tissue inhibitor of metalloproteinases-1/erythroid potentiating activity. J. Biol. Chem. **266:** 2017–2020.
8. LANGDON, C., J. LEITH, F. SMITH & C.D. RICHARDS. 1997. Oncostatin M stimulates monocyte chemoattractant protein-1 and interleukin-1–induced matrix metalloproteinase-1 production by human synovial fibroblasts in vitro. Arthritis Rheum. **40:** 2139–2146.
9. FOLKMAN, J. & P.A. D'AMORE. 1996. Blood vessel formation: what is its molecular basis? Cell **87:** 1153–1155.
10. STAHL, N., T.G. BOULTON, T. FARRUGGELLA, N.Y. IP, S. DAVIS, B.A. WITTHUHN, F.W. QUELLE, O. SILVENNOINEN, G. BARBIERI, S. PELLEGRINI, J.N. IHLE & G.D. YANCOPOULOS. 1994. Association and activation of JAK-Tyk kinases by CNTF-LIF-OSM-IL-6 beta receptor components. Science **263:** 92–95.
11. SCHEINMAN, R.I., P.C. COGSWELL, A.K. LOFQUIST, A.S. BALDWIN, JR. 1995. Role of transcriptional activation of IκBα in mediation of immunosuppression by glucocorticoids. Science **270:** 283-286.
12. AUPHAN, N., J.A. DIDONATO, C. ROSETTE, A. HELMBERG & M. KARIN. 1995. Immunosuppression by glucocorticoids: inhibition of NFκB activity through induction of IκBα synthesis. Science **270:** 286–290.

# Development and Application of a Microplate Assay Method for the Mass Screening of MMP Inhibitors

TOSHIFUMI AKIZAWA,[a,b] TAKAYUKI URATANI,[a] MOTOMI MATSUKAWA,[a] ASUKA KUNIMATSU,[a] YUKO ITO,[a] MICHIYASU ITOH,[a] YUKIO OHSHIBA,[c] MASASHI YAMADA,[c] AND MOTOHARU SEIKI[d]

[a]Department of Analytical Chemistry, Faculty of Pharmaceutical Sciences, Setsunan University, 45-1, Nagaotoge-cho, Hirakata, Osaka 573-0101, Japan

[c]Meiji Milk Products Co., Ltd., 26-11, Midori 1-chome, Sumida-ku, Tokyo 130-8502, Japan

[d]Department of Cancer Cell Research, Institute of Medical Sciences, University of Tokyo, 4-6-1, Shirokanedai, Minato-ku, Tokyo 108-8639, Japan

The matrix metalloproteinases (MMPs) are membrane-bond zinc endopeptidases and are thought to be essential for the diverse invasive processes of angiogenesis and tumor metastasis. Thus, inhibition of MMPs has generated considerable interest as a therapeutic strategy. Two approaches have been applied in a search for useful MMP inhibitors: substrate-based design of pseudopeptide derivatives, and random screening of compound libraries and natural products. A number of MMP inhibitors (MMPIs), which are mainly synthetic, have entered into Phase III clinical trials.

To find novel MMPIs that are oral, active, and substrate specific, a simple and convenient assay method is necessary. For this purpose, we developed a microplate assay method, a modification of the flow injection analysis (FIA) method.[1] Two recombinant MMPs (r-MMP-2 and r-MT1-MMP) were expressed as latent forms in *E. coli* in the manner previously used for expression of r-MMP-7.[2] After activation, MMPs were incubated with a fluorogenic substrate peptide (MOCAc-Pro-Leu-Gly-Leu-$A_2$pr(DNP)-Ala-Arg-$NH_2$). The fluorescent intensity of a proteolytic peptide (MOCAc-PropLeu-Gly) was measured every 15 min up to 2 hr with a microplate reader by monitoring at 340 nm for excitation and at 400 nm for emission. The $IC_{50}$ values were calculated from the data obtained after 2 hr of incubation. The $IC_{50}$ values of a PRCGVPD-containing peptide (BS-10; MQKPRCGVPD) and of its analogue peptides against r-MMP-7 were similar to the $IC_{50}$ values obtained by FIA. To confirm the activities of r-MMP-2 and MT1-MMP, the $IC_{50}$ values of the same peptides were measured, and a broad spectrum of inhibitory activities was observed.[3] By using this method, inhibitory activities of more than 80 test samples can be analyzed in only 4 min. For application to the development of novel MMP inhibitors, Chinese drugs and crude drugs were studied by this method. The inhibitory activities of 35 crude drugs are summarized in FIGURE 1. The results of two assays of the components of a green tea are illustrated in FIGURE 2. Only one day was required to mea-

[b]Corresponding author. Phone/fax, +81-720-66-3129; e-mail, akizawa@pharm.setsunan.ac.jp

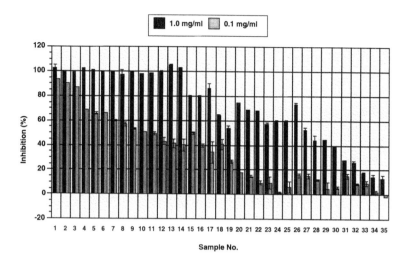

**FIGURE 1.** Inhibitory activities of 35 crude drugs against r-MMP-7.

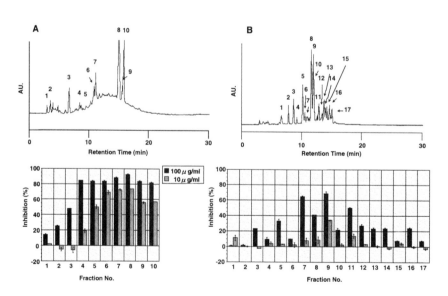

**FIGURE 2.** Results of two assays of the components of a green tea, showing the inhibitory activities of its chromatographic fractions. Separation was performed on a reversed-phase HPLC column (Capcell Pak C$_{18}$, 4.6 mm × 250 mm) for 30 min with a programmed gradient from H$_2$O to 50% CH$_3$CN containing 0.1% TFA. The chromatographic fractions were monitored by UV absorbance at 220 nm.

sure the inhibitory activities of all 17 collected compounds. It is concluded that the microplate assay is a powerful method for the mass screening of novel MMP inhibitors.

## REFERENCES

1. ITOH, M. *et al.* 1997. Flow injection analysis for measurement of activity of matrix metalloproteinase-7 (MMP-7). J. Pharm. Biomed. Anal. **15:** 1417–1426.
2. ITOH, M. *et al.* 1996. Purification and refolding of recombinant human proMMP-7 (pro-matrilysin) expressed in *Escherichia coli* and its characterization. J. Biochem. **119:** 667–673.
3. AKIZAWA, T. *et al.* 1998. Inhibitory activities of peptide-conformers originating from matrix metalloproteinase-7 against MMPs. Peptide Sci. In press.

# Heparin and Fragments Modulate the Expression of Collagen-Degrading Enzymes (Matrix Metalloproteinases 1 and 2) by Human Gingival Fibroblasts

W. HORNEBECK,[a,b] B. GOGLY,[c] G. GODEAU,[c] H. EMONARD,[a] AND B. PELLAT[c]

[a]UPRESA 6021 CNRS, Faculté de Médecine, 51, rue Cognacq Jay, 51100 Reims, France

[c]Laboratoire de Biologie et Physiopathologie Crânio-Faciale, Faculté d'Odontologie, 92120 Montrouge, France

Periodontitis is characterized by a loss of 60–70 percent of gingival collagen. Collagen degradation is catalyzed mainly by collagenases belonging to the matrix metalloproteinase (MMP) family.[1] Neutrophil collagenase (MMP-8) is detected at high levels in gingival crevicular fluid, and expression of fibroblast collagenase (MMP-1) is markedly enhanced in diseased gingiva. More recently, a major function was attributed to gelatinase A (MMP-2), an endopeptidase constitutively expressed by fibroblasts in culture, in collagenolysis of soft connective tissue.[2] Interleukin-1 (IL-1) has been detected in the gingiva of patients with periodontitis and is believed to be one of the most potent inducer of MMP-1 in human fibroblasts.[3,4] Heparin, but not dermatan and chondroitin-6 sulfate, was found to inhibit MMP-1 expression mediated by phorbol ester in arterial smooth muscle cells.[5] We investigated the influence of heparin and a low-molecular-weight heparin fragment (SR80258A), devoid of anticoagulant activity, on the constitutive and IL-1β–induced expression of collagenolytic enzymes—i.e., MMP-1 and MMP-2 by human gingival fibroblasts (HGF) in culture.[6]

## HEPARIN AND FRAGMENT INHIBIT THE CONSTITUTIVE AND IL-1β–INDUCED EXPRESSION OF MMP-1 BY HUMAN GINGIVAL FIBROBLASTS

HGF, at passage 5, were found to secrete 5.6 ng/h/$10^6$ cells of MMP-1 as determined by immunoblot assay. When cell culture medium was supplemented with heparin or fragment, MMP-1 production dropped in a quasilinear fashion. Fifty percent inhibition of enzyme amount was attained at 20 and 7 µg/ml concentrations of heparin and fragment, respectively.

IL-1β (100 U/mL) induced, on average, a 2.4-fold enhancement of MMP-1 secretion by HGF. Heparin or fragment, at increasing doses, were added one hour following cytokine supplementation, and production of MMP-1 was quantified both at protein and mRNA levels (FIG. 1). Both glycosaminoglycans (GAGs) downregulat-

[b]Address for telecommunication: Phone, +33/326 91 35 35; fax, +33/326 91 80 55.

ed the IL-1β–induced MMP-1 expression to basal or subbasal levels; heparin fragment behaved as the more potent inhibitor as compared to its original counterpart, and similar findings were obtained with fibroblasts from dermal origin.

**A**

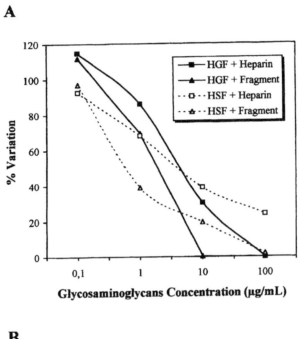

**B**

**FIGURE 1.** Downregulation of the IL-1β–induced MMP-1 in human fibroblasts (G, gingival; S, skin) by heparin and heparin fragment. **(A)** At the protein level (immunoblot assays). **(B)** At the mRNA level (Northern blot analyses). *Lane 1,* HGF without treatment (control); *lane 2,* HGF in presence of IL-1β (100 U/mL); *lane 3,* HGF in presence of IL-1β and 0.1 μg/mL heparin (1 hr later); *lane 4,* HGF in presence of IL-1β and 1 μg/mL heparin (1 hr later); *lane 5,* HGF in presence of IL-1β and 10 μg/mL heparin (1 hr later); *lane 6,* HGF in presence of IL-1β and 100 μg/mL heparin (1 hr later); *lane 7,* HGF in presence of IL-1β and 0.1 μg/mL heparin fragment (1 hr later); *lane 8,* HGF in presence of IL-1β and 1 μg/mL heparin fragment (1 hr later); *lane 9,* HGF in presence of IL-1β and 10 μg/mL heparin fragment (1 hr later); *lane 10,* HGF in presence of IL-1β and 100 μg/mL heparin fragment (1 hr later).

## HEPARIN AND FRAGMENT STIMULATE MMP-2 EXPRESSION BY HUMAN GINGIVAL FIBROBLASTS

HGF in culture secreted an average level of 12.7 ng/h/$10^6$ cells of MMP-2 as determined by gelatin quantitative zymography. Only the proform of the enzyme was evidenced in conditioned medium. Contrary to what was observed for MMP-1, supplementation of HGF culture medium with heparin or fragment stimulated proMMP-2 secretion by cells. Maximal enhancement (2.5-fold) following 36 hr of culture was observed for heparin(s) concentration equal to 10 μg/mL; at 100 μg/mL, however, heparin(s) enhancing effect on MMP-2 secretion was abolished. Here again, similar data were obtained using human skin fibroblasts.

## CONCLUDING REMARKS

In healthy tissues with high collagen degradation rates, as in gingival tissue, collagenase (MMP-1, -8) activity is believed not to be a prerequisite for collagen turnover.[2] To substantiate this assumption, the presence of collagenase in healthy periodontal ligament could not be demonstrated. On the contrary, those soft tissues contained high levels of gelatinase A, an endopeptidase also able to degrade interstitial collagen to some extent.[2] We reported that heparin, as well as a fragment devoid of anticoagulant activity, could enhance MMP-2 production by human gingival fibroblasts in culture, suggesting that such compounds could exhibit an accelerating influence on collagen turnover under normal circumstances.

Periodontitis is strongly associated with the presence of interleukins, major mediators of intense breakdown of collagenous proteins in keeping with their propensity to stimulate collagenase expression.[3,4] We showed that heparin(s) could inhibit IL-1β–induced MMP-1 expression by human gingival fibroblasts; under such investigation, we also reported that c-fos expression induced by Il-1β could also be down-regulated by heparin(s).[6] Recent data in IL-1 signaling indicated that several protein kinases—e.g., protein kinases A and C—but also members of the MAP kinase cascades[7] were involved; activation of those signaling pathways are early events occurring within minutes following binding of the cytokine to its cognate receptor. Since heparin(s) exerted its inhibitory effect on MMP-1 expression one hour following IL-1/receptor interaction, a modulatory influence of those GAGs on the above-mentioned kinases could be partly excluded. Alternatively, as previously reported, heparin(s) could interfere with the binding of AP-1 to TRE common to many MMP genes and c-fos.[5]

These preliminary investigations indicate that heparin fragments are strong modulators of collagen degradation, and, provided their mode of action on MMP-1 and -2 expression be more clearly specified, could be envisaged as useful pharmacological agents during periodontitis.

## ACKNOWLEDGMENTS

This work was supported by the Faculty of Dental Surgery (Montrouge, France) and the Centre National de la Recherche Scientifique (UPRESA 6021 CNRS, France).

## REFERENCES

1. BIRKEDAL-HANSEN, H. 1995. Proteolytic remodeling of extracellular matrix. Curr. Opin. Cell Biol. **7:** 728–735.
2. CREEMERS, L.B., I.D.C. JANSEN, A.J.P. DOCHERTY, J.J. REYNOLDS, W. BEERTSEN & V. EVERTS. 1998. Gelatinase A (MMP-2) and cysteine proteinases are essential for the degradation of collagen in soft connective tissue. Matrix Biol. **17:** 35–46.
3. HAVEMOSE-POULSEN, A. & P. HOLMSTRUP. 1997. Factors affecting IL-1-mediated collagen metabolism by fibroblasts and the pathogenesis of periodontal disease: a review of the literature. Crit. Rev. Oral Biol. Med. **8:** 217–236.
4. VAN DER ZEE, E., V. EVERTS & W. BEERTSEN. 1997. Cytokines modulate routes of collagen breakdown. Review with special emphasis on mechanisms of collagen degradation in the periodontium and the burst hypothesis of periodontal disease progression. J. Clin. Periodontol. **24:** 297–305.
5. AU, Y.P.T., K.F. MONTGOMERY & A.W. CLOWES. 1992. Heparin inhibits collagenase gene expression mediated by phorbol ester-responsive element in primate arterial smooth muscle cells. Circ. Res. **70:** 1062–1069.
6. GOGLY, B., W. HORNEBECK, N. GROULT, G. GODEAU & B. PELLAT. 1998. Influence of heparin(s) on the interleukin-1β-induced expressions of collagenase, stromelysin-1 and tissue inhibitor of metalloproteinases-1 in human gingival fibroblasts. Biochem. Pharmacol. **56:** 1447–1454.
7. LO, Y.Y.C., L. LUO, C.A.G. MCCULLOCH & T.F. CRUZ. 1998. Requirements of focal adhesions and calcium fluxes for interleukin-1-induced ERK kinase activation and c-*fos* expression in fibroblasts. J. Biol. Chem. **273:** 7059–7065.

# Inhibition of Matrix Metalloproteinases by Tea Catechins and Related Polyphenols

M. ISEMURA,[a] K. SAEKI, T. MINAMI, S. HAYAKAWA, T. KIMURA,
Y. SHOJI, AND M. SAZUKA

*School of Food and Nutritional Sciences, University of Shizuoka,
Shizuoka 422-8526, Japan*

We have reported that green tea infusion inhibited *in vitro* invasion and *in vivo* metastasis of murine tumor cells.[1] Since matrix metalloproteinases (MMPs) play roles in these processes, we examined whether green tea catechins inhibit MMPs from the murine tumor cells.[2] Mouse Lewis lung carcinoma LL2-Lu3 cells were cultured in serum-free medium Cosmedium 001, and MMPs were partially purified from the medium by affinity chromatography with gelatin-agarose. Gelatin zymography showed that the carcinoma cells produced MMP-2 and MMP-9 as judged by their molecular masses (FIG. 1). (−)-Epicatechin gallate and (−)-epigallocatechin gallate (EGCG) inhibited in a dose-dependent manner these MMP activities (FIG. 2B), while no inhibition was observed for (+)-catechin and (−)-epicatechin at least up to 100 µM. The black tea components theaflavin and its digallate also exhibited the inhibitory activity (FIG. 2B) and bound LL2-Lu3 MMPs as EGCG did (FIG. 1). The results of affinity chromatography with EGCG, theaflavin, or theaflavin digallate immobilized on agarose (FIG. 1) indicated the direct binding of MMPs to these compounds, which is likely to cause the inhibition.

Since MMPs have been linked to tumor cell invasion, we examined if these catechin-related compounds inhibit the invasion by using a Matrigel (reconstituted basement membrane) invasion system.[1,2] The results indicated that the inhibitors of MMPs inhibited the invasion as well (FIG. 2A).

In order to examine the importance of galloyl groups, galloyl monosaccharides were tested for these activities.[3] The degrees of inhibition of LL2-Lu3 MMPs were 78.9%, 0%, and 0% by tetragalloyl glucose, digalloyl hamamelose, and (mono)galloyl α-glucoside each at 50 µM, respectively. The degrees of inhibition of Matrigel invasion of LL2-Lu3 cells were 99.2%, 72.4%, and 0% each at 50 µM, respectively. The inhibition of invasion by digalloyl hamamelose with no MMP inhibitory activity may be explained by its inhibition (by 39% at 50 µM) of cell adhesion of LL2-Lu3 cells to Matrigel.

These results suggest that previously observed inhibition of metastasis by green tea infusion[1] can be explained at least partly by inhibition by tea catechins of MMPs the activities of which are necessary for tumor cell invasion.

In a subsequent experiment, we examined the effects of EGCG on mRNA expression of MMPs. Expression of mRNAs for MMP-2, MMP-9, and MT1-MMP were assessed by the reverse transcriptase–polymerase chain reaction method, using total

---

[a]Corresponding author. Phone, +81 54 264 5531; fax, +81 54 264 5530;
e-mail, isemura@fns1.u-shizuoka-ken.ac.jp

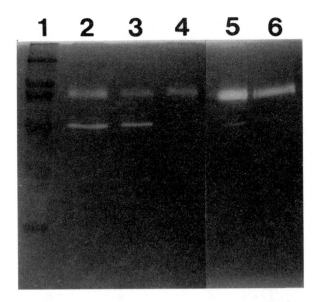

**FIGURE 1.** Zymography of MMPs from LL2-Lu3 cells. Standard proteins *(lane 1):* rabbit muscle myosin (220 kDa), *E. coli* β-galactosidase (116 kDa), rabbit muscle phosphorylase b (97 kDa), bovine serum albumin (66 kDa), chicken ovalbumin (45 kDa), and bovine carbonic anhydrase (29 kDa) *from top to bottom. Lane 2:* conditioned culture medium of LL2-Lu3 cells. *Lane 3:* gelatin-agarose–bound fraction obtained by elution with 1 M NaCl. *Lane 4:* EGCG-agarose–bound fraction obtained by elution with 1 M NaCl. *Lane 5:* theaflavin-agarose–bound fraction obtained by elution with 1 M NaCl. *Lane 6:* theaflavin digallate-agarose–bound fraction obtained by elution with 1 M NaCl. Bands with collagenase activity remain unstained after staining with Coomassie brilliant blue R-250.

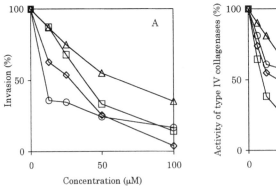
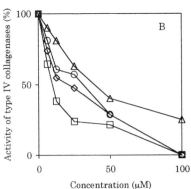

**FIGURE 2.** (**A**) Effects of gallate-containing catechins and theaflavins on Matrigel invasion by LL2-Lu3 cells. The number of invading cells with no EGCG was taken as 100%. □, EGCG; △, (–)-epicatechin gallate; ◇, theaflavin; ○, theaflavin digallate. (**B**) Effects of gallate-containing catechins and theaflavins on type IV collagenases from LL2-Lu3 cells. The activity of type IV collagenases with no EGCG was taken as 100%. □, EGCG; △, (–)-epicatechin gallate; ◇, theaflavin; ○, theaflavin digallate.

RNA from human fibrosarcoma HT-1080 cells cultured in the absence or presence of EGCG. EGCG downregulated expression of mRNAs for MMP-2 and MMP-9 to 52% and 43% of the control (without EGCG) at 4 nM, respectively; while it gave no effects on mRNA expression for MT1-MMP at least up to 40 µM. These results suggest that the low level of MMP production in LL2-Lu3 cells in the presence of EGCG may also contribute to the antimetastatic activity of green tea infusion containing EGCG as a major constituent.

## ACKNOWLEDGMENTS

This work was supported in part by the Program for Promotion of Basic Research Activities for Innovative Biosciences, by the Sasakawa Science Research Grant from the Japan Science Society, and by the Inter-institutional Research Program of the Shizuoka Prefecture for Food Functions.

## REFERENCES

1. SAZUKA, M. *et al.* 1995. Cancer Lett. **98:** 27–31.
2. SAZUKA, M. *et al.* 1997. Biosci. Biotech. Biochem. **61:** 1504–1506.
3. SAEKI, K. *et al.* 1999. Planta Med. **65:** 227–229.

# The Citrus Flavonoid Nobiletin Suppresses the Production and Gene Expression of Matrix Metalloproteinases-9/Gelatinase B in Rabbit Synovial Cells

AKIRA ITO,[a,b] JUN ISHIWA,[b] TAKASHI SATO,[b] YOSHIHIRO MIMAKI,[c] AND YUTAKA SASHIDA[c]

*Departments of [b]Biochemistry and [c]Medicinal Plant Science, School of Pharmacy, Tokyo University of Pharmacy and Life Science, Horinouchi, Hachioji, Tokyo 192-0392, Japan*

## INTRODUCTION

The destruction of connective tissue matrix components under pathological conditions such as in rheumatoid arthritis (RA) and osteoarthritis (OA) is well recognized as impairing joint functions, and matrix metalloproteinases (MMPs) are considered to play a critical role in these processes.

In articular chondrocytes and synoviocytes, interleukin 1 (IL-1) is known to induce and/or augment the production of proMMPs as well as prostaglandin (PG) $E_2$. Recently, it was also reported that PGE suppresses the production of proMMP-1 in human fibroblasts and rabbit synoviocytes along with the increase in intracellular cyclic 3′, 5′-AMP (cAMP). $PGE_2$ also suppresses the proMMP-9/progelatinase B in rabbit articular chondrocytes.[1]

Recently, flavonoids, including quercetin and nobiletin, are known to exert antiinflammatory[2] and anticancer activity,[3] *in vitro*. We have, therefore, examined the effects of citrus flavonoids on the production of proMMPs and $PGE_2$ in rabbit synovial cells.

## MATERIALS AND METHODS

Rabbit synoviocytes were prepared from Japanese white rabbits (female, weighing 1–1.5 kg) as described previously.[1] Confluent rabbit synovial cells were treated with IL-1α and/or flavonoid in serum-free DMEM containing 0.2% (w/v) lactalbumin hydrolysate (LAH) to examine the production of proMMPs and $PGE_2$.

Six flavonoids including nobiletin, sinensetin, tangeretin, and their derivatives were isolated from juice of *Citrus depressa,* Hayata (Rutaceae); the chemical structures of flavonoids are shown in FIGURE 1. The test compounds were added to the

[a]Corresponding author: Akira Ito, Ph.D., Department of Biochemistry, School of Pharmacy, Tokyo University of Pharmacy and Life Science, Horinouchi, Hachioji, Tokyo 192-0392, Japan. Phone, +81-426-76-5706; fax, +81-426-76-5734; e-mail, itoa@ps.toyaku.ac.jp

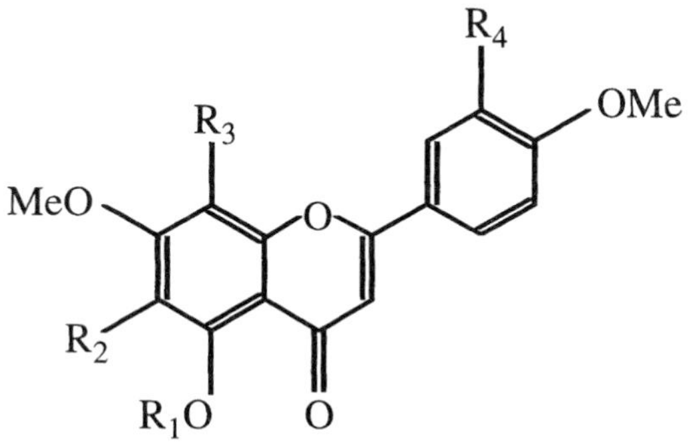

| | R$_1$ | R$_2$ | R$_3$ | R$_4$ |
|---|---|---|---|---|
| 1 5-demethylnobiletin | H | OMe | OMe | OMe |
| 2 Tangeretin | Me | OMe | OMe | H |
| 3 Nobiletin | Me | OMe | OMe | OMe |
| 4 Sinensetin | Me | OMe | H | OMe |
| 5 6-demethoxytangeretin | Me | H | OMe | H |
| 6 6-demethoxynobiletin | Me | H | OMe | OMe |

**FIGURE 1.** Structures of flavonoids isolated from juice of *Citrus depressa* (Rutaceae).

culture medium as a DMSO solution; the final DMSO concentration was 0.1% (v/v) in all cultures.

The contents of proMMPs in the culture media were monitored by gelatin-zymography and Western blotting. The expression of proMMP-9 mRNA were also determined by Northern blotting. PGE$_2$ in the culture media was assayed by radio-immunoassay.

## RESULTS AND DISCUSSION

When confluent rabbit synovial cells were cotreated with IL-1α (1 ng/ml) and the flavonoid (4–64 μM) for 24 hr, nobiletin suppressed most effectively the IL-1α–induced production of proMMP-9 along with the decrease in its mRNA expression in a dose-dependent manner. Nobiletin also inhibited the IL-1α–mediated produc-tion of proMMPs-1 and -3 to a lesser extent than that of proMMP-9. Furthermore, nobiletin effectively interfered with the IL-1α–induced production of PGE$_2$ in the synovial cells. Similar suppressive effects of nobiletin on the production of proMMPs and PGE$_2$ were also observed in rabbit chondrocytes. The above suppres-sive effects of nobiletin were not due to the cytotoxicity since nobiletin modulated neither protein nor DNA synthesis in the confluent synovial cells. In addition, nobiletin did not modulate the intracellular cAMP levels in the synovial cells, indicating that the suppression by nobiletin of the production of proMMPs was not due to the increase in the intracellular cAMP. By contrast, nobiletin suppressed the proliferation of synovial cells at the growth phase in a dose-dependent manner.

These results suggest that nobiletin is a novel antiinflammatory candidate that effectively prevents the matrix breakdown of the cartilage as well as $PGE_2$ production in osteoarthritis and/or rheumatoid arthritis.

## ACKNOWLEDGMENT

We are grateful to Dr. H. Nagase of the University of Kansas Medical Center, Kansas City, Kansas for generously providing us with proMMPs antibodies.

## REFERENCES

1. ITO, A. *et al.* 1995. Cyclooxygenase inhibitors augment the production of pro-matrix metalloproteinase 9 (progelatinase B) in rabbit articular chondrocytes. FEBS Lett. **360:** 75–79.
2. SATO, M. *et al.* 1997. Quercetin, a bioflavonoid, inhibits the induction of interleukin 8 and monocyte chemoattractant protein-1 expression by tumor necrosis factor–$\alpha$ in cultured human synovial cells. J. Rheumatol. **24:** 1680–1684.
3. CHEN, J. *et al.* 1997. Two new polymethoxylated flavones, a class compounds with potential anticancer activity, isolated from cold pressed dancy tangerine peel oil solids. J. Agric. Food Chem. **45:** 364–368.

# Gelastatins, New Inhibitors of Matrix Metalloproteinases from *Westerdykella multispora* F50733

HO-JAE LEE,[a] MYUNG-CHUL CHUNG,[a] CHOONG-HWAN LEE,[a]
HYO-KON CHUN,[a] JOON-SHICK RHEE,[b] AND YUNG-HEE KHO[a,c]

[a]*Enzyme Inhibition Research Unit,*
*Korea Research Institute of Bioscience and Biotechnology,*
*P.O. Box 115, Yusong, Taejon 305-600, Korea*

[b]*Department of Biological Sciences,*
*Korea Advanced Institute of Science and Technology,*
*373-1 Kusong, Yusong, Taejon, 305-701, Korea*

## INTRODUCTION

Matrix metalloproteinases (MMPs) have been implicated in a variety of disease states, including rheumatoid arthritis, periodontal disease, tumor invasion, and metastasis.[1] In the MMP family, gelatinases (MMP-2 and MMP-9) are unique in their ability to cleave type IV collagen, a principal structural component of basement membrane, and have been focused on as pharmacological targets. Consequently specific inhibitors of gelatinases may be of value in the therapy of cancers as well as other disease states involving tissue remodeling. We have isolated nonpeptidic inhibitors of gelatinases from fungal metabolites, designated gelastatins A and B. In this report, inhibitory activities of gelastatins against MMPs and their effects on tumor cell invasion are discussed.

## MATERIALS AND METHODS

Gelastatins A and B were isolated as inhibitors of gelatinase A (MMP-2) from the culture broth of *Westerdykella multispora* F50733.[2] Their structures are shown in FIGURE 1. As reported previously, gelastatins A and B were purified as a mixture of two stereoisomers (5*E*- and 5*Z*-form). After that, gelastatins A and B were successfully separated by HPLC with a chiral column. However, gelastatins A and B are easily converted into each other, so the mixture of gelastatins A and B were used for biological assays.

MMP-2 (gelatinase A) and MMP-9 (gelatinase B) were purchased from Boehringer Mannheim. The catalytic domain of MT1-MMP and BB-94 were generous gifts from Dr. S.E. Ryu (Korea Research Institute of Bioscience and Biotechnology), and recombinant MMP-1 (interstitial collagenase) and MMP-3 (stromelysin-1)

[c]Corresponding author. Phone, +82-42-860-4350; fax, +82-42-860-4595;
e-mail, yhkho@kribb4680.kribb.re.kr.

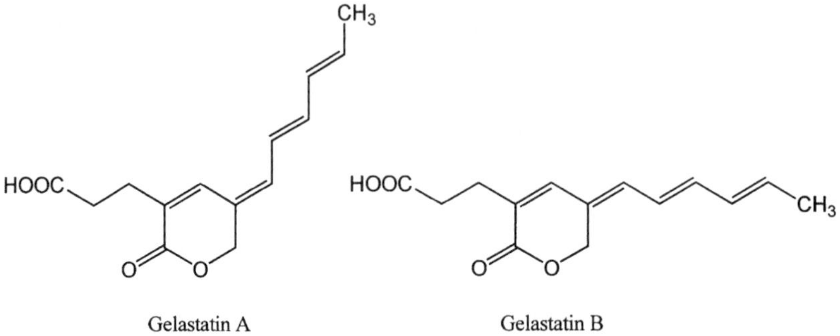

Gelastatin A                                    Gelastatin B

**FIGURE 1.** Structures of Gelastatins A and B.

were from Dr. M.Y. Kim (Hanhyo Institute of Technology, Korea). The activities of MMPs were assayed with a fluorogenic peptide, Mca-Pro-Leu-Gly-Leu-Dpa(Dnp)-Ala-Arg-$NH_2$ (Sigma M6412) as a substrate, and proMMPs were activated with 4-aminophenylmercuric acetate (APMA).[3] The hydrolysis of substrate was assessed by fluorescence with a luminescence spectrometer (Perkin Elmer LS-50B) with excitation at 325 nm and emission at 393 nm. The invasive activity of B16F10 cells was assayed in Transwell cell culture chambers with a membrane filter (Costar 3422) as described previously.[4]

## RESULTS AND DISCUSSION

The inhibitory effects of gelastatins and other MMP inhibitors were determined against MMP-2, MMP-9, MT1-MMP, MMP-1, and MMP-3. Their inhibitory activities are shown in TABLE 1. The 2:1 mixture of gelastatins A and B inhibited MMP-2, MMP-9, and MT1-MMP with an $IC_{50}$ value of 0.63, 5.29, and 6.40 μM, respectively. Gelastatins A and B exhibited rather modest activities against MMPs when compared with BB-94 but are selective inhibitors of MMP-2. The inhibition of MMP-2 by a mixture of gelastatins A and B was reversible and competitive with a $K_i$ value of $6.90 \times 10^{-7}$ M. On the other hand, gelastatins A and B did not inhibit other metalloproteinases including aminopeptidase M and thermolysin with 100 μM. In addition, the methyl esters of the mixture of gelastatins A and B with diazomethane had no inhibitory activity against MMP-2 with 100 μM. This result suggested that carboxylate of gelastatins A and B should be the zinc binding group (ZBG) of the inhibitors. Carboxylate is a less potent ZBG compared to other ZBGs such as hydroxamate, sulfhydryl, and phosphinate. Therefore, more detailed study on the structure-activity relationship will be informative in evaluating the inhibitory activities of gelastatins A and B. In order to investigate the effects of gelastatins A and B on cell invasiveness, *in vitro* tumor cell invasion assays were performed. The mixture of gelastatins A and B inhibited the invasion of reconstituted basement membrane Matrigel by B16F10 melanoma cells dose dependently with an $IC_{50}$ value

**TABLE 1. Comparison of *in vitro* inhibitory activities of gelastatins and other inhibitors**

| Compound | $IC_{50}$ $(\mu M)^a$ | | | | |
|---|---|---|---|---|---|
| | MMP-2 | MMP-9 | MT1-MMP | MMP-1 | MMP-3 |
| Gelastatins[b] | 0.63 | 5.29 | 6.40 | 15.12 | 13.94 |
| Actinonin | 0.18 | 0.23 | 1.58 | 0.84 | 0.52 |
| PLG-NHOH[c] | 1.52 | 0.68 | 7.38 | 12.55 | 2.25 |
| BB-94 | 0.003 | 0.001 | 0.004 | 0.006 | 0.012 |

[a]The concentration of the compound that resulted in 50% inhibition ($IC_{50}$) was calculated from a least-squares fit of percent inhibition and inhibitor concentration. The enzymatic activities of MMPs were measured at 37°C with a fluorometric assay. Final enzyme concentrations in the assay were between 0.2 and 10 nM depending on the enzyme and potency of the inhibitor tested. The substrate, Mca-Pro-Leu-Gly-Leu-Dpa(Dnp)-Ala-Arg-NH$_2$, was presented at a final concentration of 10 $\mu$M in all assays.
[b]2:1 mixture of gelastatins A and B.
[c]N-carbobenzoxy-Pro-Leu-Gly-hydroxamate.

of 9.50 $\mu$g/ml (38.3 $\mu$M). Among microbial metabolites, actinonin,[5] matlystatins,[6] and BE16627B[7] are reported to be the inhibitors of MMPs. Gelastatins A and B are nonpeptidic compounds to inhibit MMPs and are expected to provide the leads for the development of antimetastatic agents.

## REFERENCES

1. WOESSNER, J.F., JR. 1991. Matrix metalloproteinases and their inhibitors in connective tissue remodeling. FASEB J. **5**: 2145–2154.
2. LEE, H.J., M.C. CHUNG, C.H. LEE, B.S. YUN, H.K. CHUN & Y.H. KHO. 1997. Gelastatins A and B, new inhibitors of gelatinase A from *Westerdykella multispora* F0733. J. Antibiotics **50**: 357–359.
3. KNIGHT, C.G., F. WILLENBROCK & G. MURPHY. 1992. A novel coumarin-labelled peptide for sensitive continuous assays of the matrix metalloproteinases. FEBS Lett. **296**: 263–266.
4. SAITO, K., T. OKU, N. ATA, H. MIYASHINO, M. HATTORI & I. SAIKI. 1997. A modified and convenient method for assessing tumor cell invasion and migration and its application to screening for inhibitors. Biol. Pharm. Bull. **20**: 345–348.
5. FAUCHER, D.C., Y. LELIEVRE & T. CARTWRIGHT. 1987. An inhibitor of mammalian collagenase active at micromolar concentrations from an actinomycete culture broth. J. Antibiotics **40**: 357–359.
6. TANZAWA, K., M. ISHII, T. OGITA & K. SHIMADA. 1992. Matlystatins, new inhibitors of type IV collagenases from *Actinomadura altramentaria*. II. Biological activities. J. Antibiotics **45**: 1733–1737.
7. NAITO, K., S. NAKAJIMA, N. KANBAYASHI, A. OKUYAMA & M. GOTO. 1993. Inhibition of metalloproteinase activity of rheumatoid arthritis synovial cells by a new inhibitor [BE16627B; L-N-(N-hydroxy-2-isobutylsuccinamoyl)-seryl-L-valine]. Agents Actions **389**: 182–186.

# Human Metalloproteinase-1 (Collagenase-1) Is a Tumor Suppressor Protein p53 Target Gene

YUBO SUN,[a] YI SUN,[b] LEONOR WENGER,[a] JONI L. RUTTER,[c]
CONSTANCE E. BRINCKERHOFF,[d] AND HERMAN S. CHEUNG[a,e]

[a]Department of Medicine, University of Miami School of Medicine and Geriatric
Research, Education, and Clinical Center, VA Medical Center,
Miami, Florida 33101, USA

[b]Department of Molecular Biology, Parke-Davis Pharmaceutical Research,
Division of Warner-Lambert Company, Ann Arbor, Michigan 48105, USA

Departments of [c]Pharmacology/Toxicology and [d]Medicine,
Dartmouth Medical School, Hanover, New Hampshire 03755, USA

Matrix metalloproteinases (MMPs) are a family of secreted or transmembrane proteins that can degrade all the proteins of the extracellular matrix and have been implicated in many abnormal physiological conditions, including arthritis and cancer metastasis.[1,2] Rheumatoid arthritis (RA) is marked by the destruction of the extracellular matrix and the degradation of native collagen, which are initiated mainly by the action of one type of metalloproteinase, the collagenases: collagenase-1 (MMP-1), neutrophil collagenase (MMP-8), and a recently discovered one, collagenase-3 (MMP-13)[1,3]

The p53 tumor suppressor gene has been implicated in the malignant progression of cancers, and the mutational inactivation of p53 is the most frequent genetic alteration in human cancers. Moreover, recent studies have linked this powerful tumor suppressor to RA.[4–6] It was demonstrated that p53 protein was overexpressed in RA synovium,[4] and that mutant p53 transcripts are present.[5,6] Because p53 protein can function directly as a transcriptional regulator, we speculated that the overexpression of MMP-1 seen in cancer or RA might be associated with the inactivation of p53.

We tested this hypothesis directly by cotransfecting Saos-2 cells with a full-length hMMP1 promoter/luciferase construct[7] (hMMP-1/luci) plus wild-type (wt) and mutant p53 expression plasmids. Transcription assays revealed that wt-p53 downregulated the promoter activity of hMMP-1 in a dose-dependent fashion (FIG. 1, top). Clearly, if this repression activity was crucial to the tumor suppressor activity of wt-p53, naturally occurring p53 mutants might be defective in this function. As expected, most of tumor-derived p53 mutants examined lost this repression activity (FIG. 1, top). Northern blot assays give similar results (FIG. 1, bottom). Two lines of evidence suggest that the repression of hMMP-1 by wt-p53 is a physiologically relevant response. First of all, serum-stimulated expression of endogenous hMMP1 messenger was completely repressed by wt-p53, but not mutant p53 (FIG. 1, bottom). Second, hMMP-1/luci reporter activities were stimulated by etoposide only

[e]Address for correspondence: Dr. Herman S. Cheung, Arthritis Division (D-26), Department of Medicine, P.O. Box 016960, Miami, Florida 33101. Phone, 305/243-5735; fax, 305/243-5655; e-mail, hcheung@mednet.med.miami.edu

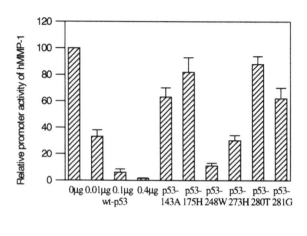

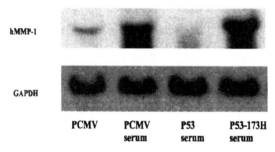

**FIGURE 1.** Modulation of the promoter activity and messenger of hMMP1 by wild-type and mutant p53. Saos-2 cells were cotransfected with varying amounts of the wild-type p53 expression plasmid pCMVp53 or 100 ng mutant p53 together with 1.5 μg of the −4327pMMP-1/luci reporter plasmid using LipofectAMINE in OPM I medium for 18 hours. Cells were then washed with PBS, and fresh McCoy's medium containing 0.5% serum (for dose-response curve experiment) or containing 8% serum (for p53 and p53 mutant) were added. After 24 hr, cells were harvested, and luciferase activity was assayed. The means ± standard errors of the means were derived from three independent transfections and assays, each run in triplicate. The parent vector without p53 coding insert was used to keep total plasmid transfected constant. Relative promoter activity was calculated by arbitrarily setting the activity of the control (cotransfected with parent vector) as 100 *(top)*. Total RNA was isolated from Saos-2 cells after p53 transfection and subjected to Northern analysis *(bottom)*. The Northern blot showed that hMMP-1 messenger was repressed by wt-p53. *Lane 1:* Saos-2 cells were transfected with pCMV (the parent vector plasmid) and incubated in McCoy's medium containing 0.5% serum. *Lane 2:* pCMV transfected cells were incubated in McCoy's medium containing 8% serum. *Lane 3:* wt-p53 expressing plasmid-transfected cells incubated in McCoy's medium containing 8% serum. *Lane 4:* mutant p53-175H transfected cells incubated in McCoy's medium containing 8% serum.

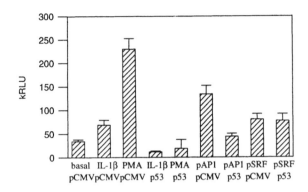

**FIGURE 2.** Repression of both basal and induced hMMP-1 promoter activity by wt-p53 *(columns 1–5)*. Saos-2 or HFF cells were transfected with −4372phMMP-1/luci (1.5 µg) together with 100 ng of pCMVp53 or pCMV as control. Half of the cells were subsequently stimulated either with PMA at a final concentration of 200 nM (Saos-2) or treated with IL-1β at a concentration of 5 ng/ml (HFF) for 24 hr; the other half were incubated in fresh medium without serum as control. Cells were lysed and protein extracts were assayed for luciferase activity. AP-1 as a mediator of p53 repression activity *(columns 6–9)*. Saos-2 cells were cotransfected with pAP1-luci or pSRF-luci together with 100 ng wt-p53 expression plasmid or pCMV as control. The cells were harvested 24 hr later, and protein extracts were assayed for luciferase activity. The means ± standard errors of the means were derived from two independent transfections and assays, each run in triplicate. Relative promoter activity was calculated by arbitrarily setting the activity of the control (cotransfected with parent vector) as 100.

in p53-negative Saos-2 cells, but not in p53-positive U2-OS cells (data not shown). Since etoposide induces both AP-1 and NF-kb,[8] two positive regulators of hMMP-1 expression, it is not surprising to observe hMMP-1 induction by etoposide in Saos-2 cells. Lack of hMMP-1 induction by etoposide in U2-OS cells, however, implies that etoposide-activated p53 executes a suppressive activity, thus diminishing the AP-1/NF-kb effect.[9]

PMA and IL-1β are known to induce hMMP-1 gene expression.[9,10] To examine whether p53 can inhibit the PMA- or IL-1β–induced hMMP-1 gene transcription, Saos-2 or human fibroblast (HFF) cells were transfected with hMMP-1/luci together with 100 ng of p53 expression plasmid or parent vector (pCMV), and subsequently stimulated with PMA or IL-1β. As shown in FIGURE 2, p53 is a strong repressor of the hMMP-1 gene at both basal and induced levels. Cotransfection of the wt-p53 expression plasmid abolished a 7-fold PMA-induced promoter activity in Saos-2 cells and a 3-fold IL-1β–induced promoter activity in HFF cells (FIG. 2). The hMMP-1 promoter activity of all deletion fragments (−4400, −3400, −2400, −1900, −512) was repressed dramatically by wt-p53, indicating that the major cis-acting elements involved in the downregulation of hMMP-1 gene expression reside within the −512/+67 of the 5′ flanking region of the hMMP-1 gene (data not shown). The fact that the hMMP1 promoter lacks p53 binding sites indicates that p53 induced hMMP1 repression is mediated by a p53-binding site-independent mechanism.

Since the AP-1 sites within the hMMP-1 promoter are crucial to both the basal and PMA- or IL-1β–induced promoter activity of hMMP-1,[9,10] it is possible that the p53 repression activity could be mediated partially through AP-1 sites. Transcription assays (Saos-2 cells were cotransfected with pAP1-luci or pSRF-luci together with 100 ng of wt-p53 expression plasmid or pCMV) indicated that the p53 repression activity on hMMP-1 promoter could indeed be mediated, at least partially, through the AP-1 sites found in the hMMP-1 promoter. The luciferase activity of pAP1-luci reporter, but not pSRF-luci reporter, was repressed substantially by wt-p53 (FIG. 2).

Taken together, the results described above show clearly that hMMP-1 is a p53 target gene. Overexpression of human wt-p53 can exert a strong inhibitory effect on the human hMMP-1 promoter, while most p53 mutants reverse this inhibition substantially. Our studies show for the first time that the extracellular matrix degradation seen in RA or in cancer cells due to high-level MMP1 might be, at least partially, related to inactivation of p53.

## REFERENCES

1. WOESSNER, J.F., JR. 1991. Matrix metalloproteinases and their inhibitors in connective tissue remodeling. FASEB J. **5:** 2145–2154.
2. CHAMBERS, A.F. & L.M. MATRISIAN. 1997. Changing views of the role of matrix metalloproteinases in metastasis. J. Nat. Cancer Inst. **89:** 1260–1270.
3. KNAUPER, V., C. LOPEZ-OTIN, B. SMITH, G. KNIGHT & G. MURPHY. 1996. Biochemical characterization of human collagenase-3. J. Biol. Chem. **271:** 1544–1550.
4. FIRESTEIN, G.S., K. NGUYEN, K.R. AUPPERLE, M. YEO, D.L. BOYLE & N.J. ZVAIFLER. 1996. Apoptosis in rheumatoid arthritis. Am. J. Pathol. **149:** 2143–2151.
5. FIRESTEIN, G.S., F. ECHEVERRI, M. YEO, N.J. ZVAIFLER & D.R. GREEN. 1997. Somatic mutations in the p53 tumor suppressor gene in rheumatoid arthritis synovium. Proc. Natl. Acad. Sci. USA **94:** 10895–10900.
6. REME, T., A. TRAVAGLIO, E. GUEYDON, L. ADLA, C. JORGENSEN & J. SANY. 1998. Mutations of the p53 tumor suppressor gene in erosive rheumatoid synovial tissue. Clin. Exp. Immunol. **111:** 353–358.
7. RUTTER, J.L., U. BENBOW, C.I. COON & C.E. BRINCKERHOFF. 1997. Cell-type specific regulation of human interstitial collagenase-1 gene expression by interleukin-1β (IL-1β) in human fibroblasts and BC-8701 breast cancer cells. J. Cell Biochem. **66:** 1–15.
8. GARCIA-BERMEJO, L., C. PEREZ, N.E. VILABOA, E.D. BLAS & P. ALLER. 1998. CAMP increasing agents attenuate the generation of apoptosis by etoposide in promonocytic leukemia cells. J. Cell Sci. **111:** 637–644.
9. GUTMAN, A. & B. WASYLK. 1990. The collagenase gene promoter contains a TPA and oncogene-responsive unit encompassing the PEA3 and AP-1 binding sites. EMBO J. **9:** 2241–2246.
10. AREND, W.P. & J.-M. DAYER. 1993. Cytokines and growth factors. *In* Textbook of Rheumatology. W.N. Kelly, E.D. Harris, Jr., S. Ruddy & C.B. Sledge, Eds.: 227–247. W.B. Saunders. Philadelphia.

# Interaction between Stromal Cells and Tumor Cells Induces Chemoresistance and Matrix Metalloproteinase Secretion

BAOQIAN ZHU, NORMAN L. BLOCK, AND BAL L. LOKESHWAR[a]

*Department of Urology (M-800), University of Miami School of Medicine, Miami, Florida 33101, USA*

## INTRODUCTION

Metastatic prostate cancer is resistant to most cytotoxic chemotherapeutic drugs.[1,2] At the cellular or molecular level the development of resistance to chemotherapeutic drugs by cancer cells is a complex phenomenon. Among the many non-mutational mechanisms that regulate chemosensitivity to anticancer drugs, cell-cell and cell-extracellular matrix (ECM) interactions are believed to be important.[3–5] Because the normal prostatic epithelial cells rely on their interactions with stromal cells and the ECM for their survival, it may be argued that stromal cells or the factors released by them may also alter the response of tumor cells to anticancer drugs.[6] Moreover, the role of tumor cells in modifying the healthy stroma to facilitate tumor cell growth, invasion, and metastasis is also just being unraveled.[7] Furthermore, it has long been known that tumors growing in different metastatic sites are not equally sensitive to cytotoxic actions of chemotherapy drugs.[8,9]

We tested two hypothesis: (1) The differences in the responses of the metastatic tumors to anticancer drugs *in vivo* is due to the interaction of tumor cells with the normal cells that surround them; and therefore, by creating these conditions *in vitro* we should be able to reproduce some of the observed differences *in vitro*. (2) Tumor cells modify the stromal cells and endothelial cells to their advantage for invasion and metastasis by inducing them to secrete matrix-degrading enzymes. These hypotheses were tested on prostate tumor cells by coculturing tumor cells with stromal cells isolated from both the primary tumor site (prostate) and a distant metastatic site (e.g., lung). The drugs used were Taxol and CMT-3 (6,deoxy, 6-dimethyl, 4-dedimethylamino tetracycline). CMT-3 is a a novel cytotoxic compound with multiple cellular targets.[10–12] The effect of tumor cells on the pattern of MMP secretion by stromal cells and endothelial cells was examined in culture-conditioned medium using species-specific enzyme immunoassays (ELISA).

[a]Address for correspondence: Bal L. Lokeshwar, Department of Urology (M-800), University of Miami School of Medicine, P.O. Box 016960, Miami, FL 33101. Phone, 305/243-6321; fax, 305/243-6893.

## RESULTS AND CONCLUSIONS

### *Effect of Stromal Cells on Cytotoxicity of Anticancer Drugs*

We tested the effect of CMT-3 and Taxol on both the human prostate tumor cell line DU145 and on the Dunning rat MAT LyLu prostate tumor cells. Tumor cells were cultured in the bottom wells of Transwell plates with 3μ filter inserts (Corning-Costar Cat. no. 3415). Cells were cultured with or without the normal cells in the inserts. The inserts included complete medium (control), or cultures of normal human prostate epithelial cells, prostate stromal cells, or dermal microvessel endothelial cells (DMVEC). Drugs were added to both top and bottom cultures two days after plating cells. Cytotoxicity was determined after 48 hours by a colorimetric assay (methyl tetrazolium bromide reduction assay, the MTT assay). The 50% inhibition dose ($IC_{50}$) for each drug was estimated for each culture condition. Cells were considered chemoresistant if the ratio between the $IC_{50}$ under coculture condition and tumor cells cultured alone was greater than one.

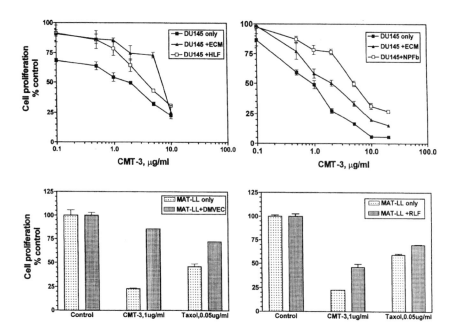

**FIGURE 1.** Effect of stromal cells and ECM on the inhibition of prostate tumor cell proliferation by CMT-3 and Taxol. Tumor cells (DU145 or MAT-LL) were cultured separately using the Costar Teranswell inserts in the presence of a human lung fibroblast cell line (HLF), rat lung fibroblasts (RLF), or on their ECM, prepared as described.[13] Both tumor cells and the fibroblasts were exposed to CMT-3 or Taxol for 48 hours. The cell proliferation (as a measure of cell viability) was quantified by the MTT assay as described.[12] Results presented are for two cell lines from three experiments (mean ± SE). Similar results were obtained for several other prostate cancer cell lines.

Tumor cells cocultured with prostate fibroblasts showed significant chemoresistance to all the three cytotoxic drugs. The resistance was dependent on both tumor cells and the normal cell combination (FIG. 1). Prostate fibroblasts and DMVEC rendered both DU145 and the MAT LyLu cells resistant to CMT-3, and Taxol by 3.0- and 1.7-fold, respectively. Tumor cells cocultured with the normal prostatic epithelial cells did not show any chemoresistance (data not shown). We next examined the effect of ECM prepared from the stromal cells on drug sensitivity of tumor cells. The results shown are for CMT-3 only, although other drugs were tested. The human prostate cancer (DU145) cells cultured on ECM prepared from both lung fibroblasts and prostate fibroblasts showed significant resistance to CMT-3–induced cytotoxicity. The $IC_{50}$ for CMT-3 was 3.7 μg/ml, 1.5 × higher than that for tumor cells cultured alone (FIG. 1A, B). Similarly, the $IC_{50}$ of CMT-3 for MAT-LyLu cells, cultured on rat lung fibroblast–derived ECM, was significantly higher ($IC_{50}$ 7.2 ± 1.2 μg/ml) than that of the same cells cultured without the ECM ($IC_{50}$ 2.5 ± 0.7 μg/ml). These findings show that it is the *stromal cell–tumor (epithelial) cell interaction,* and not the tumor–normal epithelial cell interaction, that protects both cell types from drug-induced cytotoxicity. Interestingly, cell-cell contact was not necessary to induce chemoresistance. The ECM prepared from the specific stromal components alone was also capable of inducing significant chemoresistance. Furthermore, our experiments demonstrate for the first time that conditions observed only *in vivo* can also be reproduced *in vitro,* under an appropriate organ-specific metastasis-dependent microenvironment.

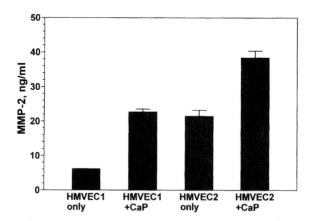

**FIGURE 2.** Tumor cells induce MMP-2 production in microvessel endothelial cells. HMVECs were cultured in the bottom well of the Transwell plates until the cells cover 50% of the growth surface. Tumor cells (MAT-LyLu) cultured in the top inserts separately were transferred to HMVEC culture wells. Culture-conditioned medium from the bottom wells was collected and assayed for MMP-2 or MMP-9 by ELISA. The assay did not detect any MMP-2 or MMP-9 produced by the MATLyLu cells, as the primary antibodies used in the assay kit were specific to human MMPs. Results presented are for MMP-2 only, as the amount of MMP-9 was below the detection limit of the assay. Results presented are pooled from two independent experiments each with triplicate wells.

## *Tumor Cells Induce MMP Secretion*

We analyzed the MMP content in the conditioned medium from cocultures. To distinguish the MMP secreted by stromal cells or endothelial cells from that of tumor cells, we used an immunoassay to analyze MMPs. The anti-MMP antibody used to detect MMP in the conditioned medium was specific to human cell–derived MMP (Biotrack human MMP ELISA kits, Oncogene Sciences/Calbiochem, Cambridge, MA). Conditioned media collected from cocultures of rat MAT LyLu prostate tumor cells and human prostate fibroblasts, human lung microvessel endothelial cells (HMVEC-1), or dermal human microvessel endothelial cells (HMVEC-2) were analyzed for both MMP-9 and MMP-2. Normal lung MVEC or dermal MVECs did not produce significant amounts of MMP-2 in our culture systems. However, there was a two- to fivefold increase in MMP-2 produced by these cells under coculture conditions (FIG. 2). Very little MMP-9 was secreted and therefore could not be quantitated using the ELISA system. These results indicate that tumor cells modulate invasive enzyme production by host cells (stromal cells). This modulation plausibly enhances the invasive and angiogenic potential of tumor cells.

## ACKNOWLEDGMENT

This work was funded by a NIH grant (1R29-CA-61038), the U.S. Army Prostate Cancer Research Program (DAMD 17-98-272), and the L. Austin Weeks Endowment (U. Miami).

## REFERENCES

1. ISMAIL, M. & L.G. GOMELLA. 1997. Current treatment of advanced prostate cancer. Tech. Urol. **3**(1): 16–24.
2. RAGHAVAN, D., B. KOCZWARA & M. JAVLE. 1997. Evolving strategies of cytotoxic chemotherapy for advanced prostate cancer. Eur. J. Cancer **33**(4): 566–574.
3. DONG, Z., R. RADINSKY, D. FAN *et al.* 1994. Organ-specific modulation of steady state *mdr* gene expression and drug resistance in murine colon cancer cells. J. Natl. Cancer Inst. **86:** 913–920.
4. GLEAVE, M., J.T. HSIEH, C. GAO *et al.* 1991. Acceleration of human prostate cancer growth in vivo by factors produced by prostate and bone fibroblasts. Cancer Res. **51:** 3753–3761.
5. PASSANITI, A., J.T. ISAACS, J.A. HANEY *et al.* 1992. Stimulation of human prostatic carcinoma tumor growth in athymic mice and control of migration in culture by extracellular matrix. Int. J. Cancer **51:** 318–324.
6. HAYWARD, S.W., M.A. ROSEN & G.R. CUNHA. 1997. Stromal-epithelial interactions in the normal and neoplastic prostate. Brit. J. Urol. **79** (Suppl 2): 18–26.
7. HEPPNER, K.J., L.M. MATRISIAN, R.A. JENSEN *et al.* 1997. Expression of most matrix metalloproteinase family members in breast cancer represents a tumor-induced host response. Am. J. Pathol. **149**(1): 273–282.
8. DONNELLI, M.G., R. RUSSO & S. GARRATTININ. 1975. Selective chemotherapy in relation to the site of tumor transplantation. Int. J. Cancer **32:** 78–86.
9. TEICHER, B.A., T.S. HERMAN, S.A.HOLDEN *et al.* 1990. Tumor resistance to alkylating agents conferred by mechanisms operative only in vivo. Science **247:** 1457–1461.
10. VAN DEN BOGERT, C., B.H.J. DONTJE, M. HOLTROP *et al.* 1986. Arrest of the proliferation of renal and prostate carcinomas of human origin by inhibition of mitochondrial protein synthesis. Cancer Res. **46:** 3283–3289.

11. SIPOS, E.P., R.J. TAMARGO, J.D. WEINGART & H. BREM. 1994. Inhibition of tumor angiogenesis. Ann. N.Y. Acad. Sci. **732:** 263–272.
12. LOKESHWAR, B.L., H.L. HOUSTON-CLARK, M.G. SELZER *et al.* 1998. Potential application of achemically modified non-antimicrobial tetracycline (CMT-3) against metastatic prostate cancer. Adv. Dent. Res. **12:** 97–102.
13. MIZUGUCHI, H., N. UTOGUCHI & T. MAYUMI. 1997. Preparation of glial extracellular matrix: a novel method to analyze glial-endothelial cell interaction. Brain Res. Protocols **1:** 339–343.

# Inhibition of Gelatinase A by Oleic Acid

H. EMONARD,[a,b] V. MARCQ,[c] C. MIRAND,[c] AND W. HORNEBECK[b]

[b]UPRESA 6021 CNRS, IFR 53 Biomolécules, Faculté de Médecine,
51, rue Cognacq Jay, 51100 Reims, France

[c]UPRESA 6013 CNRS, IFR 53 Biomolécules, Faculté de Pharmacie,
51100 Reims, France

Oleic acid (*cis*-9-octadecenoic acid, OA) inhibited the formation of lung metastases from *subcutaneous* implantation of colon carcinoma cells in mice, such an inhibition being associated with a decrease of gelatinase A (MMP-2) activity in tumor-tissue extracts.[1] We similarly showed that OA considerably reduced MMP-2 activity released into the culture medium conditioned by oncogene-transformed human bronchial epithelial cells (BZR cells).[2] The interaction between this fatty acid and MMP-2 was further analyzed.

We first confirmed the inhibitory capacity of OA towards MMP-2 activity using a fluorogenic quenching substrate, (7-methoxycoumarin-4-yl)acetyl-L-Pro-Leu-Gly-Leu-[N-3-(2,4-dinitrophenyl)-L-2,3-diaminopropionyl]-Ala-Arg-$NH_2$ (Mca-Pro-Leu-Gly-Leu-Dpa-Ala-Arg-$NH_2$).[2]

The inhibition was concentration dependent and significant at OA concentration as low as 1 μM (FIG. 1). $K_i$ was calculated using the following equation:

$$\frac{v_i}{v_0} = 1 - \frac{([E]_0 + [I]_0 + K_i) - \{([E]_0 + [I]_0 + K_i)^2 - 4[E]_0[I]_0\}^{1/2}}{2[E]_0}$$

and found equal to $4.3 \pm 0.4$ μM. A series of carboxylic acids containing long-chain alkyl groups in P1′ proved to act as potent MMP inhibitors.[3] We therefore initially hypothesized that similarly the carboxylate end group of OA could ligate zinc at the active site of MMP-2, the remaining alkyl chain of the fatty acid occupying the hydrophobic subsite of the enzyme. Since hydroxamate is a more potent bidentate ligand for zinc present in the active site of all MMPs, we synthesized the hydroxamate derivative of oleic acid (OA-Hy) aiming to improve the MMP-2 inhibitory capacity of the fatty acid. Similarly to OA, OA-Hy inhibited MMP-2 activity in a dose-dependent manner (FIG. 1). The $K_i$ value was found to be equal to $1.7 \pm 0.2$ μM. Thus, replacement of COOH with CONHOH, a more potent zinc ligand, amplified the inhibitory potential of oleic acid towards MMP-2 by a factor of only 2.5, while 200- and 500-fold enhancements have been previously reported between carboxylate and hydroxamate inhibitors for MMP-7 and MMP-1, respectively.[4,5]

Stromelysin-1 (MMP-3), as MMP-2, contains a deep S1′ hydrophobic "selectivity pocket"; we found, however, that the activity of this enzyme was only partly affected by OA. In addition to the main structural elements characteristic to all MMPs,[6] MMP-2 contains three tandem copies of a 58-amino acid residue fibronec-

[a]Corresponding author. Phone, +33/326 91 80 56; fax, +33/326 91 80 55;
e-mail, herve.emonard@univ-reims.fr

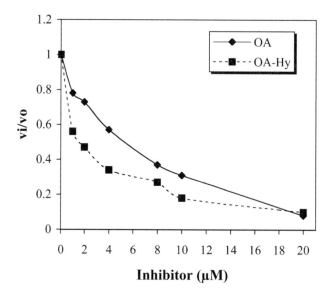

**FIGURE 1.** Inhibition of MMP-2 by oleic acid and its hydroxamate derivative. The active 62-kDa species of MMP-2 (200 pM) was assayed with increasing concentrations of OA or OA-Hy, using the fluorogenic substrate Mca-Pro-Leu-Gly-Leu-Dpa-Ala-Arg-NH$_2$.

tin type II–like module inserted at the NH$_2$-terminal boundary of the zinc binding site.[7] Those domains, located in the vicinity of the active site of MMP-2, were found to mediate the interaction of the enzyme to substrates as gelatin and elastin.[8,9] We recently found that binding of OA to MMP-2 could prevent the adsorption of the proteinase to gelatin (H. Emonard and W. Hornebeck, in preparation). As a whole, these preliminary data suggest that inhibition of MMP-2 by OA is driven by binding of the fatty acid to a hydrophobic region located within the fibronectin type II–like domains of the enzyme. The fit between the original (COOH) or modified (CONHOH) end groups of OA and zinc within the MMP-2 active site is probably far from being maximized.

The fatty acid binding ability of fibronectin-like domains containing MMPs— i.e., MMP-2 and MMP-9—could be useful for designing selective inhibitors for those enzymes.

## ACKNOWLEDGMENT

This work was supported by the Centre National de la Recherche Scientifique (UPRESA 6021 and 6013 CNRS, France).

## REFERENCES

1. Suzuki, I., M. Iigo, C. Ishikawa, T. Kuhara, M. Asamoto, T. Kunimoto, M.A. Moore, K. Yazawa, E. Araki & H. Tsuda. 1997. Inhibitory effects of oleic and docosahexaenoic acids on lung metastasis by colon-carcinoma-26 cells are associated with reduced matrix metalloproteinase-2 and -9 activities. Int. J. Cancer **73:** 607–612.
2. Polette, M., E. Huet, P. Birembaut, F.X. Maquart, W. Hornebeck & H. Emonard. 1999. Influence of oleic acid on the expression, activation, and activity of gelatinase A produced by oncogene-transformed human bronchial epithelial cells. Int. J. Cancer. **80:** 751–755.
3. Beckett, R.P. & M. Whittaker. 1998. Matrix metalloproteinase inhibitors. Exp. Opin. Ther. Patents **8:** 259–282.
4. Browner, M.F., W.W. Smith & A.L. Castelhano. 1995. Matrilysin-inhibitor complexes: common themes among metalloproteases. Biochemistry **34:** 6602–6610.
5. Grobelny, D., L. Poncz & R.E. Galardy. 1992. Inhibition of human skin fibroblast collagenase, thermolysin, and *Pseudomonas aeruginosa* elastase by peptide hydroxamic acids. Biochemistry **31:** 7152–7154.
6. Woessner, J.F. 1991. Matrix metalloproteinases and their inhibitors in connective tissue remodeling. FASEB J. **5:** 2145–2154.
7. Collier, I.E., S.M. Wilhelm, A.Z. Eisen, B.L. Marmer, G.A. Grant, J.L. Seltzer, A. Kronberger, C. He, E.A. Bauer & G.I. Goldberg. 1988. H-ras oncogene-transformed human bronchial epithelial cells (TBE-1) secrete a single metalloprotease capable of degrading basement membrane collagen. J. Biol. Chem. **263:** 6579–6587.
8. Banyai, L.B., H. Tordai & L. Patthy. 1996. Structure and domain-domain interactions of the gelatin-binding site of human 72-kilodalton type IV collagenase (gelatinase A, matrix metalloproteinase-2). J. Biol. Chem. **271:** 12003–12008.
9. Shipley, J.M., G.A.R. Doyle, C.J. Fliszar, Q.Z. Ye, L.L. Johnson, S.D. Shapiro, H.G. Welgus & R.M. Senior. 1996. The structural basis for the elastolytic activity of the 92-kDa and 72-kDa gelatinases. Role of the fibronectin type II-like repeats. J. Biol. Chem. **271:** 4335–4341.

# Attenuation of Oxidant-Induced Lung Injury by the Synthetic Matrix Metalloproteinase Inhibitor BB-3103

HUSSEIN D. FODA,[a,b,c] ELLEN E. ROLLO,[b] PETER BROWN,[d]
HEDAYATOLLAH PAKBAZ,[c] HASAN I. BERISHA,[c] SAMI I. SAID,[b,c]
AND STANLEY ZUCKER[b,c]

[b]Departments of Medicine and Research, VAMC Northport, New York, USA

[c]Department of Medicine, State University of New York at Stony Brook,
Stony Brook, New york 11794, USA

[d]British Biotechnology Ltd., Oxford, England

## INTRODUCTION

Acute, diffuse lung injury often complicates sepsis, gastric acid aspiration, extensive trauma, and other conditions. The lung endothelial and epithelial cells are the early targets of this injury leading to increased pulmonary vascular permeability and pulmonary edema. Clinically, this condition is characterized by catastrophic respiratory failure, known as the *a*dult *r*espiratory *d*istress *s*yndrome (ARDS). Despite advances in our understanding of the pathogenesis of this type of lung injury and in the management of patients with this disorder, the outcome remains grave. There is an urgent need for an effective treatment.

Recently matrix metalloproteinases (MMPs), especially gelatinases, have been implicated in the pathogenesis of acute lung injury. Gelatinase A and B and their activated forms are increased in the bronchoalveolar lavage fluid (BAL) of both animal models of acute lung injury and patients with ARDS.[1–4]

In this study we sought to investigate the protective effect of an MMP inhibitor in an experimental model of acute lung injury caused by oxygen free radicals. We examined the MMP inhibitor BB-3103, a soluble, low-molecular-weight broad spectrum inhibitor obtained from British Biotechnology Ltd., Oxford, England.

## METHODS

We used a known model of oxidant-induced lung injury.[5] Briefly, adult Sprague-Dawley rats (male, 300–350 g) were anesthetized with pentobarbital and tracheotomized. The lungs were ventilated by a Harvard respirator at tidal volume of 6.5 ml/kg. The chest was opened, and the heart was exposed but left *in situ*. Heparin (400 units) was injected into the inferior vena cava. An outflow cannula was placed

[a]Address for correspondence: Hussein D. Foda, M.D., Pulmonary and Critical Care Medicine, SUNY at Stony Brook, Health Science Center, Stony Brook, NY 11794-8172. Phone, 516/261-4400 ext. 2857; fax, 516/261-6016; e-mail, hfoda@mail.som.sunysb.edu

in the left atrium through an incision in the left ventricle, and an inflow cannula was inserted into the main pulmonary artery through the right ventricle. The pulmonary vessels were perfused with a Krebs solution containing 4% albumin equilibrated with 95% $O_2$–5% $CO_2$ at 37°C. The initial perfusion pressure was kept at 8–10 cm $H_2O$ (usually at a flow rate of 9 ml/min). Peak airway pressure ($P_{AW}$) and pulmonary arterial pressure ($P_{PA}$) were continuously monitored by pressure transducers. At the end of the experiment, 60 min later, the left lung was lavaged with 3 ml saline for measurement of bronchoalveolar lavage (BAL) fluid protein content. The right lung was dissected free and gently blotted, weighed (wet weight), and oven dried to a constant weight to determine dry weight and wet-to-dry (W/D) lung weight ratio.

The animals were divided into four groups: (1) xanthine (X) + xanthine oxidase (XO) injury: xanthine (1 mM) was added to the perfusate, followed 10 min later by XO (from buttermilk substantially free of uricase 0.1 U/ml); (2) X + XO as above with pretreatment with BB-3103 (0.5 mg/kg·min) infused into the pulmonary artery 10 min prior to the addition of XO, with the infusion continued for the balance of the experiment; (3) X + XO as above, with pretreatment with BB-3103 (1 mg/kg·min) infused as above; (4) control animals: the rat lungs were perfused with Krebs-4% albumin for 1 hour.

Substrate zymography was performed in 10% polyacrylamide gels that had been cast in the presence of 0.1% gelatin. SDS-PAGE was performed with Tris-glycine SDS in sample and running buffers as described by the manufacturer (NOVEX, San Diego, CA).

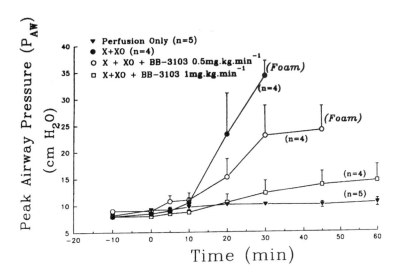

**FIGURE 1.** The MMP inhibitor BB-3103 decreased the rise in $P_{AW}$ pressure caused by X + XO and delayed the onset of lung injury and the appearance of foam in the airways.

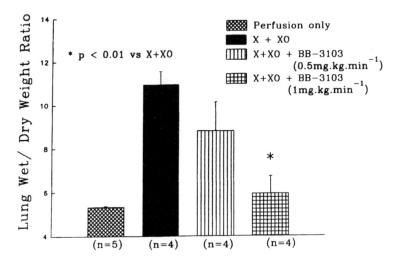

**FIGURE 2.** BB-3103 attenuates the increase in the rise in W/D weight ratio caused by X + XO. * = $p < 0.01$ vs. X + XO only and NS vs. control.

## RESULTS

With the application of X and XO the $P_{AW}$ and $P_{PA}$ increased gradually for the first 10 min then rapidly climbed for the next 20 min until foam appeared in the airway and the experiment stopped (FIG. 1). X + XO also caused an increase of the W/D lung weight ratio, an index of pulmonary edema, of 102% over the W/D weight ratio in control animals. The induction of acute lung injury was accompanied by large bands of activity at 72 kDa (progelatinase A), 62 kDa (activated gelatinase A), and 92 kDa (progelatinase B) in the BAL fluid demonstrated by zymography.

BB-3103 (1 mg/kg·min) completely prevented the rise in $P_{AW}$ and $P_{PA}$ caused by the oxygen free radicals generated by X + XO and significantly attenuated the increase in wet/dry lung weight ratio (FIGS. 1 and 2). Zymogram of the BAL fluid of the rats revealed that BB-3103 prevented the activation of gelatinase A and B caused by the oxidant injury.

## DISCUSSION

These experiments demonstrated that this nonspecific MMP inhibitor BB-3103 is an effective inhibitor of acute lung injury. This effect is most likely secondary to the inhibition of gelatinase activity; however, we cannot exclude the importance of the inhibition of other MMPs.

These results support our hypothesis that MMPs play an important role in acute lung injury and suggest that MMP inhibitors may be useful in the management of patients with ARDS.

## REFERENCES

1. CHRISTNER, P. *et al.* 1985. Am. Rev. Respir. Dis. **131:** 690–695.
2. RICOU, B. *et al.* 1996. Am. J. Respir. Crit. Care Med. **154:** 346–352.
3. TORII, K. *et al.* 1997. Am. J. Respir. Crit. Care Med. **155:** 43–46.
4. DELCLAUX, C. *et al.* 1997. Am. J. Physiol. (Lung Cell Mol. Physiol.) **272:** 442–451.
5. BERISHA, H.I. *et al.* Am. J. Physiol. (Lung Cell Mol. Physiol. 3) **259:** L151–L155.

# Influence of Putative Antiinvasive Agents on Matrix Metalloproteinase Secretion by Human Neoplastic Glia *in Vitro*

H.K. ROOPRAI,[a] A. KANDANEARATACHI, G. RUCKLIDGE, AND G.J. PILKINGTON

*Department of Neuropathology, Institute of Psychiatry, De Crespigny Park, London SE5 8AF, UK*

## INTRODUCTION

The major biological feature of human brain tumors that precludes successful therapy is their propensity to invade the contiguous normal brain. Invasion is a multistep process including adhesion of tumor cells to normal brain elements and to extracellular matrix (ECM) components, degradation and remodeling of the ECM by proteases (such as matrix metalloproteinases), and migration.[1] Matrix metalloproteinases (MMPs) constitute a large family of zinc-dependent endopeptidases that cooperatively degrade all components of the ECM. Much attention has been drawn to the gelatinases (MMP-2 and -9) in invasion since these enzymes can degrade the basement type IV collagen.

The use of antiinvasive agents has been considered as a possible therapeutic approach in addition to conventional surgery and radio-/chemotherapy. The aim of this study was to investigate the effects of four putative antiinvasive agents: swainsonine (a locoweed alkaloid), captopril (an angiotensin-converting enzyme inhibitor that has been used as an antihypertensive drug), and tangeretin and nobiletin (citrus flavonoids) on parameters underlying brain tumor invasion in five cell lines derived from human brain tumors.

## MATERIALS AND METHODS

Surgical samples from patients with various types of brain tumors were obtained from the neurosurgical staff at King's Hospital, London. Cells were routinely maintained in Dulbecco's modified Eagle medium (DMEM) supplemented with antibiotic/antimycotic and 10% fetal calf serum in small plastic culture flasks at 37°C, 5% $CO_2$ in a standard humidified incubator and gradually weaned down to serum-free medium. A chemosensitivity (MTT) assay was initially performed to obtain a drug-dose response.

The cells were treated with the antiinvasive agents: swainsonine at 0.3 μg/ml medium in 1% ethanol, captopril at 30 mM, tangeretin and nobiletin at 4 μg/ml in dimethyl sulphoxide (DMSO); control samples were incubated in serum-free medium

[a]Address for telecommunication: fax, +44 (0) 171 708 3895; e-mail, spkahkr@iop.kcl.ac.uk

**TABLE 1. Qualitative evaluation of MMP-2 and MMP-9 in five human brain tumor cell lines *in vitro***

| Cell line | Control | | Swainsonine | | Captopril | | Tangeretin | | Nobiletin | |
|---|---|---|---|---|---|---|---|---|---|---|
| | MMP-2 | MMP-9 | MMP-2 | MMP-9 | MMP-2 | MMP-9 | MMP-2 | MMP-9 | MMP-2 | MMP-9 |
| IPMA-E | + | + | NC | NC | ↓ | ↓ | ↑ | NC | ↓ | ↓ |
| IPSF-PA | + | + | NC | NC | ↓ | ↓ | NC | NC | NC | NC |
| IPSH-OA2 | + | + | ↑↑ | ↑↑ | ↑ | ↑ | ↓ | ↓ | ↓ | ↓ |
| IPMC-A3 | + | + | ↓ | ↓ | ↓ | ↓ | ↓ | ↓ | ↓ | ↓ |
| AMHFB-GM | + | + | ↑ | ↑ | ↑ | ↑ | ↓ | ↓ | ↓ | ↓ |

NOTE: The prefix "IP" designates the Institute of Psychiatry; "AMH" designates the Atkinson's Morley Hospital, London, UK. + = presence of MMP expression in control samples; − = absence of MMP expression in control samples; NC = no change in enzyme expression; ↑ = upregulation; ↓ = downregulation.

alone and maintained for 48 hours, whereupon they were harvested and counted. The cell-conditioned medium samples served as the source of MMP-2 and MMP-9 for zymography and were freeze-dried and reconstituted to provide a standard of $1 \times 10^6$ cells/ml of reconstituted volume. Zymogram analysis was performed by a modification of the method of Heussen and Dowdle[2] to investigate the activities of the two gelatinases MMP-2 and MMP-9. Proteolytic activity appeared as clear bands on a dark blue background.

## RESULTS

TABLE 1 represents the cumulative results for the qualitative analysis of MMP-2 and MMP-9 secretion by five brain tumor cell lines under the influence of various agents. The zymogram results showed that the four agents had a differential effect upon MMP secretion by the tumor cell lines. Taking each agent into consideration, it appears that swainsonine was least effective, as it showed downregulation of MMP-2 in three high-grade cell lines (IPAB-AO3, IPMC-A3 and IPLC-GM) and downregulation of MMP-9 in only one cell line (IPMC-A3). Captopril showed downregulation of both enzymes in two low-grade tumor cell lines (IPMA-E and IPSF-PA) and one high-grade glioma cell line (IPMC-A3). Interestingly, both citrus flavonoids, tangeretin and nobiletin, were most effective in downregulating MMP activity. Indeed, nobiletin was the best putative antiinvasive agent studied here, as it downregulated MMP expression in all but one cell line (IPSF-PA). Taking individual cell lines into consideration also reveals a differential pattern in that MMP expression by the pilocytic astrocytoma (IPSF-PA) was generally unaffected by the agents. In contrast, enzyme expression by the grade 3 astrocytoma (IPMC-A3) was down-regulated by all four agents.

## DISCUSSION

For an agent to be of any therapeutic value it should be able to downregulate MMP expression. Our data indicates that the four putative antiinvasive agents had a differential effect on MMP expression in a variety of brain tumor cells *in vitro*. Both tangeretin and nobiletin were effective downregulators of MMP-2 and MMP-9 expression in the cell lines studied, compared to swainsonine and captopril. The other interesting finding was that the effects of the four agents varied considerably depending on the cell line used such that the MMP expression by the low-grade pilocytic astrocytoma cell line was generally unaffected by the agents whereas both MMP-2 and MMP-9 expression by the high-grade glioma (IPMC-A3) was downregulated irrespective of the agent used. Moreover, our findings for the inhibitory effects of captopril on MMP expression by some of the cell lines confirms a previous report by Nakagawa *et al.*[3] In conclusion, this study has shown that swainsonine, captopril, tangeretin, and nobiletin have differential inhibitory effects on MMP expression by human brain tumor *in vitro*. Further studies are in progress to assess their effects on different ECM proteins by using motility, invasion, and adhesion assays *in vitro*.

## ACKNOWLEDGEMENTS

This work was supported by the Association for International Cancer Research. The authors are grateful to Dr. W. Widmer, Department of Citrus, Florida for providing the citrus flavonoids.

## REFERENCES

1. LIOTTA, L.A. 1986. Tumour invasion and metastasis: role of the extracellular matrix (Rhodes Memorial Award lecture). Cancer Res. 46: 1–7.
2. HEUSSEN, C. & E.B. DOWDLE. 1980. Electrophoretic analysis of plasminogen activators in polyacrylamide gels containing sodium dodecyl sulfate and copolymerised substrates. Anal. Biochem. **102:** 196–202.
3. NAKAGAWA, T., T. KABOTA, M. KABUTO & T. KODERA. 1995. Captopril inhibits glioma cell invasion in vitro: involvement of matrix metalloproteinases. Anticancer Res. **15:** 1985–1990.

# Effect of a Matrix Metalloproteinase Inhibitor (BB-1101) on Nerve Regeneration

MARIA DEMESTRE,[a,c] N.A. GREGSON,[a] K.J. SMITH,[a] K.M. MILLER,[b] A.J. GEARING,[b] AND R.A.C. HUGHES[a]

[a]Department of Clinical Neuroscience, GKT Medical School, Guy's Campus, London SE1 9RT, UK

[b]British Biotech Pharmaceuticals, Watlington Road, Oxford OX4 5LY, UK

## INTRODUCTION

Guillain-Barré syndrome (GBS) is an acute monophasic autoimmune inflammatory disease of the peripheral nervous system (PNS).[1] The disease is modeled by the experimentally induced disease experimental allergic neuritis (EAN) produced by immunization with peripheral nerve tissue. The neurological deficits in EAN and GBS are a consequence of conduction block in peripheral axons from demyelination and axonal degeneration resulting from autoimmune inflammation. Matrix metalloproteinases are involved in many pathological processes in inflammatory disorders of the nervous system including MS and GBS, as well as their experimental counterparts EAE and EAN.[2,3] The drug BB-1101 is a broad-spectrum MMP inhibitor and a potent TNF-$\alpha$ processing inhibitor. It has been shown that when given from the day of immunization it will prevent the development of EAN, and when given from the onset of symptoms it reduces intraneural inflammation and axonal degeneration.[4] Later studies have demonstrated the expression of MMPs by Schwann cells and endoneurial blood vessels in rats with EAN and also after traumatic nerve injury.[5,6] This suggests that MMPs not only play a role in inflammatory processes, but they could also participate in demyelination and axonal degeneration, and in the reparative phase following the disease episode. In studies on EAN, BB-1101 given at the onset of symptoms not only reduced the severity of the disease but also enhanced the initial recovery; but BB-1101 failed to maintain this effect.[4] These observations raised the possibility that MMP inhibition could be counterproductive in nerve regeneration. We have therefore examined the drug's effects on nerve regeneration following crush injury.

## METHODS

The left sciatic nerve of 27 male Lewis rats, body weight 200–250 g, was crushed by a standardized procedure at the midthigh level and were randomly assigned to one of three treatment groups. Group 1 were given a daily subcutaneous injection of

[c]Address for correspondence: Maria Demestre, Department of Clinical Neurosciences, GKT Medical School, Hodgkin Building, Guy's Hospital, London Bridge, London SE1 9RT, UK. Phone, 0044-171/955-4079; fax, 0044-171/378-1221; e-mail, md@umds.ac.uk

BB1101 2 mg/kg, prepared in PBS-Tween-80(Sigma); group 2 received vehicle only, PBS-Tween-80 daily; and group 3 were left uninjected. The contralateral leg was used for control measurements and to provide control tissue samples. Animals were weighed and monitored by a blinded observer for 63 days following nerve crush. Regeneration was monitored by: (a) observing the recovery of reflex toe spreading using a four-point scale; (b) electromyography (EMG) of motor activity in the extensor digitorum. Animals from each of the groups were anesthetized on selected days for EMG measurements (no animal was anesthetized more than three times before the terminal measurement). Supramaximal stimulation of the tibial nerve was applied at the ankle and the amplitude of the CMAP and the latency recorded. After the final EMG on day 63 the sciatic nerve from both legs and the flexor digitorum from some animals were removed for the measurement of axon and muscle fiber diameters. Axon and muscle area measurements were made using SigmaScan software (Jandel).

## RESULTS

All groups showed the start of recovery of toe spreading at day 16 and complete functional recovery by day 42 after nerve crush (FIG. 1). The extensor digitorum was found to become responsive to nerve stimulation from day 46 onwards (FIG. 1). The CMAP increased up to the termination of the experiment on day 63. On day 63 the mean CMAP for the BB-1101–treated group was significantly increased ($p < 0.05$, unpaired $t$-test) over that for the other two groups, but was not different from the control naive group (unpaired $t$-test, $p = 0.6341$, TABLE 1). The latency measurement was the same in all groups, suggesting that the nerve regeneration in the three groups was the same, but at day 63 was greater than for the control intact limb (see TABLE 1). After crush the axon diameter distribution was unimodal with a mode of 1.5–3 μm, irrespective of treatment, and distinct from the bimodal distribution seen in control nerve. The muscle fiber diameter distribution at 63 days after nerve crush was reduced compared to the control. However, the drug-treated group showed a marginally significant increased size over the control lesioned groups ($p < 0.05$, Kolgomoroff-Smirnov test).

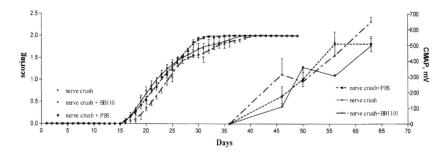

**FIGURE 1.** Recovery of the toe-spreading reflex and of CMAP amplitude of extensor digitorum, from day 0 to day 63 after nerve crush.

**TABLE 1. CMAP amplitude and latency in extensor digitorum at 63 days after nerve crush**

| Groups | n | Mean amplitude (mV) | SEM | Mean latency (ms) | SEM |
|---|---|---|---|---|---|
| Normal control | 14 | 624.2 | 31.81 | 1.99 | 0.154 |
| Nerve crush | 7 | 438.7 | 47.7 | 3.176 | 0.062 |
| Nerve crush + PBS | 9 | 509.6 | 47.7 | 3.188 | 0.050 |
| Nerve crush + BB1101 | 8 | 647.3 | 32.9 | 3.126 | 0.047 |

## DISCUSSION

The results from our study suggest that the general MMP inhibitor BB1101 neither enhances nor retards nerve repair following Wallerian degeneration. However, examination of CMAP amplitude at 63 days in a muscle most distal from the lesion site was close to the control value in drug-treated animals and significantly greater than that seen in the control lesioned groups. This correlated with a slightly increased muscle fiber diameter distribution. Skeletal muscle fibers normally atrophy on denervation, and with a long delay in reinnervation can degenerate completely. Proteases and protease inhibitors have been reported to be present at the neuromuscular junction,[7] including the MMPs 2 and 9.[8] Our results suggest that MMP inhibition following denervation may help to maintain the muscle basal lamina cues used to guide the regenerating motor nerve terminals and thus provide more efficient reinnervation. Experiments are under way to confirm the above findings and to further test the hypothesis.

## REFERENCES

1. HUGHES, R.A.C. 1990. Guillain-Barré Syndrome, Springer-Verlag. Heidelberg.
2. LIEDTKE, W., B. CANNELLA, R.J. MAZZACCARO, J.M. CLEMENTS, K.M. MILLER, K.W. WUCHERPFENNIG, A.J. GEARING & C.S. RAINE. 1998. Effective treatment of models of multiple sclerosis by matrix metalloproteinase inhibitors. Ann. Neurol. **44:** 35–46.
3. KIESEIER, B.C., J.M. CLEMENTS, H.B. PISCHEL, G.M. WELLS, K. MILLER, A.J. GEARING & H.P. HARTUNG. 1998. Matrix metalloproteinases MMP-9 and MMP-7 are expressed in experimental autoimmune neuritis and the Guillain-Barre syndrome. Ann. Neurol. **43:** 427–434.
4. REDFORD, E.J., K.J. SMITH, N.A. GREGSON, M. DAVIES, P. HUGHES, A.J. GEARING, K. MILLER & R.A.C. HUGHES. 1997. A combined inhibitor of matrix metalloproteinase activity and tumour necrosis factor-alpha processing attenuates experimental autoimmune neuritis. Brain **120:** 1895–1905.
5. HUGHES, P.M., G.M. WELLS, J.M. CLEMENTS, A.J. GEARING, E.J. REDFORD, M. DAVIES, K.J. SMITH, R.A.C. HUGHES, M.C. BROWN & K.M. MILLER. 1998. Matrix metalloproteinase expression during experimental autoimmune neuritis. Brain **121:** 481–494.

6. LA FLEUR, M., J.L. UNDERWOOD, D.A. RAPPOLEE & Z. WERB. 1996. Basement membrane and repair of injury to peripheral nerve: defining a potential role for macrophages, matrix metalloproteinases, and tissue inhibitor of metalloproteinases-1. J. Exp. Med. **184:** 2311–2326.
7. HANTAI, D., J.S. RAO & B.W. FESTOFF. 1990. Rapid neural regulation of muscle urokinase-like plasminogen activator as defined by nerve crush. Proc. Natl. Acad. Sci. USA **87:** 2926–2930.
8. KHERIF, S., M. DEHAUPAS, C. LAFUMA, M. FARDEAU & H.S. ALAMEDDINE. 1998. Matrix metalloproteinases MMP-2 and MMP-9 in denervated muscle and injured nerve. Neuropathol. Appl. Neurobiol. **24:** 309–319.

# Selective Inhibition of Collagenase-1, Gelatinase A, and Gelatinase B by Chemotherapeutic Agents

ULRIKE BENBOW,[a] RANGAN MAITRA,[b] JOSHUA W. HAMILTON,[b]
AND CONSTANCE E. BRINCKERHOFF[a,c,d]

[a]Department of Medicine, [b]Department of Pharmacology and Toxicology,
and [c]Department of Biochemistry, Dartmouth Medical School,
Hanover, New Hampshire 03755, USA

Matrix metalloproteinases (MMPs) play a crucial role in tumor cell invasion and metastasis due to their ability to digest basement membrane and extracellular matrix components and thus facilitate cell movement. MMPs, therefore, are important pharmacological targets, and a number of synthetic inhibitors have been developed.[1] In keeping with this approach we screened a panel of chemotherapeutic agents for their ability to inhibit collagenase-1, gelatinase A, and gelatinase B expression in the highly aggressive human melanoma cell line A2058. We focused on the expression of these enzymes because of their ability to digest collagen types I, III, and IV, the major components of the extracellular matrix and basement membrane.[2] Cells were treated with a variety of cytotoxic drugs such as carboplatin, mitomycin C (MMC), cisplatin, doxorubicin, and N-acetylcysteine (NAC). Among these agents, only doxorubicin specifically repressed collagenase-1 expression at both the protein and mRNA levels, with no effect on gelatinase A and B.[3] Time- and dose-dependency studies revealed that repression of collagenase-1 synthesis could be obtained at non-

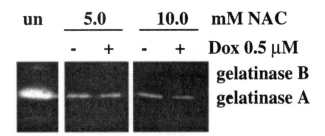

**FIGURE 1.** Effect of NAC on gelatinase A and gelatinase B expression. A2058 melanoma cells were pretreated with doxorubicin (0.5 μM) for 1 hr. At this time point, the medium was removed and replaced with new serum-free DMEM. Culture medium was harvested 18 hr later. Gelatin zymography was used to determine the expression of gelatinase A and gelatinase B. NAC was added during the incubation time (24 hr, at 37°C).

[d]Corresponding author. Phone, 603/650-1609; fax, 603/650-1128;
e-mail, constance.e.brinckerhoff@dartmouth.edu

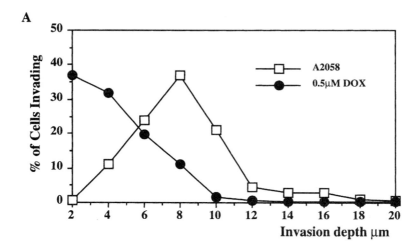

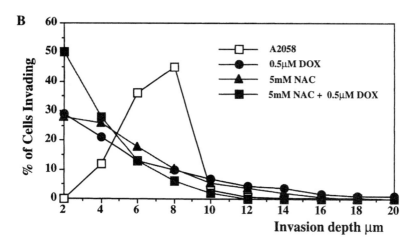

**FIGURE 2.** Quantitative invasion profiles of untreated (□) and doxorubicin- and NAC-treated A2058 cells. Melanoma cells were seeded on a collagen type I matrix (**A**) or on Matrigel® (**B**) under serum-free conditions, as described in Ref. 3. For doxorubicin treatment cells were pretreated with the drug for 1 hr. At this point the medium was replaced with serum-free DMEM in the absence or presence of 5 mM NAC. Filters were processed, and the number of cells invading the matrix layer was determined by confocal laser scanning microscopy–acquired images.

cytotoxic concentrations—i.e., lower than those normally used in cancer chemotherapy.[3,4] These low doses were noncytotoxic and did not affect proliferation in this cell line. Furthermore, use of these low doses also circumvented the induction of multidrug resistance, a condition that is observed clinically and that is associated with high concentrations of doxorubicin.[3,4] In contrast, N-acetylcysteine (NAC), a precursor of reduced glutathione, inhibited the expression of gelatinase A and gelatinsase B in this tumor cell line (FIG. 1). This inhibition occurred at the level of enzymic activity, as shown by gelatin zymography.[5]

To study the effect of doxorubicin on melanoma cell invasion we used a newly described invasion system.[3] This system allows the determination of the number of cells invading a matrix of choice. Using confocal laser scanning microscopy, images of propidium iodide–labeled cells were taken every two micrometers from the top of the matrix in the $z$-direction. For these experiments, cells were seeded on a collagen type I matrix to mimic the interstitial stroma environment. Cells were treated with concentrations of doxorubicin that inhibited collagenase-1 mRNA, but were not cytotoxic.[3] We found that about 38% of untreated cells invaded the collagen type I matrix to a depth of 8 μm. In contrast, following treatment of cells with 0.5 μM doxorubicin, only about 12% of the cells invaded to this depth (FIG. 2A). Next we used Matrigel®, a reconstituted basement membrane, to determine the effect of doxorubicin and NAC on basement membrane invasion. We found that both doxorubicin and NAC inhibited the invasion of this membrane by about 30%. In combination, low doses of doxorubicin and NAC resulted in a synergistic inhibition of Matrigel invasion by the A2058 melanoma cells (FIG. 2B).

In conclusion, even though doxorubicin is not commonly used as a therapeutic agent for malignant melanoma, the selective inhibition of collagenase-1 by low noncytotoxic doses indicates that this agent may be clinically useful. Our results suggest a novel application for combination therapy using doxorubicin and N-acetylcysteine with respect to malignant melanoma, due to their ability to decrease the invasive potential of tumor cells that are otherwise unaffected by this drug.

## REFERENCES

1. RASMUSSEN, H. & P.P. McCANN. 1997. Matrix metalloproteinase inhibition as a novel anticancer strategy: a review with specific focus on Batimastat and Marimastat. Pharmacol. Ther. **75:** 69–77.
2. MATRISIAN, L.M. 1992. The matrix-degrading metalloproteinases. Bioassays **14:** 455–463.
3. BENBOW, U. et al. 1999. Selective modulation of collagenase-1 gene expression by the chemotherapeutic agent doxorubicin. Clin. Cancer Res. **5:** 203–208.
4. IHNAT, M.A. et al. 1997. Suppression of p-glycoprotein expression and multidrug resistance by DNA crosslinking agents. Clin. Cancer Res. **3:** 1339–1346.
5. ALBINI, A. et al. 1995. Inhibition of invasion, gelatinase activity, tumor take and metastasis of malignant cells by N-acetylcysteine. Int. J. Cancer **61:** 121–129.

# Meloxicam and Indomethacin Activity on Human Matrix Metalloproteinases in Synovial Fluid

A. BARRACCHINI,[a,b] N. FRANCESCHINI,[c] G. MINISOLA,[a,d] G.C. PANTALEONI,[a] A. DI GIULIO,[c] A. ORATORE,[a] AND G. AMICOSANTE[c]

*Departments of [a]Pharmacology and [c]Enzymology, University of L'Aquila, I-67100 L'Aquila, Italy*

[d]*Rheumatology Unit, Villa Betania Hospital, Rome, Italy*

The agents primarily responsible for cartilage and bone destruction in joint diseases are active proteinases degrading collagen and proteoglycan. Under physiologic conditions, the activity of matrix-degrading enzymes is balanced by specific proteinase inhibitors. In contrast, in rheumatic synovitis, a shift of this equilibrium leads to excessive matrix degradation. Because collagenase and stromelysin play a fundamental role in the pathophysiology of rheumatic disease and because the connective tissue destruction that they cause is largely irreversible, their inhibition would seem to be of utmost importance in designing effective therapeutic strategies.[1,2] For this purpose we tested and compared the effectiveness of two nonsteroidal antiinflammatory drugs (NSAIDs), meloxicam and indomethacin, in inhibiting matrix metalloproteinase activity in 10 synovial fluid specimens from patients with osteoarthritis (OA) or rheumatoid arthritis (RA). The study was performed using a fluorometric assay and a synthetic esapeptide Mca-Pro-Leu-Gly-Leu-Dpa-Ala-Arg-$NH_2$, fluorogenic substrate for matrix metalloproteinases (MMPs).[3] The increased fluorescence, following the removal of the $NH_2$ terminal group, after cleavage of the probe by MMPs, was monitored at 393 nm and related to the enzyme activity. Under the same experimental conditions the interaction of NSAIDs with MMPs *ex vivo* was measured. The value of enzyme inhibition, at each drug concentration, was expressed as percent of residual activity with respect to the control.[4] The residual activity was plotted against drug concentration and the $IC_{50}$ values graphically determined for each compound. The residual enzyme activity using meloxicam was measured in the range of 10–100 μM and an $IC_{50}$ value for each synovial fluid estimated. When indomethacin was used as inhibitor at concentrations higher than 100 μM, the enzyme activity was not affected at all. The enzyme inhibition was clearly related to the meloxicam concentration, while the same results were not obtained using high indomethacin concentrations, demonstrating a competitive inhibition model by the former. In particular, meloxicam showed an $IC_{50}$ on human MMP activity from synovial fluids ranging from 13 to 18 μM (TABLE 1). Although the $IC_{50}$ of meloxicam is higher than the therapeutic concentration, the elimination half-life in the synovial

[b]Address for correspondence: Dr.ssa A. Barracchini, Chair of Pharmacology, Department of Internal Medicine, University of L'Aquila, Loc. Coppito—67100 L'Aquila, Italy. Fax, + 39/0862-432858; e-mail, amicosante@aquila.infn.it

**TABLE 1.** *Ex vivo* **anti-MMPs: effect and effective concentration of indomethacin and meloxicam**

| Drug | $IC_{50}{}^a$ | | Therapeutic plasma concentration | Therapeutic synovial fluid concentration | Distribution volume | $T_{1/2}{}^b$ |
|------|------|------|------|------|------|------|
| | (μM) | (mg/l) | (mg/l) | (mg/l) | ($V_d$ [l/Kg]) | (hours) |
| Indomethacin | — | — | 0.3 | 0.3 | $0.29 \pm 0.04$ | 3 |
| Meloxicam | 13–18 | 4.6–6.3 | 1.0–2.0 | 0.5–1.0 | 10.7 | 20 |

[a]Amount of drug that inhibited the *ex vivo* MMP activity by 50%.
[b]Drug elimination half-time.

fluid varies little with respect to the plasma concentration; and its inhibiting effect, even if only partial, is protracted over the time.[5] The extrapolation of our data in the *in vivo* conditions may be the subject of debate, but the inhibitory effect of meloxicam at concentrations achievable in the synovial fluids of patients receiving long-term therapy might be an interesting finding.

Better understanding of the pathogenesis of rheumatic diseases leads to important advances in therapy: MMP inhibition, like that exerted by meloxicam, could be an interesting area of development. Further studies on the interaction between NSAIDs and MMPs from synovial fluids could help to predict long-term effects of NSAIDs on the outcome of the most common rheumatic diseases (OA and RA).

## REFERENCES

1. CAWSTON, T.E. 1996. Metalloproteinase inhibitors and the prevention of connective tissue breakdown. Pharmacol. Ther. **70:** 163–182.
2. VINCENTI, M.P., I.M. CLARK & C.E. BRINCKERHOFF. 1994. Using inhibitors of metalloproteinases to treat arthritis. Arthritis Rheum. **8:** 1115–1126.
3. KNIGHT, C.G., F. WILLENBROCK & G. MURPHY. 1992. A novel coumarin-labelled peptide for sensitive continuous assays of the matrix metalloproteinases. FEBS Lett. **296:** 263–266.
4. DI GIULIO, A., A. BARRACCHINI, L. CELLAI, C. MARTUCCIO, G. AMICOSANTE, A. ORATORE & G.C. PANTALEONI. 1996. Rifamycins as inhibitors of collagenase activity: their possible pharmacological role in collagen degradative diseases. Int. J. Pharm. **144:** 27–35.
5. BROOKS, P. & G. LEE. 1998 The therapy of rheumatic diseases. *In* Textbook of Clinical Rheumatology. H.S. Howe & P.H. Feng, Eds.: 389–417. National Arthritis Foundation. Singapore.

# Non-antimicrobial and Antimicrobial Tetracyclines Inhibit IL-6 Expression in Murine Osteoblasts

KEITH L. KIRKWOOD,[a,b,f] LORNE M. GOLUB,[c] AND PETER G. BRADFORD[b,d,e]

[a]Department of Periodontics and [b]Center for Molecular Mechanisms of Diseases and Aging, State University of New York at Buffalo, Buffalo, New York 14214, USA

[c]Department of Oral Biology and Pathology, State University of New York at Stony Brook, Stony Brook, New York 11794, USA

[d]Department of Pharmacology and Toxicology and [e]Department of Oral Biology, State University of New York at Buffalo, Buffalo, New York 14214, USA

IL-6 regulation in osteoblasts has been the focus of several groups, since this cytokine appears to contribute toward the pathophysiological state of postmenopausal osteopenia.[1] IL-6 is the major cytokine regulated by sex steroids as compared to other cytokines, such as IL-1β, IL-11, and TNF-α, that are not regulated by sex steroids.[2] Previous results from our laboratory have indicated that intracellular calcium is a critical determinant of the osteoblast secretory capacity.[3,4] Since tetracycline is well known for its ability to bind divalent cations, such as calcium and zinc, and to affect intracellular calcium concentrations,[5] we evaluated the abilities of doxycycline and chemically modified tetracyclines, which lack antimicrobial activity, to affect osteoblast IL-6 secretion from MC3T3-EI osteoblastic cells.

## METHODS

### IL-6 Secretion Measurements

MC3T3-EI cells were cultured in 6-well tissue culture dishes under standard tissue culture conditions. Test cells were pretreated for 18–24 hr with doxycycline, CMTs (1–50 μg), or vehicle (DMSO 10 μg/ml). IL-1β was added to culture media at physiological relevant concentrations (10–20 pM), and the cells continued to incubate for an additional 18 hr. Time course experiments as well as dose response experiments with CMTs were conducted in the same manner. Mouse IL-6 secretion was measured by a capture ELISA kit (Quantikine, R&D Systems) following the manufacturer's protocol. Quantities of IL-6 were expressed in pg/ml of cultured supernatant.

[f]Address for correspondence: Keith L. Kirkwood, D.D.S. Ph.D., State University of New York at Buffalo, Department of Periodontics, 215 Squire Hall, 3435 Main Street, Buffalo, New York 14214-3000. Phone, 716/829-3845; fax, 716/837-7623; e-mail, klkirk@acsu.buffalo.edu

### Northern Blot Analysis of Osteoblast Gene Expression

For detection of mRNA from control and CMT-treated MC3T3-EI osteoblastic cells, we employed Northern blot using methods as described previously.[3] Rat IL-6 cDNA clone was obtained as a gift from Dr. J. Goldie, McMaster University, Hamilton, Ontario. Dr. R. Franceschi, University of Michigan, kindly provided the mouse type 1($\alpha$1) procollagen cDNA probe. After hybridization with the appropriate probes, membranes were washed at high stringency, and the hybridized probe was quantitated using Bio-Rad's phosphoimager system and Molecular Analyst software version 1.5.

## RESULTS AND DISCUSSION

### Chemically Modified Tetracyclines Inhibit IL-6 Secretion in MC3T3-E1 Cells

To determine if IL-6 secretion can be regulated by doxycycline or the other CMTs, IL-6 secretion from MC3T3-El cells was measured using a capture ELISA kit specific for mouse IL-6. Results shown in FIGURE 1A indicate that CMT-8 (10 $\mu$g/ml) can inhibit IL-6 secretion from MC3T3-EI cells when stimulated by IL-1$\beta$ (10 ng/ml). CMT-8 decreased IL-6 secretion by ~49% compared to cells pre-

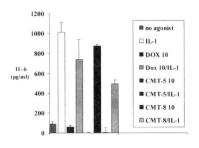
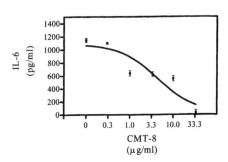

**FIGURE 1.** *(Left)* The effects of chemically modified tetracyclines on IL-1$\beta$–induced IL-6 secretion in osteoblasts. MC3T3-E1 cells were pretreated for 18 hours with doxycycline or CMTs at 10 $\mu$g/ml. Fresh media was added with the pretreatment conditions ± IL-1$\beta$ (12.5 ng/ml). Cells continued to incubate for an additional 18 hours, and cells supernatants were harvested. IL-6 secretion was measured using an IL-6 Quantikine Kit (R&D Systems). Quantitation of IL-6 was made using linear regression with a standard curve. Each measurement was made in duplicate, and the mean of two different experiments was used for analysis with standard deviations. The results are considered statistically significant ($p = 0.005$). *(Right)* CMT-8 dose-response in IL-1$\beta$ treated MC3T3-E1 osteoblasts. MC3T3-E1 cells were pretreated with CMT-8 at the following concentrations: 0, 0.3, 1.0, 3.3, 10.0, and 30.0 $\mu$g/ml. Cells were subsequently treated with IL-1$\beta$ (12.5 ng/ml) after a change in media and addition of pretreatment conditions. IL-6 measurements were made as described. Means of two separate experiments performed in duplicates ± standard deviations of the mean. The $IC_{50}$ was calculated to be 4.4 $\mu$g/ml.

treated with vehicle. CMT-5 decreased IL-6 secretion by ~13%. Doxycycline decreased IL-1β–induced IL-6 secretion by ~33%. The results shown are the means of two separate experiments performed in duplicate. Standard deviations are shown; these results are considered statistically significant ($p = 0.005$). Inhibition of IL-1β–stimulated IL-6 secretion occurred in a dose-dependent manner with pharmacologically relevant concentrations of this drug (1–10 µg/ml) reducing IL-6 by about 50% as shown in FIGURE 1A. Pretreatment with CMT-8 was reduced to 8 hours in these experiments. FIGURE 1B illustrates the dose-response effect with decreasing amounts of IL-6 secretion when increasing the amount of CMT-8. The $IC_{50}$ for CMT-8 inhibition of IL-1β–induced IL-6 secretion was experimentally determined to be 4.4 µg/ml.

### CMT-8 Decreases Steady State IL-6 mRNA Levels in MC3T3-E1 Cells

Northern blot analysis was used to determine IL-6 mRNA expression in IL-1β–treated cells with or without pretreatment with doxycycline or CMTs. IL-6 mRNA species of 1.2–2.4 kb were detected in MC3T3-EI cells (FIGURE 2). As shown previously in these cells, IL-1β increases steady state mRNA levels. When cells are pretreated with doxycycline or CMTs or treated after IL-1β stimulation (data not shown), CMT-8 decreases the IL-6 mRNA steady state levels in these cells. In contrast, neither doxycline nor CMT-5 were able to decrease IL-6 mRNA steady state levels (FIG. 2). When levels of GAPDH were used to normalize the data, an approximate 50% reduction in IL-6 mRNA with CMT-8 treatment is seen compared to

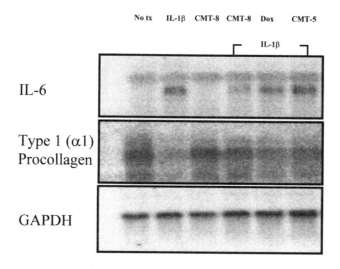

**FIGURE 2.** MC3T3-E1 cells were cultured in the presence or absence of CMTs of doxycycline and then exposed to IL-1β (12.5 ng/ml) for 18 hours. Total RNA was purified, separated on agarose gels, blotted onto nylon membranes, and suquentially hybridized with [32-P]-labeled cDNA probes specific for rat IL-6, mouse type 1(α1) procollagen, and rat GAPDH

IL-1β treated cells ($n = 2$). Additionally, CMT-8 had no effect on constitutive type I procollagen α1(I) gene expression in MC3T3-E1 cells. IL-1β decreased type 1 procollagen α1(I) gene expression by 54% in these cells. This effect was somewhat inhibited by CMT-8 (34%) and to a greater extent by doxycycline (76%).

The results of the present study suggest an additional mechanism of CMTs, inhibition of IL-6 gene expression, and secretion from osteoblasts. Our data indicate that IL-1β–stimulated expression and secretion of IL-6 in MC3T3-E1 osteoblasts can be inhibited by CMT-8 in a dose-dependent manner, and the effect of CMT-8 is consistent with decrease of IL-6 gene expression. These results support the role for chemically modified tetracyclines to provide anabolic effects in bone. In this paper, CMT-8 and, to a lesser extent, doxycycline inhibited the potentially catabolic effects of IL-1β, suggesting a novel molecular mechanism of these drugs with the treatment of metabolic bone diseases, such as osteoporosis.

## REFERENCES

1. HOROWITZ, M.C. 1993. Cytokines and estrogen in bone: antiosteoporotic effects. Science 260: 626–627.
2. MANOLAGAS, S.C. 1995. Role of cytokines in bone resorption. Bone 17: 63S–67S.
3. KIRKWOOD, K., R. DZIAK & P. BRADFORD. 1996. Inositol trisphosphate receptor gene expression and hormonal regulation in osteoblast-like cell lines and primary osteoblastic cell cultures. J. Bone Miner. Res. 11: 1877.
4. KIRKWOOD, K., M. DRAGON, K. HOMICK & P. BRADFORD. 1997. Cloning and characterization of the type I inositol 1,4,5-trisphosphate receptor gene promoter: regulation by 17β-estradiol in osteoblast. J. Biol. Chem. 272: 22425–22431.
5. DONAHUE, H.J., K. IIJIMA, M.S. GOLIGORSKY, C.T. RUBIN & B.R. RIFKIN. 1992. Regulation of cytoplasmic calcium concentration in tetracycline-treated osteoclasts. J. Bone Miner. Res. 7: 1313.
6. LITTLEWOOD, A.J., J. RUSSELL, G.R. HARVEY, D.E. HUGHES, R.G.G. RUSSELL & M. GOWEN. 1991. The modulation of the expression of IL-6 and its receptor in human osteoblasts. Endocrinology 129: 1513–1520.

# CMT-8/Clodronate Combination Therapy Synergistically Inhibits Alveolar Bone Loss in LPS-Induced Periodontitis

A. LLAVANERAS,[a,b] L.M. GOLUB,[a] B.R. RIFKIN,[a] P. HEIKKILÄ,[c] T. SORSA,[c] O. TERONEN,[c] T. SALO,[d] Y. LIU,[a] M.E. RYAN,[a] AND N.S. RAMAMURTHY[a,e]

[a]Department of Oral Biology and Pathology, School of Dental Medicine, Health Sciences Center, State University of New York at Stony Brook, Stony Brook, NY 11794-8702, USA

[b]School of Dental Medicine, Central University of Venezuela, Caracas, Venezuela

[c]Department of Oral and Maxillofacial Surgery, University of Helsinki, Helsinki, Finland

[d]Departments of Oral Surgery and Pathology, University of Oulu, Oulu, Finland

Earlier studies have reported that a chemically modified doxycycline (CMT-8) administered as a monotherapy inhibited alveolar bone loss in a lipopolysaccharides-(LPS) injected rat periodontitis model.[1,2] Teronen et al. reported that clodronate, a bisphosphonate, also inhibited the activity of MMP-8, the predominant collagenase in inflamed gingival tissue and in gingival crevicular fluid in human adult periodontitis.[2] In the present study the effects of the combination of subtherapeutic levels of chemically modified doxycycline (CMT-8) and a bisphosphonate (clodronate) were investigated in a LPS-induced periodontal disease model. This model, described here, was published by Ramamurthy et al.[1] and involves the injection of E. coli endotoxin into the gingival tissues. Injection of endotoxin into the gingiva produces marked inflammation in the periodontium, pathologically elevated levels of tissue-destructive matrix metalloproteinases (MMPs), leading to severe alveolar bone resorption and bone loss around the affected teeth.

## METHODS

Thirty adult male Sprague-Dawley rats (350–375 g) were distributed into the following experimental groups: saline-injected group; endotoxin (E) injected group (10μg/10μl); E+CMT-8 (1 mg/day by oral gavage in 2% carboxymethyl cellulose) E+clodronate (by a single subcutaneous injection of 1 mg of clodronate), or E+CMT-8+Clod combination. A dose response study using 0.5, 1.0, and 2 mg of either clodronate or CMT-8 indicated that 0.5 mg was ineffective, 1.0 mg /rat was suboptimal, and 2.0 mg/rat was optimally effective in this model.

Twenty-four hours before the start of the treatment, the rats were injected into the anterior gingiva and into the palatal interdental gingiva between first and second up-

[e]Corresponding author. Fax, 516/623-9705.

per molars with 10 µl of endotoxin (1 mg/ml) or saline solution. The LPS injections were repeated every other day to complete the three injections.[3]

On day 7, all rats were anesthetized, blood samples were collected by cardiac puncture, and the rats were euthanized. Tooth mobility was measured, and gingival tissues were dissected and extracted for functional and immunological MMP analysis. The jaws were defleshed for bone loss determination. All procedures described here have been previously published by Ramamurthy et al.[1]

## RESULTS

### Tooth Mobility

Tooth mobility, a reflection of severity of inflammation and collagen loss in the periodontal tissues plus alveolar bone loss, was measured using a forceps to move the tooth, according to the following scoring system: 0 = no movement; 1 = slight movement (vestibule-palatal); 2 = medium movement (vestibule-palatal); 3 = severe movement (vertical mobility in and out of socket).

The saline injected rats showed a mobility score of 1.2 compared to the LPS-injected group, which showed a score of 2.8; both the CMT-8 alone group and the clod alone group showed mobility scores of 2.2. The combination treated rats showed synergistically reduced mobility, below the saline-injected controls, with a score of 0.5 ($p < 0.05$ compared to LPS or LPS + monotherapy alone).

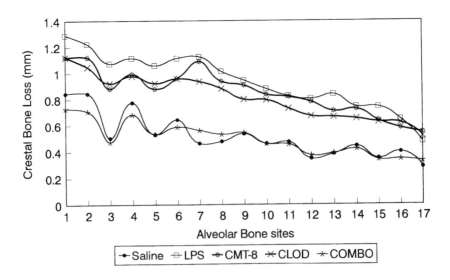

**FIGURE 1.** Effects of CMT-8 and clodronate combination therapy on LPS-induced alvelor bone loss in rats.

### Alveolar Bone Loss

Based on computer-assisted morphometric analysis of the defleshed rat jaws as shown in FIGURE 1, the rats in the LPS-injected group showed greater alveolar bone loss in all 17 sites in half-maxillas compared to the saline-injected group, especially at site 7, the site of LPS injection, which showed a 140% increase in bone loss. Treating the LPS-injected rats with either CMT-8 alone or clodronate alone reduced the bone loss slightly. In sharp contrast, the combination therapy essentially "normalized" the alveolar bone loss, induced by *E. coli* endotoxin (similar to saline injection).

### MMP-Activities

Collagenase and gelatinase activity in the partially purified gingival extracts was determined from the different treatment groups. Collagenase activity was assessed using [$^3$H-methyl] collagen as substrate and separating the intact $\alpha$ collagen components from the $\alpha^A$ collagen breakdown fragments by a combination of SDS-PAGE and fluorography. The fluorograms were then scanned with a laser densitometer to calculate collagenase activity as the percent conversion of $\alpha$ to $\alpha^A$ components. Collagenase generated three of four breakdown products of collagen($\alpha^A$) by SDS-PAGE. These collagen components and fragments were visualized by autoradiography; their levels were calculated by scanning the fluorograms with a laser densitom-

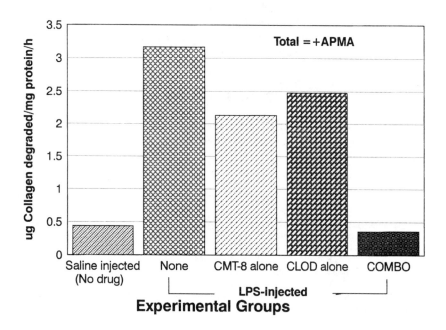

**FIGURE 2.** Effects of CMT-8 and clodronate combination therapy on LPS-induced gingival collagenase in rats.

eter. The data in FIGURE 2 represent the total collagenase activity in the gingival tissues after activating the Pro-collagenase by the addition of 1.2 mM amino phenyl mercuric acetate (APMA). FIGURE 2 shows that injecting the gingiva with LPS dramatically increased collagenase activity by 700% compared to saline, which is consistent with collagenase being responsible for the degradation of type-1 collagen present in the gingiva and in bone matrix. Treating the LPS-injected rats with either CMT-8 monotherapy or clodronate monotherapy reduced the collagenase activity by 22–31%. In contrast, and like the effect on tooth mobility and alveolar bone loss, the combination therapy synergistically reduced the collagenase by 91% down to the levels seen in the saline-injected control gingival tissues. Gelatinase activity measured in the gingival extract showed the same pattern of change as collagenase for different groups of rats; combination therapy again produced a synergistic reduction in the excessive activity of this MMP in the gingival tissue of LPS-injected rats. Zymography and Western blot analysis using specific antisera for MMPs showed different molecular species of gelatinase (MMP-2 and -9) and collagenase (MMPS-8 and -13) in the gingival tissues. It is noteworthy that all of these assays demonstrated synergistic downregulation and inhibition of the MMP activities due to combination therapy.

## CONCLUSION

LPS injection into the gingival tissue increased tooth mobility and periodontal bone loss associated with increased MMP expression and activation in the periodontium. CMT-8 and clodronate combination each at subtherapeutic doses were shown to be potent inhibitors of the complex cascade of periodontal breakdown including alveolar bone loss, mediated by MMPs and serine proteinases acting in cascade.

## ACKNOWLEDGMENTS

This study was supported by NIDR Grants DE-03987 and DE-09576, Collagenex, the Center for Biotechnology, SUNY Stony Brook, the Finnish Dental Society, and the Academy of Finland.

## REFERENCES

1. RAMAMURTHY, N.S., L.M. GOLUB, A.J. GWINNET, T. SALO, Y. DING & T. SORSA. 1998. *In vivo* and *in vitro* inhibition of matrix metalloproteinases including MMP-13 by several chemically modified tetracyclines (CMTs). *In* Biological Mechanisms of Tooth Eruption, Resorption and Replacement by Implants. Z. Davidovitch & J. Mah, Eds.: 271–277. Harvard Society for the Advancement of Orthodontics. Boston.
2. TERONEN, O., Y.T. KONTTINEN, C. LINDQUIST, T. SALO, T. INGMAN, A. LAUHIO, Y. DING, S. SANTAVIRTA & T. SORSA. 1997. Human neutrophil collagenase MMP-8 in periimplant sulcus fluid and inhibition by clodronate. J. Dent. Res. **76:** 1529–1537.
3. RAMAMURTHY, N.S., R.A. GREENWALD, M. SCHNEIR & L.M. GOLUB. 1985. The effect of alloxan diabetes on prolyl and lysyl hydroxylase activity in uninflamed and inflamed rat gingiva. Arch. Oral Biol. **30:** 679–683.

# MMP Inhibition by Chemically Modified Tetracycline-3 (CMT-3) in Equine Pulmonary Epithelial Lining Fluid

PÄIVI MAISI,[a,c] MARKUS KIILI,[a] SAARA M. RAULO,[a] EMMA PIRILÄ,[b] AND TIMO SORSA[b]

[a]Department of Clinical Veterinary Sciences, Faculty of Veterinary Medicine, University of Helsinki, 00014 Helsinki, Finland

[b]Institute of Dentistry, University of Helsinki, 00014 Helsinki, Finland

The imbalance of matrix metalloproteinases (MMPs) and tissue inhibitors of metalloproteinases seems to be important in chronic respiratory tract disease tissue destruction. We have found elevated levels of tissue-destructive MMPs, gelatinases

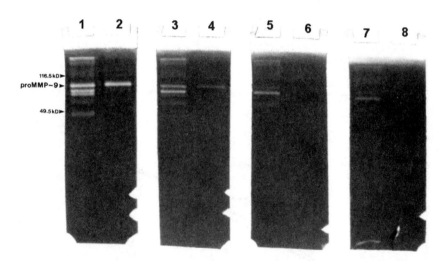

**FIGURE 1.** Inhibitory effect of CMT-3 (150 μM, 50 μM, 25 μM) on tracheal epithelial lining fluid (ELF, 0.04 μl) and MMP-9 (4 ng) gelatin degradation level as evaluated by gelatin SDS-PAGE zymography. *Lane 1* contains ELF, *lane 2* contains MMP-9, *lane 3* contains ELF incubated with 25 μM CMT-3, *lane 4* contains MMP-9 incubated with 25 μM CMT-3, *lane 5* contains ELF incubated with 50 μM CMT-3, *lane 6* contains MMP-9 incubated with 50 μM CMT-3, *lane 7* contains ELF incubated with 150 μM CMT-3, and *lane 8* contains MMP-9 incubated with 150 μM CMT-3. CMT-3 was used both in preincubation (1 hr, 37°C) and in washing and incubation (16 hr, 37°C) solutions of zymography.

[c]Address for correspondence: Päivi Maisi, Department of Clinical Veterinary Sciences, Faculty of Veterinary Medicine, Helsinki University, Box 57, 00014 Helsinki, Finland. Phone, 358-9-708 49623; fax, 358-9-708 49670; e-mail, paivi.maisi@helsinki.fi

and interstitial collagenases, to be present in pulmonary epithelial lining fluid (ELF) in equine chronic obstructive pulmonary disease (COPD)[1,2] due to poor-quality hay and/or mold antigens. The elevation is detected only locally in the lung but not on the blood level.[1] COPD is a progressive disease leading to chronic cough, bronchoconstriction, lowered oxygen levels in arterial blood, and impaired breathing with pronounced exercise intolerance.

Chemically modified tetracycline-3 (CMT-3) is known to inhibit both MMP-9 and MMP-8.[3] We studied whether the increased metalloproteinase activity in ELF from COPD horses could be inhibited by CMT-3. In parallel the inhibitory capacity on human recombinant MMP-9 was determined. A method to test inhibitory capacity on gelatin zymography was developed. Western blot analysis was used to identify MMP-9 and to detect effects of CMT-3 on structure and potential fragmentation of native MMP-9 and MMP-8.

ELF (0.04 μl), in parallel with human recombinant MMP-9 (4 ng), was preincubated without and with CMT-3 (150 μM, 50 μM, 25 μM) for 1 hour at 37°C prior to zymography or Western blotting. Zymography was performed as described,[4] except that after electrophoresis the lanes on gels were separated by cutting, and the different gel pieces were washed and incubated separately with CMT-3 (150 μM, 50 μM, 25 μM) and part in parallel without CMT-3.

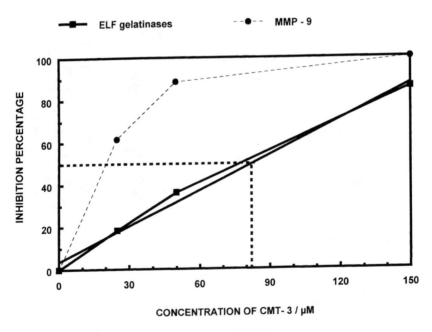

**FIGURE 2.** Inhibition percentage of tracheal epithelial lining fluid (ELF, 0.04 μl) and MMP-9 (4 ng) gelatin degradation by CMT-3 (150 μM, 50 μM, 25 μM). ELF and MMP-9 total gelatinolytic activity determined by an image analysis and processing system (Cream®, Kem-En-Tec, Copenhagen, Denmark). $IC_{50}$ for ELF indicated by *dashed line.*

No inhibition of gelatinolytic activity in zymography was detected if CMT was used only in preincubation. When CMT-3 was additionally used in washing and incubation solutions of zymography, 86% inhibition of ELF gelatinolytic activity and 100% inhibition of human recombinant MMP-9 gelatinolytic activity was achieved with 150 μM CMT-3 (FIG. 1). $IC_{50}$ was found to be 20–90 μM, lower for human recombinant MMP-9, and higher for equine COPD ELF (FIG. 2). CMT treatment of ELF and human recombinant MMP-9 as analyzed by Western blotting and zymography did not show any inducible fragmentation of MMP-8 or MMP-9 by the inhibitor used.

CMT-3 has the potential to act as a tissue-protective drug in equine COPD. The direct inhibition of CMT appeared to be reversible, while CMT-3 addition into washing and incubation solutions was required in order to achieve inhibition in gelatin zymography. This inhibitory effect was not due to CMT-3–induced fragmentation of MMP-9 or MMP-8.

## REFERENCES

1. RAULO, S.M. & P. MAISI. 1998. Gelatinolytic activity in tracheal epithelial lining fluid and in blood from horses with chronic obstructive pulmonary disease. Am. J. Vet. Res. **59:** 818–823.
2. KOIVUNEN, A.-L. *et al.* 1997. Collagenolytic activity and its sensitivity to doxycyclin inhibition of tracheal aspirates of horses with chronic obstructive pulmonary disease. Acta Vet. Scand. **38:** 9–16.
3. SORSA, T. *et al.* 1998. Functional sites of chemically modified tetracyclines: inhibition of the oxidative activation of human neutrophil and chicken osteoclast pro-matrix metalloproteinases. J. Rheumatol. **25:** 975–982.
4. SEPPER, R. *et al.* 1994. Gelatinolytic and type IV collagenolytic activity in bronchiectasis. Chest **106:** 1129–1133.

# CMT-3, a Chemically Modified Tetracycline, Inhibits Bony Metastases and Delays the Development of Paraplegia in a Rat Model of Prostate Cancer

MARIE G. SELZER, BAOQIAN ZHU, NORMAN L. BLOCK, AND BAL L. LOKESHWAR[a]

*Department of Urology (M-800), University of Miami, Miami, Florida 33101, USA*

## INTRODUCTION

In 1998 an estimated 186,500 Americans will be diagnosed with prostate cancer.[1] At the time of diagnosis approximately 50% of prostate cancer patients will have some form of extraprostatic disease—e.g., metastasis to lymph node, lung, and bone.[2] Carcinoma of the prostate is, by far, the most common of the neoplasms that produce osteoblastic metastases.[3] These metastases result from tumor cell invasion into the bony matrix by the enzymatic degradation of collagen and other matrix components.[4] Key enzymes in collagen degradation are the matrix metalloproteinases (MMPs), secreted by tumor and stromal cells, during tumor-induced bone remodeling.[5] Existing therapies to control this painful disease are only palliative, not curative.[6] We reported, some years ago, that primary cultures of human prostate tissue secrete high levels of activated MMP-2 and MMP-9, but low to undetectable amounts of their natural endogenous inhibitors (TIMPs).[7] In this study we examined whether CMT-3 (6-demethyl, 6-deoxy, 4-dedimethylamino tetracycline) a nontoxic, non-antimicrobial, orally bioavailable MMP inhibitor, with a strong affinity to bone,[8] could inhibit prostate cancer skeletal metastasis.

## MATERIALS AND METHODS

Male Copenhagen rats, 90–100 days old, were purchased from Harlan Labs, Indianapolis, IN. CMT-3 was a gift from CollaGenex Pharmaceuticals Inc., Newtown, PA. Tissue culture reagents and supplies were from Life Technologies, Inc., Gaithersberg, MD. All other reagents were from Sigma Chemical Corp, St. Louis, MO.

[a]Address for correspondence: Bal L. Lokeshwar, Department of Urology (M-800), University of Miami School of Medicine, P.O. Box 016960, Miami, FL 33101. Phone, 305/243-6321; fax, 305/243-6893.

## Cell Cultures

An androgen-independent, highly metastatic rat prostate tumor model, the Dunning MAT LyLu, was obtained from Dr. John T. Isaacs, Johns Hopkins Oncology Center, Baltimore, MD. MAT LyLu cells were cultured in RPMI containing 10% fetal bovine serum, gentamicin (0.2 mg/L) and dexamethasone (0.25 μM). Cultures were routinely tested and found negative for common mycoplasma. Cultures were frequently innoculated into rats for tumor production and were found to retain their stable phenotype, capable of producing tumor nodules at cell concentrations ≥10,000 cells/site.

## Induction of Skeletal Metastasis

Tumors metastatic to lumbar bones were induced in Copenhagen rats by the method of Geldof and Rao.[9] Rats were anesthetized by ketamine injection and the lower abdomen was shaved and cleaned with alcohol. A small midline incision was made and the inferior vena cava isolated; a small surgical "bulldog" clamp was placed on the vena cava, followed immediately by injection of $5 \times 10^4$ Dunning MAT LyLu cells into the lateral tail vein. The vena cava was clamped for a maximum of 5 minutes. The clamp was then removed and the incision closed in two layers with 3-0 silk. The animals withstood the procedure well and usually recovered within two hours, without any noticeable discomfort. Postoperative posture—e.g., conduct, food and water intake—was similar to that observed prior to the procedure.

## Agent Administration

CMT-3 (40 mg/kg) was administered by oral gavage daily for up to 4 weeks with no sign of systemic toxicity. Four groups of six animals were treated with CMT-3 as follows: group I were predosed daily for 7 days; group II were predosed daily for 2 days; group III were dosed one day postimplant; group IV (control) were given 2% carboxymethyl cellulose in water (vehicle) only. Treatment in all groups continued until the animals were euthanized. The criterion for euthanasia was either complete hind limb paralysis or acute respiratory distress. At necropsy the lungs, lumbar vertebrae, and femurs were excised. Marrow plugs, from the femurs of paralyzed rats, were collected and cultured in MAT LyLu culture medium. The lungs and vertebrae were fixed in Bouin's fixing fluid and formalin, respectively.

## RESULTS

Most animals developed acute pulmonary distress with or without paraplegia starting from 12 days after tumor cell injection, and therefore were euthanized. There was a significant delay in the development of pulmonary morbidity and subsequently an increase in survival in rats treated with CMT-3. For example, the median survival was increased by 29.2% (group I, CMT-3 predosed for 7 days), 22.7% (group II, CMT-3 predosed 2 days), and 10.5% (group III, CMT-3 dosed from one day after tumor cell injection) (FIG. 1, top). Upon necropsy, induction of hematogenous metastases, by tail vein injection of tumor cells, was evident as visible pleural tumors, and in the control group the lung surface was completely covered with tumor

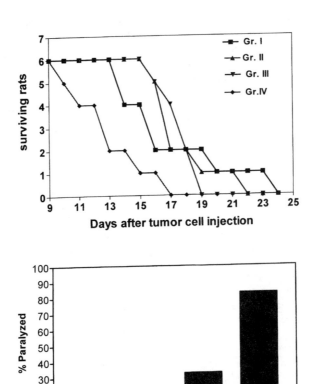

**FIGURE 1.** *(Top)* Oral administration of CMT-3 prolongs survival of rats injected with Dunning MAT LyLu tumor. Tumors were induced and treated with CMT-3 or vehicle alone as described. Animals with acute pulmonary distress or paraplegia were euthanized on the day indicated in the figure. Results are from a single experiment with six animals per group *(Bottom)* Tumor-induced paralysis in the Dunning MAT LyLu tumor-bearing rats. Tumor growth in the vertebral and lumbar bones were induced by iv injection of $5 \times 10^4$ MAT LyLu cells with simultaneous vena cava clamping. All animals survived surgery. Groups of rats were gavaged with CMT-3 (40 mg/kg) starting from 7 days before tumor cell injection or as indicated in the figure. Animals that could not stand on hind limbs or that dragged their limbs during forward motion were considered paraplegic and were euthanized.

foci. A majority of the animals (83%) in the control group (IV) also developed paraplegia at about the same time as they developed pulmonary distress. Tumor cells were recovered from the marrow plugs in about 50% of the animals that had developed paraplegia, and none from nonparaplegic animals. There was no evidence of tumor in any other organ. There was a significant and remarkable reduction in the number of animals that developed paraplegia before they developed pulmonary distress, an end point in our experiment. Compared to the 83% of the animals that developed paraplegia in the control group, only 17% developed paraplegia in groups I and II, and 33% in group III (FIG. 1, bottom). Furthermore, one animal in group IV also developed enlargement of the urinary bladder due to urinary retention.

## DISCUSSION

An experimentally induced bone metastasis model, using the MAT LyLu tumor cells, provided the opportunity to test the potential of an anti-MMP drug to affect prostate tumor metastatis to bone. This complemented our earlier study, where we demonstrated the potential of CMT-3 against tumor metastasis to lungs. Our observation of a significant reduction in the size and number of pleural tumors in those animals treated with CMT-3 demonstrates the principle that prostate cancer metastasis is associated with increased MMP activity. An increase in longevity and the delayed onset, or total lack, of paraplegia demonstrate that bony metastasis can be countered with an administration of strong inhibitor of MMPs, such as CMT-3. CMT-3 could potentially be a potent new, site-directed (bone), orally administrable, and safe drug against prostate cancer metastatic bone, a highly painful and debilitating stage of the malignant disease.

## ACKNOWLEDGMENTS

This work is funded by grants from the National Institutes of Health (1R29-CA 61038), the U.S. Army Prostate Cancer Research program (DAMD 17-98-272), and the L. Austin Weeks Endowment (University of Miami).

## REFERENCES

1. LANDIS, S.H., T. MURRAY, S. BOLDEN & P.A., WINGO. 1998. Cancer statistics, 1998. CA Cancer J. Clinicians **48**(1): 6–29.
2. JACOBS, S.C. 1983. Spread of prostate cancer to bone. Urology **21**: 337–344.
3. GLOTZMAN, D. 1997. Mechanisms of the development of osteoblastic metastases. Cancer **80**: 1581–1587.
4. QUAX, P.H.A., A.C.W. BART, J.A. SCHALKEN & J.H. VERHEIJEN. 1997. Plasminogen activator and matrix metalloproteinase production and extracellular matrix degradation by rat prostate cancer cells in vitro: correlation with metastatic behavior in vivo. Prostate **32**: 196–204.
5. STETLER-STEVENSON, W.G., S. AZNAVOORIAN & L.A. LIOTTA. 1993. Tumor cell interactions with the extracellular matrix during invasion and metastasis. Ann. Rev. Cell Biol. **9**: 541–573.

6. MAHLER, C. & L. DENIS. 1992. Management of relapsing disease in prostate cancer. Cancer **70:** 329–334.
7. LOKESHWAR, B.L., M.G. SELZER, N.L. BLOCK & Z. GUNJA-SMITH. 1993. Secretion of matrix metalloproteinases and their inhibitors (TIMPs) by human prostate in explant cultures: reduced tissue inhibitor of metalloproteinase secretion by malignant tissues. Cancer Res. **53:** 4493–4498.
8. SASAKI, T., N.S. RAMAMURTHY & L.M. GOLUB. 1994. Bone cells and matrix bind chemically modified non-antimicrobial tetracycline. Bone **15:** 373–375.
9. GELDOF, A.A. & B.R. RAO. 1990. Prostate tumor (R3327) skeletal metastasis. Prostate **16:** 279–290.
10. LOKESHWAR, B.L., M.G. SELZER, N.L. BLOCK & L.M. GOLUB. 1997. Inhibition of tumor growth and metastasis by a non-antimicrobial tetracycline analogue in a prostate cancer model. Proc. Amer. Assoc. Cancer Res. **38:** 2868a.
11. LOKESHWAR, B., S. DUDAK, M. SELZER, N. BLOCK & L. GOLUB. 1996. Novel therapies for metastatic prostate cancer: chemically modified tetracycline. *In* Therapeutic strategies in Molecular Medicine, Miami Biotechnology Short Report 7: Advances in Gene Technology. W.J. Whelan *et al.* Eds. Oxford University Press. London.
12. LOKESHWAR, B.L., M.G. SELZER, N.L. BLOCK & L. M. GOLUB. 1997. COL-3: a modified non-antimicrobial tetracycline decreases prostate tumor growth and metastasis. J. Dent. Res. **76:** 3481A.

# Collagenase-2 and -3 Are Inhibited by Doxycycline in the Chronically Inflamed Lung in Bronchiectasis

RUTH SEPPER,[a,e] KAIU PRIKK,[b] TAINA TERVAHARTIALA,[c] YRJÖ T. KONTTINEN,[c] PÄIVI MAISI,[c] CARLOS LÓPES-OTÍN,[d] AND TIMO SORSA[c]

[a]Institute of Experimental and Clinical Medicine, Tallinn, Estonia

[b]Lung Clinic, University of Tartu, Estonia

[c]Dental and Veterinary Clinics, University of Helsinki, Finland

[d]University of Oviedo, Oviedo, Spain

## INTRODUCTION

Bronchiectasis (BE) is an inflammatory lung disease characterized by an extremely vigorous tissue-destructive course that ends up with irreversible structural and functional discord of the whole lung. The etiology of BE is diverse, however; the development of BE lung tissue destruction is initiated by cooperative action of various serine and matrix metalloproteinases (MMPs).[1,2] We have previously shown that BE lung contains high amounts of collagenases capable of cleaving almost completely the lung fibrillar collagen matrix and basement membrane (BM) components depending on the inflammatory disease severity.[2] To study whether collagenolytic activity in BE lung can be inhibited by doxycycline, we reevaluated the origins of collagenases and tested the doxycycline inhibition capacity of collagenolytic activities in bronchoalveolar lavage fluid (BALF) from bronchiectatics and healthy controls.

## MATERIAL AND METHODS

Bronchoalveolar lavage fluid (BALF) from 33 bronchiectatics (14 male, 19 female, ages 16–42 years) and 14 healthy controls (9 male, 5 female, ages 21–24 years) was obtained via fiber-optic bronchoscope. Five 20-ml portions of 0.9% NaCl that were instilled into the middle or lingular lobe were immediately aspirated back, and the material was separated by centrifugation at 500 g for 20 min into its to cellular and supernatant parts. To show whether doxycycline has anticollagenolytic activity toward BE lung collagenases, the origins of collagenolytic burden were investigated by Western blot by the use of specific antibodies directed to MMP-1 (Amersham Int., Little Chalfont, UK), MMP-8 (kind gift from Dr. J. Michaelis,

[e]Address for correspondence: Dr. Ruth Sepper, Institute of Experimental and Clinical Medicine, Hiiu str 42, Tallinn, EE0016 Estonia. Phone, +37 256479973; fax, +37 2 6312653; e-mail, rsepper@tervis.ee

Christchurch School of Med., New Zealand), and MMP-13 (kind gift from Prof. C. Löpes-Otín) by the method described elsewhere.[2] The inhibitory capacity of doxycycline[3,4] was demonstrated by substrate-based collagenase assay followed by incubation of BALFs with 100 μM doxycycline for 48 hours at room temperature as described elsewhere.[2] The total collagenolytic activities in BALF before and after doxycycline inhibition were quantified by an image analyzing system (GS 700 Image Analyzer, BioRad, CA) calculating amounts of cleaved native type I collagen substrate by BALF collagenases.

## RESULTS AND DISCUSSION

In addition to previously identified MMP-8,[2] the MMP-13/collagenase-3 was found to contribute type I collagenolysis in BALF from bronchiectatics with clinically very severe course of the disease. Importantly, collagenase-3 existed mostly in an activated 48-kDa form in BALF. No collagenase-3 protein was found in BALF from healthy controls. Collagenases in BALF from BE patients cleaved the native collagen substrate 68.7% being the highest in more severe patients; whereas collagen substrate was cleaved 2.9% by BALF-collagenases from HC ($p = 0.0001$). After incubation of BALF with 100 μM doxycycline, the collagenolytic activity diminished significantly in BALF from BE patients (mean, 8.6%, $p = 0.01$), but no corresponding effects were seen in BALF from healthy controls (mean, 1.6%, FIG. 1).

Collagen types I and III are the most abundant structural components of lung extracellular matrix (ECM), having a critical role in the maintenance of the structural and functional integrity of the lung. As demonstrated here, collagenase burden in BE lung contains active forms of collagenases-2 and -3, which evidently play significant

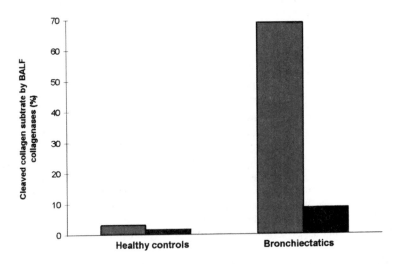

**FIGURE 1.** Cleavage amounts (mean, %) of collagen substrate by the BAL fluid collagenases from healthy controls and bronchiectatics before and after doxycycline inhibition.

roles in initiating the tissue-destructive events in the inflamed lung. The anticollagenolytic effect of doxycycline towards collagenolytic burden in BE lung *in vivo* suggests a new therapeutic approach to the treatment policy of all destructive lung disorders, especially BE.

## REFERENCES

1. SEPPER, R., Y.T. KONTTINEN, T. SORSA & H. KOSKI. 1994. Gelatinolytic and type IV collagenolytic activity in bronchiectasis. Chest **106:** 1129–1133.
2. SEPPER, R., Y.T. KONTTINEN, Y. DING, M. TAKAGI & T. SORSA. 1995. Human neutrophil collagenase (MMP-8), identified in bronchiectasis BAL fluid, correlates with severity of disease. Chest **15:** 27–34.
3. GOLUG, L.M., N.S. RAMAMURTHY, T.F. MCNAMARA, R.A. GREENWALD & B.R. RIFKIN. 1991. Tetracyclines inhibit connective tissue breakdown: new therapeutic implications for an old family of drugs. Crit. Rev. Oral Biol. Med. **2:** 297–321.
4. SORSA, T., P.L. LUKINMAA, U. WESTERLUND, T. INGMAN, Y. DING, H. TSCHESCHE, H. HELAAKOSKI & T. SALO. 1996. The expression, activation and chemotherapeutic inhibition of matrix metalloproteinase-8 (neutrophil collagenase/collagenase-2) in inflammation. *In* Biological Mechanisms of Tooth Movement and Craniofacial Adaptation. Z. Davidovich & L.A. Norton, Eds.: 317–323. Harvard Society for the Advancement of Orthodontics. Boston.

# The Effect of MMP Inhibitor Metastat on Fissure Caries Progression in Rats

L. TJÄDERHANE,[a,e] M. SULKALA,[a] T. SORSA,[b] O. TERONEN,[b] M. LARMAS,[a,c] AND T. SALO[a,c,d]

[a]Institute of Dentistry, University of Oulu, Oulu, Finland

[b]Institute of Dentistry, University of Helsinki, Helsinki, Finland

[c]Oulu University Hospital, Oulu, Finland

[d]Department of Pathology, University of Oulu, Oulu, Finland

An imbalance between activated matrix metalloproteinases (MMPs) and their endogenous tissue inhibitors (TIMPs) leads to pathologic breakdown of the bone and soft tissue extracellular matrix during periodontitis. Another collagenous tissue in the oral cavity is tooth, in which the major component is dentin. Dentin is a bonelike structure with mineralized collagen matrix consisting mainly of type I collagen. Progression of caries into dentin causes demineralization, with subsequent degradation of the collagen matrix. It has been shown *in vitro* that matrix degradation by proteases is necessary for cavity formation.[1,2] The important role of host MMPs in the degradation of this matrix has only recently been observed,[3] bringing up the question of the possible role of MMP inhibitors as therapeutic agents in the prevention of caries progression. We investigated the effect of chemically modified tetracycline-3 (Metastat™) on the progression of young Wistar rat fissure caries.

## MATERIALS, METHODS, AND RESULTS

Wistar rats, 21 days old, were weaned and randomly divided into three groups of 10 animals each. The first group received daily, five days per week, 20 mg/kg of chemically modified tetracycline COL-3 (Metastat) or its inactive analogue COL-5 (CollaGenex Pharmaceuticals Inc., Newtown, PA, USA), with gelatinous 1% carboxymethyl cellulose (CMC) as a vehicle. The control group received CMC alone. In order to get both the local and the systemic effect of the drugs, they were delivered intraorally. To induce caries, rats were fed a 41% sucrose diet and were infected with *Str. sobrinus* weekly as described elsewhere.[4] After seven weeks, dentinal caries progression was evaluated with the fluorescence method from the fissures of hemisected mandibular molars.[4] Kruskal-Wallis 1-way ANOVA and Mann-Whitney U tests were used for statistical analysis. No differences were observed in the dentinal caries progression between the groups in the first and second molars (TABLE 1). In the third

[e]Address for correspondence: Dr. Leo Tjäderhane, Institute of Dentistry, University of Oulu, P.O. Box 5281, FIN-90401 Oulu, Finland. Phone, +358 8 5375455; fax, +358 8 5375560; e-mail, LST@cc.oulu.fi

**TABLE 1.** Sizes of the caries lesions in the first and second molars of the rats

| Molar | Group | Mean ($\mu m^2$) | SD ($\mu m^2$) |
|-------|-------|------|-----|
| 1st molars | Metastat | 4172.3 | 774.9 |
| | COL-5 | 4128.3 | 944.6 |
| | CTR | 4179.3 | 1369.2 |
| 2nd molars | Metastat | 8148.8 | 1968.2 |
| | COL-5 | 6873.2 | 1227.4 |
| | CTR | 5986.3 | 893.3 |

NOTE: No statistical differences were observed between the groups. SD: standard deviation; CTR: control group.

molars, COL-3 significantly reduced caries progression in the third molars when compared to the COL-5 ($p = 0.02$) or control ($p = 0.01$) groups (FIG. 1).

## DISCUSSION

Odontoblasts, the cells responsible for the synthesis of dentin, express several MMPs, including at least gelatinases (MMP-2 and -9),[5] collagenases (MMP-1 and -8),[6] and enamelysin (MMP-20).[7] These findings explain the MMPs observed in the mineralized dentin.[8,9] As the MMPs have a key role in the degradation of demineralized dentin,[3] they also indicate that the pulp-derived MMPs may have a role in the progression of dentinal caries lesion. As almost all of the dentin formation in the third molars of the rats in this study took place during the experiment, the inhibition of the caries progression in the third molars further indicates that the inhibition of the odontoblast MMPs during the active phase of the formation of dentin organic matrix may have a role in the caries progression later on. Osteoblasts take up chem-

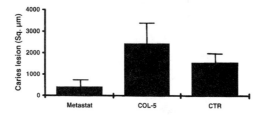

**FIGURE 1.** The area (mean ± standard deviation) of the dentinal caries lesion in the third molars of the rats. The area of caries lesions was significantly lower in the Metastat group than in the COL-5 ($p = 0.02$) or in the control ($p = 0.01$) groups.

ically modified tetracycline,[10] and the low-dose doxycycline treatment inhibits osteoblast collagenase.[11] As the odontoblasts bear a strong resemblance to osteoblasts, the mechanism behind the findings observed in the third molars may indeed be the same.

The absence of differences in the first and second molars between the groups does not necessarily mean that the carious lesions are alike. The spontaneous fluorescence reaction is apparent already after the initial phase of demineralization or even in mantle dentin, in which the structure and composition of the organic matrix as well as the mineral structure may be different.[12] Also, the first alterations in the organic matrix during the demineralization occur in the noncollagenous components.[13] Therefore, combination of different techniques to measure lesion progression, as well as the studies with root caries lesions, should be done for the further analysis of the role of MMP inhibitors in caries progression.

## ACKNOWLEDGMENTS

This study was supported financially by the Academy of Finland and the Finnish Dental Society.

## REFERENCES

1. KATZ, S. *et al.* 1987. In-vitro root surface caries studies. J. Oral Med. **42:** 40–48.
2. KAWASAKI, K. *et al.* 1997. Effects of collagenase on root demineralization. J. Dent. Res. **76:** 588–595.
3. TJÄDERHANE, L. *et al.* 1998. The activation and function of host matrix metalloproteinases in dentin matrix breakdown in caries lesion. J. Dent. Res. **77:** 1622–1629.
4. HIETALA, E-L. *et al.* 1993. Dentin caries recording with Schiff's reagent, fluorescence and back scattered electron image. J. Dent. Res. **72:** 1588–1592.
5. TJÄDERHANE, L. *et al.* 1998. A novel organ culture method to study the function of the human odontoblasts *in vitro*: gelatinase expression by odontoblasts is differentially regulated by TGF-β1. J. Dent. Res. **77:** 1488–1498.
6. PALOSAARI, H. *et al.* The expression of mesenchymal-type MMP-8 in odontoblasts and dental pulp cells is down-regulated by TGF-β1. J. Dent. Res. In press.
7. LLANO, E. *et al.* 1997. Identification and structural and functional characterization of human enamelysin (MMP-20). Biochemistry **36:** 15101–15108.
8. DAYAN, D. *et al.* 1983. A preliminary study of activation of collagenase in carious human dentine matrix. Arch. Oral Biol. **28:** 185–187.
9. FUKAE, M. *et al.* 1991. Metalloproteinases in the mineralized compartments of porcine dentine as detected by substrate-gel electrophoresis. Arch. Oral Biol. **36:** 567–573.
10. SASAKI, T. *et al.* 1994. Bone cells and matrix bind chemically modified non-antimicrobial tetracycline. Bone **15:** 373–375.
11. RAMAMURTHY, N.S. *et al.* 1993. Reactive oxygen species activate and tetracyclines inhibit rat osteoblast collagenase. J. Bone Miner. Res. **8:** 1247–1253.
12. TJÄDERHANE, L. *et al.* 1995. Mineral element analyses of carious and intact dentin by electron probe microanalyser combined with back-scattered electron image. J. Dent. Res. **74:** 1770–1774.
13. LORMÉE, P. *et al.* 1986. Morphological and histochemical aspects of carious dentine in Osborne-Mendel rats. Caries Res. **20:** 251–262.

# Tetracycline Derivative CMT-3 Inhibits Cytokine Production, Degranulation, and Proliferation in Cultured Mouse and Human Mast Cells

KARI K. EKLUND[a,c] AND TIMO SORSA[b]

[a]Institute of Biomedicine, Department of Medical Chemistry and Department of Internal Medicine, Helsinki University Hospital, Helsinki, Finland

[b]Department of Dentistry, Helsinki University, Helsinki, Finland

## INTRODUCTION

Activated mast cells (MC) produce a wide variety of inflammatory mediators such as histamine, eicosanoids, proteases, and several cytokines. Recent findings emphasize the pathogenetic role of MC not only in allergic diseases, but also in diseases associated with chronic inflammation, such as connective tissue diseases.[1]

Tetracyclines are commonly used antibiotics that have therapeutic properties other than those related to their antimicrobial activity. They have been shown to be potent inhibitors of collagenases and to have beneficial antiinflammatory effects in rheumatoid arthritis.[2,3] Chemically modified tetracyclines (CMT) are tetracycline derivatives that do not have the antimicrobial activity but have retained their other properties. We decided to study the effect of three CMTs (CMT-1, CMT-3, CMT-5) on key functions of mast cells to find out whether CMTs could also have antiallergic properties.

## MATERIALS AND METHODS

### Cell Culture

Mouse bone marrow–derived mast cells (mBMMC) were generated by culturing bone marrow cells of BALB/c mice (animal facilities of Helsinki University) for 2–5 weeks in enriched medium supplemented with 50% WEHI-3 cell (line TIB-68; ATCC, Rockville, MD) conditioned medium as a source of IL-3. Human mast cell line (HMC)-1 cells[4] were cultured in Iscove's medium supplemented with 10% FCS.

### Activation of MC

mBMMC were activated with calcium ionofore A23187 (5 mM, Sigma), and supernatants and cells were analyzed for β-hexosaminidase activity. HMC-1 cells were

[c]Address for correspondence: Kari Eklund, Institute of Biomedicine, Department of Medical Chemistry, University of Helsinki, Siltavuorenpenger 10, 00170 Helsinki, Finland.

activated with a combination of phorbol 12-myristate 13-acetate (PMA) 50 ng/ml (Sigma) and calcium ionofore A23187 0.5 µM, and cytokines were analyzed 24 hr after activation of the cells using a commercial ELISA method (R&D Systems, London, England). The viability of the cells was assessed by counting the trypan blue excluding cells 24 hr after activation. Viability was not affected by the CMTs in the concentrations used.

## RESULTS AND DISCUSSION

A clear dose-dependent inhibition of TNF-α and to a lesser degree of IL-8 production was observed in HMC-1 cells by CMT-3 but not by CMT-1 or CMT-5 (FIG. 1). In the presence of 25 µM CMT-3 75% and 40% inhibition of TNF-α and IL-8 production, respectively, was observed. The effect of CMT-3 on mBMMC degranulation, as revealed by the granule-associated β-hexosaminidase release, is shown in FIGURE 2. A clear inhibition of calcium ionofore–induced degranulation was observed in the presence of 25 µM and 50 µM CMT-3. No clear effect on degranulation could be observed by the CMT-1 or CMT-5 (data not shown).

The reason for the inhibition of degranulation and cytokine production by CMT-3 is not clear at present. Both serine and metalloproteinases have been implicated in the degranulation of mast cells. Tetracyclines are known also to have metal chelating properties, which could perhaps account for the inhibitory effect of CMT-3. To con-

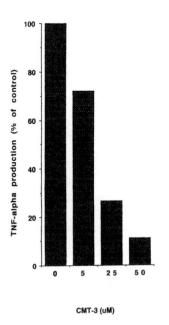

**FIGURE 1.** The effect of CMT-3 on TNF-α production in HMC-1 cells. Cells $(1 \times 10^6/\text{ml})$ were activated with PMA (50ng/ml) and calcium ionofore A12387 (0.5 µM). CMT-3 was added 1 hr before the activators. Similar results were obtained in a replicate experiment.

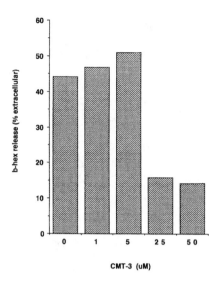

**FIGURE 2.** Degranulation of mouse bone marrow–derived mast cells ($1 \times 10^6$/ml) in the presence of CMT-3. Degranulation was induced with calcium ionofore (A12387, 5 μM), and β-hexosaminidase activity of the cells and supernatants was analyzed 20 min later. Similar results were obtained in a replicate experiment.

clude, CMT-3 inhibits very efficiently several key functions of MC and could therefore have potential use in the treatment of mast cell–related diseases such as various allergic diseases.

## REFERENCES

1. ARNASON, J. & D.G. MALONE. 1995. Role of mast cells in arthritis. Chem. Immunol. **62:** 204–238.
2. 2. KLOPPENBURG, M., F.C. BREEDVELD, J.P. TERWIEL, C. MALLEE & B.A.C. DIJKMANS. 1994. Minocycline in active rheumatoid arthritis: a double-blind placebo-controlled trial. Arthritis Rheum. **37:** 629–636.
3. TILLEY, B.C., G.S. ALARCON, S.P. HEYSE, D.E. TRENTHAM *et al.* 1995. Minocycline in rheumatoid arthritis: a 48-week, double blind, placebo controlled trial. Ann. Intern. Med. **122:** 81–89.
4. BUTTERFIELD, J.H., D. WEILER, G. DEWALD & G.J. GLEICH. 1988. Establishment of an immature mast cell line from a patient with mast cell leukemia. Leuk. Res. **2:** 345.

# TNF-α–Converting Enzyme Activity in Colonic Biopsy Specimens from Patients with Inflammatory Bowel Disease Revealed by mRNA and *in Vitro* Assay

PIA FØGH, CHRISTINA ELLERVIK, TORBEN SÆRMARK,
AND JØRN BRYNSKOV[a]

*Department of Gastroenterology C, Herlev University Hospital,
DK-2730 Herlev, Copenhagen, Denmark*

Tumor necrosis factor (TNF)-α is a proinflammatory protein that seems to play an important pathogenic role in inflammatory bowel disease (IBD) as judged from the remarkable clinical response in patients with chronically active or refractory Crohn's disease (CD) or ulcerative colitis (UC) treated with anti-TNF-α antibodies.[1,2] TNF-α is synthesized as a membrane-bound precursor protein that is subsequently processed to the mature form by cleavage of the Ala76-Val77 bond.[3,4] The soluble 17-kDa form of TNF-α is released by a membrane-anchored proteinase, recently identified and cloned as a multidomain metalloproteinase called TNF-α converting enzyme (TACE) or ADAM 17 due to structural similarities with other metzincins, particularly the family of snake venom adamalysins (ADAMs).[5] Previous Northern blot screening analyses have detected high levels of TACE mRNA in libraries of various human tissues, but apparently not in the colon.[3,6] The aim of this study was to investigate whether TACE is involved in the release of TNF-α in normal and inflamed human colonic mucosa using the more sensitive RT-PCR technique and an oligopeptide-based *in vitro* assay for enzyme activity.

## METHODS

### *RT-PCR*

Colonoscopic biopsies were obtained from 46 patients with IBD (CD, $n = 18$, UC, $n = 28$) and 20 controls. None of the patients received immunosuppressants, and only three patients were on oral prednisolone (5–10 mg/day) at the time of the study. Thirteen patients received topical treatment with a glucocorticoid or mesalazine (5-ASA) preparation; 37 patients were maintained on an oral 5-ASA–containing drug. Biopsies were collected from various segments of the colon, except for one that was taken from the terminal ileum of a patient with CD. Disease activity was graded using separate criteria for CD[7] and UC.[8] The colonoscopic biopsies were immediately frozen in liquid $N_2$ and stored at $-80°C$ for later analysis. RNA was extracted from the fro-

[a]Corresponding author: Dr. J. Brynskov, M.D., D.M.Sc. Phone, +45 4488 3625; fax, +45 4494 4056; e-mail, brynskov@dadlnet.dk

zen biopsies using Trisol™ and measured spectrophotometrically. The same amount of RNA (500 ng) from each patient sample was used to generate cDNA. RT-PCR for TACE was carried out using an annealing temperature of 62°C, 32 cycles, and primers for the human TACE sequence designed specifically for this study. The products were analyzed on agarose electrophoresis gels stained by ethidium bromide and measured with a gel scanner to determine the optical density of PCR bands. Semiquantitation of TACE was carried out using GAPDH to correct for sample variation and a placental cDNA library as an external standard.

### In vitro *Assay for TACE Activity*

Cell membranes isolated from macroscopically normal colonic mucosa were dissolved in 1% Nonidet P-40. The dissolved enzyme activity was incubated with a synthetic substrate (10 µg/ml), leupeptin and pepstatin (10 µM), soyabean inhibitor (20 U/ml), and α-1-antitrypsine (10 µg/ml) for up to 1 hour, followed by HPLC analysis of the test sample in order to detect the enzymatic activity. The peptide used was an oligopeptide containing 15 amino acids representing the processing site at the Ala76-Val77 bond[9,10] acSPLAQAVRSSSRTPSNH$_2$. Other protease inhibitors as well as a hydroxamic acid inhibitor (CH4474) were tested in separate experiments. Processing of the peptide was stopped by addition of 0.1 M HCl. HPLC analysis us-

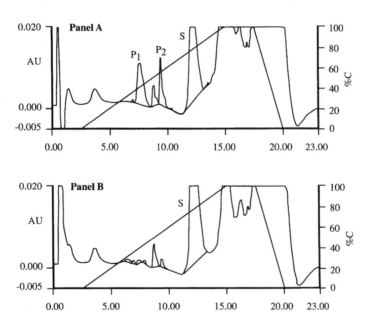

**FIGURE 1.** HPLC analysis of the breakdown of an oligopeptide of the pro-TNF-α cleavage site (S) in the presence of biopsy extracts from normal colon without (**A**) or with 5 mM EDTA (**B**). Processing of the oligopeptide to the expected products P1 and P2 was almost completely inhibited by EDTA, suggesting that a metalloproteinase is involved.

ing reverse-phase C18 columns and a linear gradient from 0 to 80% methanol in 0.1% TFA was used to quantify the processing of the TNF-$\alpha$-1–like substrate using synthetic standards for identification of breakdown products.

## RESULTS

TACE mRNA was expressed in all mucosal biopsies, except four controls and two ulcerative colitis specimens, indicating that TACE is almost ubiquitously expressed in human colonic bowel mucosa. Stratification according to clinical disease activity showed that the median TACE mRNA level was significantly higher in IBD patients with moderate/high disease activity (3.14, range 0–25.21) as compared with patients with low activity (1.52, range 0.61–4.05) or inactive disease (0.88, range 0–2.72) and controls (1.45, range 0–4.25) ($p = 0.0005$, Jonckheere-Terpstra test for trend, one tailed). The same correlations were found when TACE mRNA expression was analyzed separately in patients with Crohn's disease ($p = 0.025$) and ulcerative colitis ($p = 0.003$). No strong correlations were evident between TACE expression and endoscopic or histological activity, except for a trend towards higher levels in active as compared to inactive CD.

TACE activity in colonic mucosa was measured by processing of a synthetic analogue of the pro-TNF cleavage site. The predicted products were quantitated by HPLC analysis as shown in FIGURE 1. This breakdown was sensitive to EDTA and CH4474, but not to serine or other types of protease inhibitors, suggesting that it is caused by a metalloproteinase similar to TACE. The effects of various inhibitors on

**TABLE 1. TACE activity from normal colonic mucosa measured in the absence or presence of various protease inhibitors**

|  | Protease specificity | Final concentration | Inhibitory effect |
|---|---|---|---|
| Leupeptin | serine-, cysteine- | 10 μM | — |
| Pepstatin | aspartat-, metallo- | 10 μM | — |
| Soyabean inhibitor | serine- | 20 U/ml | — |
| α-1-Antitrypsine | trypsine- | 10 μg/ml | — |
| N-methoxysuccinyl-ala-ala-pro-val-CMK | serine- | 0.1 mg/ml | — |
| 1,10-Phenanthroline | metallo- | 5 mM | — |
| TLCK | serine- | 0.2 mg/ml | — |
| Dichloroisocoumarin | serine- | 0.1 mM | — |
| EDTA | metallo- | 5 mM | + |
| Basitrasin | lysozymal | 100 U/ml | — |
| CH4474 | TACE | 1 μg/ml | + |

NOTE: Only CH4474 and EDTA were able to inhibit the breakdown of a synthetic substrate, indicating that TACE is responsible for processing of TNF-$\alpha$ in human colonic mucosa.

the processing of the synthetic substrate are summarized in TABLE 1. TACE activity was found in 7/7 macroscopically normal mucosal colonic biopsies specimens from controls or IBD patients, thus supporting the results obtained by RT-PCR.

## CONCLUSION

The results of this study suggest that TACE is likely to be responsible for cleavage of TNF-α from its membrane-bound precursor in normal and inflamed human colonic mucosa. TACE expression is increased in patients with active IBD, suggesting that TACE inhibitors may have a therapeutic role in IBD; but further studies are needed to validate this concept, including experiments designed to locate the TACE-producing cells in human intestinal mucosa.

## REFERENCES

1. TARGAN, S.R. *et al.* 1997. A short-term study of chimeric monoclonal antibody cA2 to tumor necrosis factor α for Crohn's disease. N. Engl. J. Med. **337:** 1029–1035.
2. EVANS, R.C. *et al.* 1997. Treatment of ulcerative colitis with an engineered human anti-TNF-α antibody CDP571. Aliment Pharmacol. Ther. **11:** 1031–1035.
3. BLACK, R.A. *et al.* 1997. A metalloproteinase disintegrin that releases tumour-necrosis factor-α from cells. Nature **385:** 729–732.
4. MOSS, M.L. *et al.* 1997. Cloning of a disintegrin metalloproteinase that processes precursor tumour-necrosis factor-α. Nature **385:** 733–736.
5. MASKOS, K. *et al.* 1998. Crystal structure of the catalytic domain of human tumor necrosis factor-α–converting enzyme. Proc. Natl. Acad. Sci. USA **95:** 3408–3412.
6. PATEL, I.R. *et al.* 1998. TNF-alpha convertase enzyme from human arthritis-affected cartilage: isolation of cDNA by differential display, expression of the active enzyme, and regulation of TNF-α. J. Immunol. **160:** 4570–4579.
7. MUNKHOLM, P. *et al.* 1992. Incidence and prevalence of Crohn's disease in the county of Copenhagen, 1962–87: a sixfold increase in incidence. Scand. J. Gastroenterol. **27:** 609–614.
8. LANGHOLTZ *et al.* 1994. Course of ulcerative colitis: analysis of changes in disease activity over years. Gastroenterology **107:** 3–11.
9. MOHLER, K.M. *et al.* 1994. Protection against a lethal dose of endotoxin by an inhibitor of tumour necrosis factor processing. Nature **370:** 218–220.
10. BLACK, R.A. *et al.* 1996. Relaxed specificity of matrix metalloproteinases (MMPS) and TIMP insensitivity of tumor necrosis factor-α (TNF-α) production suggest the major TNF-α converting enzyme is not an MMP. Biochem. Biophys. Res. Commun. **225:** 400–405.

# MMPs Are IGFBP-Degrading Proteinases: Implications for Cell Proliferation and Tissue Growth

JOHN L. FOWLKES,[a,d] DELILA M. SERRA,[b] HIDEAKI NAGASE,[c] AND KATHRYN M. THRAILKILL[a]

[a]*Department of Pediatrics, University of Kentucky Medical Center, Lexington, Kentucky 40536, USA*

[b]*Department of Pediatrics, Duke University Medical Center, Durham, North Carolina, USA*

[c]*Department of Biochemistry and Molecular Biology, University of Kansas Medical Center, Kansas City, Kansas, USA*

Binding of IGF-I or IGF-II by high-affinity IGF-binding proteins (IGFBPs) sequesters IGFs, preventing their interactions with cell surface IGF receptors. Hypothetically, IGFs might be released from IGFBP/IGF complexes via limited proteolysis of the IGFBP, but not IGFs. Recent studies from our laboratory demonstrate that at least two IGFBPs, IGFBP-3 and -5, are degraded under physiologic conditions by MMPs. Although this family of zinc-dependent proteinases is generally believed to be important in the degradation and turnover of extracellular matrix molecules, both in physiologic and in pathologic conditions, our studies suggest a new role for these enzymes as regulators of IGF bioavailability and bioactivity.

In an earlier study, we demonstrated that IGFBP-3 is degraded by a zinc-dependent proteinase produced by human fibroblasts.[1] To characterize this metalloproteinase(s), we developed IGFBP-3 substrate zymography.[2,3] This method allows for the determination of the molecular mass of IGFBP-degrading proteinases, as well as their inhibitor profile. This method identified several IGFBP-3–degrading proteinases ($M_r$ 52,000–72,000) in human fibroblast–conditioned media (FIG. 1, lane 1) that were inhibited by EDTA (lane 2). Gelatin substrate zymography revealed that the same conditioned media contained several gelatinases with identical molecular masses to those that degraded IGFBP-3.[2] This was the first data suggesting that gelatinases might function as IGFBP-degrading proteinases. MMP-1, MMP-2, and MMP-3 were all detected at their expected molecular masses and were identical in size to the IGFBP-3–degrading proteinases detected by zymography. These and other data[2] strongly suggest that MMPs were involved in IGFBP-3 degradation in human fibroblast cultures.

Because bone cells produce MMPs, we hypothesized that MMPs might account for the cation-dependent, IGFBP-5–degrading proteinase activity produced by murine (MC3T3-E1) osteoblasts.[4] IGFBP-5 substrate zymography identified IGFBP-5–

[d]Address for correspondence: John L. Fowlkes, J464 Kentucky Clinic, 740 South Limestone, Lexington, Kentucky 40536-0284. Phone, 606/323-5404; fax, 606/323-8179; e-mail, jlfowlk@pop.uky.edu

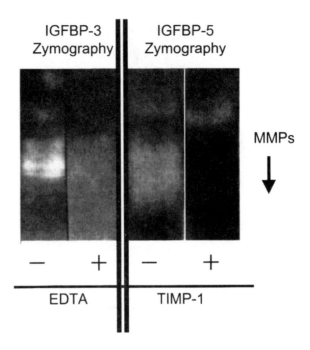

**FIGURE 1.** IGFBP-substrate zymography of IGFBP-3–degrading proteinases in human fibroblast–conditioned medium *(lanes 1 and 2)* and IGFBP-5–degrading proteinases in murine osteoblast–conditioned medium *(lanes 3 and 4)*. Lytic areas represent proteinase activity. For details of techniques, see Refs. 2–4.

degrading proteinases with $M_r$ 52,000–72,000 and $M_r$ 97,000 (FIG. 1, lane 3). The 52–72 kDa proteinases were inhibited by EDTA and TIMP-1, confirming their identities as members of the MMP family. In contrast, the 97-kDa proteinase was partially inhibited by PMSF, but not by EDTA or TIMP-1, suggesting that the 97-kDa proteinase is a serine proteinase. Immunoprecipitation studies showed that only antisera to human MMP-1 and human MMP-2 immunoprecipitated IGFBP-5–degrading proteinases ($M_r$ 52 kDa and 69/72 kDa, respectively).[4] Together, these studies suggest that murine MMPs antigenically related to human MMP-1 and MMP-2 are involved in IGFBP-5 degradation in MC3T3-E1 cultures.[4]

To clarify the overall contribution of MMPs to IGFBP-3 and IGFBP-5 degradation, IGFBP-3 or radiolabeled IGFBP-5 were incubated with conditioned medium from human fibroblasts or murine bone cells, respectively. The addition of TIMP-1 inhibited IGFBP-3 proteolysis by >80%; and, as shown in FIGURE 2, IGFBP-5 degradation was inhibited by >85% (compare lanes 1 and 2), suggesting MMPs are the primary proteinases involved in the degradation of IGFBP-3 and IGFBP-5 in these cell types.[2,4]

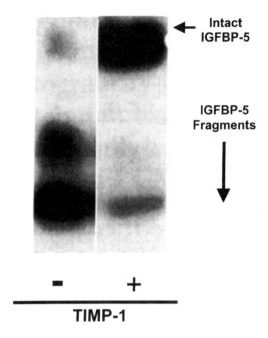

**FIGURE 2.** Inhibition of IGFBP degradation by TIMP-1. Radiolabeled IGFBP-5 *(lanes 1 and 2)* were incubated with murine osteoblast–conditioned medium in the absence *(lane 1)* or presence *(lane 2)* of TIMP-1. For details of techniques, see Refs. 2 and 4.

For MMPs to regulate IGF action, they must first be capable of releasing functioning IGFs from IGF/IGFBP complexes. To investigate the effects of MMP hydrolysis on IGF bioavailability, MMP-3, the most potent IGFBP-3–degrading MMP,[2] has been tested for its ability to digest IGF-I/IGFBP-3 complexes (our unpublished data). IGF-I/IGFBP-3 complexes digested by MMP-3 result in significant fragmentation of IGFBP-3, while IGFs remain essentially intact. Furthermore, MMP-3 degradation of IGF-I/IGFBP-3 results directly in increased IGF-I bioavailability as evidenced by (1) phosphorylation of cell surface type 1 IGF receptors, and (2) an increase in cellular proliferation. The specificity of this effect has been confirmed by the findings that TIMP-1 or an antibody to the type 1 IGF receptor mitigates these effects.

Both IGFs and MMPs have been independently implicated in a variety of processes including tumor cell growth and invasion, morphogenesis, trophoblast growth and invasion, cartilage and bone repair and turnover, wound healing, and angiogenesis. Thus, the interaction of these two systems may play a critical role in the delicate balance between biologic and pathologic growth, remodeling, and tissue destruction.

## REFERENCES

1. FOWLKES, J.L. 1994. Degradation of insulin-like growth factor (IGF)–binding protein-3 (IGFBP-3) by a metal-dependent protease produced by human fibroblasts. Endocrine **2**: 63–68.
2. FOWLKES, J.L., J.J. ENGHILD, K. SUZUKI & H. NAGASE. 1994. Matrix metalloproteinases degrade insulin-like growth factor–binding protein-3 in dermal fibroblast cultures. J. Biol. Chem. **269**: 25742–25746.
3. FOWLKES, J.L., K.M. THRAILKILL, D.M. SERRA & H. NAGASE. 1997. Insulin-like growth factor binding protein (IGFBP) substrate zymography: a new tool to identify and characterize IGFBP-degrading proteinases. Endocrine **7**: 33–36.
4. THRAILKILL, K.M., L.D. QUARLES, H. NAGASE, K. SUZUKI, D.M. SERRA & J.L. FOWLKES. 1995. Characterization of insulin-like growth factor–binding protein-5–degrading proteases produced throughout murine osteoblast differentiation. Endocrinology **136**: 3527–3533.

# Phosphonate Inhibitors of Adamalysin II and Matrix Metalloproteinases

C. GALLINA,[a] E. GAVUZZO,[b] C. GIORDANO,[c] B. GORINI,[c] F. MAZZA,[b,d,f]
M. PAGLIALUNGA-PARADISI,[c] G. PANINI,[c] G. POCHETTI,[b] AND V. POLITI.[e]

[a]Istitituto di Scienze del Farmaco, Università G. d'Annunzio, Chieti, Italy

[b]Istituto di Strutturistica Chimica, CNR, Rome, Italy

[c]Centro di Chimica del Farmaco, CNR, Università La Sapienza, Rome, Italy

[d]Dipartimento di Chimica, Università di L'Aquila, L'Aquila, Italy

[e]Polifarma Research Center, Rome, Italy

The aberrant regulation of matrix metalloproteinases (MMPs) has been implicated in a number of disease states characterized by unwanted degradation of the extracellular matrix, including tumor invasion and metastasis, arthritis, bone destruction, and multiple sclerosis. These enzymes therefore represent attractive targets for inhibitor design and drug development. The majority of MMP inhibitors contain a zinc binding function (ZBF) linked to a substrate-like pseudopeptide unit including two $sp^3$ carbons ($\alpha$ = hydrophobic chain) or three $sp^3$ atoms (Z = $CH_2$ or NH) between the ZBF and the first peptide bond:[1]

ZBF-CH($\alpha$)-CH(P$'_1$)-CO-NH....    ZBF-CH($\alpha$)-Z-CH(P$'_1$)-CO-NH....
(Class I inhibitors)    (Class II inhibitors)

Adamalysin II is a snake venom Zn-dependent proteinase whose active site presents substantial structural similarities with MMPs and TNF-$\alpha$ converting enzyme.[2] We have recently determined the mode of binding of the peptidomimetic inhibitor 1a (FIG. 1) in the active site of adamalysin-II by crystal structure solution of the complex.[3] The phosphonate group binds to the catalytic zinc ion, and the peptide backbone occupies the primed region. Contrary to inhibitors of class I and II, 1a has only one $sp^3$ carbon between the ZBF and the first peptide bond and fits at the primed region of the active site adopting a retrobinding mode. Moreover, 1a shows moderate inhibiting activities also against two representative MMPs (TABLE 1). Molecular modeling studies, based on the X-ray structure of this complex,[3] were therefore undertaken in search of new analogs more potent and selective against the snake venom peptidase and potentially useful as new models of inhibitors against MMPs and TNF-$\alpha$ converting enzyme.

The analogs 1b–1g, containing pentatomic heterocyclic rings replacing furan, were prepared by coupling the appropriate N-acyl L-leucine with L-phosphotryptophan diethyl ester by the mixed anhydride method, followed by removal of protecting groups. Preparation of 2a, which is the monobenzyl ester of 1a, required the independent synthesis of D,L-phosphotryptophan dibenzyl ester. Contrary to the di-

[f]Address for correspondence: Fernando Mazza, Istituto di Strutturistica Chimica, CNR, C.P. 10, 00016 Monterotondo Stazione, Rome, Italy. Phone, 0039-06-90625142; fax, 0039-06-90673630; e-mail, mazza@isc.mlib.cnr.it

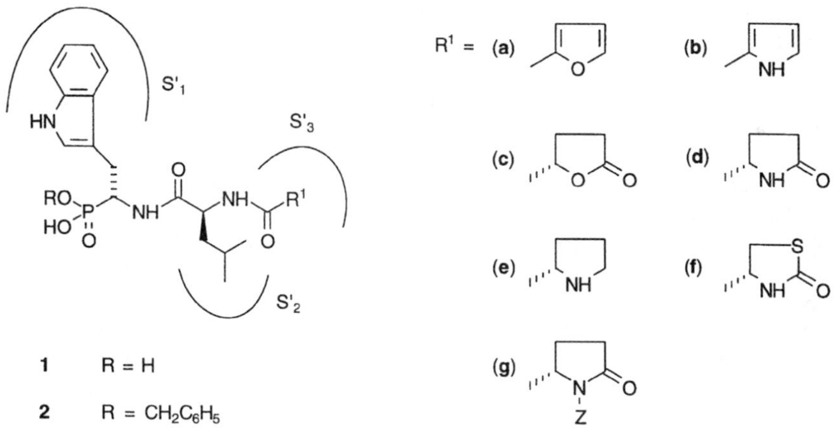

**FIGURE 1.** New analogs of pseudopeptide phosphonate inhibitors and schematic representation of their subsite of binding.

ethyl analog, this intermediate could not be resolved into single enantiomers. Acylation of the racemic dibenzyl ester with furan-2-carbonyl-L-leucine, followed by selective debenzylation, gave the desired monobenzyl ester 2a as a mixture of diastereoisomers.

The new compounds were tested for inhibitory activities against adamalysin II, MMP-2, MMP-9, MMP-8, and MMP-3 by evaluating the residual enzyme activity by continuous fluorimetric assay. Phosphonate inhibitor 1a, adopting the retrobind-

**TABLE 1. Inhibition of adamalysin II, gelatinase A, gelatinase B, human neutrophil collagenase, and stromelysin I by inhibitors containing phosphonic zinc binding groups**

| | $IC_{50}$ (µM) or % inhibition at 50 µM | | | | |
|---|---|---|---|---|---|
| Inhibitor | Adam. II | MMP-2 | MMP-9 | MMP-8 | MMP-3 |
| 1a | 0.4 | 30% | n.i. | 15% | n.i. |
| 1b | 0.2 | 9.0 | 9% | 37% | 90 |
| 1c | 0.7 | n.i. | n.i. | 15% | n.i. |
| 1d | 1.0 | 23% | 14% | 39% | 6% |
| 1e | 22 | 11 | 5.0 | 3.2 | 20 |
| 1f | 0.5 | 40 | n.i. | 20% | 6% |
| 1g | 1.0 | n.i. | n.i. | n.i. | 28% |
| 2a | 7.2 | n.i. | n.i. | 8% | n.i. |

ABBREVIATIONS: Adam. II = adamalysin II; MMP-2 = gelatinase A; MMP-9 = gelatinase B; MMP-8 = human neutrophil collagenase; MMP-3 = stromelysin I; n.i. = not inhibiting at 50 µM.

ing mode, is about 100-fold more potent than its carboxylate analog against adamalysin II, as previously reported.[3] All analogs 1b–1g, except 1e, substantially retain the inhibiting activity of 1a against adamalysin II but show lower or no activity against MMPs. The L-prolyl derivative 1e, containing a pyrrolidine ring whose amino group is probably protonated in the complex with the enzyme, exhibits an interesting increase of affinity against MMP-8 and MMP-9, whereas its potency against adamalysin II decreases. Preliminary results of crystallographic studies show that 1e binds to the unprimed region of the MMP-8 active site (paper in preparation with M. Pieper and H. Tschesche, University of Bielefeld, Germany). Another example of left-hand-side inhibitor (Pro-Leu-Gly-NHOH) has been previously observed.[4] The 2a benzyl group can partially occupy the S1 hydrophobic subsite of adamalysin II (modeling); its unexpected decrease of affinity could be due to weakening of the phosphonate-zinc interaction provoked by accommodation of the bulky benzyl group at the $S_1$ subsite. Studies on phosphonate inhibitors combining binding interactions both at primed and unprimed regions of MMPs are continuing.

## ACKNOWLEDGMENTS

Financial support by CNR, Prog. Fin. Biotecnologie, and Cofin. MURST 97 CFSIB is acknowledged.

## REFERENCES

1. BABINE, E. & S.L. BENDER. 1997. Chem. Rev. **97:** 1359–1472.
2. MASKOS, K., C. FERNANDEZ-CATALAN, R. HUBER, G.P. BOURENKOV, H. BARTUNIK, G.A. ELLESTAD, P. REDDY, M.F. WOLFSON, C.T. RAUCH, B.J. CASTNER, R. DAVIS, H.R.G. CLARKE, M. PETERSEN, J.N. FITZNER, D.P. CERRETTI, C.J. MARCH, R.J. PAXTON, R.A. BLACK & W. BODE. 1998. Proc. Natl. Acad. Sci. USA **95:** 3408–3412.
3. CIRILLI, M., C. GALLINA, E. GAVUZZO, C. GIORDANO, F.X. GOMIS-RÜTH, B. GORINI, L. KRESS, F. MAZZA, M. PAGLIALUNGA-PARADISI, G. POCHETTI & V. POLITI. 1997. FEBS Lett. **418:** 319–322.
4. GRAMS, F., P. REINEMER, J.C. POWERS, T. KLEINE, M. PIEPER, H. TSCHESCHE, R. HUBER & W. BODE. 1995. Eur. J. Biochem. **228:** 830–841.

# Roles of MT1-MMP in the Regulation of Cell Surface Proteolysis

SARA MONEA,[a,c] BRIGETTE ROBERTS,[a] STUART G. MARCUS,[a]
PETER SHAMAMIAN,[a] AND PAOLO MIGNATTI[a,b]

[a]Department of Surgery, S. Arthur Localio Laboratory for General Surgery Research, and
[b]Department of Cell Biology, New York University School of Medicine,
New York, New York 10016, USA

Membrane type 1 matrix metalloproteinase (MT1-MMP) has been implicated as a physiological activator of progelatinase A (MMP-2) and a cell surface receptor for the tissue inhibitor of metalloproteinases-2 (TIMP-2)/MMP-2 complex.[1] MMP-2 also binds to $\alpha_v\beta_3$ integrin on the cell membrane.[2] We have previously shown that plasmin can activate MMP-2 on the cell surface.[3] We have now characterized the relative contribution of $\alpha_v\beta_3$ integrin and TIMP-2 to plasmin-mediated activation of proMMP-2.

Clones of human HT1080 fibrosarcoma cells transfected with MT1-MMP cDNA, its antisense cDNA, or the vector alone were characterized for MT1-MMP expression, MMP-2 activation, and for expression of $\alpha_v\beta_3$ integrin and TIMP-2. Consistent with previous reports, multiple forms of MT1-MMP with $M_r$s 63 kDa, 60 kDa, 58 kDa, and 43 kDa were detected by Western blotting with antibody to the intracellular domain of MT1-MMP[4] (FIG. 1). Vector-transfected or nontransfected cells constitutively expressed the 60-kDa and 58-kDa, active forms of MT1-MMP.[4] The levels of these MT1-MMP forms were considerably reduced in antisense cDNA transfectants. MT1-MMP transfectants expressed 60-kDa and 58-kDa MT1-MMP, in addition to high levels of the 43-kDa form of MT1-MMP.[4] Although control, vector-transfected cells expressed active MT1-MMP, they did not activate MMP-2. Similarly, antisense cDNA transfectants secreted MMP-2 only in its inactive, 72-kDa form. In contrast, active forms of MMP-2 with $M_r$s 68/66 kDa and 64/62 kDa could be detected in the conditioned medium of MT1-MMP cDNA-transfected cells (FIG. 1).

MT1-MMP transfectants expressed very low amounts of soluble and cell-associated TIMP-2 and levels of $\alpha_v$ and $\beta_3$ integrin chains dramatically lower than vector- or antisense cDNA–transfected cells (FIG. 1). The difference in TIMP-2 levels was abolished when the cells were grown in the presence of the MMP inhibitor marimastat (FIG. 2), implicating MMP activity in the observed downregulation of TIMP-2. However, marimastat had no effect on $\alpha_v\beta_3$ integrin levels, indicating that they may reflect differences in gene expression. The inhibitor blocked proMMP-2 activation by MT1-MMP transfectants, showing that the concentration used was indeed effective (FIG. 2).

[c]Address for correspondence: Dr. Sara Monea, Department of Cell Biology, New York University School of Medicine, 550 First Avenue, New York, New York 10016. Phone, 212/263-1478; fax, 212/263-0147; e-mail, moneas01@popmail.med.nyu.edu

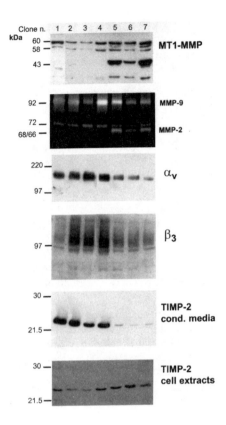

**FIGURE 1.** Characterization of MT1-MMP, gelatinases, $\alpha_v\beta_3$ integrin, and TIMP-2 in clones of HT1080 cells transfected with either vector alone *(clones 1, 4)* or with MT1-MMP antisense *(clones 2, 3)* or sense cDNA *(clones 5–7)*. Confluent cells were grown for 16 hr in serum-free medium and lysed with Triton X-100 0.5% (v/v) in Tris-HCl 0.1 M, pH 8.1. Eighty μg of Triton X-100 cell extracts was analyzed by Western blotting with antibodies to MT1-MMP, to the $\alpha_v$ or $\beta_3$ integrin chains, or to TIMP-2. Concentrated conditioned medium was analyzed by gelatin zymography and Western blotting with antibody to TIMP-2. Molecular masses are shown in kDa *on the left of each panel*. These experiments were repeated three times with comparable results.

Gelatin zymography of cell extracts (not shown) showed MMP-2 associated with cells transfected with vector- or MT1-MMP cDNA-transfected cells. In contrast, no MMP-2 was associated with extracts of cells transfected with MT1-MMP antisense cDNA.[5] Because these cells express high levels of $\alpha_v\beta_3$ integrin, whereas MT1-MMP transfectants have very low $\alpha_v\beta_3$ integrin levels (FIG. 1), this finding indicates MT1-MMP as a major cell surface binding site for MMP-2. Addition of plasmin(ogen) to clones of HT1080 cells that possess cell-associated MMP-2 resulted in proMMP-2 activation. In contrast, plasmin did not activate proMMP-2 with

**+ MARIMASTAT**

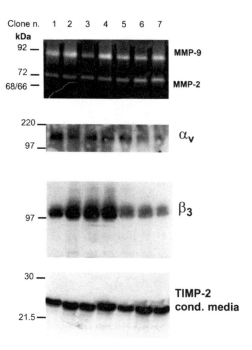

**FIGURE 2.** Characterization of gelatinases, $\alpha_v\beta_3$ integrin and TIMP-2 in clones of HT1080 cells transfected with either vector alone *(clones 1, 4)* or with MT1-MMP antisense *(clones 2, 3)* or sense cDNA *(clones 5–7)*. Confluent cells were grown for 16 hr in serum-free medium containing marimastat 10 μM. Conditioned medium was analyzed by gelatin zymography and cell extracts by Western blotting with antibodies to the $\alpha_v$ or $\beta_3$ integrin chains, and to TIMP-2 as described in the legend to FIGURE 1.

clones of antisense HT1080 cell transfectants that expressed no MT1-MMP and had no cell-bound MMP-2. The effect of plasmin(ogen) was blocked by inhibitors of plasmin but not of metalloproteinases, indicating that plasmin-mediated activation requires pro-MMP-2 binding to, but not the catalytic activity of, MT1-MMP.[5]

These results show several features of the roles of MT1-MMP in the regulation of extracellular proteolysis: (1) expression of active MT1-MMP is necessary but not sufficient for proMMP-2 activation; (2) plasmin can cooperate with MT1-MMP to activate proMMP-2; (3) cells expressing MT1-MMP can bind MMP-2 on the cell surface, although they do not express $\alpha_v\beta_3$ integrin; (4) MT1-MMP may modulate TIMP-2 levels through a proteolysis-dependent mechanism, and $\alpha_v\beta_3$ integrin levels through proteolysis-independent mechanism(s).

## REFERENCES

1. SATO, H., Y. OKADA & M. SEIKI. 1997. Membrane-type matrix metalloproteinases (MT-MMPs) in cell invasion. Thromb. Haemost. **78:** 497–500.
2. BROOKS, P.C., S. STRÖMBLAD, L.C. SANDERS, T.L. VON SCHALSCHA, R.T. AIMES, W.G. STETLER STEVENSON, J.P. QUIGLEY & D.A. CHERESH. 1996. Localization of matrix metalloproteinase MMP-2 to the surface of invasive cells by interaction with integrin alpha v beta 3. Cell **85:** 683–693.
3. MAZZIERI, R., L. MASIERO, L. ZANETTA, S. MONEA, M. ONISTO, S. GARBISA & P. MIGNATTI. 1997. Control of type IV collagenase activity by components of the urokinase-plasmin system: a regulatory mechanism with cell-bound reactants. EMBO J. **16:** 2319–2332.
4. LEHTI, K., J. LOHI, H. VALTANEN & J. KESKI-OJA. 1998. Proteolytic processing of membrane-type-1 matrix metalloproteinase is associated with gelatinase A activation at the cell surface. Biochem. J. **334:** 345–353.
5. MONEA, S., K. LEHTI, J. KESKI-OJA & P. MIGNATTI. Membrane-type 1 matrix metalloproteinase is required for plasmin-mediated activation of gelatinase A. Submitted for publication.

# Transient Increase of Intracellular cAMP by Heat Shock Initiates the Suppression of MT1-MMP Production in Tumor Cells

YASUNOBU SAWAJI,[a] TAKASHI SATO,[a] MOTOHARU SEIKI,[b] AND AKIRA ITO[a,c]

[a]Department of Biochemistry, School of Pharmacy, Tokyo University of Pharmacy and Life Science, Hachioji, Tokyo 192-0392, Japan

[b]Department of Cancer Cell Research, Institute of Medical Science, University of Tokyo, Minato, Tokyo 108-0071, Japan

## INTRODUCTION

Membrane type 1 matrix metalloproteinase (MT1-MMP) is identified as an *in vivo* activator of progelatinase A/proMMP-2 on cell surface, and the activated MMP-2 as well as MT1-MMP degrades extracellular matrices to increase tumor invasiveness. On the other hand, hyperthermia is one of the therapeutic strategies for preventing tumor progression by inhibiting tumor growth.[1] We found the evidence that the heat shock preferentially suppresses the gene expression and production of MT1-MMP in human tumor cell lines. To clarify the suppressive mechanism, we examined the involvement of 3´, 5´-cyclic AMP (cAMP) in the regulation of MT1-MMP production in human fibrosarcoma HT-1080 cells and normal human fibroblasts, since cAMP is known as a suppressor of proMMPs production in human fibroblasts[2] and is increased by heat shock in human epidermoid carcinoma A431 cells.[3]

## MATERIALS AND METHODS

Confluent HT-1080 cells and human uterine cervical fibroblasts (HUCF) were treated with heat shock at 42°C or forskolin for 4 hr in 0.2% (w/v) lactalbumin hydrolysate/MEM and then incubated for further 24 hr at 37°C following replacement of the fresh same medium with or without 10 µg/ml of concanavalin A (Con A). The production and gene expression of MT1-MMP and tissue inhibitor–2 of metalloproteinases (TIMP-2) were analyzed by Western blot analysis and Northern blot analysis, respectively. Activation of proMMP-2 in the culture medium was monitored by gelatin zymography. Intracellular cAMP level was measured using enzyme immunoassay. Invasion assay was performed using transwell chambers with Matrigel-coated membrane.

[c]Address for correspondence: Akira Ito, Ph.D., Department of Biochemistry, School of Pharmacy, Tokyo University of Pharmacy and Life Science, 1432-1 Horinouchi, Hachioji, Tokyo 192-0392, Japan. Phone, (81) 426-76-5706; fax, (81) 426-76-5734; e-mail, itoa@ps.toyaku.ac.jp

## RESULTS AND DISCUSSION

Heat shock suppressed the gene expression and production of MT1-MMP and proMMP-2 activation in the Con A–treated or untreated HT-1080 cells, but not in the Con A–treated HUCF. Heat shock transiently increased intracellular cAMP level within 20 min in HT-1080 cells (FIG. 1) but not in HUCF. The transient increase of intracellular cAMP by forskolin (50 μM) or dibutyryl cAMP (3 mM) mimicked the suppressive action of heat shock on the MT1-MMP production and proMMP-2 activation in HT-1080 cells (FIG. 1, inset). Furthermore, heat shock and forskolin inhibited *in vitro* invasive activity of HT-1080 cells. On the other hand, TIMP-2 production was augmented in the heat shocked and forskolin-treated HT-1080 cells without alteration of its mRNA level. However, the production and gene expression

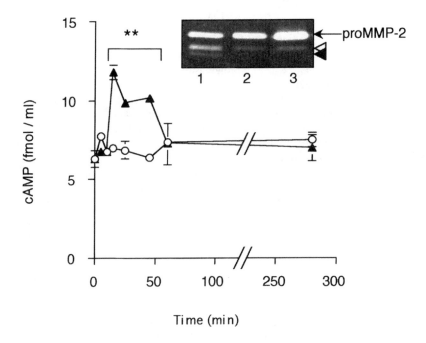

**FIGURE 1.** Heat shock and forskolin increase intracellular cAMP level and inhibit proMMP-2 activation in HT-1080 cells. Confluent HT-1080 cells at the 25th passage were pretreated with IBMX (1 mM) for 30 min, and then incubated at 37 or 42°C for indicated periods following replacement of fresh serum-free medium. The intracellular cAMP level was measured by enzyme immunoassay. The data show the mean ± SD of triplicate wells at each point. **, significantly different from each control ($p < 0.01$). *Open circles,* control cells; *closed triangles,* heat shocked cells. *Inset*: Confluent HT-1080 cells at the 26th passage were treated with or without heat shock or forskolin (50 μM) for 4 hr and then maintained for further 24 hr at 37°C following replacement of fresh serum-free medium. After the incubation, the harvested culture medium was subjected to gelatin zymography. *Lane 1,* control cells; *lane 2,* heat shocked cells; *lane 3,* forskolin-treated cells. *Open triangle,* 64-kDa intermediate MMP-2; *closed triangle,* 62-kDa active MMP-2.

of TIMP-2 in HUCF was not modified by heat shock and forskolin, suggesting that alteration of TIMP-2 level in the culture medium might be associated with the down-regulation of MT1-MMP production in HT-1080 cells.

We conclude that transient increase of intracellular cAMP by heat shock initiates the suppression of MT1-MMP production and the subsequent inhibition of proMMP-2 activation in HT-1080 cells, and that the involvement of cAMP in the regulation of MT1-MMP production may be different between tumor and normal cells. Furthermore, these results suggest that heat shock shows potential as a therapeutic strategy to prevent tumor invasiveness.

## REFERENCES

1. URANO, M. *et al.* 1983. Effect of whole-body hyperthermia on cell survival, metastasis frequency, and host immunity in moderately and weakly immunogenic murine tumors. Cancer Res. **43:** 1039–1043.
2. TAKAHASHI, S. *et al.* 1991. Cyclic adenosine 3′,5′-monophosphate suppresses interleukin 1–induced synthesis of matrix metalloproteinases but not of tissue inhibitor of metalloproteinases in human uterine cervical fibroblasts. J. Biol. Chem. **266:** 19894–19899.
3. KIANG, J.G., Y.Y. WU & M.C. LIN. 1991. Heat treatment induces an increase in intracellular cyclic AMP content in human epidermoid A-431 cells. Biochem. J. **276:** 683–689.

# The 9-kDa N-Terminal Propeptide Domain of MT1-MMP Is Required for the Activation of Progelatinase A

JIAN CAO,[a] MICHELLE DREWS,[b] HSI M. LEE,[c] CATHLEEN CONNER,[b] WADIE F. BAHOU,[a] AND STANLEY ZUCKER[a,b,d]

[a]Department of Medicine, School of Medicine and Dentistry, State University of New York at Stony Brook, Stony Brook, New York 11794, USA

[b]Department of Veterans Affairs Medical Center, Northport, New York 11768, USA

[c]Department of Oral Biology, School of Medicine and Dentistry, State University of New York at Stony Brook, Stony Brook, New York 11794, USA

MT1-MMP has a paired basic amino acid cleaving enzyme recognition motif (RRKR) sandwiched between the propeptide and catalytic domains of the latent molecule.[1] Our previous report demonstrated that furin cleaves secreted MT1-MMP lacking the transmembrane domain extracellularly at the RRKR consensus sequence but not intracellularly, and further indicated that furin is not required for the function of MT1-MMP in terms of progelatinase A activation.[2] Based on this finding, we hypothesized that a component of or the entire N-terminal propeptide domain of MT1-MMP participates in the process of MT1-MMP–induced progelatinase A activation. In the current study we have examined structure-functional relationships within the propeptide domain of MT1-MMP.

Employing recombinant DNA technique, we have constructed an entire N-terminal propeptide domain–deleted MT1-MMP (MTΔpro), which theoretically represents an activated form of MT1-MMP. By Western blotting, expressed mutant protein migrated predictably based on the truncated length of the propeptide domain; MTΔpro appeared as a protein band of 53 kDa, while MT1-MMP was detected as a 63-kDa protein. In contrast to the activation of progelatinase A induced by MT1-MMP (monitored by gelatin zymogram), transfection of COS-1 cells with the MTΔpro failed to result in progelatinase A activation. These data are consistent with our hypothesis that conformational effects induced by the plasma membrane provide functional activity to membrane-bound MT1-MMP without cleavage of the molecule.[2] By transfection of mutant MT1-MMP cDNAs into COS-1 cells, we have demonstrated that deletion of the entire N-terminal propeptide sequence of MT1-MMP resulted in loss of [125]I-TIMP-2 binding activity. This defect was not due to failure of insertion of MT1-MMP protein into COS-1 plasma membranes as demonstrated by surface biotinylation.

We then examined whether the prodomain of other secretory MMPs could function in this capacity. A substituted mutation in the prodomain of MT1-MMP was constructed. Since the prodomain of collagenase-3 (which classically is responsible

[d]Address for correspondence: Dr. Stanley Zucker, Mail Code 151, VA Medical Center, Northport, NY 11768. Phone, 516/261-4400, ext. 2861; fax, 516/544-5317.

for maintaining latency of this MMP) was determined to share minimal identity with that of MT1-MMP, the signal and prodomain of cDNA in MT1-MMP was replaced by the homologous cDNA of collagenase-3 (Col-3/MT). As demonstrated by gelatin zymogram, Col-3/MT–transfected COS-1 cells did not induce progelatinase A activation; and [125]I-TIMP-2 binding was not demonstrated in COS-1 cells transfected with the same plasmid.

To clarify which part of the propeptide domain of MT1-MMP is essential for the unique property of this protein, a deletion mutant of MT1-MMP lacking the pro-domain from Ser[34] to Arg[51] was constructed (MTΔ34–51). This mutation was expressed in COS-1 cells transiently transfected with the plasmid; no progelatinase A activation and [125]I-TIMP-2 binding were noted in MTΔ34–51 transfected COS-1 cells. This data strengthens the concept that the initial portion of the N-terminal propeptide domain of MT1-MMP is necessary for MT1-MMP–induced progelatinase A activation.

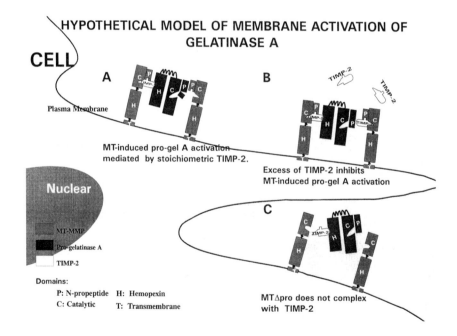

**FIGURE 1.** MT1-MMP, TIMP-2, and gelatinase A interactions. MT1-MMP is expressed at the cell surface in transfected COS-1 cell system. TIMP-2 will bind through its N-terminal domain to the MT1-MMP catalytic domain but can still bind progelatinase A by C-terminal domain interactions. Under stoichiometric conditions, a second free MT1-MMP cleaves the propeptide domain of gelatinase A leading to activation of the latent enzyme (**A**). If there is an excess of TIMP-2, all MT1-MMP molecules in this system are occupied by TIMP-2 (**B**). Prodomain-deleted MT1-MMP changes the conformation of the enzyme at the cell surface; therefore, triplex molecules cannot be formed on the cell surface (**C**).

To further examine the role of the prodomain of membrane-bound MT1-MMP, we have inserted the cDNA encoding the propeptide sequence of MT1-MMP ($MT_{1-109}$) in an expression vector. Cotransfection of COS-1 cells with both MTΔpro cDNA and $MT_{1-109}$ cDNA resulted in reconstitution of MT1-MMP function; cotransfected cells activated recombinant progelatinase A. These experiments suggest that the isolated prodomain of MT1-MMP is capable of binding to MTΔpro in the plasma membrane and thereby reconstituting of MT1-MMP.

Taken together, our data indicate that the entire propeptide domain of MT1-MMP plays an essential role in TIMP-2 binding and subsequent activation of progelatinase A. FIGURE 1 shows our hypothesized model to explain the mechanism of membrane activation of gelatinase A. MT1-MMP is expressed at the cell surface through a C-terminal transmembrane domain. TIMP-2 will bind through its N-terminal domain to the catalytic domain of MT1-MMP,[3] which is in a conformation associated with prodomain of MT1-MMP. TIMP-2 can still bind progelatinase A by C-terminal domain interactions. Under stoichiometric conditions, a second free MT1-MMP molecule will cleave the N-terminal propeptide of gelatinase A to initiate the activation of gelatinase A. If there is an excess of TIMP-2 in the system, no free MT-MMP is available; hence no gelatinase A activation ensues. N-terminal deletion of MT1-MMP changes the conformation of the mutant protein at the cell surface, so that this molecule is unable to bind to TIMP-2. Therefore, the triplex of MT1-MMP, TIMP-2, and progelatinae A on the cell surface cannot be formed.

## REFERENCES

1. SATO, H., T. TAKINO, Y. OKADA et al. 1994. A matrix metalloproteinase expressed on the surface of invasive tumor cells. Nature **370:** 61–65.
2. CAO, J., A. REHEMTULLA, W. BAHOU et al. 1996. Membrane type matrix metalloproteinase 1 activates pro-gelatinase A without furin cleavage of the N-terminal domain. J. Biol. Chem. **271:** 30174–30180.
3. ZUCKER, S., M. DREWS, C. CONNER et al. 1998. Tissue inhibitor of metalloproteinase-2 binds to the catalytic domain of the cell surface receptor, membrane type 1 metalloproteinase 1. J. Biol. Chem. **273:** 1216–1222.

# Cell Type–Specific Involvement of Furin in Membrane Type 1 Matrix Metalloproteinase– Mediated Progelatinase A Activation

TAKASHI SATO,[a,c] TAKAYUKI KONDO,[a] MOTOHARU SEIKI,[b] AND AKIRA ITO[a]

[a]Department of Biochemistry, School of Pharmacy, Tokyo University of Pharmacy and Life Science, Horinouchi, Hachioji, Tokyo 192-0392, Japan

[b]Department of Cancer Cell Research, Institute of Medical Science, University of Tokyo, Shirokanedai, Minato, Tokyo 108-0071, Japan

## INTRODUCTION

Membrane type 1 matrix metalloproteinase (MT1-MMP) is synthesized as a pro-form (proMT1-MMP) and then expressed as an active form on the cell surface. It has been considered that the activation of proMT1-MMP is accomplished by an intracellular processing protease furin because MT1-MMP possesses a furin recognition sequence R-R-K-R in the propeptide.[1] Thus, it is suggested that furin might be a key enzyme to control MT1-MMP function such as progelatinase A/proMMP-2 activation. However, it is not fully understood how furin production is regulated in the MT1-MMP–producing cells and whether furin is substantially required for MT1-MMP–mediated proMMP-2 activation. We investigated the involvement of furin in the MT1-MMP–mediated proMMP-2 activation process using three types of cell lines: human fibrosarcoma HT-1080, human uterine cervical fibroblasts (HUCF), and rabbit dermal fibroblasts (RDF).

## MATERIALS AND METHODS

The confluent HT-1080, HUCF, and RDF were treated with concanavalin A (Con A, 10 μg/ml), furin sense and antisense oligonucleotides (F-S and F-AS, respectively),[2] and/or the specific and poor inhibitors for furin; decanoyl (Dec)-R-V-K-R-$CH_2$Cl and Dec-R-A-I-R-$CH_2$Cl,[3] respectively, in 0.2% (w/v) lactalbumin hydrolysate/MEM or DMEM. The furin inhibitors were kindly provided by Fuji Chemical Industries, Ltd. (Takaoka, Japan). ProMMP-2 activation and MT1-MMP production were monitored by gelatin zymography and Western blotting, respectively. The gene expression of furin was monitored by a quantitative reverse transcriptase–polymerase chain reaction (RT-PCR). Intracellular furin activity was measured using $t$-butoxycarbonyl (Boc)-R-V-R-R-4-methyl-coumaryl-7-amide (MCA) as a substrate.

[c]Address for correspondence: Takashi Sato, Ph.D., Department of Biochemistry, School of Pharmacy, Tokyo University of Pharmacy and Life Science, 1432-1 Horinouchi, Hachioji, Tokyo 192-0392, Japan. Phone, (81) 426-76-5728; fax, (81) 426-76-5734; e-mail: satotak@ps.toyaku.ac.jp

## RESULTS AND DISCUSSION

An advanced activation of proMMP-2 was observed in Con A–treated HT-1080, HUCF, and RDF, which was correlated with increasing MT1-MMP production. A quantitative RT-PCR demonstrated that furin mRNA was detectable in all cell lines tested, and Con A did not influence its gene expression. Furthermore, an intracellular furin activity was constitutively detectable in all cell lines, and also the activity was not altered by the Con A treatment. These results suggest that furin activity is constitutively expressed and, unlike MT1-MMP, the production is not regulated by Con A.

When HUCF and RDF were treated with a specific inhibitor for furin, Dec-R-V-K-R-CH$_2$Cl (100 µM), the Con A–induced proMMP-2 activation was inhibited in HUCF but not in RDF (FIG. 1). However, a poor inhibitor for furin, Dec-R-A-I-R-CH$_2$Cl (100 µM), did not influence the advanced proMMP-2 activation in both cell lines. On the other hand, when HUCF and RDF were treated with F-AS or F-S together with Con A, F-AS (40 µM) specifically inhibited the Con A–induced activation of proMMP-2 in HUCF but not in RDF (FIG. 2), while F-AS suppressed the furin gene expression in both HUCF and RDF. These results suggest that the intracellular activation of proMT1-MMP in RDF might be independent of furin.

We conclude that the involvement of furin in the MT1-MMP–mediated proMMP-2 activation depends entirely upon the cell species, and that other types of furin-like convertase may participate in the processing of proMT1-MMP to augment proMMP-2 activation in RDF.

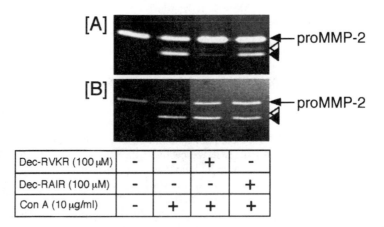

**FIGURE 1.** Effect of furin inhibitors on Con A–induced proMMP-2 activation. Confluent human uterine cervical fibroblasts (HUCF) **(A)** and rabbit dermal fibroblasts (RDF) **(B)** were pretreated with a specific inhibitor for furin, Dec-R-V-K-R-CH$_2$Cl (Dec-RVKR, 100 µM), or a poor inhibitor for furin, Dec-R-A-I-R-CH$_2$Cl (Dec-RAIR, 100 µM), for 6 hr and then treated with Con A (10 µg/ml) together with the inhibitors for 18 hr. The harvested culture media were subjected to gelatin zymography. *Open triangles,* 64-kDa intermediate MMP-2; *closed triangles,* 62-kDa active MMP-2.

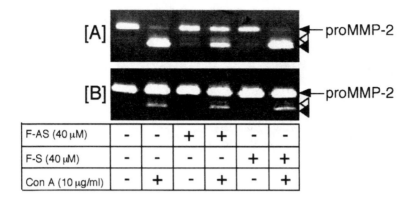

**FIGURE 2.** Effect of furin sense and antisense oligonucleotide on Con A–induced proMMP-2 activation. Confluent human uterine cervical fibroblasts (HUCF) **(A)** and rabbit dermal fibroblasts (RDF) **(B)** were pretreated with furin antisense (F-AS, 40 µM) or sense (F-S, 40 µM) oligonucleotide for 6 hr and then treated with Con A (10 µg/ml) together with F-AS or F-S for 18 hr. The harvested culture media were subjected to gelatin zymography. *Open triangles,* 64-kDa intermediate MMP-2; *closed triangles,* 62-kDa active MMP-2.

# REFERENCES

1. SATO, H. *et al.* 1994. A matrix metalloproteinase expressed on the surface of invasive tumors cells. Nature (London) **370:** 61–65.
2. VAN DEN OUWELAND, A.M.W. *et al.* 1990. Structural homology between the human furin gene product and the subtilisin-like protease encoded by yeast KEX2. Nucleic Acids Res. **18:** 664.
3. ANGLIKER, H. *et al.* 1993. The synthesis of inhibitors for processing proteinases and their action on the Kex2 proteinase of yeast. Biochem. J. **293:** 75–81.

# High Fructose–Fed Rats: A Model of Glomerulosclerosis Involving The Renin-Angiotensin System and Renal Gelatinases

P. ZAOUI,[a] E. ROSSINI, N. PINEL, D. CORDONNIER, S. HALIMI, AND F. MOREL

*GREPI UJF, LBSO, Nephrology, Diabetology, Cell Pathology, Centre Hospitalier Universitaire de Grenoble, Grenoble, France*

## BACKGROUND

The use of fructose oral diets as an easy source of sugar substitute could represent at first sight an interesting carbohydrate (CH) intake, especially for diabetic patients, with weak hyperglycemic effect and allegedly reduced secondary insulin secretion. In the long run, metabolic side effects proceed from an inoperant metabolic control by fructose-6-phosphate kinase, induction of glycogen-synthase, and glycerol-3-phosphate/VLDL pathways, and may favor the development of progressive insulin resistance, hyperuricemia, and hypertriglyceridemia, which links fructose-rich diets with the clinical dilemma of insulin resistance, systemic hypertension, and vascular renal lesions suspected in human metabolic syndrome X even without established NIDDM.

## AIM OF THE STUDY

In a prospective observational study, we intended to assess the combined effects of aging, diet modifications, and pharmacological interventions on progressive vascular and renal scarring lesions in rats fed for a prolonged period of time with a standardized supplemented high fructose-content diet.

## MATERIAL AND METHODS

A cohort of 42 Wistar rats (mean weaning weight 50 g) was allocated to three dietary intervention protocols: group C (control, $n = 12$) was fed with a standard chow diet (30% CH); group F (fructose, $n = 21$) was fed with a 58% fructose-enriched diet; group FM (fructose-metformin, $n = 9$) received simultaneously the fructose diet and an oral stimulus of insulin secretion (metformin). Animals were followed up during aging by monitoring glucose and lipid profiles and assessing basal and angiotensin II–stimulated renal function. Glucose tolerance was evaluated at two months of age

[a]Address for correspondence: Philippe Azoui, M.D., Ph.D., Centre Hospitalier Universitaire de Grenoble, Practicien Hospitalier/Service de Nephrologie, Laboratoire d'Enzymologie, BP 217, 38043 Grenoble Cedex, France. Phone, 33-04-76765523; fax, 33-04-76765263; e-mail, Philippe.Zaoui@ujf-grenoble.fr

by an euglycemic clamp technique in awake rats. At the end of the study, animals (one pair per cohort) were transferred to metabolic cages in order to obtain renal and hemodynamic assessments (cuff-tail BP, creatinine and albumin excretion rates) and to monitor the effect of acute angiotensin II infusion.[1]

In preliminary experiments, fructose-fed and control animals were infused for 1 to 4 hours with angiotensin II (50 ng/ml/hr) using jugular catheters. Urinary albumin excretion rate, kidney and isolated glomeruli metalloprotease content[2] (quantitative normalized zymography), and renal histology were compared to baseline levels in animals submitted to acute angiotensin infusion either with or without angiotensin II type 1 receptor blockade (valsartan 40 mg/d in drinking water) during the 24 hr preceding the angiotensin infusion. Sacrifice was performed in all animals after 11 to 17 months of continuous diet in order to evaluate organomegaly, target tissue histology, and protein content.

## RESULTS

### Animal Morphology and Vascular Risk

Glucose intolerance observed after two months in group F was normalized by metformin (group FM). Final weights in group C ($707 \pm 82$ g), F ($791 \pm 121$ g), and FM ($727 \pm 106$ g) were noticeable but not significantly different and occurred without fasting hyperglycemia. Hypertriglyceridemia increased to $3.6 \pm 1.2$ g/l in group F vs. $2.4 \pm 1.9$ g/l in group C ($p < 0.01$) and was normalized in group FM. Systolic blood pressure remained normal in all three groups below $150 \pm 20$ mm Hg. Organomegaly of kidneys and liver in group F was associated with a significant increase of abdominal and epididymal fat content and with macroscopic lipid deposits on aortic bifurcation. Nevertheless, retinal and vascular fat remained within the normal limits without constituting atherosclerosis or retinopathy in all animals after 11 and 17 months. Metformin did not change significantly these morphological findings.

### Renal Findings

An elevated proteinuria in the nephrotic range with abnormal urinary albumin excretion rate ($>1.5 \pm 0.6$ g/l) without renal insufficiency (normal creatininuria) suggested a fixed glomerular hyperfiltration state as observed in early nephroangiosclerosis[3] and diabetic nephropathy.[4,5] Urinary and renal gelatinases were assessed by gelatin zymography scanning, and increased levels of collagen-degrading enzymes were observed in the F and FM groups. Renal lesions were focal and characterized by glomerular mesangium thickening without typical nodular deposits of diabetic glomerulosclerosis or vascular hyalinosis. Areas of ischemic tubulointerstitial fibrosis and inflammatory infiltrates were observed in all animals and were potentially reduced in metformin-treated rats.

Urinary albumin excretion rates, creatininuria, and renal gelatinases were acutely increased by angiotensin infusion and significantly more in HFF than in control animals, while valsartan blockade was able to control the acute hyperfiltration episode. Renal histology of HFF treated with angio-2 infusions showed more pronounced subcortical infarcted zones than in infused control rats. These results suggest an in-

# Renal Gelatinases

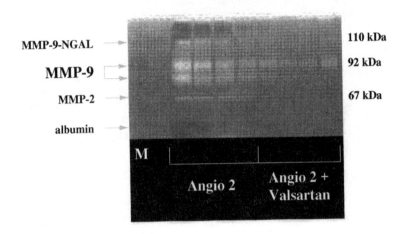

**FIGURE 1.** Equal volumes of urine samples from fructose-fed rats were loaded for electrophoresis under nondenaturating conditions together with molecular weight markers *(lane M)* on a 10% polyacrylamide gel containing 1 mg/ml gelatin. After migration was completed, the gel was incubated overnight at 37°C in Hanks' buffer containing calcium and magnesium salts in order to activate metalloproteinases and to degrade the gelatin matrix at sites where gelatinases had migrated. The gel was stained with Coomassie brilliant blue, then rapidly destained in acetic acid. Bands were photographed and quantitated by scanning densitometry. *Lanes 1 to 4:* urine samples from fructose-fed rats infused with angiotensin II. *Lanes 5 to 8:* urine samples from fructose-fed rats infused with angiotensin II after pretreatment with angiotensin II type 1 receptor blocker (valsartan). Urinary albumin was observed as a *dark band* at 55 kDa below four major degradative bands: MMP-9 (gelatinase B) complexed with N-GAL at 110 kDa; latent MMP-9 at 95 kDa; active MMP-9 at 88 kDa; meprin at 67 kDa. Active MMP-9 and albumin were predominantly observed in angiotensin II-treated animals, while animals pretreated with valsartan and infused with angiotensin II excreted lower levels of MMP and albumin as observed in control fructose-fed rats, which remained significantly higher than in standard diet–fed rats (data not shown).

appropriate response of the renin-angiotensin axis in HFF rats. Whether these findings are due to glomerular hemodynamic changes, growth factor effect[6] of angiotensin II or leukocyte margination inside kidneys remains to be elucidated.

## CONCLUSION

This model of aging rats chronically submitted to a fructose-rich diet appears promising in defining and differentiating the effects of carbohydrate alternate pathways, in observing *in vivo* the consequences of insulin resistance and glomerular hyperfiltration outside or before installed lesions of systemic hypertension, atherosclerosis, or hyperglycemic milieu[4] and in characterizing the functional and/ or tissular impact of interventions on insulin secretion or on the renin-angiotensin cascade.

## REFERENCES

1. ANDERSON, S.S., Y. KIM & E.C. TSILIBARY. 1994. Effects of matrix glycation on mesangial cell adhesion, spreading and proliferation. Kidney Int. **46:** 1359–1367.
2. IKEDA, T. & T. HOSHINO. 1996. An inhibition of urinary albumin excretion by protease inhibitor in streptozotocin-diabetic rats. Nephron **74:** 709–712.
3. BAYLIS, C. 1994. Age-dependent glomerular damage in the rat. Dissociation between glomerular injury and both glomerular hypertension and hypertrophy. Male gender as a primary risk factor. J. Clin. Invest. **94:** 1823–1829.
4. ADLER, S. 1994. Structure-function relationships associated with extracellular matrix alterations in diabetic glomerulopathy. J. Am. Soc. Nephrol. **5:** 1165–1172.
5. ANDERSON, P.M., X.Y. ZHANG, J. TIAN, J.D. CORREALE, X.P. XI, D. YANG, K. GRAF, R.E. LAW & W.A. HSUEH. 1996. Insulin and angiotensin II are additive in stimulating TGF-β1 and matrix mRNAs in mesangial cells. Kidney Int. **50:** 745–753.
6. DOI, T., L.J. STRIKER, C. QUAIFE, F.G. CONTI, R. PALMITER, R. BEHRINGER, R. BRINSTER & G.E. STRIKER. 1988. Progressive glomerulosclerosis develops in transgenic mice chronically expressing growth hormone and growth hormone releasing factor but not in those expressing insulin-like growth factor-1. Am. J. Pathol. **131:** 398–403.

# Wound Healing in Aged Normal and Ovariectomized Rats: Effects of Chemically Modified Doxycycline (CMT-8) on MMP Expression and Collagen Synthesis

N.S. RAMAMURTHY,[a] S.A. McCLAIN,[a] E. PIRILA,[b] P. MAISI,[b] T. SALO,[c] A. KUCINE,[a] T. SORSA,[b] F. VISHRAM,[a] AND L.M. GOLUB[a]

[a]Department of Oral Biology and Pathology, School of Dental Medicine, University Hospital and Medical Center, SUNY at Stony Brook, Stony Brook, New York, NY 11794-8702, USA

[b]University of Helsinki, Helsinki, Finland

[c]University of Oulu, Oulu, Finland

## INTRODUCTION

Skin as well as bone is an organ dependent on estrogen. Skin thickness has been correlated with bone density and hormonal parameters.[1] Recently Ashcroft et al.[2] reported a marked delay in repair of acute incisional wounds in ovariectomized young female rats. The effect was reversed by topical estrogen, suggesting that both the rate and quality of wound healing can be modulated. Tetracycline and its non-antimicrobial derivative, CMT-8, are able to inhibit bone loss in ovariectomized rats (for review see Sasaki et al.[3]). A recent study by Ramamurthy et al.[4] has shown enhanced wound healing in diabetic rats with topically applied CMT-2. In this study we describe the effect of oral administration of CMT-8 or estrogen on the healing of skin wounds in aged female rats, 120 days post ovariectomy (OVX).

## METHODS

Six-month-old adult female Sprague-Dawley rats (body weight approx. 350g) were either sham operated or bilaterally ovariectomized. After 120 days the sham-operated rats were divided into normal, normal + CMT-8, OVX + placebo, and OVX + CMT-8 groups. The rats were anesthetized with a mixture of ketamine and xylazine, the dorsal skin was shaved, and, using a 6-mm diameter circular biopsy punch, eight full-thickness skin wounds were made in the dorsal thorax and were allowed to heal by secondary intention. CMT-8 (15 mg/kg bodyweight) was administered daily by oral gavage in 1 ml of 2% carboxymethyl cellulose (CMC). Placebo rats received 1 ml of 2% CMC. On day 7 after the wounds were created, the animals were anesthetized, blood samples were collected, and the skin containing four wounds was excised for histology. For biochemical analysis the granulation tissue was removed by 3-mm punch biopsy from three wounds and frozen at $-70°C$ until analysis. One remaining wound tissue was removed under sterile condition for in *situ* hybridization and fixed in 4% sterile paraformaldehyde in PBS at 4°C, treated with 0.1% diethylpyrocarbonate (DEPC), and frozen in OTC mounting medium (Sakwa Finetech Co., Torrance, CA). Frozen sections were cut and processed for *in situ* hybridization using a 95-bp SphI-fragment of human MMP8 cDNA,[5] which was ligated into the pGEM-3Z(+) vector and linearized with a suitable enzyme. A ribo-probe transcription kit (Boehringer) was used for transcription according to the manufacturers instructions.

### MMP Activities

The amount of collagenase and gelatinase in the partially purified wound tissue extracts was determined from the different treatment groups using Western blot with antisera against MMP-8, MMP-9, and MMP-13.

For the hydroxyproline analysis, a disposable 3-mm biopsy punch was used to remove the central core of the wound tissue, the samples were hydrolyzed in 6N HCl (106°C, 24 hr), and then assayed by Stegmenn's colorimetric assay.[4]

### RESULTS

The wound collagen content (hydroxyproline) in the 120 days post OVX rats was ($p < 0.05$) reduced by 50% of normal or normal + CMT-8. Treating the OVX rats with CMT-8 or estrogen significantly increased the collagen deposition in the wound tissue to that of normal animals (FIG. 1). At time 0 the skin collagen content in the

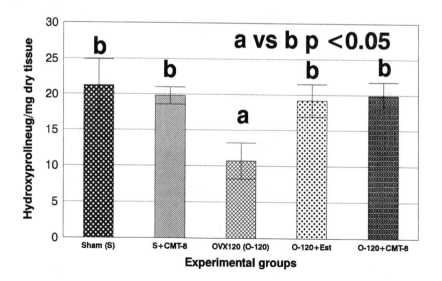

**FIGURE 1.** Effects of CMT-8 treatment on wound collagen content in ovariectomized rats.

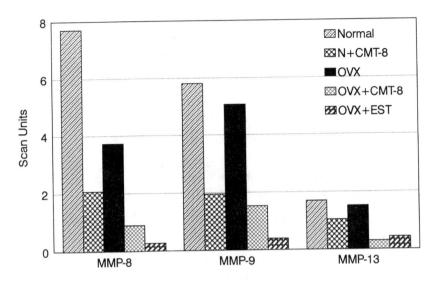

**FIGURE 2.** Immunoreactive MMPs in the wound tissues of normal and ovariectomized rats treated with CMT-8.

OVX rats was reduced by 35% compared to normal (data not shown). *In situ* hybridization showed that MMP-8 was clearly expressed in the normal and OVX rat wounds. CMT-8 or estrogen treatment reduced the MMP-8 expression in fibroblasts. Western blot for MMPs also showed a reduced expression of MMP-8, -9, and -13 in the wound tissues from OVX + CMT-8– and OVX + Est–treated groups (FIG. 2).

## CONCLUSION

The skin wound healing in OVX rats demonstrated reduced collagen synthesis and decreased MMP-8 expression. The reduced MMPs and collagen in the healing OVX rats' wounds may be due to an overall reduction in protein synthesis. Daily administration of CMT-8 or estrogen to the OVX rats increased the collagen content and downregulated the MMP-8 expression, thereby strengthening the wound.

## ACKNOWLEDGMENTS

This study was supported by NIDR Grant DE-03987, Collagenex, the Finnish Dental Society, and the Academy of Finland.

## REFERENCES

1. GRUBER, D., M. SATOR, P. FRIGIO, W. KNOGLER & J.C. HUBER. 1995. Correlation of skinfold thickness with bone density and estradiol, FSH and prolactin level in a sample of 231 women. Wien. Klin. Wochenschr. **107:** 622–625.
2. ASHCRAFT, G.S., J. DODSWORTH, E. VAN BOXTEL, R.W. TARNUZZER, M.A. HORAN, G.S. SCHULTZ & M.W. FERGUSON. 1997. Estrogen accelerates cutaneous wound healing associated with an increase in TGF-beta levels. Nat. Med. **3:** 1209–1215.
3. SASAKI, T., N.S. RAMAMURTHY & L.M. GOLUB. 1998. Long-term therapy with a new chemically modified tetracycline (CMT-8) inhibits bone loss in femurs of ovariectomized rats. Adv. Dent. Res. **12:** 76–81.
4. RAMAMURTHY, N.S., A.J. KUCINE, S.A. MCLEAN, T.F. MCNAMARA & L.M. GOLUB. 1998. Topically applied CMT-2 enhances wound healing in Streptozotocin diabetic rat skin. Adv. Dent. Res. **12:** 144–148.
5. HASTY, K.A., T.F. POURMOTABBED, G.I. GOLDBERG, J.P. THOMPSON, D.G. SPINELLA, R.M. STEVENS & C.L. MAINARDI. 1990. Human nuetrophil collagenase: a distinct gene product with homology to other matrix metalloproteinases. J. Biol. Chem. **265:** 11421–11424.

# Therapeutic Options in Small Abdominal Aneurysms: The Role of *in Vitro* Studies

M.M. THOMPSON,[a] J.R. BOYLE, M. CROWTHER, S. GOODALL, A. WILLS, I.M. LOFTUS, AND P.R.F. BELL

*Department of Surgery, University of Leicester, Leicester LE2 7LX, UK*

## INTRODUCTION

Abdominal aneurysms are characterized by increased expression and production of metalloproteinases (MMPs) within the arterial wall, degradation of the extracellular matrix—particularly elastin—and a diffuse inflammatory infiltrate. These changes have been described in a rodent model, which uses instillation of pancreatic elastase into an isolated aortic segment to initiate aneurysm formation. To further isolate cellular interactions during aneurysm formation, we have developed an *in vitro* model of aneurysmal disease, which uses aortic organ cultures exposed to a brief pulse of elastase to stimulate MMP production and elastin degradation.

## IN VITRO ANEURYSM MODEL

Porcine thoracic organ cultures were pulsed with pancreatic elastase (100 units/ml) for a period of 24 hr, prior to culture in standard medium for up to 14 days. The extent of matrix degradation was determined by stereologic estimation of elastin, collagen, and smooth muscle concentration. MMP production within the tissue was quantified by substrate zymography, immunoblotting, and densitometry.

Although all of the exogenous elastase was removed at 24 hr, there was a time-dependent degradation of elastin within the organ cultures that was not confined to the initial period of elastase exposure. Additionally there was a similar increase in MMP-2 and -9 production within the aortic tissue as demonstrated by zymography.[1]

The model can be utilized in the presence or absence of macrophages,[2] and mirrors the pathophysiological effects within established arterial aneurysms. The model has been utilized to investigate the effect of pharmacologic agents on the aortic wall.

## EFFECT OF DOXYCYCLINE AND AMLODIPINE IN A MODEL OF ANEURYSMAL DISEASE

The *in vitro* model of aneurysmal disease was utilized to investigate the effects of both doxycycline (a nonspecific MMP inhibitor) and amlodipine (a dihydropyridine calcium antagonist) on elastin degradation and MMP production within the arterial

[a]Address for correspondence: Mr. M.M. Thompson, M.D., F.R.C.S., Department of Surgery, Clinical Sciences Building, Leicester Royal Infirmary, Leicester LE2 7LX, UK. Phone, 44 -116-2523252; fax, 44-116-2523179; e-mail, MattT11@aol.com

organ cultures. Our original hypothesis was that doxycycline might reduce elastin degradation as demonstrated previously in animal models, while amlodipine might accelerate aneurysm formation through enhanced MMP-2 proteolytic capacity of vascular smooth muscle cells.

## EXPERIMENTAL DESIGN/METHODS

Porcine aortic segments were preincubated in pancreatic elastase (100 units/ml) for 24 hr, prior to culture in standard conditions for 13 days with both 1 and 10 mg/l doxycycline. Control segments were cultured both without doxycycline and without elastase. Similarly, in a different experiment, aortic segments were again incubated with elastase for 24 hr, prior to culture for 6 days in either 10 or 100 µg/l amlodipine. Controls were as previously described.

At the termination of culture MMPs were extracted from the tissue and quantified by a combination of gel enzymography and immunoblotting. The volume fractions of elastin, collagen, and smooth muscle cells were determined by stereological analysis of EVG-stained sections.

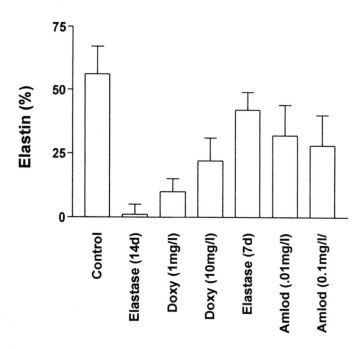

**FIGURE 1.** Graph illustrating elastin content (%) of aortic organ cultures preincubated with elastase for 24 hr, then cultured for 13 days or 7 days in normal medium (elastase 14 d; elastase 7 d), or medium supplemented with doxycycline (Doxy) or amlodipine (Amlod). Values are medians with interquartile ranges.

## RESULTS—DOXYCYCLINE

Stereological analysis demonstrated that there was a significant preservation of elastin in the elastase-exposed aortic segments treated with the highest dose of doxycycline ($p < 0.001$, Wilcoxon; FIG. 1). Densitometric analysis of substrate zymograms revealed a reduction in MMP-9 ($p = 0.04$, Wilcoxon) in the doxycycline-treated cultures when compared to the elastase-treated control (FIG. 2).[3]

## RESULTS—AMLODIPINE

In contradistinction to the previous experiment, stereological analysis of the amlodipine-treated sections demonstrated acceleration of elastin degradation in comparison to elastase-treated controls ($p < 0.05$, Wilcoxon). This enhanced elastin degradation was associated with increased MMP-9 and MMP-2.[4]

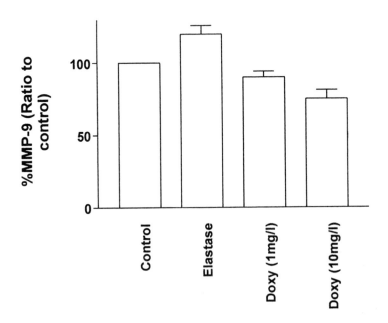

**FIGURE 2.** Graph illustrating MMP-9 production from aortic segments cultured in the presence of elastase for 24 hr and then in standard conditions (elastase) or medium supplemented with doxycycline (Doxy) for 13 days. A segment of aorta cultured in absence of elastase or doxycycline is shown as a control. MMP-9 production is given as a ratio to that of control tissue. Values are medians with interquartile ranges.

## CONCLUSIONS

The *in vitro* model presented has potential applications in understanding early events in aneurysm pathogenesis as well as in investigation of potential therapeutic agents. In this model, doxycycline, administered in a therapeutic concentration, significantly inhibited both MMP production and elastin degradation, and may thus have a role in clinical trials to prevent small aneurysm growth. Conversely, amlodipine accelerated elastin degradation via an MMP-dependent pathway. Clinical data are required to investigate the effect of calcium antagonists on aneurysm development.

## REFERENCES.

1. WILLS, A., M.M. THOMPSON, M. CROWTHER, N.P. BRINDLE, A. NASIM, R.D. SAYERS *et al.* 1996. Elastase-induced matrix degradation in arterial organ cultures: an in vitro model of aneurysmal disease. J. Vasc. Surg. **24:** 667–679.
2. THOMPSON, M.M., A. WILLS, E. MCDERMOTT, M. CROWTHER, N. BRINDLE & P.R.F. BELL. 1996. An *in vitro* model of aneurysmal disease: effect of leukocyte infiltration and shear stress on MMP production within the arterial wall. Ann. N.Y. Acad. Sci. **800:** 270–273.
3. BOYLE, J.R., E. MCDERMOTT, M. CROWTHER, A. WILLS, P.R.F. BELL & M.M. THOMPSON. 1998. Doxycycline inhibits elastin degradation and reduces metalloproteinase activity in a model of aneurysmal disease. J. Vasc. Surg. **27:** 354–361.
4. BOYLE, J.R., I.M. LOFTUS, S. GOODALL, M. CROWTHER, P.R.F. BELL & M.M. THOMPSON. 1998. Amlodipine potentiates metalloproteinase activity and accelerated elastin degradation in a model of aneurysmal disease. Eur. J. Vasc. Endovasc.Surg. In press.

# Inhibition of the Metalloproteinase Domain of Mouse TACE

AUGUSTIN AMOUR,[a,d] MIKE HUTTON,[a] VERA KNÄUPER,[a]
PATRICK M. SLOCOMBE,[b] AILSA WEBSTER,[b] MICHAEL BUTLER,[a]
C. GRAHAM KNIGHT,[c] BRYAN J. SMITH,[b] ANDREW J.P. DOCHERTY,[b]
AND GILLIAN MURPHY[a]

[a]School of Biological Sciences, University of East Anglia, Norwich, NR4 7TJ, UK

[b]Celltech Therapeutics Ltd., Slough, SL1 4EN, UK

[c]Department of Biochemistry, University of Cambridge, Cambridge CB2 1QW, UK

TNF-$\alpha$ converting enzyme (TACE/ADAM-17) is a type I membrane-bound metalloproteinase that processes the type II membrane-bound cytokine proTNF-$\alpha$ into soluble TNF-$\alpha$.[1,2] Because TNF-$\alpha$ is a major mediator of diseases such as rheumatoid arthritis and sepsis, TACE may represent an important target for the design of specific inhibitors with pharmacological applications. However, concern remains as to whether TACE is involved in the shedding of membrane proteins other than proTNF-$\alpha$.[3] Recent studies have shown that hydroxamate MMP inhibitors and TIMP-3, but not TIMP-1 or -2, can modulate cell shedding of proTNF-$\alpha$, L-selectin, and interleukin-6 receptor.[4,5] This suggests that an enzyme, possibly TACE or very similar enzymes, are required for those shedding events.

In a previous report, we described the purification of the metalloproteinase domain of mouse TACE (rTACE) expressed in a stable mouse myeloma (NS0)–transfected cell line and identified TIMP-3 as a candidate physiological inhibitor of TACE.[6] Here, we investigated (i) the proteolytic activity of rTACE and (ii) the inhibitory effects of recombinant wild-type and mutant forms of TIMP-3 refolded from inclusion bodies after expression in *E. coli*. Unless stated otherwise, rTACE (20 µg/ml) was incubated with the target protein (1 mg/ml) for 16 hr at 37°C. Our first attempts were to see whether TACE could play a role in extracellular matrix (ECM) degradation. However, rTACE did not degrade collagens I and IV, laminin, or fibronectin as observed by SDS-PAGE. Furthermore, no increased activation of proMMP-2 (7 µg/ml) or proMMP-9 (6 µg/ml) in the presence of rTACE (4 µg/ml) was detected by gelatin zymography after 16-hr incubations at 25°C (rTACE was not active on gelatin zymograms). Those preliminary results suggest that TACE is not involved in the degradation of the ECM proteins tested and does not indirectly participate in ECM degradation by activating MMP-2 or MMP-9.

We then examined the effect of the general proteinase inhibitor alpha-2-macroglobulin ($\alpha_2$M) on rTACE. rTACE partially cleaved $\alpha_2$M (seen at 180 kDa on reduced SDS-PAGE) into fragments of 100 and 80 kDa, consistent with cleavage in the bait region. After $\alpha_2$M incubation, rTACE remained active towards a human proTNF-$\alpha$–based peptidic substrate Mca-Ser-Pro-Leu-Ala-Gln-Ala-Val-Arg-Ser-Ser-Ser-Arg-Lys(Dnp)-NH$_2$ (QF45)[6] and was inhibited by the hydroxamate MMP-

[d]Corresponding author. Fax, +441603 592250; e-mail, A.Amour@uea.ac.uk

inhibitor BB-94, but not by TIMP-3. Those observations could suggest that rTACE becomes entrapped subsequent to cleavage of the $\alpha_2$M bait region. This then prevents rTACE from further association with TIMP-3 but not with the low molecular weight compounds QF45 or BB94. Experiments were repeated using rTACE-Fc fusion protein.[6] In contrast to the rTACE data, preincubation with $\alpha_2$M did not prevent TIMP-3 inhibition of rTACE-Fc, suggesting that, in this case, $\alpha_2$M entrapment is less efficient, probably due to the size of the fusion protein (150 kDa instead of 43 kDa for rTACE). Similar results have been obtained when comparing the inhibitory effect of $\alpha_2$M on a family of metalloproteinases related to the mammalian ADAMs, the snake venom hemorrhagic metalloproteinases.[7]

We then analyzed the effect of rTACE on nonspecific protein substrates: bovine myelin basic protein (MBP), calf thymus histone (Lys-rich fraction), and human fibrinogen. As previously described with ADAM-10,[8] histone and MBP were good substrates of rTACE with proteolytic activity occurring at multiple sites as judged by the appearance of several bands of different sizes (FIG. 1). The fragment A$\alpha$ of fibrinogen, known to be a good substrate for snake venom hemorrhagic metalloproteinases, was also cleaved by rTACE (FIG. 1). Two of the cleavage sites of MBP corresponded to F42-F43 and F88-F89 (-SLGRF↓FGSDR- and -PVVHF↓FKNIVT-, respectively). MADM cleaved MBP at P73-Q74.[8]

We previously reported that rTACE was inhibited by TIMP-3 but not by TIMP-1, -2, and -4.[6] In an attempt to understand the basis of the inhibition of rTACE by TIMP-3, we have initiated a new project involving rapid and large-scale production

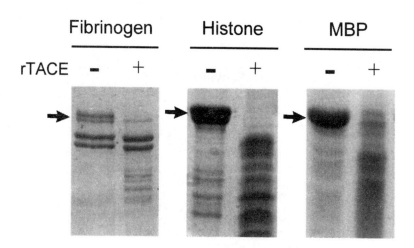

**FIGURE 1.** Coomassie-stained SDS-PAGE of samples containing 1 mg/ml of myelin basic protein (MBP), histone, or fibrinogen after 2 hr incubation at 37°C alone (−) or in presence of 10 μg/ml of rTACE (+). Fibrinogen, histone, and MBP were run under reducing conditions on 6.7, 10, and 13.3% gels, respectively. The proteolytic sensitive A$\alpha$ chain of fibrinogen degraded by rTACE was already partially cleaved prior to assay with rTACE (see *upper doublet of lane on left*).

**TABLE 1. Inhibition by refolded wild-type and mutant TIMP-3**

| Inhibitor | rTACE | MMP-2 |
|---|---|---|
| TIMP-3 | 166 | < 6 |
| N-TIMP-3 | 542,000 | 180 |
| N-TIMP-2/C-TIMP-3 | N.I. | < 10 |

NOTE: All values expressed as apparent $K_i$ (pM).

of various forms of TIMP-3 in *E. coli*. Our long-term aim is the design of new forms of TIMP-3 that may selectively inhibit TACE and could thus represent useful therapeutic agents for pathological events in which TNF-$\alpha$ secretion is involved. When compared with TIMP-3 expressed in NS0 cells, TIMP-3 purified and refolded from inclusion bodies after expression in *E. coli* was as efficient in inhibiting rTACE as TIMP-3 expressed in NS0 cells.[6] However, the N-terminal domain of TIMP-3 alone (N-TIMP-3) was 3000-fold less inhibitory than full-length TIMP-3 (TABLE 1). This important decrease in inhibitory potency indicates that the C-terminal domain of TIMP-3 (C-TIMP-3) is required for inhibition of rTACE. To ascertain the importance of C-TIMP-3 in rTACE inhibition, we produced a chimeric TIMP consisting of N-TIMP-2 fused to C-TIMP-3 (N-TIMP-2/C-TIMP-3). This chimeric TIMP did not inhibit rTACE, even though it was functional towards MMP-2 (TABLE 1). Taken together, these results indicate that both N-TIMP-3 and C-TIMP-3 are required for rTACE inhibition. On the basis of these results, our future efforts will concentrate on the design of mutant forms of full-length TIMP-3 and on the analysis of their effects on TACE inhibition.

## ACKNOWLEDGMENT

This work was funded by the Arthritis Research Campaign (UK).

## REFERENCES

1. BLACK, R.A., C.T. RAUCH, C.J. KOZLOSKY *et al.* 1997. A metalloproteinase disintegrin that releases tumour-necrosis factor–alpha from cells. Nature **385:** 729–733.
2. MOSS, M.L., S.L. JIN, M.E. MILLA *et al.* 1997. Cloning of a disintegrin metalloproteinase that processes precursor tumour-necrosis factor–alpha. Nature **385:** 733–736.
3. BUXBAUM, J.D., K.-N. LIU, Y. LUO *et al.* 1998. Evidence that tumor necrosis factor-alpha converting enzyme is involved in regulated alpha-secretase cleavage of the Alzheimer amyloid protein precursor. J. Biol. Chem. **273:** 27765–27767.
4. BORLAND, G., G. MURPHY & A. AGER. Tissue inhibitor of Metalloproteinases (TIMP)-3 inhibits shedding of L-selectin from leukocytes. J. Biol. Chem. **274:** 2810–2815.
5. HARGREAVES, P.G., F. WANG, J. ANTCLIFF *et al.* 1998. Human myeloma cells shed the interleukin-6 receptor: inhibition by tissue inhibitor of metalloproteinase-3 and a hydroxamate-based metalloproteinase inhibitor. Br. J. Haematol. **101:** 694–702.

6. AMOUR, A., P.M. SLOCOMBE, A. WEBSTER *et al.* 1998. TNF-alpha converting enzyme (TACE) is inhibited by TIMP-3. FEBS Lett. **435:** 39–44.
7. BARAMOVA, E.N, J.D. SHANNON, J.B. BJARNASON & J.W. FOX. 1990. Interaction of hemorrhagic metalloproteinases with human $\alpha_2$ -macroglobulin. Biochemistry **29:** 1069–1074.
8. CHANTRY, A., N.A. GREGSON & P. GLYNN. 1989. A novel metalloproteinase associated with brain myelin membranes. Isolation and characterization. J. Biol. Chem. **264:** 21603–21607.

# The Effects of Sustained Elevated Levels of Circulating Tissue Inhibitor of Metalloproteinases–1 on the Development of Breast Cancer in Mice

TODD B. BUCK,[a,d] H. YOSHIJI,[b] STEVEN R. HARRIS,[c] OPAL R. BUNCE,[d] AND UNNUR P. THORGEIRSSON[a,e]

[a]Tumor Biology and Carcinogenesis Section, Laboratory of Cellular Carcinogenesis and Tumor Promotion, Division of Basic Sciences, National Cancer Institute, National Institutes of Health, Bethesda, Maryland 20892, USA

[b]The Third Department of Internal Medicine, Nara Medical University, Kashihara, Nara, Japan

[c]Department of Pharmacology, Pikeville College School of Osteopathic Medicine, Pikeville, Kentucky, USA

[d]Department of Pharmaceutical and Biomedical Sciences, College of Pharmacy, University of Georgia, Athens, Georgia, USA

Under normal physiological conditions, extracellular matrix (ECM) degradation is tightly regulated by coordinated expression of matrix metalloproteinases (MMPs) and their naturally occurring inhibitors, tissue inhibitor of metalloproteinases (TIMPs). During some pathological processes such as rheumatoid arthritis and tumorigenesis, the balance between proteolysis and conservation of the extracellular matrix shifts towards degradation, due to an upregulation of MMPs in both the neoplastic and normal adjacent tissues.[1] However, elevated serum levels of TIMP-1 have also been demonstrated in several types of cancer, including prostate[2] and bladder.[3] In order to determine the role that elevated serum levels of TIMP-1 may have on tumor development, we generated a transgenic mouse line expressing human TIMP-1 (hTIMP-1) under the direction of the mouse albumin promoter. This resulted in a very tightly regulated liver-specific expression of the transgene, and subsequent release into the bloodstream. As a result, we were able to obtain very high serum levels of biologically active hTIMP-1 in these transgenic mice.

We measured hTIMP-1 levels in the serum of the transgenic mice at various ages, using an hTIMP-1–specific ELISA kit (Calbiochem). Briefly, whole blood was collected and spun down at $4000 \times g$ for 15 minutes to obtain serum. Equal volumes of serum were analyzed in triplicate according to the manufacturer's instructions. We observed the highest concentrations of hTIMP-1 in two-week-old homozygous

[e]Address for correspondence: Unnur P. Thorgeirsson, National Institutes of Health, National Cancer Institute, Division of Basic Sciences, Laboratory of Cellular Carcinogenesis and Tumor Promotion, Tumor Biology and Carcinogenesis Section, Building 37, Room 2D02, Bethesda, MD 20892. Phone, 301/496-1982; fax, 301/402-0153; e-mail, thorgeiu@dc37a.nci.nih.gov

**TABLE 1.** Serum concentrations of transgenic animals at various ages, as measured by ELISA

| Age | Serum concentration |
|---|---|
| 2 weeks | 816.5 ± 71.0 ng/ml |
| 6 weeks | 523.0 ± 66.8 ng/ml |
| 10 weeks | 474.6 ± 56.7 ng/ml |
| 20 weeks | 518.1 ± 97.2 ng/ml |
| 30 weeks | 420.4 ± 143 ng/ml |

animals (816.5 ± 71.0 ng/ml), while at six weeks of age the levels dropped 36% and were roughly stable for all other subsequent ages tested (TABLE 1).

In order to assess biological activity of the hTIMP-1 transgenic protein against MMP-2 activity, we performed a reverse zymogram as previously described[4] on the serum collected from wild-type and transgenic animals at various ages. To ensure that equal amounts of protein were added to the reverse zymogram, a protein assay (Pierce) was performed on all samples. Densitometric analysis of the inhibitory activity was performed, and levels were normalized to TIMP-4 densitometric levels, which remained unchanged between wild-type and transgenic animals. At two weeks of age, the average normalized value for inhibition of MMP-2 activity in the transgenic animals was 1.681 optical density units (OD), while at six weeks of age the value was 0.9217 OD. MMP-2 inhibitory activity of serum from age-matched wild-type animals were roughly equal at both ages, with an average value of 0.3437 OD.

To assess the impact of biologically active hTIMP-1 on the development of breast cancer in mice, we treated transgenic and wild-type animals with medroxyprogesterone acetate (MPA) and 7,12-dimethylbenz[a]anthracene (DMBA) as previously described.[5] Briefly, slow-release MPA pellets (Innovative Research of America) were subcutaneously implanted interscapularly at six weeks of age. Beginning at nine weeks of age, DMBA (Sigma) dissolved in corn oil (Sigma) at a concentration of 50 mg/kg was administered intragastrically once a week for five consecutive weeks.

Tumor development was monitored weekly in each animal by palpation for 30 weeks post-DMBA dosing. All mice were euthanized for humane reasons after the tumors reached ~2.0 cm in diameter, or at the end of the study. Tumors were then excised, and part was fixed in formalin, while the remaining portion was snap frozen in liquid nitrogen for RNA isolation.

Tumor incidence in the wild-type animals was 83.3% (10/12), while in the transgenics it was 25.0% (4/16) (FIG. 1). The time for tumor development in the wild-type animals was 19.8 ± 3.6 weeks after the last DMBA dose, while in the transgenics it was 19.2 ± 1.8 weeks. Latency was recorded as the length of time required to first detect a distinct mass in the mammary glands of the animals. The multiplicity (tumors per tumor-bearing mouse) in the transgenics was 1.0 ± 0.0, whereas the wild-type had an average of 2.1 ± 0.23 tumors per animal (FIG. 1).

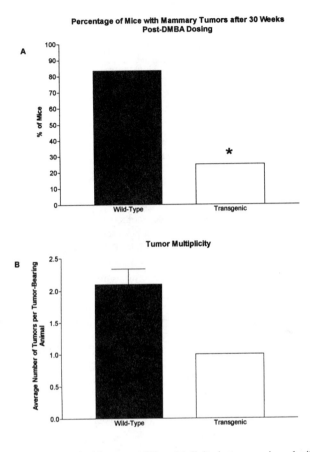

**FIGURE 1.** (**A**) Tumor incidence and (**B**) multiplicity in transgenic and wild-type mice 30 weeks post-DMBA treatment.

There is increasing evidence that TIMP-1 is upregulated during the tumorigenesis process, but whether this is a host-mediated response or a tumor-driven phenomenon remains unclear. In order to determine whether overexpression of TIMP-1 is beneficial to host survival or tumor growth, we generated a transgenic mouse line with high levels of circulating biologically active hTIMP-1. Compared to the wild-type animals, the TIMP-1 transgenics had a 58.3% decrease in the number of animals with tumors. Furthermore, there were over 80% fewer total tumors in the mammary glands of the TIMP-1 transgenics versus the wild-type animals. However, tumors that did manage to grow in the presence of high circulating levels of TIMP-1 did not have a prolonged latency. Our findings indicate that TIMP-1 may not only inhibit tumor growth, but it may actually block the tumorigenesis cascade at a very early stage.

## REFERENCES

1. CRAWFORD, H.C. *et al.* 1994. Tumor and stromal expression of matrix metalloproteinases and their role in tumor progression. Invasion Metastasis **14:** 234–245.
2. BAKER, T. *et al.* 1994. Serum metalloproteinases and their inhibitors: markers for malignant potential. Br. J. Cancer. **70:** 506–512.
3. NARUO, S. *et al.* 1994. Serum levels of a tissue inhibitor of metalloproteinases-1 (TIMP-1) in bladder cancer patients. Int. J. Urol. **1:** 228–231.
4. OLIVER, G.W. *et al.* 1997. Quantitative reverse zymography: analysis of picogram amounts of metalloproteinase inhibitors using gelatinase A and B reverse zymograms. Anal. Biochem. **244:** 161–166.
5. ALDAZ, C.M. *et al.* 1996. Medroxyprogesterone acetate accelerates the development and increases the incidence of mouse mammary tumors induced by dimethylbenzanthracene. Carcinogenesis **17**(9): 2069–2072.

# Macrophage-Specific Expression of Human Collagenase (MMP-1) in Transgenic Mice

VINCENT LEMAÎTRE,[a] TIMOTHY K. O'BYRNE,[a] SEEMA S. DALAL,
ALAN R. TALL, AND JEANINE M. D'ARMIENTO[b]

*Department of Medicine, College of Physicians and Surgeons of Columbia University,
New York, New York 10032, USA*

The extracellular matrix (ECM) is essential for maintaining the integrity of all tissues, with alterations in its structural components leading to a variety of pathological conditions. The ECM is maintained by an intricate balance between the synthesis and degradation of its structural proteins. The continuous turnover and remodeling of the ECM involve a family of proteolytic enzymes called matrix metalloproteinases (MMPs).[1] Some MMPs are secreted by macrophages in various inflammatory processes and are believed to be critical in the progression of the diseases. Among these macrophage-secreted MMPs, interstitial collagenase (MMP-1) specifically cleaves the major fibrillar collagens of the ECM at physiologic temperature and pH.[2]

To investigate the *in vivo* role of MMP-1 in disease, we decided to generate a transgenic mouse that would specifically express this human enzyme in their macrophages. The mouse is a particularly appropriate animal model to study the function of MMP-1 because it does not contain a homologue of this human collagenase. Therefore, a transgenic mouse model has been generated that harbors the human MMP-1 gene under the control of the scavenger receptor promoter-enhancer A (SREP). The SREP allows targeted expression of genes specifically in differentiated macrophages of transgenic mice.[3]

The SREP/MMP-1 transgene, represented in FIGURE 1A, was generated through removal of all sequences 5′ to the translational initiation codon of the full-length human MMP-1 gene.[4] A *Sma*I site, immediately preceding the translational initiation codon of the MMP-1 gene, was introduced by *in vitro* mutagenesis. The *Sma*I-*Sal*I MMP-1 genomic fragment (9.3 kb) was ligated downstream of the SREP sequence of 4.5 kb containing a newly inserted *Sma*I site (the human SREP was obtained from Dr. C. Glass, University of California, San Diego). The SREP/MMP-1 gene junction was sequenced to confirm that the translation start site and the reading frame of the MMP-1 gene were intact, and that Kozak's rules for translational efficiency were maintained.[5]

The 13.8 kb transgene was isolated from the cloning plasmid by *Not*I-*Sal*I digestion, purified by CsCl gradient centrifugation,[6] and microinjected into fertilized mouse eggs (F1[C57Bl/6 × CBA/J] × F1 [C57Bl/6 × CBA/J]). Ninety-five live newborn mice were obtained, of which eight carried the transgene. These eight mice were mated with wild-type mice to establish transgenic lines through Southern blot

[a]These authors contributed equally to this work

[b]Address for correspondence: Dr. Jeanine D'Armiento, College of Physicians and Surgeons, Department of Medicine, PH8-101, 622 West 168th Street, New York, NY 10032. Phone, 212/305 3745; fax, 212/305 5052; e-mail, jmd12@columbia.edu

**A.**

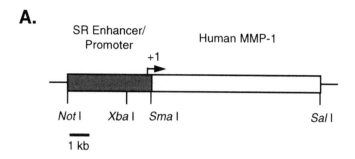

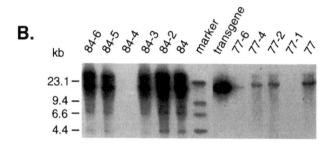

**FIGURE 1.** Generation of transgenic mice expressing human MMP-1 under the control of the *scavenger receptor enhancer/promoter* (SREP). (**A**) Construct of the SREP/MMP-1 transgene. The transcription initiation (+1) occurs 46 bp upstream from the *Sma*I site within the promoter. (**B**) Southern blot analysis of MMP-1 founder and F1 progeny mice. DNA from mice were analyzed using the transgene as the probe. Mice 84 and 77 are original founders. Mice 84-6, -5, -3, -2, 77-6, -4, and -2 are F1 progenies that contain the transgene. Mice 84-4 and 77-1 do not contain the transgene. Marker: [32]P-labeled Lambda *Hin*dIII marker. Transgene: original microinjected DNA fragment.

analysis. Tail DNA of 3-week-old pups was cleaved with *Hin*dIII, separated on a 0.6% agarose gel, and blotted onto nitrocellulose filter for Southern blot analysis.[7] The [32]P-labeled probe ($5 \times 10^6$ cpm) was generated by nick-translation, with the SREP/MMP-1 construct used as a template. FIGURE 1B shows a Southern blot performed on lines 77 and 84, confirming single integrations of the construct.

Quantitative levels of MMP-1 protein from transgenic mouse macrophages were measured by ELISA. Thioglycolate-induced peritoneal macrophages were incubated in DMEM serumless culture media for 48 hours. The media was collected and concentrated approximately 12-fold. Protein concentrations were determined by a BCA Protein Assay (Pierce). The standards and 40 micrograms of protein from each sample were assayed on a two-site ELISA "sandwich" format (Biotrack, Amersham) using monoclonal anti-human MMP-1 antibodies. TABLE 1 shows that macrophages of both transgenic lines 77 and 80 expressed MMP-1; the highest expression was observed in line 77. Collagenolytic activity of the activated transgenic protein was demonstrated through a diffuse fibrillar collagenase assay[8] (data not shown).

**TABLE 1. ELISA of concentrated macrophages culture media for quantification of MMP-1 expression**

| Mouse line | MMP-1 | |
| --- | --- | --- |
| # | Genotype | Protein (ng/ml) |
| 77-7 | Wt | ND |
| 77-9 | Wt | ND |
| 77-13 | Tg | 135.2 |
| 77-15 | Tg | 111.3 |
| 77-17 | Tg | 128.8 |
| 80-3 | Wt | ND |
| 80-8 | Wt | ND |
| 80-9 | Tg | 25.5 |
| 80-10 | Tg | 16.2 |
| 80-11 | Tg | 17.5 |

NOTE: Results are presented in ng of MMP-1 per ml of concentrated media. Transgenic (Tg) line 77 expressed the highest levels of human MMP-1. Culture media from macrophages of wild type (Wt) littermates had no detectable (ND) levels of MMP-1.

In conclusion, we have generated a transgenic mouse model that specifically expresses human MMP-1 mRNA and protein in its tissue macrophages. The affect of MMP-1 expression on inflammatory processes will be analyzed after direct comparisons with the wild-type mice, which do not normally express MMP-1. This unique animal model will give new insights into the pathophysiological involvement of MMP-1 in the complex mechanisms of extracellular matrix remodeling during human inflammatory diseases.

## REFERENCES

1. BIRKEDAL-HANSEN, H. 1995. Proteolytic remodeling of extracellular matrix. Curr. Opin. Cell Biol. **7**: 728–735.
2. WELGUS, H.G., E.J. CAMPBELL, Z. BAR-SHAVIT, S.M. SENIOR & S.L. TEITELBAUM. 1985. Human alveolar macrophages produce a fibroblast-like collagenase and collagenase inhibitor. J. Clin. Invest. **76**: 219–224.
3. HORVAI, A., W. PALINSKI, H. WU, K. MOULTON, K. KALLA & C.K. GLASS. 1995. Scavenger receptor A gene regulatory elements target gene expression to macrophages and to foam cells of atherosclerotic lesions. Proc. Natl. Acad. Sci. USA **92**: 5391–5395.
4. D'ARMIENTO, J., S.S. DALAL, Y. OKADA, R.A. BERG & K. CHADA. 1992. Collagenase expression in the lungs of transgenic mice causes pulmonary emphysema. Cell **71**: 955–961.
5. KOZAK, M. 1986. Point mutations define a sequence flanking the AUG initiator codon that modulates translation by eukaryotic chromosomes. Cell **44**: 283–292.

6. HOGAN, B., R. BENDDINGTON, F. COSTANTINI & E. LACY. 1994. Manipulating the mouse embryo: a laboratory manual, 2nd edit. Cold Spring Harbor Laboratory. New York.
7. SAMBROOK, J., E.F. FRITSCH & T. MANIATIS. 1989. Molecular Cloning: A Laboratory Manual, 2nd edit. Cold Spring Harbor Laboratory. New York.
8. CAWSTON, T.E. & A. BARRETT. 1979. A rapid and reproducible assay for collagenase using [1-[14]C]acetylated collagen. Anal. Biochem. **99:** 340–345.

# TIMP-4 Is Regulated by Vascular Injury in Rats

CLARE M. DOLLERY,[a,e] JEAN R. McEWAN,[a] MINGSHENG WANG,[b]
QINGXIANG AMY SANG,[c] YILIANG E. LIU,[b] AND Y. ERIC SHI[d]

[a]The Hatter Institute, University College London Hospitals, London, UK

Departments of [b]Paediatrics and [d]Pathology, Long Island Jewish Hospital,
Albert Einstein College of Medicine, New Hyde Park, New York 11040, USA

[c]Department of Chemistry, Florida State University, Tallahassee, Florida 32306, USA

The role of basement membrane–degrading matrix metalloproteinases (MMPs) in enabling vascular smooth muscle cell migration after vascular injury has been established in a number of animal models of restenosis.[1–5] In contrast, the role of their native inhibitors, the tissue inhibitors of matrix metalloproteinases (TIMPs), has remained unproved despite frequent coregulation of matrix metalloproteinases and TIMPs in other disease states.[1,6] We have investigated the time course of expression and localization of TIMP-4 in rat carotid arteries 6 hours, 24 hours, 3 days, 7 days, and 14 days after balloon injury by *in situ* hybridization, immunohistochemistry, and Western blot analysis. TIMP-4 protein was present in the adventitia of injured carotid arteries from 24 hours after injury. At 7 and 14 days after injury widespread immunostaining for TIMP-4 was seen throughout the neointima, media, and adventitia of injured arteries. Western blot analysis confirmed the quantitative increase in TIMP-4 protein at 7 and 14 days. *In situ* hybridization detected increased expression of TIMP-4 as early as 24 hours after injury and a marked induction in neointimal cells 7 days after injury. We then studied the effect of TIMP-4 protein on the migration of smooth muscle cells through a matrix-coated membrane *in vitro* and demonstrated a 53% reduction in invasion of rat vascular smooth muscle cells. This data and the temporal relationship between the upregulation of TIMP-4, its accumulation, and the onset of collagen deposition suggest an important role for TIMP-4 in the proteolytic balance in the vasculature in controlling both smooth muscle migration and collagen accumulation in the injured arterial wall.

## REFERENCES

1. WEBB, K.E., A.M. HENNEY, S. ANGLIN, S.E. HUMPHRIES & J. McEWAN. 1997. The expression of matrix metalloproteinases and their inhibitor TIMP-1 in the rat carotid artery carotid following balloon injury. Arterioscler. Thromb. Vasc. Biol. **17:** 1837–1844.

[e]Address for correspondence: Dr. Clare M. Dollery M.R.C.P., Ph.D., Department of Cardiology, 4th Floor, Jules Thorn Building, The Middlesex Hospital, Mortimer Street, London W1N8AA, UK. Phone, 44 171 636-8333; fax, 44 171 380-9415; e-mail, c.dollery@med.ucl.ac.uk.

2. BENDECK, M.P., N. ZEMPO, A.W. CLOWES, R.E. GALARDY & M.A. REIDY. 1994. Smooth muscle cell migration and matrix metalloproteinase expression after arterial injury in the rat. Circ. Res. **75:** 539–545.
3. STRAUSS, B.H., R.J. CHISHOLM, F.W. KEELEY, A.I. GOTLIEB, R.A. LOGAN & P.W. ARMSTRONG. 1994. Extracellular matrix remodeling after balloon angioplasty injury in a rabbit model of restenosis. Circ. Res. **75:** 650–658.
4. SOUTHGATE, K.M., M. FISHER, A.P. BANNING, V.J. THURSTON, A.H. BAKER, R.P. FABUNMI, P.H. GROVES, M. DAVIES & A.C. NEWBY. 1996. Upregulation of basement membrane-degrading metalloproteinase secretion after balloon injury of pig carotid arteries. Circ. Res. **79:** 1177–1187.
5. ZEMPO, N., R.D. KENAGY, Y.P. AU, M. BENDECK, M.M. CLOWES, M.A. REIDY & A.W. CLOWES. 1994. Matrix metalloproteinases of vascular wall cells are increased in balloon-injured rat carotid artery. J. Vasc. Surg. **20:** 209–217.
6. HASENSTAB, D., R. FOROUGH & A.W. CLOWES. 1997. Plasminogen activator inhibitor type 1 and tissue inhibitor of metalloproteinases-2 increase after arterial injury in rats. Circ. Res. **80:** 490–496.

# *In Vivo* Adenoviral Gene Transfer of TIMP-1 after Vascular Injury Reduces Neointimal Formation

CLARE M. DOLLERY,[a,e] STEVEN E. HUMPHRIES,[b] ALAN McCLELLAND,[c] DAVID S. LATCHMAN,[d] AND JEAN R. McEWAN[a]

[a]*The Hatter Institute, University College London Hospitals, London, UK.*

[b]*Division of Cardiovascular Genetics, Department of Medicine, University College London, London, UK.*

[c]*Genetic Therapy Inc., Gaithersburg, Maryland 20882, USA*

[d]*Department of Molecular Pathology, University College London Medical School, London, UK*

Cell migration across matrix boundaries is a major component in the formation of neointimal hyperplasia following vascular injury and is dependent on the alteration of the proteolytic balance within the arterial wall towards matrix breakdown.[1–5] This change is mediated in part by the matrix metalloproteinases (MMPs) and may be modified by their natural inhibitors, the tissue inhibitors of metalloproteinases (TIMPs).[1,3–5] In this study an increase in expression of biologically active and immunoreactive TIMP-1 was seen *in vitro* in response to infection of rat smooth muscle cells (SMCs) with Av1.TIMP-1 (an adenoviral vector containing the human TIMP-1 cDNA). Infection of rat SMCs with Av1.TIMP-1 reduced migration *in vitro* by 27% compared to control virus-infected cells ($37.6 \pm 4.34$ vs. $51 \pm 5.01$, $p < 0.05$, $n = 12$). The adenoviral vector was delivered to the rat carotid artery following balloon injury; and four days later immunoreactive protein was identified, and migration of smooth muscle cells was reduced by 60% ($5.2 \pm 0.5$ vs. $12.8 \pm 1.5$ cells per section, $p < 0.05$, $n = 5$). Neointimal area 14 days after injury showed a 30% reduction in the animals receiving the Av1.TIMP-1 virus compared with controls ($0.09 \pm 0.01$ vs. $0.14 \pm 0.01$ mm$^2$, $p = 0.02$, $n = 14$). These studies support the hypothesis that the response to arterial balloon injury involves MMP-dependent smooth muscle cell migration, which can be attenuated *in vivo* by the transmural expression of TIMP-1 using adenoviral gene transfer.

## REFERENCES

1. BENDECK, M.P., N. ZEMPO, A.W. CLOWES, R.E. GALARDY & M.A. REIDY. 1994. Smooth muscle cell migration and matrix metalloproteinase expression after arterial injury in the rat. Circ. Res. **75:** 539–545.

[e]Address for correspondence: Dr. Clare M. Dollery, M.R.C.P., Ph.D., Department of Cardiology, 4th Floor, Jules Thorn Building, The Middlesex Hospital, Mortimer Street, London, W1N8AA, UK. Phone, 44 171 636-8333; fax, 44 171 380-9415; e-mail, c.dollery@ucl.ac.uk

2. BENDECK, M.P., C. IRVIN & M.A. REIDY. 1996. Inhibition of matrix metalloproteinase activity inhibits smooth muscle cell migration but not neointimal thickening after arterial injury. Circ. Res. **78:** 38–43.
3. STRAUSS, B.H., R.J. CHISHOLM, F.W. KEELEY, A.I. GOTLIEB, R.A. LOGAN & P.W. ARMSTRONG. 1994. Extracellular matrix remodeling after balloon angioplasty injury in a rabbit model of restenosis. Circ. Res. **75:** 650–658.
4. SOUTHGATE, K.M., M. FISHER, A.P. BANNING, V.J. THURSTON, A.H. BAKER, R.P. FABUNMI, P.H. GROVES, M. DAVIES & A.C. NEWBY. 1996. Upregulation of basement membrane-degrading metalloproteinase secretion after balloon injury of pig carotid arteries. Circ. Res. **79:** 1177–1187.
5. WEBB, K.E., A.M. HENNEY, S. ANGLIN, S.E. HUMPHRIES & J. MCEWAN. 1997. The expression of matrix metalloproteinases and their inhibitor TIMP-1 in the rat carotid artery carotid following balloon injury. Arterioscler. Thromb. Vasc. Biol. **17:** 1837–1844.

# Paradoxical Stimulation of Matrix Metalloproteinase–9 Expression in HT1080 Cells by a Broad-Spectrum Hydroxamate-Based Matrix Metalloproteinase Inhibitor

E. MAQUOI,[a] C. MUNAUT,[a] A. COLIGE,[b] C. LAMBERT,[b] F. FRANKENNE,[a] A. NOËL,[a] F. GRAMS,[c] H.-W. KRELL,[d] AND J.-M. FOIDART[a,e]

[a]Laboratoire de Biologie des Tumeurs et du Développement, Université de Liège, Tour de Pathologie (B23), Sart Tilman, B-4000 Liège, Belgium

[b]Laboratoire de Biologie des Tissus Conjonctifs, Université de Liège, Tour de Pathologie (B23), Sart Tilman B-4000 Liège, Belgium

[c]Boehringer Mannheim, GmbH, Mannheim, Germany

[d]Boehringer Mannheim, GmbH, Penzberg, Germany

## INTRODUCTION

Due to their unusual ability to cleave and degrade a variety of ECM components, including triple-helical type IV collagen, two members of the family of matrix metalloproteinases (MMPs), MMP-2 and MMP-9, are thought to play critical roles during tumor invasion and metastasis. The proteolytic activity of mature MMPs is modulated by a group of specific inhibitors, the tissue inhibitors of metalloproteinases (TIMPs). Excessive ECM degradation observed during tumor invasion and metastasis frequently correlates with an excess of active MMPs. Therefore, adding exogenous inhibitors was contemplated for anticancer therapy. Indeed, synthetic MMP inhibitors have been demonstrated to reduce tumor invasion and metastasis in several *in vitro* and *in vivo* models.[1]

Here, we show that some of these synthetic inhibitors paradoxically enhance the production of MMP-9 even as they inhibit the activity of others (MMP-2, MT1-MMP).

## RESULTS

### GI Reduced the Invasion of HT1080 Cells through Type IV Collagen

We first demonstrated, using Transwell chambers coated with type IV collagen,[2] that GI129471 (GI, 1 µM), an hydroxamate-based inhibitor,[3] reduces HT1080 cell invasion up to about 20% of that of the vehicle-treated cells (data not shown). In contrast, neither the ability of the cells to migrate through uncoated filters nor their metabolism were affected by GI, indicating that the inhibition of HT1080 invasion was most probably mediated through reduced proteolytic activity. Gelatin zymography

[e]Corresponding author.

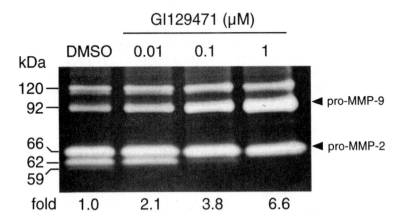

**FIGURE 1.** GI modulates MMP-9 secretion in HT1080 cells. Conditioned media obtained from cells cultured for 48 hr with GI (0–1 µM) were analyzed by gelatin zymography. The relative gelatinolytic activity corresponding to MMP-9 was evaluated by scanning densitometry. A value of 1 was arbitrarily given to the control condition. Relative stimulation (fold) of MMP-9 secretion is shown.

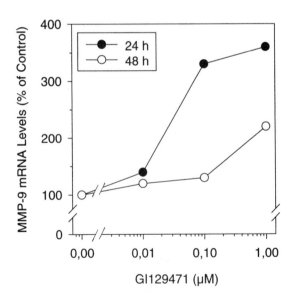

**FIGURE 2.** GI increases MMP-9 mRNA levels in HT1080 cells. MMP-9 mRNA levels were quantified by quantitative RT-PCR and densitometric analysis of autoradiographies of the gels. All results were corrected by densitometric data obtained for the 28S rRNA. A value of 1 was arbitrarily attributed to the relative level of MMP-9 mRNA measured in control condition.

of conditioned media from the chemoinvasion assays revealed that HT1080 essentially secreted pro-MMP-2 and pro-MMP-9. In addition, two activated forms of MMP-2 and a 120-kDa gelatinase activity (corresponding to a complex between TIMP-1 and the pro-MMP-9) were also detected (FIG. 1). Upon treatment with GI, the MMP-2 activated forms were undetectable, indicating the inhibition of MTl-MMP activity. Unexpectedly, the levels of both pro-MMP-9 and 120-kDa complex were strongly increased.

## GI Stimulated MMP-9 Production by HT1080 Cells

To further investigate the effect of GI on MMP-9 secretion, HT1080 cells were treated for 48 hours with graded GI concentrations (0–1 μM). GI dose-dependently stimulated the secretion of pro-MMP-9 with a maximal 6.6-fold stimulation at 1 μM (FIG. 1). In contrast, pro-MMP-2 processing was dose-dependently inhibited. On the other hand, neither rTIMP-1 nor rTIMP-2 (0.01–1 μM), the physiological MMP inhibitors, significantly modulated MMP-9 secretion (data not shown).

Using quantitative RT-PCR, we have shown that GI (0–1 μM) treatment increases the MMP-9 mRNA level (FIG. 2). The effect of GI was dose dependent up to the concentration of 1 μM. Northern blot analysis showed that GI does not alter MMP-1, MMP-2, MT1-MMP, TIMP-1, or TIMP-2 mRNA levels (data not shown).

## GI Stimulates the Transcriptional Activity of the MMP-9 Promoter

In order to assess the effect of GI on MMP-9 transcription, HT1080 cells were transiently transfected with pNA7.8 reporter vector. This construct was made up of the murine MMP-9 promoter (−7745 to +1 bp) linked to the β-galactosidase reporter gene. In the presence of GI, a fourfold stimulation of β-galactosidase activity was noted after 48 hours (data not shown). These data clearly demonstrated that the GI-stimulated MMP-9 synthesis was mediated through an enhanced transcription of the MMP-9 gene.

## CONCLUSIONS

Our findings clearly demonstrate that GI129471, a broad-spectrum hydroxamate-based MMP inhibitor, could, in addition to its classical inhibitory activities, specifically upregulate both transcription and synthesis of pro-MMP-9. Our data thus emphasize the need to carefully investigate the potential side effects of such unspecific broad-spectrum MMP inhibitors.

## REFERENCES

1. TALBOT, D.C. & P.D. BROWN. 1996. Experimental and clinical studies on the use of matrix metalloproteinase inhibitors for the treatment of cancer. Eur. J. Cancer, **14:** 2528–2533.
2. MAQUOI, E., A. NOÉL, F. FRANKENNE, H. ANGLIKER, G. MURPHY & J.M. FOIDART. 1998. Inhibition of matrix metalloproteinase 2 maturation and HT1080 invasiveness by a synthetic furin inhibitor. FEBS Lett. **424:** 262–266.
3. CAMPION, C., J.P. DICKENS & M.J. CRIMMIN. PCT Patent Application, WO 90/05719, 1990.

# Identification of the TIMP-2 Binding Site on the Gelatinase A Hemopexin C-Domain by Site-Directed Mutagenesis and the Yeast Two-Hybrid System

CHRISTOPHER M. OVERALL,[a,c] ANGELA E. KING,[a] HEATHER F. BIGG,[a]
ANGUS McQUIBBAN,[a] JULIET ATHERSTONE,[a] DOUGLAS K. SAM,[a]
ALDRICH D. ONG,[a] TIM T.Y. LAU,[a] U. MARGARETHA WALLON,[a]
YVES A. DeCLERCK,[b] AND ERIC TAM[a]

[a]Faculty of Dentistry and the Department of Biochemistry and Molecular Biology,
Faculty of Medicine, University of British Columbia, Vancouver,
British Columbia, Canada V6T 1Z3

[b]Division of Hematology-Oncology, Childrens Hospital Los Angeles and
University of Southern California, Los Angeles, California, USA

Concanavalin A,[1] PMA,[2] and collagen[3] can induce the cellular activation of the matrix metalloproteinase (MMP) gelatinase A (MMP-2). In this process, gelatinase A utilizes tissue inhibitor of metalloproteinases-2 (TIMP-2) as an adapter to dock with membrane-type MMP (MT-MMP) on the cell membrane.[4] Here, the $NH_2$-domain of TIMP-2 binds to the active site of the MT-MMP with the TIMP-2 COOH-domain thought to bind to the gelatinase A COOH terminal hemopexin domain (C-domain). A second, uncomplexed, furin-activated MT-MMP then initiates gelatinase A activation by cleavage of the prodomain. Activation is completed by autolytic intra- and intermolecular gelatinase A cleavages (reviewed in Ref. 4). These interactions in the quaternary activation complex are specifically required since progelatinase A cell surface binding to $\beta_1$ integrin-bound collagen via the triple fibronectin type II repeats of the enzyme is not only insufficient for activation, but in fact protects progelatinase A from activation.[5]

In this study we sought to identify the TIMP-2 binding site on the gelatinase A C-domain in molecular detail. First, we confirmed binding of the recombinant (r) gelatinase A C-domain protein[6] with full-length TIMP-2 (FIG. 1A). Notably, TIMP-2 did not bind to recombinant C-domain of MT1-MMP. Next, the binding of recombinant TIMP-2 $NH_2$-domain protein[7] with the gelatinase A rC-domain protein[6] was studied. By both solid-phase assays and affinity chromatography, a weak binding interaction was demonstrated between these two domains that was disrupted by 0.3–0.5 M NaCl. In contrast, TIMP-1 $NH_2$-domain protein did not bind. However, it is known that gelatinase A activation does not occur using TIMP-2 $NH_2$-domain protein in place of full-length TIMP-2.[8] This indicates that the TIMP-2 C-

[c]Address for correspondence: Dr. C.M. Overall, University of British Columbia, 2199 Wesbrook Mall, Vancouver, B.C., Canada V6T 1Z3. Phone, 604/822-2958; fax, 604/822-3562; e-mail, overall@interchange.ubc.ca

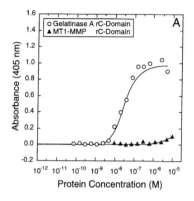

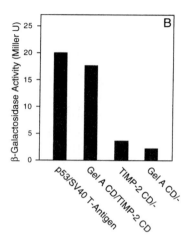

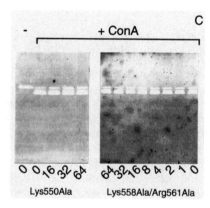

**FIGURE 1.** *Panel A,* Gelatinase A rC-domain ($7.67 \times 10^{-11}$–$5.03 \times 10^{-6}$ M) and MT1-MMP rC-domain ($9.77 \times 10^{-9}$–$1.00 \times 10^{-5}$ M) in PBS, pH 7.4, were incubated in microtiter plates coated with full-length TIMP-2. Bound protein was quantitated by ELISA using a specific antibody against the His$_6$ fusion peptide. *Panel B,* β-galactosidase activity was measured in yeast cultures cotransformed with plasmids coding for Gal4 DNA binding or activator proteins fused with the following targets as indicated: p53/SV40 T-antigen (positive control); gelatinase A C-domain/TIMP-2 C-domain; TIMP-2 C-domain alone; gelatinase A C-domain alone. *Panel C,* Zymograms of cell culture medium from human fibroblasts treated or not with concanavalin A (± as indicated) with the gelatinase A C-domain mutants in pmol/100 μL as indicated. Lys550Ala exhibits a weak reduction in TIMP-2 binding and retains gelatinase A activation blocking ability as evident by the increase in amounts of the latent *(upper)* gelatinase band relative to the lower activated band, with increasing concentrations of the mutant protein. Lys558Ala/Arg561Ala did not compete for activation at any concentration (up to 256 pmol, not shown). Here, the ratio of latent to active gelatinase A bands was unaltered by this domain.

domain interaction is pivotal for gelatinase A activation, but this interaction has yet to be directly demonstrated. To do so, we utilized the yeast two-hybrid system. Shuttle/expression vectors containing the gelatinase A and TIMP-2 C-domain cDNAs ligated to the Gal4 DNA binding and activation domains were cotransformed into a yeast strain with two reporter genes (lacZ and HIS3). The TIMP-2 and gelatinase A C-domains fused to Gal4 DNA binding and activator domains, respectively, showed a strong positive interaction as evidenced by high LacZ levels (FIG. 1B) and robust yeast colony growth on -His plates. In contrast, either plasmid alone did not support -His growth or exhibit β-galactosidase activity in control cells. This direct demonstration of binding is significant because hitherto binding of the gelatinase A C-domain to the TIMP-2 C-domain has only been indirectly inferred by domain deletion experiments. The additional novel interaction between the gelatinase A C-domain with the TIMP-2 NH$_2$-domain may stabilize the bound TIMP-2.

We next sought to identify by oligonucleotide–directed mutagenesis sites on the gelatinase A C-domain that interacted with the potential binding site (QEFLDIEDP) on the TIMP-2 COOH-terminal anionic tail.[8] The overall shape of the gelatinase A C-domain is a squat cylinder comprised of four β-sheets, each representing a hemopexin module and each forming a blade of the four-bladed β-propeller structure. Unlike all other MMPs, the gelatinase A C-domain is rich in basic residues that lie predominantly in hemopexin modules III and IV. The 3-D arrangement of the four antiparallel β-strands that comprise each β-sheet brings into spatial juxtaposition basic residues that are distantly spaced in the sequence. Thus, concatamers of basic residues are formed from adjacent strands, on β-blade III in particular, that pack lysine residues together to generate cationic clusters on the top face and outer rim of the C-domain. We hypothesized that the hemopexin module III and IV cationic clusters bind the highly negatively charged COOH-terminal peptide tail of TIMP-2. Twelve single, double, and triple site-directed mutants were generated that replaced basic residues in each of these clusters (FIG. 2). The recombinant proteins were expressed and purified from *E. coli* inclusion bodies, as described before for the wild-type domain.[6] Electrospray mass spectrometry confirmed NH$_2$-terminal methionine processing and that the correct amino acid substitution had been made.

Using a solid-phase assay,[9] the wild-type rC-domain bound human TIMP-2 with an apparent $K_d$ of $0.6 \times 10^{-8}$ M. The cationic cluster Lys$^{566}$/Lys$^{567}$/Lys$^{568}$ on hemopexin module III was not essential for TIMP-2 binding since the double mutant protein Lys566/568Ala showed a binding pattern essentially identical to that of the wild-type rC-domain. This was confirmed by affinity chromatography, where TIMP-2 was found to bind tightly to the mutant protein in an interaction that was not disrupted by 1.0 M NaCl or 10% DMSO. Moreover, TIMP-2 binding was not blocked by preincubation of the wild-type rC-domain with an affinity-purified antibody raised against a peptide encompassing this lysine cluster (NH$_2$-RYNEVKKK-MDPG-COOH). However, the role of this triple lysine cluster is enigmatic. Although not appearing to directly influence TIMP-2 binding, combining Lys566/568Ala with Lys550Ala as a triple mutant (Lys550/566/568Ala) synergistically further reduced the mild disruption in TIMP-2 binding produced by Lys550Ala alone. Moreover, this effect was not unique to position 550. The Lys617Ala mutation located on the outer β-strand of hemopexin module IV reduced TIMP-2 binding by nearly an order of magnitude (apparent $K_d$ $0.8 \times 10^{-7}$ M). Despite the lack of an effect on TIMP-2 binding by the alanine mutations at positions 566 and 568, when combined with the effective Lys617Ala substitution (Lys566/568/617Ala), TIMP-2 interaction was reduced to undetectable levels. The importance of position 617 was also highlighted by the Lys610Thr/Lys617Ala mutation, which reduced TIMP-2 binding to undetectable levels both in the plate assay and by affinity chromatography. That Lys566/568Ala did not globally destabilize the domain and so account for these synergistic results was shown by analysis of another triple mutant, Lys566/568/575Ala. These additional substitutions did not alter the TIMP-2 interaction observed for the single alanine point mutation at position 575 alone. Indeed, for both mutants (Lys575Ala and Lys566/568/575Ala), TIMP-2 binding was essentially unaltered from the wild-type rC-domain. We infer that $Lys^{566}/Lys^{567}/Lys^{568}$ is likely to be on the edge of the TIMP-2 binding site, thereby accounting for the small direct contribution to TIMP-2 binding. Nonetheless, this cluster appears to be involved in TIMP-2 binding in some manner, due to its interesting property of synergistically destabilizing TIMP-2 binding when mutations here are combined with effective mutations.

An interesting differential effect on TIMP-2 binding was observed for each of the three lysines in the cationic cluster at positions 547, 549, and 550. Alanine substitution at $Lys^{547}$ raised the apparent $K_d$ of TIMP-2 interaction by approximately an order of magnitude. The Lys550Ala mutation in this cluster showed a slight increase in $K_d$, whereas the Lys549Ala mutation showed essentially unaltered TIMP-2 binding. Indeed, the relative importance of Lys547 in this cluster was shown by the triple mutation of all three residues. Lys547/549/550Ala exhibited TIMP-2 binding essentially unaltered from that seen to Lys547Ala alone.

The differential importance of individual lysines in module III for TIMP-2 binding was further demonstrated around position 558. The mutant Lys575Ala had no apparent effect on TIMP-2 interaction. $Lys^{558}$ is topographically adjacent to $Lys^{575}$ in the C-domain, but located on the underside of the domain away from the important $Lys^{547}$ residue. Therefore, when Lys558Ala was combined with Arg561Ala (Lys558Ala/Arg561Ala) and almost a total loss of TIMP-2 binding was observed, this indicated a more important role in the TIMP-2 interaction for $Arg^{561}$ rather than $Lys^{558}$.

Examination of the position of the mutants on the 3-D structure of the gelatinase A C-domain together with the binding data locates the TIMP-2 binding site to the upper surface of the C-domain towards the outer rim, at the junction of hemopexin

---

**FIGURE 2.** Location of the TIMP-2 binding site on the human gelatinase A C-domain. The spacefill graphic of the C-domain was made using coordinates under Protein Data Base accession identifier 1RTG. The top of the pseudosymmetry axis with the central $Ca^{2+}$ ion on the upper face of the domain is to the *top*. The location of hemopexin modules I to IV is indicated (HxI–HxIV). The orientation of the *lower panel* is indicated by the position of the hemopexin modules marked and the upper and lower surfaces as indicated. The side of module I would be located further around the rim of the domain past module IV at the bottom of the model. Thus, this image represents a tilted side view of the C-domain. All lysine and arginine residues are colored *black*. From the results obtained an estimate of the boundary of the TIMP-2 binding site is shown by the *dotted white line*. Mutation sites directly influencing TIMP-2 interaction are within the *dotted white line* and labeled in *white*. In the top panel, $Lys^{610}$, $Lys^{617}$ and $Arg^{561}$ (labeled in *black*) also contribute to the TIMP-2 binding site. A weaker interaction surface is between $Lys^{550}$ and $Lys^{556-568}$.

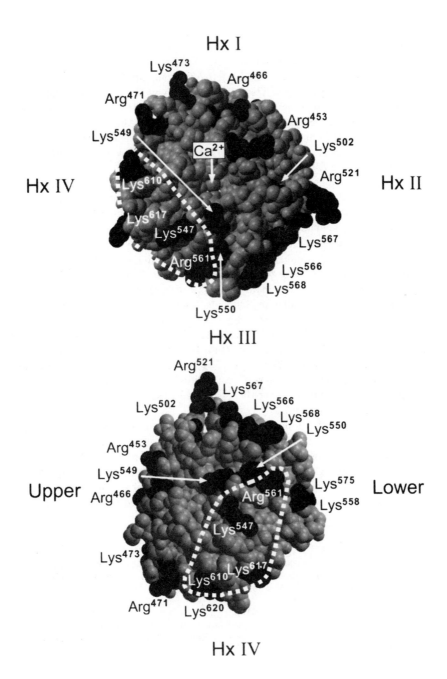

modules III and IV (FIG. 2). Obviously it is not just the basic residues that are important for TIMP-2 interaction. The surrounding residues will also contribute to the free energy of association. Of note, the strong dipole moment in the gelatinase A C-domain produced by the unique clustering of positive-charged residues to hemopexin modules III and IV is likely to favor correct orientation of gelatinase A with respect to TIMP-2 prior to binding. In the active enzyme, this dipole moment "action at a distance" may thereby favor TIMP-2 interaction with the C-domain binding site rather than with the active site. By analogy with the position of hemopexin modules III and IV of porcine collagenase,[10] we predict the bound TIMP-2 lies away from the active site, opposite from the fibronectin-like collagen binding domain of gelatinase A. The implications of such an orientation are that it may prevent autoinhibition *in cis* of the gelatinase A catalytic domain by the bound TIMP-2. (To distinguish between autoinhibition and inhibition of other MMPs, we term these types of inhibition as *in cis* and *in trans*, respectively). This orientation would also allow the inhibitory $NH_2$-domain of the bound TIMP-2 to interact with the MT-MMP active site, leaving the prodomain accessible for activation cleavage by a second MT-MMP.

To determine whether reduced TIMP-2 binding resulting from mutations in the gelatinase A C-domain could affect gelatinase A activation, we utilized competition experiments. We have previously shown that addition of rC-domain to concanavalin A–treated cells reduces endogenous gelatinase A activation in a concentration-dependent manner.[4] In similar experiments, addition of mutant C-domains that showed minor or no evidence of reduced TIMP-2 binding (Lys549Ala, Lys550Ala, Lys566/568Ala, and Lys566/568/575Ala) reduced gelatinase A activation in an identical manner to the wild-type rC-domain (FIG. 1C). However, mutants displaying reduced TIMP-2 binding (Lys547Ala, Lys617Ala) were less effective in competing for gelatinase A activation by the cell surface activation complexes. Indeed, the TIMP-2 binding knockout mutations (Lys610/617Ala and Lys558Ala/Arg561Ala) did not block gelatinase A activation.

Thus, the recombinant domains and mutants generated here have proved to be incisive tools in directly demonstrating the importance of both the TIMP-2 C- and $NH_2$-domains in binding the gelatinase A C-domain, in mapping the TIMP-2 binding site of the gelatinase A C-domain, and in confirming the importance of the TIMP-2 interaction for gelatinase A activation. The mutant proteins generated and their TIMP-2 binding properties have recently been fully described in Overall *et al.*[11]

## REFERENCES

1. OVERALL, C.M. & J. SODEK. 1990. Concanavalin A produces a matrix degradative phenotype in human fibroblasts: induction and endogenous activation of collagenase, 72-kDa gelatinase, and PUMP-1 is accompanied by the suppression of TIMP. J. Biol. Chem. 265: 21141–21151.
2. BROWN, P.D., A.T. LEVY, I.M.K. MARGULIES, L.A. LIOTTA & W.G. STETLERSTEVENSON. 1990. Independent expression and processing of Mr 72,000 type IV collagenase and interstitial collagenase in human tumorigenic cell lines. Cancer Res. 50: 6184–6191.
3. GILLES, C., M. POLETTE, P. BIREMBAUT & E.W. THOMPSON. 1997. Collagen type I-induced MT1-MMP expression and MMP-2 activation: implications in the metastatic progression of breast carcinoma. Lab. Invest. 76: 651–660.

4. OVERALL, C.M., U.M. WALLON, B. STEFFENSEN, Y. DECLERK, H. TSCHESCHE & R. ABBEY. 1999. Substrate and TIMP interactions with human gelatinase A recombinant COOH-terminal hemopexin-like and fibronectin type II-like domains: both the N- and C-domains of TIMP-2 bind the C-domain of gelatinase A. *In* Inhibitors of Matrix Metalloproteinases in Development and Disease. D. Edwards, S. Hawkes & R. Kokha, Eds. Gordon and Breach. Amsterdam. In press.

5. STEFFENSEN, B., H.F. BIGG & C.M. OVERALL. 1998. The involvement of the fibronectin Type II–like modules of human gelatinase A in cell surface localization and activation. J. Biol. Chem. **273:** 20622–20628.

6. WALLON, U.M. & C.M. OVERALL. 1997. The COOH-terminal hemopexin-like domain of human gelatinase A (MMP-2) requires $Ca^{2+}$ for fibronectin and heparin binding: binding properties of recombinant gelatinase A C-domain to extracellular matrix and basement membrane components. J. Biol. Chem. **272:** 7473–7481.

7. DECLERCK, Y.A., T. YEAN, H.S. LU, J. TING & K.E. LANGLEY. 1991. Inhibition of autoproteolytic activation of the interstitial procollagenase by recombinant metalloproteinase inhibitor TIMP-2. J. Biol. Chem. **266:** 3893–3899.

8. WILLENBROCK, F., T. CRABBE, P.M. SLOCOMBE, C.W. SUTTON, A.J.P. DOCHERTY, M.I. COCKETT, M. O'SHEA, K. BROCKLEHURST, I.R. PHILLIPS & G. MURPHY. 1993. The activity of the tissue inhibitors of metalloproteinases is regulated by C-terminal domain interactions: a kinetic analysis of the inhibition of gelatinase A. Biochemistry **32:** 4330–4337.

9. BIGG, H.F., Y.E. SHI, Y.E. LIU, B. STEFFENSEN & C.M. OVERALL. 1997. Specific, high affinity binding of TIMP-4 to the COOH terminal hemopexin-like domain of human gelatinase A: TIMP-4 binds progelatinase A and the COOH terminal domain in a similar manner to TIMP-2. J. Biol. Chem. **272:** 15496–15500.

10. LI, J., P. BRICK, M.C. O'HARE, T. SKARZYNSKI, L.F. LLOYD, V.A. CURRY, I.M. CLARK, H.F. BIGG, B.L. HAZLEMAN, T.E. CAWSTON & D.M. BLOW. 1995. Structure of full-length porcine synovial collagenase reveals a C-terminal domain containing a calcium-linked, four bladed beta-propeller. Structure **3:** 541–549.

11. OVERALL, C.M., A.E. KING, D.K. SAM, A.D. ONG, T.T.Y. LAU, U.M. WALLON, Y.A. DECLERCK & J. ATHERSTONE. 1999. Identification of the TIMP-2 binding site on the hemopexin carboxyl domain of human gelatinase A by site-directed mutagenesis. The hierarchical role in binding TIMP-2 of the unique cationic clusters of hemopexin modules III and IV. J. Biol. Chem. **274:** 4421–4429.

# Index of Contributors